Foundations and Fundamentals in Human-Computer Interaction

This book serves as a foundation to the field of HCI, equipping readers with the necessary knowledge and skills to engage in this field.

This book:

- Discusses human functionalities and characteristics relevant to interaction, including sensory perception, attention and memory, language and communication, emotions, decision-making, as well as mental models, human error, and human actions.
- Explores the evolution of HCI design approaches and the role of social and organizational psychology in HCI.
- Discusses key concepts and societal aspects of interactive technologies, such as user acceptance, ethics, privacy, and trust.
- Covers the historical background, contributing disciplines, essential concepts, and theories within the domain.

This book will appeal to individuals interested in Human-Computer Interaction research and applications.

Foundations and Fundamentals in Human-Computer Interaction

Edited by
Constantine Stephanidis and Gavriel Salvendy

CRC Press
Taylor & Francis Group
Boca Raton London New York

CRC Press is an imprint of the
Taylor & Francis Group, an **informa** business

First edition published 2025
by CRC Press
2385 NW Executive Center Drive, Suite 320, Boca Raton FL 33431

and by CRC Press
4 Park Square, Milton Park, Abingdon, Oxon, OX14 4RN

CRC Press is an imprint of Taylor & Francis Group, LLC

© 2025 selection and editorial matter, Constantine Stephanidis and Gavriel Salvendy; individual chapters, the contributors

ISBN: 978-1-032-36992-1 (hbk)
ISBN: 978-1-032-80034-9 (pbk)
ISBN: 978-1-003-49510-9 (ebk)

DOI: 10.1201/9781003495109

Typeset in Times
by codeMantra

Contents

Preface

Human-Computer Interaction (HCI) is a multidisciplinary field exhibiting increasing significance in the modern world, shaping our interactions with technology and transforming our daily lives. HCI plays a critical role in bridging the gap between human capabilities and technological breakthroughs. As technology becomes progressively intertwined with our daily lives, it plays a significant role in enhancing human interaction with advanced information technologies.

We are pleased to present six books in the series titled 'HUMAN-COMPUTER INTERACTION: FOUNDATIONS AND ADVANCES', which provide extensive coverage of the HCI field. Collectively, this book series addresses the theoretical and application aspects of HCI and thus provides important resources for HCI researchers, practitioners, and students. Each book addresses theoretical and practical aspects of HCI, as well as state-of-the-art technological advancements, aiming to illuminate how innovations reshape this field, redefine user interactions, and elevate the overall user experience. The six book titles in the series are:

- Foundations and Fundamentals in Human-Computer Interaction
- Designing for Usability, Inclusion and Sustainability in Human-Computer Interaction
- User Experience Methods and Tools in Human-Computer Interaction
- Interaction Techniques and Technologies in Human-Computer Interaction
- Human-Computer Interaction in Intelligent Environments
- Human-Computer Interaction in Various Application Domains

Within this series, readers will discover a wealth of information encompassing the foundational elements, the established and emerging domains, the cutting-edge technological advancements, as well as invaluable insights into key concepts, challenges, and the latest and most important research and practices. Each book provides a thorough examination and a standalone source of information.

To further enhance this exploration, the books also showcase the practical applications of HCI approaches and methodologies, illustrating their utility. Insightful figures, tables, and references supplement the above, while an in-depth discussion and future directions in each chapter aim to engage readers in spearheading their research and application endeavors in the field of HCI.

This book serves as a foundation for the field of HCI. It covers the historical background, contributing disciplines, essential concepts and theories within the domain. This book also delves into human functionalities and characteristics relevant to interaction, including sensory perception, attention and memory, language and communication, emotions, decision-making, as well as mental models, human error, and human actions. Additionally, it explores the evolution of HCI design approaches and the role of social and organizational psychology in HCI, it discusses the key HCI concept of user acceptance of interactive technologies and addresses societal aspects such as ethics, privacy, and trust.

This book contains 14 chapters, written by a total of 25 authors, with 73 figures, 20 tables, and 2,789 references for documenting and providing supporting data to the presented and discussed information.

Constantine Stephanidis and Gavriel Salvendy
November 2023

Editors

Constantine Stephanidis is Professor at the Department of Computer Science of the University of Crete, past Director of the Institute of Computer Science of FORTH and Founding Head of its Human-Computer Interaction (HCI) Laboratory and its Ambient Intelligence Program. He has been the Principal Investigator for over 180 European Commission, National and Industry funded projects. He is the founding editor of the International Journal *Universal Access in the Information Society*, co-editor of the *International Journal of Human-Computer Interaction* and General Chair of the HCI International Conference. He is the President of the Council for Research and Innovation of the Region of Crete and the President of the Hellenic National Accessibility Authority.

Gavriel Salvendy is University Distinguished Professor at the College of Engineering and Computer Science at the University of Central Florida and Founding President of the Academy of Science, Engineering, and Medicine of Florida. He is Professor Emeritus of Industrial Engineering at Purdue University and Chair Professor Emeritus and Founding Head of the Department of Industrial Engineering at Tsinghua University, Beijing, P.R. China. He is a member of the National Academy of Engineering and the recipient of the John Fritz Medal, which is frequently referred to as the Nobel Prize in Engineering. He is the Founding Editor of the *International Journal of Human-Computer Interaction* and the Founder of the International Conference on Human-Computer Interaction – now in its 40th year.

Contributors

Simone Diniz Junqueira Barbosa
Department of Informatics
PUC-Rio
Rio de Janeiro, Brazil

Gabriela Fernández Castillo
Department of Psychological Sciences
Rice University
Houston, Texas

Andrina Granić
Department of Computer Science
Faculty of Science
University of Split
Split, Croatia

Don Harris
Faculty of Engineering, Environment &
    Computing
Coventry University
Coventry, United Kingdom

Ya-Hsin Hung
Department of Psychological Sciences
Purdue University
West Lafayette, Indiana

Victor Kaptelinin
Department of Informatics
Umeå University
Umeå, Sweden

Maha Khalid
Department of Psychological Sciences
Rice University
Houston, Texas

Bart P. Knijnenburg
School of Computing
Clemson University
Clemson, South Carolina

David Lamas
School of Digital Technologies
Tallinn University
Tallinn, Estonia

Zhizhong Li
Department of Industrial Engineering
Tsinghua University
Beijing, P.R. China

Johanna Löchner
Department of Child and Adolescent
    Psychiatry and Psychotherapy, Tuebingen
    Center for Mental Health
University of Tübingen
Tübingen, Germany
and
German Center for Mental Health (DZPG)
Munich, Germany

Nathan J. McNeese
School of Computing
Clemson University
Clemson, South Carolina

Vinícius Carvalho Pereira
Modern Languages Department
Instituto de Linguagens
Universidade Federal de Mato Grosso
Cuiabá, Brazil

Robert W. Proctor
Department of Psychological Sciences
Purdue University
West Lafayette, Indiana

Janet C. Read
School of Engineering and Computing
University of Central Lancashire
Lancashire, United Kingdom

Marco C. Rozendaal
Industrial Design Engineering
Delft University of Technology
Delft, The Netherlands

Eduardo Salas
Department of Psychological Sciences
Rice University
Houston, Texas

Björn W. Schuller
GLAM
Imperial College London
London, United Kingdom
CHI - Chair of Health Informatics
MRI, Technische Universität München (TUM)
Munich, Germany
Munich Data Science Institute (MDSI)
Munich, Germany
Munich Center for Machine Learning (MCML)
Munich, Germany

Manrong She
Department of Industrial Engineering
Tsinghua University
Beijing, P.R. China

Constantine Stephanidis
University of Crete
Computer Science Department and
Foundation for Research and Technology -
  Hellas (FORTH)
Institute of Computer Science
Heraklion, Crete, Greece

Vladimir Tomberg
School of Digital Technologies
Tallinn University
Tallinn, Estonia

Kim-Phuong L. Vu
Department of Psychology
California State University Long Beach
Long Beach, California

Zijian Yin
Department of Industrial Engineering
Tsinghua University
Beijing, P.R. China

Jia Zhou
College of Management Science and Real
  Estate
Institute of Neuroergonomics and Management
  Innovation
Chongqing University
Chongqing, P.R. China

1 The HCI Discipline
Past, Present and Future

Constantine Stephanidis

1.1 INTRODUCTION

In the dynamic domain of human-computer interaction (HCI), gaining insight into its evolutionary journey, current status, and emerging trends is imperative for keeping abreast of the swiftly evolving technological landscape. This introductory chapter explores the HCI domain by delving into its historical foundations, contemporary relevance, and future potential. It encompasses the core discipline of HCI and its fundamental components: the human, the computer, and the intricate interaction that binds them. It also examines the theoretical underpinnings and practical dimensions of HCI, including the frameworks, guidelines, standards, and methodologies that drive its development.

This chapter unveils the latest trends and developments likely to revolutionize the way we engage with technology and outlines the challenges and opportunities that emerge on the horizon, shedding light on how HCI intersects with global challenges, from sustainability to inclusivity. In other words, this chapter aims to illuminate the path ahead by underscoring the paramount significance of HCI in the context of the continuously and radically changing technological landscape, where human-machine relationships are increasingly becoming an integral part of our daily lives.

1.1.1 THE HCI DISCIPLINE

HCI is a multidisciplinary field dedicated to the close examination, design, and assessment of interactions between humans and computers. It encompasses a wide array of aspects related to how humans engage with and utilize technology, with the aim of enhancing the usability, efficiency, and overall user experience of digital systems, applications, and devices (Shneiderman & Plaisant, 2010). At the heart of HCI lies the principle of user-centered design (UCD). This entails comprehending the requirements, behaviors, preferences, and constraints of users to create technology that is intuitive, effective, and satisfying for them (Abras et al., 2004). To accomplish this, HCI draws on a variety of disciplines, including computer science, psychology, design, sociology, anthropology, and many more (Carroll, 2003). This comprehensive approach fosters a holistic comprehension of the human-computer relationship. Additionally, a profound attention to human factors, which encompass cognitive, physical, and social aspects of human behavior, is integral to HCI (Salvendy, 2012). Understanding these factors and the knowledge foundations that inform HCI design is critical in designing technology that aligns with human capabilities and limitations (see also Chapter 11 'HCI Design Perspectives, Trends and Approaches', Book 'FOUNDATIONS AND FUNDAMENTALS IN HUMAN-COMPUTER INTERACTION').

Given these considerations, HCI initially aimed to develop technology primarily with usability in mind. Usability pertains to how easily and efficiently users can complete tasks and attain their objectives using a specific system or interface. It can be assessed through techniques like user testing and expert evaluations (Lewis, 2012; Shneiderman & Plaisant, 2010). Today, HCI extends its focus to creating positive User Experiences (UX) that go beyond mere usability and encompass the emotional and psychological dimensions of technology usage (Benyon, 2019). A positive user experience leads to user satisfaction and loyalty (Bilgihan, 2016). At the same time, accessibility

DOI: 10.1201/9781003495109-1

has become a prominent concern, recognizing that technology has evolved as an integral part of people's lives. This requires designing interfaces and interactions that are accessible and usable by individuals with various impairments (Stephanidis, 2009a, 2013, 2021).

From a practical standpoint, HCI professionals concentrate on crafting user-technology interactions (e.g., the design of buttons, menus, touch gestures, voice commands, and other interaction methods), structuring and organizing information within digital systems to facilitate user navigation (Frauenberger, 2019). Prior to software engineers embarking on system development, HCI designers create prototypes or mockups of digital interfaces to test and refine the design (See also Chapter 3 'Collaboration between HCI Design and Software Engineering', Book 'DESIGNING FOR USABILITY, INCLUSION AND SUSTAINABILITY IN HUMAN-COMPUTER INTERACTION', Chapter 14 'Prototyping Techniques for UX Research' and Chapter 15 'UX Design Tools', Book 'USER EXPERIENCE METHODS AND TOOLS IN HUMAN-COMPUTER INTERACTION'). This iterative process aims to guarantee that the final product is user-friendly and efficient (Issa & Isaias, 2022). From the earliest stages, extensive user research (e.g., surveys, interviews, and observation) is employed to gain insights into user behavior, preferences, and pain points (Baxter et al., 2015). Furthermore, HCI addresses ethical concerns linked to technology, such as privacy, security, and the societal impact of technology. Ethical considerations are particularly relevant in the design of systems that manage sensitive data or influence user behavior (Frauenberger et al., 2017; Owen et al., 2020).

HCI emerged in the mid-1980s, but in today's digital era, it has become indispensable in the design of digital systems and devices that are integral to our daily lives. The digital age has seen HCI ascend to paramount importance, given the pervasive integration of technology into everyday activities (Madakam et al., 2015). This integration encompasses an array of technologies, including smartphones (Montag et al., 2015), tablets (Radesky et al., 2020; Ryan & Lewis, 2017), websites, software applications, wearable devices (Ferreira et al., 2021), smart appliances (Nord et al., 2019), and newer technologies such as virtual reality and artificial intelligence. HCI serves as the driving force behind ensuring that these technologies are not merely functional but also user-centric and accessible. In this digital age, user satisfaction is intricately tied to the success of technology, and HCI research and design methodologies aim to create enjoyable and satisfying user experiences.

In parallel, HCI propels technological innovation by perpetually seeking ways to enhance the quality of interaction between humans and machines, fostering the development of new technologies and paradigms, such as touchscreens, voice recognition, and eXtended Reality (XR). It also adapts and evolves design principles to meet the demands of rapidly changing digital environments (Stephanidis et al., 2019).

In a nutshell, HCI aims to ensure that we can fully harness the potential of digital tools and systems while minimizing user frustration and access barriers. As technology continues to evolve, HCI remains at the forefront of creating meaningful and seamless interactions between humans and computers, and more recently, with their environments (Alavi et al., 2019; Harper, 2019).

1.1.2 THE HUMAN

Understanding the human is a prerequisite for the design of technological systems that are perceivable, understandable, and operable. It encompasses several crucial aspects such as human sensation, perception, attention, memory, information processing, mental models, language, communication, decision-making, problem solving, error, and emotions.

Sensation and perception are fundamental in HCI (see also Chapter 2 'Sensation and Perception', Book 'FOUNDATIONS AND FUNDAMENTALS IN HUMAN-COMPUTER INTERACTION'), influencing how users interpret information presented through digital interfaces, as well as impacting elements such as color, layout, and auditory feedback (Ruhland et al., 2015). Multiple senses, including vision, hearing, and touch, are relevant in HCI, especially in multimodal interfaces and interactions with advanced technology like robots and autonomous vehicles (Raats et al., 2020).

Understanding human attention, memory, and information processing is crucial in HCI (see also Chapter 3 'Human Information Processing, Attention and Memory: An Overview for HCI', Book 'FOUNDATIONS AND FUNDAMENTALS IN HUMAN-COMPUTER INTERACTION'). It is necessary to present information in a way that aligns with human cognitive processes and minimizes cognitive overload (Bakker & Niemantsverdriet, 2016; Roda, 2011). Addressing the limitations of human attention, designing to avoid cognitive overload, and recognizing the role of attention in tasks like visual search are essential. Memory, including sensory memory, working memory, and long-term memory, is integral to HCI, influencing how users interact with systems and remember information (Monk, 2014).

Correspondingly, language and communication play essential roles in HCI, involving linguistic and communicative models in various activities and stages (see also Chapter 4 'Language and Communication in HCI', Book 'FOUNDATIONS AND FUNDAMENTALS IN HUMAN-COMPUTER INTERACTION'). As technology becomes more opaque, designing languages to communicate with advanced technological systems is becoming a challenge (Seffah, 2015). HCI involves multiple languages that mediate communication between users, systems, and designers (Munteanu et al., 2013). Users interact with systems through the user interface language, which specifies system state, user actions, and goal achievement. Users themselves employ language to interact with the interface and express their intentions, while some designers offer scripting languages for end users to customize software behavior according to their needs (De Souza, 2005). End-user development introduces communication about the system, as users extend its functionality (Santos & Villela, 2019). Designers also employ various languages at different levels of abstraction, ranging from describing user-oriented qualities to specifying interaction mechanisms during design (i.e., define the system's user interface language, through which the state of the system is expressed, what the user can or should do, and how users will be able to express their intentions in order to achieve their goals) (Hartson & Pyla, 2018; Lowgren & Stolterman, 2007).

Emotions are crucial in HCI as they impact user interactions with technology (see also Chapter 8 'Affect and Emotions', Book 'FOUNDATIONS AND FUNDAMENTALS IN HUMAN-COMPUTER INTERACTION'), with several relevant research thrusts in the field. Emotional design focuses on creating products that evoke specific emotional responses from users through elements such as color schemes, visual aesthetics, and sound design (Jeon, 2017). "Cuteness engineering" is a subset of emotional design that is targeted to enhance user engagement (Marcus et al., 2017). Designing for emotions can improve engagement, usability, and decision-making (Plass & Kaplan, 2016). Consideration of cultural and ethical factors is essential in implementing emotional design, as it affects user experiences. Affective computing, a multidisciplinary field, seeks to imbue machines with the ability to interpret, respond to, and simulate human emotions. This capability has applications ranging from virtual assistants providing emotional support to enhancing gaming (Denisova et al., 2021) or learning experiences (Plass & Kaplan, 2016).

Another central component of HCI is mental models (see also Chapter 5 'Mental Models', Book 'FOUNDATIONS AND FUNDAMENTALS IN HUMAN-COMPUTER INTERACTION'), representing users' perceptions and understanding of how a system works (Xie et al., 2017). Ensuring that a system's functionality aligns with users' mental models is essential for a positive user experience. Differences between the system's conceptual model and the user's mental model can lead to interaction problems. On the contrary, properly leveraging mental models enhances collaboration among practitioners, reveals the motivations behind behaviors, and improves interaction performance.

Human error is a significant concern in various domains (see also Chapter 6 'Human Error', Book 'FOUNDATIONS AND FUNDAMENTALS IN HUMAN-COMPUTER INTERACTION'), including safety-critical fields, healthcare, and consumer product usage (Woods et al., 2017). Designing interfaces that minimize the likelihood and impact of user errors is crucial for user satisfaction and overall system performance (Issa & Isaias, 2022). Human errors can lead to accidents, safety issues, and user dissatisfaction. Understanding the various factors contributing to human errors, like performance-shaping factors, is essential for error prevention and control.

In the interaction context, decision-making and problem-solving are fundamental, as they are also in human behavior in general (See also Chapter 10 'Decision-making and Problem Solving', Book 'FOUNDATIONS AND FUNDAMENTALS IN HUMAN-COMPUTER INTERACTION'). The design of interfaces that support informed decision-making and provide tools and feedback for problem-solving is essential (Wang et al., 2023). The study of decision-making paradigms, such as normative, naturalistic, and process models, is integral to enhancing the quality of decisions and providing decision-aiding tools (Cai et al., 2019).

When interacting with a system, humans, informed by the aforementioned mechanisms, proceed to accomplish concrete actions (See also Chapter 7 'Human Actions', Book 'FOUNDATIONS AND FUNDAMENTALS IN HUMAN-COMPUTER INTERACTION'). HCI has evolved to recognize that human actions are more than just tasks; they involve meaningful interactions and experiences (Clemmensen et al., 2016), a shift that acknowledges the contextual and subjective aspects of human-computer interactions. Future HCI research is expected to explore how technology transforms human agency, address the role of emerging non-human actors, and consider global implications for design, focusing on ethics, responsibility, sustainability, and inclusiveness.

These aspects collectively shape the field of HCI, ensuring that technology interfaces are designed to optimize human capabilities and provide an effective, efficient, and enjoyable user experience, catering to the needs of all users. Particular emphasis is given to comprehending the characteristics and requirements of diverse target user groups, including children, older adults, and individuals at risk of exclusion (See also Chapter 11 'HCI Design for Children' and Chapter 12 'HCI Design for Older Adults', Book 'DESIGNING FOR USABILITY, INCLUSION AND SUSTAINABILITY IN HUMAN-COMPUTER INTERACTION').

1.1.3   THE COMPUTER

The evolution of computer-based interaction devices is a fascinating journey (Campbell-Kelly et al., 2023) that began with the early computers, which were initially programmed using wires to connect various function units such as adders, multipliers, and dividers. This manual and mechanical approach was labor-intensive and lacking efficiency. The concept of interactive devices, as we know them today, began developing in later generations of computers. Early computers relied on mechanical and manual methods for input and programming, with operators connecting one unit's outputs to another's inputs to perform computations.

A significant advancement came with the introduction of punch cards as a storage medium for computer programs. This innovation simplified the programming process, making it easier than manually moving wires on a plugboard. Punch cards were portable and allowed for easy re-execution of programs, enhancing the efficiency of computing systems. Some early computers like ENIAC (Haigh et al., 2016) even featured rudimentary output mechanisms, including punched cards and panels of lights for visual indications of operational status and intermediate results. However, during this era, interactions between humans and computers were unidirectional, with humans providing commands and machines executing them. Meanwhile, pioneers were already contemplating more interactive uses of computer technology, and the concept of an interactive machine started to emerge. The modern concept of computer interaction devices truly took shape in the 1960s, when the need for human end effectors, such as fingers, hands, eyes, and more, to control input devices became apparent. These effectors are the body parts used to interact with input devices and deliver information, while computer sensors recognize and process these inputs.

The most versatile and dexterous end effectors for interaction are the human hands and fingers (MacKenzie, 2003). The majority of input devices have been designed to respond to hand and finger movements, including switches, buttons, knobs, sliders, keyboards, mice, trackballs, touchscreens, styluses, pens, brushes, and joysticks (see also Chapter 3 'Input, Output and Interaction Devices', Book 'INTERACTION TECHNIQUES AND TECHNOLOGIES IN HUMAN-COMPUTER INTERACTION'). Other body parts such as the eyes, mouth, and even brain activity, have found use in input devices such as eye-trackers, mouth-operated interfaces, brain-computer interfaces, and voice and auditory interfaces, where sound serves as a medium for transferring information (Freeman et al., 2017; Hasan & Yu, 2017; Zander et al., 2014). Notably, some input devices rely on broader body-level movements, moving away from hand manipulations. These devices involve motion detectors, mid-air hand-tracking sensors, and locomotion platforms that respond to body movements or gestures. These developments cater to various interaction preferences and applications (see also Chapter 1 'Interaction Styles', Book 'INTERACTION TECHNIQUES AND TECHNOLOGIES IN HUMAN-COMPUTER INTERACTION').

In response to input devices, output devices aim to create human sensory experiences. They bridge the gap between digital creations and our innate senses, enhancing our interactions with the digital world. Sight is the most prominent sense, and screen devices, including curved and foldable screens, projectors, and electronic ink displays, provide visual feedback. Virtual reality (VR) and augmented reality (AR) glasses offer immersive visual experiences (Ashtari et al., 2020; Boletsis & Cedergren, 2019).

Hearing, our sense of sound, can convey a wide range of information (Chamberlain et al., 2017). Sound feedback serves as cues and alerts in video games, enhances immersion in VR experiences, and enables hands-free interaction with computer systems through speech feedback and voice commands. Speakers, headphones, earphones, piezoelectric transducers, and other sound output devices deliver audio information effectively (see also Chapter 8 'Sound-based Interaction', Book 'INTERACTION TECHNIQUES AND TECHNOLOGIES IN HUMAN-COMPUTER INTERACTION').

Touch, or the haptic system (Bermejo & Hui, 2021; Dangxiao et al., 2019), allows us to perceive tactile sensations such as pressure, texture, temperature, and vibration. Tactile feedback devices generate vibrations, force feedback devices provide macro physical sensations, and electrical muscle stimulation (EMS) stimulates muscle contractions (see also Chapter 5 'Haptics and Tactile Interaction with Computers', Book 'INTERACTION TECHNIQUES AND TECHNOLOGIES IN HUMAN-COMPUTER INTERACTION'). Additionally, non-contact tactile feedback and thermal feedback devices are being explored to enrich haptic interactions.

Smell and taste feedback devices are in the early stages of development due to their inherent complexity. Replicating and controlling smells and tastes accurately pose significant challenges, but ongoing research aims to further explore these uncharted territories in human-computer interaction.

The diversity of input and output devices in HCI highlights the field's continuous evolution, accommodating a wide range of interaction needs and preferences. Whether through touchpads, pointing devices, mid-air hand tracking, screens, projectors, headphones, or emerging technologies like non-contact haptic feedback and EMS, HCI continues to evolve, enhancing the way humans interact with digital systems (Karpov & Yusupov, 2018). While the conventional setup involves users sitting in front of a screen using a mouse and keyboard, newer approaches leverage physical movement. These methods make use of locomotion and sensorimotor cues to enhance the interaction experience. This allows users to interact without the need for a mouse and keyboard, opening a world of possibilities for more engaging and immersive interactions.

Large displays, in this context, support collaboration (see also Chapter 4 'Large and Multi-screen Displays, Proxemics and Interaction in Public Spaces', Book 'INTERACTION TECHNIQUES AND TECHNOLOGIES IN HUMAN-COMPUTER INTERACTION'). They enable multiple users to interact together in front of the same system, viewing, discussing, and editing content in a shared environment. Another critical aspect of large displays is their ability to provide immersive experiences (Shneiderman et al., 2016), which allow users to reference their own bodies within the experience, encouraging also group interactions and discussions, proving powerful for industrial applications. While it may be challenging to fulfill both the broad field of view and high resolution simultaneously, large displays offer a compelling alternative to immersive experiences.

Moreover, interaction with large displays has evolved to accommodate various interaction modalities suitable for different users and contexts (Claes & Vande Moere, 2015). Interaction modalities range from long-distance interactions using devices or gestures to close-proximity touch interactions. With stereoscopic displays and multi-touch capabilities, research has explored touch metaphors that function both on or near the touch surface and for three-dimensional content presented at varying distances from the display surface. In addition to touch, other modalities include mid-air gestures, controllers, and systems that use presence, pose, and gaze to determine user proximity and attention, initiating interactions based on the user's actions and intentions.

1.1.4 THE INTERACTION

In the domain of HCI, the focus is on the intricate relationship between humans and technology, exploring how people interact with computers and digital systems. HCI involves the study, design, and optimization of interfaces that bridge human cognitive processes and sensory perceptions with computer functionalities. It encompasses various input methods, from traditional ones like keyboards and mice to technologies like touchscreens and voice recognition.

The interaction journey commences with information visualization, placing a significant emphasis on crafting user-friendly visuals to enrich usability, enhance the user experience, and foster inclusivity (see also Chapter 7 'Information Visualization', Book 'DESIGNING FOR USABILITY, INCLUSION AND SUSTAINABILITY IN HUMAN-COMPUTER INTERACTION'). Particular attention is also paid to information architecture and navigation design (see also Chapter 6 'Information Architecture and Navigation Design', Book 'DESIGNING FOR USABILITY, INCLUSION AND SUSTAINABILITY IN HUMAN-COMPUTER INTERACTION'). Overall, information visualization aims to deliver compelling storytelling experiences (Boy et al., 2015) and revolves around core design principles, considering the target audience, managing data intricacies, and adapting to diverse application contexts.

Furthermore, aesthetics in the realm of HCI hold immense importance, transcending mere visual allure to encompass emotional, sensory, and interactive dimensions of design (Miniukovich & Marchese, 2020). Aesthetic interfaces extend their impact beyond aesthetics, significantly shaping user perceptions and interactions with digital systems (see also Chapter 5 'Aesthetics in Design', Book 'DESIGNING FOR USABILITY, INCLUSION AND SUSTAINABILITY IN HUMAN-COMPUTER INTERACTION'). They wield the power to boost visual appeal while simultaneously influencing user trust, credibility, usability, preference, and overall impressions. Remarkably, users swiftly form judgments based on aesthetics, which thereby exert a profound influence on their holistic digital experience.

Moving forward to the realization of interaction, traditional techniques provide users with foundational methods for engaging with computers and digital systems. These techniques have adapted and endured, shaping user experiences and catering to evolving technological landscapes. Keyboard input, a versatile mainstay, allows efficient text entry and commands, especially valuable for programming and command-line interfaces (CLIs). The mouse, in tandem with Graphical User Interfaces (GUIs), revolutionized navigation through pointing, clicking, and dragging (Dix, 2017). Touch input, introduced with touchscreen devices, offers a direct and intuitive interaction with content. Menu-based interfaces create structured hierarchies for accessing functions and information, while CLIs remain a powerful choice for experts who value precision. Buttons and icons in GUIs serve as intuitive visual cues for interacting with software. Scrollbars facilitate content navigation, checkboxes, and radio buttons enable selection and setting choices, and dropdown lists simplify user decisions. Text fields are versatile for text and data input, while dialog boxes manage messages and user input effectively. Hyperlinks empower seamless web navigation. Drag-and-drop actions streamline tasks like file management.

These traditional techniques, alongside contemporary approaches, compose the toolkit for user engagement in HCI, ensuring a rich and multifaceted interaction landscape that continues to adapt to the ever-changing digital world. While emerging technologies and novel methods continually enrich the HCI domain, these well-established traditional techniques retain their significance in modern computing. Yet, the integration of contemporary approaches progressively enhances the user experience, delivering a holistic and intuitive interaction framework for users engaging with computers and digital systems.

Gestures provide natural and intuitive ways for users to interact with computer systems (Vuletic et al., 2019). They leverage users' ability to convey intentions and manipulate objects through bodily movements. Proper user interface and interaction design are essential for devices, applications, and systems that detect, recognize, and interpret gestures accurately (see also Chapter 6 'Gesture-based Interaction', Book 'INTERACTION TECHNIQUES AND TECHNOLOGIES IN HUMAN-COMPUTER INTERACTION'). Various gesture taxonomies contribute to interactions with computer systems, drawing from research in motor control, psycholinguistics, and HCI. The adoption of gestures is motivated by their deep connection to language and their effectiveness in conveying information and facilitating interactions, making them increasingly relevant in touch-screens, mobile devices, wearables, and immersive environments.

Voice user interfaces (VUI) enable users to interact with technology using spoken language or vocal commands, but present distinct challenges due to their reliance on audio interactions without visual displays (Murad et al., 2019). These interfaces rely on speech recognition, synthetic speech production, and dialog management technologies to guide users efficiently through tasks (see also Chapter 7 'Voice Interaction Design', Book 'INTERACTION TECHNIQUES AND TECHNOLOGIES IN HUMAN-COMPUTER INTERACTION'). When considering voice interaction in an application, it is vital to analyze user capabilities, task requirements, and motivations. VUIs are suitable for scenarios where auditory acuity is primary, but designers should avoid them in contexts where noise, hearing-impaired users, or other limitations pose challenges.

Wearable technologies encompass devices worn on or within the user's body, ranging from smart clothing and accessories to implanted devices (Francés-Morcillo et al., 2020). These wearables provide information or operate as closed-loop systems. Usability is critical for wearable acceptance, considering customization, accuracy, and comfort. Privacy concerns and impacts on social interactions should also be considered. Wearables have applications in various fields, from healthcare and education to manufacturing, offering mobility and accessibility for diverse user demographics (see also Chapter 10 'Human Interaction with Wearable Technologies', Book 'INTERACTION TECHNIQUES AND TECHNOLOGIES IN HUMAN-COMPUTER INTERACTION'). Brain-computer interfaces, both wearable and implantable, enable thought-controlled interactions with external devices.

To address the diverse range of existing interaction paradigms and promote digital accessibility, while accommodating a broad spectrum of user capabilities, multimodal interfaces have emerged (see also Chapter 2 'Multimodal Interaction', Book 'INTERACTION TECHNIQUES AND TECHNOLOGIES IN HUMAN-COMPUTER INTERACTION'). These interfaces empower users to engage through multiple sensory channels or modalities, expanding the ways they interact with digital systems. Within multimodal interfaces, users have the flexibility to select their preferred input or output modes, seamlessly transitioning between them to suit the demands of different tasks and physical environments. This versatility not only enhances accessibility but also unlocks potent and expressive options, making for a richer and more adaptable interaction experience.

In the realm of HCI, animation serves as a complex instrumental affordance (Chevalier et al., 2016) with three primary functions. It draws attention, depicts spatial changes over time, and provides expressiveness, conveying meaning and facilitating user understanding. Animations can be used in various interface categories, aiming for various effects such as maintaining context continuity, enhancing the user experience, encoding data relationships, supporting visual discourse and storytelling, and serving as a teaching aid (see also Chapter 9 'Animation and Motion in UX Design', Book 'INTERACTION TECHNIQUES AND TECHNOLOGIES IN HUMAN-COMPUTER INTERACTION'). However, they can pose challenges in processing complex animated content, affecting perceptual, attentional, and cognitive processes. Moreover, animations can elicit emotional responses, influenced by the movement of animated elements and the content's emotional context.

Finally, help and user support is a very important component of the design of a technological system or application (Chilana et al., 2015). It covers various forms of support, from addressing typing errors to offering comprehensive task explanations. Support mechanisms include online documentation, tutorials, animated demonstrations, command-based assistants, intelligent wizards, offline help centers, chatbots, and live chat (see also Chapter 17 'Help and User Support', Book 'USER EXPERIENCE METHODS AND TOOLS IN HUMAN-COMPUTER INTERACTION'). As technology evolves, the way users access help and support has shifted from traditional paper manuals to electronic systems integrated into graphical interfaces and online resources. Demographic factors such as age or disability need to be considered when emphasizing the role of help and support in creating inclusive systems. Additionally, help and support can guide users toward eco-friendly behaviors and sustainable interactions. Different types of help and support systems cater to various user needs and design considerations, focusing on improving user experience with a system, fostering learnability, efficiency in use, and user satisfaction.

1.2   A HISTORICAL PERSPECTIVE ON HCI

Human-Computer Interaction is a relatively new field compared to other computing-related disciplines. Most reports on its history (e.g., Carroll, 2010; Grudin, 2012) agree that HCI became a distinct field during the late 1970s and early 1980s, but there is no single event that can mark the beginning of HCI. The main reason for this is that HCI is a multidisciplinary field, a crossroad of disciplines where computing science, design (as an engineering discipline and as an art), ergonomics, psychology, social sciences, and more meet to investigate how humans interact with computers and provide guidance on how to do that well with ease, efficiently, and even joyfully.

This section investigates the various disciplines that eventually converged into the goal of studying the human, the computer, and their interactions. It is not meant to be an exhaustive historical account of the major events or people that shaped the field. There are some wonderful, detailed works that serve this purpose, which the reader is encouraged to explore; Shackel's (1997) 'Human-Computer Interaction – Whence and Whither?' paper is a good place to start. Brad Myers also did a concise and detailed review of research, applications, and tools that helped shape HCI (Myers, 1998). Grudin's A Moving Target – The Evolution of HCI (2012) goes into detail about events, people, and disciplines that shaped the field. He revisited the topic in 2017 (and again in 2022) book "From Tool to Partner", which is probably the most comprehensive treatise on the subject (Grudin, 2012, 2017, 2022). Finally, some readers may find interesting the approach of Stephanidis et al. (2012), looking at the evolution of interactivity.

This section's focus is to outline chronologically the evolution of the HCI discipline, looking back at its origins and its constant augmentation through the influence of technological developments and the introduction, by pioneers, of new perspectives from other disciplines, until it became a truly distinct discipline.

When the first electronic computers were constructed in the 1940s, they were massive machines, with thousands of tubes, switches, and knobs, initially operated by engineers and used for military purposes (cypher and code-breaking during the war and for performing complex mathematical calculations). They were obviously designed, and their operators interacted with them, before the concept of human-computer interaction even existed. Back then, as already discussed in Section 1.1.3, the interaction was limited to pushing and pulling cables, flipping switches and switching tubes. Tracing their evolution briefly, some major highlights we can point to, in terms of vision of how we can interact with computers, are Vannevar Bush's concept of the Memex (Bush, 1945), which in turn influenced Sutherland's Sketchpad (first GUI, 1963) and Engelbart's oNLine System, aka NLS, most famously known for its demo in 1968, known as The Mother of All Demos – which might just be the perfect candidate for ground zero of Human-Computer Interaction, despite the previous disclaimer. That event unveiled to the world in a single demonstration the mouse (invented 4 years prior), interactive text and hypertext, email, video conferencing, and teleconferencing; in short, it featured many of the concepts that we use today. More notably, the paper describing Stanford's research center "for augmenting human intellect" refers to man-computer interaction (Engelbart & English, 1968). But that demonstration did not only show what computers could do. It also, unknowingly perhaps, demonstrated in retrospect the need for an HCI discipline, as it was notoriously difficult to use, which ultimately led to its failure. Researchers and engineers at the time would have liked computers to be more usable, if possible, but certainly that was not critical. After all, most users of these systems were scientists and engineers, and it was expected that they would need an expert to operate them. The driving force was still technological – what more could these machines do? How easy it is to make them do things, at least then (1940s to early 1970s) was not the primary focus. That goal came from people who brought perspectives from different disciplines, other than computer science.

If we had to choose two basic ingredients for the formulation of the early HCI discipline besides computer science, then it would have to be Human Factors and Ergonomics (HFE) and Psychology. HFE is the closest thing to an origin discipline of HCI. Some might argue that humankind started studying ergonomics the moment one of our ancient ancestors decided to make their primitive tools more efficient to use. Certainly, by the time of Ancient Greece, Hippocrates, the father of medicine himself, had written a treatise on how to organize the surgeon's workplace, how to arrange tools, and more (Marmaras et al., 1999), clearly applying ergonomic principles that still hold true today.

It is generally accepted, however, that HFE became a rigorous discipline during World War II when it was necessary to systematically study airplane cockpits to make them safer. In 1949, in Britain, the Ergonomics Research Society was formed. In the USA, where the term "Human Factors" is favored, the Human Factors Society was formed in 1957 (renamed the Human Factors and Ergonomics Society in 1994). Two years later, in 1959, the International Ergonomics Association was established. HFE pioneers understood they had to study "Human Factors", starting with the physiological aspects that influence the way humans operate and work and gradually including the cognitive and behavioral aspects as well. Today, HFE is defined as a "unique and independent discipline that focuses on the nature of human-artifact interactions, viewed from the unified perspective of the science, engineering, design, technology, and management of human-compatible systems, including a variety of natural and artificial products, processes, and living environments" (Karwowski, 2005).

When computers were invented, it was only natural that part of HFE's research would be dedicated to the new machines, and it is not uncommon to view HCI as an offshoot of HFE. However, that would be a limiting view of the discipline. HFE is the origin of core HCI goals such as usability, as well as methods such as task analysis, observing users to help them work more efficiently and safely and research findings that apply to user interfaces, such as Fitt's Law. However, HFE lacked the theoretical and explanatory power to describe the complexity of human-computer interactions in a holistic way. A significant part of that theoretical perspective was provided by the field of Psychology.

Jakob Nielsen points to 1945 and Bell Labs as a significant moment, when "the line between traditional human factors and what we might call 'user-experience', aimed at human-centered design of interactive systems" was crossed, when they hired John E. Karlin, a psychologist, to design telephone systems. "By the 1950s, Bell Labs definitely did UX work, in particular on the design of the touchtone keypad" (Nielsen, 2017). Bell Labs is cited as one of the most important research labs in the history of HCI research, alongside the famous Xerox Palo Alto Research Center (PARC), established in 1970. Xerox PARC is arguably the first laboratory that recognized the importance of the psychological factor in studying human-computer interactions. There is the famous 1971 AIP Memo 1 by Allan Newell (Newell, 1971; Schirvar, 2023), which was a proposal for a Psychological Research Unit, "… a psychological laboratory within a computer science oriented industrial research laboratory […] Xerox Research Laboratory". This laboratory would become a reality 3 years later. Newell correctly argued that computer science ought to be more symmetric relative to the human and the machine, and that more work was due on the human side. In that memo, he explicitly mentions "information processing psychology" as the key to understand and even calculate human behavior. During this time, borrowing from psychology, concepts such as mental models or the characteristics of human memory, how humans process information (and how fast), or why they make errors, informed the design of systems.

The most influential work that emerged from the marriage between cognitive psychology and human factors, as well as computer science, or as Shackel argues "the first major attempt to formulate some theoretical bases for the field of HCI" (Shackel, 1997) was the work by Card et al. (1983). It is interesting to note the background of the authors. Card, with a background in psychology, was working in Human Factors at Xerox, and Newell, already a well-respected researcher in artificial intelligence, took an interest in studying human behavior (Grudin, 2012; McCarthy, 1988). Their work gave us the first models of user performance with interactive systems, first the keystroke-level model (Card et al., 1980), followed by the cognitive model GOMS (Goals, Operators, Methods, and Selection rules) in 1983, presented in the seminal book 'The Psychology of Human-Computer Interaction' by Card et al. (1983), incidentally, the first time "Human-Computer Interaction" appeared on a book title and partially responsible for the popularization of the term.

However, Xerox PARC is more famous for the pioneering work in GUIs with the Xerox Alto (1973) and the Xerox Star (1981), the former being the first system to pull together all the elements of the modern Graphical User Interface and the latter the first system to use a fully integrated desktop metaphor and application suite. It would take two more years until the (commercially unsuccessful) Apple Lisa introduced the first menu bar and window controls and another year for the (commercially very successful) Apple Macintosh to break the WIMP (Windows, Icons, Menus, Pointers) paradigm into the mainstream until Microsoft Windows (especially Windows 95) made it ubiquitous.

As Newell argued for a more balanced focus between machines and humans, in 1986 the seminal book 'User-Centred System Design' (Draper and Norman, 1986) was released, arguing in favor of placing the user at the heart of any design effort. User-centred design, later renamed to Human-Centred Design, has become the defining principle of human-computer interaction, later expressed in the concept of the user experience. Human-centered design advocates the iterative design and development of technological products, involving four main phases, namely understanding and specification of the context of use, specification of user requirements, design of solutions to meet user requirements, and evaluation of the designs (ISO, 2019). Don Norman also authored in 1988 the 'Psychology of Everyday Things' (later renamed the 'Design of Everyday Things'), which basically argued for designing for usability and introduced many design principles and concepts that are still relevant today.

When personal computers became available in the late 1970s and early 1980s, it became a real need to design them in a way that ordinary people, not engineers or technology enthusiasts, would be able to use them (Carroll, 2010). It was already noted in the literature (Bennett, 1979; Shackel, 1997) that "users are no longer mainly computer professionals but are mostly discretionary users". Nielsen (2017) highlighted that "The PC revolution of the 1980s put extra pressure on the computer industry to improve the usability of its products. Before personal computing, there had definitely been a need for usability, which is why UX personnel grew by a factor of 100 during the mainframe era. But mainframes were enterprise computers, meaning that those who used the computer were not the ones making the purchase decision. So, during that early era, the computer industry had little incentive to produce high-quality user interfaces. With personal computers, the user and the buyer were the same, so the user experience directly impacted purchase decisions".

In 1993, Mosaic, the first graphical web browser, was released marking the start of the World Wide Web (WWW) explosion, transforming it from a researcher's tool available to the few, into an exciting tool for the masses. It also brought with it the first real concentrated effort toward UX. Unlike before, when the customer first paid and then experienced the product, with the internet, the user experience came first – and if all went well, then the customer paid. Companies had to invest more in the user experience because poor UX meant poor sales.

Parallel to the WWW boom came the mass proliferation of mobile phones. A whole new set of UX challenges emerged with the small screen and minimal keypad that fueled HCI research until 2007, when the iPhone was released, and smartphones dominated the market to this day. It is the combination of the internet and the smartphone that spearheaded another era of computing, where interaction was no longer confined to the desk, but spread everywhere.

1.3   CURRENT STATE OF HCI

Since its early days until now, HCI has substantially grown as a field and user interfaces have acquired a dominant role in our everyday life. HCI research and practice have undergone a substantial transformation in recent years, fueled by technological advances and the ever-expanding horizons of the discipline. The emergence and widespread adoption of novel technologies, including mobile devices, wearables, VR, AR, and Internet of Things (IoT) devices, have significantly influenced HCI. These innovations have diversified interaction methods and contexts, leading researchers to concentrate on crafting user experiences for these cutting-edge platforms. Furthermore, the concepts of Ubiquitous or Pervasive Computing (Krumm, 2018) and Ambient Intelligence (Bibri, 2015) have assumed central roles in HCI. These paradigms envision computing seamlessly woven into our daily lives, enabling interaction across a multitude of environments. Consequently, HCI is actively exploring ways to create adaptive and context-aware interfaces that can seamlessly transition between devices and locations. The omnipresence of technology in our daily lives has thrust UX Design into the spotlight within HCI (Hartson & Pyla, 2018). Researchers and practitioners now place a strong emphasis on fashioning delightful and memorable user experiences.

AI and Machine Learning have also become integral components of HCI (Harper, 2019). These technologies are harnessed to construct intelligent user interfaces, deliver personalized recommendations, and discern emotional states, as well as to design compelling UX (see also Chapter 5 'AI in HCI Design and User Experience', Book 'HUMAN-COMPUTER INTERACTION IN INTELLIGENT ENVIRONMENTS'). AI-driven chatbots and virtual assistants are increasingly prominent in HCI research and application. Moreover, HCI researchers are delving into how humans and AI systems can effectively collaborate, including the development of interfaces that facilitate transparent communication and shared decision-making (see also Chapter 11 'Human-Agent Teaming', Book 'HUMAN-COMPUTER INTERACTION IN INTELLIGENT ENVIRONMENTS').

Overall, the landscape of HCI has expanded to encompass a broad spectrum of technologies and societal concerns, with a focus on UCD, inclusivity, privacy, sustainability, and the ethical utilization of technology. The field retains its significant role in shaping the ways in which individuals engage with technology in our ever-evolving digital world, all the while maintaining a vigilant eye on the social and ethical repercussions of technological advancement. This section performs a review of the field as it stands today, introducing horizontal aspects, elaborating on the role of standards and guidelines, and presenting fundamental HCI methods. It also carries out an overview of contemporary interaction paradigms and discusses the role of the field in today's society.

1.3.1   Horizontal HCI Aspects

Despite the notable progress of the field, and advancements across various application domains, the fundamental principle that serves as a foundational component of HCI remains its focus on the human users, how they experience their interaction with a technological system, how they can benefit from this interaction, and what can be done to improve it. As a result, HCI is laid upon a series of core pillars that govern and steer research and practice in the field, which include usability, user experience, user acceptance, human-centered design, design for all and accessibility, sustainability, privacy, trust, and ethics.

Usability constitutes a major concept in HCI, defined as the extent to which a product can be used by target users to achieve specified goals with effectiveness, efficiency, and satisfaction (ISO, 1998,

2018), exhibiting qualities such as learnability, memorability, and error tolerance (Nielsen, 1994c). There are plenty of methods reported in the literature to design for and evaluate the usability of a technological product, as these have been discussed earlier in this chapter. In the 2000s, along with the evolution of technology and the emergence of novel interaction paradigms, the need for extending the scope of usability to encompass additional attributes became evident, and the concept of User Experience (UX) was introduced. Despite its widespread acceptance, UX was not precisely defined initially (Hassenzahl & Tractinsky, 2006; Law et al., 2008; Zimmermann, 2008). A clarifying definition of UX was provided by ISO 9241-210:2010 (ISO, 2010), which was later revised by ISO 9241-210:2019 (ISO, 2019) and described UX as a person's perceptions and responses resulting from the use and/or anticipated use of a product, system or service, and includes all the users' emotions, beliefs, preferences, perceptions, physical and psychological responses, behaviors and accomplishments that occur before, during, and after use (See also Chapter 1 'Usability and User Experience', Book 'DESIGNING FOR USABILITY, INCLUSION AND SUSTAINABILITY IN HUMAN-COMPUTER INTERACTION'). Several methods have been proposed toward measuring UX, encompassing traditional methods and extending them to include additional concepts such as affect and emotion, playfulness and fun, aesthetics, and hedonic quality (Costello & Edmonds, 2007; Desmet et al., 2001; Hassenzahl, 2004; Lavie & Tractinsky, 2004; Read & MacFarlane, 2006). As additional interaction paradigms became widespread, digital transformation took place, and new ubiquitous and intelligent computing environments emerged, the need for further UX design processes and evaluation tools became evident (See also Chapter 1 'Design for Intelligent Environments', Book 'HUMAN-COMPUTER INTERACTION IN INTELLIGENT ENVIRONMENTS'). As a result, several research efforts have focused on providing novel comprehensive UX evaluation frameworks (Gervasi et al., 2020; Metsis et al., 2008; Ntoa et al., 2021; Scholtz & Consolvo, 2004) and appropriate tools (Bateman et al., 2009; Büschel et al., 2021; Ntoa et al., 2019; Stefanidi at al., 2022).

The focus on humans is a core objective in HCI, initially introduced by the notion of UCD (Draper & Norman, 1986), which later evolved to Human-Centered Design (ISO, 2019). In this context of strong human focus, digital accessibility and inclusivity became a major concern (Stephanidis, 2009b; Emiliani, et al., 2011). The former underscores the imperative of making digital systems accessible to individuals with disabilities and older people, driving HCI researchers to devise technologies and guidelines to enhance accessibility across websites, apps, and digital platforms (See also Chapter 13 'Digital Accessibility for Users with Disabilities', Book 'DESIGNING FOR USABILITY, INCLUSION AND SUSTAINABILITY IN HUMAN-COMPUTER INTERACTION'). In the realm of inclusivity, the concepts of *Design for All* and *Universal Access* were established, advocating the need for proactively taking into account the needs of diverse user groups and applying appropriate methods, principles and tools, to ensure that Information and Communication Technologies' products are accessible by anyone, anywhere, and through any technological platform, supporting diverse user tasks (Stephanidis, 1995, 2021). Universal Access, therefore, mandates the digital accessibility of technological products and services by all potential target users, including persons with disabilities, older adults, persons from various cultural backgrounds, and generally anyone at risk of exclusion. Today, as technology pervades every aspect of daily life, access to technology encompasses not only access to information but also to crucial services and goods such as healthcare, education, employment, and justice. Therefore, providing every individual, including those with disabilities, with equal access to technological products and services is a fundamental human right that is globally recognized and protected through constitutional provisions and international agreements such as the United Nations Convention on the Rights of Persons with Disabilities (UN General Assembly, 2006; Stephanidis, 2023). Considering that digital accessibility is an ongoing pursuit that is challenging even in well-established contexts, such as websites, it becomes evident that additional efforts are required to methodologically approach and streamline it in novel contexts, such as intelligent environments (Margetis et al., 2012; Ntoa et al., 2022).

Also motivated by a strong human focus, sustainable HCI was established as a research field encompassing work on promoting a sustainable way of living and environmental awareness (Hansson et al., 2021). In this regard, HCI is well-aligned with the United Nations Sustainable Design Goals (SDGs) (United Nations, 2015), promoting not only ecological awareness but also inclusiveness and reduced inequalities, health and well-being, quality education, industry and innovation, sustainable cities and communities, etc. Considering that according to monitoring reports for the progress on SDGs, humanity is far behind achieving those goals (United Nations, 2023), the HCI discipline is called upon to further promote multidisciplinary work in this area.

Technological advancements have brought a notable shift in the interactivity of people with digital technologies, with reports highlighting that with a global population of 7.9 billion as of 2022, 5.3 million people are using the internet,[1] the number of smartphone subscriptions significantly exceeds the number of people on the planet,[2] whereas 90% of the world's data was generated between 2019 and 2023.[3] It is therefore indisputable that data privacy constitutes one of the most important concerns in the current technological landscape. Several streams within HCI are relevant and decisive in preserving the user's data privacy putting them in control of their data, in conformance with legal directives (European Union, 2016), including design considerations, understanding of how people interact with and through systems, understanding users themselves, and shaping next generation architectures (Ackerman & Mainwaring, 2005; Easwara Moorthy & Vu, 2015; Knijnenburg, 2017; Zheng et al., 2018). Researchers explore strategies to strike a harmonious balance between personalization and privacy, often adhering to privacy-by-design principles (See also Chapter 13 'Privacy and Trust in HCI', Book 'FOUNDATIONS AND FUNDAMENTALS IN HUMAN-COMPUTER INTERACTION'). In this regard, ethical design becomes a key concern in the development of contemporary and future technologies, acting as a means to 'materialize morality' (Verbeek, 2006) through value-sensitive design (Friedman & Hendry, 2019). Ethics in HCI (see also Chapter 14 'Ethics in HCI', Book 'FOUNDATIONS AND FUNDAMENTALS IN HUMAN-COMPUTER INTERACTION') extend to other dimensions beyond design, encompassing aspects of research with human participants, as well as responsible research and innovation (Bruckman, 2014; Frauenberger et al., 2017).

It is important to highlight that the above-mentioned qualities of a technological system foster user trust, and ultimately lead to their acceptance (see also Chapter 12 'User Acceptance of Interactive Technologies', Book 'FOUNDATIONS AND FUNDAMENTALS IN HUMAN-COMPUTER INTERACTION'). Trust is a psychological concept that can be defined as the willingness of an individual to depend on another party, because of the characteristics of the other party, and is of uttermost importance in situations in which uncertainty exists or undesirable outcomes may arise (Mcknight et al., 2011). Overall, trust is a complex and dynamic phenomenon that may depend on various factors, such as one's personality and experience with technology, the reputation of the service provider, as well as perceived privacy and security, information quality, and benevolence (Beldad et al., 2010. In this regard, qualities of a system such as the ones discussed earlier in this section, exhibiting consistently, can foster user trust. Ultimately, trust constitutes one of the key qualities that affect the perceived ease of use and usefulness of a product, which in turn determines the user acceptance of technology, as evidenced by numerous relevant models (Ghazizadeh et al., 2012; Kamal et al., 2020; Wu et al., 2011; Vorm & Combs, 2022).

1.3.2 STANDARDS AND GUIDELINES

Throughout the entire history of the HCI field, the need for designing usable and easy-to-use systems has been eminent. Aiming to provide guidance on how to design and evaluate usable interactive systems, numerous HCI guidelines have been proposed in the literature and several standards have been established, acting as a valuable resource for researchers, designers, and developers (See also Chapter 4 'HCI Standards and Guidelines', Book 'DESIGNING FOR USABILITY, INCLUSION AND SUSTAINABILITY IN HUMAN-COMPUTER INTERACTION').

Standards are usually associated with precise technical specifications, and while several HCI standards have adopted the approach of providing detailed specifications of the nature of the user interface, standard user interfaces face the peril of becoming obsolete with the rapid technological evolution (Bevan, 1995). As a result, the majority of international HCI standards have focused on describing principles that need to be adhered to in order to produce interfaces that meet user needs and are appropriate for the designated tasks. Standards are developed by technical committees involving groups of experts from all over the world and come in many forms, such as international, national, military or governmental, commercial and industrial, independent, or project-based (Buie, 1999). A standard typically includes information on the underlying rationale and principles, examples, exceptions if applicable, references, and in some cases compliance instructions (ibid.). The list of standards is long, and the standards to use should be carefully selected based on the objectives of the system being developed. Overall, using HCI standards in product development encompasses several steps, with a prominent one being the compliance assessment phase. Several compliance and conformance evaluation techniques have been reported in the literature, including content analysis, documented evidence, observation, analytical evaluation, and empirical evaluation (Reed et al., 1999). In most cases, the standards themselves are not legally binding; however, specific laws and regulations may enforce adherence to specific standards, such as the Directive (EU) 2016/2102 on the accessibility of the websites and mobile applications of public sector bodies, mandating that all public sector bodies in the EU comply with the harmonized standard EN 301 549, on the 'Accessibility requirements suitable for public procurement of Information and Communication Technologies (ICT) products and services in Europe'. HCI standards have often been criticized for being vague (Çakir, 1997), not useful, and limiting creativity (Buie, 1999); however, it is important to acknowledge that they can guarantee the high quality of a product and give evidence of its qualities such as usability, efficiency, effectiveness, or accessibility.

In contrast to the authoritative nature of standards, guidelines constitute advisory rules aiming to assist designers in following design principles, as these have been formulated by HCI theories. Guidelines may pertain to general HCI principles or be focused on specific technological domains, target user groups or even corporations for achieving consistency within their products. Well-established design guidelines that are general and can be applied to the design of any interactive product or service include Shneiderman's Eight Golden Rules (Shneiderman, 2016), Norman's Seven Principles (Norman, 2013), Nielsen's Ten Usability Heuristics (Nielsen, 1994a), as well as the design rules by Dix, Finlay, Abowd and Beale (2003). Considering the interaction challenges and intricacies entailed by different application domains, concrete guidelines have been developed with the objective of providing additional recommendations and support to designers. Indicatively, guidelines for web design (e.g., Instone, 1997; Bevan & Spinhof, 2007), e-commerce (e.g., Fang & Salvendy, 2003; Maguire, 2023), games (e.g., Tekinbas & Zimmerman, 2003; Mueller & Isbister, 2014), or educational applications (e.g., Hirsh-Pasek et al., 2015; Linehan et al., 2011) are reported in the literature. Furthermore, following the technological evolution and the proliferation of various technologies, targeted design guidelines have been proposed to address the novel interaction aspects entailed, with notable examples of guidelines for mobile design (e.g., Häkkilä & Mäntyjärvi, 2006), extended reality design (e.g., Vi et al., 2019), and AI design (e.g., Amershi et al., 2019; Shneiderman, 2020a). The appearance of novel interaction techniques has also resulted in the proposal of pertinent guidelines, such as for example guidelines for touch-based interaction (e.g., Gorlewicz et al., 2020), gesture-based interaction (McAweeney et al., 2018), haptics (Van Erp, 2002), and speech-based interaction (Murad et al., 2018). Other sets of standards reflect the field's movement from the creation of systems to fulfill organizational requirements that were primarily utilized by workers to systems that support a wider variety of activities for a broader audience, encompassing older adults (Kurniawan & Zaphiris, 2005; Loureiro & Rodrigues, 2014), children (McKnight & Cassidy, 2012), cross-cultural audiences (Alexander et al., 2017), and persons with disabilities (Ballantyne et al., 2018; Kirkpatrick et al., 2018). Finally, guidelines developed by specific industries are available and should be taken into consideration to pursue uniformity in the appearance and experience of products tailored for a particular company or products' suite. It is evident that the HCI research community has produced numerous guidelines, driven by the need to offer tangible instructions in response to fundamental HCI theories, as well as to address the interaction challenges that arise as new technologies come into play. The challenge that is currently at stake is how to handle this large body of information so as to effectively select the resources that are relevant and useful for the design endeavor one is embarking on.

1.3.3 HCI Methods

HCI researchers and practitioners have a plethora of methods and techniques at their disposal for designing, implementing, and evaluating interactive applications and systems (See also Chapter 2 'HCI Design Processes', Book 'DESIGNING FOR USABILITY, INCLUSION AND SUSTAINABILITY IN HUMAN-COMPUTER INTERACTION'). As these methods are not mutually exclusive and each has its strengths and weaknesses, they can often be used in combination to enhance the comprehension of the user needs that the system should aim to fulfill and ultimately the usability and UX of the designed system. Some of the methods can be used in multiple stages in the UCD cycle iterations, but with a different purpose each time. For example, interviews and survey questionnaires can be conducted in the user understanding and user requirements definition stage to gather information regarding user needs, attitudes, and preferences toward a technology or a system, but they can also be conducted in the evaluation stage to collect information regarding the usability or UX of an existing system and application or a new system or application prototype. In all cases, participatory design is actively pursued, including experts and representative end users in the design process (See also Chapter 10 'Collaborative and Participatory Design', Book 'USER EXPERIENCE METHODS AND TOOLS IN HUMAN-COMPUTER INTERACTION'). Contemporary approaches employ crowdsourcing in HCI research as a means of participatory design (See also Chapter 11 'Crowdsourcing in UX Research', Book 'USER EXPERIENCE METHODS AND TOOLS IN HUMAN-COMPUTER INTERACTION').

Overall, we can roughly group HCI methods into four main categories: inquiry methods, ideation and design methods, inspection methods, and modeling methods. Examples of methods for each of the categories are provided next.

1.3.3.1 Inquiry Methods

In the early stages of a design cycle, i.e., the user requirements and problem definition stages, the focus is on collecting as much information as possible regarding the target users of the application or system, the environment in which it is going to be used, and the context of its use (See also Chapter 1 'Requirements Engineering and User Needs Analysis', Book 'USER EXPERIENCE METHODS AND TOOLS IN HUMAN-COMPUTER INTERACTION'). Only by deeply understanding and empathizing with the users' true needs, motivations, frustrations, preferences, and attitudes, will the design team be able to produce solutions that are both functional and usable. One of the best and simplest ways to find out about the user's needs and preferences is by asking them. In HCI, there are many inquiry methods and techniques available for the team to choose from (See also Chapter 2 'Ethnography, User Observation and Interviews' and Chapter 6 'Designing and Analyzing Questionnaires and Surveys', Book 'USER EXPERIENCE METHODS AND TOOLS IN HUMAN-COMPUTER INTERACTION').

The two most common and effective inquiry research methods are interviews and questionnaires. They both can be used in the early design stages of the design cycle to collect data regarding the users and stakeholders' needs and preferences toward a specific technology or an existing application or system, but they can also be used in later stages of the design cycle, i.e., the evaluation stage, as a post-study method to collect data regarding the usability and UX of the inspected application or system. Interviews produce qualitative data and can have a structured format, i.e., the interviewer asks a pre-defined set of questions and does not deviate from these, or a semi-structured format, i.e., the interviewer uses a set of pre-defined questions as a guide but may ask additional questions as needed and depending on the flow of the discussion. Questionnaires, on the other hand, can produce both qualitative and quantitative data depending on the format of the questions, open-ended versus closed-ended questions, respectively. Specialized questionnaires are also very often used in the evaluation stages to measure usability and user-experience aspects of the interactive system, as described further down. Both methods have their benefits and challenges and are often used in conjunction, as they are complementary to each other.

Another important inquiry method that is also used in the explorative early design stages is contextual inquiry, an ethnographic type of approach developed by Karen Holtzblatt and Hugh Beyer (1997), in which the HCI practitioner observes the user while using the system or application in their natural environment as they would normally do, and makes targeted inquiries to draw an in-depth understanding of how processes are done and why (Salazar, 2020). In this method, the researcher or practitioner in a sense assumes the role of the apprentice and the user the role of the master craftsman. This method is very powerful and may produce more accurate data than an interview because the researcher becomes the witness of how the user uses an application or system to accomplish their tasks and does not rely on the accuracy of the user's recollection.[4]

Focus groups are another qualitative research method that has its roots in marketing research, where it is used to gather information regarding people's attitudes and opinions on products, services, etc. (Rosenbaum et al., 2002). In this method, the moderator of the session guides a group discussion of usually up to ten individuals who have been selected to share and discuss their perspectives on a given subject. In HCI, focus groups can be used for a variety of purposes at all stages of the design and development application or system cycle, such as for exploring users' attitudes and motivations toward the use of a technology or system, generating design ideas, usability purposes, i.e., validating ease of use and design aspects (Rosenbaum et al., 2002). Special attention needs to be given when using this technique to avoid common pitfalls such as the groupthink effect (participant's desire for group consensus prevents them from presenting alternative thoughts, critiques, or expressing an unpopular opinion), uneven communication dynamics (domineering versus more timid participants), and difficulties in interpreting the results (Rosenbaum et al., 2002). Particular attention needs to be paid as well to cater to the needs of all participants under a Universal Access perspective (Antona et al., 2009).

1.3.3.2 Design and Ideation Techniques

In the middle stages of the design cycle, the design team shifts its focus to the production of ideas and designs for solutions that will support the defined user needs and preferences. There is a plethora of available HCI methods and techniques to facilitate the design team in this process, the most common ones are discussed next (See also Chapter 3 'Ideation, Focus Groups and Brainstorming' and Chapter 4 'Personas, Scenarios, Journey Maps and Storyboarding', Book 'USER EXPERIENCE METHODS AND TOOLS IN HUMAN-COMPUTER INTERACTION').

Brainstorming, which was originated by Alex Osborn, an advertising executive, as a method for creative problem-solving (Osborn, 1953), is a widely used ideation technique. It involves having the participants come up with as many ideas for designs or solutions about a well-defined problem, topic, product, application, or service. The focus of the technique is on the number of ideas and not on the quality. The session is moderated by a person who ensures that the participants follow the rules and the provided time limits and guides the process along. In a traditional brainstorming session, each participant writes their ideas on separate sticky notes, and at the end, each person takes turns to state their idea. The ideas are not judged, criticized, or discussed during the session in order to promote creativity and encourage participants to freely express their ideas with no fear of rejection. The ideas are then aggregated on a board where they can then be further processed and organized into meaningful clusters based on the drawn associations. Brainstorming can be conducted internally with members of the design and development team, but can also be expanded to include representative end users and other stakeholders such as in the context of participatory design.

In this stage of the design cycle, the design team may also benefit from constructing personas for the application or system they are designing. Personas are fictitious characters that serve as archetypal representations of an application or solution's target users. A persona describes a typical target user, i.e., who they are, what is their background, what are their expectations, goals, motivations, frustrations, and needs. Although they are fictitious, their information is based on real user data collected in the earlier design stages. The personification of the target user allows the design team to think and refer to their specific needs in a more concrete and relatable way throughout the design stage. Furthermore, they can be used in combination with other methods such as journey maps, user scenarios, storyboarding, etc., to further facilitate ideation and design activities.

Storyboarding is a form of prototype technique that involves visually plotting out elements of a design concept from start to finish. It is an easy and quick way to see how an idea can unfold. It can be done for fragments of the design or User Interface (UI) concept or entire solutions. It does not have to be elaborate or of high-fidelity, a simple paper and pen setup is sufficient, although there are free online tools and templates available to help generate storyboards.

Another UX technique that also allows the team to visualize the process and the series of steps a user has to make to accomplish a goal using a system, application, or product is Journey maps.[5] The main ingredients of a journey map are: (1) the actor, or in our case the user, whose journey we are describing. The description of the actor may align with a persona constructed earlier; (2) a specific user scenario that aims to fulfill a specific goal, expectation, or need of the user; (3) the journey stops, i.e., a description of each step along the way; (4) the thoughts, feelings, motivations as the user moves along the way; (5) opportunities. Journey maps are very useful because they create a vision of how the user experience is going to be shared among the entire team.

1.3.3.3 Inspection Methods

Usability inspection (or evaluation) is one of the most crucial components of HCI and plays an integral part in all the main design methodology approaches, i.e., iterative design, user-centred design, design thinking, etc. (See also Chapter 7 'Inspection Methods for Usability Evaluation', Book 'USER EXPERIENCE METHODS AND TOOLS IN HUMAN-COMPUTER INTERACTION'). Usability testing is a generic name for a set of methods that involve having an evaluator examine or inspect usability-related parameters (such as ease of use, effectiveness, efficiency, learnability, and satisfaction) of a UI of a system or application. In general, usability inspection methods in HCI can be split into two major categories, expert-based and user-based inspections. In expert-based inspections, an HCI specialist or domain expert inspects the system with the mindset of the target user and tries to find any areas that may cause problems or difficulties to the user. The most common expert-based inspection method is Heuristics analysis, which uses a set of ten design principles or heuristics as a base for the inspection (Nielsen, 1994b). Another common expert-based method is cognitive walkthroughs, in which the evaluator performs typical user tasks and tries to identify problematic areas in the design or the functionality by answering four questions: (1) Does the user know what to do? Is it the correct action? (2) Will the user notice that the correct action is available? Is the action visible? Will users recognize it? (3) Will the user associate the correct action with the effect to be achieved? The action may be visible, but will the user understand it? (4) If the correct action is performed, will the user see that progress is being made toward the solution of the task? Is there system feedback to inform the user of progress? Will they see it? Will they understand it? (Cockton et al., 2009).

In user-based evaluations, the application or system is evaluated by a selected sample of representative end users (See also Chapter 8 'Designing, Conducting, Analyzing and Reporting Usability Testing Experiments', Book 'USER EXPERIENCE METHODS AND TOOLS IN HUMAN-COMPUTER INTERACTION'). In this type of evaluation, the evaluator gives the user a pre-determined set of typical tasks to complete and records how easy it was for them to complete the task. Very often, in this type of evaluation, the Think Aloud protocol (Lewis, 1982) is applied, in which the user is asked to verbalize their inner thoughts while trying to complete each task in order for the evaluator to capture the user's thinking process and to discover what they genuinely think of the interface design.

Typical usability metrics aiming to measure usability in relation to user's performance on a given task include time-on-task, task completion success rate, number of errors, etc. Additionally, there are a few post-study or after-task standardized questionnaires that measure user-perceived usability, such as the System Usability Satisfaction (SUS) questionnaire, the After-Task Questionnaire (ASQ) (Lewis, 1995), Post-Study System Usability Questionnaire (PSSUQ) (Lewis, 1992), the Usability Metric for User Experience (UMUX) (Finstad, 2010), the User Experience Questionnaire – long and short version – (UEQ) (Hinderks et al., 2019), the Attrakdiff questionnaire, etc. The UEQ and Attrakdiff questionnaires measure the overall user experience in terms of both pragmatic and hedonic aspects, such as attractiveness, novelty, stimulation, etc.

1.3.3.4   Modeling Methods

Modeling techniques aim at creating representations of integral parts of the interaction, thus promoting more detailed and well-informed design. Task analysis models the tasks that users perform with the system being designed, thus enhancing the understanding of user workflows and the on-time identification of potential usability issues (see also Chapter 5 'Task Analysis and Modeling', Book 'USER EXPERIENCE METHODS AND TOOLS IN HUMAN-COMPUTER INTERACTION'). Focused on users, user modeling techniques aim at representing individual users as a whole, or as a selected part of their characteristics, such as demographics and roles, attitudes and opinions, goals and intentions, functional characteristics, and emotional attributes (See also Chapter 12 'Responsible User Modeling', Book 'USER EXPERIENCE METHODS AND TOOLS IN HUMAN-COMPUTER INTERACTION'). Furthermore, digital representations of the human body fall within the scope of Digital Human Modeling and are used to simulate and analyze human movements, ergonomics, and interactions in various environments and contexts (see also Chapter 13 'Digital Human Modelling' Book 'USER EXPERIENCE METHODS AND TOOLS IN HUMAN-COMPUTER INTERACTION').

1.3.4   Interaction Paradigms

1.3.4.1   Desktop Computing

In the dynamic landscape of ICTs, the way users interact with their devices has seen remarkable transformation. Desktop computing, as well as web-based applications, are no exception. These platforms embrace a variety of interaction paradigms meticulously designed to provide users with intuitive, efficient, and responsive experiences (See also Chapter 8 'Web Design', Book 'DESIGNING FOR USABILITY, INCLUSION AND SUSTAINABILITY IN HUMAN-COMPUTER INTERACTION').

Traditional input methods, such as keyboards and mice, continue to serve as the fundamental elements of both desktop computing and web interactions. These paradigms undergo continuous refinement aimed at enhancing efficiency and overall user experience. Notably, ongoing advancements include the development of ergonomic keyboard designs and customizable mouse inputs, which cater to the diverse needs and preferences of users (Lee et al., 2012).

Moreover, the scope of interaction methods has expanded beyond the confines of mobile devices, with touch and gestures making their mark in the domain of desktop computers and web applications (Farhadi-Niaki et al., 2013). This evolution has been particularly pronounced in the context of hybrid devices and touchscreen-equipped desktop monitors, introducing an additional dimension of versatility to user engagement. As a result, users can navigate and interact with content in a more intuitive and fluid manner, further blurring the lines between traditional desktop computing and the dynamic world of web-based interactions. Gesture recognition has continued to develop, enabling users to interact with desktop computers through hand and body movements. Devices like the Leap Motion[6] controller and depth-sensing cameras allow gesture-based interactions.

In creative sectors, paradigms such as stylus and pen input have become indispensable tools (Chen et al., 2019). Artists, designers, and note-takers benefit greatly from these paradigms, appreciating their precision and natural feel.

Voice recognition is another popular paradigm in both desktop and web environments. Users can perform tasks, initiate searches, and control functions through natural language, with voice assistants and recognition technologies driving this paradigm. Developments in the domain of Natural language processing (NLP) have led to the emergence of voice assistants and chatbots that enable users to communicate with devices and software through natural language, profoundly altering the landscape of information retrieval and service access on the web (Witte & Gitzinger, 2008).

With the growing prevalence of multiple monitors in desktop computing, we have witnessed the evolution of interaction paradigms tailored to accommodate extended display setups (Gallagher et al., 2021). New tools and techniques facilitate content management across screens, providing a seamless experience for users within a broader workspace.

Eye-tracking technology is primarily employed in specialized fields such as usability research, market research, and assistive technology (Jacob & Karn, 2003). It can be used to analyze how users visually interact with on-screen elements, and it has applications in gaming and certain accessibility features (See also Chapter 9 'Eye-tracking and Physiological Measurements for UX Evaluation', Book 'USER EXPERIENCE METHODS AND TOOLS IN HUMAN-COMPUTER INTERACTION'). However, it is not a standard method for general desktop interactions.

1.3.4.2 Mobile Interaction

The credits for the first touchscreen mobile phone go to IBM's Simon Personal Communicator, a bulky device with a screen that combined the functionalities of a mobile phone and a personal digital assistant. It came with pre-installed applications such as fax, email client, address book, calendar, calculator, and notepad. Its interface was operated with a stylus and less easily with a finger (Dainow, 2017). But it was not until Apple's introduction of the first iPhone in 2007, a touchscreen device with full internet connectivity, a wide LCD screen for video viewing, a camera, and a list of built-in applications, that the landscape for what is now known as the modern mobile era truly changed. Apart from the mobile device, Apple also developed a special operating system, the iOS, and was the first to provide a Software Development Kit (SDK) with custom software languages and tools to independent developers to build their own applications and created the App store for them to distribute their applications. Apple's first iPhone and its App store became the catalyst for the widespread adoption of smartphones and the subsequent mobile revolution. Google followed in 2005, by buying Android, a small start-up that was developing a generic operating system for smartphones, and in 2008 the first Android smartphones appeared. It also provided the Android SDK for software development in JAVA and introduced the Google Store for Android apps (Dainow, 2017). Since then, mobile technology has advanced at an incredibly fast pace, and nowadays people depend on this multi-tool for just about everything, other than just making and receiving calls, such as sending messages, reading emails, surfing the Web, going on social media, watching movies and videos, taking high-quality pictures and videos, navigating around the world, playing games, and many more.

The mobile device revolution has brought on new interaction paradigms, but also many challenges in the field of HCI (See also Chapter 9 'Mobile Design', Book 'DESIGNING FOR USABILITY, INCLUSION AND SUSTAINABILITY IN HUMAN-COMPUTER INTERACTION'). The first shift in the interaction paradigm was brought on by the dominance of the touchscreen over any other interaction mode. Nowadays, all mobile devices, i.e., smartphones, tablets, PDAs, and wearable devices, use touchscreen interface technology for user interaction. Touchscreen technology was used even before the first iPhone, but Apple perfected the user experience for it. Over the years, this mode of interfacing has evolved to become highly responsive and intuitive, allowing a plethora of gestures such as tapping, pinching, spreading, swiping, dragging, etc., to support multiple commands such as selecting an option, activating a button, scrolling, swapping screens, zooming in and out, panning, etc. Pretty much anything that can be done with a mouse and a keyboard can now be done with the use of the finger. However, UI design for this mode of interaction has its own challenges and requires different design guidelines than those for designing for the Web. Designers now have to consider tap targets and gestures and ensure that there is appropriate spacing and placement of the interactive elements on the UI to make it easy to navigate with fingers. Furthermore, due to the limited screen real estate and the limited mental processing power of the user, content and features must be prioritized, and the designers must decide what to hide behind menus or in separate screens and what to display up front. Another great challenge that designers face is the wide range of screen size and resolutions that their application needs to be displayed on. As browsing the Web over a smartphone is nowadays an essential activity for people, websites must be responsive and flexible to adjust to various screen resolutions without compromising the user experience. Lastly, native mobile applications have the additional complexity of portability, their ability to operate on different platforms, iOS or Android. Developers must decide whether they are going to develop for a specific platform or for all platforms and select the appropriate framework tools accordingly.

The future for mobile applications seems even more promising. High-speed mobile internet connections (5G and 6G), the integration of mobile technology with the IoT, advancements in mobile sensors such as gyroscope, accelerometer, and location-based services, the emergence of wearable devices like smartwatches and fitness trackers, the integration of AR and VR capabilities in the mobile platforms are just some of the technological trends that are going to have a great impact on the mobile devices in the near future. These technologies have already created the foundation for amazing mobile applications, but they are going to further improve the way we interact with the world around us. At the same time, today, mobile users are more experienced than ever and have high expectations from a mobile application. They expect apps to be fast-loading, easy to use, and provide a positive user experience, making usability and user experience essential aspects of the design and not mere nice-to-have features. As applications require from users more sensitive personal information, data privacy, ethics, and safety also become increasingly important, designers and developers must place enormous attention to these matters as they move forward in this field.

1.3.4.3   Ubiquitous Computing

Ubiquitous computing has as its goal to enhance computer use not only by making many computers available throughout the physical environment but also by making them effectively invisible to the user (Weiser, 1993). The idea of ubiquitous computing was first thought by Mark Weiser in 1988 at the Computer Science Lab at Xerox PARC. He envisioned a person able to interact with hundreds of computers at a time, each invisibly embedded in the environment (embedded in walls, tabletops, and everyday objects) and wirelessly communicating with each other (Weiser, 1993). Ubiquitous computing is the third wave of computing (Alcañiz & Rey, 2005). The first wave was *many people, one computer*, and the second wave was the era of *one person, one computer* while the third wave was the era of *many computers per person*.

Ubiquitous computing assumes a large number of *invisible* small computers embedded into the environment and interacting with mobile users (Erickson, 1995). Users experience the world through devices they wear (e.g., medical monitoring systems), they carry (e.g., personal communicators that integrate mobile phones), devices that are implanted in vehicles or public spaces (e.g., car and public space information systems), and devices integrated into the architectural environment (e.g., interactive walls and furniture).

The *Disappearing Computer* initiative had as an overall goal to explore how everyday life can be supported and enhanced through the use of collections of interacting artifacts. Together, these artifacts form new people-friendly environments in which the *computer-as-we-know-it* has no role (Streitz & Nixon, 2005). In the context of ubiquitous computing two types of *disappearance* have been defined, namely physical and mental disappearance (Ermes et al., 2008):

- *Physical disappearance*: Miniaturization of computer parts that allows convenient and easy integration into other artifacts, mostly into *close-to-the-body* objects so that the result can fit in a hand. These parts could be integrated into clothing or even implanted in the body. Features usually associated with a computer are not visible anymore and the interaction happens via the compound artifact in which the computer parts disappear.
- *Mental disappearance*: Computers become *invisible* to the *mental* eye of the users. This can happen by embedding computers or parts of computers in the architectural environment (walls, doors, etc.) or in furniture (tables, chairs, etc.). Traditional everyday objects can yet interact, communicate, and cooperate with humans due to the multifunctional character they get. An interactive table, for instance, still has the standard table form, though it offers additional functions to the users.

Ubiquitous computing is reflected in many very similar concepts, such as *pervasive computing, ambient intelligence,* and the *internet of things* (Friedewald & Raabe, 2011). In practice, the difference between these terms is of a rather academic nature: common to all is the goal of assisting people as well as a continuous optimization and promotion of economic and social processes by numerous microprocessors and sensors integrated into the environment. The following features characterize ubiquitous computing (Gabriel et al., 2006)

- Decentralization or modularity of the systems and their comprehensive networking.
- Embedding of the computer hardware and software in other equipment and objects of daily use, mobile support for the user through information services anywhere and anytime.
- Context awareness and adaptation of the system to current information requirements.
- Automatic recognition and autonomous processing of repetitive tasks without user intervention.

These characteristics, coupled with systems' invisibility, present significant challenges in creating new human-computer interaction methods. Over time, ubiquitous computing has the potential to influence every aspect of our lives (Aarts & Encarnação, 2006). In home spaces, ubiquitous computing systems aim to enhance comfort and improve energy efficiency (Amazon Echo, Google Home, and smart thermostats have become commonplace in many households). From refrigerators to light bulbs, more and more objects are now connected to the internet. This connectivity allows for more intelligent decision-making and automation in various scenarios. Wearable devices, such as smartwatches and AR glasses, have become mainstream since they offer a blend of health monitoring, communication, and entertainment features. Designing intuitive human-computer interaction techniques remains a big challenge given the *invisible* nature of many of these ubiquitous computing systems.

1.3.4.4   IoT, Smart Cities and Urban Interfaces

Within the context of Smart Cities and Urban Interfaces in the era of the IoT, interactions revolve around paradigms aimed at improving urban living, optimizing resource allocation, enhancing sustainability, and fostering active community engagement (Ahmad et al., 2022; Khatoun & Zeadally, 2016). These interactions are meticulously designed to harness the potential of IoT, where a network of interconnected devices, equipped with sensors and seamless connectivity, collects data on diverse aspects of city life (see also Chapter 6 'Interacting with the Internet of Things', Book 'HUMAN-COMPUTER INTERACTION IN INTELLIGENT ENVIRONMENTS'). This valuable dataset, often referred to as "smart data," serves as the bedrock for well-informed decision-making within the city's administrative framework. Urban planners and local authorities place their trust in this data to fine-tune resource allocation, streamline services, and ultimately uplift the quality of life for city residents.

In smart cities, residents are not passive recipients but active stakeholders. Urban interfaces and digital platforms should be crafted with user experience in mind, ensuring accessibility, intuitiveness, and responsiveness to the diverse needs of the community (see also Chapter 13 'Urban Interfaces', Book 'HUMAN-COMPUTER INTERACTION IN INTELLIGENT ENVIRONMENTS'). Through interactive interfaces, residents are encouraged to provide real-time feedback, report issues, and engage in the urban planning process (Ismagilova et al., 2019; Javed et al., 2022; Leonidis et al., 2022; Syed et al., 2021). This real-time engagement fosters a sense of co-ownership in the city's development.

In a smart city, various types of interactions between humans and technologically augmented surroundings play a crucial role in creating efficient, sustainable, and user-friendly urban environments. These interactions include: (1) digital services and mobile apps (Hou et al., 2020) through which residents and visitors can access a wide range of urban services (e.g., public transport, issue reporting, accessing information about local events); (2) smart mobility Interfaces (Paiva et al., 2021) enabling people to interact with various transportation options (e.g., public transit systems, bike-sharing programs, electric vehicle charging stations); (3) environmental sensing and feedback interfaces that provide residents with information about environmental conditions, such as air quality and noise levels, allowing them to make informed decisions about their daily activities and provide their own feedback on environmental concerns (Alías & Alsina-Pagès, 2019; Malche et al., 2019); (4) smart home and building controls (Alam et al., 2020) through which residents can control various aspects of their homes, such as lighting, heating, and security, enhancing energy efficiency and user comfort; (5) public kiosks that provide access to information about city services, events, and directions (Ilhami et al., 2022); (6) interactive signage offering real-time information on public transportation schedules, local news, and emergency alerts (Toh et al., 2020); (7) AR and VR technologies that provide immersive experiences for educational and informational purposes (e.g., virtual city tours, historical recreations, and real-time data overlays); (8) voice assistants and smart speakers integrated into urban environments, allowing residents to access information and control smart devices using voice commands; (9) interactive public displays in public spaces enabling users to interact with information or services mainly through gestures and touch, and more (Parker et al., 2018).

These several types of (HCI) interactions contribute to making smart cities and urban interfaces more user-centric, efficient, and responsive to the needs of their residents and visitors. Thus, they enhance convenience, access to services, and overall quality of life within the urban environment.

1.3.4.5 Automotive UIs

In the world of automotive UIs, interaction paradigms have evolved significantly to enhance both usability and safety for drivers and passengers (See also Chapter 9 'Automotive User Interfaces', Book 'HUMAN-COMPUTER INTERACTION IN INTELLIGENT ENVIRONMENTS'). These paradigms are crucial in providing a seamless and intuitive in-car experience.

Traditional physical controls, such as buttons, knobs, and switches, still play a vital role in the automotive UI. They are essential for immediate, tactile access to core functions such as climate control and audio adjustments. These physical controls ensure drivers can make critical adjustments without diverting their attention from the road (Meixner & Müller, 2017).

Nowadays, touchscreens have become central in many modern vehicles, offering a versatile interface for a wide range of functions, from navigation and entertainment to climate control. However, their use while driving demands careful design to reduce distractions and cognitive load, prioritizing safety (Rümelin & Butz, 2013). Physical dials and touchpads allow users to interact with on-screen elements without directly touching the screen. These controls aim to reduce smudging and improve precision.

Voice commands have gained prominence as voice recognition systems have become more sophisticated (Braun et al., 2019). They allow drivers to perform tasks hands-free, reducing the need for manual or visual interactions with the UI. Voice assistants like Apple CarPlay[7] and Android Auto[8] can understand natural language, making interactions more intuitive.

Gesture control, found in high-end vehicles, enables drivers to interact with the infotainment system through hand gestures (Zengeler et al., 2018). For instance, making a circular motion with your hand can adjust the volume. Furthermore, Head-Up Displays (HUD) project critical information directly onto the windshield, reducing the need for drivers to look away from the road. They can display navigation directions, speed, and other essential data, enhancing situational awareness.

Haptic feedback, such as vibrating seats or steering wheels, provides tactile cues for notifications, like lane departure warnings or incoming calls. These cues offer feedback without demanding visual attention, further enhancing safety (Telpaz et al., 2015).

Many modern automotive UIs combine multiple interaction methods, offering flexibility for users. For example, drivers might use voice commands for navigation while relying on physical controls for audio adjustments. Furthermore, recent advancements have introduced biometric authentication methods, including facial recognition, which not only enhance security but also enable personalized settings and preferences (Villa et al., 2018). These innovative features not only contribute to a more secure driving experience but also make interactions with the UI more tailored and user-friendly.

The development of automotive UIs balances safety, user experience, and technological advancement. Designers continually explore innovative interaction paradigms to make the in-car experience safer and more user-friendly. As self-driving technology advances, the role of interaction paradigms in automotive UIs will continue to evolve to meet the needs of both drivers and passengers.

1.3.4.6 Ambient Intelligence and Ambient Assisted Living

In the realm of smart and Intelligent environments (IEs), interaction techniques play a pivotal role in ensuring seamless and user-friendly interactions between individuals and the integrated technology that surrounds them (Stephanidis et al., 2021). These environments, often driven by the IoT and a network of sensors, are designed to be responsive and context-aware, making everyday tasks more convenient and efficient.

One of the most prevalent methods of interaction in IEs is voice and NLP. Devices like smart speakers and virtual assistants use NLP to comprehend and respond to natural language requests. This empowers users to control various aspects of their environment with simple voice commands, such as adjusting lighting, thermostats, and other connected devices (Isyanto et al., 2020; Raj et al., 2019; Stefanidi et al., 2019). Virtual assistants and chatbots are making their presence felt in recent years as well. These AI-powered entities provide information, answer questions, and execute tasks, making the interaction experience more conversational and dynamic (Baby et al., 2017).

Gesture recognition offers an intriguing technique that empowers users to engage with devices or initiate actions through hand or body movements (Roberge et al., 2022). For instance, certain mid-air gestures, including tilting the palm, pinching fingers, and swiping the hand, enable users to swiftly and naturally execute specific actions. These gestures can include adjusting volume, navigating through lists, zooming in or out, or toggling lights on and off, among other possibilities. The recognition of user posture and body movements is also invaluable in the context of Ambient Assisted Living (AAL) environments (Sanchez-Comas et al., 2020). In AAL, fall detection systems are critical for ensuring the safety of elderly individuals (See also Chapter 12 'Ambient Assisted Living Solutions', Book 'HUMAN-COMPUTER INTERACTION IN INTELLIGENT ENVIRONMENTS'). These systems have the capability to automatically detect falls and trigger emergency response mechanisms, such as alerting caregivers or medical services, thereby enhancing the well-being and security of those in need.

Certain IEs are specifically engineered to provide automated and context-aware interactions. They harness an array of sensors to discern user presence, preferences, and requirements. These sensors enable automatic adaptations in lighting, temperature, and other environmental variables, all contingent on real-time user behavior and context. Additionally, advanced computer vision techniques permit implicit interaction with the IE by enabling object detection, adding another layer of user-friendliness and responsiveness to the environment (Leonidis et al., 2019). Interaction with physical objects is another input of modality often encountered in such environments (See also Chapter 11 'Tangible and Embodied Interaction', Book 'INTERACTION TECHNIQUES AND

TECHNOLOGIES IN HUMAN-COMPUTER INTERACTION'). Such interactions capitalize on the simplicity and intuitiveness of activities in the physical world, offering seamless integration in the intelligent environment and the development of new interaction paradigms (Gupta & Tanenbaum, 2019; Margetis et al., 2019a; Zuckerman et al., 2016).

Touchscreens and displays integrated into walls, mirrors, or appliances within smart environments offer intuitive and direct control interfaces (Leonidis et al., 2021a; Sun et al., 2018; Zhang et al., 2018). These interfaces enable users to manage various functions, access information, and use apps within the environment with ease.

Wearable devices, particularly smartwatches and similar wearables, function as convenient control hubs within smart environments. Users can effortlessly oversee and fine-tune settings on their interconnected devices directly from their wrists. This approach streamlines interaction, ensuring users remain well-informed and in control. Yet, the utility of these devices extends beyond merely controlling the environment. They play a pivotal role in improving the quality of life and advancing AAL (Adami et al., 2021; Liu et al., 2016; Nath & Thapliyal, 2021). For instance, applications designed for wearables aid users in managing stress levels and monitoring their sleep patterns, enhancing their overall well-being (Leonidis et al., 2021b). AAL environments often utilize wearables and other sensors for health monitoring, such as heart rate, blood pressure, and motion detection. Users can interact with these systems through wearable devices or dedicated AAL interfaces to keep track of their vital signs and health status.

Mobile apps are a familiar and accessible means of control within IEs (de Oliveira et al., 2016). Users can effortlessly adjust settings, monitor devices, and receive real-time notifications on their smartphones or tablets. This ensures that they can stay connected and in control even when they're away from the physical space. Remote control and automation are key features that allow users to manage smart environments even when they are not physically present. This can be done through smartphones, voice commands, or other remote-control mechanisms.

AR and VR are also making strides in enhancing interaction within IEs (Mahroo et al., 2019). AR technologies can overlay digital information in the real world, while VR can provide immersive experiences. For instance, AR glasses provide users with real-time information about objects and spaces in their immediate surroundings. Additionally, they offer the capability to remotely control devices, allowing users to manage them without the need for physical proximity or direct physical touch.

Intelligent Environments often integrate biometric sensors, such as fingerprint recognition and facial recognition, to enhance security and personalize the user experience (Ghazali & Zakaria, 2018). These biometric methods not only secure access control but also tailor the environment to individual preferences and settings.

Moreover, the deployment of machine learning and predictive analytics is elevating the quality of interaction within these environments. These systems analyze user behavior and preferences over time, enabling predictive actions (Chaurasia & Jain, 2019). For instance, the environment can automatically adjust settings based on historical user behavior, resulting in a more intuitive and personalized experience.

Finally, it is important to note that in the majority of IEs employ multimodal interaction. For example, a user might use voice commands in conjunction with gesture recognition to create a more fluid and dynamic interaction experience. These interaction techniques within Intelligent Environments are in a constant state of evolution, driven by advancements in technology and the pursuit of making everyday interactions more efficient and user-friendly. As IEs continue to develop, users can expect increasingly seamless and intuitive ways to interact with the expanding array of interconnected devices and systems that define intelligent spaces.

1.3.4.7   Extended Reality

The notion of eXtended Reality (XR) was created as a reference umbrella term for all the immersive technology facets included in the spectrum of the Real–Virtual world continuum (Mann et al., 2018; Margetis et al., 2019b). The hype of XR technologies emerged with the improvement of the hardware needed to support such technologies, which was a result of getting big companies such as Apple, Microsoft, Google, and Meta into the game, aspiring to create new market trends. Although the anticipated breakthrough of using XR technologies as a mainstream means of interaction between reality and virtuality has not materialized yet, it is a common aspiration that such a future is very near.

The idea of augmenting the real world with digital information and enabling a blended interaction between physical and virtual objects and environments is rooted in the late 1990s when Ishii and Ulmer presented their idea of "Tangible Bits" (Ishii & Ullmer, 1997). Their concept was to create technology-augmented environments enabling the blending of digital information with physical objects, advancing the natural interaction of humans with their environment, beyond the conventional interaction paradigm of point and click on a desktop computer (Kim et al., 2017). Since then, numerous research studies have been conducted in this direction, evolving to AR that aims to augment users' environment with digital information, entailing several ways of direct or implicit interaction paradigms, as well as novel means of displaying digital information almost upon any physical object or surface (See also Chapter 13 'Interaction Design for Augmented, Virtual and Extended Reality Environments', Book 'INTERACTION TECHNIQUES AND TECHNOLOGIES IN HUMAN-COMPUTER INTERACTION').

On parallel routes with AR, VR was embraced by a vast research community. The idea of escaping to other fictitious or virtual worlds wearing a piece of technological equipment intrigued people's imagination and was expressed eloquently through literacy and art for almost a century. For example, the first Head-Up Display was described in Weinbaum's short story "Pygmalion's Spectacles" in 1935.[1] However, technology became mature only in the 2010s with the appearance of the ground-breaking headset Oculus Rift, which sparked the interest of key market technological players as the new big thing for their customers. VR is at the center of research interest in many different domains, including education, health and well-being, cultural heritage, entertainment, and many more, because of the immersive affordances it provides, such as embodiment, navigability, sense-ability, interactivity, and create-ability (Dincelli & Yayla, 2022), liberating users from physical limitations of the real world.

As VR and AR evolved, the idea of blurring the borders between virtuality and reality toward unifying them into one common world has emerged and was illustrated through the concept of Mixed Reality (MR). However, current MR research efforts tend to improve the user's sense of reality rather than intertwine with VR. In this regard, MR applications can be considered in line with the AR end of the virtuality continuum. On the other hand, the abundance of technologies and the advancements in the fields of computer graphics, human-computer interaction, machine learning, computer vision, etc., makes today's reality the possibility of blending of the real and virtual worlds into one. A conceptual framework enabling such potential is discussed by Margetis et al. (2021), introducing the concept of "true mediated reality", which harnesses the technological elements needed to create realistic interactive virtual artifacts that can be employed in the real world as they were parts of it. Furthermore, the authors discuss how these virtual elements can directly affect user interaction and deliver a realistic User Experience.

In a nutshell, eXtended Reality (XR) technologies are becoming increasingly prevalent (Ratcliffe et al., 2021; Xu et al., 2023) and their role in HCI is evolving and becoming vital in many applications, like driverless vehicles and wearable health devices, and are integrated with AI, 5G networks and the IoT, progressing toward the ultimate target of a unique "virtual/real" universe.

1.3.4.8 Human–Robot Interaction

While human–robot interaction is not a new topic in HCI, this area has witnessed significant progress in recent years, mainly due to advances in AI which have widened the possibilities of applying and deploying robots in many everyday life contexts beyond industry. Typical examples are robots deployed in hospitals, hotels, restaurants, classrooms, and homes. This has driven research targeted at understanding the user experience and acceptance of robots in various application contexts

(See also Chapter 10 'Human-Robot Interaction', Book 'HUMAN-COMPUTER INTERACTION IN INTELLIGENT ENVIRONMENTS').

Relevant robot's characteristics that inform the user experience are the robot's degree of autonomy and decision-making policies (e.g., Furlough et al., 2021), the robot's anthropomorphism (e.g., Onnasch & Roesler, 2021; Fox & Gambino, 2021), as well as the robot's personality (e.g., Robert et al., 2020) and social skills, such as, for example, approaching distance (e.g., Joosse et al., 2021), ability of understanding and expressing emotions (e.g., Stock-Homburg, 2022), and communication style (Babel et al., 2021). These characteristics contribute to shaping trust, which is considered by many authors as the main factor determining the user acceptance of robots in both work and everyday life tasks (See also Chapter 10 'Human-Robot Interaction', Book 'HUMAN-COMPUTER INTERACTION IN INTELLIGENT ENVIRONMENTS'). On the other hand, a significant inhibiting factor is fear of negative effects of robots, including job loss, dehumanization, privacy infringement, and malfunctions (Makarius et al., 2020; Ruijten et al., 2019). Robots' acceptance may also be dependent on cultural issues (Bröhl et al., 2019).

A specific application domain that has been widely investigated recently is the use of assistive robots for the aging at home or in care environments. Research in this direction has addressed the extent to which robots may be effective in helping elderly people in various tasks related to medication and health care (e.g., Christoforou et al., 2020), exercising (e.g. Martinez-Martin & Cazorla, 2019), cognitive stimulation (e.g., Coşar et al., 2020), risk detection and prevention (e.g., Olde Keizer et al., 2019), personal mobility (e.g., Moustris et al., 2021), everyday living tasks such as cooking (e.g., Perotti & Strutz, 2022), entertainment (e.g., Gupta et al., 2021) and socialization (e.g., D'Onofrio et al., 2019), thus contributing to prolonging independent living and relieving caregivers' burden (Fiorini et al., 2021). Various studies have pointed out that elderly people are generally favorable to the presence of robots in their lives, recognizing their usefulness, finding them pleasant to use and sometimes even fun, but also struggling with their technical unreliability (e.g., Gasteiger et al., 2022). However, the uncanny valley effect can also arise with humanoid robots (Esposito et al., 2020).

Other research issues of recent interest in HRI include the human-centred design of robots' (Prati et al., 2021), interaction adaptation and personalization (Funk et al., 2020), as well as multimodal UIs for robots (Su et al., 2023).

1.3.4.9 Adaptive and Intelligent UIs

UI adaptation has been defined as the modification of a software system's UI in order to satisfy specific requirements, such as the needs and preferences of a particular user or a group of users (Abrahão et al., 2021). Adaptation may fall into various categories depending on how it is performed, by whom and according to which parameters (See also Chapter 2 'User Interface Adaptation and Design for All', Book 'HUMAN-COMPUTER INTERACTION IN INTELLIGENT ENVIRONMENTS'). For example, adaptability refers to the end-user's ability to adapt the UI, whereas adaptivity or self-adaptation refers to the system's ability to autonomously perform UI adaptation (Bouzit et al., 2017). Personalization, on the other hand, is considered a particular form of adaptivity, usually affecting the UI contents, based on user data such as personal traits and interactive behavior (Fan & Poole, 2006). Intelligent User Interfaces (IUIs) aim to improve the efficiency, effectiveness, and naturalness of HCI by representing, reasoning, and acting on models of the user, domain, task, discourse, and media (Maybury, 1998). The term intelligent UI is often used along with various adaptation-related terms. Many intelligent interfaces can be described as adaptive interfaces, though not all adaptive interfaces are intelligent.

Initially, UI adaptation was mainly targeted to benefit end users by optimizing various aspects of the end-user's experience (e.g., increase effectiveness and efficiency, or improve the user's subjective satisfaction (Abrahão et al., 2021). Subsequently, the concept was revisited when mobile devices appeared in the market. In such context, UI adaptation was meant to cater to the development of UIs capable of displaying themselves on devices with various screen sizes and resolutions, to seamlessly migrate from device to device supporting the continuity of the user experience, and to improve mobile interaction (Eisenstein et al., 2000). Additionally, over the years, UI adaptation has been investigated as a technical approach to achieve accessibility and Design for All. However, despite research progress, the adoption of adaptation-based and IUIs has been limited, mainly due to difficulties in designing effective adaptations and the lack of efficient development methods and tools (Höök, 2000; Hartmann, 2009).

However, today adaptive and intelligent UIs are acquiring renewed interest thanks to the use of AI and ML algorithms and tools, which bring the promise of contributing to reduce the above bottlenecks. IUIs mostly utilize artificial neural networks, instance-based algorithms, classification, NLP, and probabilistic algorithms (Brdnik et al., 2022; Ferraro & Giacalone, 2022). Emerging AI-based solutions hold the promise of reducing the effort needed to create the expansive realm of possible adaptations, through carefully designed decision-making mechanisms producing optimal solutions to given problems in real time. Machine Learning techniques can be used to monitor the adaptation process over time, learn what the good adaptations are, or which are preferred by the end-user, and recommend them (Bouzit et al., 2017; Todi et al., 2021). Another approach combines ontology modeling and reasoning with combinatorial optimization to decide what information to present, when to present it, where to visualize it in the display – and how, taking into consideration contextual factors as well as placement constraints (Stefanidi et al., 2022). Recent trends also emphasize the integration of traditional "white box" models, and in particular user models, with ML "black box" models, by feeding classical models with users' data obtained through ML techniques (Mezhoudi & Vanderdonckt, 2021). A representative example is a model-based reinforcement learning approach which plans a sequence of adaptation steps (instead of a one-shot adaptation) and exploits a model to assess the cost/benefit ratio (Todi et al., 2021).

1.3.4.10   Affective Computing

Affective computing, as introduced by Picard (2000), refers to the capacity of computers to perceive, comprehend, and respond appropriately to human emotions, thereby tailoring human-computer interactions accordingly. It seeks to imbue emotional intelligence into these interactions, allowing systems to adapt their responses or suggest calming activities when they detect a user's frustration or stress. This enhances the overall quality and UX of interactions. The applications of affective computing are vast and encompass diverse fields, including mental health monitoring, customer service, gaming, education, and marketing (See also Chapter 12 'Emotion Recognition and Affective Computing'), Book 'INTERACTION TECHNIQUES AND TECHNOLOGIES IN HUMAN-COMPUTER INTERACTION'). As an illustration, in the context of mental health monitoring, affective computing proves invaluable. It can identify signs of depression or anxiety and offer customized interventions and support to those in need (Zucco et al., 2017). This not only underscores technology's potential to enhance well-being but also emphasizes the human-centered approach inherent in the field of affective computing.

To realize affective computing, systems leverage a variety of sensors and data sources to discern human emotions. These sources include facial expression analysis, voice analysis, physiological data (such as heart rate and skin conductance), and text analysis. The choice of sensors and their integration into devices, such as cameras, microphones, or wearables, depends on the specific deployment context (Wang et al., 2022). Advanced algorithms play a pivotal role in analyzing data acquired from these sensors, drawing from machine learning and deep learning techniques to recognize patterns associated with different emotional states. This capability forms the foundation of emotionally intelligent systems. Beyond adapting interactions, some affective computing systems enable computers to express emotions, often through avatars or virtual agents. This mode of expression fosters a sense of connection between users and technology, enhancing the overall interaction (Rosenthal-von der Pütten et al., 2018). Emphasizing personalization, a cornerstone of affective computing, these systems meticulously adjust the user experience based on detected emotions. For instance, a virtual assistant can offer varied responses or activities depending on whether the user is feeling happy, sad, or stressed. This personalized approach elevates the quality and relevance of interactions, demonstrating the value of emotional awareness in computing systems.

Nevertheless, affective computing faces challenges (Daily et al., 2017). These challenges encompass concerns about the accuracy of emotion detection, the influence of cross-cultural differences in emotional expression, and the interpretability of AI-driven emotional analysis. Addressing these challenges is crucial for the ongoing advancement of affective computing, ensuring emotionally aware and culturally sensitive interactions. Ethical, privacy, and security concerns also arise as they underpin the collection and analysis of emotional data (Beavers & Slattery, 2017). With the integration of AI and machine learning technologies, the future of affective computing holds the potential to revolutionize various fields, including mental health, human-computer interaction, and education. As systems become increasingly sophisticated, they open doors to new possibilities in multiple domains, enriching human-computer interactions even further.

1.3.4.11   Brain-Computer Interfaces

Brain-Computer Interfaces (BCIs) refer to devices that allow a direct and sometimes bidirectional communication pathway between the brain's electrical activity and an external software or hardware device, most commonly a computer or robotic limb, without the involvement of muscular stimulation (Saha et al., 2021). In a sense, a BCI facilitates the acquisition, manipulation, analysis, and translation of brain signals into human actions independent of the peripheral nerves or muscles, to control external devices or applications (Maiseli et al., 2023). The challenge in this form of interaction lies in the way the device acquires the brain signals. This can be achieved in two main ways, the non-invasive way and the invasive way (See also Chapter 14 'Non-Invasive Brain Computer Interfaces', Book 'INTERACTION TECHNIQUES AND TECHNOLOGIES IN HUMAN-COMPUTER INTERACTION'). The most commonly used non-invasive method of acquiring brain signals is Electroencephalography (EEG)-based and it involves wearing a device on the scalp with a set of electrical sensors that serve as two-way communication channels between a patient's brain and a machine. This method is preferred due to its low-cost, user-friendliness, and portability (Rashid et al., 2020). However, due to the fact that the skull blocks a lot of the electrical signal, the acquired data may get distorted and less accurate.

The invasive method involves surgically implanting the electrodes under the scalp and directly onto the gray matter for direct communication with the brain signals. Although this method acquires a signal that is "stronger" and of higher amplitude and thus produces more accurate data than the non-invasive method, it is far more complicated and highly risky for the patient. This method is reserved only for very severe neuromuscular and other conditions (Kawala-Sterniuk et al., 2021).

Non-invasive BCIs offer several advantages over their invasive counterparts which make them more applicable to HCI and everyday life applications. Most non-invasive BCIs are EEG-based and have applications varying from gaming to rehabilitation via various external control devices, such as, among others, wheelchairs, robotic arms, and video displays (Kawala-Sterniuk et al., 2021).

The term BCI was coined by Jacques Vidal, a professor at UCLA, who was the first to describe this type of interfacing in 1973 through the results of his scientific research on using EEG signals to control external objects (Vidal, 1973). Since then, a great deal of research has been done on BCIs as researchers continue to explore the potential of this technology in various domains, such as rehabilitative medicine and assistive technologies for people with motor impairments, paralysis, and other neurological disabilities, robotics, as well as education, gaming, entertainment, diagnosing and evaluating brain diseases (Song et al., 2021), general health and well-being, etc.

Continuous advancements in health tech research and machine learning technologies have kept the excitement for this technology alive and growing, with the BCI market currently reaching $1.74 billion and being forecasted to reach up to $6.2 billion by the end of the 2020 decade.[9] The big challenge from this point forward is for researchers to find ways to increase the accuracy of non-invasive BCIs to the point where paralyzed patients will be able to control their environment or robotic limbs using their own thoughts.

1.3.4.12 Conversational Interfaces

Conversational user interfaces (CUIs) are software applications that use natural language as the main form of interaction with humans, through speech or text. A wide range of synonyms or near synonyms is present in the literature, including conversational agents, chatbots, and personal assistants (See also Chapter 7 'Conversational Agents', Book 'HUMAN-COMPUTER INTERACTION IN INTELLIGENT ENVIRONMENTS'). The idea of communicating with computers in natural language, thus mimicking human communication, emerged as early as the 1950s, but started acquiring increasing popularity in the 2000s, with the integration of conversational agents into instant messaging tools, the creation of smart personal voice assistants built into smartphones or home speakers (e.g., Apple Siri, IBM Watson, Google Assistant, Microsoft Cortana, and Amazon Alexa), as well as the appearance of chatbots integrated in social media platforms (Adamopoulou & Moussiades, 2020). Recently, advances in machine learning allowed the training of large-scale language models, considerably improving the performance of conversational agents. The most prominent example is ChatGPT, well-known for its impressive abilities in a variety of contexts, including text generation, programming, and language translation (van Dis et al., 2023).

Recently, besides AI and language model developments, the focus of investigation in CUIs has widened to include interaction and user experience issues. In this context, recent research has addressed various aspects of the evaluation of conversational user interfaces, for example by comparing available UX questionnaires and their suitability to assess CUIs (Kocaballi et al., 2019), or by developing dedicated usability scales (Borsci et al., 2022). Other efforts have assessed, on the one hand, user experience differences between conversational and graphical UIs (Manojlović & Kumarswamy, 2021), and on the other hand, the specificity and peculiarities of CUIs with respect to human conversation, building on existing psychology, linguistics, and sociology theories and approaches (Cowan et al., 2023). Under this perspective, key principles of user experience pertaining to conversational interface design have also been investigated (Rossouw & Smuts, 2023). An important specificity dimension of CUIs is the linguistic one. In Sugisaki and Bleiker (2020) linguistic principles for natural language conversation and high-level usability heuristics are combined to derive a set of 53 checkpoints for text-based CUIs (chatbots). Such checkpoints provide both guidelines for the design and criteria for the evaluation of chatbots.

Another aspect of CUIs that has recently attracted research interest is their accessibility for people with various types of disabilities (Lister et al., 2020). Sign language CUIs are emerging to facilitate deaf users (Glasser et al., 2020). Multimodal interaction in the context of CUIs is also being investigated in order to support more natural and human-like forms of interaction (Schaffer & Reithinger, 2019). In this context, various types of agent embodiments and non-verbal communication cues are also aspects of face-to-face communication that could be implemented in CUIs to make them more human-like (Foster, 2019).

Finally, as the use of CUIs is expanding in various application domains and contexts of use, issues related to privacy become relevant from a user experience perspective (Sohn et al., 2019; Brüggemeier & Lalone, 2022), leading to the elaboration and testing of privacy strategies (Leschanowsky et al., 2023).

1.3.4.13 Artificial Intelligence

The proliferation of AI nowadays has led to becoming an integral part of our everyday life. We are using AI to make decisions in many different facets of our daily activities ranging from trivial tasks, such as selecting music of our interest, to serious aspects, such as the discovery of cancerous tumors. Although AI has the potential to improve our lives, there are concerns about its long-term sustainability for human benefit. A fundamental concern while creating AI systems is how to make autonomous intelligence behave and act in a way that benefits people and societies, upholds human rights, and makes moral and ethical decisions. The debates about models for human-centered, reliable, and equitable AI are very clear reflections of this major concern.

Recently, great awareness has been raised among the research community and policymakers about the importance of designing and implementing human-centered AI systems, putting the human not only at the center of the design process, but continuously in the loop during the entire implementation lifecycle of AI systems (See also Chapter 4 'Human-Centered AI', Book 'HUMAN-COMPUTER INTERACTION IN INTELLIGENT ENVIRONMENTS'). Shneiderman (2020b), suggested fundamental principles and a framework for human-centered AI toward creating reliable, safe, and trustworthy systems. He discusses the strategies for achieving such AI systems and elaborates on how the suggested framework describes different design objectives and the path to high levels of human control and high levels of automation. Shneiderman (2020b) uses concrete examples to showcase how his human-centered AI framework can help stakeholders to decide when rapid automated actions should be controlled by computers; when human mastery for control should come first; and when there are risks associated with too much automation or too much human control.

Garibay et al. (2023) define "six grand challenges" regarding human-centered AI, putting in place the directions that should be considered for the data collection and curation, the AI model design and implementation, and its deployment and use. They contemplate the need for responsible design of AI, privacy safeguarding, the existence of design and evaluation frameworks, human-AI interaction, human well-being, as well as governance, and independent oversight. For each one of these challenges, the authors provide recommendations that researchers, developers, business leaders and policymakers should take into account for ensuring a human-centered approach to AI. In alignment with the aforementioned principles and directions, Margetis et al. (2021) suggested a conceptual framework for the human-centered design of AI. They illustrate concrete approaches and tools that can be used during the implementation lifecycle of AI systems, rooted in already well-established research domains that delve into realizing and understanding AI, such as explainable AI, cognitive computing, visual predictive analytics, interactive machine learning, and federated learning. Hence, the suggested framework tools can inherently address human-centered AI challenges, since they use fundamental principles and practices of AI in a human-centered manner.

Although a lot of work has focused so far on unraveling paths toward human-centered AI, we have seen only the tip of the iceberg. In the years to come, research is expected to intensify, with new caveats surfacing and new challenges being addressed in the context of a continuously evolving AI field.

1.3.5 HCI AND THE SOCIETY

HCI in the 1980s and most of the 1990s was a term used mainly in research and reflected in a handful of commercial positions, starting in 1995 with Don Norman's position in Apple, in the role of "User Experience Architect". Since then, HCI has gradually become part of corporate culture and today the prefix "UX" is found in many job titles.

As modern HCI researchers and practitioners, we find ourselves in a world where our chosen path is no longer an obscure discipline, but one that is rapidly gaining recognition and has real impact in the world, albeit under a different name, i.e., UX. According to NN/g's 2019 report Rosala and Krause (2019) and Nielsen (2017), there are now ~1 million UX experts in the world and by 2050 that number may reach 100 million.

Young people today can choose to have a career in UX, something non-existent just 20 years ago. This means that the findings, research and ultimately practice of HCI have proven to be so effective as to be adopted in the market as significant roles in the production process, especially digital products.

The NN/g 2019 report presents their global survey that discovered 134 unique job titles related to UX. Half of those contained the word "designer" (like UX designer, Interaction Designer, UI Designer, etc.) It is a bit discouraging to observe that still there is not an agreed-upon job title to identify effectively HCI experts ("UX Designer" is the most popular) in the workplace, but it is very good to know that the science behind Human-Computer Interaction is recognized as essential to any product development. The adoption of UX specialists in the workplace has been noted in the same report, and between 2007 and 2019 (the time between the two editions of the report), there was a significant uptake of these careers in places outside North America and Northern Europe. Nielsen argues that the maturity level of UX practice is still low but steadily rising, and new technologies will only increase the demand for UX expertise (NNGroup, 2021). It is anticipated that this vital part of any development team and process will continue to spread globally, as to reach the estimated 100 million experts by 2050. Alongside people versed in HCI joining the workforce comes the mindset of the human-centred designer and design thinking. More and more corporations want to plan for UX in their product and adopt some sort of iterative process in their development model. This is a slowly moving phenomenon outside North America and Europe, but it is happening and influences the adoption of human-centric values (including accessibility guidelines) worldwide. This is important because the modern technological landscape is one that sees a proliferation of computers in the environment on an unprecedented scale. Alongside the myriad of desktop computers, there are now billions of smartphones, tablets, and wearables, automobiles equipped with screens and connections to the internet. More and more objects are connected, new media and new capabilities of intelligent agents make the intelligent environment a reality; in short, computers are quite literally everywhere, and they are connected. All this has had an immediate effect on multiple domains, and more is expected in the very near future, especially with XR technologies and AI. In simpler terms, modern society is constantly interacting with computers and soon it will be doing so in new and occasionally uncharted ways. It is up to HCI researchers and UX professionals to make sure that this transition to the digital age happens smoothly and does not exclude any members of the society, including the rapidly growing aging population.

As already discussed, HCI can be viewed as the discipline that focuses on the needs of the user and thus leverages the knowledge and research of other disciplines into solutions. One look into some of the most active domains of research will reveal this interdisciplinarity and how HCI is now a central factor in ongoing efforts.

Arguably, one of the most affected domains is Social Computing, which has expanded widely (See also Chapter 3 'Social Computing', Book 'HUMAN-COMPUTER INTERACTION IN VARIOUS APPLICATION DOMAINS'). There is a huge heterogeneity in topics of social computing studies, the most common themes being users, mediated interaction, collaborative work, social networking, and social media. All of them combine several different disciplines into one shared research agenda. Collaborative work or Computer-Supported Cooperative Work (CSCW) as is the traditional field name, has seen a tremendous rise in interest (See also Chapter 4 'Computer Supported Cooperative Work (CSCW), Teleconferencing and Remote Working', Book 'HUMAN-COMPUTER INTERACTION IN VARIOUS APPLICATION DOMAINS'), not only due to new technological opportunities (such as large, interactive screens and VR environments) but because of the COVID-19 pandemic that necessitated remote working and teleconferencing, which have persisted even after the end of the restrictions (Vargas Llave et al., 2022).

One other obvious example of different disciplines coming together is the way research on traditional business/marketing domains, such as customer service and consumer behavior, has crossed over with HCI research on design as a fundamental factor of both. There is no doubt that HCI is now important for business (Stephanidis et al., 2019), UX being a primary consideration for any modern enterprise, including online commerce via regular channels (websites or apps) but certainly not limited there. Conversational agents are another technology that is increasingly used for customer service, having reached a maturity that was previously found in sci-fi movies (i.e., naturally conversing with an intelligent computer), bringing together AI, Voice interfaces, and Business (See also Chapter 8 'Artificial Intelligence for Customer Services', Book 'HUMAN-COMPUTER INTERACTION IN INTELLIGENT ENVIRONMENTS'). In this new digital landscape, privacy, and security have become extremely important, and therefore usable security is, and is expected to be for the foreseeable future, a primary research field, no longer being confined to safe algorithms and encryption but also considering human behavior as well (See also Chapter 3 'Human Behavior in Cybersecurity Privacy and Trust', Book 'HUMAN-COMPUTER INTERACTION IN INTELLIGENT ENVIRONMENTS').

Radical transformations have also taken place in businesses and organizations, with HCI research and practice guiding and retrofitting breakthroughs. Social and organizational psychology have informed the design of technology to support productivity, satisfaction, and organizational culture (See also Chapter 9 'Applying the Science of Social and Organizational Psychology to HCI', Book 'FOUNDATIONS AND FUNDAMENTALS IN HUMAN-COMPUTER INTERACTION'). At the same time, the wide adoption of technology has paved the way for digital transformation and the field has had a significant impact on modern business processes (See also Chapter 11 'HCI in Business and Organizations: Digital Transformation with HCI, Metaverse and AI Technologies', Book 'HUMAN-COMPUTER INTERACTION IN VARIOUS APPLICATION DOMAINS'). Such a business field that has enhanced and prospered is that of commerce, which has become electronic, mobile, and omni-channel, applying design guidelines and principles and researching consumer behavior and customer experience to further advance and evolve (See also Chapter 10 'HCI in e-Commerce', Book 'HUMAN-COMPUTER INTERACTION IN VARIOUS APPLICATION DOMAINS').

The Service industries have also benefited much from HCI research, particularly when it comes to digitalization and servitization (See also Chapter 13 'HCI in Service Industries', Book 'HUMAN-COMPUTER INTERACTION IN VARIOUS APPLICATION DOMAINS'), in the form of smart services that have influenced value creation, work progress and consumer behavior (Paluch, 2017). The manufacturing factory is also being transformed into the digital factory, where digital information systems support workers, enhance their capabilities or even substitute them in the context of automation (See also Chapter 16 'HCI in the Manufacturing Industries', Book 'HUMAN-COMPUTER INTERACTION IN VARIOUS APPLICATION DOMAINS').

Older, more "traditional" domains such as aviation, which, as seen, is arguably the foundational domain for HFE, continue to be a very important HCI research domain (See also Chapter 17 'HCI in Aviation', Book 'HUMAN-COMPUTER INTERACTION IN VARIOUS APPLICATION DOMAINS'). Automation has transformed the cockpit by reducing the number of individuals required in large aircraft (from three to typically two nowadays), with efforts to reduce it to one or even none. Already, unmanned aerial systems (UAS), including vehicles, are a reality, requiring remote operators or even being fully automated. These advancements have been facilitated by HCI research, methods, and tools, which have allowed this progression to take place while maintaining the highest level of safety. Similarly, since the introduction of internet-enabled media stations in modern cars has become quite common, as well as the appearance of new self-driven vehicles, HCI in the automotive context has been one of the most active research areas over the past years and will continue to be so in the future, tackling issues such as driver monitoring and fatigue detection, interaction with complex systems while driving (hands free or gaze free), automation and decision issues of the self-driving car and more.

The proliferation of wearables and the ability to monitor human vitals has resulted in explosive growth in e-Health and well-being applications, which are generating a massive amount of data (See also Chapter 1 'HCI in e-Health' and Chapter 2 'Well-Being in a Digital Age', Book 'HUMAN-COMPUTER INTERACTION IN VARIOUS APPLICATION DOMAINS'). In healthcare, for example, the field has experienced substantial growth, with the creation of medical apps, wearable health monitoring devices, and telemedicine platforms (Blandford, 2019). These innovations seek to improve patients' health outcomes, enhance user experience, and facilitate interactions between healthcare professionals and patients. In the health and well-being context, but also beyond it, technology is often used as a persuasion medium, aiming to motivate people to achieve important behavior-change goals (See also Chapter 10 'Persuasive Interfaces', Book 'DESIGNING FOR USABILITY, INCLUSION AND SUSTAINABILITY IN HUMAN-COMPUTER INTERACTION').

More data is being produced by patient electronic health records, medical imaging, and clinical trial data. HCI research is looking into the visualization of the big data produced and ways for health professionals, caretakers, and patients to interact with them to gain insights into health care, disease prevention, and treatment. Big data is also used in numerous application domains and requires particular consideration in terms of visualization (See also Chapter 14 'Big Data Visualization and Visual Analytics', Book 'HUMAN-COMPUTER INTERACTION IN VARIOUS APPLICATION DOMAINS'). A common visualization approach, taking advantage of big data, and aiming to allow humans to comprehend and easily assimilate large volumes of information is that of dashboards, providing summative overviews and graphical data representations (See also Chapter 15 'Dashboard Design', Book 'HUMAN-COMPUTER INTERACTION IN VARIOUS APPLICATION DOMAINS').

The flourishing domain of educational technology has captured significant attention from HCI researchers (Scanlon et al., 2015; Van Mechelen et al., 2023). Their efforts focus on designing e-learning platforms, educational games, and interactive learning environments to elevate student engagement and learning outcomes (see also Chapter 5 'Adaptive Learning, Training and Education', Book 'HUMAN-COMPUTER INTERACTION IN VARIOUS APPLICATION DOMAINS').

Another exciting domain is that of gaming, where researchers are investigating the Player Experience (a particular instance of user experience) and what makes a game engaging, pleasurable and thus keeping the player motivated (See also Chapter 6 'Player Experience', Book 'HUMAN-COMPUTER INTERACTION IN VARIOUS APPLICATION DOMAINS'). These insights are then exploited into making any other domain application engaging, including learning applications; Serious Games and Gamification are recent research areas that have seen tremendous growth in the past years (See also Chapter 16 'Gamification Design', Book 'USER EXPERIENCE METHODS AND TOOLS IN HUMAN-COMPUTER INTERACTION'). While learning through play is an ancient concept, today the consolidation of HCI, Teaching, Learning and Gaming principles has produced a considerable corpus of knowledge to turn learning objectives into engaging games (See also Chapter 7 'Emerging Technologies and Frameworks for Serious Games', Book 'HUMAN-COMPUTER INTERACTION IN VARIOUS APPLICATION DOMAINS').

Culture is another field where the impact of HCI is clearly manifested. In this regard, engaging and intuitive experiences are designed in the context of entertainment and interactive media, taking advantage of novel technologies toward immersive and gamified interactions (See also Chapter 8 'Entertainment and Interactive Media', Book 'HUMAN-COMPUTER INTERACTION IN VARIOUS APPLICATION DOMAINS'). Interaction with cultural heritage has also been transformed with novel interactive technologies and the contributions of HCI to promoting, democratizing and sustaining tangible and intangible cultural heritage (See also Chapter 9 'Digital Presentation and Interaction with Cultural Heritage', Book 'HUMAN-COMPUTER INTERACTION IN VARIOUS APPLICATION DOMAINS'). Meanwhile, in a globalized world, cross-cultural and global HCI research assumes an essential role (Kyriakoullis & Zaphiris, 2016). Scholars are investigating how cultural variances influence user interactions and endeavor to design for diverse, worldwide user bases (See also Chapter 14 'Cross-cultural Design', Book 'DESIGNING FOR USABILITY, INCLUSION AND SUSTAINABILITY IN HUMAN-COMPUTER INTERACTION').

Sustainability is yet another pressing concern addressed by HCI, with a focus on creating eco-friendly solutions, such as energy-efficient interfaces and sustainable design practices (see also Chapter 14 'Sustainability and Citizen Science', Book 'HUMAN-COMPUTER INTERACTION IN INTELLIGENT ENVIRONMENTS').

Furthermore, the new digital landscape has brought e-Government into the forefront, with more countries offering digital services that change the quality of life of citizens (See also Chapter 12 'HCI in e-Government and e-Democracy', Book 'HUMAN-COMPUTER INTERACTION IN VARIOUS APPLICATION DOMAINS')..However, while it has been recognized that a good user experience with a company's product is vital to customer loyalty and trust, this principle has been largely ignored by the state and civil services globally. More to the point, involving the user in the design process is a core principle of HCD. In the e-Government context, this translates to involving the citizen in the decision-making process in a more direct and impactful way, essentially becoming more democratic by definition. It remains to be seen how much HCI, HCD and UX principles will impact how society governs itself.

1.4   FUTURE DIRECTIONS IN HCI

Throughout this chapter, the origins and historical evolution of the HCI field have been discussed. It must be acknowledged that the field has come a long way: it started from an era when emphasis was paid to the functionality provided by software with minimal attention to the interaction between human operators and computers, and pioneering concepts for their time have emerged; nowadays, there is a clear demand to design computer systems, applications, and services that are driven by the needs and requirements of the intended users, ensuring that end users are actively involved in their design and evaluation. Along with the marked technological evolution, HCI has substantially grown into a well-recognized multidisciplinary field, acknowledged for its substantial contributions and societal impact. Table 1.1 provides a timeline of this technological evolution.

Looking into the future, we cannot but be certain that HCI will further flourish in the years to come. However, what exactly will its role be in the future? Although we cannot predict the future, we can nevertheless formulate informed expectations based on challenges as they have emerged today from the current technological evolution and our knowledge of the recent past. A key factor so far has been the strong user focus of the field, which has today shifted further to a human focus, motivated by the implicit acknowledgment that technology should not focus only on its users, but humans in general; in other words, in the years to come, HCI is expected to expand its focus to society and also the entire humanity, advocating not only Human-Centered Design, but also Society-Centered Design and Humanity-Centered Design. HCI will be therefore the discipline that will study the design of technology and the phenomena that pertain to the interaction and communication of humans with technology and their implications at an individual, societal, and humanity level.

Considering the remarkable evolution of technology in the past few decades, new interaction paradigms are expected to emerge, as a result of further technological advancements. Natural language interaction adopted at large with everyday devices, deformable and shape-changing interfaces, human-building interaction, inbodied interaction, biodesign and biological HCI are just a few of the trends already reported in the literature (Andres et al., 2020; Becerik-Gerber et al., 2022; Boem & Troiano, 2019; Gough et al., 2020; Pataranutaporn et al., 2018; Santos et al., 2020). Technologies that will enter our daily lives include home robotics, self-driving cars, and delivery drones (Gajendar, 2017).

At the same time, areas of current research in HCI are expected to further evolve. Advances in affective computing have brought us to the cusp of a new age whereupon the computational systems we interact with are emotionally aware and responsive to our needs, exhibiting a learning potential that can truly revolutionize affective and empathic computing (Bosch et al., 2022; Han et al., 2021). The field of automotive interaction has also marked astonishing progress as vehicles become ever more intelligent and connected, expanding human-automotive interaction and cooperation to address scenarios of co-driving, assisted and automated driving, and in-car user experiences beyond driving; also, research is ongoing on ethical aspects, human behavior, inclusive design, vehicle personality, behavior, and transparency (Biondi et al., 2019; Detjen et al., 2022; Lilis et al., 2019; Liu & Tan, 2022). Human-Robot interaction constitutes already an extensive and diverse field with efforts classified as human-supervised robots for routine tasks, remote control of robots for hazardous situations or inaccessible environments, automated vehicles, and social robots with applications in numerous domains (Sheridan, 2016). The field, however, is still young and is expected to further advance, with research suggesting that future endeavors will focus on emotional and affective robots, neuroscience-informed human-robot interaction, safety, harmonic co-existence and collaboration, as well as ethics and social impact (Cross et al., 2019; de Graaf, 2016; Henschel et al., 2020; Ostrowski et al., 2022; Spezialetti et al., 2020; Zacharaki et al., 2020).

TABLE 1.1
Technological Breakthroughs and HCI Milestones from 1940 until 2024

1940–1950	• 1941: Isaac Asimov describes the "Three Rules of Robotics" in his novel "I, Robot" • 1945: Vannevar Bush – As We May Think introduces the Memex • 1945 – 1950s: Bell Labs hires a psychologist to design telephone systems, UX work on touchstone keypads • 1957: Human Factors Society is established in the US • 1950: Alan Turing's "Computing Machinery and Intelligence" article is published, introducing the concept of the Turing test to the wide public
1950–1960	• 1958: ARPA (later renamed to DARPA) founded • 1959: Shackel publishes "Ergonomics for a Computer" • 1960: Licklider – "Man-Computer Symbiosis" is published • 1960: Engelbart – "Special Considerations of the Individual as a User, Generator and Retriever of Information" is published
1960–1970	• 1963: First GUI (Sunderland's Sketchpad) • 1966: Joseph Weizenbaum creates Eliza, the first natural language processing conversational agent • 1968: The Mother of All Demos by Engelbart is presented • 1968: Ivan Sutherland presents the first Head-Mounted Display "The Sword of Damocles" • 1969-1971: ARPANET, precursor to the Internet is established • 1970: Xerox Palo Alto Research Center (PARC) is established
1970–1980	• 1971: Allen Newell writes the AIP Memo 1 that pushes to utilize psychology in Xerox Research to understand the nature of the human user • 1973: Xerox Alto – first modern GUI is presented • 1975: VIDEOPLACE, the first interactive VR platform is introduced by Myron Krueger • 1980: Keystroke-level model by Card, Moran and Newell is published
1980–1990	• 1981–1984: Xerox Star ('81), IBM PC ('81), Apple Lisa ('83), Apple Macintosh ('84) released; the personal computer era and WYSIWYG GUIs • 1982: "Human Factors in Computing Systems" conference in Maryland • 1982: The Special Interest Group on Computer-Human Interaction (SIGCHI) is formed • 1983: First CHI conference • 1983: The Psychology of Human-Computer Interaction published • 1984: First HCI International Conference • 1986: User-Centred Design by Donald Norman and Stephen Draper is published • 1986: The "Super Cockpit" flight simulator is being developed, allowing pilots wearing a helmet to control the aircraft using gestures, speech and eye movements • 1988: The Psychology (later Design) of Everyday Things by D. Norman is published • 1988: Mark Weiser coins the term "Ubiquitous Computing" to describe interaction with invisible devices embedded in the built environment • 1989: The World Wide Web is invented by Tim Berners-Lee
1990–2000	• 1990: The term "augmented reality" is coined by Tom Caudell to define the integration of computer-generated images into the real world • 1991: World Wide Web opens to the public • 1992: Louis Rosenburg creates the first AR system at the US Air Force Research Laboratory, a virtual flying training for Air Force pilots • 1993: The first graphical WWW browser, Mosaic, is released • 1993: The first smartphone, the IBM Simon, is released • 1993: Jakob Nielsen publishes the seminal textbook 'Usability Engineering' • 1994: Amazon.com is founded, marking the beginning of e-Commerce • 1995: Windows 95 makes the WIMP paradigm ubiquitous • 1995: Constantine Stephanidis introduces the concept of 'User Interfaces for All' • 1995: Don Norman introduces the term 'User Experience' • 1995: Navlab5, the first self-driving car is successfully piloted from Pittsburgh to San Diego • 1995: Hiroshi Ishii pioneers the Tangible User Interface through his paper "Tangible Bits: Towards Seamless Interfaces between People, Bits and Atoms"

(Continued)

TABLE 1.1 (*Continued*)

Technological Breakthroughs and HCI Milestones from 1940 until 2024

	• 1997: Kismet, an expressive robot head, is designed by Cynthia Breazeal at the MIT Media Lab, marking a significant step in social robotics
	• 1998: The term "ambient intelligence" is coined by the European Commission's Information Society Technologies Advisory Group (ISTAG) to describe a vision of the future where technology would be embedded in our surroundings and would adapt to our needs and preferences
	• 1999: The term "Web 2.0" is coined by Darcy DiNucci, marking a new era for more interactive online experiences
	• 1999: Human-centered design is established as a standard through ISO 13407:1999 "Human-centred design processes for interactive systems"
	• 1999: W3C publishes the Web Content Accessibility Guidelines, a set of recommendations for making web content more accessible to people with disabilities
	• 1999: The term "Internet of Things" is coined by Kevin Ashton to describe a system where the Internet is connected to the physical world via ubiquitous sensors
	• 1999: Sony introduces AIBO a four-leg robot dog showcasing the potential for robots in entertainment and companionship
2000–2010	• 2000: The first AR game is launched, named 'AR Quake', with players wearing a head mounted display and a backpack containing a computer and gyroscope
	• 2000: Honda introduces ASIMO, the first humanoid robot that can run, walk, communicate with humans, recognize faces, environment, voices and posture
	• 2003: MySpace is launched, mainstreaming social networking
	• 2007: iPhone is released – smartphones as we know them dominate and the modern era begins
	• 2008: Google releases the Android operating system
	• 2010: Apple releases the iPad, the first successful tablet, essentially introducing that category of devices into the mainstream
	• 2010: The definition of UX is included in the ISO standard ISO 9241-210:2010
2010–2025	• 2011: Apple releases Siri, an intelligent personal assistant generating responses and taking actions in response to voice requests
	• 2011: The term responsive web design is established by Ethan Marcotte, to reflect the need for flexible HTML design supported by appropriate stylesheets (CSS) addressing the need for web pages that adapt optimally to the viewing medium
	• 2012: Palmer Luckey designs the first prototype of the Oculus Rift headset and Oculus Touch controllers
	• 2013: Release of Google Glass, a wearable AR technology in the form of smart glasses
	• 2014: "Pepper", the first personal humanoid robot that can read emotions is introduced by SoftBank robotics
	• 2015: AR becomes popular in social networks through filters introduced by SnapChat
	• 2015: Tesla introduces the "Autopilot" feature that enables hands-free control for highway and freeway driving, to be soon followed by automated driving features in models of other automobile companies, such as Volvo, Honda, and Mercedes
	• 2015: Amazon launches its voice assistant, Alexa, integrated into the Echo smart speaker, that can interact with users using natural language
	• 2015: The United Nations establish the 2030 Agenda for Sustainable Development with 17 Sustainable Development Goals at its core
	• 2016: Pokemon Go, a popular game, is launched, mainstreaming AR technology
	• 2016: Microsoft releases HoloLens, a wearable AR device, allowing users to scan their environment and create their own AR experiences
	• 2018: Generative AI is made available to the wide public through ChatGPT, a Large Language Model (LLM) released by OpenAI
	• 2021: OpenAI introduces DALL-E, a generative AI system creating images from text prompts
	• 2023: Generative AI efforts expand with systems becoming readily available to end users for different purposes (e.g., Google Bard, Microsoft Copilot, MidJourney image generation platform, OpenAI Codex, Adobe Firefly, etc.)

Furthermore, technological domains that have not yet reached their full potential will further mature and provide new perspectives, innovative interaction paradigms and expand to a wide range of application domains. BCIs have already marked a notable evolution from invasive to non-invasive approaches and a remarkable shift from their application in the medical domain to their incorporation in consumer applications (Stegman et al., 2020); however, a future in which we can effectively and efficiently control our computers with our thoughts is pending major technological advancements. XR technologies constituted the next big thing in the 2000s; however, they failed to match the predictions for truly revolutionizing our interactions and are currently facing low popularity (Muñoz-Saavedra et al., 2020). At the same time, hardware advances hold a promising potential for their revitalization (Xiong et al., 2021). In the future landscape, discourse about unified realities, i.e., virtuality and reality, is anticipated to emerge along with novel experiences in the reality-virtuality continuum (Margetis et al., 2020; Skarbez et al., 2021; Steinicke & Wolf, 2020). At the same time, the concept of the Metaverse still remains unclear, leaving room for further exploration, refinement and substantial advancements, joint research efforts, and technological innovations, toward becoming the next generation internet (Cheng et al., 2022). Looking into the future prospects of most of the aforementioned technologies, a common denominator that has the potential to propel their progress and revolutionize their achievements is AI. AI has already pervaded everyday life, becoming a tool that can be used by everyone in various application domains. HCI and AI are closely connected and can be mutually beneficial, with HCI on the one hand catering to the design of AI technologies, and AI on the other hand supporting existing and new HCI practices.

In this complicated landscape, it is evident that there are ample opportunities for HCI not only to prosper as a field, but more importantly to truly serve humanity. Considering the autonomy and intelligence exhibited by the technology, as well as its omnipresence, HCI has a substantial role to play in averting dark scenarios and dystopian realities and fostering meaningful and ethical symbiosis and interaction with intelligent technologies of the future, nurturing well-being, health, eudaimonia, inclusivity, prosperity, and democracy (Stephanidis et al., 2019). More specifically, when it comes to the design and development of AI technologies, HCI is called upon to promote Human-Centered AI, which is centered on human well-being, is designed responsibly, respects privacy, follows human-centered design principles, is subject to appropriate governance and oversight and interacts with individuals while respecting human's cognitive capacities (Garibay et al., 2023). Furthermore, a major issue that needs to be addressed is algorithmic bias against individuals or entire social groups, including persons with disabilities, older adults and vulnerable individuals, advancing the agenda toward 'Human-Centered Universally Designed AI' (Stephanidis, 2023).

In this light, the discussion on HCI and ethics is not only timely but also critical. Looking at the contemporary technological landscape, it is clear that besides the immense potential of technology, frightening risks are now at stake. This is already acknowledged and an apparent agreement that AI should be 'ethical' has been achieved, while there is a global consensus on five key ethical principles, namely transparency, justice and fairness, non-maleficence, responsibility, and privacy (Jobin et al., 2019). In this regard, numerous documents on AI ethics, policy, and governance have been produced, having an impact on relevant ethics research efforts, on public discourse, as well as on informing policies, regulations, and strategies (Schiff et al., 2020). Yet, AI ethics is profoundly failing in many cases, lacking a reinforcement mechanism, being often faced as a marketing strategy or extraneous to technical concerns, and in all cases an unbinding framework beyond the technical community (Hagendorff, 2020). HCI, being a field that builds upon and cultivates the collaboration between designers and software engineers, can contribute to instilling ethical guidelines and supporting AI designers and engineers in adopting them, by adapting existing methods and tools. Furthermore, aiming to address a major concern, that of keeping humans in control of the technology evolution and always in the loop, knowledge, methodologies, and findings from

Human-Centered Design can be further evolved and applied to address the escalating challenges of designing contemporary and future technologies.

Such significant responsibilities cannot be attributed generally to the entire field nor do they merely constitute abstract wishful thinking. In practice, these responsibilities fall on the shoulders of researchers and practitioners themselves. As a result, the issue of HCI education is of paramount importance. Looking ahead, HCI education should not pertain only to good design principles and UX design, but should also expand to an in-depth understanding of diverse methodologies, state-of-the-art and emerging technologies, as well as ethical considerations, in order to prepare future HCI researchers and practitioners to address the challenges of the ever-evolving technological landscape. The issue of multidisciplinarity is also brought to the forefront, mandating that experts from multiple fields collaborate to design and deliver technologies of the future.

In conclusion, HCI has played a critical role in the technological evolution, shifting the focus of technology development from merely functional aspects to human-centered views, emphasizing that the design should focus on humans and their needs, cultivating a relationship of trust between people and technology, and therefore leading to the acceptance of technology by people.

Today, the maturity of technology is remarkable, and further rapid leaps are anticipated, leading to new interaction paradigms, escalated engagement, and continuous human collaboration with technology. At the same time, potential risks have grown in scope, laying out threatening consequences upon their materialization. At this turning point, we are very hopeful that HCI can positively contribute to a brighter future of human, societal, and planetary prosperity and well-being, by promoting human values and ethical design, putting forward well-established principles and guidelines and unfolding its research potential toward innovating practical contributions for the design of the future technological world.

NOTES

1. https://www.statista.com/statistics/273018/number-of-internet-users-worldwide/.
2. https://www.statista.com/statistics/262950/global-mobile-subscriptions-since-1993/ .
3. https://www.statista.com/statistics/871513/worldwide-data-created/.
4. https://research-methodology.net/research-methods/qualitative-research/interviews/.
5. https://www.nngroup.com/articles/journey-mapping-101/.
6. https://leap2.ultraleap.com/leap-motion-controller-2/.
7. https://www.apple.com/ios/carplay/.
8. https://www.android.com/auto/.
9. https://builtin.com/hardware/brain-computer-interface-bci.

REFERENCES

Aarts, E., & Encarnação, J. (2006). Into ambient intelligence. In Aarts, E. H. L. (ed.) *True Visions: The Emergence of Ambient Intelligence* (pp. 1–16). Berlin, Heidelberg: Springer.

Abrahão, S., Insfran, E., Sluÿters, A., & Vanderdonckt, J. (2021). Model-based intelligent user interface adaptation: Challenges and future directions. *Software and Systems Modeling*, 20(5), 1335–1349.

Abras, C., Maloney-Krichmar, D., & Preece, J. (2004). User-centered design. In Bainbridge, W. (ed.), *Encyclopedia of Human-Computer Interaction* (Vol. 37, pp. 445–456). Thousand Oaks, CA: Sage Publications.

Ackerman, M. S., & Mainwaring, S. D. (2005). Privacy issues and human-computer interaction. *Computer*, 27(5), 19–26.

Adami, I., Foukarakis, M., Ntoa, S., Partarakis, N., Stefanakis, N., Koutras, G., ... & Stephanidis, C. (2021). Monitoring health parameters of elders to support independent living and improve their quality of life. *Sensors*, 21(2), 517.

Adamopoulou, E., & Moussiades, L. (2020). Chatbots: History, technology, and applications. *Machine Learning with Applications*, 2, 100006. https://doi.org/10.1016/j.mlwa.2020.100006.

Ahmad, K., Maabreh, M., Ghaly, M., Khan, K., Qadir, J., & Al-Fuqaha, A. (2022). Developing future human-centered smart cities: Critical analysis of smart city security, data management, and ethical challenges. *Computer Science Review*, 43, 100452.

Alam, T., Salem, A. A., Alsharif, A. O., & Alhejaili, A. M. (2020). Smart home automation towards the development of smart cities. *APTIKOM Journal on Computer Science and Information Technologies*, 5(1), 152–159.

Alavi, H. S., Churchill, E. F., Wiberg, M., Lalanne, D., Dalsgaard, P., Fatah gen Schieck, A., & Rogers, Y. (2019). Introduction to human-building interaction (HBI) interfacing HCI with architecture and urban design. *ACM Transactions on Computer-Human Interaction (TOCHI)*, 26(2), 1–10.

Alcañiz, M., & Rey, B. (2005). New technologies for ambient intelligence. *Ambient Intelligence*, 3, 3–15.

Alexander, R., Murray, D., & Thompson, N. (2017). Cross-cultural web design guidelines. In *Proceedings of the 14th International Web for All Conference, Perth Western Australia Australia April 2 - 4, 2017* (pp. 1–4). New York: Association for Computing Machinery.

Alías, F., & Alsina-Pagès, R. M. (2019). Review of wireless acoustic sensor networks for environmental noise monitoring in smart cities. *Journal of Sensors*, 2019, 13.

Amershi, S., Weld, D., Vorvoreanu, M., Fourney, A., Nushi, B., Collisson, P., ... & Horvitz, E. (2019). Guidelines for human-AI interaction. In *Proceedings of the 2019 CHI Conference on Human Factors in Computing Systems*, Glasgow Scotland UK, May 4–9, 2019 (pp. 1–13). New York: Association for Computing Machinery.

Andres, J., Schraefel, M. C., Patibanda, R., & Mueller, F. F. (2020). Future inBodied: A framework for inbodied interaction design. In *Proceedings of the Fourteenth International Conference on Tangible, Embedded, and Embodied Interaction*, Sydney NSW Australia, February 9–12, 2020 (pp. 885–888). New York: Association for Computing Machinery.

Antona, M., Ntoa, S., Adami, I., & Stephanidis, C. (2009). User requirements elicitation for universal access. In Stephanidis, C. (ed.), *The Universal Access Handbook* (pp. 15-1–15-14). Boca Raton, FL: Taylor & Francis (ISBN: 978-0-8058-6280-5, 1.034 pages).

Ashtari, N., Bunt, A., McGrenere, J., Nebeling, M., & Chilana, P. K. (2020). Creating augmented and virtual reality applications: Current practices, challenges, and opportunities. In *Proceedings of the 2020 CHI Conference on Human Factors in Computing Systems* Honolulu HI USA, April 25–30, 2020 (pp. 1–13). New York: Association for Computing Machinery.

Babel, F., Kraus, J., Miller, L., Kraus, M., Wagner, N., Minker, W., & Baumann, M. (2021). Small talk with a robot? The impact of dialog content, talk initiative, and gaze behavior of a social robot on trust, acceptance, and proximity. *International Journal of Social Robotics* 13, 1485–1498. https://doi.org/10.1007/s12369-020-00730-0.

Baby, C. J., Khan, F. A., & Swathi, J. N. (2017). Home automation using IoT and a chatbot using natural language processing. In I-PACT, & IEEE Staff (eds.), *2017 Innovations in Power and Advanced Computing Technologies (i-PACT)* (pp. 1–6). New York: IEEE.

Bakker, S., & Niemantsverdriet, K. (2016). The interaction-attention continuum: Considering various levels of human attention in interaction design. *International Journal of Design*, 10(2), 1–14.

Ballantyne, M., Jha, A., Jacobsen, A., Hawker, J. S., & El-Glaly, Y. N. (2018). Study of accessibility guidelines of mobile applications. In *Proceedings of the 17th International Conference on Mobile and Ubiquitous Multimedia*, Cairo Egypt, November 25–28, 2018 (pp. 305–315). New York: Association for Computing Machinery.

Bateman, S., Gutwin, C., Osgood, N., & McCalla, G. (2009). Interactive usability instrumentation. In *Proceedings of the 1st ACM SIGCHI Symposium on Engineering Interactive Computing Systems* (pp. 45–54).

Baxter, K., Courage, C., & Caine, K. (2015). *Understanding Your Users: A Practical Guide to User Research Methods*. Burlington, MA: Morgan Kaufmann.

Beavers, A. F., & Slattery, J. P. (2017). On the moral implications and restrictions surrounding affective computing. In Jeon, M. (ed.), *Emotions and Affect in Human Factors and Human-Computer Interaction* (pp. 143–161). Amsterdam: Elsevier.

Becerik-Gerber, B., Lucas, G., Aryal, A., Awada, M., Berges, M., Billington, S. L., ... & Zhao, J. (2022). Ten questions concerning human-building interaction research for improving the quality of life. *Building and Environment*, 226, 109681.

Beldad, A., De Jong, M., & Steehouder, M. (2010). How shall I trust the faceless and the intangible? A literature review on the antecedents of online trust. *Computers in Human Behavior*, 26(5), 857–869.

Bennett, J. L. (1979). The commercial impact of usability in interactive systems. *Man-Computer Communication, Infotech State-of-the-Art*, 2, 1–17.

Benyon, D. (2019). *Designing User Experience*. London: Pearson UK.

Bermejo, C., & Hui, P. (2021). A survey on haptic technologies for mobile augmented reality. *ACM Computing Surveys (CSUR)*, 54(9), 1–35.

Bevan, N. (1995). Human-computer interaction standards. In Meister, D. (ed.), *Advances in Human Factors/ Ergonomics* (Vol. 20, pp. 885–890). Amsterdam: Elsevier.

Bevan, N., & Spinhof, L. (2007). Are guidelines and standards for web usability comprehensive? In *International Conference on Human-Computer Interaction*, Beijing, China, July 22–27, 2007 (pp. 407–419). Berlin, Heidelberg: Springer.

Bibri, S. E. (2015). The Shaping of Ambient Intelligence and the Internet of Things. Berlin, Heidelberg: Springer.

Bilgihan, A. (2016). Gen Y customer loyalty in online shopping: An integrated model of trust, user experience and branding. *Computers in Human Behavior*, 61, 103–113.

Biondi, F., Alvarez, I., & Jeong, K. A. (2019). Human-vehicle cooperation in automated driving: A multidisciplinary review and appraisal. *International Journal of Human-Computer Interaction*, 35(11), 932–946.

Blandford, A. (2019). HCI for health and wellbeing: Challenges and opportunities. *International Journal of Human-Computer Studies*, 131, 41–51.

Boem, A., & Troiano, G. M. (2019). Non-rigid HCI: A review of deformable interfaces and input. In *Proceedings of the 2019 on Designing Interactive Systems Conference*, San Diego CA USA, June 23–28, 2019 (pp. 885–906). New York: Association for Computing Machinery.

Boletsis, C., & Cedergren, J. E. (2019). VR locomotion in the new era of virtual reality: An empirical comparison of prevalent techniques. *Advances in Human-Computer Interaction*, 15 pages, 2019.

Borsci, S., Malizia, A., Schmettow, M., Van Der Velde, F., Tariverdiyeva, G., Balaji, D., & Chamberlain, A. (2022). The Chatbot usability scale: The design and pilot of a usability scale for interaction with AI-based conversational agents. *Personal and Ubiquitous Computing*, 26, 95–119.

Bosch, E., Bethge, D., Klosterkamp, M., & Kosch, T. (2022). Empathic technologies shaping innovative interaction: Future directions of affective computing. In *Adjunct Proceedings of the 2022 Nordic Human-Computer Interaction Conference*, Aarhus Denmark, October 8–12, 2022 (pp. 1–3). New York: Association for Computing Machinery.

Bouzit, S., Calvary, G., Coutaz, J., Chêne, D., Petit, E., & Vanderdonckt, J. (2017, May). The PDA-LPA design space for user interface adaptation. In *2017 11th International Conference on Research Challenges in Information Science (RCIS)*, Brighton, UK, May 10-12, 2017 (pp. 353–364). IEEE.

Boy, J., Detienne, F., & Fekete, J.-D. (2015). Storytelling in information visualizations: Does it engage users to explore data? *Proceedings of the 33rd Annual ACM Conference on Human Factors in Computing Systems*, Seoul Republic of Korea, April 18–23, 2015 (pp. 1449–1458). New York: Association for Computing Machinery.

Braun, M., Mainz, A., Chadowitz, R., Pfleging, B., & Alt, F. (2019). At your service: Designing voice assistant personalities to improve automotive user interfaces. In *Proceedings of the 2019 CHI Conference on Human Factors in Computing Systems*, Glasgow Scotland UK, May 4–9, 2019 (pp. 1–11). New York: Association for Computing Machinery.

Brdnik, S., Hericko, T., & Šumak, B. (2022). Intelligent user interfaces and their evaluation: A systematic mapping study. *Sensors*, 22(15), 5830.

Bröhl, C., Nelles, J., Brandl, C., Mertens, A., & Nitsch, V. (2019). Human-robot collaboration acceptance model: Development and comparison for Germany, Japan, China and the USA. *International Journal of Social Robotics,* 11, 709–726. https://doi.org/10.1007/s12369-019-00593-0.

Bruckman, A. (2014). Research ethics and HCI. In Olson, J. S., & Kellogg, W. A. (ed.), *Ways of Knowing in HCI* (pp. 449–468). Berlin, Heidelberg: Springer Science & Busines.,

Brüggemeier, B., & Lalone, P. (2022). Perceptions and reactions to conversational privacy initiated by a conversational user interface. *Computer Speech & Language*, 71, 101269.

Buie, E. (1999). HCI standards: A mixed blessing. *Interactions*, 6(2), 36–42.

Büschel, W., Lehmann, A., & Dachselt, R. (2021). Miria: A mixed reality toolkit for the in-situ visualization and analysis of spatio-temporal interaction data. In *Proceedings of the 2021 CHI Conference on Human Factors in Computing Systems*, Yokohama Japan, May 8–13, 2021 (pp. 1–15). New York: Association for Computing Machinery.

Bush, V. (1945). As we may think. *The Atlantic Monthly*, 176(1), 101–108.

Cai, C. J., Reif, E., Hegde, N., Hipp, J., Kim, B., Smilkov, D., Wattenberg, M., Viegas, F., Corrado, G. S., Stumpe, M. C., & Terry, M. (2019). Human-centered tools for coping with imperfect algorithms during medical decision-making. *Proceedings of the 2019 CHI Conference on Human Factors in Computing Systems*, Glasgow Scotland UK, May 4–9, 2019 (pp. 1–14). New York: Association for Computing Machinery.

Çakir, A. (1997). International ergonomic HCI standards. In Helander, M. G., Landauer, T. K., & Prabhu, P. V. (eds.), *Handbook of Human-Computer Interaction* (pp. 407–420). North-Holland: Elsevier.

Campbell-Kelly, M., Aspray, W. F., Yost, J. R., Tinn, H., & Díaz, G. C. (2023). *Computer: A History of the Information Machine*. New York: Taylor & Francis.

Card, S. K., Moran, T. P., & Newell, A. (1980). The keystroke-level model for user performance time with interactive systems. *Communications of the ACM*, 23(7), 396–410.

Card, S. K., Moran, T. P., & Newell, A. (1983). *The Psychology of Human-Computer Interaction*. Boca Raton, FL: CRC Press.

Carroll, J. M. (2003). *HCI Models, Theories, and Frameworks: Toward a Multidisciplinary Science*. Amsterdam: Elsevier.

Carroll, J. M. (2010). Conceptualizing a possible discipline of human-computer interaction. *Interacting with Computers*, 22(1), 3–12.

Chamberlain, A., Bødker, M., Hazzard, A., McGookin, D., De Roure, D., Willcox, P., & Papangelis, K. (2017). Audio technology and mobile human computer interaction: From space and place, to social media, music, composition and creation. *International Journal of Mobile Human Computer Interaction (IJMHCI)*, 9(4), 25–40.

Chaurasia, T., & Jain, P. K. (2019). Enhanced smart home automation system based on Internet of Things. In *2019 Third International Conference on I-SMAC (IoT in Social, Mobile, Analytics and Cloud) (I-SMAC)*, Palladam, India, December 12–14, 2019 (pp. 709–713). IEEE.

Chen, Y. T., Hsu, C. H., Chung, C. H., Wang, Y. S., & Babu, S. V. (2019). iVRNote: Design, creation and evaluation of an interactive note-taking interface for study and reflection in VR learning environments. In *2019 IEEE Conference on Virtual Reality and 3D User Interfaces (VR)*, Osaka, Japan, March 23–27, 2019 (pp. 172–180). IEEE.

Cheng, R., Wu, N., Chen, S., & Han, B. (2022). Will metaverse be nextg internet? Vision, hype, and reality. *IEEE Network*, 36(5), 197–204.

Chevalier, F., Riche, N. H., Plaisant, C., Chalbi, A., & Hurter, C. (2016). Animations 25 years later: New roles and opportunities. *Proceedings of the International Working Conference on Advanced Visual Interfaces*, Bari Italy, June 7–10, 2016 (pp. 280–287). New York: Association for Computing Machinery.

Chilana, P. K., Ko, A. J., & Wobbrock, J. (2015). From user-centered to adoption-centered design: A case study of an HCI research innovation becoming a product. *Proceedings of the 33rd Annual ACM Conference on Human Factors in Computing Systems*, Seoul Republic of Korea, April 18–23, 2015 (pp. 1749–1758). New York: Association for Computing Machinery.

Christoforou, E. G., Panayides, A. S., Avgousti, S., Masouras, P., & Pattichis, C. S. (2020). An overview of Assistive Robotics and Technologies for elderly care. In Henriques, J., Neves, N., & de Carvalho, P. (eds.), *XV Mediterranean Conference on Medical and Biological Engineering and Computing - MEDICON 2019*, Coimbra, Portugal, September 26–28, 2019. *IFMBE Proceedings* (Vol. 76). Cham: Springer. https://doi.org/10.1007/978-3-030-31635-8_118.

Claes, S., & Vande Moere, A. (2015). The role of tangible interaction in exploring information on public visualization displays. *Proceedings of the 4th International Symposium on Pervasive Displays*, Saarbruecken Germany, June 10–12, 2015 (pp. 201–207). New York: Association for Computing Machinery.

Clemmensen, T., Kaptelinin, V., & Nardi, B. (2016). Making HCI theory work: An analysis of the use of activity theory in HCI research. *Behaviour & Information Technology*, 35(8), 608–627.

Cockton, G., Woolrych, A., & Lavery, D. (2009). Inspection-based evaluations. In Sears, A. and Jacko, J. (Eds.) *Human-Computer Interaction* (pp. 289–308). Boca Raton, FL: CRC Press.

Coşar, S., Fernandez-Carmona, M., Agrigoroaie, R., Pages, J., Ferland, F., Zhao, F., … & Tapus, A. (2020). ENRICHME: Perception and interaction of an assistive robot for the elderly at home. *International Journal of Social Robotics*, 12, 779–805.

Costello, B., & Edmonds, E. (2007). A study in play, pleasure and interaction design. In *Proceedings of the 2007 Conference on Designing Pleasurable Products and Interfaces*, Helsinki Finland, August 22–25, 2007 (pp. 76–91). New York: Association for Computing Machinery..

Cowan, B. R., Clark, L., Candello, H., & Tsai, J. (2023). Introduction to this special issue: Guiding the conversation: New theory and design perspectives for conversational user interfaces. *Human-Computer Interaction*, 38(3–4), 159–167. DOI: 10.1080/07370024.2022.2161905.

Cross, E. S., Hortensius, R., & Wykowska, A. (2019). From social brains to social robots: Applying neurocognitive insights to human-robot interaction. *Philosophical Transactions of the Royal Society B*, 374(1771), 20180024.

D'Onofrio, G., Fiorini, L., Hoshino, H., Matsumori, A., Okabe, Y., Tsukamoto, M., … & Sancarlo, D. (2019). Assistive robots for socialization in elderly people: Results pertaining to the needs of the users. *Aging Clinical and Experimental Research*, 31, 1313–1329.

Daily, S. B., James, M. T., Cherry, D., Porter III, J. J., Darnell, S. S., Isaac, J., & Roy, T. (2017). Affective computing: Historical foundations, current applications, and future trends. In Jeon, M. (ed.), *Emotions and Affect in Human Factors and Human-Computer Interaction* (pp. 213–231). Cambridge, MA: Academic Press.

Dainow, E. (2017). *A Concise History of Computers, Smartphones and the Internet*. Ernie Dainow.

Dangxiao, W., Yuan, G., Shiyi, L., Zhang, Y., Weiliang, X., & Jing, X. (2019). Haptic display for virtual reality: Progress and challenges. *Virtual Reality & Intelligent Hardware*, 1(2), 136–162.

De Graaf, M. M. (2016). An ethical evaluation of human-robot relationships. *International Journal of Social Robotics*, 8, 589–598.

De Oliveira, G. A. A., de Bettio, R. W., & Freire, A. P. (2016, October). Accessibility of the smart home for users with visual disabilities: An evaluation of open source mobile applications for home automation. In *Proceedings of the 15th Brazilian Symposium on Human Factors in Computing Systems* (pp. 1–10).

De Souza, C. S. (2005). Semiotic engineering: Bringing designers and users together at interaction time. *Interacting with Computers*, 17(3), 317–341.

Denisova, A., Bopp, J. A., Nguyen, T. D., & Mekler, E. D. (2021). "Whatever the emotional experience, it's up to them": Insights from designers of emotionally impactful games. *Proceedings of the 2021 CHI Conference on Human Factors in Computing Systems*, Yokohama Japan, May 8–13, 2021 (pp. 1–9). New York: Association for Computing Machinery.

Desmet, P., Overbeeke, K., & Tax, S. (2001). Designing products with added emotional value: Development and application of an approach for research through design. *The Design Journal*, 4(1), 32–47.

Detjen, H., Schneegass, S., Geisler, S., Kun, A., & Sundar, V. (2022). An emergent design framework for accessible and inclusive future mobility. In *Proceedings of the 14th International Conference on Automotive User Interfaces and Interactive Vehicular Applications*, Seoul Republic of Korea, September 17–20, 2022 (pp. 1–12). New York: Association for Computing Machinery.

Dincelli, E., & Yayla, A. (2022). Immersive virtual reality in the age of the Metaverse: A hybrid-narrative review based on the technology affordance perspective. *The Journal of Strategic Information Systems*, 31(2), 101717.

Dix, A. (2017). Human-computer interaction, foundations and new paradigms. *Journal of Visual Languages & Computing*, 42, 122–134.

Dix, A., Finlay, J., Abowd, G. D., & Beale, R. (2003). *Human-Computer Interaction*. London: Pearson Education. ISBN 9780130461094.

Draper, S. W., & Norman, D. A. (1986). *User Centered System Design: New Perspectives on Human-Computer Interaction*. Hillsdale, NJ: L. Erlbaum Associates.

Easwara Moorthy, A., & Vu, K. P. L. (2015). Privacy concerns for use of voice activated personal assistant in the public space. *International Journal of Human-Computer Interaction*, 31(4), 307–335.

Eisenstein, J., Vanderdonckt, J., & Puerta, A. (2000). Adapting to mobile contexts with user-interface modeling. *Proceedings Third IEEE Workshop on Mobile Computing Systems and Applications*, Los Alamitos, CA (pp. 83–92). DOI: 10.1109/MCSA.2000.895384.

Emiliani, P. L., Stephanidis, C., & Vanderheiden, G. (2011). Technology and inclusion-past, present and foreseeable future. *Technology and Disability*, 23(3), 101–114.

Engelbart, D. C., & English, W. K. (1968). A research center for augmenting human intellect. In *Proceedings of the December 9-11, 1968, Fall Joint Computer Conference, Part I*, San Francisco California, December 9–11, 1968 (pp. 395–410). New York: Association for Computing Machinery.

Erickson, T. D. (1995). Working with interface metaphors. In Baecke, R. M. (ed.), *Readings in Human-Computer Interaction* (pp. 147–151). Amsterdam: Elsevier.

Ermes, M., Pärkkä, J., Mäntyjärvi, J., & Korhonen, I. (2008). Detection of daily activities and sports with wearable sensors in controlled and uncontrolled conditions. *IEEE Transactions on Information Technology in Biomedicine*, 12(1), 20–26.

Esposito, A., Esposito, A. M., Troncone, A., Maldonato, M. N., Vogel, C., Bourbakis, N., & Cordasco, G. (2020). Seniors' appreciation of humanoid robots. In Esposito, A., Faundez-Zanuy, M., Morabito, F., & Pasero, E. (eds.), *Neural Approaches to Dynamics of Signal Exchanges: Smart Innovation, Systems and Technologies* (Vol. 151). Singapore: Springer. https://doi.org/10.1007/978-981-13-8950-4_30.

European Union (2016). Regulation (EU) 2016/679 of the European Parliament and of the Council of 27 April 2016 on the protection of natural persons with regard to the processing of personal data and on the free movement of such data, and repealing Directive 95/46/EC (General Data Protection Regulation). https://eur-lex.europa.eu/eli/reg/2016/679/oj.

Fan, H., & Poole, M. (2006). What is personalization? Perspectives on the design and implementation of personalization in information systems. *Journal of Organizational Computing and Electronic Commerce*, 16, 179–202. 10.1207/s15327744joce1603&4_2.

Fang, X., & Salvendy, G. (2003). Customer-centered rules for design of e-commerce web sites. *Communications of the ACM*, 46(12), 332–336.

Farhadi-Niaki, F., Etemad, S. A., & Arya, A. (2013). Design and usability analysis of gesture-based control for common desktop tasks. In *Human-Computer Interaction. Interaction Modalities and Techniques: 15th International Conference, HCI International 2013*, *Proceedings, Part IV 15*, Las Vegas, NV, July 21-26, 2013 (pp. 215–224). Berlin Heidelberg: Springer.

Ferraro, A., & Giacalone, M. (2022). A review about machine and deep learning approaches for intelligent user interfaces. In Barolli, L., Hussain, F., & Enokido, T. (eds.), *Advanced Information Networking and Applications. AINA 2022. Lecture Notes in Networks and Systems* (Vol. 451). Cham: Springer. https://doi.org/10.1007/978-3-030-99619-2_9.

Ferreira, J. J., Fernandes, C. I., Rammal, H. G., & Veiga, P. M. (2021). Wearable technology and consumer interaction: A systematic review and research agenda. *Computers in Human Behavior*, 118, 106710.

Finstad, K. (2010). The usability metric for user experience. *Interacting with Computers*, 22(5), 323–327.

Fiorini, L., De Mul, M., Fabbricotti, I., Limosani, R., Vitanza, A., D'Onofrio, G., Tsui, M., Sancarlo, D., Giuliani, F., Greco, A., Guiot, D., Senges, E., & Cavallo, F. (2021). Assistive robots to improve the independent living of older persons: Results from a needs study. *Disability and Rehabilitation: Assistive Technology*, 16(1), 92–102. DOI: 10.1080/17483107.2019.1642392.

Foster, M. E. (2019). Face-to-face conversation: Why embodiment matters for conversational user interfaces. In *Proceedings of the 1st International Conference on Conversational User Interfaces,* Dublin Ireland, August 22–23, 2019 (pp. 1–3). New York: Association for Computing Machinery.

Fox, J., & Gambino, A. (2021). Relationship development with humanoid social robots: Applying interpersonal theories to human-robot interaction. *Cyberpsychology, Behavior, and Social Networking* 24(5), 294–299.

Francés-Morcillo, L., Morer-Camo, P., Rodríguez-Ferradas, M. I., & Cazón-Martín, A. (2020). Wearable design requirements identification and evaluation. *Sensors*, 20(9), 2599.

Frauenberger, C., Rauhala, M., & Fitzpatrick, G. (2017). In-action ethics. *Interacting with Computers*, 29(2), 220–236.

Frauenberger, C. (2019). Entanglement HCI the next wave?. *ACM Transactions on Computer-Human Interaction (TOCHI)*, 27(1), 1–27.

Freeman, E., Wilson, G., Vo, D.-B., Ng, A., Politis, I., & Brewster, S. (2017). Multimodal feedback in HCI: Haptics, non-speech audio, and their applications. In Oviatt, S., Schuller, B., Cohen, P., Sonntag, D., & Potamianos, G. (eds.), *The Handbook of Multimodal-Multisensor Interfaces: Foundations, User Modeling, and Common Modality Combinations-Volume* 1 (pp. 277–317). San Rafael, CA: Morgan & Claypool.

Friedewald, M., & Raabe, O. (2011). Ubiquitous computing: An overview of technology impacts. *Telematics and Informatics*, 28(2), 55–65. https://doi.org/10.1016/j.tele.2010.09.001.

Friedman, B., & Hendry, D. G. (2019). *Value Sensitive Design: Shaping Technology with Moral Imagination*. Cambridge, MA: MIT Press.

Funk, M., Rosen, P. H., & Wischniewski, S. (2020). Human-centered HRI design-the more individual the better? *Workshop on Behavioral Patterns and Interaction Modelling for Personalized Human-Robot Interaction*.

Furlough, C., Stokes, T., & Gillan, D. J. (2021). Attributing blame to robots: I. The influence of robot autonomy. *Human Factors*, 63(4), 592–602. https://doi.org/10.1177/0018720819880641.

Gabriel, P., Bovenschulte, M., Hartmann, E., Groß, W., Strese, H., Bayarou, K., Haisch, M., Mattheß, M., Brune, C., & Strauss, H. (2006). *Pervasive Computing: Trends and Impacts*. Ingelheim: SecuMedia.

Gajendar, U. (2017). Notes on the future of interaction design. *Interactions*, 24(5), 46–51.

Gallagher, K. M., Cameron, L., De Carvalho, D., & Boule, M. (2021). Does using multiple computer monitors for office tasks affect user experience? A systematic review. *Human Factors*, 63(3), 433–449.

Garibay, O., Winslow, B., Andolina, S., Antona, M., Bodenschatz, A., Coursaris, C., … & Xu, W. (2023). Six human-centered artificial intelligence grand challenges. *International Journal of Human-Computer Interaction*, 39(3), 391–437.

Gasteiger, N., Ahn, H. S., Fok, C., Lim, J., Lee, C., MacDonald, B. A., … & Broadbent, E. (2022). Older adults' experiences and perceptions of living with Bomy, an assistive dailycare robot: A qualitative study. *Assistive Technology*, 34(4), 487–497.

Gervasi, R., Mastrogiacomo, L., & Franceschini, F. (2020). A conceptual framework to evaluate human-robot collaboration. *The International Journal of Advanced Manufacturing Technology*, 108, 841–865.

Ghazali, T. K., & Zakaria, N. H. (2018). Security, comfort, healthcare, and energy saving: A review on biometric factors for smart home environment. *Journal of Computers*, 29(1), 189–208.

Ghazizadeh, M., Lee, J. D., & Boyle, L. N. (2012). Extending the technology acceptance model to assess automation. *Cognition, Technology & Work*, 14, 39–49.

Glasser, A., Mande, V., & Huenerfauth, M. (2020). Accessibility for deaf and hard of hearing users: Sign language conversational user interfaces. In *Proceedings of the 2nd Conference on Conversational User Interfaces*, Bilbao Spain, July 22–24, 2020 (pp. 1–3). New York: Association for Computing Machinery.

Gorlewicz, J. L., Tennison, J. L., Uesbeck, P. M., Richard, M. E., Palani, H. P., Stefik, A., … & Giudice, N. A. (2020). Design guidelines and recommendations for multimodal, touchscreen-based graphics. *ACM Transactions on Accessible Computing (TACCESS)*, 13(3), 1–30.

Gough, P., Pschetz, L., Ahmadpour, N., Hepburn, L. A., Cooper, C., Ramirez-Figueroa, C., & Catts, O. (2020). The nature of biodesigned systems: Directions for HCI. In *Companion Publication of the 2020 ACM Designing Interactive Systems Conference*, Eindhoven Netherlands July 6–10, 2020 (pp. 389–392). New York: Association for Computing Machinery.

Grudin, J. (2012). Introduction: A moving target: The evolution of human-computer interaction. In Jacko, D. (ed.), *Human Computer Interaction Handbook* (pp. xxvii–lxi). Boca Raton, FL: CRC Press.

Grudin, J. (2017). *From Tool to Partner: The Evolution of Human-Computer Interaction*. San Rafael, CA: Morgan & Claypool Publishers.

Grudin, J. (2022). *From Tool to Partner: The Evolution of Human-Computer Interaction*. Berlin, Germany: Springer Nature.

Gupta, A., Bridges, N., & Kamino, W. (2021). Musically assistive robot for the elderly in isolation. In *Companion of the 2021 ACM/IEEE International Conference on Human-Robot Interaction*, Boulder CO USA March 8–11, 2021 (pp. 620–621). New York: Association for Computing Machinery.

Gupta, S., & Tanenbaum, T. J. (2019). Shiva's Rangoli: Tangible interactive storytelling in ambient environments. In *Companion Publication of the 2019 on Designing Interactive Systems Conference 2019 Companion*, San Diego CA USA June 23–28, 2019 (pp. 29–32). New York: Association for Computing Machinery.

Hagendorff, T. (2020). The ethics of AI ethics: An evaluation of guidelines. *Minds and Machines*, 30(1), 99–120.

Haigh, T., Priestley, P. M., & Rope, C. (2016). *ENIAC in Action: Making and Remaking the Modern Computer*. Cambridge, MA: MIT Press.

Häkkilä, J., & Mäntyjärvi, J. (2006). Developing design guidelines for context-aware mobile applications. In *Proceedings of the 3rd International Conference on Mobile Technology, Applications & Systems*, Bangkok, Thailand, October 25–27, 2006 (pp. 24-es). New York: Association for Computing Machinery

Han, J., Zhang, Z., Pantic, M., & Schuller, B. (2021). Internet of emotional people: Towards continual affective computing cross cultures via audiovisual signals. *Future Generation Computer Systems*, 114, 294–306.

Hansson, L. Å. E. J., Cerratto Pargman, T., & Pargman, D. S. (2021). A decade of sustainable HCI: Connecting SHCI to the sustainable development goals. In *Proceedings of the 2021 CHI Conference on Human Factors in Computing Systems*, Yokohama Japan, May 8–13, 2021 (pp. 1–19). New York: Association for Computing Machinery.

Harper, R. H. (2019). The role of HCI in the age of AI. *International Journal of Human-Computer Interaction*, 35(15), 1331–1344.

Hartmann, M. (2009). Challenges in developing user-adaptive intelligent user interfaces. In *LWA* (pp. ABIS–6).

Hartson, R., & Pyla, P. S. (2018). *The UX Book: Agile UX Design for a Quality User Experience*. Burlington, MA: Morgan Kaufmann.

Hasan, M. S., & Yu, H. (2017). Innovative developments in HCI and future trends. *International Journal of Automation and Computing*, 14, 10–20.

Hassenzahl, M. (2004). The interplay of beauty, goodness, and usability in interactive products. *Human-Computer Interaction*, 19(4), 319–349.

Hassenzahl, M., & Tractinsky, N. (2006). User experience: A research agenda. *Behaviour & Information Technology*, 25(2), 91–97.

Henschel, A., Hortensius, R., & Cross, E. S. (2020). Social cognition in the age of human-robot interaction. *Trends in Neurosciences*, 43(6), 373–384.

Hinderks, A., Meiners, A. L., Domínguez Mayo, F. J., & Thomaschewski, J. (2019). Interpreting the results from the user experience questionnaire (UEQ) using importance-performance analysis (IPA). In *WEBIST 2019: 15th International Conference on Web Information Systems and Technologies (2019)*, Vienna, Austria, September 18–20, 2019 (pp. 388–395). ScitePress Digital Library.

Hirsh-Pasek, K., Zosh, J. M., Golinkoff, R. M., Gray, J. H., Robb, M. B., & Kaufman, J. (2015). Putting education in "educational" apps: Lessons from the science of learning. *Psychological Science in the Public Interest*, 16(1), 3–34.

Holtzblatt, K., & Beyer, H. (1997). *Contextual Design: Defining Customer-Centered Systems*. Amsterdam: Elsevier.

Höök, K. (2000). Steps to take before intelligent user interfaces become real. *Interacting with Computers*, 12(4), 409–426.

Hou, J., Arpan, L., Wu, Y., Feiock, R., Ozguven, E., & Arghandeh, R. (2020). The road toward smart cities: A study of citizens' acceptance of mobile applications for city services. *Energies*, 13(10), 2496.

Ilhami, R., Endah Marlovia, E. M., & Achmad, W. (2022). Smart government policy implementation for smart city concept realization. *International Journal of Health Sciences Scopus Coverage Years: From 2021 to Present*, 8379–8389.

Instone, K. (1997). Usability heuristics for the Web. *Retrieved April 8, 2024, from http://instone.org/heuristics*

International Organization for Standardization (ISO) (1998). ISO 9241-11:1998 Ergonomic requirements for office work with visual display terminals (VDTs) Part 11: Guidance on usability

International Organization for Standardization (ISO) (2010). ISO 9241-210:2010 Ergonomics of human-system interaction - Part 210: Human-centred design for interactive systems.

International Organization for Standardization (ISO) (2018). ISO 9241-11:2018 Ergonomics of human-system interaction - Part 11: Usability: Definitions and concepts.

International Organization for Standardization (ISO) (2019). ISO 9241-210:2019 Ergonomics of human-system interaction - Part 210: Human-centred design for interactive systems.

Ishii, H., & Ullmer, B. (1997). Tangible bits: Towards seamless interfaces between people, bits and atoms. In *Proceedings of the ACM SIGCHI Conference on Human Factors in Computing Systems*, Atlanta Georgia USA, March 22–27, 1997 (pp. 234–241). New York: Association for Computing Machinery.

Ismagilova, E., Hughes, L., Dwivedi, Y. K., & Raman, K. R. (2019). Smart cities: Advances in research-An information systems perspective. *International Journal of Information Management*, 47, 88–100.

Issa, T., & Isaias, P. (2022). Usability and human-computer interaction (HCI). In Issa, T., & Isaias, P. (eds.), *Sustainable Design: HCI, Usability and Environmental Concerns* (pp. 23–40). Berlin, Heidelberg: Springer.

Isyanto, H., Arifin, A. S., & Suryanegara, M. (2020). Design and implementation of IoT-based smart home voice commands for disabled people using Google Assistant. In *2020 International Conference on Smart Technology and Applications (ICoSTA)*, Surabaya, Indonesia, February 20, 2020 (pp. 1–6). IEEE.

Jacob, R. J., & Karn, K. S. (2003). Eye tracking in human-computer interaction and usability research: Ready to deliver the promises. In Radach, R., Hyona, J., & Deubel, H. (eds.), *The Mind's Eye* (pp. 573–605). North-Holland: Elsevier.

Javed, A. R., Shahzad, F., ur Rehman, S., Zikria, Y. B., Razzak, I., Jalil, Z., & Xu, G. (2022). Future smart cities: Requirements, emerging technologies, applications, challenges, and future aspects. *Cities*, 129, 103794.

Jeon, M. (2017). Emotions and affect in human factors and human-computer interaction: Taxonomy, theories, approaches, and methods. In Jeon, M. (ed.), *Emotions and Affect in Human Factors and Human-Computer Interaction* (pp. 3–26). Cambridge, MA: Academic Press.

Jobin, A., Ienca, M., & Vayena, E. (2019). The global landscape of AI ethics guidelines. *Nature Machine Intelligence*, 1(9), 389–399.

Joosse, M., Lohse, M., Berkel, N. V., Sardar, A., & Evers, V. (2021). Making appearances: How robots should approach people. *ACM Transactions on Human-Robot Interaction (THRI)*, 10(1), 1–24.

Kamal, S. A., Shafiq, M., & Kakria, P. (2020). Investigating acceptance of telemedicine services through an extended technology acceptance model (TAM). *Technology in Society*, 60, 101212.

Karpov, A., & Yusupov, R. (2018). Multimodal interfaces of human-computer interaction. *Herald of the Russian Academy of Sciences*, 88, 67–74.

Karwowski, W. (2005). Ergonomics and human factors: The paradigms for science, engineering, design, technology, and management of human-compatible systems. *Ergonomics*, 48(5), 436–463.

Kawala-Sterniuk, A., Browarska, N., Al-Bakri, A., Pelc, M., Zygarlicki, J., Sidikova, M., ... & Gorzelanczyk, E. J. (2021). Summary of over fifty years with brain-computer interfaces: A review. *Brain Sciences*, 11(1), 43.

Khatoun, R., & Zeadally, S. (2016). Smart cities: Concepts, architectures, research opportunities. *Communications of the ACM*, 59(8), 46–57.

Kim, S. K., Kang, S. J., Choi, Y. J., Choi, M. H., & Hong, M. (2017). Augmented-reality survey: From concept to application. *KSII Transactions on Internet & Information Systems*, 11(2), 982–1004.

Kirkpatrick, A., Connor, O. J., Campbell, A., & Cooper, M. (2018). Web content accessibility guidelines (WCAG) 2.1. W3C. https://www.w3.org/TR/WCAG21/.

Knijnenburg, B. P. (2017). Privacy? I Can't Even! Making a case for user-tailored privacy. *IEEE Security & Privacy*, 15(4), 62–67.

Kocaballi, A. B., Laranjo, L., & Coiera, E. (2019). Understanding and measuring user experience in conversational interfaces. *Interacting with Computers*, 31(2), 192–207.

Krumm, J. (2018). *Ubiquitous Computing Fundamentals*. Boca Raton, FL: CRC Press.

Kurniawan, S., & Zaphiris, P. (2005). Derived web design guidelines for older people. In *Proceedings of the 7th International ACM SIGACCESS Conference on Computers and Accessibility*, Baltimore MD USA October 9–12, 2005 (pp. 129–135). New York: Association for Computing Machinery.

Kyriakoullis, L., & Zaphiris, P. (2016). Culture and HCI: A review of recent cultural studies in HCI and social networks. *Universal Access in the Information Society*, 15, 629–642.

Lavie, T., & Tractinsky, N. (2004). Assessing dimensions of perceived visual aesthetics of web sites. *International Journal of Human-Computer Studies*, 60(3), 269–298.

Law, E., Roto, V., Vermeeren, A. P., Kort, J., & Hassenzahl, M. (2008). Towards a shared definition of user experience. In *CHI'08 Extended Abstracts on Human Factors in Computing Systems* (pp. 2395–2398). ACM.

Lee, B., Isenberg, P., Riche, N. H., & Carpendale, S. (2012). Beyond mouse and keyboard: Expanding design considerations for information visualization interactions. *IEEE Transactions on Visualization and Computer Graphics*, 18(12), 2689–2698.

Leonidis, A., Korozi, M., Antona, M., & Stephanidis, C. (2022). Interaction in smart cities. In Duffy, V. G., Ziefle, M., Rau, P.-L. P., & Tseng, M. M. (eds.), *Human-Automation Interaction: Mobile Computing* (pp. 513–564). Berlin, Heidelberg: Springer.

Leonidis, A., Korozi, M., Kouroumalis, V., Adamakis, E., Milathianakis, D., & Stephanidis, C. (2021a, June). Going beyond second screens: Applications for the multi-display intelligent living room. In *ACM International Conference on Interactive Media Experiences*, Virtual Event USA June 21–23, 2021 (pp. 187–193). New York: Association for Computing Machinery

Leonidis, A., Korozi, M., Kouroumalis, V., Poutouris, E., Stefanidi, E., Arampatzis, D., ... & Antona, M. (2019). Ambient intelligence in the living room. *Sensors*, 19(22), 5011.

Leonidis, A., Korozi, M., Sykianaki, E., Tsolakou, E., Kouroumalis, V., Ioannidi, D., ... & Stephanidis, C. (2021b). Improving stress management and sleep hygiene in intelligent homes. *Sensors*, 21(7), 2398.

Leschanowsky, A., Popp, B., & Peters, N. (2023). Privacy strategies for conversational AI and their influence on users' perceptions and decision-making. In *Proceedings of the 2023 European Symposium on Usable Security* (pp. 296–311).

Lewis, C. H. (1982). Using the "thinking aloud" method in cognitive interface design. Technical report, RC-9265, IBM.

Lewis, J. R. (1992). Psychometric evaluation of the post-study system usability questionnaire: The PSSUQ. In *Proceedings of the Human Factors Society Annual Meeting* (Vol. 36, No. 16, pp. 1259–1260). Los Angeles, CA: Sage Publications.

Lewis, J. R. (1995). IBM computer usability satisfaction questionnaires: psychometric evaluation and instructions for use. *International Journal of Human-Computer Interaction*, 7(1), 57–78.

Lewis, J. R. (2012). Usability testing. In Salvendy, G., & Karwowski, W. (eds.), *Handbook of Human Factors and Ergonomics* (pp. 1267–1312). Hoboken, NJ: John Wiley & Sons.

Lilis, Y., Zidianakis, E., Partarakis, N., Ntoa, S., & Stephanidis, C. (2019). A framework for personalised HMI interaction in ADAS systems. In *Proceedings of the 5th International Conference on Vehicle Technology and Intelligent Transport Systems (VEHITS 2019)* Heraklion Crete Greece, May 3–5, 2019 (pp. 586–593). SCITEPRESS – Science and Technology Publications, Lda.

Linehan, C., Kirman, B., Lawson, S., & Chan, G. (2011). Practical, appropriate, empirically-validated guidelines for designing educational games. In *Proceedings of the SIGCHI Conference on Human Factors in Computing Systems*, Vancouver BC Canada, May 7–12, 2011 (pp. 1979–1988). New York: Association for Computing Machinery.

Lister, K., Coughlan, T., Iniesto, F., Freear, N., & Devine, P. (2020, April). Accessible conversational user interfaces: Considerations for design. In *Proceedings of the 17th International Web for All Conference*, Taipei Taiwan, April 20–21, 2020 (pp. 1–11). New York: Association for Computing Machinery.

Liu, A., & Tan, H. (2022). Research on the trend of automotive user experience. In Rau, P. L. P. (ed.), *HCII 2022: Cross-Cultural Design: Product and Service Design, Mobility and Automotive Design, Cities, Urban Areas, and Intelligent Environments Design*. Lecture Notes in Computer Science (Vol. 13314). Cham: Springer. https://doi.org/10.1007/978-3-031-06053-3_13.

Liu, L., Stroulia, E., Nikolaidis, I., Miguel-Cruz, A., & Rincon, A. R. (2016). Smart homes and home health monitoring technologies for older adults: A systematic review. *International Journal of Medical Informatics*, 91, 44–59.

Loureiro, B., & Rodrigues, R. (2014). Design guidelines and design recommendations of multi-touch interfaces for elders. *Proceedings of the ACHI* (pp. 41–47).

Lowgren, J., & Stolterman, E. (2007). *Thoughtful Interaction Design: A Design Perspective on Information Technology*. Cambridge, MA: MIT Press.

MacKenzie, I. S. (2003). Motor behaviour models for human-computer interaction. In Carroll, J. M., & Carroll, J. M. (eds.), *HCI Models, Theories, and Frameworks: Toward a Multidisciplinary Science* (pp. 27–54). Amsterdam: Elsevier Science & Technology Books.

Madakam, S., Ramasamy, R., & Tirupathi, S. (2015). Internet of Things (IoT): A literature review. *Journal of Computer and Communications*, 3(05), 164.

Maguire, M. (2023). A review of usability guidelines for E-commerce website design. In Marcus, A., Rosenzweig, E., & Soares, M. M. (eds.), *HCII 2023: Design, User Experience, and Usability*, Lecture Notes in Computer Science (Vol. 14032). Cham: Springer. https://doi.org/10.1007/978-3-031-3570 2-2_3.

Mahroo, A., Greci, L., & Sacco, M. (2019). HoloHome: An augmented reality framework to manage the smart home. In *Augmented Reality, Virtual Reality, and Computer Graphics: 6th International Conference, AVR 2019, Proceedings, Part II*, Santa Maria al Bagno, Italy, June 24–27, 2019 (pp. 137–145). Berlin, Heidelberg: Springer International Publishing.

Maiseli, B., Abdalla, A. T., Massawe, L. V., Mbise, M., Mkocha, K., Nassor, N. A., … & Kimambo, S. (2023). Brain-computer interface: Trend, challenges, and threats. *Brain Informatics*, 10(1), 20.

Makarius, E. E., Mukherjee, D., Fox, J. D., & Fox, A. K. (2020). Rising with the machines: A sociotechnical framework for bringing artificial intelligence into the organization. *Journal of Business Research*, 120, 262–273.

Malche, T., Maheshwary, P., & Kumar, R. (2019). Environmental monitoring system for smart city based on secure Internet of Things (IoT) architecture. *Wireless Personal Communications*, 107(4), 2143–2172.

Mann, S., Havens, J. C., Iorio, J., Yuan, Y., & Furness, T. (2018). All reality: Values, taxonomy, and continuum, for virtual, augmented, eXtended/MiXed (X), Mediated (X, Y), and multimediated reality/intelligence. *Presented at the AWE 2018*.

Manojlović, S., & Kumarswamy, S. (2021). To chat or not to chat? Assessing how smartness, personalisation and efficiency differ in conversational and graphical user interfaces. In Arai, K. (ed.), *Advances in Information and Communication. FICC 2021: Advances in Intelligent Systems and Computing* (Vol. 1364). Cham: Springer. https://doi.org/10.1007/978-3-030-73103-8_68.

Marcus, A., Kurosu, M., Ma, X., & Hashizume, A. (2017). *Cuteness Engineering: Designing Adorable Products and Services*. Berlin, Heidelberg: Springer.

Margetis, G., Antona, M., Ntoa, S., & Stephanidis, C. (2012). Towards accessibility in ambient intelligence environments. In *Ambient Intelligence: Third International Joint Conference, AMI 2012*, Pisa, Italy, November 13–15, 2012 (pp. 328–337). Berlin Heidelberg: Springer.

Margetis, G., Apostolakis, K. C., Ntoa, S., Papagiannakis, G., & Stephanidis, C. (2020). X-reality museums: Unifying the virtual and real world towards realistic virtual museums. *Applied Sciences*, 11(1), 338.

Margetis, G., Ntoa, S., Antona, M., & Stephanidis, C. (2019a). Augmenting natural interaction with physical paper in ambient intelligence environments. *Multimedia Tools and Applications*, 78, 13387–13433.

Margetis, G., Ntoa, S., Antona, M., & Stephanidis, C. (2021). Human-centered design of artificial intelligence. In Salvendy, G., & Karwowski, W. (eds.), *Handbook of Human Factors and Ergonomics* (pp. 1085–1106). Hoboken, NJ: John Wiley & Sons.

Margetis, G., Papagiannakis, G., & Stephanidis, C. (2019b). Realistic natural interaction with virtual statues in x-reality environments. *The International Archives of the Photogrammetry, Remote Sensing and Spatial Information Sciences*, 42, 801–808.

Marmaras, N., Poulakakis, G., & Papakostopoulos, V. (1999). Ergonomic design in ancient Greece. *Applied Ergonomics*, 30(4), 361–368. DOI: 10.1016/S0003-6870(98)00050-7.

Martinez-Martin, E., & Cazorla, M. (2019). A socially assistive robot for elderly exercise promotion. *IEEE Access*, 7, 75515–75529.

Maybury, M. (1998). Intelligent user interfaces: An introduction. In *Proceedings of the 4th International Conference on Intelligent User Interfaces*, Los Angeles California USA, January 5–8, 1999 (pp. 3–4). New York: Association for Computing Machinery.

McAweeney, E., Zhang, H., & Nebeling, M. (2018). User-driven design principles for gesture representations. In *Proceedings of the 2018 CHI Conference on Human Factors in Computing Systems*, Montreal QC Canada, April 21–26, 2018 (pp. 1–13). New York: Association for Computing Machinery.

McCarthy, J. (1988). BP Bloomfield, the question of artificial intelligence: Philosophical and sociological perspectives. *Annals of the History of Computing*, 10(3), 224–229.

Mcknight, D. H., Carter, M., Thatcher, J. B., & Clay, P. F. (2011). Trust in a specific technology: An investigation of its components and measures. *ACM Transactions on Management Information Systems (TMIS)*, 2(2), 1–25.

McKnight, L., & Cassidy, B. (2012). Children's interaction with mobile touch-screen devices: Experiences and guidelines for design. In Lumsden, J. (ed.), *Social and Organizational Impacts of Emerging Mobile Devices: Evaluating Use* (pp. 72–89). Hershey, PA: IGI Global.

Meixner, G., & Müller, C. (2017). *Automotive User Interfaces*. Cham, Switzerland: Springer.

Metsis, V., Le, Z., Lei, Y., & Makedon, F. (2008). Towards an evaluation framework for assistive environments. In *Proceedings of the 1st International Conference on Pervasive Technologies Related to Assistive Environments*, Athens Greece, July 16–18, 2008 (pp. 1–8). New York: Association for Computing Machinery.

Mezhoudi, N., & Vanderdonckt, J. (2021). Toward a task-driven intelligent GUI adaptation by mixed-initiative. *International Journal of Human-Computer Interaction*, 37(5), 445–458.

Miniukovich, A., & Marchese, M. (2020). Relationship between visual complexity and aesthetics of webpages. *Proceedings of the 2020 CHI Conference on Human Factors in Computing Systems*, Honolulu HI USA, April 25–30, 2020 (pp. 1–13). New York: Association for Computing Machinery.

Monk, A. F. (2014). *Fundamentals of Human-Computer Interaction*. Cambridge, MA: Academic Press.

Montag, C., Blaszkiewicz, K., Sariyska, R., Lachmann, B., Andone, I., Trendafilov, B., Eibes, M., & Markowetz, A. (2015). Smartphone usage in the 21st century: Who is active on WhatsApp? *BMC Research Notes*, 8(1), 1–6.

Moustris, G., Kardaris, N., Tsiami, A., Chalvatzaki, G., Koutras, P., Dometios, A., ... & Mavridis, P. (2021). The i-walk lightweight assistive rollator: First evaluation study. *Frontiers in Robotics and AI*, 8, 677542.

Mueller, F., & Isbister, K. (2014). Movement-based game guidelines. In *Proceedings of the Sigchi Conference on Human Factors in Computing Systems*, Toronto Ontario Canada, 26 April 2014- 1 May 2014 (pp. 2191–2200). New York: Association for Computing Machinery.

Muñoz-Saavedra, L., Miró-Amarante, L., & Domínguez-Morales, M. (2020). Augmented and virtual reality evolution and future tendency. *Applied Sciences*, 10(1), 322.

Munteanu, C., Jones, M., Oviatt, S., Brewster, S., Penn, G., Whittaker, S., Rajput, N., & Nanavati, A. (2013). We need to talk: HCI and the delicate topic of spoken language interaction. In *CHI'13 Extended Abstracts on Human Factors in Computing Systems* (pp. 2459–2464).

Murad, C., Munteanu, C., Clark, L., & Cowan, B. R. (2018). Design guidelines for hands-free speech interaction. In *Proceedings of the 20th International Conference on Human-Computer Interaction with Mobile Devices and Services Adjunct*, Barcelona Spain, September 3–6, 2018 (pp. 269–276). New York: Association for Computing Machinery.

Murad, C., Munteanu, C., Cowan, B. R., & Clark, L. (2019). Revolution or evolution? Speech interaction and HCI design guidelines. *IEEE Pervasive Computing*, 18(2), 33–45.

Myers, B. A. (1998). A brief history of human-computer interaction technology. *Interactions*, 5(2), 44–54.

Nath, R. K., & Thapliyal, H. (2021). Wearable health monitoring system for older adults in a smart home environment. In *2021 IEEE Computer Society Annual Symposium on VLSI (ISVLSI)* (pp. 390–395). IEEE.

Newell, A. (1971). AIP memo 1: Notes on a proposal for a psychological research unit, October 1974 reproduction of a memo from January 1971, Carnegie Mellon University Library Allen Newell Collection (subsequently ANC), Box 91, Folder 6326, p. 1.

Nielsen, J. (1994a). Enhancing the explanatory power of usability heuristics. In *Proceedings of the SIGCHI Conference on Human Factors in Computing Systems* Boston Massachusetts USA, April 24–28, 1994 (pp. 152–158). New York: Association for Computing Machinery.

Nielsen, J. (1994b). Heuristic evaluation. In Nielsen, J., & Mack, R. L. (eds.), *Usability Inspection Methods* (pp. 25–62). New York: John Wiley & Sons.

Nielsen, J. (1994c). *Usability Engineering*. Amsterdam: Elsevier.

Nielsen, J. (2017). A 100-year view of user experience. Nielsen Norman Group. Available online: https://www.nngroup.com/articles/100-years-ux/.

NNGroup (2021). *UX 2050* (Jakob Nielsen keynote) [Video]. YouTube. https://www.youtube.com/watch?v=y4J6JR9jEUk.

Nord, J. H., Koohang, A., & Paliszkiewicz, J. (2019). The Internet of Things: Review and theoretical framework. *Expert Systems with Applications*, 133, 97–108.

Norman, D. (2013). The design of everyday things. In Norman, D. (ed.), *Revised and Expanded Edition*. New York: Basic Books. ISBN 978-0-465-06710-7.

Ntoa, S., Margetis, G., Antona, M., & Stephanidis, C. (2019). UXAmI observer: An automated user experience evaluation tool for ambient intelligence environments. In *Intelligent Systems and Applications: Proceedings of the 2018 Intelligent Systems Conference (IntelliSys)*, London UK, September 6–7, 2018 (Vol. 1, pp. 1350–1370). Cham: Springer International Publishing.

Ntoa, S., Margetis, G., Antona, M., & Stephanidis, C. (2021). User experience evaluation in intelligent environments: A comprehensive framework. *Technologies*, 9(2), 41.

Ntoa, S., Margetis, G., Antona, M., & Stephanidis, C. (2022). Digital accessibility in intelligent environments. In Duffy, V. G., Lehto, M., Yih, Y., & Proctor, R. W. (eds.), *Human-Automation Interaction: Manufacturing, Services and User Experience* (pp. 453–475). Cham: Springer International Publishing.

Olde Keizer, R. A., van Velsen, L., Moncharmont, M., Riche, B., Ammour, N., Del Signore, S., ... & N'Dja, A. (2019). Using socially assistive robots for monitoring and preventing frailty among older adults: A study on usability and user experience challenges. *Health and Technology*, 9, 595–605.

Onnasch, L., & Roesler, E. (2021). A taxonomy to structure and analyze human-robot interaction. *International Journal of Social Robotics*, 13(4), 833–849.

Osborn, A. F. (1953). *Applied Imagination; Principles and Procedures of Creative Thinking*. New York: Scribner.

Ostrowski, A. K., Walker, R., Das, M., Yang, M., Breazea, C., Park, H. W., & Verma, A. (2022). Ethics, equity, & justice in human-robot interaction: A review and future directions. In *2022 31st IEEE International Conference on Robot and Human Interactive Communication (RO-MAN)*, Napoli Italy, 29 August 2022 – 02 September 2022 (pp. 969–976). IEEE.

Owen, R., Macnaghten, P., & Stilgoe, J. (2020). Responsible research and innovation: From science in society to science for society, with society. In Marchant, G. E., & Wallach, W. (eds.), *Emerging Technologies* (pp. 117–126). New York: Routledge.

Paiva, S., Ahad, M. A., Tripathi, G., Feroz, N., & Casalino, G. (2021). Enabling technologies for urban smart mobility: Recent trends, opportunities and challenges. *Sensors*, 21(6), 2143.

Paluch, S. (2017). Smart services: Analyse von strategischen und operativen Auswirkungen. In Bruhn, M., & Hadwich, K. (eds.), *Dienstleistungen 4.0* (pp.161–182). Berlin, Heidelberg: Springer Gabler. https://doi.org/10.1007/978-3-658-17552-8_7.

Parker, C., Tomitsch, M., & Kay, J. (2018). Does the public still look at public displays? A field observation of public displays in the wild. *Proceedings of the ACM on Interactive, Mobile, Wearable and Ubiquitous Technologies*, 2(2), 1–24.

Pataranutaporn, P., Ingalls, T., & Finn, E. (2018). Biological HCI: Towards integrative interfaces between people, computer, and biological materials. In *Extended Abstracts of the 2018 CHI Conference on Human Factors in Computing Systems*, Montreal QC Canada, April 21–26, 2018 (pp. 1–6). New York: Association for Computing Machinery.

Perotti, L., & Strutz, N. (2022). Evaluation and intention to use the interactive robotic kitchen system AuRorA in older adults. *Zeitschrift für Gerontologie und Geriatrie*, 56(7):580–586.

Picard, R. W. (2000). *Affective Computing*. Cambridge, MA: MIT Press.

Plass, J. L., & Kaplan, U. (2016). Emotional design in digital media for learning. In Tettegah, S. Y., & Gartmeier, M. (eds.), *Emotions, Technology, Design, and Learning* (pp. 131–161). Amsterdam: Elsevier.

Prati, E., Peruzzini, M., Pellicciari, M., & Raffaeli, R. (2021). How to include user experience in the design of human-robot interaction. *Robotics and Computer-Integrated Manufacturing*, 68, 102072.

Raats, K., Fors, V., & Pink, S. (2020). Trusting autonomous vehicles: An interdisciplinary approach. *Transportation Research Interdisciplinary Perspectives*, 7, 100201.

Radesky, J. S., Weeks, H. M., Ball, R., Schaller, A., Yeo, S., Durnez, J., Tamayo-Rios, M., Epstein, M., Kirkorian, H., & Coyne, S. (2020). Young children's use of smartphones and tablets. *Pediatrics*, 146(1), e20193518.

Raj, V., Chandran, A., & RS, A. P. (2019). IoT based smart home using multiple language voice commands. In *2019 2nd International Conference on Intelligent Computing, Instrumentation and Control Technologies (ICICICT)*, Kammur India, July 05–06, 2019 (Vol. 1, pp. 1595–1599). IEEE.

Rashid, M., Sulaiman, N., Mustafa, M., Khatun, S., Bari, B. S., & Hasan, M. J. (2020). Recent trends and open challenges in EEG based brain-computer interface systems. In *ECCE2019: Proceedings of the 5th International Conference on Electrical, Control & Computer Engineering*, Kuantan, Pahang, Malaysia, 29th July 2019 (pp. 367–378). Singapore: Springer.

Ratcliffe, J., Soave, F., Hoover, M., Ortega, F. R., Bryan-Kinns, N., Tokarchuk, L., & Farkhatdinov, I. (2021). Remote XR studies: Exploring three key challenges of remote XR experimentation. *Extended Abstracts of the 2021 CHI Conference on Human Factors in Computing Systems*, Yokohama Japan, May 8–13, 2021 (pp. 1–4). New York: Association for Computing Machinery.

Read, J. C., & MacFarlane, S. (2006). Using the fun toolkit and other survey methods to gather opinions in child computer interaction. In *Proceedings of the 2006 Conference on Interaction Design and Children*, Tampere Finland, June 7–9, 2006 (pp. 81–88). New York: Association for Computing Machinery.

Reed, P., Holdaway, K., Isensee, S., Buie, E., Fox, J., Williams, J., & Lund, A. (1999). User interface guidelines and standards: Progress, issues, and prospects. *Interacting with Computers*, 12(2), 119–142.

Roberge, A., Bouchard, B., Maître, J., & Gaboury, S. (2022). Hand gestures identification for fine-grained human activity recognition in smart homes. *Procedia Computer Science*, 201, 32–39.

Robert Jr, L. P., Alahmad, R., Esterwood, C., Kim, S., You, S., & Zhang, Q. (2020). A review of personality in human-robot interactions. *Foundations and Trends(r) in Information Systems*, 4(2), 107–212.

Roda, C. (2011). Human attention and its implications for human-computer interaction. *Human Attention in Digital Environments*, 1, 11–62.

Rosala, M., & Krause, R. (2019). *User Experience Careers: What a Career in UX Looks Like Today*. California: Nielsen Norman Group.

Rosenbaum, S., Cockton, G., Coyne, K., Muller, M., & Rauch, T. (2002). Focus groups in HCI: Wealth of information or waste of resources? In *CHI'02 Extended Abstracts on Human Factors in Computing Systems* (pp. 702–703).

Rosenthal-von der Pütten, A. M., Krämer, N. C., & Herrmann, J. (2018). The effects of humanlike and robot-specific affective nonverbal behavior on perception, emotion, and behavior. *International Journal of Social Robotics*, 10, 569–582.

Rossouw, A., & Smuts, H. (2023). Key principles pertinent to user experience design for conversational user interfaces: A conceptual learning model. In Huang, Y. M., & Rocha, T. (eds.), *ICITL 2023: Innovative Technologies and Learning*. Lecture Notes in Computer Science (Vol. 14099). Cham: Springer. https://doi.org/10.1007/978-3-031-40113-8_17.

Ruhland, K., Peters, C. E., Andrist, S., Badler, J. B., Badler, N. I., Gleicher, M., Mutlu, B., & McDonnell, R. (2015). A review of eye gaze in virtual agents, social robotics and HCI: Behaviour generation, user interaction and perception. *Computer Graphics Forum*, 34(6), 299–326.

Ruijten, P. A., Haans, A., Ham, J., & Midden, C. J. (2019). Perceived human-likeness of social robots: Testing the Rasch model as a method for measuring anthropomorphism. *International Journal of Social Robotics*, 11, 477–494.

Rümelin, S., & Butz, A. (2013). How to make large touch screens usable while driving. In *Proceedings of the 5th International Conference on Automotive User Interfaces and Interactive Vehicular Applications*, Eindhoven Netherlands, October 28–30, 2013 (pp. 48–55). New York: Association for Computing Machinery.

Ryan, C. L., & Lewis, J. M. (2017). Computer and internet use in the United States: 2016. US Department of Commerce, Economics and Statistics Administration, USA.

Saha, S., Mamun, K. A., Ahmed, K., Mostafa, R., Naik, G. R., Darvishi, S., Khandoker, A. H., & Baumert, M. (2021). Progress in brain computer interface: Challenges and opportunities. *Frontiers in Systems Neuroscience*, 15, 578875. DOI: 10.3389/fnsys.2021.578875.

Salazar, K., 2020. Article: Contextual inquiry: Inspire design by observing and interviewing users in their context. Accessed online October 2023 at: https://www.nngroup.com/articles/contextual-inquiry/.

Salvendy, G. (2012). *Handbook of Human Factors and Ergonomics*. Hoboken, NJ: John Wiley & Sons.

Sanchez-Comas, A., Synnes, K., & Hallberg, J. (2020). Hardware for recognition of human activities: A review of smart home and AAL related technologies. *Sensors*, 20(15), 4227.

Santos, M., & Villela, M. L. B. (2019). Characterizing end-user development solutions: A systematic literature review. *Human-Computer Interaction. Perspectives on Design: Thematic Area, HCI 2019, Held as Part of the 21st HCI International Conference, HCII 2019, Proceedings, Part I*, Orlando, FL, July 26–31, 2019 (pp. 194–209). Springer, Cham.

Santos, R., Abreu, J., Beça, P., Rodrigues, A., & Fernandes, S. (2020). Voice interaction on TV: Analysis of natural language interaction models and recommendations for voice user interfaces. *Multimedia Tools and Applications*, 79, 35689–35716.

Scanlon, E., McAndrew, P., & O'Shea, T. (2015). Designing for educational technology to enhance the experience of learners in distance education: How open educational resources, learning design and MOOCs are influencing learning. *Journal of Interactive Media in Education*, 2015(1), article no. 6, 1–9.

Schaffer, S., & Reithinger, N. (2019). Conversation is multimodal: Thus conversational user interfaces should be as well. In *Proceedings of the 1st International Conference on Conversational User Interfaces*, Dublin Ireland, August 22–23, 2019 (pp. 1–3). New York: Association for Computing Machinery.

Schiff, D., Biddle, J., Borenstein, J., & Laas, K. (2020). What's next for ai ethics, policy, and governance? A global overview. In *Proceedings of the AAAI/ACM Conference on AI, Ethics, and Society*, New York NY USA, February 7–9, 2020 (pp. 153–158). New York: Association for Computing Machinery.

Schirvar, S. (2023). Machinery for managers: Secretaries, psychologists, and 'human-computer interaction', 1973-1983. *BJHS Themes*, 1–14. DOI:10.1017/bjt.2023.10.

Scholtz, J., & Consolvo, S. (2004). Toward a framework for evaluating ubiquitous computing applications. *IEEE Pervasive Computing*, 3(2), 82–88.

Seffah, A. (2015). *Patterns of HCI Design and HCI Design of Patterns: Bridging HCI Design and Model-Driven Software Engineering*. Berlin, Heidelberg: Springer.

Shackel, B. (1997). Human-computer interaction-Whence and whither? *Journal of the American Society for Information Science*, 48(11), 970–986.

Sheridan, T. B. (2016). Human-robot interaction: Status and challenges. *Human Factors*, 58(4), 525–532.

Shneiderman, B. (2016). Guidelines, Principles, and Theories. In Shneiderman, B., Plaisant, C., Cohen, M., Jacobs, S., & Elmqvist, N. (eds.), *Designing the User Interface: Strategies for Effective Human-Computer Interaction*, 6th Ed (pp. 80–121). London: Pearson.

Shneiderman, B. (2020a). Bridging the gap between ethics and practice: Guidelines for reliable, safe, and trustworthy human-centered AI systems. *ACM Transactions on Interactive Intelligent Systems (TiiS)*, 10(4), 1–31.

Shneiderman, B. (2020b). Human-centered artificial intelligence: Reliable, safe & trustworthy. *International Journal of Human-Computer Interaction*, 36(6), 495–504.

Shneiderman, B., & Plaisant, C. (2010). *Designing the User Interface: Strategies for Effective Human-Computer Interaction*. London: Pearson Education India.

Shneiderman, B., Plaisant, C., Cohen, M., Jacobs, S., Elmqvist, N., & Diakopoulos, N. (2016). Grand challenges for HCI researchers. *Interactions*, 23(5), 24–25.

Skarbez, R., Smith, M., & Whitton, M. C. (2021). Revisiting Milgram and Kishino's reality-virtuality continuum. *Frontiers in Virtual Reality*, 2, 647997.

Sohn, S., Braunschweig, T. U., & Sohn, S. (2019). Can conversational user interfaces be harmful? The undesirable effects on privacy concern. In *ICIS 2019 Proceedings*.

Song, Z., Fang, T., Ma, J., Zhang, Y., Le, S., Zhan, G., ... & Kang, X. (2021). Evaluation and diagnosis of brain diseases based on non-invasive BCI. In *2021 9th International Winter Conference on Brain-Computer Interface (BCI)*, Gangwon, Korea (South), February 22–24, 2021 (pp. 1–6). IEEE.

Spezialetti, M., Placidi, G., & Rossi, S. (2020). Emotion recognition for human-robot interaction: Recent advances and future perspectives. *Frontiers in Robotics and AI*, 9, 145.

Stefanidi, E., Foukarakis, M., Arampatzis, D., Korozi, M., Leonidis, A., & Antona, M. (2019). ParlAmI: A multimodal approach for programming intelligent environments. *Technologies*, 7(1), 11.

Stefanidi, H., Leonidis, A., Korozi, M., & Papagiannakis, G. (2022). The ARgus designer: Supporting experts while conducting user studies of AR/MR applications. *2022 IEEE International Symposium on Mixed and Augmented Reality Adjunct (ISMAR-Adjunct)*, 885–890.

Stefanidi, Z., Margetis, G., Ntoa, S., & Papagiannakis, G. (2022). Real-time adaptation of context-aware intelligent user interfaces, for enhanced situational awareness. *IEEE Access*, 10, 23367–23393.

Stegman, P., Crawford, C. S., Andujar, M., Nijholt, A., & Gilbert, J. E. (2020). Brain-computer interface software: A review and discussion. *IEEE Transactions on Human-Machine Systems*, 50(2), 101–115.

Steinicke, F., & Wolf, K. (2020). New digital realities: Blending our reality with virtuality. *i-com*, 19(2), 61–65. https://doi.org/10.1515/icom-2020-0014.

Stephanidis, C. (1995). Towards user interfaces for all: Some critical issues. Parallel session "user interfaces for all: Everybody, everywhere, and anytime". In Anzai, Y., Ogawa, K., & Mori, H. (eds.), *Symbiosis of Human and Artifact - Future Computing and Design for Human-Computer Interaction [Volume 1 of the Proceedings of the 6th International Conference on Human-Computer Interaction (HCI International '95)]*, Tokyo, Japan, 9–14 July (pp. 137–142). Elsevier Science.

Stephanidis, C. (2009a). eAccessibility. In Liu, L., & Özsu, M. T. (eds.), *Encyclopedia of Database Systems* (pp. 955–958). Berlin, Heidelberg: Springer-Verlag.

Stephanidis, C. (2009b). *The Universal Access Handbook*. Boca Raton, FL: CRC Press.

Stephanidis, C. (2013). Design 4 all. In Soegaard, M., & Dam, R. F. (eds.), *The Encyclopedia of Human-Computer Interaction*, 2nd Ed. Aarhus, DK: The Interaction Design Foundation. Retrieved April 8, 2024 from: https://www.interaction-design.org/literature/book/the-encyclopedia-of-human-computer-interaction-2nd-ed/design-4-all

Stephanidis, C. (2021). Design for all in digital technologies. In Salvendy, G., & Karwowski, W. (eds.), *Handbook of Human Factors and Ergonomics* (pp. 1187–1215). Hoboken, NJ: John Wiley & Sons.

Stephanidis, C. (2023). Paradigm shifts towards an inclusive society: From the desktop to human-centered artificial intelligence. In *Proceedings of the 2nd International Conference of the ACM Greek SIGCHI Chapter*, Athens Greece, September 27–28, 2023 (pp. 1–4). New York: Association for Computing Machinery.

Stephanidis, C., Antona, M., & Ntoa, S. (2021). Human factors in ambient intelligence environments. In Salvendy, G., & Karwowski, W. (eds.), *Handbook of Human Factors and Ergonomics* (pp. 1058–1084). Hoboken, NJ: John Wiley & Sons.

Stephanidis, C., Kouroumalis, V., & Antona, M. (2012). Interactivity: Evolution and emerging trends. In Salvendy, G., & Karwowski, W. (eds.), *Handbook of Human Factors and Ergonomics* (pp. 1374–1406). Hoboken, NJ: John Wiley & Sons.

Stephanidis, C., Salvendy, G., Antona, M., Chen, J. Y. C., Dong, J., Duffy, V. G., ... & Zhou, J. (2019). Seven HCI grand challenges. *International Journal of Human-Computer Interaction*, 35(14), 1229–1269. https://doi.org/10.1080/10447318.2019.1619259.

Stock-Homburg, R. (2022). Survey of emotions in human-robot interactions: Perspectives from robotic psychology on 20 years of research. *International Journal of Social Robotics*, 14(2), 389–411.

Streitz, N., & Nixon, P. (2005). The disappearing computer. *Communications of the ACM*. 48(3), 32–35.

Su, H., Qi, W., Chen, J., Yang, C., Sandoval, J., & Laribi, M. A. (2023). Recent advancements in multimodal human-robot interaction. *Frontiers in Neurorobotics*, 17, 1084000.

Sugisaki, K., & Bleiker, A. (2020). Usability guidelines and evaluation criteria for conversational user interfaces: A heuristic and linguistic approach. In *Proceedings of Mensch und Computer 2020* (pp. 309–319).

Sun, Y., Geng, L., & Dan, K. (2018). Design of smart mirror based on Raspberry Pi. In *2018 International Conference on Intelligent Transportation, Big Data & Smart City (ICITBS)*, Xiamen, China, January 25–26, 2018 (pp. 77–80). IEEE.

Sutherland, I. E. (1963). Sketchpad: A man-machine graphical communication system. In *Proceedings of the May 21-23, 1963, Spring Joint Computer Conference* (pp. 329–346).

Syed, A. S., Sierra-Sosa, D., Kumar, A., & Elmaghraby, A. (2021). IoT in smart cities: A survey of technologies, practices and challenges. *Smart Cities*, 4(2), 429–475.

Tekinbas, K. S., & Zimmerman, E. (2003). *Rules of Play: Game Design Fundamentals*. Cambridge, MA: MIT Press.

Telpaz, A., Rhindress, B., Zelman, I., & Tsimhoni, O. (2015). Haptic seat for automated driving: Preparing the driver to take control effectively. In *Proceedings of the 7th International Conference on Automotive User Interfaces and Interactive Vehicular Applications*, Nottingham United Kingdom, September 1–3, 2015 (pp. 23–30). New York: Association for Computing Machinery.

Todi, K., Bailly, G., Leiva, L., & Oulasvirta, A. (2021, May). Adapting user interfaces with model-based reinforcement learning. In *Proceedings of the 2021 CHI Conference on Human Factors in Computing Systems*, Yokohama Japan, May 8–13, 2021 (pp. 1–13). New York: Association for Computing Machinery.

Toh, C. K., Sanguesa, J. A., Cano, J. C., & Martinez, F. J. (2020). Advances in smart roads for future smart cities. *Proceedings of the Royal Society A*, 476(2233), 20190439.

UN General Assembly (2006). Convention on the rights of persons with disabilities. *Ga Res* 61, 106.

United Nations (2015). Transforming our world: The 2030 Agenda for Sustainable Development. https://sdgs.un.org/publications/transforming-our-world-2030-agenda-sustainable-development-17981.

United Nations (2023). The sustainable development goals report. https://unstats.un.org/sdgs/report/2023/.

van Dis, E. A. M., Bollen, J., Zuidema, W., van Rooij, R., & Bockting, C. L. (2023). ChatGPT: Five priorities for research. *Nature*, 614(7947), 224–226. https://doi.org/10.1038/d41586-023-00288-7.

Van Erp, J. B. (2002). Guidelines for the use of vibro-tactile displays in human computer interaction. In *Proceedings of Eurohaptics* (Vol. 2002, pp. 18–22).

Van Mechelen, M., Smith, R. C., Schaper, M.-M., Tamashiro, M., Bilstrup, K.-E., Lunding, M., Graves Petersen, M., & Sejer Iversen, O. (2023). Emerging technologies in K-12 education: A future HCI research agenda. *ACM Transactions on Computer-Human Interaction*, 30(3), 1–40.

Vargas Llave, O., Hurley, J., Peruffo, E., Rodriguez Contreras, R., Adăscăliței, D., Botey Gaude, L., … & Vacas-Soriano, C. (2022). The rise in telework: Impact on working conditions and regulations.

Verbeek, P. P. (2006). Materializing morality: Design ethics and technological mediation. *Science, Technology, & Human Values*, 31(3), 361–380.

Vi, S., da Silva, T. S., & Maurer, F. (2019). User experience guidelines for designing HMD extended reality applications. In *Human-Computer Interaction-INTERACT 2019: 17th IFIP TC 13 International Conference, Proceedings, Part IV 17*, Paphos, Cyprus, September 2–6, 2019 (pp. 319–341). Berlin, Heidelberg: Springer International Publishing.

Vidal, J. J. (1973). Toward direct brain-computer communication. *Annual Review of Biophysics and Bioengineering*, 2(1), 157–180.

Villa, M., Gofman, M., & Mitra, S. (2018). Survey of biometric techniques for automotive applications. In Latifi, S. (ed.), *Information Technology-New Generations: 15th International Conference on Information Technology* Al-Ain, AbuDhabi, November 14–15, 2023 (pp. 475–481). Berlin, Heidelberg: Springer International Publishing.

Vorm, E. S., & Combs, D. J. (2022). Integrating transparency, trust, and acceptance: The intelligent systems technology acceptance model (ISTAM). *International Journal of Human-Computer Interaction*, 38(18-20), 1828–1845.

Vuletic, T., Duffy, A., Hay, L., McTeague, C., Campbell, G., & Grealy, M. (2019). Systematic literature review of hand gestures used in human computer interaction interfaces. *International Journal of Human-Computer Studies*, 129, 74–94.

Wang, R., Bush-Evans, R., Arden-Close, E., Bolat, E., McAlaney, J., Hodge, S., Thomas, S., & Phalp, K. (2023). Transparency in persuasive technology, immersive technology, and online marketing: Facilitating users' informed decision making and practical implications. *Computers in Human Behavior*, 139, 107545.

Wang, Y., Song, W., Tao, W., Liotta, A., Yang, D., Li, X., Gao, S., Sun, Y., Ge, W., & Zhang, W. (2022). A systematic review on affective computing: Emotion models, databases, and recent advances. *Information Fusion*, 83, 19–52.

Weiser, M. (1993). Hot topics-ubiquitous computing. *Computer*, 26(10), 71–72.

Witte, R., & Gitzinger, T. (2008). Semantic Assistants-user-centric natural language processing services for desktop clients. In *Asian Semantic Web Conference*, Bangkok, Thailand, December 8–11, 2008 (pp. 360–374). Berlin, Heidelberg: Springer.

Woods, D., Dekker, S., Cook, R., Johannesen, L., & Sarter, N. (2017). *Behind Human Error*. Boca Raton, FL: CRC Press.

Wu, K., Zhao, Y., Zhu, Q., Tan, X., & Zheng, H. (2011). A meta-analysis of the impact of trust on technology acceptance model: Investigation of moderating influence of subject and context type. *International Journal of Information Management*, 31(6), 572–581.

Xie, B., Zhou, J., & Wang, H. (2017). How influential are mental models on interaction performance? Exploring the gap between users' and designers' mental models through a new quantitative method. *Advances in Human-Computer Interaction*, 2017, 14.

Xiong, J., Hsiang, E. L., He, Z., Zhan, T., & Wu, S. T. (2021). Augmented reality and virtual reality displays: Emerging technologies and future perspectives. *Light: Science & Applications*, 10(1), 216.

Xu, W., Dainoff, M. J., Ge, L., & Gao, Z. (2023). Transitioning to human interaction with AI systems: New challenges and opportunities for HCI professionals to enable human-centered AI. *International Journal of Human-Computer Interaction*, 39(3), 494–518.

Zacharaki, A., Kostavelis, I., Gasteratos, A., & Dokas, I. (2020). Safety bounds in human robot interaction: A survey. *Safety Science*, 127, 104667.

Zander, T. O., Brönstrup, J., Lorenz, R., & Krol, L. R. (2014). Towards BCI-based implicit control in human-computer interaction. In Fairclough, S. H., & Gilleade, K. (eds.), *Advances in Physiological Computing* (pp. 67–90). Berlin, Heidelberg: Springer Science & Business Media.

Zengeler, N., Kopinski, T., & Handmann, U. (2018). Hand gesture recognition in automotive human-machine interaction using depth cameras. *Sensors*, 19(1), 59.

Zhang, Y., Yang, C., Hudson, S. E., Harrison, C., & Sample, A. (2018). Wall++ room-scale interactive and context-aware sensing. In *Proceedings of the 2018 CHI Conference on Human Factors in Computing Systems*, Montreal QC Canada, April 21–26, 2018 (pp. 1–15). New York: Association for Computing Machinery.

Zheng, S., Apthorpe, N., Chetty, M., & Feamster, N. (2018). User perceptions of smart home IoT privacy. *Proceedings of the ACM on Human-Computer Interaction*, 2(CSCW), 1–20.

Zimmermann, P. G. (2008). Beyond usability: Measuring aspects of user experience, Doctoral dissertation, ETH.

Zucco, C., Calabrese, B., & Cannataro, M. (2017). Sentiment analysis and affective computing for depression monitoring. *2017 IEEE International Conference on Bioinformatics and Biomedicine (BIBM)* Kansas City, MO, USA, November 13–16, 2017 (pp. 1988–1995). IEEE.

Zuckerman, O., Gal, T., Keren-Capelovitch, T., Karsovsky, T., Gal-Oz, A., & Weiss, P. L. T. (2016, February). DataSpoon: Overcoming design challenges in tangible and embedded assistive technologies. In *Proceedings of the TEI'16: Tenth International Conference on Tangible, Embedded, and Embodied Interaction*, Eindhoven Netherlands, February 14–17, 2016 (pp. 30–37). New York: Association for Computing Machinery.

2 Sensation and Perception in HCI

Robert W. Proctor and Ya-Hsin Hung

2.1 INTRODUCTION

The fields of Human-Computer Interaction (HCI) and Human Factors and Ergonomics (HFE) have been concerned with the interface between the human and the non-human artifact from their beginning (Johnson & Kobler, 1963). Information must be conveyed by the computer to the human via the human's senses regardless of whether the human is explicitly interacting directly with the computer (as in text or data entry), only occasionally setting input parameters (as in programming a thermostat), or receiving alerts that need to be understood and acted on. Humans' perceptions of their interactions with the computerized systems and their ability to see, hear, and understand the information provided are essential to proper functioning of the systems. Likewise, in driverless vehicles that must interact with other humans on the roadway – human drivers, pedestrians, and bicyclists – the automation needs not only to be able to perceive the humans through sensors and infer their intentions but also to communicate its intentions to them. Moreover, wearable devices monitor health information about patients that must be communicated to healthcare providers. As computers have become more ubiquitous, artificial intelligence (AI) more omnipresent, and "eXtended Reality" (XR) technologies more immersive, the role of all aspects of perception in HCI has increased.

These examples should make clear the significance of sensation and perception for many aspects of HCI. The goal of this chapter is to promote understanding of basic concepts by reviewing empirical findings and principles that are relevant to HCI researchers and designers.

2.2 HISTORICAL ISSUES IN THE STUDY OF SENSATION AND PERCEPTION

Fortunately, philosophers, neuroscientists, psychologists, computer scientists, and others have studied sensation and perception for centuries, with many advances in understanding made over the years (Boring, 1942; Fon, 2021; Pastore, 1971). Through their treatments of whether sensation should be distinguished from perception, whether emphasis should be on phenomenal experience or mechanisms, nativism versus empiricism, and relation between perception and action, the researchers have provided a foundation of theory and methods for contemporary knowledge and research.

2.2.1 NATIVISM VERSUS EMPIRICISM

Nativism versus empiricism has been debated for many centuries (Gordon & Slater, 1998; Pastore, 1971). The issue is whether perception is based mainly on innate abilities or on acquired skills. Early representatives of the empiricist approach, which emphasizes acquired skills, were John Locke and George Berkeley. Berkeley's (1709) book, *An Essay Towards a New Theory of Vision*, was particularly influential. He argued that the distance of objects cannot be perceived directly because the retinal image is two-dimensional. Experience, primarily through the sense of touch, provides a basis for the acquisition of associations with the sensations produced by vision.

René Descarte was the primary advocate of nativism in 17th century philosophy. His general viewpoint was that experience and thinking were sources of ideas, as were "ideas created in the mind by God" (Gordon & Slater, 1998, p. 77), which did not depend on experience. Immanuel Kant

DOI: 10.1201/9781003495109-2

took the view that space and time are not perceived directly, and people's awareness of them is based on a priori intuitions. This structuring of how humans are able to perceive fits with the nativist view that much perception must be innate.

The division between perceptual psychologists who favored empiricism and those who favored nativism has continued throughout the development of experimental psychology. Hermann von Helmholtz, Donald Hebb, and Richard Gregory are among those who advocated empiricism, with an emphasis on inferential processes, whereas the Gestalt psychologists (e.g., Wolfgang Köhler, 1940; Kurt Koffka, 1940) and James J. Gibson (1979) are classified as nativists. More generally, the information-processing approach to perception (Haber, 1969) is aligned with empiricism, whereas the ecological approach spawned by Gibson is associated with nativism. Gordon and Slater (1998) emphasize, though, "Few researchers and theorists have ever maintained an absolutely extreme position regarding innate versus acquired factors in perception" (p. 74). This chapter depicts contributions made by both groups.

2.2.2 DISTINCTION BETWEEN SENSATION AND PERCEPTION

Wolfe et al. (2021) describe sensation as the ability to detect a stimulus and possibly to convert it into a personal experience, whereas perception provides meaning and purpose to the sensations. This is a straightforward distinction that can be used to help organize knowledge and research in the area into psychological effects associated with sensory physiology from effects associated primarily with from brain processes. However, the value of this distinction for psychological theory has been debated for many years. For example, Titchener (1919) considered that the senses provided elements that were associated through experience to produce percepts. In contrast, Köhler (1940), Koffka (1940), and Gibson (1979) focused exclusively on perception and not sensation. Graham (1958), a noted vision researcher, likewise concluded, "The terms sensation and perception probably do not refer to two different operationally specifiable processes" (p. 76).

2.2.3 RELATION TO ACTION

Perception and action have typically been treated as separate processes that do not interact. Gibson (1979) emphasized the relation of perception and action in his concepts of direct perception (that is, unmediated by representations) and affordances, which are what the environment "*offers* to the animal, what it *provides* or *furnishes*, either for good or ill" (p. 127). Several alternative theories of affordances have been proposed within Gibson's ecological psychology approach, attributing them to the environment or relations between the abilities of people (and other species) and features of the environment (Chemero, 2003). The concept of affordances has also been incorporated within some representational, information-processing approaches (Caligiore et al., 2010) and with reference to design of products for human use (Norman, 2013). These various uses require that readers attend closely to the intended meaning whenever the term "affordance" is used (Chong & Proctor, 2020).

In 1982, Mishkin and Ungerleider provided evidence distinguishing a ventral stream of projections from the primary visual cortex that goes to the inferotemporal cortex from a dorsal stream that goes to the posterior parietal cortex (see Figure 2.1). They concluded that the former is involved in pattern identification and the latter in spatial perception. An influential study by Goodale and Milner (1992) summarized evidence suggesting that the dorsal pathway is actually involved in visuomotor action selection and control. They deemed it the "how" pathway as opposed to the "what" ventral pathway. This pathway division underlies the distinction between an ambient mode of visual processing (which guides locomotion) and a focal mode (required for detailed pattern recognition), as applied in HFE (Horrey & Wickens, 2004; Leibowitz & Post, 1982). Current views of the dorsal and ventral pathways maintain the division of function but conceive of the pathways as interacting in the control of complex perceptual-motor behavior such as object-guided hand movements (Van Polanen & Davare, 2015).

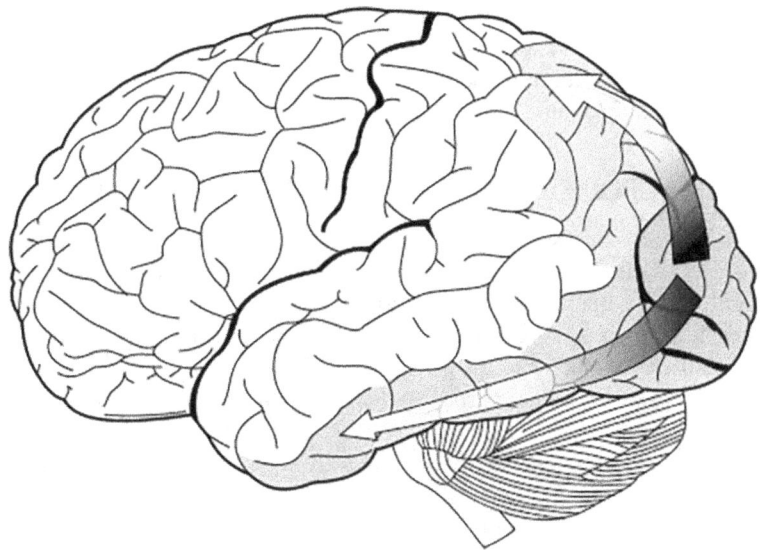

FIGURE 2.1 Dorsal (upper) and ventral (lower) pathways emanating from the primary visual cortex. (By Selket, https://commons.wikimedia.org/w/index.php?curid=1679336.)

Another line of research emphasizing a relation between perception and action is that initiated by Prinz (1997), who proposed that perceived events and planned actions are represented in a common coding system. The most impactful development of this approach is the Theory of Event Coding (TEC; Hommel et al., 2001), which provides a framework for perception and action based on the concept of event files (integrated networks of perceptual-motor feature codes). Research based on TEC has emphasized the nature of the codes shared between perception and action and the control processes that operate on the codes (Hommel, 2019). This common coding framework relies on ideomotor theory, according to which actions are initiated and controlled by anticipating the feedback that will occur as a consequence of the action (Shin et al., 2010; Stock & Stock, 2004).

2.3 WHAT IS MEANT BY "PERCEPTION"?

The first thing to note about the term "perception" is that it is used in many different ways by various authors. The most fundamental use is to refer to information that comes in through the senses, as in visual perception and auditory perception (Mortensen et al., 2009). In the human information-processing approach, perception is often distinguished from response selection and response initiation/control to emphasize processes that act primarily on the sensory input to enable representation and identification of stimuli (Proctor & Van Zandt, 2018; Sanders, 1979; see Chapter 3, this book). Issues examined are the ways in which the sensory input influences the time and accuracy for detection and identification of stimuli, as well as the performance of specific tasks (Kösem & Van Wassenhove, 2012). Perception is also often used to refer to a person's experience of the external world (Cohen et al., 2016), for which it is typically assumed that the person can describe verbally. However, the common reference to "perception without awareness" implies that some perception can occur that is not reportable (Kihlstrom, 1996; Merikle et al., 2001).

Moving beyond these more conventional approaches that link perception to the sensory systems, the term is used within the context of HCI to refer to a wide variety of experiences that are also roughly equated with conscious awareness. These experiences include, among others, perceptions of privacy and usability (Wu et al., 2021), perception of smart speaker-based voice assistants (Kim & Choudhury, 2021), perception of game design elements (Denden et al., 2021), perceptions of computer iconography (Ali et al., 2021), perception of intimacy in human and human-agent interactions

(Potdevin et al., 2020), and risk perception (Byrne et al., 2016). For example, do users perceive a website or application to be easy and pleasant to use? These uses of the term perception extend beyond the immediate information provided by the senses to judgments that involve memory and, in many cases, emotional reactions.

2.4 THEORETICAL PERSPECTIVES IN THE STUDY OF PERCEPTION

What is called the constructivist approach, for which von Helmholtz (1867) is regarded as the foremost proponent, treats perception as dependent on mediational processes that intervene between stimulation and perception (Norman, 2002). Perception, in this case, is regarded as indirect. The constructivist view of perception is linked closely to the human information-processing approach in general in its emphasis on mediating representations. In contrast, the ecological approach treats perception as a single process, that is, as direct and immediate (Norman, 2002). As Gibson (1979) notes in the Introduction to *The Ecological Approach to Visual Perception*, in his 1950 book (Gibson, 1950), *The Perception of the Visual World*, "my explanation of vision was then based on the retinal image, whereas it is now based on what I call the ambient optical array" (p. 1). He elaborates this point, saying, "I shall suggest that natural vision depends on the eyes in the head on a body supported by the ground, the brain being only the central organ of a complete visual system" (p. 1).

Neuroscience has played an ever-increasing role in understanding sensation and perception. As one example, we have already noted the role of such studies in the distinction between dorsal and ventral streams. Another well-known example is that of mirror neurons (Rizzolatti & Sinigaglia, 2016), discovered in Macaque monkeys, that "fire" both for certain actions and when viewing similar actions performed by another monkey or person. Mirror neurons have been found in several regions of the motor pathway, including primary, premotor, and parietal areas (Ossmy & Mukamel, 2018) and are thought to play a role in relating perception and action.

The computational approach to human perception uses computer simulation models to enhance understanding of perceptual processes. The research can be organized around Marr's (1982) three levels of analysis: computational theory, representation and algorithm, and hardware implementation (p. 25). The computational level involves the specification of the problem that the visual system has to solve in creating a 3D percept. The representation and algorithm level refers to the process that transforms the initial input representation into an output representation. The final, hardware implementation involves the neural implementation for living creatures and the physical implementation for machines. Much of the modeling in the computational approach uses neural networks that rely on distributed computation (Wechsler, 2014). The computational approach has been applied sufficiently broadly to vision that it served as the basis for a textbook (Frisby & Stone, 2010). Computational modeling has also been advocated for social perception, for which Brooks et al. (2021) reviewed studies using multivariate pattern analysis of functional magnetic resonance imaging data to investigate how the brain processes and represents social information.

Gestalt psychologists of the 20th century took a qualitative approach toward developing principles of perceptual organization. These principles are widely accepted and included in all textbooks on perception and most introductory psychology texts. Rosenholtz et al. (2009) stated, "Perhaps the most important aspect of human vision for design is perceptual organization" (p. 1331). Computer vision models have been developed for perceptual grouping, but Rosenholtz et al. concluded that they were inappropriate for HCI design. Therefore, they developed a perceptual grouping algorithm that represents an image in a feature space and divides it into coherent clusters. Rosenholtz et al. demonstrated that the model captures design rules and visual percepts in classic displays.

The last theoretical approach to mention is that of Egon Brunswik (1952, 1956). To understand human-environment relations, Brunswik proposed three key concepts: *probabilistic*

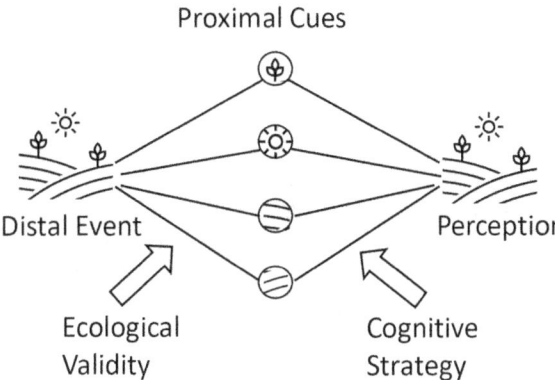

FIGURE 2.2 Depiction of Brunswik's lens model, with the fanning out of proximal cues from the distal event, the focusing lens of the perceiver, and the final perception.

functionalism, the *lens model*, and *representative design* (Dhami et al., 2004; Hammond, 1966). Probabilistic functionalism theory focuses on the probabilistic coupling of the human and environment as a system. The core idea is depicted by the lens model, in which environmental cues are processed through a lens to address the environment's uncertainties (see Figure 2.2). For Brunswik, the primary aim of psychological research is to discover probabilistic laws that describe an organism's adaptation, in terms of distal objects or events, to the causal texture of its environment. Brunswik (1952) proposed to measure the ecological or predictive validity by the correlation of the distal event with proximal cues as weighted by the perceiver's cognitive strategy, or expectations.

Besides the conceptual framework of probabilistic functionalism and the lens model, Brunswik (1952, 1956) also developed methodological procedures called representative design to guide how to conduct experiments. With representative design methods, the researcher tries to retain all of the relevant environmental variables in the design and to identify their effects on decisions after the fact through correlational statistical analyses. Potential applications of this approach are described in more detail in Proctor and Xiong (2020).

2.5 MEASURING PERCEPTION

Perceptual measurement is among the earliest concerns in experimental psychology. The methods that have been developed are of use not only for the study of perception but also for a variety of other processes. Pelli and Farell (1995) emphasize this point, stating, "Psychophysical methods are the tools for measuring perception and performance. These tools are used to reveal basic perceptual processes, to assess observer performance, and to specify the required characteristics of a display" (p. 29.1).

2.5.1 PSYCHOPHYSICAL THRESHOLDS

Sensory threshold methods are intended to determine the minimum amount of stimulation needed to detect the presence of a stimulus (absolute threshold) or the minimum amount of difference from a baseline, or standard, stimulus to detect a difference between the two (difference threshold). Methods for finding thresholds date to Ernst Weber (1834/1996) and Gustav Fechner (1860), and the impact more generally of Fechner's work is noted by Gescheider (1997): "Fechner's work, in providing methods and theory for the measurement of sensation, gave psychology basic tools for the study of mind" (p. ix). Psychophysical methods, both classic and more recent, continue to provide such tools.

Thresholds can be measured using a variety of methods (Gescheider, 1997; Kingdom & Prins, 2016). Well-known ones include the method of limits and the method of constant stimuli. With the former, the observer receives runs of increasing or decreasing magnitude of the stimulus relative to the null or standard and has to respond "yes" if the stimulus is detectable or different, or "no" if it is not. The threshold on each run is the estimated value between the switch from "no" to "yes", or vice versa. These are then averaged to get the threshold value. A variation of the method of limits is the staircase method, in which the threshold is bracketed by switching the direction of change when the response changes (e.g., from ascending to descending when the response changes from "no" to "yes"). For the method of constant stimuli, also called the yes-no method (Kingdom & Prins, 2016), the stimulus magnitudes are presented in random order, to which the participant must give a "yes" or "no" response. With this method, the threshold is estimated as the boundary between a predominance of "yes" or "no" responses.

2.5.2 Signal-Detection Methods and Theory

One of the most useful psychophysical methods is that of signal detection (Hautus et al., 2022), which has a long history of use in HFE (Swets, 1996). In a signal-detection task, an event is classified by the researcher as a signal, and the observer's task is to detect whether the signal is present in some trials but not in others (noise trials). The proportion of trials on which the observer correctly detects the signal is called the hit rate, and the proportion of trials on which it is falsely detected is called the false-alarm rate. A high hit rate with a low false-alarm rate indicates high sensitivity, whereas an overall tendency to say "yes" for both trial types indicates a liberal response bias and a tendency to say "no" a conservative bias.

Signal-detection theory assumes that the response on each trial is a function of two operations, encoding and decision. On a trial, the observer samples the information presented and decides whether this information is sufficient to warrant a "yes" response. The sample of information is assumed to provide a value along a continuum of evidence regarding the likelihood that the signal was present. The noise trials form a probability distribution along the continuum, as do the signal trials. The decision that must be made on each trial is whether the event is from the signal or noise distribution. The observer is presumed to adopt a criterion of evidence, above which she responds "yes" and below which she responds "no".

In the simplest form, the distributions are assumed to be of normal and equal variance, in which case a measure of detectability, d', can be calculated, as well as a measure of response bias, C (for criterion; Hautus et al., 2022). The d' measure represents the difference in the means for the signal and noise distributions in standard deviation units. It is found by converting the hit and false-alarm rates to standard normal scores and obtaining the difference. A d' of 0 indicates no detectability, whereas a d' of 3.0 or greater indicates essentially perfect detectability. C is calculated by summing the standardized values of the hit and false-alarm rates, and dividing by two, with a $C=0$ indicating no response bias. Positive values of C indicate a bias toward "no" responses, whereas negative values indicate a bias toward *yes* responses, with the absolute value indicating the magnitude of bias. There are numerous alternative measures of detectability and bias based on different assumptions and theories, and many task variations to which they can be applied (see Hautus et al., 2022).

Signal-detection analyses are particularly useful because they can be used for any task that can be depicted in terms of binary discriminations, not just sensation and perception. Parasuraman and Wisdom (1985) emphasized the value of signal-detection theory for HCI research, using the specific example of computer assistance for human-computer monitoring (Sorkin & Woods, 1985). Also, Martin et al. (2018) have demonstrated the utility of signal-detection theory for modeling user reactions to phishing attacks. More generally, Wixted (2020), who has applied signal-detection theory to eyewitness testimony, underscores that signal-detection theory has "had an enormous impact on basic and applied science alike" (p. 201).

2.5.3 REACTION TIME

Reaction time (RT) is a sensitive measure that is widely used to study all aspects of human-infor-mation processing. RT methods are described in Chapter 3 of this book; therefore, we only touch on the use in studies of perception in this chapter.

Time to detect a stimulus can be measured in simple reaction tasks in which there is only one possible stimulus and one response that the participant is to make immediately upon detecting the onset of the stimulus. Simple RT is a basic measure of processing speed because it does not require identification of the stimulus (only detection) or selection of a response (there is only one response to be made when the stimulus is detected). Teichner (1954) summarized many of the established phenomena, including:

- For visual stimuli, the greater the spatial extent of the stimulus, the faster the speed of responding;
- For all senses, RT is a negatively accelerated decreasing function of stimulus intensity up to a maximum intensity;
- Stimulation of more than one sensory modality simultaneously yields shorter RT than stimulation of only one modality.

Woods et al. (2015) conducted two large-sample experiments with adults across the life-span and found that mean stimulus-detection time did not differ significantly as a function of gender and edu-cation level and differed little with age. Thus, detection reactions show little variation as a function of demographic variables.

Recognition time can be investigated in go/no-go reaction tasks, which require responding to one stimulus but not others. Although this type of task requires recognition of which stimulus has occurred, the RT does not include the time to determine which response to make or effector to use because there is only one response. Choice-reaction tasks with more than one response engage response selection and initiation processes in addition, but methods can be used to isolate influences on stimulus iden-tification (Ashby & Townsend, 1980; Sternberg, 1969). For example, Broadbent and Gregory (1962) found that RTs were considerably longer in choice-reaction tasks than in go/no-go tasks when the map-pings of verbal, auditory stimuli or tactile stimuli to responses were incompatible (say the alternative word in response to a verbal stimulus; press a finger on the opposite hand to one that was stimulated). In contrast, when the mappings were compatible (make the response corresponding to the stimulus), which should minimize response selection, there was no difference between go/no-go and choice RTs.

RT is most often measured with response buttons or other input devices that are separate from the human performer, requiring that the performer not move around. This restriction can be enforced not only in the laboratory but also in a driving simulator for which response buttons are placed on the wheel (Xiong & Proctor, 2016). For naturalistic contexts in which the movement of the person is not restricted, such as with vehicle drivers and aircraft pilots, a wearable system has been developed and validated for use in a variety of contexts that allow simple and recognition RT to be measured (Abbasi-Kesbi et al., 2017).

As a final note, whenever one is measuring RT, there is also concern about the error rate because response speed can be traded off with accuracy (Heitz, 2014). As described in more detail in Chapter 3, models that conceive of information accumulating over time until a response threshold is reached provide a good way to analyze RT and accuracy in a task (Ratcliff & Smith, 2004).

2.5.4 MAGNITUDE ESTIMATION

Psychophysical scaling involves relating the physical magnitude of stimuli to their perceived magni-tude (Marks & Gescheider, 2002). The earliest work of this type was conducted by Fechner (1860), who derived psychophysical scales from the assumption that every successive difference threshold provides a constant increase in the perceived magnitude of the sensation. From this assumption, the perceived magnitude of a particular stimulus could be specified as the number of difference thresholds above the absolute threshold, yielding a logarithmic function. This type of method for deriving a scale is called *indirect* because the observer is never asked to specify the magnitude.

Magnitude methods of psychophysical scaling, in which observers judge the subjective magnitudes of stimuli, also originated in the 1800s (Merkel, 1888) but flourished in the beginning of the latter half of the 20th century. The rise of magnitude methods for deriving scales was due primarily to the work of Stevens (1953, 1956, 1975); they are *direct* in the sense that the observer is asked to judge the magnitudes of the stimuli. In one version of the method, the observer is given a magnitude as a "modulus" and told to judge the ratio of each other stimulus relative to the modulus (e.g., if it is twice as loud, assign a rating of 2). In another version, the participant selects a value to assign to the first stimulus and then judges the magnitudes of the other stimuli relative to that value. Stevens showed that magnitude estimation methods for various senses yield a power function, which is called Stevens's law:

$$S = kI^n,$$

where S = sensation magnitude, k = a constant, and I = physical intensity raised to the power n.

Magnitude estimates are subject to several biases. One way to minimize the influence of the biases is to pair magnitude estimation with magnitude production. With this latter method, the observer is presented with numerical values for stimulus magnitudes and is to adjust a stimulus to produce the corresponding sensory magnitude. Averaging the scales derived from magnitude estimation and production tends to offset biases of the two methods and may result in the most accurate scale (Hellman & Zwislocki, 1963). Another possible way to address biases is to adopt a Bayesian framework, according to which previous experience (prior) is combined with noisy sensory input, weighted according to their relative uncertainties (Petzschner et al., 2015). This combination of knowledge and sensory input yields biased magnitude judgments whenever the prior value differs from the current physical stimulus magnitude.

Magnitude estimation is applicable to many HCI problems. For example, West et al. (2001) asked observers to estimate the smoothness of videos presented at different frame rates. They found that the relation between frame rate and smoothness perception could be described by a power function for which $n = 0.68$. As another example, Wall and Harwin (2000) had participants provide magnitude estimates of the roughness of "virtual gratings" produced by a PHANToM haptic interface device and a standard visual display unit. Under haptic perception, the perceived roughness was judged to decrease with increased grating period, although with visual exploration the exact relation was less well defined.

2.5.5   SUBJECTIVE RATINGS OF EXPERIENCE

With magnitude estimation procedures, there is typically a physical stimulus that is varied to obtain the psychophysical scale. However, it is also common to ask participants to rate some aspect of their experience that cannot be matched to a specific physical variable on a subjective 5- or 7-point Likert scale, ranging from strongly disagree to strongly agree, or a Likert-type scale that is labeled in some other manner. An example is self-perceived driver ratings, in which participants rate several driving abilities as poor, fair, good, or very good (Huang et al., 2022; MacDonald et al., 2008). Although the scale is called "perceived driver ratings", the ratings are judgments that can be influenced by many factors other than the intended construct of self-perceived driving ability. This point is illustrated by a study of older adults' views on their abilities and driving performance conducted by Freund et al. (2005), for which their conclusion was that "Older drivers assign high ratings to their driving performance, even in the presence of suspected skill decline" (p. 613). Self-perceived driving ability is treated correctly as only one possible interpretation of the rating data.

Subjective ratings of mental workload are a common part of studies in HFE and HCI (Hancock et al., 2021). However, the validity of subjective measures can be questioned because they sometimes diverge from objective behavioral and physiological measures of mental workload (Matthews et al., 2020). One possible issue with the subjective measures was identified by Moore and Picou (2018). They found that participants were likely to substitute an easier question when asked to rate mental effort and often used perceived performance as a heuristic for assessing mental effort. Hancock et al. provide a list of potential problems that may threaten the validity of subjective workload measures. The main point is that subjective measures of mental workload, as with all subjective ratings, need to be interpreted with caution.

In HCI, perception from users' evaluations software or hardware is usually measured through questionnaires. The self-report subjective measurement is more cost- and time-efficient than most physiological or behavioral methods. However, a limitation similar to that identified by Moore and Picou (2018) for subjective measures of mental workload has been noted. Ben-Bassat et al. (2006) had participants perform using four versions of a data-entry application created by manipulating the application's usability and aesthetics. Five groups of participants were created as a function of evaluation method (questionnaire alone or auction and questionnaire), monetary incentive (present or absent), and experience in using the system (present or absent). Users' evaluations of usability were influenced by the aesthetics of the system but not by their experience with the system or monetary performance incentives. In contrast, auction bids (which involved the use of the systems) were affected only by the objective performance levels that could be attained with the alternative systems.

A final, general concern is the nature of perceptual awareness and how it relates to performance. Awareness must be measured by subjective report, but how awareness maps onto the reports is debatable, as mentioned earlier. One of the major advances in this regard is the Perceptual Awareness Scale introduced by Ramsøy and Overgaard (2004), which has become a widely used measure of consciousness. Rather than just distinguishing consciousness and unconsciousness, the scale distinguishes four categories: clear experience, almost clear experience, brief glimpse (experience that there was something presented but no awareness of identity or properties), and no experience (Overgaard & Sandberg, 2021).

The matter of whether the scale measures perceptual awareness and the extent to which it can be used to support various views of consciousness has received considerable discussion (Michel, 2019; Overgaard & Sandberg, 2021). One feature of the scale that is relevant for research HCI community is the distinction between no experience and the low level of experience called brief glimpse. This latter category can be thought of as detecting the occurrence of a stimulus without any experience relating to what it is. As emphasized by Overgaard and Sandberg (2021), this distinction between no awareness and low-level awareness can be used to help answer the question of whether perception can occur without any awareness.

2.6 VISUAL PERCEPTION

Displaying information in a way that humans will be able to detect and identify stimuli, and comprehend their meaning, is fundamental to many areas of HCI. As an example, if visual stimuli are outside the visual spectrum (electromagnetic wavelengths of 400–780 nm), they will have no influence on the sensory receptors and cannot be perceived. The senses influence perception in many other ways, some of which we review here, starting with vision.

2.6.1 SENSORY PROCESSES

Light is generated in the environment by the sun during the day and by artificial lighting at any time of day. Most light reaching the eyes is reflected by objects, although people are also exposed regularly to light generated by traffic signals and other light sources. Use of display screens of various sizes for desktop computers, laptops, and smart phones is ubiquitous in HCI, as are concerns about their adequacy (Liu et al., 2020). The photoreceptors that convert light energy into neural signals are the rods and cones, located on the retina at the back of the eye's interior (Kolb, 2012). The light stimulation must be in focus on the retina for the environment to be perceived clearly, as anyone who wears corrective lenses knows. Most of the focusing is accomplished by the cornea (the transparent frontal part of the eye), which is appropriate for focusing a distant object. Through the process of accommodation, the lens provides additional focusing power when fixating on a closer object (Koretz & Handelman, 1988).

Pupil size affects the amount of light entering the eye (Spector, 1990), being smaller in bright light (2–4 mm diameter) than in the dark (4–8 mm). The pupils also constrict when the eyes are focused on a near object. Bright backgrounds and the constriction of the pupil have three consequences for vision (Mathôt & Ivanov, 2019). A bright background gives rise to light scatter over a large area of the retina, reducing perceived image contrast. Small pupils reduce the amount of light falling on the retina, which also negatively affects perception. Most importantly, small pupils lead to less optical distortion, thus producing a higher quality image and increased visual acuity (Liang & Williams, 1997). This relation between pupil size and acuity is likely a major cause of the perceptual consequence that dark font on a light background is easier to read than light font on a dark background (Buchner et al., 2009; Dobres et al., 2017).

Not only does a small pupil size in well-lit conditions act to enable acuity and identification in central vision, but so do many other aspects of the visual system. The fovea is a small, pinhead size region of the retina that covers <0.1% of the visual field (Intoy & Rucci, 2020) and contains only cones, the photoreceptors that underlie daylight (photopic) vision and color perception. Several layers of neurons and blood vessels intervene between the incoming light and the retina, but these are pulled away from the fovea, allowing the image to be less noisy than at more peripheral locations (Kolb, 2012). The foveal photoreceptors have a one-to-one link to the ganglion cells whose axons make up the optic nerve, whereas the receptive fields of the ganglion cells get larger at more peripheral locations. The parvocellular pathway associated with central vision has properties that maximize acuity, whereas the magnocellular pathway has properties that maximize light detection and sensitivity to motion (Masri et al., 2020). Along the sensory pathways and into the visual cortex, the regions involved in visual perception show over-representation of the fovea (Azzopardi & Cowey, 1996): The proportion of neurons devoted to the fovea is much higher than the proportion of foveal receptors. From the primary visual cortex, the ventral stream devoted mainly to pattern recognition is distinct from the dorsal stream devoted mainly to spatial perception and action.

2.6.2 BASIC PERCEPTUAL ATTRIBUTES

Visual acuity is often measured by a Snellen eye chart for which the observer must identify letters of various sizes (Azzam & Ronquillo, 2022). In a standard vision test, the observer is fixating directly on the letters to be reported, which consequently fall on the fovea. Around this area is a region called the useful field of view (Wood & Owsley, 2014), which is the region from which information relevant to a particular task can be processed efficiently without head or eye movements. Because this region of high acuity is small, HCI designers need to know where a user is fixating, by using a cue to draw attention to a critical region or using an eye-tracker to monitor saccadic eye movements from one fixation to another (Carter & Luke, 2020).

Contrast sensitivity is another measure of acuity that can provide more information about functional vision than a standard eye chart can (Ginsburg, 2003). It can be measured with many stimuli, but the use of sine-wave gratings of different spatial frequencies allows a contrast sensitivity function to be obtained. Greatest sensitivity in an adult human is typically about 3–4 cycles/degree of visual angle (a measure of image size), with less sensitivity at lower (global properties) and higher (detailed properties) spatial frequencies, respectively. Because the visual system uses functionally independent channels to process different ranges of size relevant to different levels of shape and detail, contrast sensitivity functions provide detailed information about a person's functional vision

(Barten, 1999). Barten (2003) provides a practical formula for human contrast sensitivity that incorporates orientation angle and surround luminance for use in design of displayed images and evaluation of image quality.

Color perception is also fundamental to many aspects of humans' lives, including HCI (Sokolova & Fernández-Caballero, 2015). Three dimensions of color experience are customarily distinguished: hue (chroma), saturation, and brightness/lightness. Hue corresponds to the use of the term color in everyday language. It is mainly a function of the dominant wavelength of the light energy, with spectral hues ranging from violet and blue at short wavelengths to red at long wavelengths. The term "colorfulness" is used to describe the perceptual attribute of hue intensity. Saturation refers to the purity of the hue, with a stronger hue said to be more highly saturated than a less strong one. Brightness refers to the perceived amount of luminance emitted by a light source or reflected by a surface; it varies from dim to bright. Lightness is the reflectance of a stimulus of neutral hue along a white-to-black dimension. Detailed models exist for precise specification of these color properties (Hellwig & Fairchild, 2022).

Of importance for HCI is color mixing (Albers, 2013). An obvious example is the color display screens used for computer monitors and color televisions, for which the colors are produced from Red, Green, and Blue (RGB) pixels. When mixing light sources in this way, the hue follows additive mixing principles. Any spectral hue can be created by a combination of the RGB pixels; saturation can be varied by altering the relative intensities of the three pixels, and brightness by the total intensity. Complementary color combinations, which include red and green, and yellow and blue, tend to cancel each other when mixed but to create a strong contrast when in adjoining regions. Also, a hue in a surrounding region will tend to induce the complementary hue in a region that would typically be seen as gray or white. The relevance of complementary colors is emphasized by Pridmore (2011), who distinguishes 40 different roles for complementary colors.

For pigments, color is determined by the proportion of illuminating light reflected across various regions of the spectrum. A red object reflects light primarily in the long wavelength region, a blue object reflects light primarily in the short wavelength region, and so on. Pigment mixtures follow a subtractive color mixing principle as, for example, in computer color-printing (Gilbert & Haeberli, 2007). Each successive pigment that is mixed subtracts light from different regions of the spectrum, yielding a black neutral color when multiple pigments are mixed. A black computer case and monitor border appear black because they do not reflect much of the light present on them. People show relatively good lightness constancy in that the black computer case appears black under different levels of illumination, even though the amount of light it reflects varies as a function of the amount of illumination (Murray, 2021).

2.6.3 Perceptual Organization

Principles of perceptual organization have been emphasized in psychology and visual design since being identified in the first half of the 20th century by Gestalt psychologists (Wagemans et al., 2012a,b). The principles, which apply not only to vision but also to the other sensory modalities, are essential to HCI because of the many contexts in which information displays are crucial. The Gestalt psychologists proposed what is called the law of Prägnanz, which essentially entails that the perceptual system tends to consolidate the input into the simplest, or best, possible organization. Note that, although intuitive, this definition is unclear (Guberman, 2017). Most well-known are their principles of perceptual grouping and figure-ground organization (Wagemans et al., 2012a). The grouping principles include the following (see Figure 2.3):

- *Proximity*: Elements that are close together in space tend to be grouped together;
- *Similarity*: Elements of similar color, size, or orientation tend to be grouped together;
- *Common fate*: Elements that with common motion tend to be grouped together;
- *Symmetry*: Lines and curves that are symmetric tend to be grouped together;

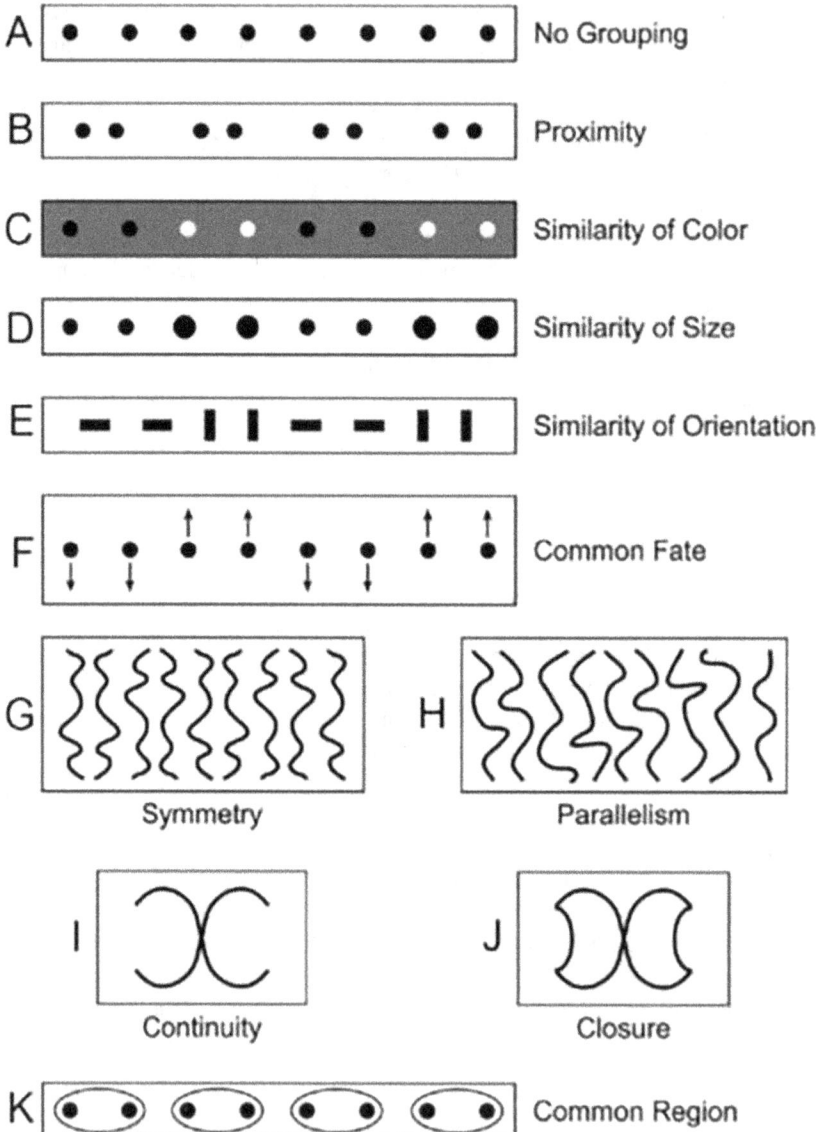

FIGURE 2.3 Illustration of several grouping principles. (From "Perceptual Organization in Vision," by S. E. Palmer, in *Stevens' Handbook of Experimental Psychology: Vol. 1. Sensation and Perception* (p. 183), ed. by H. Pashler, 2002, New York: Wiley. Copyright 2002 by John Wiley & Sons. Reprinted with permission.)

- *Parallelism*: Lines and curves that are parallel tend to be grouped together;
- *Continuity*: Contours tend to be perceived as continuous rather than changing abruptly;
- *Closure*: Gaps in images will tend to be filled-in to complete the shape or form;
- *Common region*: Elements included within the boundary of a region tend to be grouped together.

Common region is particularly important for the design of visual representations and displays in HCI, such as user interface and information visualization, as it can be used effectively to separate elements that should be considered together from ones that should not.

Wagemans et al. (2012a, p. 1188) summarized the effect of the grouping principles on perception as follows:

Perhaps the most parsimonious view consistent with the known facts is that grouping principles operate at multiple levels. It seems most likely that provisional grouping takes place at each stage of processing, possibly with feedback from higher levels to lower ones, until a final, conscious experience arises of a grouping that is consistent with the perceived structure of the 3-D environment.

Figure-ground organization refers to how shared borders between areas of the visual field are organized into a figure with shape against a background that has no perceived shape. The principles of figure-ground organization (Peterson, 2014) include the following, the first four of which are illustrated in Figure 2.4:

- *Convexity*: Regions with convex contours tend to be seen as figure;
- *Symmetry*: In at least some circumstances, globally symmetric regions are seen as figures;
- *Small area:* A region with a small area tends to be perceived as a figure rather the one with larger area;
- *Surroundedness*: A small area tends to be seen as a figure and a larger surrounding area as ground;
- *Lower region*: The lower region below a horizontal border tends to be seen as a figure;
- *Top-bottom polarity*: Regions that are narrower at the top than at the bottom tend to be seen as figures compared to regions with the opposite spatial relation.

The above principles, and others not described here, are based on image properties. However, research has shown that higher-level properties including attention, intention, and past experience also influence what is perceived as a figure, for which Peterson (2014) and Wagemans et al. (2012a) can be consulted.

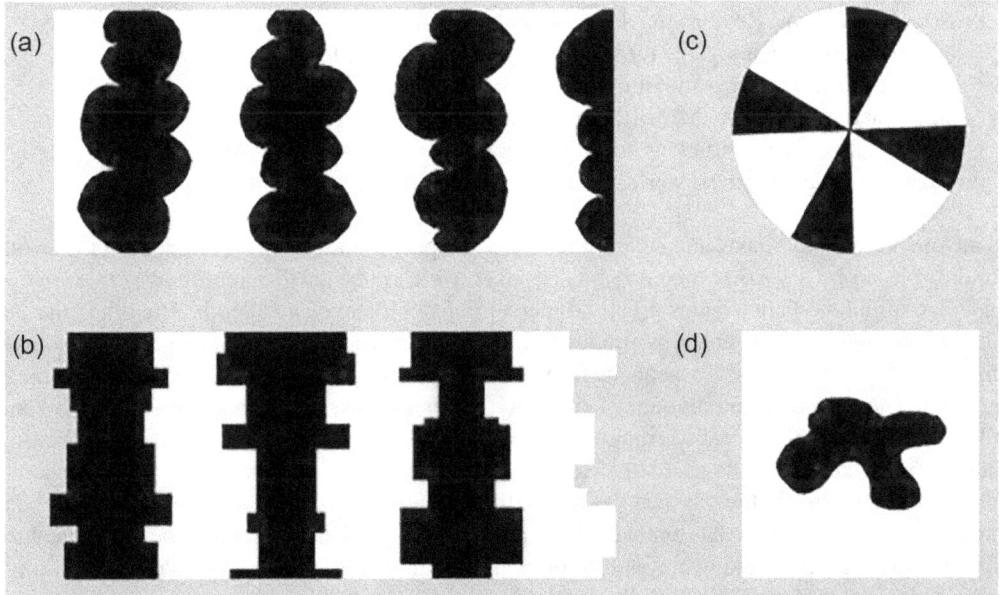

FIGURE 2.4 Cues affecting figure-ground perception: (a) convexity, (b) symmetry, (c) small area, (d) enclosure (surroundedness). (From Kubovy Lab Website (2023). Visual Perceptual Organization. https://uva. theopenscholar.com/kubovy-lab/visual-perceptual-organization.)

2.6.4 PERCEPTION OF SPACE

People perceive objects in three dimensions and not just two, even though the retinal image is 2D. Figure-ground organization illustrates this fact as even 2D black-and-white static images are perceived as a figure against a background. Much other information about distance and depth comes from monocular cues of the type involved in static displays on a computer monitor and in motion pictures that do not use special 3D methods of presentation. Perhaps surprising to many individuals is that all 50 states in the U.S. allow persons with monocular vision to drive (Whelan, 2021), though possibly with some restrictions. Thus, although the horizontal field of vision is reduced with only one eye, the monocular cues are sufficient to support relatively precise spatial functioning in the physical world.

Monocular cues include the following:

- *Occlusion*: When one image appears to cover another, this is a cue that the former is from an object that is in front of the latter;
- *Vertical distance from the horizon line*: For images below the line, a lower position implies an object that is closer than an object whose image is at a higher position; for images above the line, a higher image implies the object is closer than a lower one;
- *Linear perspective*: Actual or virtual perspective lines that converge in the middle at the horizon imply that the objects are receding in the distance;
- *Aerial perspective*: An image that is bluer and less clear than another, due to having to pass through particles in the air, is a cue that it is from an object that is farther away than the other;
- *Relative size*: An image that is larger appears to be from a closer object than an image size that is smaller;
- *Texture gradient*: A constant texture converging toward the horizon with progressively smaller elements implies a ground surface extending into the distance;
- *Familiar size*: An image of a familiar object that is relatively small is a cue that it is farther away;
- *Attached shadow*: Regions that have a darker "shadow" at the bottom tend to be perceived as farther away than regions for which the shadow is at the top;
- *Motion parallax*: For a moving person, images from objects far away will move more slowly across the scene than will those from objects that are closer;
- *Optical flow*: The continuous transformation pattern produced by forward or backward movement in the physical world.

As anyone knows who has watched a 3D movie with special glasses or moved about in a virtual world displayed by a VR/AR mixed reality headset, which presents different images to each eye, binocular stimulation contributes greatly to perceptions of distance and depth. The binocular cue is called retinal disparity, and it is produced in natural, physical environments by the two eyes being at different locations. Consequently, the left eye receives a slightly different image than the right eye. The direction and amount of disparity varies relative to a hypothetical curved plane, called the horopter, that passes through the location of an object on which a person is fixating. Thus, the cue is relative to fixation. In VR, the monocular cues not only need to match what someone would experience in the physical world, but the retinal disparity cues must as well. It has been noted that one limitation of 3D stereoscopic displays is that the distances for accommodation of the lens and convergence of the two eyes cannot be altered to match what they would be when fixating on objects at different physical distances (Reichelt et al., 2010). This mismatch may produce discomfort and fatigue.

People typically show good size constancy, which is that they perceive an object as being a constant size whether it is near or far away, even though the image size varies. This requires accurate perception of distance, as was demonstrated in a classic study by Holway and Boring (1941). They found that participants could match the size of a circular at different distances down a hallway when full depth cues were present (even when viewing was monocular), but as depth cues were progressively reduced, the match was based increasingly on retinal image size rather than object size. Moreover, size illusions occur when depth cues are misleading, with the Ames room (Ames, 1952) being the most famous example. The monocular cues in such a room are made to appear as if the far wall is crossing directly in front of the observer, when it is in receding in depth. When the room is viewed monocularly, eliminating binocular depth cues, the same size object will look much smaller when in the far corner than in the near corner because the cues suggest that the two images are coming from the same distance. Hafri et al. (2022) showed that perceived distance can alter memory for scene boundaries. They used a technique called tilt shift, which perceived distance in a photograph by blurring images located in the distance and caused an object (a train in their example) to produce fake miniaturization (causing the train to appear to be a model train). Participants who performed a scene-memory task misremembered fake-miniaturized views as farther away. Hafri et al. (2022) concluded that the memories were moderated by the spatial scale at which the image was viewed.

A distinction is commonly made between what is called peripersonal space (PPS), a region of space surrounding the body, and extra-personal space, a region that is farther from the body. The idea is that PPS is processed specifically for actions because objects within it are in reach of the individual. Hunley and Lourenco (2018) pointed out numerous empirical and theoretical issues involving the distinction between peri- and extra-personal space but arrived at the following definition: "PPS is a network of body-part-centered representations responsible for the coordination of actions toward, and in avoidance of, objects and other living entities, including people" (p. 14). They concluded that this network is multimodal, combining information from vision, audition, and other sensory modalities when relevant. Serino (2019) goes farther in situating PPS in a fronto-parietal network with neurons that integrate bodily tactile stimuli with visual or auditory information from external objects that are close to the body. Serino's view is that the resulting representation is not just multisensory but multisensory-motor, as is that of Bufacchi and Iannetti (2018).

2.6.5 MOTION PERCEPTION

Motion perception is fundamental to many online and offline activities. It is critical for most formats of animation, including movies, cartoons, and video games, or any interactive interface (Lasseter, 1997). In those contexts, the animators and designers create illusions of movement for the audience. As with all kinds of perception, the processing underlying motion perception is complex, and there are 17 or more cortical regions that respond better to stimuli that are moving than to ones that are stationary (Park & Tadin, 2018; Sunaert et al., 1999). Movement of an image across the visual field while the observer is stationary is sufficient to trigger neurons in the primary visual cortex (motion detectors) that detect changes in position on the retina in particular directions (Hubel & Wiesel, 1962). Such detectors can also account for apparent motion that is perceived from successive presentations of static stimuli (as in "moving" signs and video display screens) and to some extent for speed perception (Park & Tadin, 2018).

However, in most environments, there are many objects and surfaces, some moving in various directions and others stationary. Moreover, changes in position on the retina can be caused by movement of the observer, and the observer can also track a moving object, maintaining a roughly foveal position across time as the movement occurs. Consequently, more global motion analysis is needed that includes mechanisms of motion integration and segmentation. A well-established phenomenon is that of motion adaptation (Mather et al., 1998). When watching motion in a particular direction for a period of several seconds or longer, sensitivity to visual stimuli moving in that same direction will be decreased. Moreover, a stationary object will appear to move in the opposite direction. This is often evident when watching the credits for a movie scroll upward for several minutes, after which any stationary content presented on the screen seems to move downward. Motion adaptation has been found for neurons in the primary visual cortex (V1) and middle temporal area (MT), a region in the ventral visual pathway also known as V5 (Nishida et al., 2018). Weaker motion after-effects occur with movement durations as short as 640 ms, and evidence from ERPs suggests that short-adaptation effect involves different neural mechanisms than the long-adaptation effect (Akyuz et al., 2020).

The final topic we consider with regard to motion perception is biases in perception of estimating the final location of a moving stimulus. Representational momentum refers to a forward shift, or overestimation, along the motion trajectory, whereas representational gravity refers to a downward shift consistent with gravitational force (Merz, 2022). Merz et al. (2022) have proposed a speed prior account to explain these and other systematic biases in perception of moving stimuli, according to which the perceived location is a combination of the sensory input and prior expectations concerning stimulus speed.

2.6.6 PERCEPTION OF OBJECTS AND RECOGNITION OF PATTERNS

Visual perception allows people to make sense of their surroundings and to identify and understand the immediate objects. Among the processes included under the term object perception are those that integrate some elements in the visual input and separate them from others (i.e., the perceptual organization processes described earlier). However, the brain uses various additional processes to recognize and understand the objects (Peterson, 2001). It assigns shape and 3D structure to some of the element groupings. Visual cues, including the edges and contours, are used to perceive the object's shape; the object's size, orientation, and spatial position are also considered. The information arriving through the visual senses is matched against that for previously seen objects represented in memory, which permits their recognition. Other processes control the manner in which attention is focused on the shapes.

Pattern recognition refers more specifically to the processes associated with identifying the objects based on contact with memory (Tarr, 2000). Objects can be recognized at different categorical levels. The *entry level* (Jolicoeur et al., 1984) refers to that of the name that is generated most rapidly for a certain object, but recognition can also occur at superordinate and subordinate levels, depending on the task.

Two particular types of pattern-recognition warrant special mention. The first is reading, which is a pattern-recognition skill that literate persons possess to at least some extent (Rayner et al., 2012). Reading involves not only recognizing the meanings of individual words but also being able to obtain the meanings of sentences, paragraphs, and entire books. Readers can make inferences beyond the literal meanings of text by applying information they already know, and they can create mental representations of the material that may include visual images. Reading begins with extracting the visual information from the page, which can occur mainly from the fovea. However, more global information to direct eye movements can be gathered from the surrounding parafoveal region (Rayner, 1998). It is generally accepted that reading proceeds by way of feature detection, and tracking of the successive eye fixations of a reader can provide good evidence as to how the reader is proceeding through text.

Another special area is that of face recognition. Studies using fMRI found that certain areas of the inferotemporal cortex respond more for face recognition than for recognition of other objects (Sergent et al., 1992). However, other studies have found evidence that it is not faces per se but familiar faces that people can recognize easily (Young & Burton, 2017). Using magnetoencephalography to measure the time course of neural responses to faces, Dobs et al. (2019) found that the gender and age information, which can be based on coarse visual analysis, emerged <100 ms after onset of a face, with evidence of identity information occurring shortly thereafter. Their results also indicated that identity and gender representations were enhanced for familiar faces compared to unfamiliar ones, which they interpreted as suggesting that the benefit for recognizing familiar faces results in part from early feed-forward processing mechanisms.

2.7 AUDITORY PERCEPTION

Because audition is more temporally sensitive than vision but less spatially sensitive, it has different roles in HCI. The majority of displays utilize visual representations, while the auditory representations are often secondary and used to draw attention. Nevertheless, auditory displays are essential for many purposes. These include the creation of warnings and notifications, implementation of multimodal and virtual environments, and provision of an alternative display modality for visually impaired users.

2.7.1 Sensory Processes

Sound is generated in the environment by a mechanical disturbance that causes motion of the molecules in a medium, typically the air. The sound stimulus creates deviations from the standard air pressure level, which propagates outward in all directions from the source as a sound wave at 340 m/s. The wave frequency (or, inversely, wavelength) and amplitude are the primary determinants of the perceptual features of pitch and loudness, respectively. High-frequency waves attenuate quickly and are effective for providing information at short distances, whereas low-frequency waves transmit for longer distances. Sound level is typically specified in decibels:

$$L_p = 20 \log \left(p/p_{ref} \right) \text{dB},$$

where p is sound pressure and p_{ref} is 20 μPa.

The simplest soundwave is a sinusoidal function, as generated by a vibrating tuning fork. Young adult humans are sensitive to frequencies of ~20 Hz (very low pitch) to 20 kHz (very high pitch), with the upper limit for older adults being 15–17 kHz. Soundwaves are typically more complex than this, combining the moment-to-moment contributions of many frequencies, multiple sources, and reflections, including those produced by the head and body. Because of this additive property, the phase relationship between waves is crucial. Two 1,000-Hz tones that are in phase will add to produce a higher amplitude 1,000-Hz tone, whereas two that are 180° out of phase at the same amplitude will cancel each other and produce no sound. Active noise cancellation is used in noise-canceling headphones, for which a microphone detects the source noise sounds relative to which the headphones generate an opposing cancellation wave that reduces the amplitude of the noise wave (Liebich et al., 2018).

Soundwaves enter a person's ear through the outer part (the pinna) and travel through the ear canal, causing movement of the eardrum (tympanic membrane) where the canal terminates (Gelfand, 2018). This movement in turn causes motion in the ossicles (malleus, incus, and stapes), the lever system of small bones located in the middle ear. The footplate of the stapes is attached to the oval window, a membrane at the inner ear, which transfers the stapes' movement to the inner ear. The part of the inner ear most relevant to hearing is the cochlea, a fluid-filled, snail-shaped structure, that contains the Organ of Corti. Motion of the oval window produces a traveling-wave motion in the fluid, causing movement of the basilar membrane, which runs the length of the cochlea.

A sensory signal is generated by the bending of cilia of the hair cells, arrayed along the basilar membrane. A distinction is made between inner and outer hair cells, of which there are one and three rows, respectively (Corey et al., 2017). The inner hair cells are those most directly involved in converting the wave motion of the inner ear into neural signals, whereas the outer hair cells act to amplify the cochlear movement at low-intensity levels. Because the basilar membrane's stiffness increases from the base at the oval window end to the apex at the far end, its peak motion varies systematically as a function of tone frequency, being nearest to the oval window for high-frequency sounds and nearest to the apex for low-frequency sounds.

The inner hair cells provide input to type-I spiral ganglion neurons, which are the majority of afferent fibers within the auditory nerve (Heil & Peterson, 2015). These fibers fire spontaneously at a low rate without stimulation, and code frequency by the origin of the nerve fiber in the cochlea and the timing, or frequency, of its action potentials (Pickles, 2015). The characteristic frequency to which a fiber is most sensitive varies systematically from high frequencies for fibers that receive input from the base end of cochlea to low frequencies for fibers that receive their input from the apex end. Phase locking the responses of auditory nerve fibers occurs for low-frequency sounds, which allows the frequency of firing to code tone frequency for the low frequencies. However, phase locking does not occur for frequencies of 5 kHz or more, meaning that location coding (which neurons are signaling strongest) is most critical at higher frequencies.

The central auditory nervous system is more complex than the visual system, though there are multiple parallel systems, beginning with the auditory nerve, some that maintain temporal information and others that code patterns of activity measured over populations of neurons (Pickles, 2015). At the cochlear nucleus complex (CNC), the auditory nerve signals are directed into three primary pathways (Malmierca & Hackett, 2010). The CNC shows tonotopic organization; also of note, it receives input from descending pathways in several other brain areas. The tonotopic organization is retained in the nuclei to which the CNC projects. After passing through other regions, the auditory inputs converge at the inferior colliculus. From there, the fibers go to the medial geniculate body in the thalamus. The pathways then go to the primary auditory cortex in the temporal lobe, which also has a tonotopic organization.

That area is only one of more than 30 cortical areas involved in auditory processing. As with vision, ventral (what) and dorsal (where/how) auditory streams can be distinguished (Rolls et al., 2022). The ventral stream is involved in semantic processing of objects and faces and is thought to play a significant role in speech perception (DeWitt & Rauschecker, 2012). The dorsal stream is involved primarily in spatial actions. Rauschecker (2018) has further suggested that the dorsal route serves the function of providing an internal model for sensorimotor coordination in space and time.

2.7.2　Basic Perceptual Attributes

In addition to being influenced by sound frequency, loudness is affected by many other physical properties, including duration, spectral complexity, and frequency (Florentine, 2011), the latter of which we consider here. People are differentially sensitive to sound across the auditory spectrum. This fact is captured by equal loudness contours, which were first published by Fletcher and Munson (1933) and later updated by Robinson and Dadson (1956), the ISO (2003), and Suzuki and Takeshima (2004; see Figure 2.5). The bottom curve plots the absolute threshold as a function of tone frequency and shows the greatest sensitivity between 3,000 and 4,000 Hz. The most notable characteristic of the curves is that sensitivity decreases substantially for low-frequency sounds below about 400 Hz at low-intensity levels but not so much at higher frequency levels. Tones above about 6,000 Hz show a lesser decrease in sensitivity that does not depend on the intensity level. Because the contours in the studies cited above were obtained for participants with unimpaired hearing, they are not applicable for those with hearing limitations. Consequently, Schlittenlacher and Moore (2020) developed a method for the determination of equal loudness contours that allows them to be obtained for an individual much more rapidly than with standard threshold methods and is useful for fitting persons with hearing aids.

Tone pitch is typically described as a dimension of height, and it is primarily a function of the tone frequency. However, as with loudness, it is influenced by other factors (Houtsma, 1997). One of the more interesting findings is that for harmonics (integer multiples of the fundamental tone frequency), the pitch is determined by the fundamental frequency even when it is not present (the missing fundamental phenomenon) or detectable (Winkler et al., 1997). Other factors that can affect

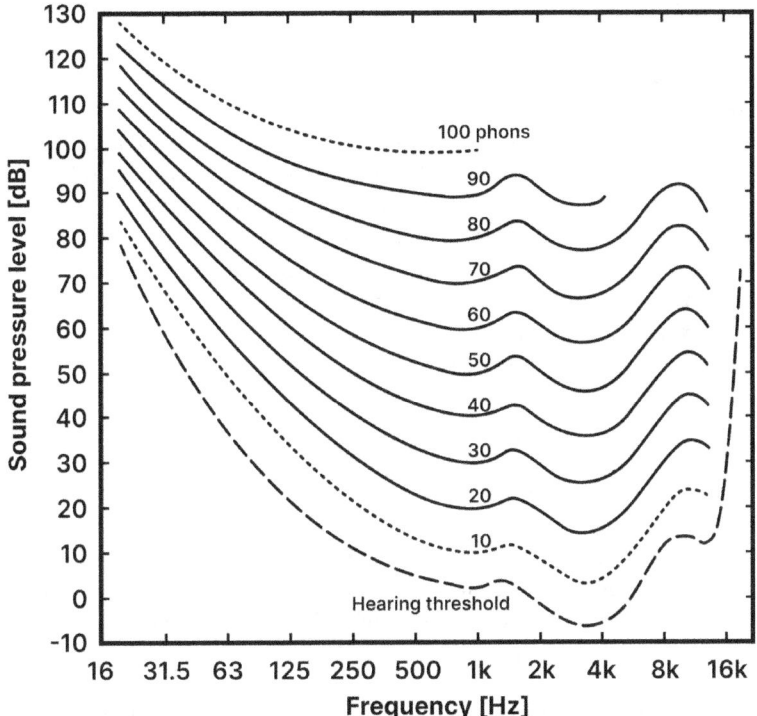

FIGURE 2.5 Equal loudness contours of Suzuki and Takeshima (2004). (From Suzuki, Y. & Takeshima, H. (2004). Equal-loudness-level contours for pure tones. *Journal of the Acoustical Society of America, 116*, 918–933. Reprinted with permission.)

tone pitch include amplitude and duration. Timbre refers to the differences in perceived sound of the same note played on different musical instruments at the same intensity. It is determined primarily by the spectral composition of the harmonics, phase relations between the harmonics, and the temporal envelope of the instrument sound. Sounds with different harmonics can be distinguished, as can sounds of different pitch and loudness (Houtsma, 1997).

2.7.3 PERCEPTUAL ORGANIZATION

Auditory perception faces several issues that are relatively distinct from those in visual perception (Denham & Winkler, 2015). Because it is more temporally based than vision, with the properties of sound stimuli unfolding across time, significant information-processing demands on a perceiver are created. Extracting the information conveyed by the sounds typically requires integration across several timescales. Moreover, because many objects generate sounds only intermittently, association of temporally separated events is required.

Effective use of grouping principles by designers can be used to facilitate users' perception of auditory displays. The principles, which are similar to those for visual perceptual organization, can segregate the acoustic stimuli into distinct auditory streams or fuse them into a single stream (Denham & Winkler, 2015; Moore & Gockel, 2012). Similarity is a major organizational principle: Sounds tend to group into distinct auditory streams when they are dissimilar from each other in frequency, timbre, or location but to fuse into a single stream when these features are similar. Continuity, or good continuation, and common fate are also significant factors, particularly for sequential grouping of sounds (Denham & Winkler, 2015). The latter factor is called "temporal coherence" in the auditory perception literature, and it has been argued that attention to one temporally coherent feature is necessary to bind other temporally coherent features into a single stream

(Shamma et al., 2011). Neural indicators of temporal coherence in the cortex show up as early as 115 ms after stimulus onset and last approximately to 185 ms with passive listening and longer (up to 265 ms) with active listening (O'Sullivan et al., 2015). This temporal coherence-based scene analysis may start as early as the cochlear nucleus (Viswanathan et al., 2022).

The grouping factors described so far can be characterized as bottom-up, that is, as coming from the senses. Auditory scene analysis is also influenced by top-down, or schema-based, factors including attentional mechanisms and learned rules (Sutojo et al., 2020). The top-down analysis can produce closure or completion of the auditory stimulus. The power of such closure is illustrated by the phonemic restoration effect. Warren et al. (1972) showed that listeners were unaware when listening to a sentence that a phoneme was missing and had been replaced by a cough. Also, they could not indicate exactly where the cough occurred in the sentence, indicating that the cough was grouped separately from the sentence stream.

2.7.4 PERCEPTION OF SPACE

Sound localization occurs in three dimensions: azimuth, height, and distance. Localization of a sound along the azimuth depends on three types of cues (Risoud et al., 2018), two of which, interaural time differences and level (intensity) differences, respectively, are based on having two ears. The time and intensity differences vary systematically as sound location is changed in successive steps from straight ahead (0°) to straight behind (180°), with the difference increasingly favoring the near ear until 90° (or 270°), after which it decreases toward zero. As illustrated in Figure 2.6, interaural time differences function most effectively at sound frequencies below about 1,000 Hz, whereas interaural level differences are accurate mainly for frequencies above 3,000 Hz. Because these are the primary cues for localization around the azimuth, localization accuracy is less accurate for tones with frequencies in the range of 1,000 and 3,000 Hz than for those outside of that range.

The additional cue type is the head-related transfer function (spectral cues generated by the pinna, trunk, head, and shoulders), and it does not require two ears. The spectral cues generated by the head-related transfer function are also strongest for higher frequency sounds, which contributes to localization being less accurate for low-frequency sounds than for high-frequency sounds. Because the spectral cues of the head-related transfer function are the main ones available for height differences, sound localization on the vertical dimension is less accurate than on the azimuth. Distance localization also depends primarily on monaural cues, with time and intensity differences from an initial sound and the reverberant sound of its reflection being of importance (Risoud et al., 2018).

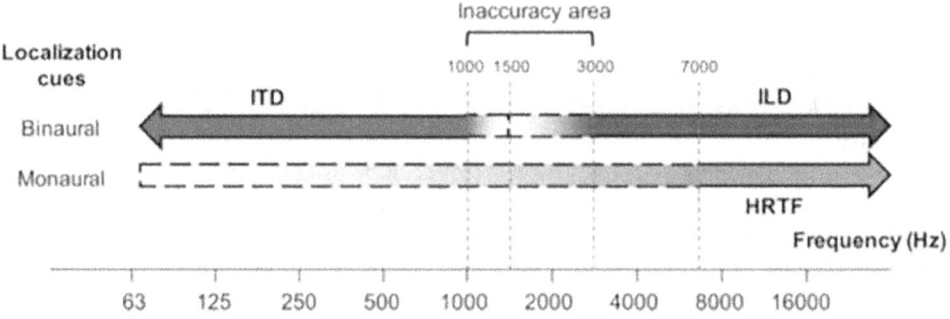

FIGURE 2.6 Representation of sound localization cues according to stimulus frequency. ITD, interaural time difference; ILD, interaural level difference; HRTF, head-related transfer function. (From Risoud, M., Hanson, J. N., Gauvrit, F., Renard, C., Lemesre, P. E., Bonne, N. X., & Vincent, C. (2018). Sound source localization. *European Annals of Otorhinolaryngology, Head and Neck Diseases, 135*(4), 259–264.)

2.7.5 Motion Perception

Our description of visual motion perception emphasized that multiple objects and the perceiver are often moving concurrently in the physical world. Thus, sorting out the various sources of change is key to being able to act in the world. The same is true for auditory motion perception. To understand the implications of an emergency vehicle's siren when driving, for example, it is essential to perceive not just its location and distance at that moment but whether the vehicle is moving toward or away from the listener and the speed with which it is doing so. The motion cues for direction and distance are crucial, but, as with vision, the changes can be brought about by movement of the sound source in the environment or of the listener.

The minimum audible movement needed to detect motion is an increasing function of velocity and a decreasing function of signal bandwidth (Carlile & Leung, 2016). A combination of cues produces the strongest effects, and changes in velocity are affected by a prior state. Many mechanisms, including prior state, prediction, dynamic updating, and attentional modulation, likely play roles in the process of distinguishing object motion from self-motion.

2.7.6 Perception of Objects and Recognition of Patterns

The primary purpose of human auditory perception, like vision, is to gather information about objects or events in the world. From an evolutionary perspective, during the early stages of human development, auditory stimuli were generated primarily by creatures and non-living physical events in the immediate environment. However, the ability to transmit and create sounds has developed over time to include warnings generated electronically, recordings of sounds of all types, and generation of artificial speech messages. As described earlier, auditory object perception relies on bottom-up processing of sensory signals, and top-down cognitive processing constraints to group the sounds into meaningful streams (Brefczynski-Lewis & Lewis, 2017). Bizley and Cohen (2013) note that the key to such grouping "is the idea of predictability: the auditory system must generate some sort of prediction from current and previously present sounds to build a model of what is likely to occur next" (p. 701). This process provides a basis for recognizing what the sound source might be (Yost, 2013). A definition of an auditory object, thus, is a perceptual representation corresponding to a sound that can be attributed to a particular source (Bizley & Cohen, 2013), and the object may span multiple auditory events as they develop across time.

Brefczynski-Lewis and Lewis (2017) distinguished three categories of sound sources (see Figure 2.7). The first is action sounds associated with living things, which are sounds created by a human or non-human animal interacting with the physical environment. The second is action sounds produced by non-living things, including environmental sounds such as the wind blowing through the leaves and mechanical sounds made by powered vehicles. The final category is that of vocalizations of living things, for which subcategories are human speech and animal vocal sounds

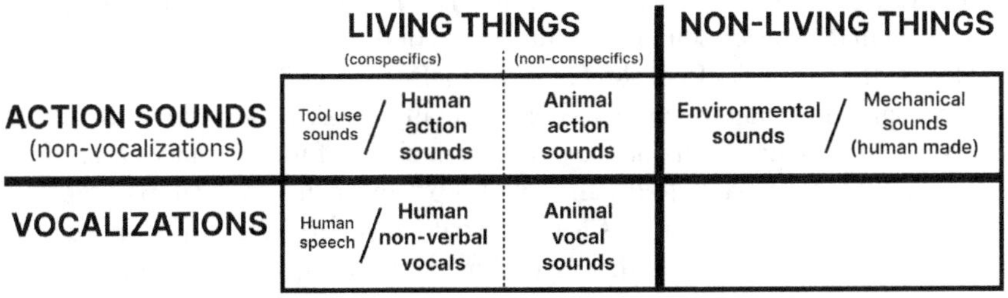

	LIVING THINGS		NON-LIVING THINGS	
	(conspecifics)	(non-conspecifics)		
ACTION SOUNDS (non-vocalizations)	Tool use sounds / Human action sounds	Animal action sounds	Environmental sounds / Mechanical sounds (human made)	
VOCALIZATIONS	Human speech / Human non-verbal vocals	Animal vocal sounds		

FIGURE 2.7 Brefczynski-Lewis and Lewis's (2017) categorization of different types of sounds. (Reprinted from Brefczynski-Lewis, J. A. & Lewis, J. W. (2017). Auditory object perception: A neurobiological model and prospective review. *Neuropsychologia, 105*, 223–242. Reprinted with permission.)

(e.g., howls of a dog). The authors present evidence, mostly derived from human neuroimaging studies, to suggest that these different sound types are processed in parallel, hierarchical pathways.

Human speech perception has been a topic of significant research for many years (Pardo et al., 2021). Speech is typically produced with the intent to communicate to someone who has to perceive the spoken message. Phonemes, the smallest possible units of speech sounds, are typically assumed to be the primitive elements of speech perception (Fowler & Iskarous, 2013). However, there has been much disagreement as to how the phonemes are extracted from the speech signal. The disagreement arises because, as a consequence of co-articulation in phoneme production, there is no obvious invariant sound pattern corresponding to a particular phoneme across contexts and no clear signals for segmentation of the speech stream.

One implication of these limitations of the speech signal is that context is very important in speech perception. As Stilp (2020) notes, stimulus variability becomes less problematic when it is realized that the perceiver is processing the speech signal relative to the perceptual context and not just basing it on the immediate stimulus values. Accounts differ as to whether they link the broader context to cues about the articulatory gestures used to produce the speech sounds (Liberman et al., 1967) or to more general auditory processing mechanisms (Rysling et al., 2019). Stilp (2020) reviewed the influences of acoustic contexts that preceded or followed the target item and were adjacent to or temporally removed from it. He concluded that acoustic context acts to disambiguate speech by magnifying differences between context and the target stimulus, which facilitates speech perception.

Although Brefczynski-Lewis and Lewis (2017) did not include music among their categories of sound types, they suggested that music may be a fourth category. Music and musical chords have several distinctive aspects, particularly with regard to its ability to generate emotions, and likely involve a distinct type of processing. Koelsch et al. (2019) emphasize the role of prediction in musical experience. They distinguish first-order predictions of perceptual content from second-order predictions involving the precision of the first-order predictions and emphasize that the listener plays an active role in the process.

2.8   MULTIMODAL PERCEPTION AND MULTIMODAL INTERFACES

We have emphasized visual and auditory perception because of their prevalence in HCI. However, it should be noted that the other senses, including the skin senses and the vestibular sense, are also relevant. Moreover, in most physical and cyber environments, events produce stimulation in multiple modalities (Bertelson & De Gelder, 2004).

2.8.1   MULTIMODAL PERCEPTION

One example of multimodal perception that has been widely studied in recent years is speech (Rosenblum & Dorsi, 2021). The vocalizations of the speaker are accompanied by movements of the mouth that are visible to a listener in many contexts. Thus, the visual stimulation provides input that is correlated with the speech sounds. An illustration of the multimodal nature of speech perception is provided by the McGurk effect (McGurk & MacDonald, 1976): When people hear a spoken utterance at the same time that they see a person visually speaking a different utterance, what is heard is affected by the visual input and may not match either of the utterances. For example, with repeated utterances of the syllable [ba] paired with lip movements for [ga], participants reported hearing [da]. This effect is compelling, as is the ventriloquism effect in which speech is perceived as coming from an inanimate puppet rather than from the ventriloquist who is producing the speech (Lavan et al., 2022). Another case of multimodal interaction is synesthesia, which a stimulus in one modality generates a sensory experience in another sensory modality. For instance, colored hearing synesthetic perceptions are connections between sounds and color sensations (Marks, 1975). Strong synesthesia can be distinguished from weak synesthesia (Martino & Marks, 2001). In the former, which is relatively rare, a person experiences vivid imagery in one sensory modality in response to stimulation in another. In the latter, which is more common, a person experiences cross-sensory correspondences that can be reported verbally and influence the performance of information-processing tasks.

Rosenblum and Dorsi (2021) provide a good summary of the current view of multimodal perception. They state, "The brain is now thought to be largely designed around multisensory input, with most major sensory areas showing crossmodal modulation. Behaviorally, research has shown that even our seemingly unimodal experiences are continuously influenced by crossmodal input" (p. 44).

2.8.2   MULTIMODAL INTERFACES AND APPLICATIONS IN HCI

Multimodal perception is becoming an increasingly vital part of HCI. For example, Seinfeld et al. (2021, p. 426) advocate a central role of user representations, with the view that temporal and spatial alignment of multisensory feedback leads to a coherent and meaningful unitary perception. They give the example of the holistic percept of the user that occurs when a virtual object touches a virtual avatar while haptic feedback is provided at the same body location.

This view of multisensory interaction relates closely to the concept of tangible interfaces (see book #4 'Interaction Techniques and Technologies in Human-Computer Interaction', Chapter 11), which allow users to interact with a computer or other device through physical objects, rather than through a traditional input device such as a keyboard or mouse. For example, de Bérigny et al. (2014) developed an interactive installation artwork called *Reefs on the Edge*, which allows the audience to explore threatened coral reef systems through a tangible user interface. They use this example to make the case that such interfaces can serve as educational and entertainment tools in environmental education. Tangible interfaces use touch as the primary modality for direct manipulation and haptic feedback, with vision also playing a key role in allowing users to see and understand the effects of their actions. Sound can be exploited to provide additional information and feedback. Other modalities like smell and taste are less commonly used but can be employed in special cases (Lai & Cao, 2019; Ranasinghe et al., 2017).

2.8.3   XR, VR, AND AR TECHNOLOGIES

The term eXtended reality (XR) has come to be used as an umbrella term for three technologies (Kaplan et al., 2021): virtual reality (VR), augmented reality (AR), and mixed reality (MR; Kaplan et al., 2021). For VR, visual perception is used to create a realistic and immersive experience in a virtual world that can respond to the user's movements and actions. The idea is that the virtual environment provides similar cues to reality, which enables the user to interact with the virtual environment in an intuitive way. For AR, the virtual objects are overlaid on top of the physical environment. This adds the complexity that the virtual environment must not only provide a satisfactory perceptual experience but also be joined with the physical world. Additionally, the complexities associated with the user coordinating the virtual and physical aspects of the task must be accommodated. This is also the case for MR, which can be thought of as a combination of VR and AR in which the physical and virtual worlds are merged seamlessly. In all XR technologies, human perception, particularly vision, is critical. Because spatial perception is fundamental to many tasks humans perform in the physical world, such as navigation, object recognition, and spatial reasoning, it is particularly essential in all forms of XR. Other requirements relating to vision include field of view, angular resolution, dynamic range, and correct depth cues (Xiong et al., 2021).

Tactile perception involves the experience of touch, pressure, and texture through sensory receptors embedded in the skin. It is essential for communicating aspects of the physical world (such as, navigating on an uneven roadway) and, in conjunction with kinesthetic feedback associated with a person's own physical effort, for tasks such as grasping objects. In XR applications, tactile perception can be used to encode or recreate the sensation of touch and texture, allowing users to interact with virtual objects as if they were real. For example, Oh et al. (2021) developed a hand-worn VR device with motion sensors and haptic feedback that uses soft, stretchable, lightweight sensors and heater sheets. The sensors and vibrators embedded in the sheets measure finger movements and provide vibro-haptic feedback, whereas the heater sheet provides thermo-haptic feedback. The device allows users to feel contact and discriminate materials with different temperatures. As another example, Kim et al. (2019) developed a training system for the Humanitude care technique for dementia patients using AR technology and a tactile sensor. This system superimposes a 3D model of a patient's face onto a soft doll and uses the tactile sensor to measure touch skills to allow the system to realize sensing and interaction simultaneously. These techniques that utilize characteristics of human perception can enhance the virtual environment to be more realistic and help users to better understand and interact with it.

2.8.4   ROBOTS AND AUTONOMOUS VEHICLES

Finally, robots and autonomous vehicles are interacting increasingly with humans, and these interactions, by their nature, use multiple modalities. For the interactions to be effective and safe, it is valuable to understand the cues humans use for similar interactions with other humans and incorporate them into the actions of the non-human actors. This is a premise underlying advocacy of human-robot and human-automation teaming (Amin et al., 2020; Calhoun, 2022) and the claim that human use of AI should be conceptualized as "collaboration" instead of "interactions" (Xu, 2020), although the considerations extend beyond perceptual ones. In human-robot interaction, the notion of the "uncanny valley" has been widely discussed since being proposed by Mori (Mori, MacDorman, & Kageki, 2012. In at least some contexts, when robots become too human-like, they become unsettling or even repulsive to humans. An issue is, what cues are present in the interactions that cause the negative reactions, and how they can be eliminated (Stepp Jr., 2022).

For autonomous vehicles to interact safely with human-driven vehicles in mixed driving conditions, it is necessary not only for the vehicles to be able to "perceive" the human drivers' intentions but also for the human drivers and pedestrians to perceive the intentions of the autonomous vehicles (Petersen & DeLuccia, 2022). This requires understanding the various visual and auditory cues that may be signaled between human drivers to convey their intents. As one example, Hudson et al. (2019) had participants wear a headset that put them as a pedestrian in VR environment with a four-way intersection that had two-way stop signs. They were to decide whether it was safer to cross in front of autonomous vehicles that signaled their intentions in different ways. Participants indicated that they understood that the word "walk" and silhouette visual features signaling someone walking meant that they could cross the road safely in front of the vehicle. They also understood that an upraised hand or a stop sign meant that it was unsafe to cross the road. The participants showed evidence of uncertainty about relying on the signal of the automated vehicle's intention when the vehicle contained a passenger in the "driver's" seat who was looking away and appeared to be distracted. In other words, the participants did not fully trust the signal from the vehicle when other visual cues implied that the passenger was not paying attention. Studies like these illustrate how the interaction between visual perception and other modalities can be studied in the area of autonomous vehicles.

2.9   CONCLUSION

HCI requires communication between the human user and the machine. One key aspect of this communication process is displaying the information that is necessary for the user, or desired by the user, in a way that will be efficiently perceived. With the increasing autonomy of cyber-physical systems, perception takes on a more prominent role since controlling actions of the user are minimized. Instead, the user engages more often in monitoring the system and occasionally responding to alerts. In XR environments, the actions of the user are more natural and intimately linked to the perceptual information, which also increases the importance of intuitive displays of information.

In this chapter, we presented an overview of some of the essentials of sensation and perception that are relevant in many HCI contexts. The principles, phenomena, and theories we covered should be of use to researchers, designers, and students. We put most emphasis on visual perception and auditory perception due to their prominence in HCI, but our coverage of these domains has only touched on salient topics related to contemporary issues. Much current research implies that when computer technology is implemented outside of the laboratory, perception is influenced by concurrent inputs from multiple sensory modalities, intended actions, and the anticipated feedback, as well as knowledge and prior experience. We encourage all persons who are interested in human interactions with computer technology to use our coverage as a starting point for delving into the details of topics in sensation and perception as they apply to specific design contexts.

REFERENCES

Abbasi-Kesbi, R., Memarzadeh-Tehran, H., & Deen, M. J. (2017). Technique to estimate human reaction time based on visual perception. *Healthcare Technology Letters*, *4*(2), 73–77.

Akyuz, S., Pavan, A., Kaya, U., & Kafaligonul, H. (2020). Short-and long-term forms of neural adaptation: An ERP investigation of dynamic motion aftereffects. *Cortex*, *125*, 122–134.

Albers, J. (2013). *Interaction of Color*. New Haven, CT: Yale University Press.

Ali, A. X., McAweeney, E., & Wobbrock, J. O. (2021). Anachronism by design: Understanding young adults' perceptions of computer iconography. *International Journal of Human-Computer Studies*, *151*, 102599.

Ames, A. (1952). *The Ames Demonstrations in Perception*. New York: Hafner Publishing Company, pp 1–130.

Amin, F. M., Rezayati, M., van de Venn, H. W., & Karimpour, H. (2020). A mixed-perception approach for safe human-robot collaboration in industrial automation. *Sensors*, *20*(21), 6347.

Ashby, F. G., & Townsend, J. T. (1980). Decomposing the reaction time distribution: Pure insertion and selective influence revisited. *Journal of Mathematical Psychology*, *21*(2), 93–123.

Azzam, D., & Ronquillo, Y. (2022). Snellen chart. In *StatPearls* [Internet]. St. Petersburg, FL: StatPearls Publishing. https://www.ncbi.nlm.nih.gov/books/NBK558961/

Azzopardi, P., & Cowey, A. (1996). The overrepresentation of the fovea and adjacent retina in the striate cortex and dorsal lateral geniculate nucleus of the macaque monkey. *Neuroscience*, *72*(3), 627–639.

Barten, P. G. J. (1999). *Contrast Sensitivity of the Human Eye and Its Effects on Image Quality*. Bellingham, WA: SPIE Press.

Barten, P. G. J. (2003). Formula for the contrast sensitivity of the human eye. *Proceedings of the Society for Photo-Optical Instrumentation Engineers*. Vol. 5294, Image Quality and System Performance.

Ben-Bassat, T., Meyer, J., & Tractinsky, N. (2006). Economic and subjective measures of the perceived value of aesthetics and usability. *ACM Transactions on Computer-Human Interaction (TOCHI)*, *13*(2), 210–234.

Berkeley, G. (1709). An essay towards a new theory of vision. Jeremy Pepyat, Bookseller. (Reprinted with minor changes for clarity by J. Bennett, 2017, https://earlymoderntexts.com/assets/pdfs/berkeley1709.pdf)

Bertelson, P., & De Gelder, B. (2004). The psychology of multimodal perception. In C. Spence, & J. Driver (Ed.), *Crossmodal Space and Crossmodal Attention* (pp. 141–177). Oxford: Oxford University Press.

Bizley, J. K., & Cohen, Y. E. (2013). The what, where and how of auditory-object perception. *Nature Reviews Neuroscience*, *14*(10), 693–707.

Boring, E. G. (1942). *Sensation and Perception in the History of Experimental Psychology*. New York: Appleton-Century.

Brefczynski-Lewis, J. A., & Lewis, J. W. (2017). Auditory object perception: A neurobiological model and prospective review. *Neuropsychologia*, *105*, 223–242.

Broadbent, D. E., & Gregory, M. (1962). Donders' B- and C-reactions and S-R compatibility. *Journal of Experimental Psychology*, *63*(6), 575–578.

Brooks, J. A., Stolier, R. M., & Freeman, J. B. (2021). Computational approaches to the neuroscience of social perception. *Social Cognitive and Affective Neuroscience*, *16*(8), 827–837.

Brunswik, E. (1952). The conceptual framework of psychology. In O. Neurath (Ed.), *International Encyclopedia of Unified Science* (Vol. 1, No. 10, pp. 656–760). Chicago, IL: University of Chicago Press.

Brunswik, E. (1956). *Perception and the Representative Design: Design of Psychological Experiments*. Berkeley, CA: University of California Press.

Buchner, A., Mayr, S., & Brandt, M. (2009). The advantage of positive text-background polarity is due to high display luminance. *Ergonomics 52*(7), 882–886.

Bufacchi, R. J., & Iannetti, G. D. (2018). An action field theory of peripersonal space. *Trends in Cognitive Sciences*, *22*(12), 1076–1090.

Byrne, Z. S., Dvorak, K. J., Peters, J. M., Ray, I., Howe, A., & Sanchez, D. (2016). From the user's perspective: Perceptions of risk relative to benefit associated with using the Internet. *Computers in Human Behavior*, *59*, 456–468.

Calhoun, G. (2022). Adaptable (not adaptive) automation: Forefront of human-automation teaming. *Human Factors*, *64*(2), 269–277.

Caligiore, D., Borghi, A. M., Parisi, D., & Baldassarre, G. (2010). TRoPICALS: A computational embodied neuroscience model of compatibility effects. *Psychological Review, 117*, 1188–1228.

Carlile, S., & Leung, J. (2016). The perception of auditory motion. *Trends in Hearing*, *20*, 1–19.

Carter, B. T., & Luke, S. G. (2020). Best practices in eye tracking research. *International Journal of Psychophysiology*, *155*, 49–62.

Chemero, A. (2003). An outline of a theory of affordances. *Ecological Psychology, 15*(2), 181–195.

Chong, I., & Proctor, R. W. (2020). On the evolution of a radical concept: Affordances according to Gibson and their subsequent use and development. *Perspectives on Psychological Science*, *15*(1), 117–132.

Cohen, M. A., Dennett, D. C., & Kanwisher, N. (2016). What is the bandwidth of perceptual experience? *Trends in Cognitive Sciences*, *20*(5), 324–335.

Corey, D. P., Maoiléidigh, D. Ó., & Ashmore, J. F. (2017). Mechanical transduction processes in the hair cell. In G. A. Manley, A. W. Gummer, A. N. Popper, & R. R. Fay (Eds.), *Understanding the Cochlea* (pp. 75–111). Berlin, Heidelberg: Springer.

de Bérigny, C., Gough, P., Faleh, M., & Woolsey, E. (2014). Tangible user interface design for climate change education in interactive installation art. *Leonardo*, *47*(5), 451–456.

Denden, M., Tlili, A., Essalmi, F., Jemni, M., Chen, N.-S., & Burgos, D. (2021). Effects of gender and personality differences on students' perception of game design elements in educational gamification. *International Journal of Human-Computer Studies*, *154*, 102674.

Denham, S. L., & Winkler, I. (2015). Auditory perceptual organisation. In D. Jaeger, & R. Jung (Eds.), *Encyclopedia of Computational Neuroscience* (2nd ed., pp. 277–288). Berlin, Heidelberg: Springer.

DeWitt, I., & Rauschecker, J. P. (2012). Phoneme and word recognition in the auditory ventral stream. *Proceedings of the National Academy of Sciences of the United States of America, 109*(8), E505–eE514.

Dhami, M. K., Hertwig, R., & Hoffrage, U. (2004). The role of representative design in an ecological approach to cognition. *Psychological Bulletin, 130*(6), 959–988.

Dobres, J., Chahine, N., & Reimer, B. (2017). Effects of ambient illumination, contrast polarity, and letter size on text legibility under glance-like reading. *Applied Ergonomics, 60*, 68–73.

Dobs, K., Isik, L., Pantazis, D., & Kanwisher, N. (2019). How face perception unfolds over time. *Nature Communications, 10*(1), 1–10.

Fechner, G. T. (1860). Element der Psychophysik [*Elements of Psychophysics*]. Leipzig: Breitkopf & Harterl.

Fletcher, H., & Munson, W. A. (1933). Loudness, its definition, measurement and calculation. *Journal of the Acoustical Society of America, 5*, 82–108.

Florentine, M. (2011). Loudness. In M. Florentine, et al. (Eds.), Springer *Handbook of Auditory Research* (Vol. 37, pp. 1–15). Berlin, Heidelberg: Springer Science+Business Media.

Fon, U. M. (2021). Brief history of perception. *Research Catalogue*. https://www.researchcatalogue.net/view/1025487/1025488/0/0.

Fowler, C. A., & Iskarous, K. (2013). Speech production and perception. In A. F. Healy, & R. W. Proctor (Eds.), *Experimental Psychology* (2nd ed., pp. 236–263). Volume 4 in I. B. Weiner (Editor-in-Chief), *Handbook of Psychology*. Hoboken, NJ: John Wiley.

Freund, B., Colgrove, L. A., Burke, B. L., & McLeod, R. (2005). Self-rated driving performance among elderly drivers referred for driving evaluation. *Accident Analysis & Prevention*, *37*(4), 613–618.

Frisby, J. P., & Stone, J. V. (2010). *Seeing: The Computational Approach to Biological Vision*. Cambridge, MA: MIT Press.

Gelfand, S. A. (2018). *Hearing: An Introduction to Psychological and Physiological Acoustics* (6th ed.). Boca Raton, FL: CRC Press.

Gescheider, G. A. (1997). *Psychophysics: The Fundamentals*. Mahwah, NJ: Lawrence Erlbaum.

Gibson, J. J. (1950). *The Perception of the Visual World*. Boston, MA: Houghton Mifflin.

Gibson, J. J. (1979). *The Ecological Approach to Visual Perception*. Boston, MA: Houghton Mifflin.

Gilbert, P. A., & Haeberli, W. (2007). Experiments on subtractive color mixing with a spectrophotometer. *American Journal of Physics*, *75*(4), 313–319.

Ginsburg, A. P. (2003). Contrast sensitivity and functional vision. *International Ophthalmology Clinics*, *43*(2), 5–15.

Goodale, M. A., & Milner, A. D. (1992). Separate visual pathways for perception and action. *Trends in Neurosciences, 15*(1), 20–25.

Gordon, G., & Slater, A. (1998). Nativisim and empiricism: The history of two ideas. In A. Slater (Ed.), *Perceptual Development: Visual, Auditory and Speech Perception in Infancy* (pp. 73–103). London: Psychology Press.

Graham, C. H. (1958). Sensation and perception in an objective psychology. *Psychological Review, 65*(2), 65–76. https://doi.org/10.1037/h0046960

Guberman, S. (2017). Gestalt theory rearranged: Back to Wertheimer. *Frontiers in Psychology*, *8*, 1782.

Haber, R. N. (Ed.). (1969). *Information-Processing Approaches to Visual Perception*. New York: Holt, Rinehart and Winston.

Hafri, A., Wadhwa, S., & Bonner, M. F. (2022). Perceived distance alters memory for scene boundaries. *Psychological Science, 33*(12), 2040–2058.

Hammond, K. R. (Ed.) (1966). *The Psychology of Egon Brunswik*. New York: Holt, Rinehart and Winston.

Hancock, G. M., Longo, L., Young, M. S., & Hancock, P. A. (2021). Mental workload. In G. Salvendy, & W. Karwowski (Eds.), *Handbook of Human Factors and Ergonomics* (pp. 203–226). Hoboken, NJ: John Wiley & Sons.

Hautus, M. J., Macmillan, N. A., & Creelman, C. D. (2022). *Detection Theory: A User's Guide*. London: Routledge.

Heil, P., & Peterson, A. J. (2015). Basic response properties of auditory nerve fibers: A review. *Cell Tissue Research, 361*, 129–158.

Heitz, R. P. (2014). The speed-accuracy tradeoff: History, physiology, methodology, and behavior. *Frontiers in Neuroscience*, *8*, 150.

Hellman, R. P., & Zwislocki, J. (1963). Monaural loudness function at 1000 cps and interaural summation. *Journal of the Acoustical Society of America*, *35*(6), 856–865.

Hellwig, L., & Fairchild, M. D. (2022). Brightness, lightness, colorfulness, and Chroma in CIECAM02 and CAM16. *Color Research & Application*, *47*, 1083–1095.

Holway, A. H., & Boring, E. G. (1941). Determinants of apparent visual size with distance variant. *American Journal of Psychology*, *54*(1), 21–37.

Hommel, B. (2019). Theory of event coding (TEC) V2.0: Representing and controlling perception and action. *Attention, Perception, & Psychophysics*, *81*(7), 2139–2154.

Hommel, B., Müsseler, J., Aschersleben, G., & Prinz, W. (2001). The theory of event coding (TEC): A framework for perception and action planning. *Behavioral and Brain Sciences*, *24*(5), 849–878.

Horrey, W. J., & Wickens, C. D. (2004, September). Focal and ambient visual contributions and driver visual scanning in lane keeping and hazard detection. In *Proceedings of the Human Factors and Ergonomics Society Annual Meeting* (Vol. 48, pp. 2325–2329). Los Angeles, CA: SAGE Publications

Houtsma, A. J. (1997). Pitch and timbre: Definition, meaning and use. *Journal of New Music Research*, *26*(2), 104–115.

Huang, G., Hung, Y. H., Proctor, R. W., & Pitts, B. J. (2022). Age is more than just a number: The relationship among age, non-chronological age factors, self-perceived driving abilities, and autonomous vehicle acceptance. *Accident Analysis & Prevention*, *178*, 106850.

Hubel, D. H., & Wiesel, T. N. (1962). Receptive fields, binocular interaction and functional architecture in the cat's visual cortex. *Journal of Physiology*, *160*, 106–154.

Hudson, C. R., Deb, S., Carruth, D. W., McGinley, J., & Frey, D. (2019). Pedestrian perception of autonomous vehicles with external interacting features. In *Advances in Human Factors and Systems Interaction: Proceedings of the AHFE 2018 International Conference on Human Factors and Systems Interaction* (pp. 33–39). Springer International Publishing.

Hunley, S. B., & Lourenco, S. F. (2018). What is peripersonal space? An examination of unresolved empirical issues and emerging findings. *Wiley Interdisciplinary Reviews: Cognitive Science*, *9*(6), e1472.

Intoy, J., & Rucci, M. (2020). Finely tuned eye movements enhance visual acuity. *Nature Communications*, *11*(1), 1–11.

ISO (2003). ISO 226: Acoustics - Normal equal-loudness-level contours. International Organization for Standardization Geneva, Switzerland.

Johnson, D. L., & Kobler, A. L. (1963). Man-computer interface study. Technical report, University of Washington, Department of Electrical Engineering.

Jolicoeur, P., Gluck, M., & Kosslyn, S. M. (1984). Pictures and names: Making the connection. *Cognitive Psychology, 16*, 243–275.

Kaplan, A. D., Cruit, J., Endsley, M., Beers, S. M., Sawyer, B. D., & Hancock, P. A. (2021). The effects of virtual reality, augmented reality, and mixed reality as training enhancement methods: A meta-analysis. *Human Factors*, *63*(4), 706–726.

Kihlstrom, J. F. (1996). Perception without awareness of what is perceived, learning without awareness of what is learned. In M. Velmans (Ed.), *The Science of Consciousness: Psychological, Neuropsychological, and Clinical Reviews* (pp. 23–46). London: Routledge.

Kim, L. H., Castillo, P., Follmer, S., & Israr, A. (2019). VPS tactile display: Tactile information transfer of vibration, pressure, and shear. *Proceedings of the ACM on Interactive, Mobile, Wearable and Ubiquitous Technologies*, *3*(2), 1–17.

Kim, S., & Choudhury, A. (2021). Exploring older adults' perception and use of smart speaker-based voice assistants: A longitudinal study. *Computers in Human Behavior*, *124*, 106914.

Kingdom, F. A. A., & Prins, N. (2016). *Psychophysics: A Practical Introduction*. Cambridge, MA: Academic Press.

Koelsch, S., Vuust, P., & Friston, K. (2019). Predictive processes and the peculiar case of music. *Trends in Cognitive Sciences*, *23*(1), 63–77.

Koffka, K. (1940). *Principles of Gestalt Psychology*. New York: Liveright.

Köhler, W. (1940). *Dynamics in Psychology*. New York: Liveright.

Kolb, H. (2012). Simple anatomy of the retina. In H. Kolb, R. Nelson, & E. Fernandez (Eds.), *Webvision: The Organization of the Retina and Visual System*. Salt Lake City, UT: University of Utah Health Sciences Center.

Koretz, J. F., & Handelman, G. H. (1988). How the human eye focuses. *Scientific American*, *259*(1), 92–99.

Kösem, A., & Van Wassenhove, V. (2012). Temporal structure in audiovisual sensory selection. *PLoS One*, *7*(7), e40936.

Lai, M.-K., & Cao, Y. Y. (2019). Designing interactive olfactory experience in real context and applications. In *TEI'19: Proceedings of the Thirteenth International Conference on Tangible, Embedded, and Embodied Interaction* (pp. 703–706). ACM.

Lasseter, J. (1998). Principles of traditional animation applied to 3D computer animation. In *Seminal Graphics: Pioneering Efforts that Shaped the Field* (pp. 263–272).

Lavan, N., Chan, W. Y., Zhuang, Y., Mareschal, I., & Shergill, S. S. (2022). Direct eye gaze enhances the ventriloquism effect. *Attention, Perception, & Psychophysics*, *84*, 2293–2302.

Leibowitz, H. W., & Post, R. B. (1982). The two modes of processing concept and some implications. In J. Beck (Ed.), *Organization and Representation in Perception* (pp. 343–363). Mahwah, NJ: Lawrence Erlbaum

Liang, J., & Williams, D. R. (1997). Aberrations and retinal image quality of the normal human eye. *JOSA A*, *14*(11), 2873–2883.

Liberman, A. M., Cooper, F. S., Shankweiler, D. P., & Studdert-Kennedy, M. (1967). Perception of the speech code. *Psychological Review*, *74*(6), 431–461.

Liebich, S., Fabry, J., Jax, P., & Vary, P. (2018). Signal processing challenges for active noise cancellation headphones. In *Speech Communication; 13th ITG-Symposium* (pp. 1–5). VDE.

Liu, Z., Chen, C., Wang, J., Huang, Y., Hu, J., & Wang, Q. (2020, September). Owl eyes: Spotting UI display issues via visual understanding. In *2020 35th IEEE/ACM International Conference on Automated Software Engineering (ASE)* (pp. 398–409).

MacDonald, L., Myers, A. M., & Blanchard, R. A. (2008). Correspondence among older drivers' perceptions, abilities, and behaviors. *Topics in Geriatric Rehabilitation*, *24*(3), 239–252.

Malmierca, M. S., & Hackett, T. A. (2010). Structural organization of the ascending auditory pathway. In A. Rees, & A. R. Palmer (Eds.), *The Oxford Handbook of Auditory Science: The Auditory Brain* (pp. 9–41). Oxford: Oxford University Press.

Marks, L. E. (1975). On colored-hearing synesthesia: Cross-modal translations of sensory dimensions. *Psychological Bulletin*, *82*(3), 303–331.

Marks, L. E., & Gescheider, G. A. (2002). Psychophysical scaling. In H. Pashler, & J. Wixted (Eds.), *Stevens' Handbook of Experimental Psychology: Volume 4: Methodology in Experimental Psychology* (3rd ed., pp. 91–138). Hoboken, NJ: Wiley.

Marr, D. (1982). *Vision*. New York: W. H. Freeman.

Martin, J., Dubé, C., & Coovert, M. D. (2018). Signal detection theory (SDT) is effective for modeling user behavior toward phishing and spear-phishing attacks. *Human Factors*, *60*(8), 1179–1191.

Martino, G., & Marks, L. E. (2001). Synesthesia: Strong and weak. *Current Directions in Psychological Science*, *10*(2), 61–65.

Masri, R. A., Grünert, U., & Martin, P. R. (2020). Analysis of parvocellular and magnocellular visual pathways in human retina. *Journal of Neuroscience*, *40*(42), 8132–8148.

Mather, G., Verstraten, F. A. J., & Anstis, S. (1998). *The Motion Aftereffect: A Modern Perspective*. Cambridge, MA: MIT Press.

Mathôt, S., & Ivanov, Y. (2019). The effect of pupil size and peripheral brightness on detection and discrimination performance. *PeerJ, 7*, e8220. doi: 10.7717/peerj.8220.

Matthews, G., De Winter, J., & Hancock, P. A. (2020). What do subjective workload scales really measure? Operational and representational solutions to divergence of workload measures. *Theoretical Issues in Ergonomics Science, 21*(4), 369–396.

McGurk, H., & MacDonald, J. (1976). Hearing lips and seeing voices. *Nature, 264*, 746–748.

Merikle, P. M., Smilek, D., & Eastwood, J. D. (2001). Perception without awareness: Perspectives from cognitive psychology. *Cognition, 79*(2), 115–134.

Merkel, J. (1888). Die Abhängigkeit zwischen Reiz und Empfindung. *Philosophische Studien, 4*, 541–594.

Merz, S. (2022). Motion perception investigated inside and outside of the laboratory: Comparable performances for the representational momentum and representational gravity phenomena. *Experimental Psychology, 69*(2), 61–74.

Merz, S., Soballa, P., Spence, C., & Frings, C. (2022). The speed prior account: A new theory to explain multiple phenomena regarding dynamic information. *Journal of Experimental Psychology: General, 151*(10), 2418–2436.

Michel, M. (2019). The mismeasure of consciousness: A problem of coordination for the perceptual awareness scale. *Philosophy of Science, 86*(5), 1239–1249.

Mishkin, M., & Ungerleider, L. G. (1982). Contribution of striate inputs to the visuospatial functions of parieto-preoccipital cortex in monkeys. *Behavioural Brain Research, 6*(1), 57–77.

Moore, B. C., & Gockel, H. E. (2012). Properties of auditory stream formation. *Philosophical Transactions of the Royal Society B: Biological Sciences, 367*(1591), 919–931.

Moore, T. M., & Picou, E. M. (2018). A potential bias in subjective ratings of mental effort. *Journal of Speech, Language, and Hearing Research, 61*(9), 2405–2421.

Mori, M., MacDorman, K. F., & Kageki, N. (2012). The Uncanny Valley [From the Field]. In *IEEE Robotics & Automation Magazine*, (vol. 19, no. 2, pp. 98–100). doi: 10.1109/MRA.2012.2192811.

Mortensen, D. H., Bech, S., Begault, D. R., & Adelstein, B. D. (2009). The relative importance of visual, auditory, and haptic information for the user's experience of mechanical switches. *Perception, 38*(10), 1560–1571.

Murray, R. F. (2021). Lightness perception in complex scenes. *Annual Review of Vision Science, 7*, 417–436.

Nishida, S. Y., Kawabe, T., Sawayama, M., & Fukiage, T. (2018). Motion perception: From detection to interpretation. *Annual Review of Vision Science, 4*, 501–523.

Norman, D. A. (2013). *The Design of Everyday Things: Revised and Expanded Edition*. New York: Basic Books.

Norman, J. (2002). Two visual systems and two theories of perception: An attempt to reconcile the constructivist and ecological approaches. *Behavioral and Brain Sciences, 25*(1), 73–96.

O'Sullivan, J. A., Shamma, S. A., & Lalor, E. C. (2015). Evidence for neural computations of temporal coherence in an auditory scene and their enhancement during active listening. *Journal of Neuroscience, 35*(18), 7256–7263.

Oh, J., Kim, S., Lee, S., Jeong, S., Ko, S. H., & Bae, J. (2021). A liquid metal based multimodal sensor and haptic feedback device for thermal and tactile sensation generation in virtual reality. *Advanced Functional Materials, 31*(39), 2007772.

Ossmy, O., & Mukamel, R. (2018). Perception as a route for motor skill learning: Perspectives from neuroscience. *Neuroscience, 382*, 144–153.

Overgaard, M., & Sandberg, K. (2021). The perceptual awareness scale-recent controversies and debates. *Neuroscience of Consciousness, 2021*(1), niab044.

Parasuraman, R., & Wisdom, G. (1985). The use of signal detection theory in research on human-computer interaction. In *Proceedings of the Human Factors Society Annual Meeting* (Vol. 29, No. 1, pp. 33–37). Los Angeles, CA: SAGE Publications.

Pardo, J. S., Nygaard, L. C., Remez, R. E. & Pisoni, D. B. (Eds.). (2021). *The Handbook of Speech Perception* (2nd ed.). Hoboken, NJ: John Wiley.

Park, W. J., & Tadin, D. (2018). Motion perception. In J. T. Wixted, & J. Serences (Eds.), *Stevens' Handbook of Experimental Psychology and Cognitive Neuroscience, Sensation, Perception, and Attention* (pp. 415–487). Hoboken, NJ: John Wiley.

Pastore, N. (1971). *Selective History of Theories of Visual Perception 1650-1950*. Oxford: Oxford University Press.

Pelli, D. G., & Farell, B. (1995). Psychophysical methods. In M. Bass, E. W. Van Stryland, D. R. Williams, & W. L. Wolfe (Eds.), *Handbook of Optics* (2nd ed., Vol. 1, pp. 29.1–29.13). New York: McGraw-Hill.

Petersen, C. M., & DeLucia, P. R. (2022, September). Perception of intention in traffic environments: A systematic review. In *Proceedings of the Human Factors and Ergonomics Society Annual Meeting* (Vol. 66, No. 1, pp. 953–957). Los Angeles, CA: SAGE Publications.

Peterson, M. A. (2001). Object perception. In E. B. Goldstein (Ed.), *Blackwell Handbook of Sensation and Perception* (pp. 168–203). Oxford: Blackwell.

Peterson, M. A. (2014). Low-level and high-level contributions to figure-ground organization. In J. Wagemans (Ed.), *The Oxford Handbook of Perceptual Organization* (pp. 259–280). Oxford: Oxford University Press.

Petzschner, F. H., Glasauer, S., & Stephan, K. E. (2015). A Bayesian perspective on magnitude estimation. *Trends in Cognitive Sciences*, *19*(5), 285–293.

Pickles, J. O. (2015). Auditory pathways: Anatomy and physiology. In G. G. Celesia, & G. Hickok (Eds.), *Handbook of Clinical Neurology: The Human Auditory System* (Vol. 129, pp. 3–25). Amsterdam, the Netherlands: Elsevier.

Potdevin, D., Sabouret, N., & Clavel, C. (2020). Intimacy perception: Does the artificial or human nature of the interlocutor matter? *International Journal of Human-Computer Studies*, *142*. https://doi.org/10.1016/j.ijhcs.2020.102464.

Pridmore, R. W. (2011). Complementary colors theory of color vision: Physiology, color mixture, color constancy and color perception. *Color Research & Application*, *36*(6), 394–412.

Prinz, W. (1997). Perception and action planning. *European Journal of Cognitive Psychology*, *9*(2), 129–154.

Proctor, R. W., & Van Zandt, T. (2018). *Human Factors in Simple and Complex Systems* (3rd ed.). Boca Raton, FL: CRC Press.

Proctor, R. W., & Xiong, A. (2020). From small-scale experiments to big data: Challenges and opportunities for experimental psychologists. In S. E. Woo, L. Tay, & R. W. Proctor (Eds.), *Big Data in Psychological Research* (pp. 35–57). Washington, DC: American Psychological Association.

Ramsøy, T. Z., & Overgaard, M. (2004). Introspection and subliminal perception. *Phenomenology and the Cognitive Sciences*, *3*(1), 1–23.

Ranasinghe, N., Jain, P., Karwita, S., & Do, E.-Y. (2017). Virtual lemonade: Let's teleport your lemonade! *TEI'17: Proceedings of the Eleventh International Conference on Tangible, Embedded, and Embodied Interaction* (pp. 183–190).

Ratcliff, R., & Smith, P. L. (2004). A comparison of sequential sampling models for two-choice reaction time. *Psychological Review*, *111*(2), 333–367.

Rauschecker, J. P. (2018). Where, when, and how: Are they all sensorimotor? Towards a unified view of the dorsal pathway in vision and audition. *Cortex*, *98*, 262–268.

Rayner, K. (1998). Eye movements in reading and information processing: 20 years of research. *Psychological Bulletin*, *124*(3), 372–422.

Rayner, K., Pollatsek, A., Ashby, J., & Clifton Jr, C. (2012). *Psychology of Reading*. London: Psychology Press.

Reichelt, S., Häussler, R., Fütterer, G., & Leister, N. (2010). Depth cues in human visual perception and their realization in 3D displays. *Proceedings of SPIE 7690, Three-Dimensional Imaging, Visualization, and Display 2010 and Display Technologies and Applications for Defense, Security, and Avionics IV* (Vol. 7690, pp. 92–103). SpIE.

Risoud, M., Hanson, J. N., Gauvrit, F., Renard, C., Lemesre, P. E., Bonne, N. X., & Vincent, C. (2018). Sound source localization. *European Annals of Otorhinolaryngology, Head and Neck Diseases*, *135*(4), 259–264.

Rizzolatti, G., & Sinigaglia, C. (2016). The mirror mechanism: A basic principle of brain function. *Nature Reviews Neuroscience*, *17*(12), 757–765.

Robinson, D. W., & Dadson, R. S. (1956). A re-determination of the equal-loudness relations for pure tones. *British Journal of Applied Physics*, *7*, 166–181.

Rolls, E. T., Rauschecker, J. P., Deco, G., Huang, C. C., & Feng, J. (2022). Auditory cortical connectivity in humans. *Cerebral Cortex*, 1–21. https://doi.org/10.1093/cercor/bhac496.

Rosenblum, L. D., & Dorsi, J. (2021). Primacy of multimodal speech perception for the brain and science. In J. S. Pardo, L. C. Nygaard, R. E. Remez, & D. B. Pisoni (Eds.). *The Handbook of Speech Perception* (2nd ed., pp. 28–57). Hoboken, NJ: John Wiley.

Rosenholtz, R., Twarog, N. R., Schinkel-Bielefeld, N., & Wattenberg, M. (2009, April). An intuitive model of perceptual grouping for HCI design. In *Proceedings of the SIGCHI Conference on Human Factors in Computing Systems* (pp. 1331–1340).

Rysling, A., Jesse, A., & Kingston, J. (2019). Regressive spectral assimilation bias in speech perception. *Attention, Perception, & Psychophysics*, *81*(4), 1127–1146.

Sanders, A. F. (1979). *Elements of Human Performance*. Mahwah, NJ: Lawrence Erlbaum.

Schlittenlacher, J., & Moore, B. C. (2020). Fast estimation of equal-loudness contours using Bayesian active learning and direct scaling. *Acoustical Science and Technology, 41*(1), 358–360.

Seinfeld, S., Feuchtner, T., Maselli, A., & Müller, J. (2021). User representations in human-computer interaction. *Human-Computer Interaction, 36*(5–6), 400–438.

Sergent, J., Ohta, S., & Macdonald, B. (1992). Functional neuroanatomy of face and object processing: A positron emission tomography study. *Brain, 115*(1), 15–36.

Serino, A. (2019). Peripersonal space (PPS) as a multisensory interface between the individual and the environment, defining the space of the self. *Neuroscience & Biobehavioral Reviews, 99*, 138–159.

Shamma, S. A., Elhilali, M., & Micheyl, C. (2011). Temporal coherence and attention in auditory scene analysis. *Trends in Neurosciences, 34*(3), 114–123.

Shin, Y. K., Proctor, R. W., & Capaldi, E. J. (2010). A review of contemporary ideomotor theory. *Psychological Bulletin, 136*(6), 943–974.

Sokolova, M. V., & Fernández-Caballero, A. (2015). A review on the role of color and light in affective computing. *Applied Sciences, 5*(3), 275–293.

Sorkin, R. D., & Woods, D. D. (1985). Systems with human monitors: A signal detection analysis. *Human-Computer Interaction, 1*(1), 49–75.

Spector, R. H. (1990). The pupils. In H. K. Walker, W. D. Hall, & J. W. Hurst (Eds.), *Clinical Methods: The History, Physical, and Laboratory Examinations* (3rd ed., pp. 300–304). London: Butterworths.

Stepp Jr, D. C. (2022). Investigating variables to reduce the uncanny valley effect in human-robot interaction: A systematic literature review. Doctoral dissertation, University of Arizona Global Campus.

Sternberg, S. (1969). Memory-scanning: Mental processes revealed by reaction-time experiments. *American Scientist, 57*(4), 421–457.

Stevens, S. S. (1953). On the brightness of lights and the loudness of sounds. *Science, 118*, 576.

Stevens, S. S. (1956). The direct estimation of sensory magnitudes: Loudness. *American Journal of Psychology, 69*(1), 1–25.

Stevens, S. S. (1975). *Psychophysics: Introduction to Its Perceptual, Neural and Social Prospects*. Hoboken, NJ: Wiley.

Stilp, C. (2020). Acoustic context effects in speech perception. *Wiley Interdisciplinary Reviews: Cognitive Science, 11*(1), e1517.

Stock, A., & Stock, C. (2004). A short history of ideo-motor action. *Psychological Research, 68*(2), 176–188.

Sunaert, S., Van Hecke, P., Marchal, G., & Orban, G. A. (1999). Motion-responsive regions of the human brain. *Experimental Brain Research, 127*(4), 355–370.

Sutojo, S., Thiemann, J., Kohlrausch, A., & van de Par, S. (2020). Auditory Gestalt rules and their application. In J. Blauert, & J. Braasch (Eds.), *The Technology of Binaural Understanding* (pp. 33–59). Berlin, Heidelberg: Springer Nature.

Suzuki, Y., & Takeshima, H. (2004). Equal-loudness-level contours for pure tones. *Journal of the Acoustical Society of America, 116*, 918–933.

Swets, J. A. (1996). *Signal Detection Theory and ROC Analysis in Psychology and Diagnostics*. Mahwah, NJ: Lawrence Erlbaum.

Tarr, M. J. (2000). Visual pattern recognition. In A. E. Kazdin (Ed.), *Encyclopedia of Psychology* (Vol. 8). Washington, DC: American Psychological Association.

Teichner, W. H. (1954). Recent studies of simple reaction time. *Psychological Bulletin, 51*(2), 128–149.

Titchener, E. B. (1919). *A Text-Book of Psychology*. London: Macmillan.

Van Polanen, V., & Davare, M. (2015). Interactions between dorsal and ventral streams for controlling skilled grasp. *Neuropsychologia, 79*, 186–191.

Viswanathan, V., Shinn-Cunningham, B. G., & Heinz, M. G. (2022). Speech categorization reveals the role of early-stage temporal-coherence processing in auditory scene analysis. *Journal of Neuroscience, 42*(2), 240–254.

von Helmholtz, H. (1867/1910/1962). *Treatise on Physiological Optics* (Vol. III). J. P. C. Southall (trans. & Ed.). Mineola, NY: Dover. (Translated from the 3rd German edition, English edition 1962).

Wagemans, J., Elder, J. H., Kubovy, M., Palmer, S. E., Peterson, M. A., Singh, M., & von der Heydt, R. (2012a). A century of Gestalt psychology in visual perception: I. Perceptual grouping and figure-ground organization. *Psychological Bulletin, 138*(6), 1172–1217.

Wagemans, J., Feldman, J., Gepshtein, S., Kimchi, R., Pomerantz, J. R., Van der Helm, P. A., & Van Leeuwen, C. (2012b). A century of Gestalt psychology in visual perception: II. Conceptual and theoretical foundations. *Psychological Bulletin, 138*(6), 1218.

Wall, S. A., & Harwin, W. S. (2000, August). Interaction of visual and haptic information in simulated environments: Texture perception. In *International Workshop on Haptic Human-Computer Interaction* (pp. 108–117). Berlin, Heidelberg: Springer Berlin Heidelberg.

Warren, R., Obusek, C., & Ackroff, J. (1972). Auditory induction: Perceptual synthesis of absent sounds. *Science 176*, 1149–1151.

Weber, E. H. (1996). In D. J. Murray, H. E. Ross, & E. H. Weber (Eds.), *E.H. Weber On The Tactile Senses* (1st ed.). London: Psychology Press.

Wechsler, H. (Ed.). (2014). Neural networks for perception. In *Volume 1: Human and Machine Perception; and Volume 2: Computation, Learning, and Architectures*. Cambridge, MA: Academic Press.

West, R. L., Boring, R. L., Dillon, R. F., & Bos, J. (2001). Human-computer interaction, quality of service, and multimedia Internet broadcasting. In A. Sloane, & D. Lawrence (Eds.), *Multimedia Internet Broadcasting* (pp. 1–15). Berlin, Heidelberg: Springer.

Whelan, C. (2021, March 1). Is it safe to drive with vision in only one eye? *Heathline*. https://www.healthline.com/health/eye-health/can-you-drive-with-one-eye.

Winkler, I., Tervaniemi, M., & Näätänen, R. (1997). Two separate codes for missing-fundamental pitch in the human auditory cortex. *The Journal of the Acoustical Society of America*, 102(2), 1072–1082.

Wixted, J. T. (2020). The forgotten history of signal detection theory. *Journal of Experimental Psychology: Learning, Memory, and Cognition, 46*(2), 201–233.

Wolfe, J. M., Kluender, K. R., Leve, D. M., Bartoshuk, Herz R. S., Klatzky, R. L., & Merfeld, D. M. (2021). *Sensation and Perception* (International 6th ed.). Sunderland, MA: Sinauer Associates.

Wood, J. M., & Owsley, C. (2014). Useful field of view test. *Gerontology*, 60(4), 315–318.

Woods, D. L., Wyma, J. M., Yund, E. W., Herron, T. J., & Reed, B. (2015). Factors influencing the latency of simple reaction time. *Frontiers in Human Neuroscience*, 9, 131.

Wu, X., Nix, L. C., Brummett, A. M., Aguillon, C., Oltman, D. J., & Beer, J. M. (2021). The design, development, and evaluation of telepresence interfaces for aging adults: Investigating user perceptions of privacy and usability. *International Journal of Human-Computer Studies*, 156, 102695.

Xiong, A., & Proctor, R. W. (2016). Decreasing auditory Simon effects across reaction time distributions. *Journal of Experimental Psychology: Human Perception and Performance, 42*(1), 23–38.

Xiong, J., Hsiang, E. L., He, Z., Zhan, T., & Wu, S. T. (2021). Augmented reality and virtual reality displays: Emerging technologies and future perspectives. *Light: Science & Applications*, 10(1), 216.

Xu, W. (2020). From automation to autonomy and autonomous vehicles: Challenges and opportunities for human-computer interaction. *Interactions*, 28(1), 48–53.

Yost, W. A. (2013). Audition. In A. F. Healy, & R. W. Proctor (Eds.), *Experimental Psychology* (2nd ed., pp. 120–151). Volume 4 in I. B. Weiner (Editor-in-Chief), *Handbook of Psychology*. Hoboken, NJ: John Wiley.

Young, A. W., & Burton, A. M. (2017). Recognizing faces. *Current Directions in Psychological Science*, 26(3), 212–217.

3 Human Information Processing, Attention, and Memory
An Overview for HCI

Kim-Phuong L. Vu and Robert W. Proctor

It is natural for an applied psychology of human-computer interaction to be based theoretically on information-processing psychology.

—Card et al. (1983)

As noted by Card et al. (1983) more than 40 years ago, human-computer interaction (HCI) is essentially an information-processing task. Any system that requires human interaction with an interface needs to consider the users' goals and subgoals. For instance, when interacting with a mobile device, a user forms an intention, and then initiates the interaction by activating the device and inputting commands (manually or verbally) to communicate with the system. The device then executes its programs in response to the commands given and provides feedback that the command was executed, typically through an output that is displayed to the user. The output provided can lead to the user initiating a series of subsequent interactions until the desired task is completed. The sequence of interactions to accomplish the task can vary in terms of complexity and alternative possible sequences, all of which influence the time needed to complete the task and the types of errors that may occur along the way. During the interaction, the user is required to perceive the displayed information, understand its meaning, and select and execute responses based on the information. Thus, for the interaction between the device and the user to be efficient, the interface must be designed in accordance with the user's information-processing capabilities.

3.1 HUMAN INFORMATION-PROCESSING APPROACH

3.1.1 BACKGROUND AND ASSUMPTIONS

3.1.1.1 Rise of the Approach in the 1950s

The view of the human as a processor of information coincides with the development of relatively sophisticated technologies around the time of World War II. The rise of the human information-processing approach in psychology is closely coupled with the growth of the fields of cognitive psychology, human factors and ergonomics (HFE), and human engineering (see Proctor & Vu, 2010a; Xiong & Proctor, 2018). As part of the war efforts, experimental psychologists working on applied topics collaborated with engineers to try to make sure that the operators would be able to perform their tasks effectively (Mead, 1948). Therefore, the psychologists were exposed not only to applied problems but also to the techniques and views being developed in areas such as communications engineering (see Roscoe, 2011). Indeed, a critical impetus was Norbert Weiner's (1948) book, *Cybernetics or Control and Communication in the Animal and the Machine*. As indicated by the title, Weiner argued that control theory and communication theory were inseparable and

DOI: 10.1201/9781003495109-3

that the behavior of humans and machines should be viewed in terms of information and feedback. Cybernetics provided "a vision of the human relationship with machines, for engineers and systems theorists of various stripes" (Mindell, 2002, p. 4). Many of the concepts from control and communications engineering, for instance, the notion of transmission of information through a limited capacity communications channel, were seen as applicable to analyses of human performance.

Closely related to cybernetics was Shannon's (1948) development of information theory, which specified information in terms of uncertainty. The essential idea is captured by the formula for discrete distributions of probabilities (p) over n alternatives, for which the amount of uncertainty (and, hence, the information conveyed by an event) is $H = -\Sigma_i \, p_i \log p_i$. The information theory metric, developed in communications engineering, provided a way to specify information quantitatively that could be applied to human information processing as well. This resulted in a burst of studies applying information theory in psychological research in the 1950s (e.g., Attneave, 1954; Garner & Hake, 1951; Hick, 1952). Indeed, many of the foundational findings in experimental studies of human information processing arose out of the applied experimental and engineering psychology research generated from information theory (e.g., Fitts' law; Fitts, 1954).

Another significant factor in the rise of the human information-processing approach was the development of group research designs and statistical methods, which sprang initially from the work of Ronald A. Fisher. Fisher developed the logic of group experimental design and statistical methods including the analysis of variance in the 1920s and 1930s (Fisher, 1925, 1935), and Neyman and Pearson (1933a,b) contributed a more systematic analysis of probability and statistics within the context of decision theory. These statistical methods started to be introduced into applied experimental psychology research in the 1940s (Chapanis & Schachter, 1945; Schachter & Chapanis, 1945), but it was in the 1950s that they came to be used widely (Rucci & Tweney, 1980). The methods, incorporating probability theory, led more generally to what has been called the probabilistic revolution (Krüger et al., 1987).

Finally, we note the close link between the development of the information-processing approach and that of computers and artificial intelligence. Among the leaders in computer simulations of information processing were Allen Newell and Herbert A. Simon. In their classic book, *Human Problem Solving* (1972), they suggested, "1956 could be taken as the critical year for the development of information processing psychology" (p. 578). They cited several major events and publications at that time, grouped under the headings "The Cybernetic Revolution" (which we already described), "Linguistics" (specifically, Noam Chomsky's, 1956, work on generative linguistic systems), and "Digital Computers." Newell and Simon noted that programming languages like FORTRAN, which were just being developed, provided researchers experience working with a non-human system that interpreted symbols in much the same way that humans did. Among the critical events they described is the Dartmouth Summer Research Project on Artificial Intelligence, a 1956 summer workshop attended by several major figures working in artificial intelligence (AI), including Marvin Minsky (1961). At the meeting, Newell and Simon gave the initial presentation of The Logic Theorist (Newell & Simon, 1956), a computer program designed to perform automated reasoning. This work led to Newell and Simon's (1972) groundbreaking work on human problem solving. The link between human information processing and computer science continues today with the widespread use of computational models of psychological processes and applications of deep-learning AI to technologies intended to assist humans (Guest & Martin, 2021; Wang et al., 2020).

3.1.1.2 Fundamental Assumptions

The human information-processing approach is based on the idea that human performance, from displayed information to response, is a function of several processing stages. The nature of these stages, how they are arranged, and the factors that influence how quickly and accurately a particular stage operates, can be discovered through appropriate research methods. The most fundamental assumption of the information-processing approach is that the human is a complex system that can be analyzed in terms of subsystems and their interrelation (Posner, 1986). This point is evident in the work of researchers on attention and performance, such as Paul Fitts (1951) and Donald Broadbent (1958), who were among the first to adopt the information-processing approach in the 1950s.

The systems perspective underlies not only human information processing but also HFE and HCI, providing a direct link between the basic and applied fields (Proctor & Van Zandt, 2018; Xiong & Proctor, 2018). HFE in general, and HCI in particular, begin with the fundamental assumption that a human-machine system can be decomposed into machine and human subsystems, each of which can be analyzed further. The human information-processing approach provides the concepts, methods, and theories for analyzing the processes involved in the human subsystem. Posner (1986) stated, "Indeed, much of the impetus for the development of this kind of empirical study stems from the desire to integrate description of the human with overall systems" (p. V-6). Young et al. (2004) emphasized that the most basic distinction between three processing stages (perception, cognition, and action), as captured in a block diagram model of human information processing, is important even for understanding the dynamic interactions of an operator with a vehicle for purposes of computer-aided augmented cognition. They noted,

> This block diagram of the human is important because it not only models the flow of information and commands between the vehicle and the human, it also enables access to the internal state of the human at various parts of the process. This allows the modeling of what a cognitive measurement system might have access to (internal to the human), and how that measurement might then be used as part of a closed-loop human-machine interface system (pp. 261–262).

In the first half of the 20th century, the behaviorist approach predominated in psychology, particularly in the U.S. Within this approach, many sophisticated theories of learning and behavior were developed that differed in various details (Bower & Hilgard, 1981). However, the research and theories of the behaviorist approach tended to minimize the role of cognitive processes and were of limited value to the applied problems encountered by psychologists in World War II. The information-processing approach was adopted because it provided a way to examine topics of basic and applied concern such as attention that were relatively neglected during the behaviorist period. It continues to be the main approach in psychology, driving much of the research conducted in cognitive psychology, cognitive neuroscience, human factors, and neuroergonomics, although contributions have been made from various other approaches.

For example, the ecological approach associated with Gibson (1979) places emphasis on analyzing the perceptual information that is available in the optic array and the dynamics of this information as the individual interacts with the environment. The cybernetic view, that cognition emerges as a consequence of motor control over sensory feedback, stresses self-regulated control of perception and cognition (Smith & Henning, 2005). The situated cognition approach focuses on the need to understand behavior in specific contexts in which, for example, a computer application will be used (Kiekel & Cooke, 2011). Finally, the embodied cognition approach, according to which knowledge is acquired and processed through interactions of the body with the environment (e.g., Robinson & Thomas, 2021) has led to alternative conceptualizations of HCI. One common feature of these alternative accounts is an emphasis on the relation between perception, cognition, and action. Although in certain areas of information-processing research, action has been viewed as a final stage that does not influence the prior stages, this is not an inherent limitation of the approach. In fact, the area of research on human performance has emphasized action since the earliest applications of the information-processing approach (Fitts & Posner, 1967). Moreover, alternative approaches have not been shown to be unambiguously more successful in producing more effective guidelines for HCI design than the information processing approach.

In this chapter, we survey methods used to study human information processing and summarize major findings and the theoretical frameworks developed to explain them. We also tie the methods, findings, and theories to HCI issues to illustrate their use. Within HCI, human information-processing analyses are used in two ways. First, empirical studies evaluate the information-processing requirements of various tasks in which humans use computers. Second, computational models are developed to characterize human information processing when interacting with computers and to predict human performance with alternative interfaces.

3.2 INFORMATION-PROCESSING METHODS

Any theoretical approach makes certain presuppositions and tends to favor some methods and techniques over others. Information-processing researchers have used a variety of methods from behavioral to psychophysiological to neuroimaging measures. However, the main body of research has relied on chronometric (time-based) methods and choice-reaction tasks. There also has been a reliance on speed-accuracy methods and flow models that are often quantified through computer simulation or mathematical modeling.

3.2.1 CHRONOMETRIC METHODS

Time-based, or chronometric, methods are the most widely used tools for studying human information processing. Thus, it is not surprising that reaction time (RT) has been the main dependent measure of the information processing approach (Lachman et al., 1979). Although many other behavioral, physiological, and neural measures are used, RT still predominates because of its sensitivity and the sophisticated techniques that have been developed for analyzing and modeling RT data (Donkin & Brown, 2018). For example, fundamental methods including the additive factors and subtractive logics (Donders, 1868/1969; Sternberg, 1969) allow the examination of patterns and differences in RT to identify the information-processing stages and the timing of those stages.

The subtractive method has been used in controlled settings in the lab to estimate the durations of a variety of processes, including rates of mental rotation (approximately 12–20 ms per degree of rotation; Shepard & Metzler, 1971) and memory search (approximately 40 ms per item; Sternberg, 1969). Similar logic can be applied to situations in which on some trials a participant receives a "stop" signal during the reaction process, indicating that the response is to be stopped (Matzke et al., 2018). They estimated the durations of the stop-signal identification process and the response-mapping process at 34 and 20 ms, respectively (Van de Laar et al., 2010).

Because the subtractive method is only applicable when discrete, serial processing stages can be assumed, one might ask if it can be used in applied settings. For simple tasks, the answer is yes. For example, an application of the subtractive method to an HCI task would be to compare the time to find a target link on two Web pages that are identical except for the number of links displayed and to attribute the extra time to the additional visual search required for the more complex Web page. Although it may be more difficult to apply this logic to more complex tasks, the general idea behind the method can be a useful tool for designers to estimate how long users will need to perform cognitive tasks. For example, Saleh and Enaizan (2022) employed the subtractive method to estimate how much time users spend being productive when interacting with a mobile application. Participants were asked to complete a series of tasks in an app while responding to specific notifications that were presented during task performance. The app automatically logs the users' interactions with a series of timestamps. Productive periods were estimated by subtracting the unproductive time from the total time to perform the task. Unproductive time was defined as the duration spent by users accessing assistance features of the app plus the time that the user took to respond to a particular notification while using the app. This productivity metric can be used by designers to evaluate the usability of their app features and, along with other metrics, to estimate the amount of cognitive load that may be demanded by certain tasks.

Sternberg (1969) developed the additive factors method to allow determination of the processes involved in performing a task. He showed that two variables that affect different stages should have additive effects on RT. In contrast, two variables that affect the same stage should have interactive effects on RT. Grobelny et al. (2005) provided an application of additive factors logic to the usability of graphical icons in the design of HCI interfaces. Mode of icon array (menu or dialog box), number of icons, and difficulty of movement had additive effects on response times, implying that these variables affect different processing stages. Designers can use additive factors methods as a tool to help determine what task combinations are likely to cause more interference (i.e., impact the same processing stages) compared to others (i.e., impact different processing stages).

Both the subtractive and additive factors methods have been challenged on several grounds (Pachella, 1974). Each assumes discrete serial stages with constant output, which is difficult to justify in many situations. Also, both methods rely on analyses of RT, without consideration of error rates, which can be problematic because performance is typically not error-free and speed can be traded for accuracy. Despite these limitations, the methods have proven to be robust and useful (Sanders, 1998). For example, Salthouse (2005) noted that the process analysis approach employed in contemporary research into aging effects on cognitive abilities "has used a variety of analytical methods such as subtraction, additive factors… to partition the variance in the target variable into theoretically distinct processes" (p. 288).

3.2.2 SPEED-ACCURACY METHODS

The function relating response speed to accuracy is called the speed-accuracy tradeoff (Heitz, 2014; Pachella, 1974). For example, if the task is to scroll down a document to find an anomaly (see Figure 3.1a), the use of different input devices can yield different performances with respect to speed and accuracy. The function, illustrated in the Figure 3.1b, shows that very fast responses can be performed with low accuracy, and accuracy will increase as responding slows down. Of importance is the fact that when accuracy is high, as in most RT studies, a small increase in errors can result in a large decrease in RT. With respect to text entry on computing devices, MacKenzie and Soukoreff (2002) stated, "Clearly, both speed and accuracy must be measured and analyzed… Participants can enter text more quickly if they are willing to sacrifice accuracy" (pp. 159–160).

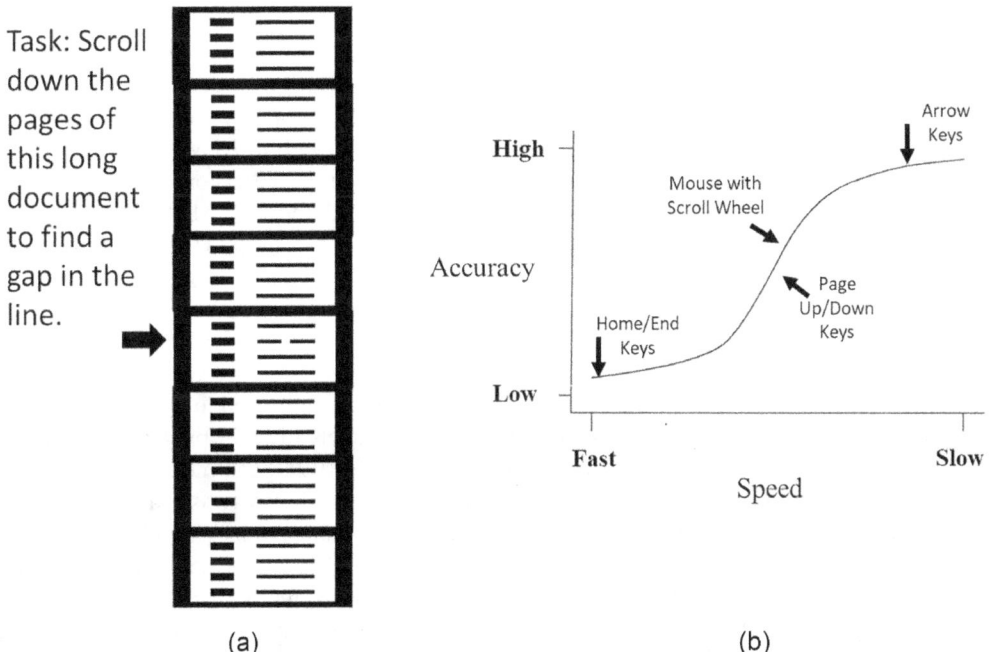

(a)   (b)

FIGURE 3.1 (a) Task of scrolling through pages of a document to find a gap in the line using four different input methods to navigate: home/end keys, page up/down keys, mouse/scroll wheel, and up/down arrow keys. (b) Speed-accuracy operating characteristic curve with where performance is estimated to be with the different navigation methods. Fastest task completion is expected to occur with the use of the home/end keys at the cost of lowest accuracy.

Nathan et al. (2022) obtained results consistent with this statement for a labeling task, where they found keyboard entry to be faster but less accurate than other methods (e.g., swiping, mouse selection, tapping).

In studies of what is called the macro speed-accuracy tradeoff, the speed-accuracy criterion is varied between blocks of trials or between subjects by using different instructions regarding the relative importance of speed versus accuracy, varying payoffs such that speed or accuracy is weighted more heavily, or imposing different response deadlines (Wickelgren, 1977). These studies have the potential to be more informative than RT studies because they can provide information about whether variables affect the intercept (time at which accuracy exceeds chance), asymptote (the maximal accuracy), and rate of ascension from the intercept to the asymptote, each of which may reflect different processes. For example, Boldini et al. (2004) obtained evidence favoring dual-process models of recognition memory over single-process models by varying the delay between a visually presented test word and a signal to respond. Recognition accuracy benefited from a modality match at study and test (better performance when the study words were also visual rather than auditory) at short response-signal delays, but it benefited from deep processing during study (judging pleasantness) over shallow processing (repeating aloud each word) at long response-signal delays. Boldini et al. (2004) interpreted these results as consistent with the view that recognition judgments are based on a fast familiarity process or a slower recollection process.

In tasks requiring the search of complex visual displays, a speed emphasis may influence more than just the criterion for emitting a response. McCarley (2009) had young adults perform a simulated baggage-screening task under instructions that emphasized speed or accuracy of responding. With speed emphasis, the participants made fewer eye fixations of shorter duration than under accuracy emphasis. Reduction in accuracy was a consequence mainly of failure to fixate the target of the search rather than a failure to respond to targets that were fixated. This study illustrates how a speed-accuracy tradeoff manipulation can be of value in applied contexts.

Because in macro speed-accuracy tradeoff studies the criterion is manipulated in addition to any other variables of interest, much more data must be collected than in a typical RT study. Consequently, the use of macro speed-accuracy methods has been restricted to situations in which the speed-accuracy relation is of major concern or of apparent significant value, rather than being widely adopted as the method of choice. Also of interest and value is what is called the micro speed-accuracy tradeoff (Wickelgren, 1977), which is obtained within a given speed-accuracy instruction condition by partitioning the RT distribution into several subdivisions (from shortest to longest) and obtaining the accuracy separately for the respective subdivisions. Methods exist as well for combining RT and error rate into a single processing efficiency measure (Liesfeld & Janczyk, 2019; Mueller et al., 2020).

3.2.3 PSYCHOPHYSIOLOGICAL AND NEUROIMAGING METHODS

Psychophysiological and neuroimaging methods have been increasingly used to evaluate the implications of information-processing models and to relate the models to brain processes. This area of research is called *cognitive neuroscience* (Gazzaniga et al., 2019). Such methods can provide details regarding the nature of processing by examining physiological activity as a task is being performed. Although imaging techniques have been increasingly used, a psychophysiological method that involves the measurement of electroencephalograms (EEGs) is still widely used. EEGs are recordings of changes in brain activity as a function of time as measured from electrodes placed on the scalp (Mangun, 2014). Different frequency bands of EEG rhythms are correlated with subjective states and the processes underlying task performance.

An application of EEGs to HCI has been the development of brain-computer interfaces that allow a person to control technological devices by brain signals. Such interfaces are of value for persons with motor disabilities who are not able to communicate through standard input devices. Changes in EEGs that arise from different types of mental processing can be coded into distinct computer commands, and people can be trained to use their thoughts to control the computer's interface (e.g., Rezeika et al., 2018; Saha et al., 2021). This mode of HCI opens possibilities for persons with motor disabilities to interact with their environment and communicate with other people.

For information-processing research, EEGs that are associated with a stimulus event, known as event-related potentials (ERPs), are extremely informative (Luck, 2012). ERPs are the changes in brain activity that are elicited by an event such as stimulus presentation or response initiation. ERPs are obtained by averaging across many trials of a task to remove background EEG noise and are thought to reflect postsynaptic potentials in the brain. There are several features of the ERP that represent different aspects of processing. These features are labeled according to their polarity, positive (P) or negative (N), and their sequence or latency. The first positive (P1) and negative (N1) components are associated with early perceptual processes. They are called exogenous components because they occur in close temporal proximity to the stimulus event and have a stable latency with respect to it. Later components reflect cognitive processes and are called endogenous because they are a function of the task demands and have more variable latency than the exogenous components. One such component that has been studied extensively is P3 (or, P300), which represents postperceptual processes. When an occasional target stimulus is interspersed in a stream of standards, P3 is observed in response to targets, but not to standards.

By comparing the effects of task manipulations on various ERP components such as P3, their onset latencies, and their scalp distributions, relatively detailed inferences about the cognitive processes can be made. Solís-Marcos and Kircher (2019) found the amplitudes of the N1 and P3 components in response to rare deviant stimuli were smaller when participants were multitasking than when performing a single task. For HCI, Miller et al. (2011) found that the amplitudes of the N1, P2, and P3 components to the probed stimuli were inversely related to the level of videogame difficulty. Their interpretation of this finding was that smaller amplitudes indicate higher workload. Another measure that has been used in studies of human information processing is the lateralized readiness potential (LRP; Eimer, 1998; Schurger et al., 2021). LRP can be recorded in choice-reaction tasks that require a response with the left or right hand. It is a measure of differential activation of the lateral motor areas of the visual cortex that occurs shortly before and during the execution of a response. The asymmetric activation favors the motor area contralateral to the hand making the response because this is the area that controls the hand. LRP has been obtained in situations in which no overt response is ever executed, allowing it to be used as an index of covert, partial response activation. LRP is thus a measure of the difference in activity from the two sides of the brain that can be used as an indicator of covert reaction tendencies, to determine whether a response has been prepared even when it is not actually executed. It can also be used to determine whether the effects of a variable are prior to or subsequent to response preparation.

EEG-based measurements often do not have the spatial resolution needed to provide precise information about the brain structures that produce the recorded activity. Thus, neuroimaging methods are often used to determine brain regions associated with human information processing. Early methods tend to rely on positron-emission tomography (PET) scan, whereas more recent studies use functional magnetic resonance imaging (fMRI), and transcranial Doppler sonography (TDS). All these imaging methods measure changes in blood flow associated with neuronal activity in different regions of the brain (Lytaev, 2022). The temporal resolution of fMRI has improved drastically over the years, and high-end scanners provide both spatial and temporal resolution (Ghuman & Martin, 2019).

In an imaging study, typically both control and experimental tasks are performed, and the functional neuroanatomy of the cognitive processes is derived by subtracting the image during the control task from that during the experimental task. This subtractive method of neuroimaging analysis has the same limitations as that for reaction-time analysis (Sartori & Umiltà, 2000). However, multivariate pattern analysis methods can detect distributed activity patterns that subtractive and additive factor methods cannot, and multiple data analysis techniques can be used in a complementary manner to support specific conclusions (Wright, 2018).

The application of cognitive neuroscience to HFE and HCI has been advocated under the heading of neuroergonomics (Ayaz & Dehais, 2021; Parasuraman, 2003). Neuroergonomics has the goal of using knowledge of the relation between brain function and human performance to design interfaces and computerized systems that are sensitive to brain function with the intent of increasing the efficiency and safety of human-machine systems.

3.2.4 Sample Empirical Studies on Information Processing Stages

Information processing can be studied using choice-reaction tasks in which each stimulus is assigned to a unique response to examine the effects of variables on three stages of processing (see Figure 3.2): stimulus identification, response selection, and response execution (Proctor & Vu, 2023). The stimulus-identification stage involves perceptual processes that are primarily dependent on stimulus properties. The response-selection stage concerns cognitive processes involved in choosing which response to make to each stimulus. Response execution refers to motor processing involved in programming and execution of motor responses. Of relevance to HCI, several principles and laws capture how human information processing is influenced by the task environment and task demands.

3.2.4.1 Stimulus Identification

The speed with which stimuli can be identified depends on several factors. As stimulus contrast, or intensity, decreases, RT increases. For example, Miles and Proctor (2009) had participants make left and right keypress responses to the non-spatial or spatial feature of centrally presented location words. The discriminability of the spatial feature of the word, or of both the spatial and non-spatial features, was manipulated. When the spatial feature of the word was task-irrelevant, decreasing the discriminability of this feature reduced the typical benefit for correspondence of the word meaning with the keypress response to the relevant stimulus feature. This correspondence benefit was restored when the discriminability of both the task-relevant and task-irrelevant features was reduced together, slowing the processing of both the relevant and irrelevant information. These results suggest that the reduction of discriminability slows the processing of the perceptual information but does not alter the response-selection processes that operate on that information.

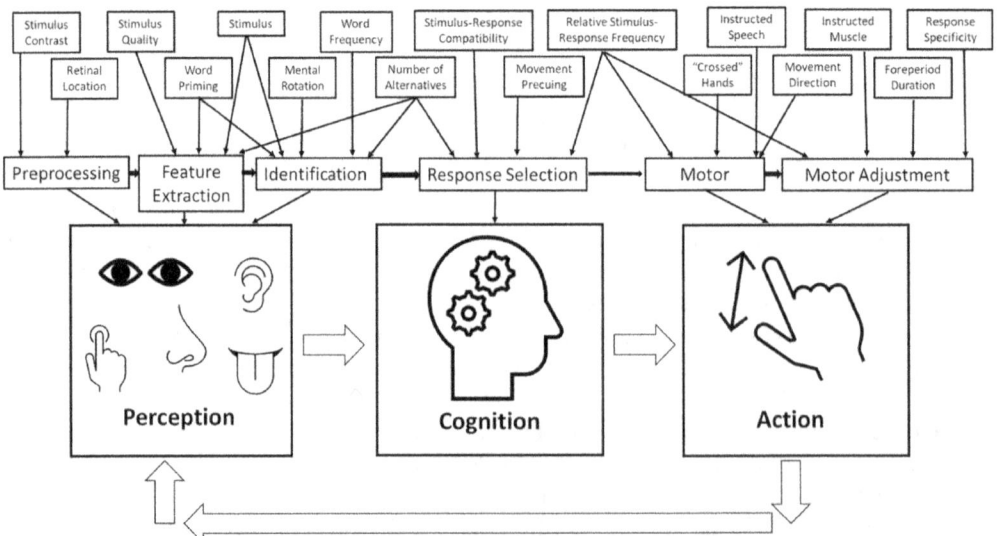

FIGURE 3.2 Information-processing stages and variables that affect them, based on Sanders' (1998) taxonomy.

Feature extraction involves lower-level perceptual processing based on area V1 (the primary visual cortex) and other early visual cortical areas. Stimulus discriminability, word priming, and stimulus quality affect the feature extraction process. For example, manipulations of stimulus quality such as superimposing a grid slow RT, presumably by creating difficulty for the extraction of features. Identification itself is influenced by word frequency and mental rotation. The latter refers to that when a stimulus is rotated from the upright position, the time it takes to identify the stimulus increases as an approximately linear function of angular deviation from upright (Shepard & Metzler, 1971). This increase in identification time is presumed to reflect a normalization process by which the image is mentally rotated in a continuous manner to the upright position.

3.2.4.2 Response Selection

Response selection refers to those processes involved in determining what response to make to a particular stimulus. It is affected by the number of alternatives, stimulus-response compatibility, and precuing (providing advance information about a forthcoming event). The influence of the number of alternatives on performance is captured by Hick's law, or the Hick-Hyman law, which states that RT increases as a logarithmic function of the number of stimulus-response alternatives (Hick, 1952; Hyman, 1953). For N equally likely alternatives, the following formula is used:

$$RT = a + b \log_2 N, \tag{3.1}$$

where a is the base processing time and b is the amount that RT increases with increases in N. The slope of the Hick-Hyman function is influenced by many factors (Proctor & Schneider, 2018). For example, the slope decreases as subjects become practiced at a task (Teichner & Krebs, 1974) and as response-selection becomes more efficient by using more compatible stimulus-response mappings (Proctor & Vu, 2006). Yablonski (2020) lists Hick's law as one of ten laws of user experience (UX) design, but Liu et al. (2020) argue that it is of relatively little value compared to other factors that influence RT in computing systems. One limitation in its relevance is that the slope of the function is negligible when a single effector (e.g., the index finger of the dominant hand) is used to make all responses (Wright et al., 2019), which is a typical mode of responding on touch screens.

Stimulus-response compatibility effects are differences in speed and accuracy of responding as a function of how natural, or compatible, the relation between stimuli and responses is, and will be described in more detail in the section on *Congruity Effects*. Of importance to HCI is that stimulus-response compatibility effects occur for older as well as younger adults, with older adults typically showing larger compatibility effects that cannot be attributed entirely to general slowing (Proctor et al., 2005). Moreover, because the older adults' response times increase disproportionally as a function of uncertainty, they benefit more from a precue that indicates which of two tasks will be performed or reduces the number of possible stimulus and response alternatives (Vu & Proctor, 2008). Implications of these findings are that older adults' performance will suffer more from incompatibility in designs, but this cost can be minimized by design strategies that limit the amount of information that must be processed.

Designers also need to be aware of the potential problems created by various types of incompatibility between display and response elements because their influences are not always obvious (Payne, 1995; Tlauka, 2004; Vu & Proctor, 2003). Fortunately, Vu and Proctor (2003) showed that providing participants with relatively little practice with the stimulus-response mappings can improve the estimates of relative compatibility. Thus, designers can get a better feel for the relative compatibility of alternative arrangements by performing tasks that use the various mappings. However, after the designer selects a few arrangements that would seem to yield good performance, more thorough usability testing of the remaining arrangements on groups of users needs to be performed.

3.2.4.3 Response Initiation and Execution

Motor programming refers to the specification of the physical response that is to be made. This process is affected by variables such as relative stimulus-response frequency and movement direction. One factor that influences this stage is movement complexity. Even in a simple RT task, the longer the sequence of movements that is to be made upon the occurrence of a stimulus, the longer is the RT to initiate the sequence (Henry & Rogers, 1960; Sternberg et al., 1978). This difference has been found to persist across warning for periods of 0–2,000 ms (Shin & Proctor, 2018) and when responding to a loud startle stimulus (which reduces RT overall) that occurs randomly on a small proportion of trials (Maslovat et al., 2014). Klapp and Maslovat (2020) concluded that these and other results are due to an inability to prepare in advance of movement initiation the code that controls the timing of the onsets for elements making up the response. Although the abstract goal for a response can be specified in advance, programming of the action's timing cannot be completed until just prior to initiation of the response.

One of the most well-known relations for response execution is Fitts' Law (Fitts, 1954; MacKenzie, 2018), which describes the time to make aimed movements to a target location:

$$\text{Movement time} = a + b\log_2\left(2D/W\right), \tag{3.2}$$

where a and b are constants, D is distance to the target, and W is target width. However, there are slightly different versions of the law. According to Fitts' law, movement time is a direct function of distance and an inverse function of target width. Fitts' law has been found to provide an accurate description of movement time in many situations, although alternatives have been proposed for certain situations. One factor that contributes to the increase in movement time as the index of difficulty increases is the need to make a corrective sub-movement based on feedback in order to hit the target location (Meyer et al., 1988).

The importance of Fitts' law for HCI was illustrated in its 50th anniversary by a special issue of the *International Journal of Human-Computer Studies* devoted to Fitts' original study. In the preface to the issue, the editors, Guiard and Beaudouin-Lafon (2004) stated, "What has come to be known as Fitts' law has proven highly applicable in Human-Computer Interaction (HCI), making it possible to predict reliably the minimum time for a person in a pointing task to reach a specified target" (p. 747).

Fitts' law has several implications for HCI. First, input devices for HCI can be selected based on the slope of the function, b, where the lower the slope, the shorter the movement times. Second, any reduction in the index of difficulty should decrease the time for movements. Card et al. (1978) found this to be the case for the selection of text using a computer mouse to point and click. However, Gillan et al. (1992) noted that designers should be cautious when applying Fitts' law to HCI because factors other than distance and target size play a role when using a mouse. Specifically, they proposed that the critical factors in pointing and dragging are different than those in pointing and clicking. Gillan et al. (1992) found that, for a text-selection task, both point-click and point-drag movement times can be accounted for by Fitts' law. For point-click sequences, as in the study by Card et al. (1978), the diagonal distance across the text object, rather than the horizontal distance, provided the best fit for pointing time. However, for point-drag, the vertical distance of the text provided the best fit. The reason why the horizontal distance is inconsequential for point-drag is that the cursor must be positioned at the beginning of the string for the point-drag sequence. Thus, task requirements should be taken into account before applying Fitts' law to determine input time for an interface design.

3.2.5 Information-Processing Models

It is common to assume that the processing between stimuli and responses consists of a series of discrete stages for which the output for one stage serves as the input for the next, as Donders and Sternberg assumed. This assumption is made for the classic Model Human Processor

(Card et al., 1983), which was applied to HCI to provide approximate estimates of task completion time. Current cognitive architectures, intended to accommodate fundamental aspects of processing across the full range of perception, cognition, and action allow for considerable parallel processing, although the processing is sequential in the sense of cognitive and physical actions being taken at discrete time intervals of about 50 ms (Laird et al., 2017).

Sequential sampling models are widely used to model both RT and accuracy, and consequently, the tradeoff between them (Forstmann et al., 2016; Ratcliff & Smith, 2004). They typically distinguish decision processes, which are the primary concern of the modeling, from the nondecision processes of perception and motor control (van den Bergh et al., 2020). The decision process is based on a dynamic model of signal detection (see Chapter 2, this book) in which decisions are based on a series of samples from the probability distributions rather than a single sample. Each sample is classified as favoring one alternative or another, and this information is fed into a decision mechanism in which gradual accumulation of the information occurs until a response threshold is reached, at which time that response is made.

The specific models differ in several respects, but the most widely used model is the Diffusion Decision Model (DDM; Forstmann et al., 2016; Ratcliff, 1978). In the model, evidence starts accumulating over time at the onset of a stimulus toward one of two response boundaries in a probabilistic manner. The main parameters are (1) *drift rate*, which represents how quickly the evidence will build up to one or the other boundary; (2) *boundary separation*, for which larger separation lengthens RT but decreases error rate; (3) *starting point*, which can be biased toward one response boundary of the other; (4) *nondecision time*, which includes the time for all additional processes unrelated to the decision process. The DDM can fit not only the mean RT and error rate of conditions but also the RT distribution. The model includes additional parameters for variability across trials in drift rate, starting point, and nondecision time, which allow it to fit known relations for RT on trials for which the response is correct as opposed to incorrect.

Variations of the DDM have been developed for specific purposes. Of interest is a variation for tasks in which an irrelevant stimulus attribute can produce conflict when it signals a response that is different from that specified by the relevant stimulus attribute. Conflict tasks of this include the widely studied Stroop-color naming task, Simon task, and Eriksen flanker task, which produce, respectively, the Stroop effect (longer RT to name a color when it is displayed in a conflicting color word; Stroop, 1935), Simon effect (longer RT to make a spatial response, e.g., left or right, when the irrelevant stimulus location does not correspond with the location of the response designated by the relevant stimulus dimension; Simon, 1990), and the flanker effect (longer RT when stimuli flanking a target stimulus signal a different response than that indicated by the target stimulus; Eriksen & Eriksen, 1974). Ulrich et al. (2015) developed the Diffusion Model for Conflict Tasks (DMC), which combines separate automatic and controlled processes in the accumulation process. Controlled processing is based on the task-relevant stimulus attribute, whereas automatic processing is affected by the task-irrelevant stimulus attribute. Allowing for these two sources of activation enables the DMC to fit the RT distributions for the various tasks but, with slight modification, the patterns of trial-by-trial congruency sequence effects that occur in such tasks as well (Luo & Proctor, 2022).

Sequential sampling can be incorporated into more complete cognitive architectures to model speed and accuracy. These architectures specify the properties of various processing stages and stores, such as memory and decision processes, and provide a means for developing specific models to simulate the performance of a range of tasks. One widely used architecture of this type is ACT-R (Adaptive Control of Thought – Rational; Anderson et al., 2004), which consists of several modules with buffers that enable communication between and within the modules. Modeling in ACT-R is based on declarative knowledge (verbalizable facts) and procedural knowledge (production rules of IF → THEN structure) for the task being modeled. Current task goals determine what declarative information is activated for quick retrieval when needed by a production rule, for which a match executes the rule's actions. ACT-R has been used to model, for example, improvements in performance and retention with practice (practice and retention (Anderson et al., 1999), choices in decision-making tasks (Gonzalez et al., 2003), and alternative interface designs (Ritter et al., 2019).

Queuing network models, developed mainly within human factors, have also been successful at modeling human performance. A queuing network has one or more components that are of limited capacity, requiring that input enters a queue to wait until the limited-capacity component becomes available (Liu, 1997). Cao and Liu (2013) integrated a queuing network (QN) with ACT-R in what is called (QN-ACTR), which combines the queuing network's strength in modeling multiple-task performance with ACT-R's strength in modeling complex cognition. Subsequent research has shown that QN-ACTR, which uses production rules to represent task-specific knowledge and skills and algorithms and parameters to represent the limitations and capacities of cognitive processing, is able to model drivers' RT to direction sign reading (Deng et al., 2019b) and the takeover time in a partially automated vehicle (Deng et al., 2019a). Thus, it provides a promising approach for dealing with multiple-task situations in interactions with technology.

3.3   ATTENTION AND MEMORY IN INFORMATION PROCESSING

Attention is a topic that has been studied since the earliest days of psychology (see Posner, 2017; Proctor & Vu, 2023, Chapter 1). Exactly how to conceive of attention, beyond the idea of attending to some events or thoughts and not to others, has been the subject of considerable debate. Moreover, there is ongoing debate as to how attention and memory are related. We start with the topic of attention because we know that attention can be specific to the processing of sensory information, as in visual attention and auditory attention, but also to the contents of working memory. That is, attention is often depicted as exerting an influence throughout the entire human information-processing system (Lee et al., 2017). Attention is also depicted to carry out many functions and to involve several distinct neural systems (Posner, 1992). The diversity of topics coming under the heading of attention has led some researchers to propose that the term be eliminated with reference to a specific functional system (Hommel et al., 2019). Regardless, when using the term attention, it should be realized that it refers to many different functions and systems related to the selection and control of information processing.

3.3.1   SELECTION AND CONTROL OF ATTENTION

Due to people's limited ability to process the massive amounts of information presented by interactions with the physical and virtual worlds, the selection of which events to attend is essential. Studies of selective attention examine factors that determine the efficiency of selection and how selection is accomplished. Selective attention plays a role in the extent to which cognitive control can be exerted.

3.3.1.1   Filtering and Processing of Irrelevant Stimuli

In an influential study, Cherry (1953) presented participants with different messages to each ear through headphones. Participants were to repeat aloud one of the two messages while ignoring the other. When subsequently asked questions about the two messages, subjects were able to accurately describe the content of the message to which they were attending but could not describe anything except physical characteristics, such as gender of the speaker, about the unattended message. This result illustrates a general property of attention: People show little awareness of or memory for stimuli or events to which they do not attend.

To account for such findings, Broadbent (1958) developed the filter theory, which assumes that the nervous system acts as a single-channel processor. According to filter theory, information is received in a pre-attentive temporary store and then is selectively filtered, based on physical features such as spatial location, to allow only one input to access the channel. Because filter theory captures that people can selectively attend to one stream of stimuli over another and that not much of an unattended stream reaches awareness, Moray (1993) pithily noted, "Broadbent's original Filter Theory…is probably both necessary and sufficient to guide the efforts of the designer" (p. 111).

Broadbent's filter theory implies that the meaning of unattended messages is not identified, as Cherry's (1953) results suggested, but later studies showed that the unattended message could be processed beyond the physical level, in at least some cases (Treisman, 1964). For example, ~30% of participants can report that their name occurred in the unattended message when it did so unexpectedly (Moray, 1959; Röer & Cowan, 2021). To accommodate the finding that the meaning of an unattended message can influence performance, Treisman (1964) offered a filter-attenuation theory, according to which an early selection filter only attenuated the information on unattended channels prior to stimulus identification, allowing identification if the stimulus is one with a low threshold, such as a person's name or an expected event. Deutsch and Deutsch (1963) proposed a late-selection theory, which states that the bottleneck occurs after the identification of all stimuli, with activation produced by the unattended ones decaying rapidly. Regardless, the general point is that the meaning of unattended events or messages can break through and influence performance in some circumstances.

Lavie et al. (2004) proposed a load theory of attention, which they claimed provides "a compelling resolution to the long-standing early and late selection debate" (pp. 351–352). Specifically, the load theory includes two selective attention mechanisms, a perceptual selection mechanism and a cognitive control mechanism. When perceptual load is high (i.e., great demands are placed on the perceptual system), the perceptual mechanism excludes irrelevant stimuli from being processed. When memory load is high, it is not possible to suppress irrelevant information at a cognitive level. In support of load theory, Lavie et al. showed that interference from distracting stimuli is reduced under conditions of high perceptual load but increased under conditions of high working memory load. Many criticisms have been leveled at load theory, including the vague nature of the concept of "load" and a possible "diluting" effect when neutral stimuli are added to a display to increase load (Murphy et al., 2016). However, one of its detractors, Benoni (2018) made the point that the theory provides good predictions about the outcomes of selectivity, even if the underlying theory is not accepted. That is, irrelevant perceptual information is more likely to intrude on performance in conditions that a designer would tend to think would have a low perceptual load.

3.3.1.2 Spotlight Model of Visual Attention

Visual attention is often likened to a spotlight (e.g., Posner et al., 1980) or zoom lens (Eriksen & St. James, 1986) that directs attention to things in its field. The spotlight of attention can be moved to different areas of space to enhance the processing of the stimuli that might be located there. The attentional spotlight is often positioned in the direction of gaze so that the attended information falls on the region of the retina with the highest acuity. This association of attention with gaze direction enables the use of eye-tracking measures – including the number of visits to a region of interest, duration of the visits, number of fixations in the region, and the length of those fixations (Borys & Plechawska-Wójcik, 2017) – as indicators of where attention is directed. Scanpaths of successive fixations while inspecting an image can be informative as to how and where attention is being directed, as can heatmaps that show the total amount of time that vision was directed to specific regions. For example, Xiong et al. (2017) used heatmap analysis to evaluate the extent to which participants attended to the informative URL in the address bar of a browser when deciding whether a web site was legitimate or fraudulent.

However, it has been known since the time of Helmholtz (1894/1968) that the spotlight of attention can be dissociated from the direction of gaze. When attention is cued to a location, processing is facilitated: detection reactions to stimulus onsets are faster (Posner et al., 1980) as is identification of a target letter or word (Eriksen & Hoffman, 1973). The zoom lens metaphor adds that attention is more concentrated when it encompasses a smaller area rather than a larger one (Eriksen & St. James, 1986). This implication is supported by research showing that when one spatial location is cued as likely to contain a target stimulus, but the stimulus is then presented at another location, items nearer to the cued location are processed more efficiently than items that are located farther away (LaBerge, 1983).

3.3.1.3   Orienting and Inhibiting Attention

One of attention's distinct functions is that of orienting (Fan & Posner, 2004). Orienting is some-what – though not entirely – under a person's control. Studies of orienting often use a precue related to the location at which a target stimulus will subsequently appear. The movement of the attentional spotlight to a location can be triggered by two types of precues, exogenous and endogenous. An endogenous cue is typically a symbol such as a central arrowhead indicating a location to which attention should be shifted intentionally. Posner et al. (1978) provided initial evidence that participants could shift attention to a location cued in this manner. In their experiments, a central left- or right-pointing arrow signaled the location to the left or right at which a target stimulus X would occur in 80% of the trials after an interval of 1 seconds or less. RT to validly cued locations showed a benefit and those to invalidly cued targets (the other 20% in which the stimulus appeared in the uncued location) showed a cost compared to a neutral cue condition. A critical finding is that this result pattern occurred even when eye movements were not made to the cued location, indicating that the attention shift can be covert. Moreover, the pattern was found for a press of a single key when a stimulus onset was detected at either location as well as for choice reactions that required selection of one of two responses corresponding to the stimulus location.

An exogenous cue is an external event such as the abrupt onset of a stimulus at a peripheral location that can involuntarily draw attention to its location. Posner (1980) provided evidence for such unintentional attention shifts by having one of two boxes to the left or right of fixation increase in brightness, after which a target stimulus appeared in one of the boxes. Even when the cue was uncorrelated with the location in which the target stimulus occurred, detection reactions to target onset were facilitated at the cued location and inhibited at the alternate location. Exogenous cues of this type produce rapid performance benefits for stimuli at the cued location, reaching a maximum ~250 ms after the cue onset, which then decreases rapidly as the cue-target interval increases. The performance benefits for endogenous arrow cues, in contrast, take longer to develop and remain evident for a longer period of time when the cues are relevant, consistent with the view that their benefits are due to intentional control of the attentional spotlight (Klein & Shore, 2000).

With exogenous cues, a phenomenon called *inhibition of return* occurs: At longer cue-target intervals beyond ~500 ms, there is a period in which performance is worse for stimuli at the cued location than for ones presented at the uncued location (Posner & Cohen, 1984). The name given to the phenomenon is that of the initial account that was proposed, that is, the return of attention to the cued location is suppressed. Despite the name, alternative accounts have been proposed (Berlucchi, 2006). Redden et al. (2021) recently provided evidence that there are two forms of inhibition of return, one of which decreases the salience of inputs when the reflexive oculomotor system is suppressed (i.e., eye movements are not to be made) and the other of which biases responding when the reflexive oculomotor system is not suppressed (i.e., when eye movements are allowed).

3.3.1.4   Congruity Effects

Stimulus-response compatibility, that is, how direct the relation is between stimuli and their assigned responses, is a major determinant of RT and accuracy (Hommel & Prinz, 1997; Proctor & Reeve, 1990: Proctor & Vu, 2006). Compatibility effects occur whenever there is dimensional overlap, or similarity, between stimuli and responses (Kornblum et al., 1990) or when strong associations of specific stimuli and responses have been created through experience (Proctor & Vu, 2010b). Of most relevance are spatial compatibility effects that occur when both the stimulus and response sets vary along physical or conceptual spatial dimensions. Fitts and Seeger (1953) provided the first demonstration of the benefit of maintaining compatibility by showing that responses were faster and more accurate for the combinations of eight-alternative stimulus and response arrays that were spatially congruent rather than incongruent. Fitts and Seeger's explanation of this result was based on a "process of information transformation or recoding in the course of a perceptual-motor activity" and assumed that "the degree of compatibility is at a maximum when recoding processes are at a minimum" (p. 199). Similar benefits for a compatible mapping of stimulus and response locations are obtained in two choice-reaction tasks, based mainly on the correspondence of the spatial stimulus and response mappings.

In recent years, more emphasis has been devoted to studying the correspondence effects of stimulus dimensions that are irrelevant to the task with either responses or another stimulus dimension. The spatial Simon effect is a spatial compatibility effect obtained when participants are responding to a non-spatial stimulus attribute but the location of the stimulus from trial-to-trial varies randomly between left and right (Simon, 1990). Even though the stimulus location is not relevant to the task, responses are faster and more accurate when the stimulus and response locations correspond. For the flanker task (Eriksen & Eriksen, 1974), one or more stimuli are assigned to each of two responses; a stimulus is the relevant target to which a response is to be made when it occurs in a known location but an irrelevant distractor when it occurs in a flanking location. The flanker effect is faster responding when the flankers signal the same response as the target than when they signal a conflicting response. This effect also seems to have its basis primarily in response activation processes, in this case due to the assignment of letters to responses as part of the task. Finally, in the Stroop color-naming task, the colors of stimuli, the color word spelled out by the stimuli, and the vocal color-name response have dimensional overlap with each other. It is not too surprising that this task is widely investigated because it yields large effects.

In the recent past, there has been considerable interest in sequential effects. A sequential effect pattern obtained across all three tasks is that the benefit for the congruent relation is larger when the prior trial was congruent than when it was incongruent (Gratton et al., 1992). The most widely accepted explanation of this congruency sequence effect is in terms of conflict adaptation (Botvinick et al., 2001; Gratton et al., 1992). The general idea is that when conflict is detected on a trial, suppression of that activation takes place. This suppression carries forward into the next trial, which results in the irrelevant dimension producing relatively little activation. Conversely, when there is no conflict on a trial, suppression does not occur, and the irrelevant dimension produces more activation on the next trial. If the conflict adaptation accounts are correct, then the congruency sequence effect reflects dynamic changes in cognitive control from one moment to another. However, alternative explanations for cognitive control have been proposed (Hommel et al., 2004; Mayr et al., 2003), and the exact explanation of the sequential effects is still a matter of debate (Schmidt, 2019).

Another common finding is that when a large proportion of the trials are incongruent, the congruency effect is eliminated or reversed. Activation/suppression accounts have also been applied to this phenomenon (Botvinick et al., 2001). Because proportion manipulations alter the relative frequencies of trials for which the preceding trial is congruent or incongruent, these proportion effects could be a consequence of the individual trial frequencies unless those are controlled. However, evidence suggests that frequency-unbiased conditions also show a proportion congruent effect (Crump et al., 2017; Torres-Quesada et al. (2014), implying that attentional control may have a proactive component based on prior probabilities.

3.3.1.5 Visual Search

Many tasks in HCI require visual search, where users need to find an object of interest on an interface. Attentional processes involved in visual search have been examined extensively in laboratory studies where participants are asked to find whether a target is present among distractors, and properties of the target and distractors are modified to examine differences in search efficiencies. When the target is distinguished from the distractors by a basic feature such as color (feature search), search times and error rate often show little increase as the number of distractors increases

(Wolfe et al., 2010). However, when two or more features must be combined to distinguish the target from distractors (conjunctive search), search time and error rate typically increase sharply as the number of distractors increases.

Treisman and Gelade (1980) developed feature integration theory to explain the results from visual search studies, where they assume that basic features of stimuli are encoded into feature maps in parallel across the visual field at a pre-attentive stage. Feature search can be based on this pre-attentive stage because a "target-present" response requires only the detection of the feature. The second stage involves focusing attention on a specific location and combining features that occupy the location into objects. Attention is required for conjunctive search, and the spotlight of attention must be moved sequentially across the search field until a target is detected or all items present have been searched. Consequently, search time increases as the number of distractors increases.

In 1989, Wolfe et al. introduced a variation of feature integration theory, called the Guided Search model. Rather than using a dichotomy of serial versus parallel search, the Guided Search model focused on the efficiency of search being guided by attention. In other words, attention can be directed to the locations or properties of items in the visual field by means of pre-attentive processes. For example, pre-attentive processes may direct attention in a stimulus-driven manner when a unique feature of items on a display pops out. Alternatively, pre-attentive processes may also direct attention in a strategic manner when the observer knows a unique property of the targeted stimulus in advance. This latter process can explain why researchers have found that participants were able to find icons of specific properties (i.e., high or low in spatial frequency) if the target was distinguishable from distractors based on that property and the users knew about this distinction (Rauschenberger et al., 2009). Thus, feature integration theory and Guided Search theory provide HCI professionals with useful frameworks for determining contexts in which search should be more or less efficient.

In another study of icon search, Huang (2008) had college-age participants search for one of four icons that differed in terms of shape (i.e., the icons itself), foreground/background color combination, and area ratio. Huang found that the search time was shortest for the icon that had a shape that was most distinctive from other icons, for foreground/background colors that had higher contrast, and for the area ratios that did not put the shape of the icon close to the edge. These findings are consistent with what would be predicted from the visual search models because all the factors leading to faster search times were related to making the icon more distinguishable and recognizable to the person. Some icons have text included to help users identify them.

Hou and Hu (2021) examined how text size and icon size influenced older adults' visual search behaviors. Participants between the ages of 57 and 71 were asked to search for a target icon, arranged in a grid, of distractors. The pictogram was small, medium, or large in size, and the accompanying text was small, medium, or large in size. Results showed that the pictogram size influenced search time, with search time being best when the pictogram was small and worst when it was large. However, the effect of pictogram size depended on the text size. For the small pictograms, the search time was shorter when it was paired with the medium or large text size. For the larger pictograms, the pairing with the larger text did not improve search times. This finding suggests that when the pictogram size was small, this group of older adult users relied on the large text to identify the icon. However, when the pictogram was large, the large text decreased search efficiency by moving the letters outside the users' visual attentional focus. Thus, designers need to consider both the features that would make an icon more distinguishable, so that it will pop-out or allow for guided search based on recognizable features within the attentional focus.

Although it is not surprising that items outside the attentional focus are not processed, even items within the field of view may not be processed. This phenomenon is sometimes referred to as "Look But Fail To See" (see e.g., Wolfe et al., 2022). For example, drivers involved in accidents with pedestrians often report looking at the location of the collision but not seeing a person there. The look but fail to see phenomenon can be a direct byproduct of normal attentional mechanisms involved in object recognition. Because attention is limited in capacity and observers move their eyes across a scene, observers only select a subset of the items in the scene they could process on each eye fixation, which Wolfe et al. refer to as the functional field of view. Items in the functional field of view can be missed if too little time is given to their processing, even though observers are looking at them. Moreover, the visual system often fills in gaps that result in observers thinking that they see more than they are actually aware of and attention can guide observers to different items in the scene. Thus, Wolfe et al. indicate that cognitive and perceptual processes can produce a state of inattentional blindness that is responsible for the look but fail to see phenomenon.

In HCI, search can also occur for items that are out of view, specifically nested in menus or other locations that may not be currently visible. For example, when searching for known items in a menu, the alphabetic and categorical organization can facilitate search (McDonald et al., 1983). Although early studies of menu design showed that users tend to prefer deep menu structures (Lee & MacGregor, 1985), broad menu structures have been found to be more beneficial for individuals with lower working memory capacity, such as older adults (Rouet et al., 2019), and under situations where working memory may be taxed, such as when voice menus are used (Commarford et al., 2008). Moreover, for complex or ambiguous situations, the use of broad menu designs may be more beneficial because users can compare items between categories (Tullis et al., 2011). Thus, designers must consider the type of search in which the user would most likely be engaged when structuring menus.

3.3.2 Divided and Sustained Attention

There are also many situations in which people try to divide attention to multitask, such as when they are attending a lecture and interacting on social media at the same time, or watching television while reading e-mail. In this section, we examine factors influencing divided and sustained attention, which is required for monitoring tasks that people have difficulty doing for an extended period.

3.3.2.1 Divided Attention and Multitasking

In many HCI tasks, the human operator must attend to multiple sources of information simultaneously, making it necessary for designers to understand how attention is allocated to different displays and events occurring in the environment. Kahneman's (1973) unitary resource model views attention as a single resource that can be divided up among various tasks in different amounts, based on task demands and voluntary allocation strategies. The model predicts that multiple tasks will produce interference when their resource demands exceed the supply that is available. Thus, two tasks can be performed concurrently without any observable performance decrement as long as the total attentional demand does not exceed the supply.

Dual-task performance is often compared to single-task performance to determine how performance is affected by having to perform more than one task. Using a performance operating characteristic (POC) curve (see Figure 3.3), task performance for Task 1 is plotted on one of the two axes and performance on Task 2 is plotted along the alternative axis. If the two tasks can be performed without interference from the other task, then dual-task performance should equal single-task performance (also known as the independence point). In most cases of dual-task performance, though, a cost of concurrence occurs, where performance on a specific task is worse even if the person allocates all their attention to that task. The use of the POC can help HCI professions determine which tasks can be performed together more efficiently by having participants devote differential amounts of attention to one task versus the other and measuring how performance is impacted. In general, it is easier to perform two tasks together when they use different stimulus or response modalities than when they use the same modalities. In addition, performance is better when one task is verbal and the other is visuospatial than when they are of the same modality. These result patterns provide the basis for multiple resource models of attention (see Wickens, 1984), according to which distinct attentional resources exist for different sensory-motor modalities and coding domains.

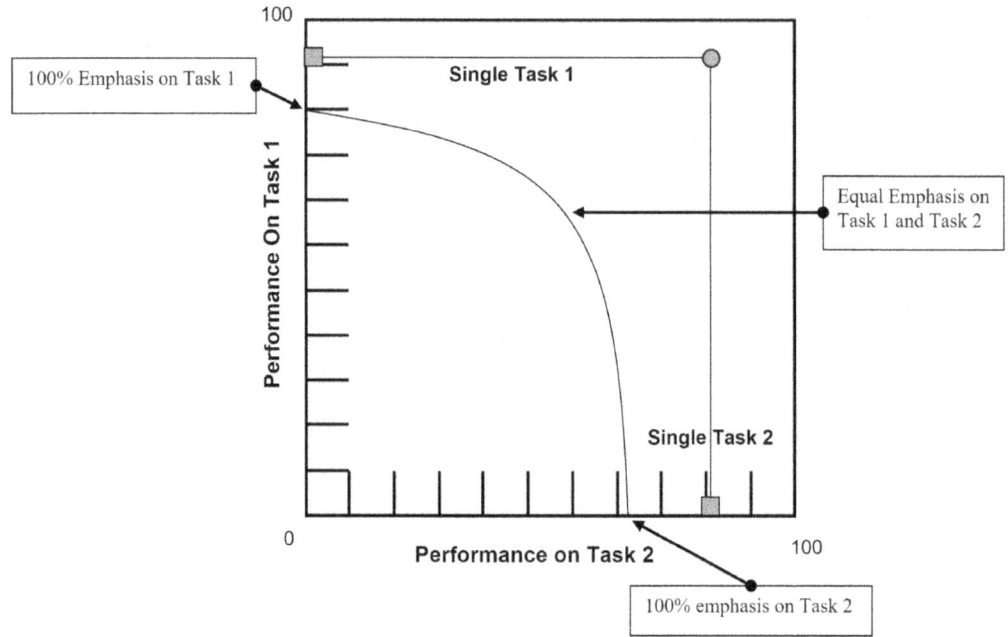

FIGURE 3.3 Illustration of a Performance Operating Characteristic (POC) curve. Task 1 performance is plotted on the Y-axis and Task 2 on the X-axis, with 0 being worst and 100 being best possible performance. Single task performance is denoted by the square data point and the circle represents the independence point. The curve is used to illustrate performance as a function of the amount of attention that is being allocated to Task 1 versus Task 2.

Multiple resource theory captures the fact that multiple-task performance is typically better when the tasks use different input-output modes than when they use the same modes. However, Thropp et al. (2004) noted that individual differences in POC curves should be examined because averaging across data may mask large individual differences.

A relevant study to HCI is that of Nijboer et al. (2016), in which participants drove on a simulated road either with or without traffic. In some conditions, the participants also performed a secondary task: listening to the radio, answering questions to a radio quiz by pressing buttons on the steering wheel, or answering questions to a quiz by reading and responding on a tablet computer. In both the no traffic and traffic scenarios, participants performed the worst when using the tablet, which would be predicted by multiple resource models. However, when the driving task was monotonous, neither listening to the radio nor answering questions to the radio quiz decreased driving performance but instead yielded better driving performance. Thus, although multiple resource models serve as a useful guideline for HCI designers, specific scenarios need to be empirically evaluated.

The role of attention in response selection has been investigated extensively using the psychological refractory period (PRP) paradigm (Pashler, 1998). In the PRP paradigm, a pair of choice-reaction tasks must be performed, and the stimulus onset asynchrony (SOA) of the second stimulus is presented at different intervals. RT for Task 2 is slowed at short SOAs, and this phenomenon is called the PRP effect. The experimental results have been interpreted with what is called locus of slack logic (Schweickert, 1978), which is an extension of additive factors logic to dual-task performance. The basic idea is that if a Task 2 variable has its effect prior to a bottleneck, that variable will have an under-additive interaction with SOA. This under-additivity occurs because, at short SOAs, the slack period during which post-bottleneck processing cannot begin can be used for continued processing for the more difficult conditions. If a Task 2 variable has its effect after the bottleneck, the effect will be additive with SOA.

The most widely accepted account of the PRP effect is the response-selection bottleneck model (Pashler, 1998). The primary evidence for this model is that perceptual variables typically have under-additive interactions with SOA, implying that their effects are prior to the bottleneck. In contrast, post-perceptual variables typically have additive effects with SOA, implying that their effects are after the bottleneck. There has been dispute as to whether there is also a bottleneck at the later stage of response initiation (De Jong, 1993), and whether the response-selection bottleneck is better characterized as a parallel processor of limited capacity that divides resources among to-be-performed tasks (Tombu & Jolicœur, 2005), and whether the apparent response-selection bottleneck is a strategy adopted by subjects to comply with task instructions (Meyer & Kieras, 1997). This latter approach is consistent with an emphasis on the executive functions of attention in the coordination and control of cognitive processes (Monsell & Driver, 2000).

A noteworthy recent account is that of Klapp et al. (2019), which is related to the explanation for response complexity effects on simple RT described earlier (Klapp & Maslovat, 2020). Klapp et al. proposed that the bottleneck in the PRP paradigm resides in a motoric process after response selection that is involved in programming the timing of response initiation. They based this hypothesis on two research findings. One is the aforementioned finding that in studies of the startle response and single-task simple (precued) RT, programming of the timing of response onsets must be delayed until immediately prior to responding (Maslovat et al., 2014; Shin & Proctor, 2018). Second, Klapp et al. (1998) found that even after training participants to tap different rhythms separately with each hand, they had difficulty tapping the two rhythms concurrently. Klapp et al. (2019) made a case that these two limitations, which are general properties of response initiation, combine to produce the PRP bottleneck at the initiation stage of processing.

In contrast to dual-task performance, where the two tasks are to be performed concurrently, task-switching refers to performing two or more tasks one at a time. When the task to be performed switches from the preceding one, a switch cost is usually obtained: RT is longer and the error rate is higher compared to when the prior task is repeated (Kiesel et al., 2010; Monsell, 2003). Switch costs can be reduced by providing advance knowledge of an up-coming task switch by way of a precue or by following a fixed sequence, suggesting that time is required for a person to adopt an appropriate task set. However, because the switch cost is often not eliminated even with sufficient time for advance preparation – a phenomenon called the residual switch cost – task switching likely involves two processes: attentional control and automatic processes of inhibition and activation from prior trials (Strobach et al., 2018). In many HCI contexts, individuals engage, or attempt to engage, in multiple tasks at the same time, which is often referred to as multitasking (Wammes et al., 2019). When multitasking, factors that influence dual-task performance and task switching must be considered, as well as other factors such as task interruptions and resumptions (Koch et al., 2018).

HCI professionals also need to be aware that media multitasking can cause interference with learning. Wammes et al. (2019) showed that media multitasking during in-person lectures results in poorer performance. In their study, students in a course were probed during the lecture to indicate what they were doing prior to the presentation of the probe question. Students could respond by indicating that they were (1) paying attention to the lecture (on task), (2) thinking about other material related to the lecture/class (elaborating on material), (3) thinking about things unrelated to the task but school-related (related mind wandering), (4) thinking about completely unrelated things (unrelated mind wandering), or (5) using a smartphone or laptop for unrelated activities (media multitasking). Results showed that the frequency of media multitasking increased as time in the lecture increased, but the frequency of engaging in elaboration of the material or mind wandering (related and unrelated) did not. Moreover, engaging in media multitasking was associated with poorer performance on quizzes compared to mind wandering during the lecture. Wammes et al. suggested that the reason why media multitasking was more detrimental to quiz performance than mind wandering is that the media draws more attention away from the lecture and demands more attention than mind wandering.

3.3.2.2 Focused and Sustained Attention

Without focused attention, people are not able to detect changes in visual stimulation that occur within their field of view or when their view is reinstated after being briefly interrupted. These phenomena are known as inattentional blindness and change blindness (Simons & Ambinder, 2005), respectively, which refer to the insensitivity that people show to detecting events and changes to which they are not attending. There are many well-known illustrations of these phenomena, including a man not realizing that the person he was talking with changed when two people carrying a door walked between them or not being aware that a gorilla danced across the screen while viewing a team passing a basketball. The probability of someone detecting a change is greatest when their attention is directed to the location of where the change occurs, and they are consciously noting what was there before and after the change.

The ability to sustain focused attention is often studied using a vigilance task, where an individual is required to monitor a display or environment to detect a critical stimulus or event. Examples of vigilance tasks include that of an airport baggage screener, who monitors displays presenting images of luggage contents to identify ones that may be restricted items, such as a weapon. Similarly, an operator of a self-driving car or a pre-programmed drone also engages in the task of monitoring the automation to ensure that it is operating safety but must be ready to intervene when the automation fails or to respond to off-nominal events that the automation is not programmed to handle. Early studies of operator performance on vigilance tasks showed that a decrement in performance on a continuous monitoring task can be observed within the first 30 minutes of a session (Mackworth, 1948).

Over the years, many researchers have examined intervention techniques to reduce the vigilance decrement. These techniques have been guided by underload and overload theories of attention. Underload theories indicate that the vigilance decrement is caused by boredom or low arousal from performing a monotonous task (Manly et al., 1999). Because individuals are not stimulated by the task, mind wandering can also occur (Thomson et al., 2015), reducing performance over time. Pattyn et al. (2008) found evidence for the underload hypothesis, with their participants showing a decrease in heart rate over time, and EEG data showed indicators of autonomic activation, both of which are suggestive of task disengagement. In contrast, overload theories indicate that the vigilance decrement results from the effort required to sustain attention to the task. This effort depletes the attentional resources needed to perform the task, causing performance to decrease over time. Warm et al. (2008) found evidence consistent with the depletion of resources in that there was a rapid decrease in the velocity of cerebral blood flow over the course of the vigil. Moreover, participants rated vigilance tasks as being unpleasant, mentally demanding, and stressful (Dillard et al., 2019). Regardless of which theory is more accurate, both attribute the vigilance decrement to the attentional resources becoming inadequate over time on a task that requires vigilance.

Another issue of concern in the research on vigilance has been whether the decrease in signal detection across time is due to a change in perceptual sensitivity or a shift in the operator's response bias. Signal detection methods can be used to determine whether the vigilance decrement is due to a decrease in the individual's sensitivity (i.e., accuracy) or a shift to a more conservative (or higher) criterion for responding that a critical event is present. McCarley and Yamani (2021) found evidence for both but indicated that the shifts in the operator's criterion accounted for the largest change in performance.

One way to conceive of the vigilance decrement is that the person is not in a state of high alertness. Alertness can be increased by presenting a neutral warning stimulus at various foreperiods prior to the target stimulus to which a response is to be made. When the foreperiod is varied across trial blocks, RT often decreases up to a foreperiod of about 150 ms, but the decrease is accompanied by an increase in error rate (Bertelson, 1967; McCormick et al., 2019; Posner et al., 1973). This result pattern implicates a shift in the response criterion to a more liberal (or lower) setting as the performer becomes prepared to respond. Han and Proctor (2022) obtained this result at a 50-ms foreperiod but found that RT decreased further at a 200-ms foreperiod without a corresponding increase in error rate. Thus, much like McCarley and Yamani, they concluded that at very brief foreperiods the warning effect is a result of response criterion adjustment but that longer foreperiods there is improved response efficiency, suggestive of a controlled preparation component.

Researchers have examined the effectiveness of a variety of interventions other than warnings to reduce the vigilance decrement. Al-Shargie et al. (2019) found the following interventions, all of which seem to increase alertness, and to reduce mean RT by 10% or more: videogame playing, transcranial and haptic stimulation, taking caffeine or modafinil (increases wakefulness), chewing gum, and listening to music. The effects of these interventions are strongest for simple vigilance tasks and may even be negative for more complex vigilance tasks that require comprehension and working memory.

3.3.3 MEMORY IN INFORMATION PROCESSING

Memory refers to the ability to recall, recognize, or reconstruct information about a previously seen stimulus or prior event. Memory can involve explicit recall of an immediately preceding event or one many years in the past, or be implicit, showing up in the performance of perceptual-motor task or as general knowledge derived from life experiences and education. There are several common classifications of memory. Episodic memory refers to accessing a specific episode such as what was the last item you searched for on the web, whereas semantic memory refers to general knowledge such as knowing how to conduct a web search. Declarative memory is a recall of verbalizable knowledge, usually measured with an explicit memory test (e.g., recall test). Procedural memory is knowledge that can be performed but not necessarily described verbally and is usually measured with implicit memory tests (e.g., sequential learning or priming test). For example, if a password is used to authenticate an identity, the recall of the password typically involves declarative memory (i.e., the user recalls the password). However, if typing patterns were used as the authentication method, then that may involve more of procedural memory (i.e., the user performs a perceptual-motor task).

Three types of memory systems are customarily distinguished: sensory memory, working memory (WM; also known as short-term or immediate memory), and long-term memory (LTM). Sensory stores refer to brief modality-specific persistence of a sensory stimulus from which information can be retrieved for one or two seconds (see Nairne, 2013). WM and LTM are the main categories by which investigations of episodic memory are classified, although there is ongoing debate about whether WM and LTM represent distinct systems (Baddeley et al., 2019) or are part of the same system (Cowan, 2017). There is evidence consistent with both conceptualizations. For example, an fMRI study by Talmi et al. (2005) found that recognition of early items in the list was accompanied by activation of areas in the brain associated with LTM, whereas recognition of recent items was not, supporting a distinction between LTM and WM. Souza et al. (2015) found that when attention is directed at individual items in a visual array, memory for those items is better compared to other items in the array, highlighting the role of WM in holding representations that are to be acted upon.

Attention and memory are clearly related, but there is debate about the way in which they are related. As noted earlier, resource models of attention are popular, viewing attention as a limited capacity resource (e.g., Kahneman, 1973) that also limits WM. In contrast, if attention is viewed as the selective process for information processing, lack of resources is not the reason why WM is limited in capacity but the selective function that attention plays in maintaining, disengaging, and coordinating processing (Oberauer, 2019). WM then helps control attentional processes by holding items or task sets that are needed to carry out task goals. Regardless of whether memory is one or multiple systems, or whether it is limited by attentional resources or a mechanism for selective attention, there is consensus that WM refers to the "mechanisms and processes that hold the mental representations currently most needed for an ongoing cognitive task available for processing" (Oberauer, 2019, p. 1). In the following sections, we emphasize WM as a distinct system because it provides a useful framework for helping designers to select alternative interface designs that reduce memory loads, an issue of importance to HCI. We also discuss the implications of views that place more stress on activation from LTM.

3.3.3.1 Working Memory

As noted, WM refers to representations that are currently being used or have recently been used and last for a short duration. For many HCI applications, items in WM are often considered to be where the focus of attention is directed. In this view, because attention is a limited resource, WM is also limited in capacity. G. A. Miller's (1956) classic article, "The Magical Number Seven Plus or Minus Two," is one of the most widely cited works for the limits of WM, and many HCI designers have found this rule to be a good heuristic. That is, designers should not design tasks that require users to hold more than seven digits or letters in WM (i.e., limit access codes and voice menus to less than seven items). HCI designers are also aware that the 7 ± 2 heuristic refers to chunks of information, and not necessarily items per se, so when possible, grouping of items into meaningful chunks can increase users' memory for those items. More recent evidence seems to indicate that the WM capacity is much lower, being three chunks when covert rehearsal is prevented (Chen & Cowan, 2009), so this is a point that HCI designers need to consider. Consistent with this latter finding, Gao et al. (2023) found that users can maintain two gesture-command associations when the associations are low in compatibility (i.e., arbitrary), but that this capacity can be increased to three to five when the gesture-command associations are high in compatibility (i.e., consistent with users' understanding of the relation).

An individual's immediate memory capacity is usually measured by a memory-span test, where the span is the number of items that can be recalled correctly, in order. When rehearsal is not prevented, the memory span for words varies as a function of word length: The number of words that can be retained decreases as word length increases (Baddeley et al., 1975). A good rule of thumb is that the capacity of immediate memory is the number of syllables that can be repeated in about 2 seconds (Schweickert & Boruff, 1986). Also, as most people are aware from personal experience, if rehearsal is prevented, such as when one is distracted by another activity, information in WM can be forgotten quickly. Studies have shown that, when rehearsal is prevented by performing a distractor task, recall of a string of letters that is within the memory span decreases to close to chance levels over a retention interval of 18 seconds (Brown, 1958; Peterson & Peterson, 1959). For HCI, the use of an operation span test (Unsworth et al., 2005) may be a more accurate indicator of an individual's WM capacity. In such a test, participants are shown a list of items that are to be remembered, but the presentation of the items is separated by a task that requires verification of a mathematical statement. This additional task places a load on WM, which reflects many environments encountered in everyday life where distractions and competing task demands are present.

Conway et al. (2001) provided evidence that participants with higher WM capacity have greater attentional control, being able to focus on or selectively attend to the relevant message. Colflesh and Conway (2007) found that individuals with higher WM capacity were also better able to divide their attention. Thus, individuals with high WM capacity have greater attentional control, being able to selectively attend or divide their attention based on task demands. In a review paper, Conway et al. (2005) indicate that WM capacity is correlated with performance on a wide range of tasks, including comprehension of written and spoken material, following oral and spatial directions, note-tasking ability, and learning of complex tasks.

The best-known model of WM is that of Baddeley and Hitch (1974), which partitions WM into three main parts: Central executive, phonological loop, and visuospatial sketchpad. The central executive is closely tied to the focus of attention, where information is being actively processed to achieve task goals. For example, the central executive is involved in performing manipulations or computations on the items being held for processing, as well as in controlling and coordinating the actions of the two other subsystems, the phonological loop and visuospatial sketchpad. The phonological loop is composed of a phonological store that is responsible for holding the to-be-remembered items and an articulatory control process that is responsible for recoding verbal items into a phonological form and rehearsal of those items. The items stored in the phonological store decay rapidly unless kept active by rehearsal from the articulatory control process. The visuospatial sketchpad retains information regarding visual and spatial information, and it is involved in mental imagery and mental rotation. The WM model has been successful in explaining several memory phenomena (Baddeley, 2003). However, the model cannot explain why memory span for visually presented material is only slightly reduced when subjects engage in concurrent articulatory suppression (such as saying the words "the" aloud repeatedly). Articulatory suppression should monopolize the phonological loop, preventing any visual items from entering it. To account for such findings, Baddeley revised the WM model to include an episodic buffer (see Figure 3.4). The buffer is a limited-capacity temporary store that can integrate information from the phonological loop, visuospatial sketchpad, and LTM. By attending to a given source of information in the episodic buffer, the central executive can create new cognitive representations that might be useful in problem solving.

Baddeley and Hitch's (1974) WM model provided a good way of characterizing different aspects of working memory that HCI designers can use when structuring tasks and designing interface to reduce WM load within a modality or to increase WM capacity by spreading demands across multiple subcomponents/modalities. However, HCI designers should note that other models of WM shifted from the characterization of distinct stores to an area of LTM that is currently being activated. For example, Engle (2002) equated WM capacity with executive control of the allocation of attention. Cowan (1999) and Oberauer (2009) developed embedded component models of WM to capture this more recent characterization. Cowan's model directly attributes WM to be the area within LTM that is currently being activated. In a review paper,

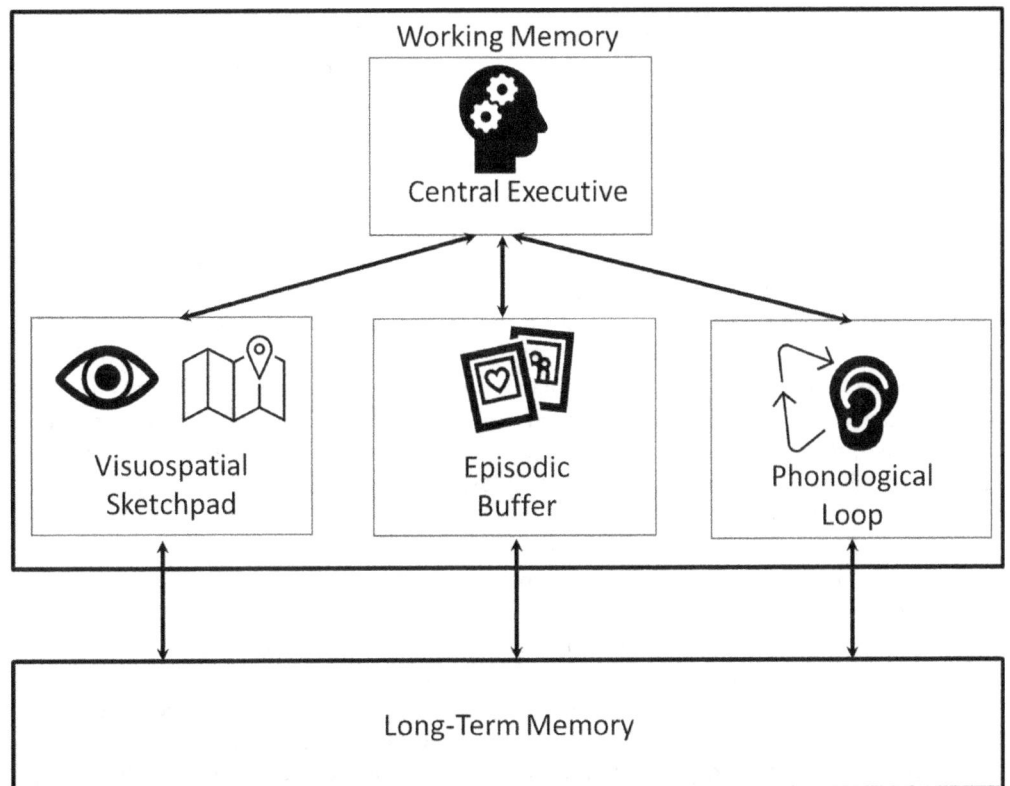

FIGURE 3.4 Depiction of Baddeley's (2000) revised working memory model.

Cowan (2001) showed that the capacity limits of WM memory were related to the number of items to which attention can be allocated at the same time. Oberauer (2009) and Oberauer and Hein (2012) extended Cowan's embedded components model to emphasize that attention acts as a selective filter that activates one chunk in an immediately available state. In other words, WM is attention to the memory representations that control what content is being activated for achieving task goals. For HCI, the implications of this characterization are that in most situations one item is the focus of attention in WM, but other items are activated to different extents based on their relations to task goals.

3.3.3.2 Long-Term Memory

LTM refers to representations that can be remembered for durations longer than those that can be attributed to WM. LTM can involve information presented minutes ago or years ago. Initially, it was thought that the probability of an item being encoded into LTM was a direct function of the amount of time that it was in WM, or how much it was rehearsed. However, Craik and Watkins (1973) showed that rehearsal by itself is not sufficient, but rather that deep-level processing of the meaning of the material is the important factor in transferring items to LTM. For example, if a multifactor authentication technique requires that a user enters an access code received on a registered device, repeating that access code over and over until the user can enter it does not lead to better recall of that code later. However, if the authentication technique requires users to select a subset of images from an array that the user selected to reflect a meaningful event then the information can be better remembered later. Oulasvirta et al. (2005) also found that participants who viewed the content area of a Web page had no better memory for the material than that guessed by a control group who had never seen the page because the participants' task was to locate links on the page and not to process the content information. Thus, if the information is not processed adequately, it cannot be recalled.

There are many LTM memory principles that are useful for HCI, several of which are summarized in Table 3.1. In addition to these LTM principles, HCI designers should be aware that events that precede or follow an event of interest can interfere with the recall of that event. The former is referred to as proactive interference, and the latter as retroactive interference. Shrestha et al. (2020) interviewed 16 expert programmers to help determine why having prior programming knowledge interfered with the learning of a new one. They found that part of the difficulty was in terms of the strategy that the programmers used to learn the new language, which was that of trying to map features of the new language to their previously learned one. Moreover, proactive interference played a role, with the programmers reporting that they had to constantly suppress old habits based on their prior knowledge or try to apply incorrect mappings. Some reported having trouble switching between mindsets. Relating to the latter point, Raissi et al. (2020) found retroactive transfer effects for which performance suffered when participants alternated between two similar layouts of a keyboard, even though they had previously learned them. One promising finding, though, is that the magnitude of the performance decrement decreased with the number of alternations between keyboards. In other words, experience or practice switching between keyboards reduced the initial cost.

For HCI, it is essential to design interfaces to support prospective memory, which is memory to perform an intended action at a later time. Prospective memory can be event-based, activity-based, or time-based (Walter & Meier, 2014). For example, remembering to stop at the grocery store when you go home from work is event-based; remembering to attach a file to an e-mail is activity-based; remembering to join an online meeting at a specific time is time-based. For older adults, forgetting intentions and planned actions characterizes at least half of the memory problems reported by healthy older adults (Hering et al., 2018). Thus, the inclusion of reminder features in applications and systems that can help users perform the correct action at the correct time will be very beneficial to users, especially more elderly users.

TABLE 3.1

Long-Term Memory Principles and Their Implications for HCI

Principle	Description	Implications for HCI
Depth of processing	Items can be processed along a continuum of shallow to deep processing. The greater the depth of processing, the better the memory (Craik & Lockhart, 1972)	Repetition is not sufficient for good memory. Relating to be remembered items to prior knowledge can improve memory
Encoding specificity	The probability that a retrieval cue results in recollection of an earlier memory is dependent on the match between encoding and retrieval (Tulving & Thomson, 1973)	Memory is often context-dependent. If the encoding and retrieval environments match, memory can be improved
Transfer appropriate processing	Deep-level processing produces better performance than shallow on standard recall or recognition memory test. However, when the memory test required judgments based on visual features; shallow processing of visual features was more beneficial (Morris et al., 1977)	Retention will be evident to the extent that the procedures engaged in during study or training are reinstated at the retention test (Healy et al., 2005)
Spaced practice is better than massed practice	Massed repetition (repeating the same item in a row) is less effective than spaced repetition (repeating the same item with one or more intervening items; Hintzman, 1974)	With spaced practice individuals have the chance of determining which study strategies are more effective and can use those strategies to benefit memory (Bahrick & Hall, 2005)
Generation effect	Recall is better when participants have to generate the items that they need to remember rather than just studying them (Slamecka & Graf, 1978)	Have users generate passwords rather than assign a password to them (Vu et al., 2007)
Testing effect	Recall is better for items that were previously tested than new items (Karpicke & Roediger, 2008)	Provide intermittent quizzes for important information
Mnemonic techniques	Connecting the items to be remembered with an established organizational structure that can facilitate later recall (Wood & Pratt, 1987; Verhaeghen & Marcoen, 1996)	Passwords generated from a mnemonic technique can be more memorable (Vu et al., 2007)

Finally, HCI designers can develop applications to help users develop prospective memory skills. Chan et al. (2019) used a smartphone app to train older adults in the technique of implementation intention (Gollwitzer, 1999), where they verbalize an intention and visualize themselves performing the intended action. The purpose of this technique is to strengthen the association between the situational cues (event, time, or location) and intended actions. In a 12-day study, the older adults who received the training were more prompt in performing prospective memory tasks compared to the control group. These findings are promising for the design of software and apps that can improve prospective memory for older adults through training.

3.3.3.3 Automaticity and Practice

Attention and WM demands are high when a person first performs a new task. However, these demands decrease, and performance improves as the task is practiced. Because the quality of performance and attentional requirements change substantially as a function of practice, it is customary to describe performance as progressing from an initial cognitively demanding phase to a phase in which processing is automatic (Anderson, 1982; Fitts & Posner, 1967). That is, when a user first learns a task, the individual must hold task components in WM and verbalize the steps (declarative phase). As the user continues performing the task, the components become linked or connected through practice (associative phase). Finally, after extensive practice, the steps are proceduralized and performed automatically, with little or no attention (autonomous phase).

Fitts's three-stage model (1964) attributes automaticity to changes in how the information is processed across the phases. Logan (1988) proposed an instance theory of automaticity that instead assumes that the information processing requirements are not changed through practice but are just performed more efficiently. According to Logan, instances are episodes or events in which attention is directed to relevant information, which results in that information being activated in memory. Retrieval of items in WM requires more effort than LTM. Automaticity is achieved with practice because the individual can perform fast and effortless retrieval of the episode from LTM rather than through the retrieval of procedures. Thus, although models also predict a shift in reliance on WM to LTM through practice, the mechanism involved in the shift differs.

Both the three-stage and instance theory of automaticity predict that the time to perform the task decreases with practice, which is consistent with findings from many tasks ranging from choice RT to solving geometry problems, with the largest benefits occurring early in practice. Newell and Rosenbloom (1981) proposed a power function to describe the changes in RT with practice:

$$RT = BN^{-\alpha}, \tag{3.3}$$

where N is the number of practice trials, B is RT on the first trial, and α is the learning rate. The defining characteristic of the power function is that the relative learning rate is a hyperbolically decreasing function of the number of trials. The power function was widely accepted for some time as the law that describes the changes in RT, but Heathcote et al. (2000) argued that an exponential function provides a better fit to individual participant data. The defining characteristic of the exponential function is that the relative learning rate is constant at all levels of practice. More recently, Evans et al. (2018) noted that the initial rate of learning in many cases is slower than either the power or exponential law suggests, and they provided evidence that the speedup with practice is best fit by a delayed exponential law. For HCI, designers can use these functions to estimate learning times and performance of groups and individuals.

It has been long established that not all tasks can be performed efficiently, even with extended practice. Schneider and Shiffrin (1977) and Shiffrin and Schneider (1977) found that when performing a memory search task, efficient search was only possible if a consistent mapping was employed, where the target items on one trial were never distractor items on another trial. When a varied mapping was employed, where target items on one trial could be distractors on subsequent trials, the search was still inefficient even with extensive practice. With a consistent mapping, participants were able to learn the target items and search for them simultaneously. In contrast, with the varied mapping, the targets on a trial must be maintained in WM and searched through in a less efficient, possibly serial, manner. HCI designers should employ consistent mappings to reap the benefits of practice and learning.

Finally, HCI professionals can look for opportunities to enhance human performance through the design of training programs and software and through the use of automation itself. For cognitive or "brain training" studies, Simons et al. (2016) concluded that the use of software or video game training is effective for developing the skills specifically being trained by the tasks, but not very effective for transfer to similar tasks and ineffective for developing general cognitive skills. Thus, designers should focus on the specific information processing skills they want to promote in the design of the training software. There is also a trend toward designing increasingly autonomous systems to assign routine or pre-determined task to automation to reduce operator workload (Vu et al., 2018). However, making the human a monitor of automation results in a vigilance task, which has been demonstrated to be a task that humans are poor at performing and find to be unpleasant (Dillard et al., 2019). Moreover, it can result in over-reliance on automation and/or a loss of situation awareness, making it difficult for the operator to take over in emergency situations (Biondi et al., 2019). There has been interest in making automation behave more like a team member in a concept known as human autonomy teaming (O'Neill et al., 2022), though there is too little evidence at present as to when or whether this is beneficial.

3.4   CONCLUSION

Because the human information-processing approach has guided much research for approximately the last 70 years, the methods, empirical databases, theories, and models are well-developed. The knowledge in this area, which we described at only a surface level in this chapter, is relevant to a range of concerns in HCI, from visual display design to the design of increasingly autonomous systems. For HCI to be effective, the interaction with the system must be compatible with human information-processing capabilities. Sophisticated computational models based on sequential information accumulation such as the DDM (Forstmann et al., 2016; Ratcliff, 1978) have been implemented that allow precise modeling of RT and accuracy in many contexts. Cognitive architectures that incorporate much of what is known about human information processing have been developed that can be applied to HCI. The use of such cognitive architectures started with the Model Human Processor (Card et al., 1983), and current, more sophisticated architectures allow the development of complete quantitative models of information processing for even complex tasks. These architectures include Adaptive Control of Thought (Anderson et al., 1997; Ritter et al., 2019), Soar (Laird, 2022; Newell, 1990), and Executive Process-Interactive Control (Kieras, 2019; Kieras & Meyer, 1997).

The human information-processing approach emphasizes using data from research studies conducted under controlled conditions in which fundamental cognitive processes and principles thought to be of broad generalizability are established. Although this emphasis typically results in the use of laboratory environment, the knowledge and principles derived from the laboratory research are relevant to applied contexts, many of which were highlighted throughout the chapter. Moreover, the information-processing approach has been used successfully to investigate human performance in many real-world tasks including distracted driving (Nijboer et al., 2016), media multitasking (Wammes et al., 2019), and monitoring increasing autonomous systems (Vu et al., 2018). Information-processing analyses and models will continue to be useful tools for understanding and predicting human behavior in general and in the domain of HCI for many years to come.

*This chapter uses the background and historical material on the subject matter from a chapter that was originally published in **Human-Computer Interaction Handbook: Fundamentals, Evolving Technologies, and Emerging Applications, Third Edition** (Jacko, J., CRC Press (2012), ISBN: 9781439829431).*

REFERENCES

Al-Shargie, F., Tariq, U., Mir, H., Alawar, H., Babiloni, F., & Al-Nashash, H. (2019). Vigilance decrement and enhancement techniques: A review. *Brain Sciences, 9*(8), 178. https://doi.org/10.3390/brainsci9080178.

Anderson, J. R. (1982). Acquisition of cognitive skill. *Psychological Review, 89*(4), 369–406.

Anderson, J. R., Bothell, D., Byrne, M. D., Douglass, S., Lebiere, C., & Qin, Y. (2004). An integrated theory of the mind. *Psychological Review, 111*, 1036–1060.

Anderson, J. R., Fincham, J. M., & Douglass, S. (1999). Practice and retention: A unifying analysis. *Journal of Experimental Psychology: Learning, Memory, and Cognition, 25*, 1120–1136.

Anderson, J. R., Matessa, M., & Lebiere, C. (1997). ACT-R: A theory of higher level cognition and its relation to visual attention. *Human-Computer Interaction, 12*, 439–462.

Attneave, F. (1954). Some informational aspects of visual perception. *Psychological Review, 61*(3), 183–193.

Ayaz, H., & Dehais, F. (2021). Neuroergonomics. In G. Salvendy, & W. Karwowski (Eds.), *Handbook of Human Factors and Ergonomics* (5th ed., pp. 816–841). Hoboken, NJ: John Wiley.

Baddeley, A. (2003). Working memory and language: An overview. *Journal of Communication Disorders, 36*, 189–208.

Baddeley, A. D., & Hitch, G. J. (1974). Working memory. In G. H. Bower (Ed.), *The Psychology of Learning and Motivation* (Vol. 8, pp. 47–89). San Diego, CA: Academic Press.

Baddeley, A. D., Hitch, G. J., & Allen, R. J. (2019). From short-term store to multicomponent working memory: The role of the modal model. *Memory & Cognition, 47*(4), 575–588.

Baddeley, A. D., Thomson, N., & Buchanan, M. (1975). Word length and the structure of short-term memory. *Journal of Verbal Learning and Behavior, 14*, 575–589.

Bahrick, H. P., & Hall, L. K. (2005). The importance of retrieval failures to long-term retention: A metacognitive explanation of the spacing effect. *Journal of Memory and Language, 52*, 566–577.

Benoni, H. (2018). Can automaticity be verified utilizing a perceptual load manipulation? *Psychonomic Bulletin & Review, 25*(6), 2037–2046.

Berlucchi, G. (2006). Inhibition of return: A phenomenon in search of a mechanism and a better name. *Cognitive Neuropsychology, 23*(7), 1065–1074. https://doi.org/10.1080/02643290600588426.

Bertelson, P. (1967). The time course of preparation. *Quarterly Journal of Experimental Psychology, 19*, 272–279.

Biondi, F., Alvarez, I., & Jeong, K. A. (2019). Human-vehicle cooperation in automated driving: A multidisciplinary review and appraisal. *International Journal of Human-Computer Interaction, 35*(11), 932–946.

Boldini, A., Russo, R., & Avons, S. E. (2004). One process is not enough! A speed-accuracy tradeoff study of recognition memory. *Psychonomic Bulletin & Review, 11*, 353–361.

Borys, M., & Plechawska-Wójcik, M. (2017). Eye-tracking metrics in perception and visual attention research. *European Journal of Medical Technologies, 3*, 11–23.

Botvinick, M. M., Braver, T. S., Barch, D. M., Carter, C. S., & Cohen, J. D. (2001). Conflict monitoring and cognitive control. *Psychological Review, 108*(3), 624–652.

Bower, G. H., & Hilgard, E. R. (1981). *Theories of Learning* (5th ed.). Hoboken, NJ: Prentice-Hall.

Broadbent, D. E. (1958). *Perception and Communication*. Oxford: Pergamon Press.

Brown, J. (1958). Some tests of the decay theory pf immediate memory. *Quarterly Journal of Experimental Psychology, 10*, 12–21.

Cao, S., & Liu, Y. (2013). Queueing network-adaptive control of thought rational (QN-ACTR): An integrated cognitive architecture for modelling complex cognitive and multi-task performance. *International Journal of Human Factors Modelling and Simulation, 4*(1), 63–86.

Card, S. K., English, W. K., & Burr, B. J. (1978). Evaluation of the mouse, rate-controlled isometric joystick, step keys, and text keys for text selection on a CRT. *Ergonomics, 21*, 601–613.

Card, S. K., Moran, T. P., & Newell, A. (1983). *The Psychology of Human-Computer Interaction*. Mahwah, NJ: Lawrence Erlbaum Associates.

Chan, S. W., Buddhika, T., Zhang, H., & Nanayakkara, S. (2019). ProspecFit: In situ evaluation of digital prospective memory training for older adults. *Proceedings of the ACM on Interactive, Mobile, Wearable and Ubiquitous Technologies, 3*(3), 1–20.

Chapanis, A., & Schachter, S. (1945). Depth perception through a P-80 canopy and through distorted glass. Memorandum report TSEAL-69S-48N. Dayton, OH: Aero Medical Laboratory.

Chen, Z., & Cowan, N. (2009). Core verbal working-memory capacity: The limit in words retained without covert articulation. *Quarterly Journal of Experimental Psychology, 62*, 1420–1429.

Cherry, E. C. (1953). Some experiments on the recognition of speech, with one and with two ears. *Journal of the Acoustical Society of America, 25*. 975–979.

Chomsky, N. (1956). Three models for the description of language. *IRE Transactions on Information Theory, 2*(3), 113–124.

Colflesh, G. J., & Conway, A. R. (2007). Individual differences in working memory capacity and divided attention in dichotic listening. *Psychonomic Bulletin & Review, 14*(4), 699–703.

Commarford, P. M., Lewis, J. R., Smither, J. A. A., & Gentzler, M. D. (2008). A comparison of broad versus deep auditory menu structures. *Human Factors, 50*(1), 77–89.

Conway, A. R., Cowan, N., & Bunting, M. F. (2001). The cocktail party phenomenon revisited: The importance of working memory capacity. *Psychonomic Bulletin & Review, 8*(2), 331–335.

Conway, A. R., Kane, M. J., Bunting, M. F., Hambrick, D. Z., Wilhelm, O., & Engle, R. W. (2005). Working memory span tasks: A methodological review and user's guide. *Psychonomic Bulletin & Review, 12*(5), 769–786.

Cowan, N. (1999). An embedded-processes model of working memory. *Models of Working Memory: Mechanisms of Active Maintenance and Executive Control, 20*(506), 1013–1019.

Cowan, N. (2001). The magical number 4 in short-term memory: A reconsideration of mental storage capacity. *Behavioral and Brain Sciences, 24*(1), 87–114. https://doi.org/10.1017/S0140525X01003922.

Cowan, N. (2017). The many faces of working memory and short-term storage. *Psychonomic Bulletin & Review, 24*(4), 1158–1170. https://doi.org/10.3758/s13423-016-1191-6.

Craik, F. I. M., & Lockhart, R. S. (1972). Levels of processing: A framework for memory research. *Journal of Verbal Learning and Verbal Behavior, 11*, 671–684.

Craik, F. I. M., & Watkins, M. J. (1973). The role of rehearsal in short-term memory. *Journal of Verbal Learning and Verbal Behavior, 12*, 599–607.

Crump, M. J., Brosowsky, N. P., & Milliken, B. (2017). Reproducing the location-based context-specific proportion congruent effect for frequency unbiased items: A reply to Hutcheon and Spieler (2016). *Quarterly Journal of Experimental Psychology, 70*(9), 1792–1807.

De Jong, R. (1993). Multiple bottlenecks in overlapping task performance. *Journal of Experimental Psychology: Human Perception and Performance, 19*, 965–980.

Deng, C., Cao, S., Wu, C., & Lyu, N. (2019a). Modeling driver take-over reaction time and emergency response time using an integrated cognitive architecture. *Transportation Research Record, 2673*(12), 380–390.

Deng, C., Cao, S., Wu, C., & Lyu, N. (2019b). Predicting drivers' direction sign reading reaction time using an integrated cognitive architecture. *IET Intelligent Transport Systems, 13*(4), 622–627.

Deutsch, J. A., & Deutsch, D. (1963). Attention: Some theoretical considerations. *Psychological Review, 70*, 80–90.

Dillard, M. B., Warm, J. S., Funke, G. J., Nelson, W. T., Finomore, V. S., McClernon, C. K., Eggemeier, F. T., Tripp, L. D., & Funke, M. E. (2019). Vigilance tasks: Unpleasant, mentally demanding, and stressful even when time flies. *Human Factors, 61*(2), 225–242. https://doi.org/10.1177/0018720818796015.

Donders, F. C. (1868/1969). On the speed of mental processes. In W. G. Koster (Ed.), *Acta Psychologica, 30, Attention and Performance II* (pp. 412–431). Amsterdam: North-Holland.

Donkin, C., & Brown, S. D. (2018). Response times and decision-making. In T. T. Wixted (Ed.-in-Chief) & E. J. Wagenmakers (Eds.), *Stevens' Handbook of Experimental Psychology and Cognitive Neuroscience* (Vol. 5, pp. 349–377). Hoboken, NJ: John Wiley.

Eimer, M. (1998). The lateralized readiness potential as an on-line measure of central response activation processes. *Behavior Research Methods, Instruments, & Computers, 30*, 146–156.

Engle, R. W. (2002). Working memory capacity as executive attention. *Current Directions in Psychological Science, 11*(1), 19–23. https://doi.org/10.1111/1467-8721.00160.

Eriksen, B. A., & Eriksen, C. W. (1974). Effects of noise letters upon the identification of a target letter in a nonsearch task. *Perception & Psychophysics, 16*(1), 143–149.

Eriksen, C. W., & Hoffman, J. E. (1973). The extent of processing of noise elements during selective encoding from visual displays. *Perception & Psychophysics, 14*(1), 155–160. https://doi.org/10.3758/BF03198630.

Eriksen, C. W., & St James, J. D. (1986). Visual attention within and around the field of focal attention: A zoom lens model. *Perception & Psychophysics, 40*(4), 225–240.

Evans, N. J., Brown, S. D., Mewhort, D. J. K., & Heathcote, A. (2018). Refining the law of practice. *Psychological Review, 125*(4), 592–605.

Fan, J., & Posner, M. (2004). Human attentional networks. *Psychiatrische Praxis, 31*(S2), 210–214.

Fisher, R. A. (1925). *Statistical Methods for Research Workers*. Edinburgh: Oliver & Boyd.

Fisher, R. A. (1935). *The Design of Experiments*. Edinburgh: Oliver & Boyd.

Fitts, P. M. (1951). Engineering psychology and equipment design. In S. S. Stevens (Ed.) *Handbook of Experimental Psychology* (pp. 1287–1340). Hoboken, NJ: Wiley.

Fitts, P. M. (1954). The information capacity of the human motor system in controlling the amplitude of movement. *Journal of Experimental Psychology, 47*, 381–391.

Fitts, P. M. (1964). Perceptual-motor skill learning. In A. W. Melton (Ed.), *Categories of Human Learning* (pp. 243–285). San Diego, CA: Academic Press. https://doi.org/10.1016/B978-1-4832-3145-7.50016-9.

Fitts, P. M., & Posner, M. I. (1967). *Human Performance*. Pacific Grove, CA: Brooks/Cole.

Fitts, P. M., & Seeger, C. M. (1953). S-R compatibility: Spatial characteristics of stimulus and response codes. *Journal of Experimental Psychology, 46*, 199–210.

Forstmann, B. U., Ratcliff, R., & Wagenmakers, E. J. (2016). Sequential sampling models in cognitive neuroscience: Advantages, applications, and extensions. *Annual Review of Psychology, 67*, 641–666.

Gao, Q., Ma, Z., Gu, Q., Li, J., & Gao, Z. (2023). Working memory capacity for gesture-command associations in gestural interaction. *International Journal of Human-Computer Interaction, 39*(15), 3045–3056.

Garner, W. R., & Hake, H. W. (1951). The amount of information in absolute judgements. *Psychological Review, 58*(6), 446–459. https://doi.org/10.1037/h0054482.

Gazzaniga, M. S., Ivry, R. B., & Mangun, G. R. (2019). *Cognitive Neuroscience: The Biology of the Mind* (5th ed.). New York: W. W. Norton.

Ghuman, A. S., & Martin, A. (2019). Dynamic neural representations: An inferential challenge for fMRI. *Trends in Cognitive Sciences, 23*(7), 534–536.

Gibson, J. J. (1979). *The Ecological Approach to Visual Perception*. Boston, MA: Houghton Mifflin.

Gillan, D. J., Holden, K., Adam, S. Rudisill, M., & Magee, L. (1992). How should Fitts' law be applied to human-computer interaction? *Interacting with Computers, 4*, 291–313.

Gollwitzer, P. M. (1999). Implementation intentions: Strong effects of simple plans. *American Psychologist, 54*, 493–503.

Gonzalez, C., Lerch, F. J., & Lebiere, C. (2003). Instance-based learning in real-time dynamic decision making. *Cognitive Science, 27*, 591–635.

Gratton, G., Coles, M. G. H., & Donchin, E. (1992). Optimizing the use of information: Strategic control of activation of responses. *Journal of Experimental Psychology: General, 121*, 480–506.

Grobelny, J., Karwowski, W., & Drury, C. (2005). Usability of graphical icons in the design of human-computer interfaces. *International Journal of Human-Computer Interaction, 18*, 167–182.

Guest, O., & Martin, A. E. (2021). How computational modeling can force theory building in psychological science. *Perspectives on Psychological Science, 16*(4), 789–802.

Guiard, Y., & Beaudouin-Lafon, M. (2004). Fitts' law 50 years later: Applications and contributions from human-computer interaction. *International Journal of Human-Computer Studies, 61*, 747–750.

Han, T., & Proctor, R. W. (2022). Effects of a neutral warning signal on spatial two-choice reactions. *Quarterly Journal of Experimental Psychology, 75*(4), 754–764.

Healy, A. F., Wohldmann, E. L., & Bourne, L. E., Jr. (2005). The procedural reinstatement principle: Studies on training, retention, and transfer. In A. F. Healy (Ed.), *Experimental Cognitive Psychology and Its Applications* (pp. 59–71). Washington, DC: American Psychological Association.

Heathcote, A., Brown, S., & Mewhort, D. J. K. (2000). The power law repealed: The case for an exponential law of practice. *Psychonomic Bulletin & Review, 7*, 185–207.

Heitz, R. P. (2014). The speed-accuracy tradeoff: History, physiology, methodology, and behavior. *Frontiers in Neuroscience, 8*, 150.

Helmholtz, H. (1968). The origin of the correct interpretation of our sensory impressions. In R. M. Warren, & R. P. Warren (Eds.), *Helmholtz on Perception: Its Physiology and Development* (pp. 249–260). Hoboken, NJ: Wiley. (Original work published 1894).

Henry, F., & Rogers, D. (1960). Increased response latency for complicated movements and a "memory drum" theory of neuromotor reaction. *Research Quarterly of the American Association for Health, Physical Education, and Recreation, 31*(3), 448–458.

Hering, A., Kliegel, M., Rendell, P. G., Craik, F. I., & Rose, N. S. (2018). Prospective memory is a key predictor of functional independence in older adults. *Journal of the International Neuropsychological Society, 24*(6), 640–645. https://doi.org/10.1017/S1355617718000152.

Hick, W. E. (1952). On the rate of gain of information. *Quarterly Journal of Experimental Psychology, 4*, 11–26.

Hintzman, D. L. (1974). Theoretical implications of the spacing effect. In R. L. Solso (Ed.), *Theories of Cognitive Psychology: The Loyola Symposium* (pp. 77–99). Mahwah, NJ: Lawrence Erlbaum Associates.

Hommel, B., & Prinz, W. (Eds.) (1997). *Theoretical Issues in Stimulus-Response Compatibility*. Amsterdam: North-Holland.

Hommel, B., Chapman, C. S., Cisek, P., Neyedli, H. F., Song, J. H., & Welsh, T. N. (2019). No one knows what attention is. *Attention, Perception, & Psychophysics, 81*(7), 2288–2303.

Hommel, B., Proctor, R. W., & Vu, K. P. L. (2004). A feature-integration account of sequential effects in the Simon task. *Psychological Research, 68*(1), 1–17. https://doi.org/10.1007/s00426-003-0132-y.

Hou, G., & Hu, Y. (2021). Designing combinations of pictogram and text Size for icons: Effects of text size, pictogram size, and familiarity on older adults' visual search performance. *Human Factors, 65*(8), 1577–1598. doi:10.1177/00187208211061938.

Huang, K. C. (2008). Effects of computer icons and figure/background area ratios and color combinations on visual search performance on an LCD monitor. *Displays, 29*(3), 237–242.

Hyman, R. (1953). Stimulus information as a determinant of reaction time. *Journal of Experimental Psychology, 45*(3), 188. Kahneman, D. (1973). *Attention and Effort*. Hoboken, NJ: Prentice Hall.

Karpicke, J. D., & Roediger, H. L., III. (2008). The critical importance of retrieval for learning. *Science, 319*, 966–968.

Kiekel, P. A., & Cooke, N. J. (2011). Human factors aspects of team cognition. In K.-P. L. Vu, & R. W. Proctor (Eds.), *Handbook of Human Factors in Web Design* (2nd ed.). Boca Raton, FL: CRC Press.

Kieras, D. E. (2019). Visual search without selective attention: A cognitive architecture account. *Topics in Cognitive Science, 11*(1), 222–239.

Kieras, D. E, & Meyer, D. E. (1997). An overview of the EPIC architecture for cognition and performance with application to human-computer interaction. *Human-Computer Interaction, 12*, 391–438.

Kiesel, A., Steinhauser, M., Wendt, M., Falkenstein, M., Jost, K., Philipp, A. M., et al. (2010). Control and interference in task switching: A review. *Psychological Bulletin, 136*, 849–874.

Klapp, S. T., & Maslovat, D. (2020). Programming of action timing cannot be completed until immediately prior to initiation of the response to be controlled. *Psychonomic Bulletin & Review, 27*(5), 821–832.

Klapp, S. T., Maslovat, D., & Jagacinski, R. J. (2019). The bottleneck of the psychological refractory period effect involves timing of response initiation rather than response selection. *Psychonomic Bulletin & Review, 26*(1), 29–47.

Klapp, S. T., Nelson, J. M., & Jagacinski, R. J. (1998). Can people tap concurrent bimanual rhythms independently? *Journal of Motor Behavior, 30*(4), 301–322.

Klein, R. M., & Shore, D. I. (2000). Relation among modes of visual orienting. In S. Monsell, & J. Driver (Eds.), *Control of Cognitive Processes: Attention and Performance XVIII* (pp. 195–208). Cambridge, MA: MIT Press.

Koch, I., Poljac, E., Müller, H., & Kiesel, A. (2018). Cognitive structure, flexibility, and plasticity in human multitasking-An integrative review of dual-task and task-switching research. *Psychological Bulletin, 144*(6), 557–583.

Kornblum, S., Hasbroucq, T., & Osman, A. (1990). Dimensional overlap: Cognitive basis for stimulus-response compatibility: A model and taxonomy. *Psychological Review, 97*, 253–270.

Krüger, L., Gigerenzer, G., & Morgan, M. S. (1987). *The Probabilistic Revolution* (Vol. 2). Cambridge, MA: MIT Press.

LaBerge, D. (1983). Spatial extent of attention to letters and words. *Journal of Experimental Psychology: Human Perception and Performance, 9*(3), 371–379. https://doi.org/10.1037/0096-1523.9.3.371.

Lachman, R., Lachman, J. L., & Butterfield, E. C. (1979). *Cognitive Psychology and Information Processing: An Introduction*. Mahwah, NJ: Lawrence Erlbaum Associates.

Laird, J. E. (2022). Introduction to soar cognitive architecture. https://arxiv.org/ftp/arxiv/papers/2205/2205.03854.pdf.

Laird, J. E., Lebiere, C., & Rosenbloom, P. S. (2017). A standard model of the mind: Toward a common computational framework across artificial intelligence, cognitive science, neuroscience, and robotics. *Ai Magazine, 38*(4), 13–26.

Lavie, N., Hirst, A., de Fockert, J. W., & Viding, E. (2004). Load theory of selective attention and cognitive control. *Journal of Experimental Psychology: General, 133*, 339–354.

Lee, E., & MacGregor, J. (1985). Minimizing user search time in menu retrieval systems. *Human Factors, 27*, 157–162.

Lee, J. D., Wickens, C. D, Liu, Y. & Boyle, L. N. (2017). *Designing for People: An Introduction to Human Factors Engineering* (3rd ed.). CreateSpace.

Lee, Y.-S. (2008). Levels-of-processing effects on conceptual automatic memory. *European Journal of Cognitive Psychology, 20*, 936–954.

Liesfeld, H. R., & Janczyk, M. (2019). Combining speed and accuracy to control for speed-accuracy trade-offs (?). *Behavior Research Methods, 51*(1), 40–60.

Liu, W., Gori, J., Rioul, O., Beaudouin-Lafon, M., & Guiard, Y. (2020, April). How relevant is Hick's Law for HCI? In *Proceedings of the 2020 CHI Conference on Human Factors in Computing Systems* (pp. 1–11), Honolulu.

Liu, Y. (1997). Queueing network modeling of human performance of concurrent spatial and verbal tasks. *IEEE Transactions on Systems, Man, and Cybernetics-Part A: Systems and Humans, 27*(2), 195–207.

Logan, G. D. (1988). Toward an instance theory of automatization. *Psychological Review, 95*(4), 492–527. https://doi.org/10.1037/0033-295X.95.4.492.

Luck, S. J. (2012). Event-related potentials. In H. Cooper, P. M. Camic, D. L. Long, A. T. Panter, D. Rindskopf, & K. J. Sher (Eds.), *APA Handbook of Research Methods in Psychology: Foundations, Planning, Measures, and Psychometrics* (Vol. 1, pp. 523–546). Washington, DC: American Psychological Association.

Luo, C., & Proctor, R. W. (2022). A diffusion model for the congruency sequence effect. *Psychonomic Bulletin & Review, 29*(6), 2034–2051.

Lytaev, S. (2022). Modern human brain neuroimaging research: Analytical assessment and neurophysiological mechanisms. In *International Conference on Human-Computer Interaction* (pp. 179–185). Berlin, Heidelberg: Springer.

MacKenzie, I. S. (2018). Fitts' law. In K. L. Norman, & J. Kirakowski (Eds.), *Handbook of Human-Computer Interaction* (Vol. 1, pp. 349–370). Hoboken, NJ: Wiley.

MacKenzie, I. S., & Soukoreff, R. W. (2002). Text entry for mobile computing: Models and methods, theory and practice. *Human-Computer Interaction, 17*, 147–198.

Mackworth, N. H. (1948). The breakdown of vigilance during prolonged visual search. *Quarterly Journal of Experimental Psychology, 1*, 6–21.

Mangun, G. R. (Ed.) (2014). *Cognitive Electrophysiology of Attention: Signals of the Mind*. San Diego, CA: Academic Press.

Manly, T., Robertson, I. H., Galloway, M., & Hawkins, K. (1999). The absent mind: Further investigations of sustained attention to response. *Neuropsychologia, 37*(6), 661–670.

Maslovat, D., Klapp, S. T., Jagacinski, R. J., & Franks, I. M. (2014). Control of response timing occurs during the simple reaction time interval but on-line for choice reaction time. *Journal of Experimental Psychology: Human Perception and Performance, 40*(5), 2005–2021.

Matzke, D., Verbruggen, F., & Logan, G. (2018). The stop-signal paradigm. In T. T. Wixted (Ed.-in-Chief) & E. J. Wagenmakers (Eds.), *Stevens' Handbook of Experimental Psychology and Cognitive Neuroscience* (Vol. 5, pp. 383–427). Hoboken, NJ: John Wiley.

Mayr, U., Awh, E., & Laurey, P. (2003). Conflict adaptation effects in the absence of executive control. *Nature Neuroscience, 6*, 450–452.

McCarley, J. S.(2009). Effects of speed-accuracy instructions on oculomotor scanning and target recognition in a simulated baggage X-ray screening task. *Ergonomics, 52*, 325–333.

McCarley, J. S., & Yamani, Y. (2021). Psychometric curves reveal three mechanisms of vigilance decrement. *Psychological Science, 32*(10), 1675–1683. https://doi.org/10.1177/09567976211007559.

McCormick, C. R., Redden, R. S., Hurst, A. J., & Klein, R. M. (2019). On the selection of endogenous and exogenous signals. *Royal Society Open Science, 6*(11), 190134.

McDonald, J. E., Stone, J. D., and Liebelt, L. S. (1983). Searching for items in menus: The effects of organization and type of target. In *Proceedings of the Human Factors Society 27th Annual Meeting* (pp. 289–338). Santa Monica, CA: Human Factors Society.

Mead, L. C. (1948). A program of human engineering. *Personnel Psychology, 1*(3), 303–317. https://doi.org/10.1111/j.1744-6570.1948.tb01310.x.

Meyer, D. E., & Kieras, D. E. (1997). A computational theory of executive cognitive processes and multiple-task performance: Part 2. Accounts of psychological refractory-period phenomena. *Psychological Review, 104*, 749–791.

Meyer, D. E., Abrams, R. A., Kornblum, S., Wright, C. E., & Smith, J. E. K. (1988). Optimality in human motor performance: Ideal control of rapid aimed movements. *Psychological Review, 86*, 340–370.

Miles, J. D., & Proctor, R. W. (2009). Reducing and restoring stimulus-response compatibility effects by decreasing the discriminability of location words. *Acta Psychologica, 130*(1), 95–102.

Miller, G. A. (1956). The magical number seven plus or minus two: Some limits on our capacity for processing information. *Psychological Review, 63*, 81–97.

Miller, M. W., Rietschel, J. C., McDonald, C. G., & Hatfield, B. D. (2011). A novel approach to the physiological measurement of mental workload. *International Journal of Psychophysiology, 80*(1), 75–78. https://doi.org/10.1016/j.ijpsycho.2011.02.003.

Mindell, D. A. (2002). *Between Human and Machine: Feedback, Control, and Computing before Cybernetics.* Baltimore, MD: Johns Hopkins University Press.

Minsky, M. (1961). Steps toward artificial intelligence. *Proceedings of the IRE, 49*(1), 8–30.

Monsell, S. (2003). Task switching. *Trends in Cognitive Sciences, 7*(3), 134–140.

Monsell, S., & Driver, J. (Eds.) (2000). *Control of Cognitive Processes: Attention and Performance XVIII.* Cambridge, MA: MIT Press.

Moray, N. (1959). Attention in dichotic listening: Affective cues and the influence of instructions. *Quarterly Journal of Experimental Psychology, 11*(1), 56–60.

Moray, N. (1993). Designing for attention. In A. D. Baddeley, & L. E. Weiskrantz (Eds.), *Attention: Selection, Awareness, and Control: A Tribute to Donald Broadbent.* Oxford: Oxford University Press.

Morris, C. D., Bransford, J. D., & Franks, J. J. (1977). Levels of processing versus transfer appropriate processing. *Journal of Verbal Learning and Verbal Behavior, 16*, 519–533.

Mueller, S. T., Alam, L., Funke, G. J., Linja, A., Ibne Mamun, T., & Smith, S. L. (2020, December). Examining methods for combining speed and accuracy in a Go/No-Go vigilance task. *Proceedings of the Human Factors and Ergonomics Society Annual Meeting, 64*(1), 1202–1206.

Murphy, G., Groeger, J. A., & Greene, C. M. (2016). Twenty years of load theory: Where are we now, and where should we go next? *Psychonomic Bulletin & Review, 23*(5), 1316–1340.

Nairne, J. S. (2013). Sensory and working memory. In A. F. Healy, & R. W. Proctor (Eds.), *Handbook of Psychology: Experimental Psychology* (Vol. 4), Editor-in-Chief: I. B. Weiner. Hoboken, NJ: Wiley.

Nathan, Y., Rosenblatt, J. D., & Bitan, Y. (2022). Effect of the user input method on response time and accuracy in a binary data labeling task. *International Journal of Human-Computer Interaction*, 1–9. https://doi.org/10.1080/10447318.2022.2128943.

Newell, A. & Rosenbloom, P. S. (1981). Mechanisms of skill acquisition and the law of practice. In J. R. Anderson (Ed.), *Cognitive Skills and Their Acquisition* (pp. 1–55). Mahwah, NJ: Lawrence Erlbaum Associates.

Newell, A. (1990). *Unified Theories of Cognition.* Cambridge, MA: Harvard University Press.

Newell, A., & Simon, H. (1956). The logic theory machine: A complex information processing system. *IRE Transactions on Information Theory, 2*(3), 61–79.

Newell, A., & Simon, H. A. (1972). *Human Problem Solving.* Hoboken, NJ: Prentice Hall.

Neyman, J., & Pearson, E. S. (1933a). On the problem of the most efficient tests of statistical hypotheses. *Philosophical Transactions of the Royal Society of London*, A *231*, 289–337. doi: 10.1098/rsta.1933. 0009.

Neyman, J., & Pearson, E. S. (1933b). The testing of statistical hypotheses in relation to probabilities a priori. *Proceedings of the Cambridge Philosophical Society, 29*, 492–510.

Nijboer, M., Borst, J. P., Van Rijn, H., & Taatgen, N. A. (2016). Driving and multitasking: The good, the bad, and the dangerous. *Frontiers in Psychology, 7*, 1718.

O'Neill, T., McNeese, N., Barron, A., & Schelble, B. (2022). Human-autonomy teaming: A review and analysis of the empirical literature. *Human Factors, 64*(5), 904–938.

Oberauer, K. (2009). Design for a working memory. In B. H. Ross (Ed.), *The Psychology of Learning and Motivation* (Vol. 51, pp. 45–100). San Diego, CA: Academic Press. https://doi.org/10.1016/S0079-7421(09)51002-X.

Oberauer, K. (2019). Working memory and attention: A conceptual analysis and review. *Journal of Cognition, 2*(1): 36, pp. 1–23.

Oberauer, K., & Hein, L. (2012). Attention to information in working memory. *Current Directions in Psychological Science, 21*(3), 164–169. https://doi.org/10.1177/0963721412444727.

Oulasvirta, A., Kärkkäinen, L., & Laarni, J. (2005). Expectations and memory in link search. *Computers in Human Behavior, 21*, 773–789.

Pachella, R. G. (1974). The interpretation of reaction time in information-processing research. In B. H. Kantowitz (Ed), *Human Information Processing: Tutorials in Performance and Cognition* (pp. 41–82). Mahwah, NJ: Lawrence Erlbaum Associates.

Parasuraman, R. (2003). Neuroergonomics: Research and practice. *Theoretical issues in ergonomics science, 4*(1–2), 5–20.

Pashler, H. (1998). *The Psychology of Attention*. Cambridge, MA: MIT Press.

Pattyn, N., Neyt, X., Henderickx, D., & Soetens, E. (2008). Psychophysiological investigation of vigilance decrement: Boredom or cognitive fatigue? *Physiology & Behavior, 93*(1–2), 369–378.

Payne, S. J. (1995). Naïve judgments of stimulus-response compatibility. *Human Factors, 37*, 495–506.

Peterson, L. R., & Peterson, M. J. (1959). Short-term retention of individual verbal items. *Journal of Experimental Psychology, 58*, 193–198.

Posner, M. I. (1980). Orienting of attention. *Quarterly Journal of Experimental Psychology, 32*(1), 3–25. https://doi.org/10.1080/00335558008248231.

Posner, M. I. (1986). Overview. In K. R. Boff, L. Kaufman, & J. P. Thomas (Eds.). *Handbook of Perception and Human Performance vol. II: Cognitive Processes and Performance* (pp. V3–V10). Hoboken, NJ: Wiley.

Posner, M. I. (1992). Attention as a cognitive and neural system. *Current Directions in Psychological Science, 1*(1), 11–14.

Posner, M. I. (Ed.). (2017). *The Psychology of Attention*. London: Routledge.

Posner, M. I., & Cohen, Y. (1984). Components of visual orienting. In H. Bouma, & D. G. Bouwhuis (Eds.) *Attention and Performance X* (pp. 531–556). Mahwah, NJ: Lawrence Erlbaum Associates.

Posner, M. I., Klein, R., Summers, J., & Buggie, S. (1973). On the selection of signals. *Memory & Cognition, 1*, 2–12.

Posner, M. I., Nissen, M. J., & Ogden, W. C. (1978). Attended and unattended processing modes: The role of set for spatial location. In H. L. Pick, Jr., & E. Saltzman (Eds.), *Modes of Perceiving and Processing Information* (pp. 137–157). Mahwah, NJ: Lawrence Erlbaum Associates.

Posner, M. I., Snyder, C. R. R., & Davidson, B. J. (1980). Attention and the detection of signals. *Journal of Experimental Psychology: General, 109*(2), 160–174.

Proctor, R. W., & Reeve, T. G. (Eds.) (1990). *Stimulus-Response Compatibility: An Integrated Perspective*. Amsterdam: North-Holland.

Proctor, R. W., & Schneider, D. W. (2018). Hick's law for choice reaction time: A review. *Quarterly Journal of Experimental Psychology, 71*(6), 1281–1299.

Proctor, R. W., & Van Zandt, T. (2018). *Human Factors in Simple and Complex Systems* (3rd ed.). Boca Raton, FL: CRC Press.

Proctor, R. W., & Vu, K.-P. L. (2006). *Stimulus-Response Compatibility: Data, Theory and Application*. Boca Raton, FL: CRC Press.

Proctor, R. W., & Vu, K.-P. L. (2010a). Cumulative knowledge and progress in human factors. *Annual Review of Psychology, 61*, 623–651.

Proctor, R. W., & Vu, K.-P. L. (2010b). Universal and culture-specific effects of display-control compatibility. *American Journal of Psychology, 123*, 425–435.

Proctor, R. W., & Vu, K.-P. L. (2023). *Attention: Selection and Control in Human Information Processing.* Washington, DC: American Psychological Association.

Proctor, R. W., Vu, K.-P. L., & Pick, D. F. (2005). Aging and response selection in spatial choice tasks. *Human Factors, 47*, 250–270.

Raissi, R., Dimara, E., Berry, J. H., Gray, W. D., & Bailly, G. (2020). Retroactive transfer phenomena in alternating user interfaces. In *Proceedings of the 2020 CHI Conference on Human Factors in Computing Systems* (pp. 1–14), Honolulu, HI.

Ratcliff, R. (1978). A theory of memory retrieval. *Psychological Review, 85*(2), 59–108.

Ratcliff, R., & Smith, P. L. (2004). A comparison of sequential sampling models for two-choice reaction time. *Psychological Review, 111*, 333–367.

Rauschenberger, R., Lin, J. J. W., Zheng, X. S., & Lafleur, C. (2009, October). Subset search for icons of different spatial frequencies. *Proceedings of the Human Factors and Ergonomics Society Annual Meeting, 53*(17), 1101–1105.

Redden, R. S., MacInnes, W. J., & Klein, R. M. (2021). Inhibition of return: An information processing theory of its natures and significance. *Cortex, 135*, 30–48. https://doi.org/10.1016/j.cortex.2020.11.009.

Rezeika, A., Benda, M., Stawicki, P., Gembler, F., Saboor, A., & Volosyak, I. (2018). Brain-computer interface spellers: A review. *Brain Sciences, 8*(4), 57.

Ritter, F. E., Tehranchi, F., & Oury, J. D. (2019). ACT-R: A cognitive architecture for modeling cognition. *Wiley Interdisciplinary Reviews: Cognitive Science, 10*(3), e1488.

Robinson, M. D., & Thomas, L. E. (Eds.). (2021). *Handbook of Embodied Psychology: Thinking, Feeling, and Acting.* Berlin, Heidelberg: Springer Nature.

Röer, J. P., & Cowan, N. (2021). A preregistered replication and extension of the cocktail party phenomenon: One's name captures attention, unexpected words do not. *Journal of Experimental Psychology: Learning, Memory, and Cognition, 47*(2), 234–242. https://doi.org/10.1037/xlm0000874.

Roscoe, S. (2011). Historical overview of human factors and ergonomics. In K.-P. L. Vu, & R. W. Proctor (Eds.), *Handbook of Human Factors in Web Design* (2nd ed.). Boca Raton, FL: CRC Press.

Rouet, J. F., Ros, C., Jégou, G., & Metta, S. (2019). Locating relevant categories in web menus: Effects of menu structure, aging and task complexity. In *Human-Centered Computing* (pp. 547–551). Boca Raton, FL: CRC Press.

Rucci, A. J., & Tweney, R. D. (1980). Analysis of variance and the "second discipline" of scientific psychology: A historical account. *Psychological Bulletin, 87*, 166–184. doi: 10.1037/0033-2909.87.1.166

Saha, S., Mamun, K. A., Ahmed, K., Mostafa, R., Naik, G. R., Darvishi, S., ... & Baumert, M. (2021). Progress in brain computer interface: Challenges and opportunities. *Frontiers in Systems Neuroscience, 15*, 578875.

Saleh, A. M., & Enaizan, O. (2022). COG Tool: An automated cognitive measurement of workload for mobile event logging. *International Journal of Interactive Mobile Technologies, 16*(20), 143–161.

Salthouse, T. A. (2005). From description to explanation in cognitive aging. In R. J. Sternberg, & J. E. Pretz (Eds.), *Cognition & Intelligence: Identifying the Mechanisms of the Mind* (pp. 288–305). Cambridge: Cambridge University Press.

Sanders, A. F. (1998). *Elements of Human Performance.* Mahwah, NJ: Lawrence Erlbaum Associates.

Sartori, G., & Umiltà, C. (2000). How to avoid the fallacies of cognitive subtraction in brain imaging. *Brain and Language, 74*(2), 191–212.

Schachter, S., & Chapanis, A. (1945). Distortion in glass and its effect on depth perception. Memorandum Report No. TSEAL-695-48B. Dayton, OH: Aero Medical Laboratory.

Schmidt, J. R. (2019). Evidence against conflict monitoring and adaptation: An updated review. *Psychonomic Bulletin & Review, 26*(3), 753–771.

Schneider, W., & Shiffrin, R. M. (1977). Controlled and automatic human information processing: I. Detection, search, and attention. *Psychological Review, 84*(1), 1–66. https://doi.org/10.1037/0033-295X.84.1.1.

Schurger, A., Pak, J., & Roskies, A. L. (2021). What is the readiness potential?. *Trends in Cognitive Sciences, 25*(7), 558–570.

Schweickert, R. (1978). A critical path generalization of the additive factor method: Analysis of a Stroop task. *Journal of Mathematical Psychology, 18*, 105–139.

Schweickert, R., & Boruff, B. (1986). Short-term memory capacity: Magic number or magic spell? *Journal of Experimental Psychology: Learning, Memory, and Cognition, 12*, 419–425.

Shannon, C. E. (1948). A mathematical theory of communication. *The Bell System Technical Journal, 27*(3), 379–423.

Shepard, R. N., & Metzler, J. (1971). Mental rotation of three-dimensional objects. *Science, 171*, 701–703.

Shiffrin, R. M., & Schneider, W. (1977). Controlled and automatic human information processing: II. Perceptual learning, automatic attending, and a general theory. *Psychological Review*, *84*(2), 127–190.

Shin, Y. K., & Proctor, R. W. (2018). Evidence for distinct steps in response preparation from a delayed response paradigm. *Acta Psychologica, 191*, 42–51.

Shrestha, N., Botta, C., Barik, T., & Parnin, C. (2020, October). Here we go again: Why is it difficult for developers to learn another programming language?. In IEEE Staff (Ed.), *2020 IEEE/ACM 42nd International Conference on Software Engineering (ICSE)* (pp. 691–701). New York: IEEE.

Simon, J. R. (1990). The effects of an irrelevant directional cue on human information processing. In R. W. Proctor, & T. G. Reeve (Eds.), *Stimulus-Response Compatibility: An Integrated Perspective* (pp. 31–86). Amsterdam: North-Holland.

Simons, D. J., & Ambinder, M. S. (2005). Change blindness: Theory and consequences. *Current Directions in Psychological Science, 14*, 44–48.

Simons, D. J., Boot, W. R., Charness, N., Gathercole, S. E., Chabris, C. F., Hambrick, D. Z., & Stine-Morrow, E. A. (2016). Do "brain-training" programs work? *Psychological Science in the Public Interest, 17*(3), 103–186.

Slamecka, N. J., & Graf, P. (1978). The generation effect: Delineation of a phenomenon. *Journal of Experimental Psychology: Human Learning and Memory, 4*, 592–604.

Smith, T. J., & Henning, R. A. (2005). Cybernetics of augmented cognition as an alternative to information processing. *Proceedings of the 1st International Conference on Augmented Cognition*, Las Vegas, NV.

Solís-Marcos, I., & Kircher, K. (2019). Event-related potentials as indices of mental workload while using an in-vehicle information system. *Cognition Technology and Work, 21*(1), 55–67.

Souza, A. S., Rerko, L., & Oberauer, K. (2015). Refreshing memory traces: Thinking of an item improves retrieval from visual working memory. *Annals of the New York Academy of Sciences, 1339*(1), 20–31.

Sternberg, S. (1969). The discovery of processing stages: Extensions of Donders' method. *Acta Psychologica, 30*, 276–315.

Sternberg, S., Monsell, S, Knoll, R. L., & Wright, C. E. (1978). The latency and duration of rapid movement sequences. In G. E. Stelmach (Ed.), *Information Processing in Motor Control and Learning*. New York: Academic Press.

Strobach, T., Wendt, M., & Janczyk, M. (2018). Multitasking: Executive functioning in dual-task and task switching situations. *Frontiers in Psychology, 9*, 108.

Stroop, J. R. (1935). Studies of interference in serial verbal reactions. *Journal of Experimental Psychology, 18*(6), 643–662.

Talmi, D., Grady, C. L., Goshen-Gottstein, Y., & Moscovitch, M. (2005). Neuroimaging the serial position curve: A test of single-store versus dual-store models. *Psychological Science, 16*, 716–723.

Teichner, W. H., & Krebs, M. J. (1974). Laws of visual choice reaction time. *Psychological Review, 81*, 75–98.

Thomson, D. R., Besner, D., & Smilek, D. (2015). A resource-control account of sustained attention: Evidence from mind-wandering and vigilance paradigms. *Perspectives on Psychological Science, 10*(1), 82–96.

Thropp, J. E., Szalma, J. L., & Hancock, P. A. (2004). Performance operating characteristics for spatial and temporal discriminations: Common or separate capacities? *Proceedings of the Human Factors and Ergonomics Society Annual Meeting, 48*(16), 1880–1884.

Tlauka, M. (2004). Display-control compatibility: The relationship between performance and judgments of performance. *Ergonomics, 47*(3), 281–295.

Tombu, M., & Jolicœur, P. (2005). Testing the predictions of the central capacity sharing model. *Journal of Experimental Psychology: Human Perception and Performance, 31*, 790–802.

Torres-Quesada, M., Lupiáñez, J., Milliken, B., & Funes, M. J. (2014). Gradual proportion congruent effects in the absence of sequential congruent effects. *Acta Psychologica, 149*, 78–86.

Treisman, A. M. (1964). Selective attention in man. *British Medical Bulletin, 20*, 12–16.

Treisman, A. M., & Gelade, G. (1980). A feature-integration theory of attention. *Cognitive Psychology, 12*, 97–136.

Tullis, T. S., Tranquada, F. J., & Siegel, M. J.. (2011). Presentation of information. In K.-P. L. Vu, & R. W. Proctor (Eds.), *Handbook of Human Factors in Web Design* (2nd ed.). Boca Raton, FL: CRC Press.

Tulving, E., & Thomson, D. M. (1973). Encoding specificity and retrieval processes in episodic memory. *Psychological Review, 80*(5), 352–373.

Ulrich, R., Schröter, H., Leuthold, H., & Birngruber, T. (2015). Automatic and controlled stimulus processing in conflict tasks: Superimposed diffusion processes and delta functions. *Cognitive Psychology, 78*, 148–174.

Unsworth, N., Heitz, R. P., Schrock, J. C., & Engle, R. W. (2005). An automated version of the operation span task. *Behavior Research Methods, 37*(3), 498–505.

van de Laar, M. C., van den Wildenberg, W. P. M., van Boxtel, G. J. M., & van der Molen, M. W. (2010). Processing of global and selective stop signals application of Donders' subtraction method to stop-signal task performance. *Experimental Psychology, 57*, 149–159.

van den Bergh, D., Tuerlinckx, F., & Verdonck, S. (2020). DstarM: An R package for analyzing two-choice reaction time data with the D∗M method. *Behavior Research Methods, 52*(2), 521–543.

Verhaeghen, P., & Marcoen, A. (1996). On the mechanisms of plasticity in young and older adults after instruction in the method of loci: Evidence for an amplification model. *Psychology & Aging, 11*, 164–178.

Vu, K.-P. L, & Proctor, R. W. (2003). Naïve and experienced judgments of stimulus-response compatibility: Implications for interface design. *Ergonomics, 46*, 169–187.

Vu, K.-P. L., & Proctor, R. W. (2008). Age differences in response selection for pure and mixed stimulus-response mappings and tasks. *Acta Psychologica, 129*, 49–60.

Vu, K.-P. L., Lachter, J., Battiste, V., & Strybel, T. Z. (2018). Single pilot operations in domestic commercial aviation. *Human Factors, 60*, 775–762.

Vu, K.-P. L., Proctor, R. W., Bhargav-Spanzel, A., Tai, B.-L., Cook, J., & Schultz, E. E. (2007). Improving password security and memorability to protect personal and organizational information. *International Journal of Human-Computer Studies, 65*, 744–757.

Walter, S., & Meier, B. (2014). How important is importance for prospective memory? A review. *Frontiers in Psychology, 5*, 657.

Wammes, J. D., Ralph, B. C., Mills, C., Bosch, N., Duncan, T. L., & Smilek, D. (2019). Disengagement during lectures: Media multitasking and mind wandering in university classrooms. *Computers & Education, 132*, 76–89.

Wang, D., Churchill, E., Maes, P., Fan, X., Shneiderman, B., Shi, Y., & Wang, Q. (2020). From human-human collaboration to Human-AI collaboration: Designing AI systems that can work together with people. In *Extended Abstracts of the 2020 CHI Conference on Human Factors in Computing Systems* (pp. 1–6). Honolulu, HI: ACM.

Warm, J. S., Parasuraman, R., & Matthews, G. (2008). Vigilance requires hard mental work and is stressful. *Human Factors, 50*, 433–441.

Weiner, N. (1948). *Cybernetics or Control and Communication in the Animal and the Machine*. Hoboken, NJ: John Wiley.

Wickelgren, W. A. (1977). Speed-accuracy tradeoff and information processing dynamics. *Acta Psychologica, 41*, 67–85.

Wickens, C. D. (1984). Processing resources in attention. In R. Parasuraman and D. R. Daives (Eds.), *Varieties of Attention* (pp. 63–102). San Diego, CA: Academic Press.

Wolfe, J. M., Cave, K. R., & Franzel, S. L. (1989). Guided search: An alternative to the feature integration model for visual search. *Journal of Experimental Psychology: Human Perception and Performance, 15*(3), 419–433.

Wolfe, J. M., Kosovicheva, A., & Wolfe, B. (2022). Normal blindness: When we look but fail to see. *Trends in Cognitive Sciences, 26*(9), 809–819.

Wolfe, J. M., Palmer, E. M., & Horowitz, T. S. (2010). Reaction time distributions constrain models of visual search. *Vision Research, 50*(14), 1304–1311.

Wood, L. E., & Pratt, J. D. (1987). Pegword mnemonic as an aid to memory in the elderly: A comparison of four age groups. *Educational Gerontology, 13*, 325–339.

Wright, C. E., Marino, V. F., Chubb, C., & Mann, D. (2019). A model of the uncertainty effects in choice reaction time that includes a major contribution from effector selection. *Psychological Review, 126*(4), 550–577.

Wright, J. (2018). The analysis of data and the evidential scope of neuroimaging results. *The British Journal for the Philosophy of Science, 69*, 1179–1203.

Xiong, A., & Proctor, R. W. (2018). Information processing: The language and analytical tools for cognitive psychology in the information age. *Frontiers in Psychology: Cognition, 9*, 1270.

Xiong, A., Proctor, R. W., Yang, W., & Li, N. (2017). Is domain highlighting actually helpful in identifying phishing web pages? *Human Factors, 59*(4), 640–660.

Yablonski, J. (2020). *Laws of UX: Using Psychology to Design Better Products & Services*. Sebastopol, CA: O'Reilly Media.

Young, P. M., Clegg, B. A., & Smith, C. A. P. (2004). Dynamic models of augmented cognition. *International Journal of Human-Computer Interaction, 17*, 259–273.

4 Language and Communication in HCI

Simone Diniz Junqueira Barbosa and Vinícius Carvalho Pereira

4.1 INTRODUCTION

The entry for "language", in A Dictionary of Linguistics and Phonetics (Crystal, 2008), maps out some of the most common uses of this noun, ranging from everyday contexts to its meanings in different academic areas, such as Linguistics, Semiotics, Computer Science, etc. In its most general sense, "language" is a mass noun that involves the human capacity to communicate through signs; in its most specific sense, it is a countable noun that can refer to specific languages (English, Portuguese, American Sign Language, etc.). Besides those meanings, there are others that fall out of the scope of our discussion here, such as those that refer to specific domain areas (scientific language, legal language, medical language, etc.), to the varieties of social groups (children's language, LGBTQIAPN+ language, etc.), or to a person's individual style of communicating (as in Shakespeare's language, Donald Trump's language, Queen Elizabeth II's language, etc.).

Due to the polysemy of the word "language", linguists are very careful in defining what concept of "language" they are using for a certain purpose. In this chapter, to introduce the discussion on language and communication in HCI, we begin with a semiotic definition of language (*a language is a set of signs with defined selective and associative rules used to convey meaning*) and refer to a pair of concepts in Structural Linguistics defined by de Saussure (2005): *langue* and *parole*. Roughly translated into English as "language", a *langue* is an abstract system composed of conventional rules and signs; a *parole* ("speech", in English), on the other hand, is a concrete oral or visual instance of the *langue* a person uses to express their thoughts. Therefore, when one is talking about the language of computer interfaces, it is fundamental to specify whether the intended meaning refers to the abstract system of signs in computer interfaces (*langue*) or to the use of signs in a specific interface (*parole*).

According to de Saussure (2005), Linguistics is a privileged area in the field of *Semiology* - that is what he called the general science of signs, although the term "Semiotics"[1] became much more common in Linguistics, Media Studies and Computer Science decades later. Hence, other sign systems may share characteristics with natural languages, such as the organization into different strata (like form and meaning), sequencing, hierarchy, and combinatory rules. Barthes (1977) used the categories of *langue* and *parole*, as well as other Structural Linguistics constructs, to analyze non-linguistic sign systems, such as fashion, food, automobilism, furniture, cinema, television, advertisement, press, etc. That is the same rationale that grounds the possibility of analyzing computer interfaces as sign systems with specific vocabulary, associative rules, and discursive conventions.

Charles Sanders Peirce (Peirce, 1867–1893 (1992), 1893–1913 (1998)) went beyond de Saussure's signifier-signified pair and defined semiosis as "an action, or influence, which is, or involves, an operation of three subjects, such as a sign, its object, and its interpretant, this tri-relative influence not being in any way resolvable into an action between pairs" (Peirce, 1893–1913 (1998), p. 411). In Peircean Semiotics, the sign is considered as "the origin of the semiosic process of interpretation" (Eco, 1986, p. 1). Peirce's Semiotics addresses signs not only pertaining to natural languages; instead, his theory postulates a broader understanding of Logics, Mathematics, visual codes, etc. as sign systems. That makes Peircean Semiotics particularly useful to discuss language

DOI: 10.1201/9781003495109-4

and communication in HCI, assuming that packing and unpacking messages coded in computer interfaces through buttons, links, text, images, etc. are semiotic processes. More on that will be presented in section "Semiotics: signs, signification, communication, and semiosis".

The next sections of this chapter are structured as follows: in Section 4.1.1, we define language, discuss its main elements, and how each level of linguistic analysis (Phonetics, Phonology, Morphology, Syntax, Semantics, and Pragmatics) has correspondents in the field of HCI. Section 4.2 defines communication and presents the two most relevant communication models for HCI: Shannon and Weaver's Mathematical Theory of Communication, and Roman Jakobson's model for communication and language functions. Section 4.3 covers some of the main topics of Pragmatics, i.e. language in use, relevant to computer interfaces: deixis, the cooperative principle of conversation, speech acts, and conversation analysis. Section 4.4 discusses key elements of Semiotics in HCI, focusing on semiotic engineering, types of signs and communicability. Section 4.5 covers languages in HCI, including design languages, languages at the user interface, and languages for end-user development. Section 4.6 presents the concluding remarks of this chapter and is followed by the references.

4.1.1 Elements of Language

As hypercomplex systems, natural languages are described and analyzed by linguists at different levels, the most common of which are the phonological, the morphological, the syntactic, the semantic, and the pragmatic. The same points of view can be adopted to describe other sign systems, such as computer interfaces, as follows.

In a natural oral language, Phonology is the study of its sound system, comprising the description of individual phonemes (abstract representation of a sound), contrastive features (what differentiates the sound /p/ from the sound /b/, for example), and the systematic phenomena of sound combination (such as the fact that the last sound of third person singular verb forms in English always sound as /s/ if the preceding sound is a voiceless consonant, as in *thinks*, *cuts* or stops, but they sound as /z/ if the preceding sound is a voiced consonant or a vowel, as in *begs*, *goes* or *loves*). A phoneme is the smallest unit of a language that can change meaning, although it does not have a meaning on its own.

In the realm of computer interfaces, the phonological[2] level of analysis can help us understand how certain buttons convey meaning purely through contrastive features. Take, for instance, Figure 4.1, with buttons from Google Docs, which provide the user with different choices for text alignment on the page:

The four possible alignments on the page are visually conveyed by iconic representations that differentiate from one another by contrastive traits: line lengths (short versus long) and distribution in the horizontal axis.

On the other hand, not all differences in sounds are phonologically contrastive in natural oral languages. Take, for example, the variations in the only vowel sound in the word /bath/ in American dialects and in British dialects. The phoneme /æ/ is performed by means of different phones (concrete manifestations of the phoneme) across these linguistic varieties, but these differences entail no change in meaning.

When it comes to computer interfaces, phonetic variations (differences in form with no impact on meaning), can also be spotted if we consider, for instance, similar icons in different systems, such as the button for "refresh" in Google Chrome, Edge and Pale Moon browsers, as can be seen in Figure 4.2. The differences in size, color, or line thickness do not impact any change in the meaning conveyed by these buttons, being mere formal variation.

FIGURE 4.1 Phonological differences in Google Docs buttons.

$$c \quad G \quad e$$

FIGURE 4.2   Phonetic variations of the refresh button across different browsers.

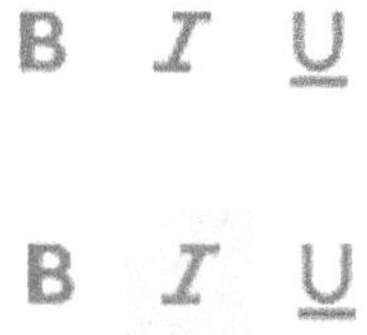

$$\textbf{B} \quad \textit{I} \quad \underline{\textsf{U}}$$

$$\textbf{B} \quad \textit{I} \quad \underline{\textsf{U}}$$

FIGURE 4.3   Morphological differences in Google Docs buttons.

Morphology is the study of word form and structure, focusing on the theoretical construct of morphemes, the minimal distinctive units of grammar (Crystal, 2008). In natural languages, morphology studies word formation and how grammatical categories such as number (singular, plural, etc.), verb aspects (simple, progressive, perfect, etc.), gender (masculine, feminine, etc.) and many others are expressed by morphemes. It is important to notice that those distinctions are not exclusive of a word or two; instead, they go across the system and show, for example, that the progressive aspect in English is regularly expressed by the opposition between -ing and null morpheme as in "doing" vs. "do", "working" vs. "work" etc.

Computer interfaces also convey meaning morphologically, that is, through systemic contrasts of meaning that arise from systemic contrasts of form. When it comes to buttons, the opposite meanings of "selected" and "not selected" may be expressed by opposite uses of color, as in "colored" and "not colored" in Google Docs buttons, as seen in Figure 4.3.

In the Linguistics of natural languages, Syntax refers to the study of the combinatory rules of words within a sentence. Every language has its own constraints on word order (e.g., that subjects often come before verbs in English), subordination (e.g., that adverbs modify verbs, adjectives, and other adverbs, but not nouns in English), agreement (e.g., that adjectives do not take plural inflections in English even if they modify plural nouns, as in "blue cars") and collocation (that a certain word can only take specific prepositions, as in "compare X **to** Y", or "compare X **with** Y", but not "compare X **for** Y" in English.

In computer interfaces, syntactic hierarchies of coordination (structures of the same level) or subordination (structures of different levels, in which there is a main structure and subordinate structures) are often expressed visually by means of alignment in the horizontal or vertical axis. In Figure 4.4, a Gmail screenshot shows that the alignment in the horizontal axis means that Updates, Forums, and Promotions are subordinate categories of Trash, whereas Trash is on the same level as (or coordinate with) Inbox, Starred, Snoozed, Important, Sent, Drafts, Categories, Chats, Scheduled, All Mail, and Spam.

In Linguistics, Semantics can be broadly understood as the study of meaning, which borrows contributions from Structural Linguistics (as in mapping out meaning relations such as synonymy, antonymy, hyperonymy, etc.), Logics (as in truth conditions, and logical consequences of statements), and Psychology (as in highlighting cognitive processes underneath linguistic phenomena, such as metaphors or conceptual blending).

In HCI, a Semantics-oriented approach sheds light on the well-known desktop metaphor (Hartson & Pyla, 2018), in which an image is drawn from a source domain very familiar to users (that of a physical office, with common objects and stationery on a desktop) to help conceptualize a more abstract domain (that of computer processes). These metaphors are both verbal and nonverbal, so that deleted files are moved to a folder called "trash", and its visual representation is often an icon for an office trash can.

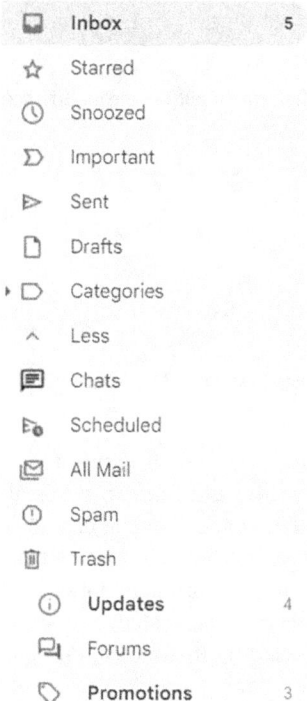

FIGURE 4.4   Syntactic organization of hierarchy in Gmail tabs.

FIGURE 4.5   Explicit and implicit information in a file manager application.

Finally, Pragmatics, the highest level linguistic stratum, is related to language in use in real contexts, especially focusing on the relation between the addresser, the addressee, and the message they get across considering the rules and constraints of different contexts of social interaction (Crystal, 2008). Among other discursive phenomena, Pragmatics studies deixis, the cooperation principle between speakers, the dynamics of conversation, speech acts, and implicatures (tacit information more or less deducible from context or utterances). As Pragmatics is, among the language levels herein presented, the one that is more often nowadays explored in the convergence between Linguistics and HCI, in this section, we will only discuss implicatures in HCI. Other pragmatic phenomena will be analyzed in further detail in the section "Language in use".

Human-human communication is composed of both explicit and implicit meanings. In daily conversations, when one says "My mobile is from the brand X, but it works fine", there are two explicit pieces of information, one in each clause: the mobile is from the brand X; the mobile works fine. However, there is an easily retrievable implicit meaning (a conventional implicature), manifested by the conjunction "but": mobiles from the brand X usually do not work fine. Other kinds of implicit meanings, such as irony, are harder to spot and depend on the previous knowledge of the interlocutor for a successful interpretation.

Likewise, human-computer communication is also made up of explicit and implicit meanings (Liang et al., 2019). Figure 4.5 shows part of a screenshot from a file manager application, where it is explicitly communicated that the user can cut, copy, rename, share, or delete a certain file, but not paste it. It is not said though, but implicitly meant and retrievable due to the conventions of the desktop metaphor, that only after cutting or copying a file will the user be able to paste it somewhere else.

In interfaces that do not use such widely known metaphors or schemes, implicit information may be harder to elicit, as in Figure 4.6, where the user is explicitly told they can click on five different buttons at the bottom of the screen from an app store, but not how different the processes triggered by the "Games" and "Arcade" button will be, since both words refer to the same domain and the icons seem to be perfectly interchangeable. The implicit difference, that the monetizing models are different (pay per purchase in "Games" vs. pay for subscription in "Arcade"), is nowhere to be inferred just from the signs on the interface and depends on the knowledge the user has about how this app store works.

4.2 COMMUNICATION

Although often used interchangeably in our daily lives, "language" and "communication" are different concepts from a technical perspective: whereas a language is a set of signs, communication refers to "the transmission and reception of information (a 'message') between a source and a receiver using a signaling system" (Crystal, 2008). That system can be a language in case of human communication, but not necessarily: human beings also communicate through colors (as in traffic light communication), flags (as in maritime communication), whistles (as in sports), etc., and communication can also refer to messages transmitted between non-human entities like animals, software, or machines in general, which do not use natural languages.

Due to the broadness of the concept of communication, very different theoretical models have been developed to describe it. Below we describe two paradigmatic models that are more often referred to in the realm of HCI: Shannon and Weaver's Mathematical Theory of Communication, and Jakobson's Language Functions Model.

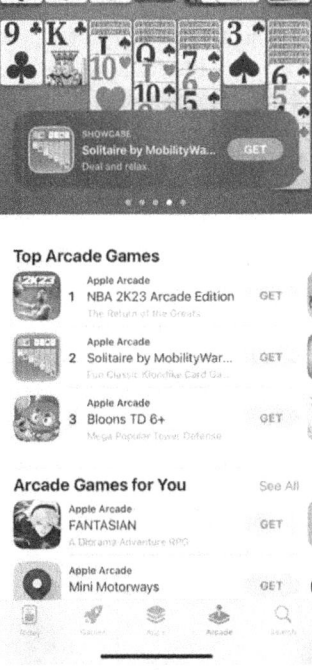

FIGURE 4.6 Explicit and implicit information in an app store.

4.2.1   THEORIES AND MODELS OF COMMUNICATION

4.2.1.1   Shannon and Weaver's Mathematical Theory of Communication

With the goal of improving the efficiency of telecommunication systems, Shannon (1948) defined a mathematical theory of communication involving the transmission, reception, and processing of information. Shannon and Weaver (1949) recognized three levels of communication problems:

- technical, concerned with how accurately symbols can be transmitted (as signals);
- semantic, concerned with how the transmitted symbols convey the desired meaning (i.e., the relation between the receiver's interpreted meaning and the sender's intended meaning); and
- effectiveness, concerned with how effectively the received meaning affects conduct in the desired way.

They focused on the technical problem of signal transmission, as proposed by Shannon (1948) in his mathematical theory of communication. Figure 4.7 presents Shannon's model of a communication system.

The information source defines a message, which is changed (encoded) into a signal by a transmitter through a coding process. The transmitter sends the signal over a physical communication channel to the receiver. In its turn, the receiver changes (decodes) the transmitted signal back into a message and conveys it to the destination. During the signal transmission, unintended interferences and changes to the signal (i.e., noise) can be introduced by different sources, which can result in information loss.

In face-to-face oral communication, the information source is a person's brain, the transmitter is their voice mechanism, the signal is varying sound pressure, the communication channel is the air, the receiver is another person's ear, and the destination is that person's brain. In computer-mediated communication, the transmitter and receiver are usually interactive software, and the communication channel is the internet.

With their mathematical model, Shannon and Weaver were concerned with measuring the amount of information in a message and the capacity of a communication channel (i.e., the rate in which it can convey information), and with defining the characteristics of an efficient coding process. Regarding noise, they were interested in defining its characteristics, in assessing how it affects the accuracy of the message received at the destination, and in finding ways to minimize or even eliminate its undesirable effects, especially by resorting to redundancy. They also investigated the differences between the transmission of a continuous signal (e.g., oral speech, music) and of a discrete signal (e.g., written text).

It is noteworthy that they were not concerned with the meaning of the message. Their notion of information is related to the number of alternative messages that can be transmitted. The amount of information is then defined as the logarithm of the number of available choices. For instance, if there are only two messages, A and B, only one unit of information (bit) is necessary: 0 would signal A, and 1 would signal B.

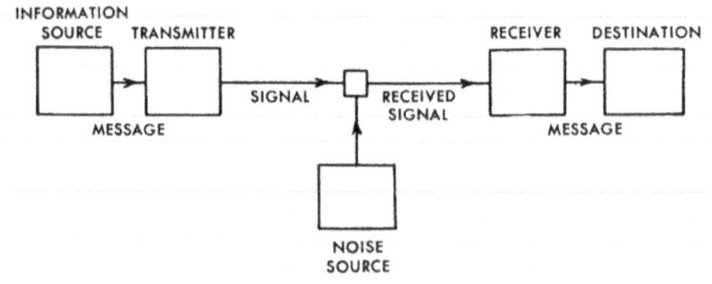

FIGURE 4.7   Schematic diagram of a general communication system (Shannon & Weaver, 1949, p. 34).

4.2.1.2 Jakobson's Language Functions

When asked to relate poetics to linguistics, Roman Jakobson (1960) discussed six functions of language, based on a model of verbal communication that bears some superficial resemblance to Shannon's schema. As depicted in Figure 4.8, the factors in Jakobson's model are addresser, message, addressee, context, csode, and contact (seen in fuller detail, in the context of HCI, in section "Semiotics and HCI: Semiotic Engineering"):

> The addresser sends a message to the addressee. To be operative, the message requires a context referred to, seizable by the addressee, and either verbal or capable of being verbalized; a code fully, or at least partially, common to the addresser and addressee (or, in other words, to the encoder and decoder of the message); and, finally, a contact, a physical channel and psychological connection between the addresser and the addressee, enabling both of them to enter and stay in communication".
>
> *Jakobson (1960, p. 353)*

Jakobson goes on to define six functions of language, one associated with each of the factors in his model. He clarifies that messages usually fulfill more than one function, but that the messages' (verbal) structure depends primarily on a predominant function, as decided by the addresser. The language functions are the following:

- referential function: focused on the context, describing a situation, object, or mental state;
- emotive (or expressive) function: focused on the addresser, direct expression of the speaker's attitude toward what they are speaking about, tending to produce an impression of a certain emotion;
- conative function: oriented to the addressee, as in vocative or imperative sentences;
- phatic function: related to the channel, "serving to establish, to prolong, or to discontinue communication, to check whether the channel works, to attract the attention of the interlocutor or to confirm their continued attention";
- metalingual (or metalinguistic) function: associated with the code, for instance, to check whether the addresser and addressee are using the same code; and
- poetic function: focused on the message "for its own sake", on how something is said.

We can also map Jakobson's language functions onto user interfaces. The referential function is predominant when informing users about the system state or giving feedback of an operation. For instance, after a user asks to save a file, the system can respond with "File saved successfully". The expressive function is often used to persuade the users to taking (or not taking) an action. For instance, when a user asks to unsubscribe from a newsletter, the user interface can say "I am sorry you no longer want to hear from us", speaking for the people or institution responsible for the system. The conative function occurs in direct instructions at the user interface, such as "Provide the location and name of the file". The phatic function occurs in indications of the availability of the system to receive user input, for instance, a "ready" (or "busy") indication when the user is defining or executing a search. The metalinguistic function occurs in explanations of other user interface elements, such as "Asterisks indicate mandatory fields" on a form. Finally, the poetic function can occur in how information is formatted, for instance, data formatted as a list or as a table.

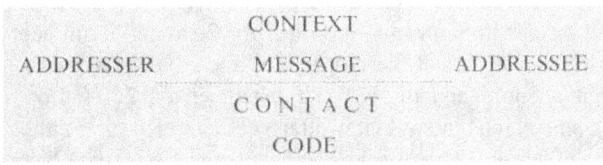

FIGURE 4.8 Jakobson's schema of the factors involved in verbal communication.

Although the examples above use verbal language, we can find examples of language functions in non-verbal user interface elements, such as icons, buttons, etc. For instance, a progress bar has a referential function, indicating how much of an operation has already been completed; a blinking caret may be viewed as having a phatic function, indicating to the user that the channel is ready for their input (conversely, an hourglass indicates that the channel is busy); and a tooltip has a metalinguistic function, explaining the meaning or behavior of the user interface element under the mouse cursor.

Being aware of the six functions allows designers to reflect not only on what they want to convey to users through the user interface but also on how they could do so to be more effective in their communication with users. For that purpose, semiotic engineering (de Souza, 2005a) adapts Jakobson's model to structure the HCI design space, as seen in section "Semiotics and HCI: Semiotic Engineering".

4.3 LANGUAGE IN USE

When talking about communication, the rules of language in use matter as much as (or maybe more than) the rules of language structure (Brisard et al., 2023). Any discussion of communication in HCI (or in any other field) must, therefore, consider not only the structural elements of languages (discussed in the section "Elements of language"), but also how they work

> From the point of view of the users, especially of the choices they make, the constraints they encounter in using language in social interaction, and the effects their use of language has on the other participants in an act of communication.

Crystal (2008)

Language in use is often approached from the perspective of Pragmatics, analyzing phenomena that can be observed both in face-to-face communication and in communication mediated by computer interfaces. In section "Elements of language", we briefly mentioned how implicatures can be observed in human-computer communication. In the following subsections, we will see how other pragmatic phenomena, such as deixis, the cooperative principle of conversation, and speech acts, affect user interaction with computer systems. Then, we will move on to see how the interaction with a system can be understood as a dialogue and also discussed from the perspective of conversation analysis, an emerging subfield of Pragmatics.

4.3.1 Deixis

Deixis is a linguistic term that refers to the indexical relation between more general words, especially adverbs and pronouns, such as "now", "here", and "he/she/they", and specific references to time, space, or person in context (Hanks, 2017; Stapleton, 2017). Differently from nouns, adjectives, and verbs, which have intrinsic context-independent meanings that can be pinned down in dictionaries, such as in the noun "computer", the adjective "computational", and the verb "to compute", adverbs and pronouns often point to temporal, spatial and/or personal references that vary in the context of real-time communication. For example, if the chair of a conference says "I am happy to see you all here today" on November 15th, in New York, the pronouns "I" and "you", and the adverbs "here" and "today" point out to entities different from the ones referred to by a high school teacher who greets their class in San Francisco on April 2 saying "I am happy to see you all here today". That is because the referents of these pronouns and adverbs are indexical.

Deixis is a relevant phenomenon in HCI (Kranstedt et al., 2004; Lopes et al., 2017; Navas Medrano et al., 2018), since signs in computer interfaces can also have either *context-independent* or *context-dependent* meaning. The former comes from signs that have stable and predictable meaning, such as scroll bars, buttons, etc., and would fall within the scope of Semantics.

On the other hand, deictic, or context-dependent meaning, is related to Pragmatics and concerns signs that make indexical references of space, time, and/or person to the moment of communication through the interface. For example, when a system has to inform when a certain file was created, changed, or deleted, it can either use signs that express time in a context-independent way (such as mentioning specific times and dates, e.g. 09:08 a.m., Nov 25, 2022, etc., or through deictic signs, e.g. 5 minutes ago, last year, today, recently, etc.). Other examples of temporal deixis can be seen in Figure 4.9, which shows that the tab "Recents" in Dropbox indexes files through context-dependent time referents (as in "47 minutes ago", or "1 hour ago"), whereas Figure 4.10 shows that other Dropbox tabs index files through context-independent time referents (as in "27/11/2020 8:04 am").

Space is also a category that is subject to deixis. Computer interfaces can use signs that refer to the user's location in context-independent and context-dependent ways, especially in locative mobile applications. For example, when the Uber app recommends recent rides for the "Where to" field, it uses context-independent signs to refer to addresses, such as street/avenue/road names and numbers. On the other hand, deictic signs are used in the "Around you" screen, where the location of the user is pinned on a map, and some elements of the neighborhood and Uber drivers passing by are spotted. That map implies a visual representation of "you are here", and "these are the things in the vicinity of where you are", pieces of information that will change according to where the user is. Figure 4.11 shows both kinds of signs relating to spatial referents in the Uber app.

One or more users can also be the referents of context-independent and context-dependent signs. For example, in popular social media platforms, such as Facebook, Instagram, and Twitter, when different users interact in fields such as feeds, stories, or individual posts, they are often identified by their profile name and photo. The referents of those signs will remain the same no matter who sees them, and where or when they do so. On the other hand, personal deixis consists in references to people (basically, designers, users, or third parties in the realm of HCI) by means of context-dependent signs. Whenever a user is addressed in an interface as an "I" (like when email services have to refer to messages sent by the user as "sent by me", or as a "you" (such as in direct orders and requests, like in "Protect your account with 2-step verification"), the referents of these pronouns will of course vary depending on who is using the system.

4.3.2   THE COOPERATIVE PRINCIPLE OF CONVERSATION

One of the main goals of communication is to have interlocutors act cooperatively and understand one another. In describing how to achieve effective communication, Paul Grice (1975) formulated the cooperative principle of conversation, phrased as follows: "Make your conversational contribution such as is required, at the stage at which it occurs, by the accepted purpose or direction of the talk exchange in which you are engaged" (Grice, 1975, p. 45). This principle, widely studied in the field of Pragmatics, and increasingly present in HCI research (Jacquet et al., 2019; Panfili et al., 2021; Setlur & Tory, 2022), assumes that people engaged in

uber.png
Opened 47 minutes ago - Downloads   ...

WhatsApp Image 2023-04-15 at 18.46.21.jpeg
Added 1 hour ago - Downloads   ...

FIGURE 4.9   Context-dependent time referents in "Recents" Dropbox tab.

Name		Who can access	Modified ↑
PDF	Introduction to Academic Writing.pdf	☆ Only you	27/11/2020 8:04 am by you

FIGURE 4.10   Context-independent time referents in other Dropbox tabs.

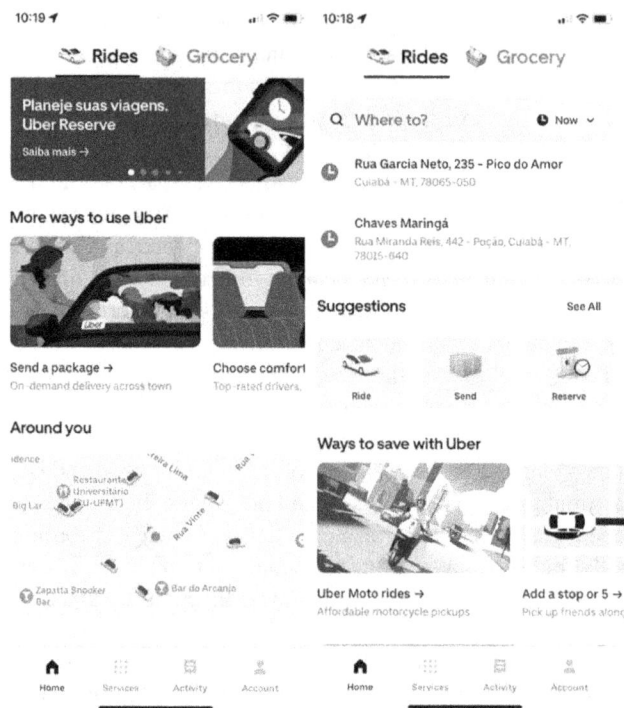

FIGURE 4.11   Context-dependent and context-independent space referents in the Uber app.

communicative interaction will do their best to get their message across, and to do so they will abide by a number of conversational conventions, or *maxims*.

> *Maxim of quantity*: Make your contribution as informative as is required (for the current purposes of the exchange). Do not make your contribution more informative than is required.
> *Maxim of quality*: Do not say what you believe to be false. Do not say that for which you lack adequate evidence.
> *Maxim of relation*: Be relevant.
> *Maxim of manner*: Be perspicuous. Avoid obscurity of expression. Avoid ambiguity. Be brief (avoid unnecessary prolixity). Be orderly.

In addition, Grice proposes that a maxim such as "Be polite" is also normally observed.

In daily conversations between humans, the cooperative principle helps explain why some dialogues with an apparent disjointed structure are still perfectly understandable, such as if speaker A yells "The doorbell", and speaker B replies "I'm in the shower". Both speakers are following the maxim of quantity when they do not give more information than the bare minimum for communicating that the doorbell is ringing, and someone must open the door (meant by speaker A), and that one of these people cannot do so because they are in the shower (speaker B). Likewise, it is the maxim of quality that leads a speaker to ask someone else what time it is and assume there is no reason why the addressee should intentionally lie about the time. Also, when someone is telling a story to a friend, they do not describe all events in minute detail, and the listener would normally assume that all relevant information is included in the report and negligible parts are left out, according to the maxim of relation. However, miscommunication might happen in case the speaker and the listener fail to agree on what parts of the story are relevant or irrelevant. In that case, they need to ask for more or less information, and, if they are co-operating, they are expected to do so politely, according to the maxim of manner.

Communication mediated by computer interfaces is also subject to the cooperation principle. Therefore, according to the maxim of quantity, when a designer is projecting an interface, they must ensure users will be presented with all the necessary signs (buttons, items from menus, checkboxes, etc.) to communicate how they should use the system to meet their goals. Nevertheless, this maxim is sometimes flouted, as when an interface redundantly presents more than one sign/UI element to communicate the same functionality, as can be seen in the Notes app on iPhones, as in Figure 4.12.

The maxim of quality is also paramount in human-computer interaction. On any file manager, when a user requests a file to be deleted, in general they cannot access the computer memory directly to check whether the file was really deleted or not. They simply assume any signs of "file successfully deleted" to be true – which is a problem when the signs on the interface come from malwares that intentionally flout the maxim of quality.

The maxim of relation is directly connected to the fact that the designer must decide on what to show and what to hide about the system considering the perspective of the user on what is relevant to them. This means foregrounding buttons and menus associated with the most popular function-alities and hiding signs that might overwhelm the user with information not needed at interaction time, such as a list of all the processes the system is running, or the memory addresses of all data involved.

Finally, the maxim of manner, concerning briefness and clarity of expression, manifests in the use of recurrent metaphors guiding the design of the interface, such as the desktop metaphor, or conventional icons such as floppy disks or stars to refer to meanings like "save" or "favorite". The fact that metaphors often consist of using concrete and well-known signs to represent abstract ideas makes them a common resource in user interfaces to effectively communicate computer processes, especially in the case of highly conventionalized metaphors. Although a floppy disk is not a faithful representation of the media where data is saved in computers today, replacing this kind of icon every time a new storage media is invented would go against the maxim of manner.

FIGURE 4.12 More than one interface element to trigger the same process on the iPhone Notes app, flouting the maxim of quantity.

Blackwell draws on philosophers who have investigated how metaphorical language helps us extract new abstract concepts, arguing that "a computer user interface might be considered to be a kind of 'literary' description; a representation created to help the user understand the abstract operation and capabilities of the computer" (Blackwell, 2006). He reminds us of common user interface elements that exemplify metaphors and analogies to objects in the real world, such as menus (analogy to a restaurant), buttons (to a control panel), carriage return (to typewriter), scroll (to papyrus), and, of course, the desktop, files and folders, and direct manipulation operations. Through prolonged exposure and use, some metaphors become *dead metaphors*, so ingrained that we lose track of their original references. Because they are so pervasive in our thought processes, metaphors are valuable design tools, not only regarding the user interface language, but also regarding how designers view users, environments, and processes.

4.3.3 SPEECH ACTS

When talking about speech acts, we are discussing how to do things with words, that is, how utterances affect the behavior of interlocutors and their surroundings. Every speech act is a communicative activity (a locutionary act), driven by the speaker's intention (the illocutionary force of the utterance) and a certain effect on listeners (the perlocutionary effect) (Crystal, 2008). As the theory of speech acts relates signs used in communication and the processes they trigger, it is quite useful in HCI (Hanna & Richards, 2019; Hassell, 2007; Morelli et al., 1990).

The most common typology of speech acts in Linguistics (and also in HCI) is the one based on (Searle, 1969), who defined that speakers perform the following five types of speech acts when using language: *declaratives* (declarations that change something in the extraverbal reality, such as in "I now declare you husband and wife"), *assertives* (referential affirmations about the world, as in "The water boiling point at atmospheric pressure is 212 °F"), *expressives* (expressions of the speaker's feelings or attitudes toward the content of their utterance, as in "I'm afraid there is no solution for your problem"), *directives* (commands or requests, as in "Please, give me a minute") and commissives (promises or vows, as in "I'll handle that for you").

The interfaces of Google Inactive Account Manager are explicitly structured as a conversation between the designer and the user, in which the former performs *directive speech acts* (commanding the user to decide when Google should consider the account inactive and to add a phone number, a contact email, and a recovery email), as seen in Figure 4.13.

In response, the user is expected to perform another *directive speech act*, by defining the inactivity time Google should respond to, and also *assertive speech acts*, by informing their phone number, contact email, and recovery email. The interface also conveys *commissive speech acts* performed by the system, when it speaks of actions that will take place in a future point of time, after these utterances are exchanged: "We will only trigger the plan you set up if you haven't used your Google Account for some time", "We'll contact you by SMS on this number", "We'll contact you on this email" and "We'll also contact you on your recovery email".

Declarative speech acts can also be performed by the user in that system in case they decide to name a third party as an heir. That decision changes the extraverbal reality of the user of Google Inactive Account Manager and their heir, who will be able to access some kinds of data from the inactive account, chosen by the owner (Figure 4.14).

4.3.4 CONVERSATION ANALYSIS

If we assume the interaction between designers and users, mediated by interfaces, is communicative, we can also address it as a kind of dialogue. Therefore, conversation analysis, as an area of Pragmatics focused on the sequential structure and coherence of conversations (Crystal, 2008), can inform HCI design (Norman & Thomas, 1991). Conversational interfaces, such as those in chatbots and chat-based artificial intelligence, are the ones that most explicitly try to mimic the dynamics of

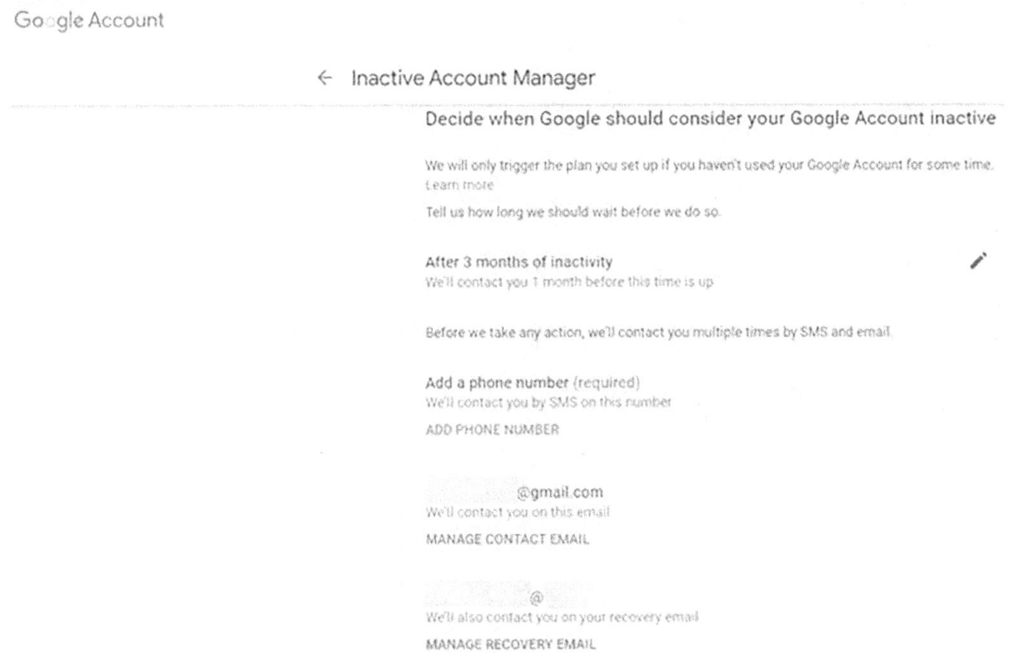

Google Account

← Inactive Account Manager

Decide when Google should consider your Google Account inactive

We will only trigger the plan you set up if you haven't used your Google Account for some time. Learn more

Tell us how long we should wait before we do so.

After 3 months of inactivity
We'll contact you 1 month before this time is up

Before we take any action, we'll contact you multiple times by SMS and email.

Add a phone number (required)
We'll contact you by SMS on this number
ADD PHONE NUMBER

@gmail.com
We'll contact you on this email
MANAGE CONTACT EMAIL

@
We'll also contact you on your recovery email
MANAGE RECOVERY EMAIL

FIGURE 4.13   Directive speech acts in Google Inactive Account Manager.

daily dialogues between human beings, but any graphic user interface works as a dialogue where the system sends messages to the user and the latter responds to the former by inputting data. In the case of collaborative systems, those conversations are increasingly complex, as they will involve more than two participants (user and system), and the communication will be at least three-fold.

Conversations are basically structured as interactions between two or more interlocutors, who take turns coordinately, focusing their cognitive attention on a common task at the same time (Dittmann, 2011). They require not only a shared symbolic system (that is, a natural language, or a set of conventions on computer interfaces), but also a shared understanding of how communication occurs in a certain culture, or a certain domain. A conversation can be carried out between hierarchically symmetrical interlocutors, who have equal power to determine its length, topic, turn-taking dynamics, etc. (like when two friends meet at a bar and talk freely), or between hierarchically asymmetrical interlocutors, one of which is responsible for determining how the dialogue will flow (as in a job interview, where the interviewer decides the whole script and the interviewee is expected to reply to prompts) (Dittmann, 2011). Human-computer interaction is mostly asymmetrical in that sense: although users can make different actions/speech acts when using a system, they are restricted to the interactive possibilities previously defined in the system design.

The basic constitutive element of conversations is the turn, i.e., what an interlocutor says while they have the floor and before they pass it on to the other interlocutor, so that they take turns in their communicative interaction. The general rule of conversation is that one interlocutor speaks at a time, although there may be quick overlaps when one of the interlocutors has to cede (Sacks et al., 1974). In human-computer interaction, the system and the user also take turns in their communication, such as when a user clicks a button and passes the turn to the system, which now must respond with the action triggered by the click. On certain occasions, the user should wait for that response before they take another action; otherwise, the system may crash or freeze. If the system response (i.e. its turn) takes a bit longer, it may display a progress bar, a loading icon, or any similar verbal message informing the user their turn is not over yet, and will only finish when this message disappears.

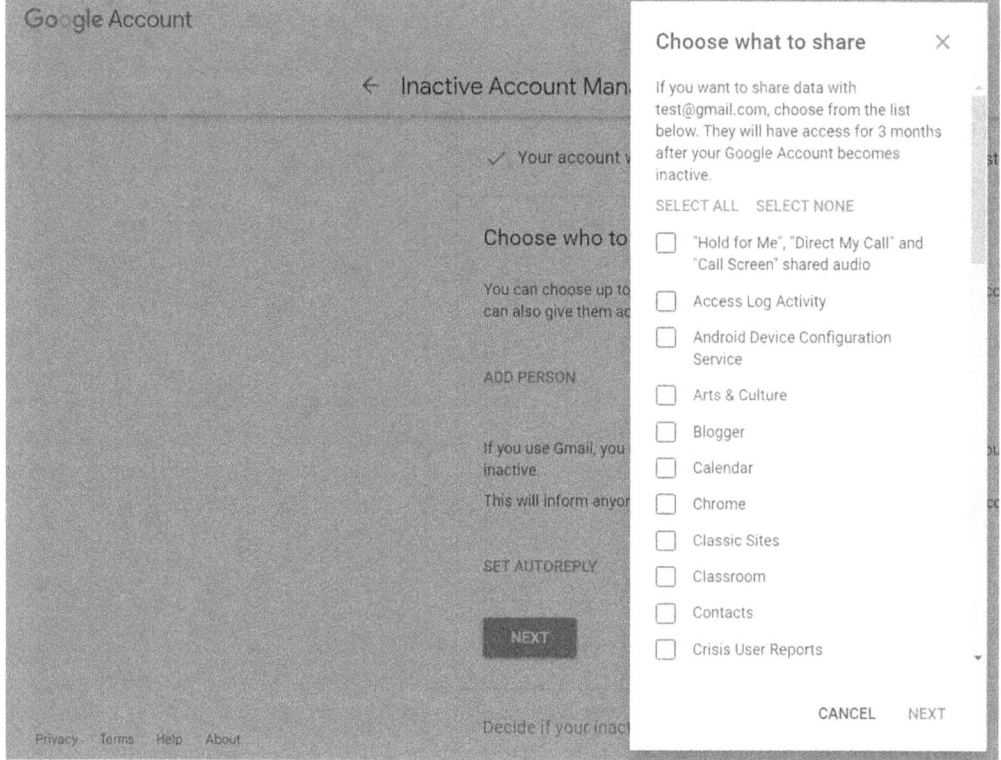

FIGURE 4.14   Declarative speech act in Google Inactive Account Manager.

However, other systems are less restrictive in that sense, allowing the user to click several buttons or menu icons before any of them gets the expected response. In that case, the user is given a longer turn and is not interrupted by the system processing their requests. Systems' usability can be augmented by their capability to interrupt user's turns with autocomplete functions, as in browsers that help users fill out recurrent fields in forms, such as name, address, credit card data, etc. A user only writes the initial values for those fields to have them automatically completed by the system before the user's turn is over.

As another general rule of conversations, long pauses between turns are supposed to be avoided, as they tend to be understood as awkward moments between human interlocutors. In communication between humans and computers, long pauses between turns may have different consequences. For security reasons, some systems automatically log off users who do not start new turns (i.e., who click on no other interactive elements at the interface) within a couple of minutes, thereby ending conversations. Others, such as typical office suites, simply remain waiting for the user to take a new turn, even if it takes long, as they are not constrained by expectations that a conversation should be finished by a certain period of time. Users, on the other hand, tend to have strong expectations of how long the turn of a system should last and get frustrated when those expectations are not met. Some end up dismissing the conversation before their interactive goal is met (shutting down a program taking too long to respond), as one might do in daily conversations where one party holds the floor for too long.

Turn interruption, a common element of human conversations manifested by phrases such as "sorry, but I have to say…", changes in intonation and pace and/or hand gestures, is also present in HCI, when a system prevents a user from finishing their turn by means of a busy or error message, thereby stating that some message cannot be sent yet, or at all. Users can also interrupt the turns of systems, either by clicking on "stop" buttons or by accessing task manager functionalities to kill ongoing processes.

Conversations commonly also feature repairs, that is, attempts to prevent or correct any mis-understanding. Those repairs can be self-initiated, when a speaker uses "I mean" to clarify their previous utterance, or other-initiated, when the interlocutor double-checks their understanding or asks for clarification, as in "What is it that you said?" In HCI, a user can self-initiate a repair of their present turn or a previous one, as when they click any "back" or "undo" button. Likewise, systems can make other-initiated repairs, as when Google search engine cannot find many results for a cer-tain string and suggests an alternative one by asking, "Did you mean X?"

The application of conversational analysis to HCI is limited, though, as human-computer interac-tion is obviously not the same as human-human interaction. For example, most computer interfaces are not sensitive to signs emitted by the user other than through mice, keyboards, or touchscreens. On the other hand, face-to-face conversations between human beings are driven not only by their words but also by very important paralinguistic features, such as speech pace, pitch, tone of voice, body language, facial expressions, etc.

4.4   SEMIOTICS: SIGNS, SIGNIFICATION, COMMUNICATION, AND SEMIOSIS

Semiotics investigates signs, signification, and communication processes (Eco, 1978). Peirce defines

> A *sign* [as] a thing which serves to convey knowledge of some other thing, which it is said to *stand for* or *represent*. This thing is called the *object* of the sign; the idea in the mind that the sign excites, which is a mental sign of the same object, is called an *interpretant* of the sign.
>
> *Peirce (1893–1913 (1998), p. 13)*

In other words, a sign is a representation that stands for something (an object) to someone. In Peircean Semiotics, the term "sign" includes "every picture, diagram, natural cry, pointing fin-ger, wink, knot in one's handkerchief, memory, dream, fancy, concept, indication, token, symp-tom, letter, numeral, word, sentence, chapter, book, library, (...)" (Peirce, 1893–1913 (1998), p. 326). A sign can exist, be connected, or refer to an existing object or event in the physical or psychological (mental) world.

According to this definition, every user interface element produced by a designer is a sign to them, and every user interface element interpreted by a user is a sign to them. According to Peirce's pragmatism, what gives significance to an object or concept are the practical effects resulting from its use or application, that is, no object or concept possesses inherent validity or importance.

For a person's mind to interpret a sign, they rely on social or cultural conventions that define a signification system (Eco, 1978, p. 4). From a design perspective, a signification process occurs every time a designer produces a sign, associating an idea or content (object) to a user interface ele-ment (the expression or representation of that content), following known conventions, using known signs in innovative or creative ways, or even inventing signs (de Souza, 2005a, p. 98).

Semiotics investigates what something means, how it means what it means, and why it means what it means (Danesi & Perron, 1999, p. 46). When designing a user interface, the designer engages not only in a signification process, producing signs in a signification system, but also in a commu-nication process, aiming at communicating their intention to the users. Figure 4.15 illustrates a sign designed to represent an operation for saving the current document, composed of a command button with the label "Save" and an icon of a floppy disk, a legacy icon commonly associated with a save operation. In producing this composite sign, the designer hopes users will interpret it as "Clicking on this button, I am able to save the current document".

Peirce calls the interpretation process *semiosis*, and its result an *interpretant* of the sign (Eco, 1978; Peirce, 1867–1893 (1992), 1893–1913 (1998)). It is noteworthy that semiosis does not necessar-ily stop at the first interpretant. It can itself become a sign which, when interpreted, generates a new interpretant, in a potentially indefinitely long and unpredictable process, in an unbounded semiosic process, as illustrated in Figure 4.16.

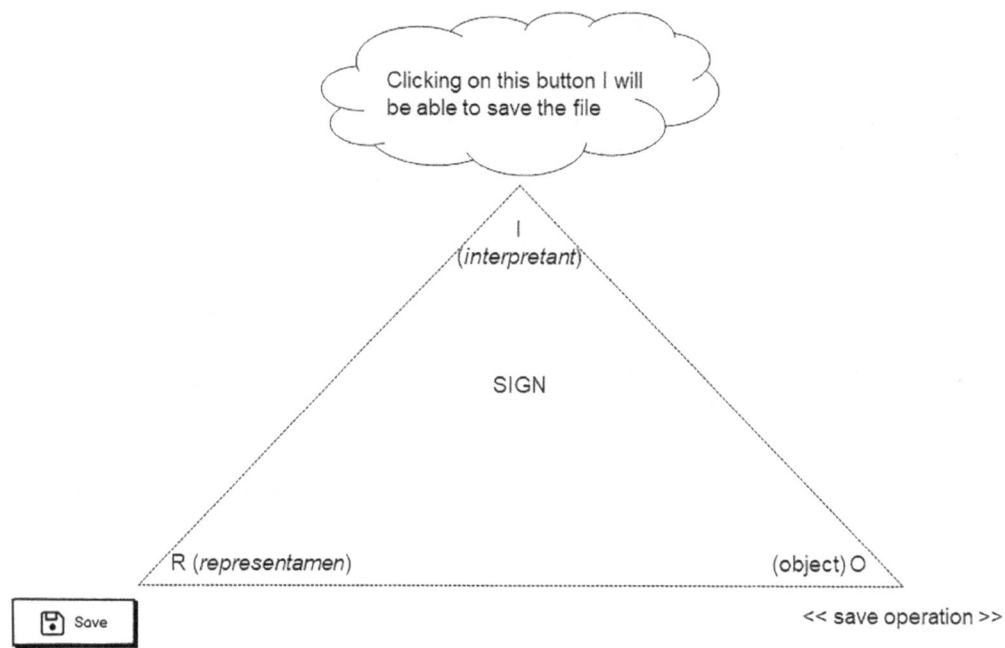

FIGURE 4.15   A sign that represents an operation of saving a document.

Although semiosis is unbounded, in practice it is usually interrupted when (1) the interpreter (e.g., user or designer) is satisfied with the generated interpretant, i.e., with the provisional meaning attributed to the sign; or (2) the interpreter no longer has time or another resource necessary to continue generating new meanings. This interruption is considered temporary, because the interpreter can resume the interpretation (semiosic) process anytime and for any reason, for instance, due to the emergence of a new piece of information, complementing or revising the previously generated meanings.

The temporary, circumstantial, and culture-dependent nature of the interpretant highlights the importance of designers, as producers of user interface signs, to study not only what the users want or need to do but also their activities and experiences, their values and expectations, as well as their culture and the environments in which they will use the interactive computational system being designed. Based on such understanding, the designer devises problem-solving strategies for the users they are aiming to support with the system and designs the user interface language through which the users will communicate with the user interface in order to achieve their goals (de Souza, 2005a,b).

4.4.1 Semiotics and HCI: Semiotic Engineering

Semiotic Engineering is an explanatory theory that characterizes HCI as a particular case of computer-mediated communication (de Souza, 2005a). According to the theory, communication occurs in two levels (Figure 4.17): (1) designer-to-user metacommunication and (2) user-system communication.

Metacommunication is the designer's communication about how users will be able to communicate (i.e., interact) with the system, through its user interface; about why they should do it in a certain way; and to what effect (de Souza, 2005a). According to semiotic engineering, the content of this metacommunication message can be paraphrased as follows (de Souza, 2005a, p. 25):

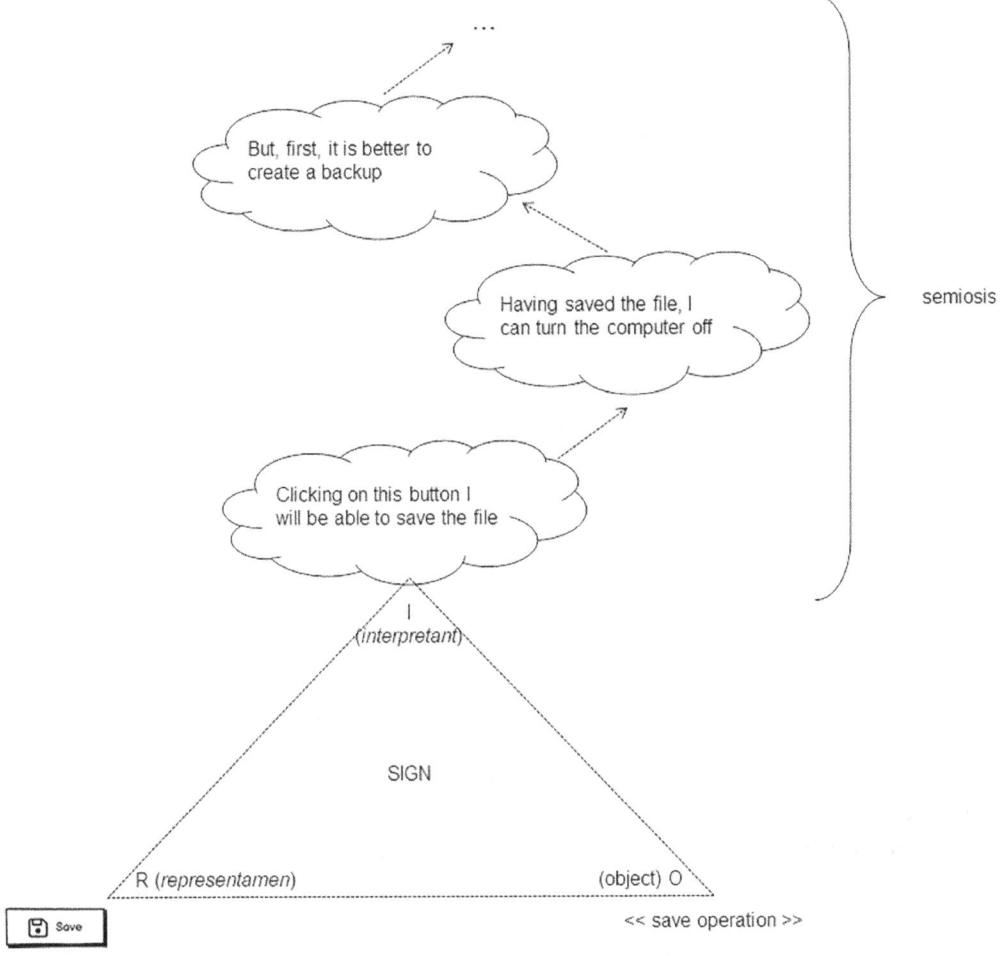

FIGURE 4.16 Example of semiosis, a chain of signs generated through interpretation processes.

> Here is my understanding of who you are, what I've learned you want or need to do, in which preferred ways, and why. This is the system that I have therefore designed for you, and this is the way you can or should use it in order to fulfill a range of purposes that fall within this vision.

Analyzing the metacommunication message, we see that it can be divided into two parts, one resulting from the analysis or research activity, and another related to the design activity (Table 4.1).

During the analysis/research activity, the designer investigates the users, their activities, their environment, the application domain, and competing systems. Based on this analysis, during the design activity, the designer conceives their vision about how to satisfy the users' wants and needs, as well as how the users, their activities, and their environment may or should change with the introduction of the system being designed.

Semiotic Engineering organizes the HCI design space based on Jakobson's (1960) model (Figure 4.18), briefly described in the section "Jakobson's language functions": "A sender transmits a message to a receiver through a channel. The message is expressed in a code and refers to a context. In communication, sender and receiver are alternate roles taken on by interlocutors" (de Souza, 2005a, p. 65).

In order for the communication to be successful, the sender must carefully choose an expression for the content they aim to communicate, using a code that the receiver is able to interpret and, in the case of HCI, that the system is able to process.

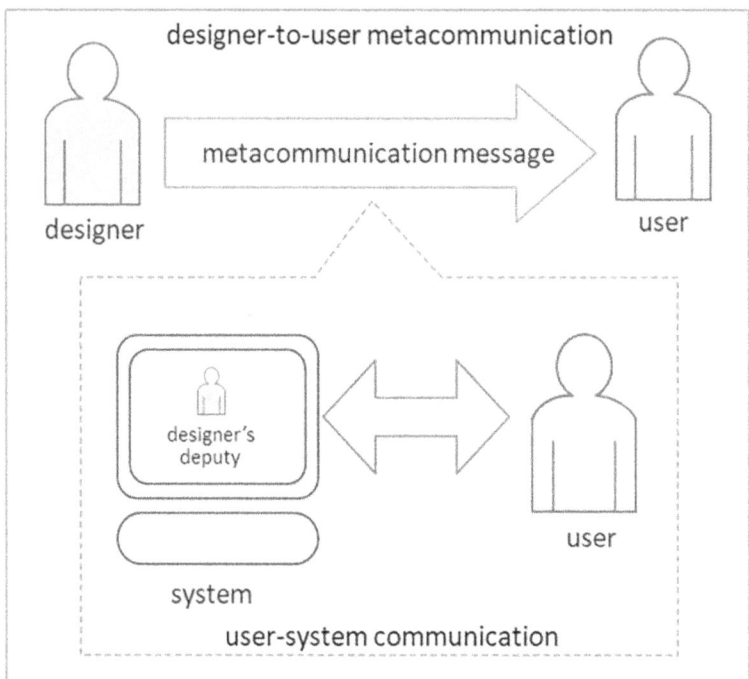

FIGURE 4.17   Designer-to-user metacommunication and user-system communication (interaction).

TABLE 4.1

Metacommunication Message Associated to Analysis and Design Activities

Analysis	Here is my understanding of who you are, what I've learned you want or need to do, in which preferred ways, and why
Design	This is the system that I have therefore designed for you, and this is the way you can or should use it in order to fulfill a range of purposes that fall within this vision

Semiotic Engineering asks the designer to answer the following questions about each element of the model (de Souza, 2005a, p. 87):

- Who is the sender (designer)? What aspects of their own constraints, motivations, beliefs, and preferences should be communicated to the user for the benefit of metacommunication?
- Who is the receiver (users)? What aspects of the user's constraints, motivations, beliefs, and preferences, as conceived by the designer, should be communicated to the actual users to support their role as the system's interlocutor?
- What is the context of communication? What elements of the user's expected spectrum of interactive contexts (psychological, sociocultural, technological, physical, etc.) must be processed by the system's semiotic computations, and how?
- What is the communication code? What computable codes may or must be used to support effective metacommunication? Should this include alternative, complementary, and redundant codes?
- What is the channel? What channels of communication are available for designer-to-user metacommunication and how they should or could be used?
- What is the message? What does the designer want to tell users, and to what effect i.e., what is the designer's communicative intent?

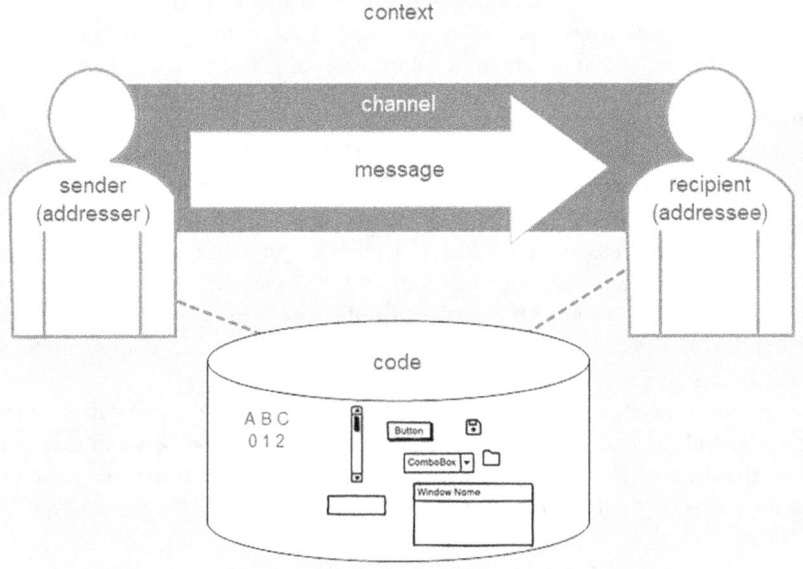

FIGURE 4.18   Semiotic Engineering's HCI design space (de Souza, 2005a, p. 65), based on Jakobson's model (1960).

To express their vision, as expressed by the metacommunication message, the designer must carefully design the user interface language in the form of computational technology, with which the users will interact in order to achieve their (designer's) communicative intent. This means that the users will be able to adequately interpret and benefit from the designed artifact, achieving their goals through the system.

Semiotic Engineering considers the user interface as the *designer's deputy*, that is, a representative of the designer crystallized within the computational system, with which the user will be able to communicate in order to achieve their goals. It is the designer's deputy role to communicate to the users the designer's metacommunication message, which "contains all the meanings and supports all meaning manipulations that the designers have rationally chosen to incorporate in the application in order to have it do what it has been designed to do" (de Souza, 2005a, p. 24). As the designer will not be physically present during the interaction for the users to speak with them, their conversation with the users is always mediated by the designer's deputy, and everything the designer's deputy will need to communicate must be fully planned at design time and implemented in the computational system.

During their interaction with the system, users will have to decode and interpret the metacommunication message, that is, interpret the user interface language – expressed in words, visual elements, interactive elements, behavior, online help, explanations, etc. – and use it to trigger system actions and respond to system behavior, all by means of its user interface.

An important aspect of the metacommunication message is that it should be written from the point of view of the designer, in first person, addressing the user in the second person, and explaining the reasoning behind each design decision. For instance, instead of stating "the system shows the product search results in a list sorted by relevance, but the user may alter the sorting order by value or number of items sold", the metacommunication message would state

> Because I (designer) want to provide the results of the product search that are more relevant to you (user), I will initially sort them taking into account the correspondence between your search query text and the product titles. However, I know that there are situations in which it is more important for you to save money or, instead, look for something more sophisticated (and hence more expensive). Moreover, I believe you like to know what is being sold more often. Therefore, I will allow you to alter the sort order of the results according to these criteria.

By phrasing the metacommunication message in the first person, the designer takes on responsibility for their decisions and runs a lower risk of delegating it to a third party, namely the system, which in fact cannot be held accountable for anything (Barbosa et al., 2021).

4.4.1.1   Types of Signs

When designing the user interface, we use three types of signs that users will interact with when communicating with the system (de Souza et al., 2006; de Souza & Leitão, 2009, p.19): static, dynamic, and metalinguistic signs.

Static signs can be interpreted independent of temporal and causal relations, only by viewing a snapshot of the user interface with the signs present at the user interface at that moment. They often express the state of the system. An example of static sign is the general layout and position of elements at the user interface (e.g., the position of navigation bars, menus and menu items, toolbar buttons, main canvas, fields and buttons on a form, etc.).

Dynamic signs, in contrast, emerge with the interaction and are associated with its temporal and causal aspects. Examples of dynamic signs are the causal association between the user clicking on a menu item and the system presenting them with a dialog box for further input, and the causal association between the user filling in one or more text fields and the system enabling a button for submitting a form.

Metalinguistic signs explain the meaning of other signs. Designers use them to explicitly communicate to users how they can interact with the system and to what effect. Examples of metalinguistic signs are instructions, warning and error messages, and tooltips.

Designers use static signs to engage users, prompting them to anticipate what will happen when users interact with them, while dynamic signs will either confirm or disconfirm what users have anticipated (de Souza & Leitão, 2009, p. 19). Metalinguistic signs tend to be used for disambiguating the other signs or explaining to users why and how to interact with them. Other HCI theories grounded on Semiotics may resort to different classifications of signs, some of which are based on the classic Peircean model of icons, indexes, and symbols, respectively grounded on the phenomena of firstness, secondness, and thirdness (de Souza et al., 2010).

4.4.1.2   Communicability

Semiotic Engineering defines an HCI quality criterion named communicability (de Souza, 2005b). The communicability of a system is its capacity to communicate to users the designers' intentions and the design logic, so that the users will attain the knowledge necessary to interact with the system.

The design logic incorporates all the decisions made throughout the design process, about: to whom the system is designed, which purposes it serves, the advantages of using it, how it works, and its general interaction principles (de Souza, 2005a; de Souza & Leitão, 2009; Prates et al., 2000). The better the system conveys the metacommunication message to its users (i.e., the higher the communicability), the better are the users' chances of using it creatively, productively, and efficiently. By understanding the design logic, users will be better equipped to take advantage of the technology and to follow problem-solving strategies that are adequate for each usage situation. For instance, we do not need to know how the mechanisms of formatting styles or automatic numbering in a text editor work, but having this knowledge we may make better use of those mechanisms, making fewer mistakes and working more efficiently.

An example of low communicability may be observed in an operation for copying multiple files, according to the following scenario:

> John decided to copy multiple files to an external disc, but gave up when, after a few files had been copied, he realized that there were still very large files to go and the operation would take too long.
> Examining the screenshots in Figure 4.19, before and after John asked to cancel the copy operation, what will the resulting state of the external disc be, if he decides to confirm the cancelation of the copy operation at this moment?

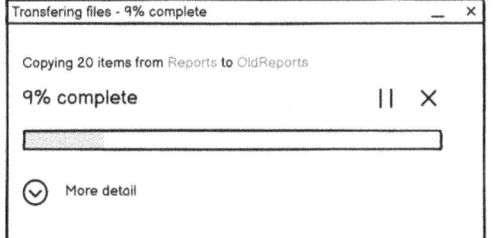
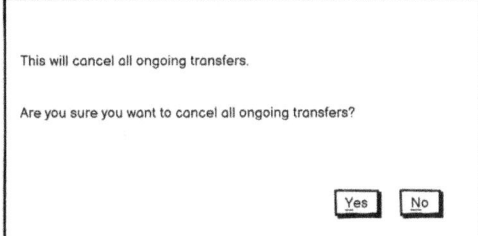

FIGURE 4.19   Before and after asking to cancel an operation for copying multiple files.

Some possible results are:

1. The operation is canceled immediately, and everything that was already copied, including the incomplete fragment of the file that is being copied at the moment, will be at the external disc;
2. The operation is canceled immediately, and everything that was already fully copied, excluding the incomplete fragment of the file that is being copied at the moment, will be at the external disc;
3. The operation will only be canceled after concluding the copy of the current file, and the external disc will contain everything that was copied entirely, including the current file;
4. The operation is canceled immediately, and none of the files that were or would be copied will be on the external disc. The files that had been copied (entirely or partially) will be removed from the external disc.

Notice that, independently of the correct answer, the only way John will know what the result would be is to proceed with canceling the operation and verify its result. The logic implemented in the future was not adequately communicated through its user interface.

In this and other cases of low communicability, if users do not understand the design logic, the interaction often becomes a process of trial and error, tedious, inefficient, or even risky. When a user interface has low communicability, the user may not always be able to anticipate the consequences of their actions, which makes them hesitate, waste time verifying the effect of those actions, and even give up exploring the user interface to increase their knowledge about the system and attain mastery in its use.

When it comes to broader strategies of system use to achieve a range of goals, failures to communicate the designer's intent to the user become even more significant. The designer needs to not only produce a coherent design logic, but also convey it adequately to users during their interactions with the system, introducing that design logic so that users will be able to not only extract value from effective usage strategies conceived by the designer, but also engage in innovative uses that take advantage of a deep understanding of the design logic (de Souza, 2005b).

In addition to explicit information about the behavior of the user interface elements (often through metalinguistic signs), designers can use analogies to increase the communicability of a system. An analogy allows the user to formulate hypotheses about the interaction with a system based on their previous interactions with existing and sometimes similar systems. However, designers must choose the analogies carefully so that the users' hypotheses about how to interact with the system are compatible with the designer's intended use, and communicate the limits of the analogy clearly so that users will not formulate hypotheses that will lead them to error or ineffective interactions.

A successful example of analogy is the use of a "play" button inspired by a media player to start a process such as backup or restore. However, including an "eject" button in the same system to exit the program could be considered as stretching the analogy too far, resulting in hesitation (e.g., for fear that a backup will be deleted), dissatisfaction (e.g., for not understanding the meaning of the interface element), or even errors (e.g., prematurely interrupting a restore operation).

It is important to note that communicability does not replace or supersede other quality criteria in HCI: the designer should always strive to remove barriers that may impede the user's interaction (accessibility) and to make interaction effortless and efficient (usability), as well as pleasing (user experience). In general, though, if a user can understand how the system works because the designer has adequately expressed their intent through the user interface (communicability), it becomes easier for the user to learn how to use it effectively (usability).

4.5 LANGUAGES IN HCI

From an HCI design perspective, language is used for defining the digital artifacts at different levels of abstractions, from defining general "use-oriented qualities" (Löwgren & Stolterman, 2007) up to fully modeling and specifying how the interaction may take place at usage time (Barbosa & de Paula, 2003; Hartson & Pyla, 2018).

At design time, design and developers use design and specification languages to conceive and specify the user interface and interaction mechanisms through which the users will be able to achieve their goals. Designers can be viewed as specifying the system's user interface language, through which they express the state of the system, what the user can or should do, and how users will be able to express their intentions in order to achieve their goals.

From a usage perspective, language is used as a means of communication between users and an interactive system (through its user interface) or between people through an interactive system (de Souza, 2005a). At interaction time, users interact with the interface, expressing themselves in the user interface language, to achieve their goals. Acknowledging that not all user goals can be anticipated at design time, in addition to the user interface language, some designers also provide scripting or end-user programming languages, through which end users can extend or adjust the software behavior to automate tasks and better match their goals.

4.5.1 DESIGN LANGUAGES

The most widely used language in HCI design is arguably natural language, especially for narrating scenarios (Carroll, 2000) and describing personas (Nielsen, 2019), supporting designers in representing specific people, situations, problems, and solutions to particular projects. Natural language is also used for describing proven solutions for common problems, in the form of design pattern languages (Borchers, 2000; Pan & Stolterman, 2013; Tidwell, 2011), coupled with images of the user interface and usage situations.

A number of design and modeling languages have been proposed over the years as attempts to bring more structure to HCI design, to ease the understanding of an application domain, record design rationale and the results of design discussions, and to promote consistency within applications and of applications and their underlying conceptual model. Languages and models support designers in conceiving and specifying various aspects of HCI design at different levels of abstraction (Hoover et al., 1991). Some examples are languages that allow designers to represent mental models (Young, 2008), conceptual models (Hartson & Pyla, 2018), task models (Annett, 2003; Diaper & Stanton, 2003; Kieras, 2004; Paternò, 2000), dialogue and interaction models (Barbosa & de Paula, 2003; Lao et al., 2009), information architecture and navigation models (Garrett, 2010), and user interface specifications (OASIS, 2009; UsiXML, 2014). Less structured notations have also been proposed to support HCI design in either verbal or visual form, such as user journeys (Lichaw, 2016) and empathy maps (Ferreira et al., 2015; Gray, 2018).

According to Löwgren and Stolterman (2007, p. 101), a core element of interaction design ability is "the development of a sense of quality and a language for articulating use-oriented qualities". In their view, design methods also serve as languages for planning and coordinating purposes, providing a common ground for actors in the design process to clearly communicate with one another.

4.5.2 Languages at the User Interface

User interface languages can be graphical, verbal (written or oral), tangible, gestural, etc., increasingly blurring the frontiers between the physical and the digital worlds (Sharp et al., 2019; Shneiderman et al., 2016). Natural languages have a special standing as and within user interface languages. While in the past they were used mostly in fixed or template form to express the system state and provide instructions to users, users are increasingly able to interact with the system by expressing their intentions in natural language, to be interpreted by the system through conversational agents with various degrees of sophistication.

The content of the user interface is expressed by combining signs from the repertoire of the user interface language the designer chooses or creates. The designer then adopts a certain language style to express the intended content with their audience in mind. In doing so, the designer is not only designing *for* their audience but also designing *their* audience (Bell, 1984; Portmann, 2022). However, differently from natural conversations, the user interface language does not usually accommodate the social variation of the users that may interact with the system through expression and stylistic variations. Adaptive user interfaces, which might change the expression and style of the user interface language to better suit the users, are still an exception and, more often than not, focus on task efficiency rather than expression and style variations.

The crystallization of the designer's vision into immutable user interfaces brings greater responsibility to them, as they should carefully design the user interface language with an explicit intent to avoid excluding certain social groups. To help designers face this challenge, Burnett et al. (2016) created the GenderMag method to evaluate software's gender inclusiveness and, more recently, the InclusiveMag design meta-method, which supports inclusivity researchers and software practitioners to design with more diversity dimensions in mind (Mendez et al., 2019).

4.5.3 Languages for End-User Development

A major goal of HCI is to empower users to achieve their goals. Regardless of how well designers and developers create software tailored to specific users, users are too diverse and their goals and environments are ever changing. To face this challenge, the area of end-user development (EUD) emerged (Barricelli et al., 2019; Lieberman et al., 2006), encompassing efforts ranging from tailoring and customizing software to end-user programming (EUP) (Nardi, 1993) and end-user software engineering (EUSE) (M. M. Burnett & Scaffidi, 2013).

When EUD involves writing code in an end-user programming language (EUPL) – either a scripting language or a full-fledged programming language – designers should clearly establish and communicate to end users the correspondences between the EUPL and the user interface language (UIL). Designers should ensure that the UIL is an *interpretive abstraction* of the EUPL, i.e., not only should end users be able to fully understand the UIL without having any knowledge of the EUPL, but the semantics of the UIL should be fully describable by the EUPL (de Souza et al., 2001). Moreover, designers should ensure the UIL and EUPL are pragmatically coupled, so that any instance of a well-formed EUPL text can always be translated to a valid combination of user interface signs at the extended UIL; in other words, the UIL and the EUPL should be *semiotically continuous* (de Souza et al., 2001).

Moving away from artificial languages, Lieberman and Liu (2006) argued that end users should be able to program computers using natural language. They discussed the feasibility of that endeavor, arguing that, differently from open-ended natural language interpretation and generation, when applied to EUD, natural language is constrained by the underlying semantic model of the program, which provides clear referents that facilitate parsing the texts provided by the end user. However, they point out some relevant challenges of their endeavor: the underlying semantic model is not static, so the language must be evolvable; users may not specify their intent fully and grammatically, nor do they think linearly and orderly at a homogeneous level of abstraction, so the interpretation mechanisms should be flexible and tolerant; and users are diverse, so the system should adapt its parsing and dialogue strategies accordingly.

Although we usually think of end-user developers as adult professionals, Monteiro et al. (2015) investigated how middle-school children engaged in EUD activities as part of computational thinking acquisition and as a means of self-expression in first-person digital discourse. In line with semiotic engineering, they define digital discourse as "the intentional expression of personally elaborated meanings in the form of a computer program whose output delivers such discourse and enables its intended effects" (Monteiro et al., 2015). The children produced games and the accompanying learning support material for their games. More distinctively, they used the SideTalk browser extension to produce interactive dialogues with which users could interact with the application as if they were talking with its designer, gaining a better understanding of the designer's vision and effectively accomplishing their goals while doing so.

4.6 CONCLUDING REMARKS AND THE ROAD AHEAD

Many languages and linguistic/communicative models are involved in HCI in several different activities and stages, ranging from languages that users actually engage with (i.e., user interface languages and end-user programming/development languages) to languages designers and developers use when conceiving, specifying, and implementing interactive software.

All these languages mediate communication between different interlocutors. At interaction time, individual users engage in user-system communication, which, from a semiotic engineering perspective, means they are conversing with the designer-to-user metacommunication message. Communication may also take place between users through the system, as in computer-mediated communication. In the case of end-user development, users engage in communication about the system, when extending it through scripting languages and other EUD mechanisms, such as macros and programming by demonstration, to name a few.

At design time, designers have a variety of languages at their disposal, both to support their own reflection on the artifact they are creating (Schon, 1984) and to support the team members' communications with one another (Brown, 2010). Finally, at implementation time, developers make use of programming languages and APIs, which influence them in terms of what and how they can express themselves and may even limit or obfuscate what they can convey (de Souza et al., 2016). As Marino eloquently demonstrated, "it is not enough to understand what code does without fully considering what it means" (Marino, 2020).

As HCI researchers and practitioners, we should be involved in the design not only of the final product but also of the languages at these various levels of abstraction, focusing on the meanings we are creating and allowing to be created and encoded into digital technologies, lest we abdicate our responsibility toward the users and social groups we should serve. To do so, borrowing concepts, theories, and methods from other areas, such as Linguistics, Communication and Media Studies, and Semiotics, helps us better understand interaction as communication from an interdisciplinary perspective, and HCI as an advantageous point of view to understand communication with computer systems, or mediated by them.

Going forward, HCI designers need to face the challenge of not only avoiding communication breakdowns but also actively engaging in communicating to the users their design vision and software meanings so they can understand those meanings and express themselves to achieve their goals through the system. However, as not all goals and uses can be anticipated, HCI designers need to deal with a significant amount of *external* uncertainty and design not only user interface languages for systems' use but also end-user development languages for systems' extensibility.

In addition, there are *internal* sources of uncertainty that designers need to deal with, i.e., uncertainties about the system behavior. In today's world, with the increasing development and adoption of opaque technologies, such as applications that make use of deep learning models, the challenge of conveying the system meanings and behaviors has acquired an unprecedented level of complexity, as the underlying semantics are inscrutable even for the systems' developers. As those models become further embedded into systems that support a wide range of activities and decisions, designers need to communicate not only known and established meanings but also their own uncertainties and lack of knowledge about the system's emergent properties and behaviors.

Future research opportunities, therefore, include investigating a wider range of communicative strategies and languages for interacting with and adapting opaque systems, while maintaining their high performance. This entails a shift from designing *for* to designing *around* system behavior.

NOTES

1 Both "Semiology" and "Semiotics" derive from the Greek word *sēmeîon,* meaning sign, token or mark.
2 Although the terms "Phonology" and "Phonetics" were associated only with sound components of oral languages in the beginning of Linguistics, they are nowadays also used to describe formal contrastive elements of systems that are not sound-based, such as sign languages, computer system interfaces, etc.

REFERENCES

Annett, J. (2003). Hierarchical task analysis. In D. Diaper & N. Stanton (Eds.), *The Handbook of Task Analysis for Human-Computer Interaction* (pp. 67–82). Mahwah, NJ: Lawrence Erlbaum.

Barbosa, S. D. J., & de Paula, M. G. (2003). Designing and evaluating interaction as conversation: A modeling language based on semiotic engineering. In J. A. Jorge, N. Jardim Nunes, & J. Falcãoe Cunha (Eds.), *Interactive Systems. Design, Specification, and Verification, DSV-IS* 2003 (pp. 16–33). Berlin, Heidelberg: Springer. https://doi.org/10.1007/978-3-540-39929-2_2.

Barbosa, S. D. J., Barbosa, G. D. J., de Souza, C. S., & Leitão, C. F. (2021). A semiotics-based epistemic tool to reason about ethical issues in digital technology design and development. *Proceedings of the 2021 ACM Conference on Fairness, Accountability, and Transparency* (pp. 363–374). https://doi.org/10.1145/3442188.3445900.

Barricelli, B. R., Cassano, F., Fogli, D., & Piccinno, A. (2019). End-user development, end-user programming and end-user software engineering: A systematic mapping study. *Journal of Systems and Software, 149,* 101–137. https://doi.org/10.1016/j.jss.2018.11.041.

Barthes, R. (1977). *Elements of Semiology.* New York: Hill and Wang.

Bell, A. (1984). Language style as audience design. *Language in Society, 13*(2), 145–204. https://doi.org/10.1017/S004740450001037X.

Blackwell, A. F. (2006). The reification of metaphor as a design tool. *ACM Transactions on Computer-Human Interaction, 13*(4), 490–530. https://doi.org/10.1145/1188816.1188820.

Borchers, J. O. (2000). A pattern approach to interaction design. *Proceedings of the 3rd Conference on Designing Interactive Systems: Processes, Practices, Methods, and Techniques* (pp. 369–378). https://doi.org/10.1145/347642.347795.

Brisard, F., Gras, P., D'hondt, S., & Vandenbroucke, M. (Eds.) (2023). *Handbook of Pragmatics.* John Benjamins Publishing Company. Retrieved April 15, 2023, from https://benjamins.com/catalog/hop.

Brown, D. (2010). *Communicating Design: Developing Web Site Documentation for Design and Planning* (2nd edition). Indianapolis, IN: New Riders Publishing. https://www.amazon.com/Communicating-Design-Developing-Documentation-Planning/dp/0321712463.

Burnett, M. M., & Scaffidi, C. (2013). End-user development. In C. Ghaoui (Ed.), *The Encyclopedia of Human-Computer Interaction* (2nd edition). Interaction Design Foundation. https://www.interaction-design.org/literature/book/the-encyclopedia-of-human-computer-interaction-2nd-ed/end-user-development.

Burnett, M., Stumpf, S., Macbeth, J., Makri, S., Beckwith, L., Kwan, I., Peters, A., & Jernigan, W. (2016). GenderMag: A method for evaluating software's gender inclusiveness. *Interacting with Computers, 28*(6), 760–787. https://doi.org/10.1093/iwc/iwv046.

Carroll, J. M. (2000). *Making Use: Scenario-Based Design of Human-Computer Interactions* (1st edition). Cambridge, MA: MIT Press.

Crystal, D. (2008). *A Dictionary of Linguistics and Phonetics* (6th edition). Oxford: Blackwell Publishing.

Danesi, M., & Perron, P. (1999). *Analyzing Cultures: An Introduction and Handbook.* Bloomington, IN: Indiana University Press.

de Saussure, F. (2005). *Course in General Linguistics (Nachdr.).* New York: McGraw-Hill.

de Souza, C. S. (2005a). *The Semiotic Engineering of Human-Computer Interaction*. Cambridge, MA: MIT Press. https://mitpress.mit.edu/books/semiotic-engineering-human-computer-interaction.

de Souza, C. S. (2005b). Semiotic engineering: Bringing designers and users together at interaction time. *Interacting with Computers*, *17*(3), 317–341. https://doi.org/10.1016/j.intcom.2005.01.007.

de Souza, C. S., Barbosa, S. D. J., & da Silva, S. R. P. (2001). Semiotic engineering principles for evaluating end-user programming environments. *Interacting with Computers*, *13*(4), 467–495. https://doi.org/10.1016/S0953-5438(00)00051-5.

de Souza, C. S., de Gusmao Cerqueira, R. F., Afonso, L. M., de Mello Brandão, R. R., & Ferreira, J. S. J. (2016). *Software Developers as Users: Semiotic Investigations in Human-Centered Software Development*. Berlin, Heidelberg: Springer International Publishing. https://www.springer.com/gp/book/9783319428291.

de Souza, C. S., & Leitão, C. F. (2009). Semiotic engineering methods for scientific research in HCI. *Synthesis Lectures on Human-Centered Informatics*, *2*(1), 1–122. https://doi.org/10.2200/S00173ED1V01Y200901HCI002.

de Souza, C. S., Leitão, C. F., Prates, R. O., & da Silva, E. J. (2006). The semiotic inspection method. *Proceedings of VII Brazilian Symposium on Human Factors in Computing Systems* (pp. 148–157). https://doi.org/10.1145/1298023.1298044.

de Souza, C. S., Leitão, C. F., Prates, R. O., Amélia Bim, S., & da Silva, E. J. (2010). Can inspection methods generate valid new knowledge in HCI? The case of semiotic inspection. *International Journal of Human-Computer Studies*, *68*(1), 22–40. https://doi.org/10.1016/j.ijhcs.2009.08.006.

Diaper, D., & Stanton, N. (Eds.). (2003). *The Handbook of Task Analysis for Human-Computer Interaction* (1st edition). Boca Raton, FL: CRC Press.

Dittmann, J. (2011). Einleitung-Was ist, zu welchen Zwecken und wie treiben wir Konversationsanalyse? In J. Dittmann (Ed.), *Arbeiten zur Konversationsanalyse* (pp. 1–43). Max Niemeyer Verlag. https://doi.org/10.1515/9783111346007.1.

Eco, U. (1978). *A Theory of Semiotics* (Illustrated edition). Bloomington, IN: Indiana University Press.

Eco, U. (1986). *Semiotics and the Philosophy of Language* (Reprint edition). Bloomington, IN: Indiana University Press.

Ferreira, B., Conte, T., & Diniz Junqueira Barbosa, S. (2015). Eliciting requirements using personas and empathy map to enhance the user experience. *2015 29th Brazilian Symposium on Software Engineering* (pp. 80–89). https://doi.org/10.1109/SBES.2015.14.

Garrett, J. J. (2010). *The Elements of User Experience: User-Centered Design for the Web and Beyond* (2nd edition). Indianapolis, IN: New Riders.

Gray, D. (2018, July 21). Updated empathy map canvas. *The XPLANE Collection*. https://medium.com/the-xplane-collection/updated-empathy-map-canvas-46df22df3c8a.

Grice, H. P. (1975). Logic and conversation. In J. P. Kimball, J. L. Morgan, & P. Cole (Eds.), *Syntax and Semantics 3: Speech Acts* (pp. 41–58). Cambridge, MA: Academic Press.

Hanks, W. F. (2017, March 29). *Deixis and Pragmatics*. Oxford Research Encyclopedia of Linguistics. https://doi.org/10.1093/acrefore/9780199384655.013.213.

Hanna, N., & Richards, D. (2019). Speech act theory as an evaluation tool for human-agent communication. *Algorithms*, *12*(4), 4. https://doi.org/10.3390/a12040079.

Hartson, R., & Pyla, P. S. (2018). *The UX Book: Agile UX Design for a Quality User Experience* (2nd edition). Burlington, MA: Morgan Kaufmann Publishers.

Hassell, L. (2007). Speech-act-theory-communication-modeling. In N. Kock (Ed.), *Encyclopedia of E-Collaboration*. Hershey, PA: IGI Global. Https://Services.Igi-Global.Com/Resolvedoi/Resolve.Aspx?Doi=10.4018/978-1-59904-000-4.Ch087.

Hoover, S. P., Rinderle, J. R., & Finger, S. (1991). Models and abstractions in design. *Design Studies*, *12*(4), 237–245. https://doi.org/10.1016/0142-694X(91)90039-Y.

Jacquet, B., Hullin, A., Baratgin, J., & Jamet, F. (2019). The impact of the Gricean Maxims of quality, quantity and manner in Chatbots. *2019 International Conference on Information and Digital Technologies (IDT)* (pp. 180–189). https://doi.org/10.1109/DT.2019.8813473.

Jakobson, R. (1960). Closing statement: Linguistics and poetics. In T. A. Sebeok (Ed.), *Style in Language* (pp. 350–377). Cambridge, MA: MIT Press.

Kieras, D. (2004). GOMS models for task analysis. In D. Diaper & N. Stanton (Eds.), *The Handbook of Task Analysis for Human-Computer Interaction* (pp. 83–116). Oxford: Taylor & Francis.

Kranstedt, A., Kühnlein, P., & Wachsmuth, I. (2004). Deixis in multimodal human computer interaction: An interdisciplinary approach. In A. Camurri & G. Volpe (Eds.), *Gesture-Based Communication in Human-Computer Interaction* (pp. 112–123). Berlin, Heidelberg: Springer. https://doi.org/10.1007/978-3-540-24598-8_11.

Lao, S., Heng, X., Zhang, G., Ling, Y., & Wang, P. (2009, September 1). A gestural interaction design model for multi-touch displays. *HCI 2009: People and Computers XXIII Celebrating People and Technology.* https://doi.org/10.14236/ewic/HCI2009.55.

Liang, C., Proft, J., Andersen, E., & Knepper, R. A. (2019). Implicit communication of actionable information in human-AI teams. *Proceedings of the 2019 CHI Conference on Human Factors in Computing Systems* (pp. 1–13). https://doi.org/10.1145/3290605.3300325.

Lichaw, D. (2016). *The User's Journey: Storymapping Products That People Love.* Brooklyn, NY: Rosenfeld Media.

Lieberman, H., & Liu, H. (2006). Feasibility studies for programming in natural language. In H. Lieberman, F. Paternò, & V. Wulf (Eds.), *End User Development* (Vol. 9, pp. 459–473). Berlin, Heidelberg: Springer Netherlands. https://doi.org/10.1007/1-4020-5386-X_20.

Lieberman, H., Paternò, F., Klann, M., & Wulf, V. (2006). End-user development: An emerging paradigm. In H. Lieberman, F. Paternò, & V. Wulf (Eds.), *End User Development* (Vol. 9, pp. 1–8). Berlin, Heidelberg: Springer Netherlands. https://doi.org/10.1007/1-4020-5386-X_1.

Lopes, A. D., Pereira, V. C., & Maciel, C. (2017). An analysis of deictic signs in computer interfaces: Contributions to the semiotic inspection method. *Journal of Visual Languages and Computing, 40*(C), 51–64. https://doi.org/10.1016/j.jvlc.2017.01.001.

Löwgren, J., & Stolterman, E. (2007). *Thoughtful Interaction Design: A Design Perspective on Information Technology* (New edition). Cambridge, MA: MIT Press. https://books.google.com/books?id=g573kw36QMUC.

Marino, M. C. (2020). *Critical Code Studies.* Cambridge, MA: MIT Press.

Mendez, C., Letaw, L., Burnett, M., Stumpf, S., Sarma, A., & Hilderbrand, C. (2019). From GenderMag to InclusiveMag: An inclusive design meta-method. *2019 IEEE Symposium on Visual Languages and Human-Centric Computing (VL/HCC)* (pp. 97–106). https://doi.org/10.1109/VLHCC.2019.8818889.

Monteiro, I.T., de Souza, C.S., & Tolmasquim, E.T. (2015). My Program, My World: Insights from 1st-Person Reflective Programming in EUD Education. In P. Díaz, V. Pipek, C. Ardito, C. Jensen, I. Aedo, & A. Boden (Eds.), *End-User Development. IS-EUD 2015. Lecture Notes in Computer Science* (Vol. 9083). Cham: Springer. https://doi.org/10.1007/978-3-319-18425-8_6

Morelli, R. A., Goethe, J. W., & Bronzino, J. D. (1990). A language/action model of human-computer communication in a Psychiatric Hospital. *Proceedings of the Annual Symposium on Computer Application in Medical Care* (pp. 574–578). Washington, DC. https://www.ncbi.nlm.nih.gov/pmc/articles/PMC2245351/

Nardi, B. A. (1993). *A Small Matter of Programming: Perspectives on End User Computing.* Cambridge, MA: MIT Press.

Navas Medrano, S., Pfeiffer, M., & Kray, C. (2018). Deictic communication across distances: Visualising remote pointing gestures on mobile devices. *Proceedings of the 32nd International BCS Human Computer Interaction Conference* (pp. 1–13). https://doi.org/10.14236/ewic/HCI2018.11.

Nielsen, L. (2019). *Personas-User Focused Design.* Berlin, Heidelberg: Springer. https://doi.org/10.1007/978-1-4471-7427-1.

Norman, M. A., & Thomas, P. J. (1991). Informing HCI design through conversation analysis. *International Journal of Man-Machine Studies, 35*(2), 235–250. https://doi.org/10.1016/S0020-7373(05)80150-6.

OASIS. (2009). User Interface Markup Language (UIML) Version 4.0. Retrieved April 12, 2023, from https://www.oasis-open.org/standard/uiml-v4-0/.

Pan, Y., & Stolterman, E. (2013). Pattern language and HCI: Expectations and experiences. *CHI'13 Extended Abstracts on Human Factors in Computing Systems* (pp. 1989–1998). https://doi.org/10.1145/2468356.2468716.

Panfili, L., Duman, S., Nave, A., Ridgeway, K. P., Eversole, N., & Sarikaya, R. (2021). Human-AI interactions through a Gricean lens. *Proceedings of the Linguistic Society of America, 6*(1), 288. https://doi.org/10.3765/plsa.v6i1.4971.

Paternò, F. (2000). *Model-Based Design and Evaluation of Interactive Applications.* Berlin, Heidelberg: Springer-Verlag. https://doi.org/10.1007/978-1-4471-0445-2.

Peirce, C. S. (1992). *The Essential Peirce, Volume 1:* Selected Philosophical Writings. Bloomington, IN: Indiana University Press.

Peirce, C. S. (1998). *The Essential Peirce, Volume 2: Selected Philosophical Writings, 1893-1913.* Bloomington, IN: Indiana University Press.

Portmann, L. (2022). Crafting an audience: UX writing, user stylization, and the symbolic violence of little texts. *Discourse, Context & Media, 48*, 100622. https://doi.org/10.1016/j.dcm.2022.100622.

Prates, R. O., de Souza, C. S., & Barbosa, S. D. J. (2000). A method for evaluating the communicability of user interfaces. *Interactions, 7*(1), 31–38. https://doi.org/10.1145/328595.328608.

Sacks, H., Schegloff, E., & Jefferson, G. (1974). A simplest systematics for the organization of turn-taking for conversation. *Language*, *50*(4), 696–735.

Schon, D. A. (1984). *The Reflective Practitioner: How Professionals Think In Action* (1st edition). New York: Basic Books.

Searle, J. R. (1969). *Speech Acts: An Essay in the Philosophy of Language*. Cambridge: Cambridge University Press.

Setlur, V., & Tory, M. (2022). How do you converse with an analytical Chatbot? Revisiting Gricean Maxims for designing analytical conversational behavior. *Proceedings of the 2022 CHI Conference on Human Factors in Computing Systems* (pp. 1–17). https://doi.org/10.1145/3491102.3501972.

Shannon, C. E. (1948). A mathematical theory of communication. *The Bell System Technical Journal*, *27*(3), 379–423. https://doi.org/10.1002/j.1538-7305.1948.tb01338.x.

Shannon, C., & Weaver, W. (1949). *The Mathematical Theory of Communication*. Champaign, IL: University of Illinois Press.

Sharp, H., Preece, J., & Rogers, Y. (2019). *Interaction Design: Beyond Human-Computer Interaction* (5th edition). Hoboken, NJ: Wiley.

Shneiderman, B., Plaisant, C., Cohen, M., Jacobs, S., Elmqvist, N., & Diakopoulos, N. (2016). *Designing the User Interface: Strategies for Effective Human-Computer Interaction* (6th edition). London: Pearson.

Stapleton, A. (2017). Deixis in modern linguistics. *Essex Student Journal*, *9*(1), 1. https://doi.org/10.5526/esj23.

Tidwell, J. (2011). *Designing Interfaces: Patterns for Effective Interaction Design* (Second edition). Sebastopol, CA: O'Reilly Media.

UsiXML (USer Interface eXtended Markup Language). (2014). Retrieved April 12, 2023, from http://www.usixml.org/.

Young, I. (2008). Mental models: Aligning design strategy with human behavior. *Ubiquity*, *2008*, 1–1. https://doi.org/10.1145/1376142.1376141.

5 Mental Models

Jia Zhou

5.1 INTRODUCTION

The mental effects of interacting with information technology (IT) remain mysterious. Mental models can be used to explore the mental states or mental processes behind user interaction behaviors. Incomplete and inaccurate mental models may influence how people process information, interpret its significance, and react reasonably to it (Endsley, 2012). Differences in users' mental models and designers' conceptual models are the root causes of interaction problems (Norman, 1988). Resolving these differences could contribute to a deeper understanding of users and may help with better designs.

Despite the prevalence of human-computer interaction (HCI), mental models are not always considered important. In fact, the role of mental models is inconclusive and changes over time. During these three phases of development, a turning point occurred when mental models became central to cognitive psychology. Subsequently, mental models have been applied in various domains and have made notable contributions.

To better understand mental models, this section explores the following questions: What are the characteristics of mental models? How do we elicit and analyze mental models in HCI research? How do we apply mental models in practice? This chapter briefly discusses previous studies on mental models and their use through a case study.

5.1.1 DEFINITION OF MENTAL MODELS

Mental model theory originated from the idea of "small-scale models" (Craik, 1943) and was first expounded by Johnson-Laird (1980) to explain how humans draw inferences (Held et al., 2006, p. 11). Craik (1943) assumed that a "small-scale model" exists within organisms' minds to explain how external reality works and how to deal with future events based on this explanation. Specifically, the internal model was developed to

> Try out various alternatives, conclude which is the best of them, react to future situations before they arise, utilize the knowledge of past events in dealing with the present and future, and in every way to react in a much fuller, safer, and more competent manner to the emergencies which face it.
>
> *Craik (1943, p. 61)*

To further describe the small-scale model within the brain, one branch of studies described the model in computational terms using computer metaphors, whereas another branch of studies described the model through mental models (Held et al., 2006). As Johnson-Laird (1975, p. 10) indicated, human beings do not behave like logicians in daily life. Instead, people create internal models of the world and make deductions based on them. Thus, mental models have been used to explore situations in which people do not make inferences according to formal logic (Johnson-Laird, 1980). Mental modeling is similar to simulated "thought experiments" (Nersessian, 1992) and could further generate how-to-do-it knowledge (Tauber & Ackermann, 1991, p. 339).

Norman (1983) applied mental models, defined as "the conceptual models in people's minds that represent their understanding of how things work," to HCI. Norman (1983) clarified that users' mental models in HCI are mental representations of the systems they interact with. In this context, mental models are specified as representations of IT, including information objects, information systems, and related processes (Zhang, 2008). In the HCI domain, mental models explain what a system is composed of, how it works, how to use it, and how it responds (Wickens & Liu, 2014).

5.1.2 The Role of Mental Models

Mental models are perceived as "a key and even a panacea for successful design and development" by many researchers and practitioners from the domain of human factors (Stanney & Salvendy, 1995; Wilson & Rutherford, 1989). In practice, when an IT product is designed or released, different or even opposing opinions always exist regarding the product. This cannot be explained merely by directly observable behaviors or user capabilities. To further explore the underlying causes and effects of interaction behaviors, mental models are a possible way to understand user diversity, going beyond users' social characteristics (such as income and career), capabilities and needs, and interaction behaviors. Mental models provide perspectives from which to infer cognitive processes.

Properly leveraging mental models can improve the user experience. First, mental models help unveil the different understandings of IT among users, designers, software engineers, human factor analysts, and cognitive psychologists (Carroll & Olson, 1987). Awareness of these differences might help improve collaboration among practitioners from various backgrounds and departments to lay a common ground for discussion. Second, mental models go beyond the limitations of attitudes, values, and beliefs and could help unveil the motivations of human behaviors (Jones et al., 2011). In other words, mental models dig deeply into why certain behaviors occur, which may apply to users with diverse characteristics. Third, mental models may influence interaction performance. Particularly when users are confronted with complex information architectures, better mental models contribute to better performance (Xie et al., 2017). Furthermore, comparing the mental models of designers and users between the mental models of experts and those of novice users would help identify directions for improvement. Despite these merits, the power of mental models to explain the human mind and behavior should not be exaggerated (Wilson & Rutherford, 1989).

5.1.3 Six Concepts Related to Mental Models

Mental models are associated with six concepts in different domains: mental representations, schemas and scripts, semantic networks, cognitive maps, predictive cognition, and conceptual models.

Mental models are occasionally used interchangeably with mental representation. In mental representation, "one entity 'stands for' another—the represented object." It is difficult for the represented objects to be infinite and not directly observed because they are intentional states (Stelmach, 2006). However, mental models are subsets of mental representations. Unlike other types of representation, mental models are abstract, representing multiple situations rather than images representing a single situation (Johonson-Laird, 1996, p. 120).

Furthermore, mental models overlap with schemas, but there is no consensus on whether they are the same. A schema refers to "the entire knowledge structure about a particular topic" (Wickens & Liu, 2014). Some researchers hold that mental models are a special category of schemas, defined as schemas of dynamic systems (Wickens & Liu, 2014, p. 117). Consequently, schemas and mental models are often used interchangeably (Sharit, 2012, p. 741). Jones et al. (2011) summarized three major differences between mental models and schemas. First, mental models are dynamic and flexible, while schemas are static, inflexible, and procedural in terms of data structure (Rutherford & Wilson, 2004). Second, mental models combine multiple schemas to deal with new situations, while schemas mainly deal with "highly regular and routine situations" (Holland et al., 1986). Third, mental models use general knowledge to generate specific knowledge about new situations, while schemas mainly concern general knowledge (Brewer, 1987).

Mental models probably rely on semantic networks (Sharit, 2012). Mental models are stored in long-term memory, mainly using associated networks to structure data. There are four ways to organize information in long-term memory: semantic networks, schemas and scripts, mental models,

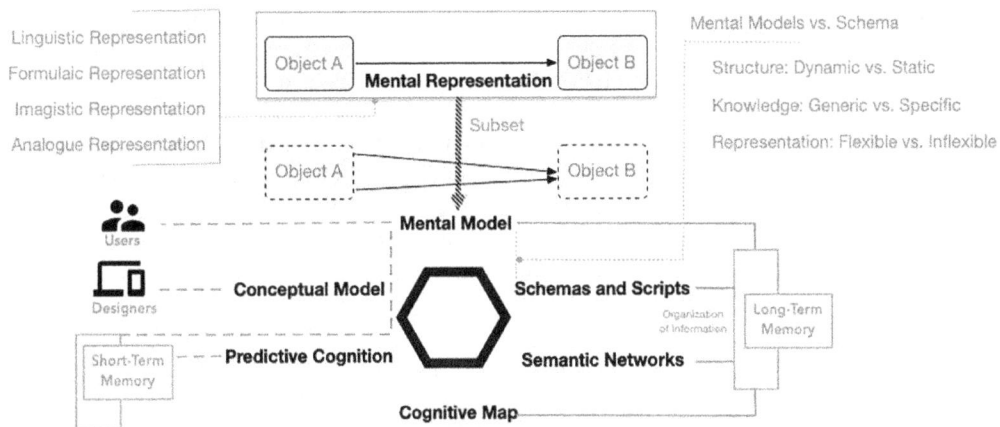

FIGURE 5.1 Mental models and six related concepts.

and cognitive maps (Wickens & Liu, 2014, p. 116). Among these, semantic networks carry much knowledge in daily life. Therefore, it is unsurprising that mental models involve semantic networks to bridge what is already known to what is yet to be known.

At first glance, mental models and cognitive maps may appear distinct. Cognitive maps are frequently used in space cognition, which was first defined by Tolman (1948) and referred to as the map in the mind of a mouse. Subsequently, cognitive maps usually refer to mental representations of physical or virtual environments during wayfinding and navigation. Although spatial information processing differs from verbal information (Ishikawa, 2020), it is not easy to separate them completely in practice. For example, when certain verbal information contains information on the spatial relationships among objects, spatial cognition is triggered, and mental models, as well as a cognitive map of the objects, are also formed (Johnson-Laird, 2010) (Figure 5.1).

Mental models are also related to the theory of predictive cognition (Ware, 2022), which emphasizes the ability of the brain to make short-term predictions (Clark, 2013). Although mental models could be interpreted as "mental simulation" (Payne, 2007, p. 66), whether mental models locate in working memory is questionable. Previous studies have found that mental models are involved in working and long-term memory (Nersessian, 2002; Jones et al., 2011). Two reasons may explain the disagreement regarding the location of the mental models. First, the nature and format of mental models do not find common ground, given that the four types of mental representations involved in previous studies could result in different operations: linguistic representation, formulaic representation, imagistic representation, and analog representation. Second, previous studies seem to put different emphases: mental models capture salient features of objects and store them in long-term memory, or mental models simulate real-world situations or imaginary situations in working memory for comprehension or reasoning tasks (Nersessian, 2002).

Conceptual models can be interpreted as mental models (Allen, 1997) and are necessary for designers to understand users and help them use physical systems (Norman, 1983). The ideal relationship between conceptual models of physical systems and users' mental models is direct and simple; however, there is often a gap between them. Conceptual models can be further divided into four types: metaphors, surrogates, task-action mappings, and propositional knowledge (Allen, 1997).

Based on these analyses, it is easier to find common ground on what mental models are than to find a clear answer to what they are not. Thus, mental models lack clear boundaries.

5.1.4 Brief History of Mental Models

The notion of mental models was not a central theme in cognitive psychology and cognitive science until the 1950s. Mental models underwent three phases, similar to the development of psychology.

The first phase involved behavioral methods used to explore mental models. Research in this phase treated the cognitive system as a black box and mainly tested the relationship between stimuli and response. During this phase, mental models were not highly valued.

The second phase involved research using cognitive psychology methods to explore the mental models. A defining case is the study of Tolman (1948), which explained rats' behaviors in mazes by proposing 'cognitive maps' in their minds. The major premise of cognitive psychology is that the human brain is a physical symbol system (Newell & Simon 1976), so cognition can be simulated through computation, where stimulus plus internal states are mapping responses plus internal states. Research in this phase mainly added methods from artificial intelligence research methods to psychological methods. In this context, mental models have mainly explored mental logic, treating human brains as computers.

The third phase involved the neuroscience method used to explore the mental models. Brain activity adds information to the understanding of human thinking and reasoning. Studies on mental models exploring brain events, particularly locations and dissociations, would help us understand cognitive processes (Stelmach, 2006, pp. 6–8, 14).

5.2   REPRESENTATIONS AND CHARACTERISTICS OF MENTAL MODELS

There are four types of mental representations: linguistic, formulaic, imagistic, and analogous. As a subset, mental models also have different formats. Specifically, mental models can be expressed in terms of simple sequences, surrogates, metaphor models, glass-box models, and network representations (Carroll & Olson, 1987).

Regarding linguistic representations, mental models make three main assumptions. First, each mental model represents a mutually exclusive possibility. Second, mental models are iconic, including not only visual images but also abstract relationships after cognitive processing. Third, mental models represent what is true rather than what is false (Johnson-Laird, 2010). These assumptions were supported by experiments aimed at understanding human reasoning based on sentential inference. Given that text is a common element in user interfaces, text comprehension is necessary to understand the system, particularly when users read instructions or onboarding tips.

The characteristics of mental models can be compared with other branches of human reasoning and mental logic. Mental logic assumes that humans are rational and that their reasoning can be represented through formal rules of inference (Johnson-Laird, 2010). However, mental models emphasize situations in which the cognitive process is not necessarily logical, static, or complete.

5.2.1   DYNAMIC REPRESENTATION

Mental models are unstable and continuously evolving. Given that users' backgrounds and experiences influence their construction of mental models, during interactions with information technologies, and as their experience grows, they modify mental models to make better explanations and predictions (Norman, 1983). In addition, there were no fixed boundaries among the mental models. Users' mental models of similar devices or operations may not be easily distinguishable (Norman, 1983).

5.2.2   INCOMPLETE REPRESENTATION

Mental models are often incomplete because people do not usually use all IT functions. Such mental models do not match designers' conceptual models (Norman, 1983). If designers' conceptual models are assumed to be correct, users' mental models are often inaccurate. The gap between designers' conceptual models and users' mental models may cause difficulties in interaction.

5.2.3 Simplified Representation

Mental models are developed to save users' mental efforts. Norman (1983) stressed that "mental models are parsimonious." Although mental planning could cause users to avoid extra physical effort, they still prefer to do physical actions (Norman, 1983).

5.3 ELICIT MENTAL MODELS

Mental models are located in minds and are, therefore, not directly observable. However, it is possible to infer people's mental models from their thoughts, attitudes, and behaviors. To this end, methods rely either on users' self-reports or researchers' guidance, follow structured frameworks and predefined activities, or use tools to aid in the elicitation process.

Methods for eliciting mental models, such as interviews and questionnaires, overlap with methods for studying HCI. This raises two important questions. The first issue concerns validity. It is necessary to check what has been interviewed or surveyed as a mental model rather than regular user needs. Furthermore, researchers usually need to interpret users' self-reported thoughts or observed behaviors, which might result in a combination of users' mental models and designers' conceptual models.

The second issue concerns dynamics. Since mental models are dynamic and flexible, they may evolve as users spend more time interacting with the system or use different system functions. For example, the System Usability Scale (SUS) requires using the system for 20–60 minutes before completing the SUS (Brooke, 1996). Therefore, it is necessary to set up interaction tasks and the usage duration of the system before eliciting mental models.

5.3.1 Interviews

The interview questions were structured according to the following framework. This framework covers the primary aspects of IT interactions. For example, interaction with the web had four-component frameworks: web space, information structure, search functions, and web interfaces (Wang et al., 2000) or components, functions, attributes, and feelings (Zhang, 2008), and then participants were interviewed through questions with phrases such as "could you describe … to me?" and "how do you think … is organized?" (Zhang, 2008).

Interviews can be integrated with other techniques, depending on the task. For example, interviews within the framework of the Critique Incident Technique were used to elicit mental models of stakeholders during care transitions, and the results were analyzed through semantic analysis (Werner et al., 2021). When reading a text, utterances automatically elicit knowledge by processing words and sentences to make deductions (Johnson-Laid, 1975, p. 51).

Klein and Baxter (2009) proposed the cognitive transformation theory, which indicates that mental models, rather than knowledge, effectively improve cognitive skills through training. To change mental models, it is necessary to connect with existing mental models, unlearn the flawed parts of mental models, and then transform the mental models through feedback. To apply cognitive transformation theory in practice, Klein and Borders (2016) further adapted the ShadowBox approach. The ShawdowBox was initially proposed by Hintze (2008). This provides a way to compare novices' mental models with experts' mental models using scenario-based decision-making procedures. The adapted ShawdowBox selected mental models, although they were not mentioned throughout the process, and the experts did not need to be present.

5.3.2 Questionnaire

Given that experience and knowledge influence people's construction of mental models, questionnaires are typically used to measure users' prior experience with IT or to test their knowledge of IT.

Regarding experience, questions such as IT usage frequency, duration, and breadth were used (Zhang, 2008). Regarding knowledge, questions are specific to IT. For example, a four-item questionnaire assessed mental models of adaptive cruise control. Based on the questionnaire scores, respondents could be classified into weak and strong mental model groups (Gaspar et al., 2020) and used in studies to quantify mental models of adaptive cruise control (Carney et al., 2022).

LaMere et al. (2020) proposed a Rich Elicitation Approach (REA) to transfer mental knowledge to written knowledge. REA integrates direct and indirect elicitation methods over five phases: preparation, direct elicitation (e.g., draw diagrams) and questionnaire, indirect elicitation (e.g., code transcripts and notes), verification and standardization, and methodology evaluation.

5.3.3   Card Sorting and Concept Mapping

Mental models can be expressed through drawings in terms of abstract concepts or concrete components (e.g., words, pictures, symbols) of information systems and the relationships among them. Participants could draw by hand or create predefined structures (Jones, et al., 2011). Furthermore, the participants could explain why they drew lines between cards and how they understood the relationships in the flowcharts (Dray et al., 2006). In addition to drawing relationships between concepts, participants were able to indicate the weight and direction of the relationships. The method of assigning a number between −1 and 1 to the relationships during drawing is termed fuzzy cognitive maps (Özesmi & Özesmi, 2004), and the directions of relationships may correspond to the click streams that can be identified through the path diagram method during interaction with IT (Xie et al., 2017).

Concept maps attempt to explicitly represent mental models (Kinchin et al., 2000). Concept maps are also related to mind mapping and spider diagramming (Atkinson et al., 2020) and cognitive mapping (Hartmeyer et al., 2018). Participants first wrote down concepts in text boxes and then drew links between them. Labeled arrows can accompany the links to clarify the association between concepts. Consequently, concept maps can contain different types of graphical structures. Kinchin et al. (2000) were originally inspired by a group of 8-year-old students who examined the reproduction of flowering plants. After analyzing how the students learned new concepts, they identified three concept map structures: spoke, chain, and net. The spoke structure features radical links between regular and core concepts, whereas direct links between regular concepts do not exist. This indicates that learning a new concept does not disturb existing concepts and that the chain structure features linear links of concepts following a logical sequence. This implies that adding a new concept is easy if it is not in the middle of the sequence and the net structure features an integrated and hierarchical network. This implies wider and more flexible knowledge because a new concept is linked to existing concepts through multiple routes. In addition to these three structures, a hierarchical structure was added as the fourth basic structure of concept maps (Hartmeyer et al., 2018).

A conceptual content cognitive map (3CM) was proposed to measure the cognitive map or general knowledge of conceptual themes. First, concepts related to a specific theme are usually chosen or generated by the participants through questionnaires or interviews. Participants then sorted and organized the concepts to explain the conceptual theme to others who did not think much about it. Next, the organized concepts were labeled and placed in envelopes. 3CMs can be conducted either in an open-ended or structured manner, depending on the breadth and depth of the research (Kearney & Kaplan, 1997).

Given that mental models are dynamic and influenced by others, some studies have elicited mental models from groups of people. The co-discovery method is used to create mental models of new information systems. After participants completed the information search tasks in pairs, they were asked to visually and verbally express their thoughts on the institutional repository work (Rieh et al., 2010). Other methods, such as Actors, Resources, Dynamics, and Interaction, offer a framework to elicit shared mental models (Etienne et al., 2011).

5.4   MENTAL MODEL ANALYSIS

The results of eliciting mental models have two main categories: qualitative results, such as interview transcripts and drawings, and quantitative results, such as questionnaires and card sorting results. There is no clear-cut line between the two categories because, after using appropriate techniques, quantitative metrics can be drawn from the qualitative results. For example, semantic networks and cognitive maps drawn by participants can be transformed into a similarity or adjacency matrix.

The goal of analyzing mental models is twofold. First, we compared the differences in mental models, either between users and designers or between individuals and all the participants. Second, it is necessary to find ideas for a better design, such as identifying the appropriate information architecture. Semantic analysis is typically used for qualitative results to achieve these goals, whereas multivariate techniques and graph theory are typically used for quantitative analysis.

5.4.1   QUALITATIVE ANALYSIS

Qualitative analytical methods have two branches. One branch is guided by theoretical frameworks that vary in their degree of rigorous adherence. Four methods belong to this branch: conversation analysis (CA), interpretative phenomenological analysis (IPA), grounded theory (GA), and narrative analysis (NA). The second branch is more explorative, with fewer constraints on existing theories. A typical example is semantic analysis, which forms the basis of qualitative analysis (Braun & Clarke, 2006).

Semantic analysis aims to identify repeated patterns in texts. The patterns include not only explicit semantic themes but also latent or interpretative themes. As a result, apart from descriptive statistics, such as word frequency, semantic analysis can further generate thematic maps showing themes and their relationships. Braun and Clarke (2006) presented six steps of semantic analysis: transcribing and familiarizing the data, coding interesting features of the data, collating codes into themes, generating a thematic map, refining and naming themes, and producing a report. Following this procedure, 15 criteria for good thematic analysis were achieved.

5.4.2   CLUSTER ANALYSIS

Before conducting the statistical analysis on card-sorting data, it is necessary to check three important issues. The first issue is about the titles of the cards. If participants did not create specific group titles for certain cards or just used labels such as "other" or "miscellaneous," these cards did not belong to a real conceptual group, and they should be separated from each other to be individual groups. The second issue is the hierarchical structure in card sorting. Participants may create hierarchical groups, which are difficult to code using a similarity matrix. In this case, their hierarchal groups could be flattened into a border group or split into smaller groups. The third issue is duplicate cards. Sometimes, participants placed a card in multiple groups. Participants were asked to choose one of the best locations (Spencer, 2009, p. 131).

During data preparation, some challenging points remain to be addressed, although Spencer (2009) provided detailed suggestions to deal with various card-sorting results in practice. For example, how to transform hierarchical groups of cards with more than three-level depth into an adjacency matrix remains unclear. A quick and simple method is to flatten the hierarchical groups into one level, which results in the loss of information conveyed in the raw data. Instead, if hierarchal structures are maintained, one question to be answered is how to code the distances between different groups and within a group of deep structures.

Card-sorting data were analyzed using five methods: k-means cluster analysis, hierarchical cluster analysis (HCA), multidimensional scaling (MDS), latent partition analysis (LPA), and 3CM.

The principles of these statistical analyses are beyond the scope of this section; however, this section briefly discusses the advantages and disadvantages of these statistical methods. K-means cluster analysis can generate different grouping results after inputting the number of groups. Therefore, it is good for exploring grouping but cannot obtain a confirmative answer. HCA can generate a dendrogram and is effective for visualizing a hierarchical tree. However, the results of HCA largely depend on the definition of the distance between two clusters (Spencer, 2009). MDS can transform the distance matrix into a map showing the proximity of points. In practice, measuring the distance between two points on a map is intuitive, and it is easy to obtain a distance matrix among the points, whereas MDS goes the opposite way from the distance matrix to the map (Ishikawa, 2020, p. 85). MDS can not only be applied to analyze the results of card sorting but is also widely used in generating cognitive maps. LPA is an alternative to MDS, where individuals' matrices of categorizations can be averaged to obtain the group's matrices. LPA is similar to the factor analysis of categorical data (Kearney & Kaplan, 1997; Wiley, 1967). The 3CM combined hierarchal clustering, LPA, and MDS in a structured manner (Kearney & Kaplan, 1997).

5.4.3 ANALYSIS OF NETWORKS AND GRAPHS

Regardless of the methods used to elicit mental models, the graphical structure of mental models may be inferred from but not limited to the following: (1) clustering structure after coding card sorting results, (2) semantic networks after semantic analysis of the text, (3) concept maps and path diagrams, (4) causal relationships after coding the sketches or drawings, and (5) networks derived from the cognitive maps.

Coding is usually required to prepare for the analysis. The goal was to identify the variables and their relationships. The coding framework may vary depending on the type of information used. For example, the participants' drawings and sketches could be divided into four categories: processing, global view, interface, and interactivity. Each category is defined based on a detailed understanding of information-system operations (Rieh et al., 2010). Other categorization methods are technical, functional, process, and connection (Zhang, 2008). The coding results are typically networks and graphs. They can be directional/unidirectional, with/without the quantified strength of the relationships, with/without feedback loops, and with/without text labels to clarify the relationships among variables.

The structure and dynamics of mental models can then be quantitatively analyzed using network analysis and graph theory. The boundary between them is blurred because graph theory is the mathematical foundation of network science (Iñiguez et al., 2020). Some studies use cognitive maps and networks interchangeably (Özesmi & Özesmi 2004). According to the graph theory, there are two major steps.

The first step is to obtain the adjacency matrices by transforming the networks or graphs or directly deriving them from the card sorting results. For each participant, adjacency matrices were derived individually. The adjacency matrices of an individual can then be combined with the group's adjacency matrices for further analysis. The row sum of the absolute values in the adjacency matrix is termed the outdegree, and the column sum is the indegree (Özesmi & Özesmi, 2004).

The second step was to compute the metrics of the networks and graphs. Common metrics include but are not limited to similarity, centrality, density, and complexity. Similarity can reflect the differences between two individuals or gaps between users and designers. The similarities between the mental models can be further divided into directionless and directional similarities. Based on graph theory, Xie et al. (2017) elicited mental models of websites through card sorting and path diagrams, the results of which were analyzed using three proposed calculation formulas for directionless and directional similarity, and then evaluated their role in interaction performance in experiments. The results indicated that directional similarity was a better predictor of task completion time and the number of clicks for searching tasks on websites than directionless similarity. This method applies to the analysis of mental models of smart TVs (Ouyang & Zhou, 2018). In the context of the shared mental models of teams, similarity can be measured using the following four indices: r_{WG}, r^*_{WG}, a_{WG}, and r_{RG}. The first three mainly consider the degree of shared common knowledge in a team, whereas the fourth considers team-specific knowledge compared to common knowledge (Biemann et al., 2014). In addition to similarity, centrality is the sum of the indegree and outdegree, and density reflects the percentage of existing connections between variables compared to all possible connections (Özesmi & Özesmi, 2004).

5.5   CASE STUDY

A case study is presented to indicate the mental models of information architecture in smartphones. A detailed description of smartphones has not been provided for intellectual property reasons. The following sections present the process of eliciting mental models and discuss decision points for improvement.

5.5.1   PROJECT BACKGROUND

Why was this application downloaded? I did not even know where it came from.
How could I go back? I just wanted to skip advertisements but was misled by shopping websites.

These two quotes are responses from older adults who participated in a series of user studies. The project aimed to tailor mobile operation systems for older adults and develop elderly-friendly standards. This goal was set because market research revealed two dominant problems among older users: phone security and disorientation.

Regarding phone security, older adults were bothered by adware installed without users' knowledge and produced never-ending pop-ups. Most older participants found it difficult to manage permissions, prevent authorized access to apps, and free up smartphone storage. As a result, the adware redirects users to advertisements and crash or brick older adults' phones.

Regarding disorientation, older adults were easily lost when lured to advertisements or external links, sliding between one screen and another, tracking a quick message from a group of notifications, and transitioning among widgets, websites, and applications. They did not know where they were, where they came from, or how they arrived at their destination in the "maze" of the limited screen of smartphones.

These two problems are associated with the mental smartphone models of older adults. To elicit mental models, card sorting and interviews were used. Taking the setting as an example, the detailed process is as follows.

5.5.2   CARD SORTING EXPERIMENT

A total of 24 older adults aged above 60 participated in card sorting. The average duration of smartphone usage was 5.5 years. The self-reported time spent using smartphones was 4.8 hours per day. This experiment used smartphones because they are popular among older Chinese adults. Given that it has a built-in simplified elderly-specific mode of mobile operating systems, a simplified version of the settings application on the phone that provides a list of suggestions and options for users to manage their phones was used in the experiment. Although simplified, the settings still contained 148 features.

To prepare cards, one decision to be made is which features should be included in card sorting. In the hierarchical structure of the settings, features are aligned from parent to child nodes. One includes only leaf nodes that do not have a child, whereas the other includes all the nodes. This study used the latter method for two reasons. First, usability testing studies in the former phase of the project indicated that disorientation is serious in the current setting. Second, the goal is not to explore new groups of features but to check the differences between the designers' conceptual model and users' mental models.

A whiteboard and A3 papers were presented to the participants as group cards. Additionally, a smartphone was used to capture photos and record videos of the card-sorting process. Furthermore, a notebook computer was used to record participants' feedback in real-time.

The participants were briefed about the experiment and signed the consent forms. Subsequently, they completed a questionnaire regarding their demographic background and a phone usage experiment. Next, the participants were trained to sort cards according to the features of familiar applications (i.e., WeChat and Taobao). They did not enter the experiment until they understood how to align the cards with parent and child features.

During the card-sorting experiment, participants were presented with 148 cards with all the features in the settings. Given that the participants did not necessarily know all the features, they were given the Not Applicable option. They could leave cards they were unfamiliar with, did not understand, or did not use off to the side. Next, participants went through each card and sorted the cards that belonged together into groups, and they could further split the groups into subgroups. They were encouraged to think aloud during the experiment, particularly when they hesitated to place the cards or change their original alignment. Participants were then asked to name the groups. They could either choose an existing card to name the group or write a new name on a blank card. If they had difficulty naming, they could simply skip the naming step. Participants were asked to explain their rationale after completing each group session.

5.5.3 Analyzing Card-Sorting Data

This project analyzed the results of card sorting without considering the direction of the relationships. This is because the goal was not to examine the interaction flow but to check the differences between the information architecture in the designers' conceptual model and that in users' mental models. The detailed procedure is as follows:

1. *Define a coding scheme*: The total number of themes depended on whether the card sorting was open, closed, or a hybrid. For a predefined number of cards in closed-card sorting, just count the total number of themes/cards. In contrast, for cards defined by participants in open/hybrid card sorting, their input should be grouped into themes, including explicit and latent themes. In this case study, cards in the phone setting application were predefined by the mobile operating system, and the total number of predefined cards was 148. Cards defined by participants who filled in blank cards for additional categories were grouped into 60 themes. Given that fewer than 5% of the participants proposed new cards, this case study chose 148 cards for further analysis.

2. *The cards are structured into spreadsheets*: The card structure was recorded for each participant. In this case study, the card-sorting results of each participant were recorded by taking a photo, and their verbal explanations were audio-recorded. The two sources of information were cross-checked and integrated into a spreadsheet. This step was necessary, especially when participants had inconsistent feedback or when they grouped certain cards incorrectly. Based on the spatial relationships among cards and the verbal feedback from each participant, "move" each group on the whiteboard/paper to a spreadsheet. The structure of each group on the spreadsheet was the same as that of the paper cards. The order of groups is flexible if they do not belong to each other. In case the participants did not name a group, its group name is "Unnamed+ID."

3. *Create an adjacency matrix for each participant*: Convert the text on cards into numbers. The n features were converted to "F_1," … "F_n." There is no consensus on whether unnamed groups should be discarded. If there are x unnamed groups, they could be ignored or

labeled from "F_{n+1}" to "F_{n+x}." This project tried both methods and built an $n \times n$ matrix and an $(n+x) \times (n+x)$ matrix. Next, the cells in the matrix are filled. This step is easy if the hierarchy depth is less than two levels but difficult if the hierarchy depth is above three levels. For a hierarchy of fewer than two levels, fill in the cells in matrix 1 if two features belong to a group; otherwise, fill in 0. However, for a hierarchy with more than three levels, there is no consensus on how to convert the leaves of different branches of a tree into numbers. Specifically, we quantified the distance between features F002 and F025 and between F011 and F050, as shown in Figure 5.2. An intuitive approach is to count the number of steps. However, these steps assume that users are on the right route among the features, as reflected in the designers' conceptual model. In fact, based on participants' feedback from the think-aloud task, they did not seem to clearly distinguish within branches of trees, but they may distinguish between branches of trees. Given that most of the participants' grouped cards were on one or two levels, this project used 0 or 1 to fill in the matrix.

4. Analyze network metrics, including similarity, centrality, density, and complexity. Given that each user has an adjacency matrix, combining the adjacency matrices of users could generate visualizations such as a colored similarity matrix, dendrograms, and network graphs, which could be derived through linear operation, clustering analysis, and network analysis, respectively. As shown in Figures 5.3 and 5.4, this project first explored highly correlated features (correlation coefficient higher than 0.6) in terms of deprograms, and then used the Newman–Girvan algorithm (Newman & Girvan, 2004) to find possible groups of features.

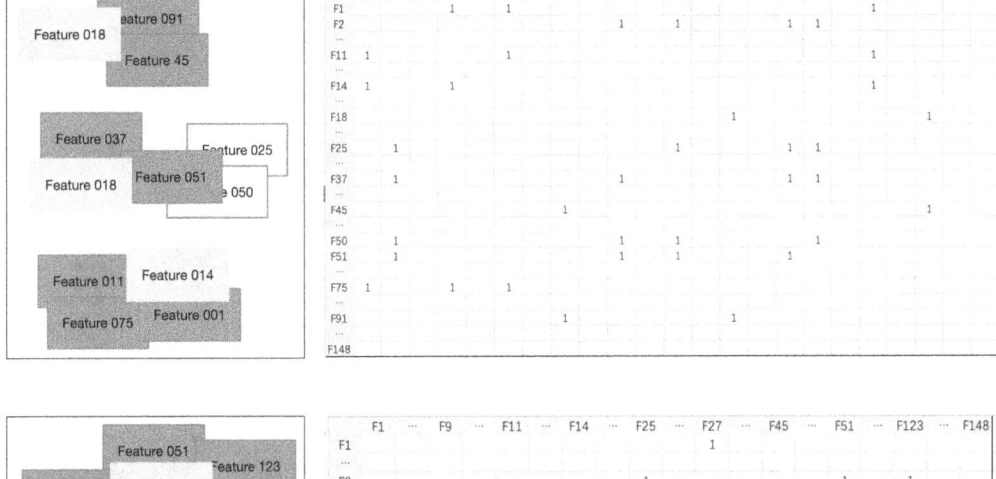

	F1	F2	⋯	F11	⋯	F14	⋯	F18	⋯	F25	⋯	F37	⋯	F45	⋯	F50	F51	⋯	F75	⋯	F91	⋯	F148
F1				1		1													1				
F2										1		1				1	1						
F11	1							1											1				
F14	1			1															1				
F18														1							1		
F25	1											1				1	1						
F37	1									1						1	1						
F45								1													1		
F50	1									1		1					1						
F51	1									1		1				1							
F75	1			1		1																	
F91								1						1									
F148																							

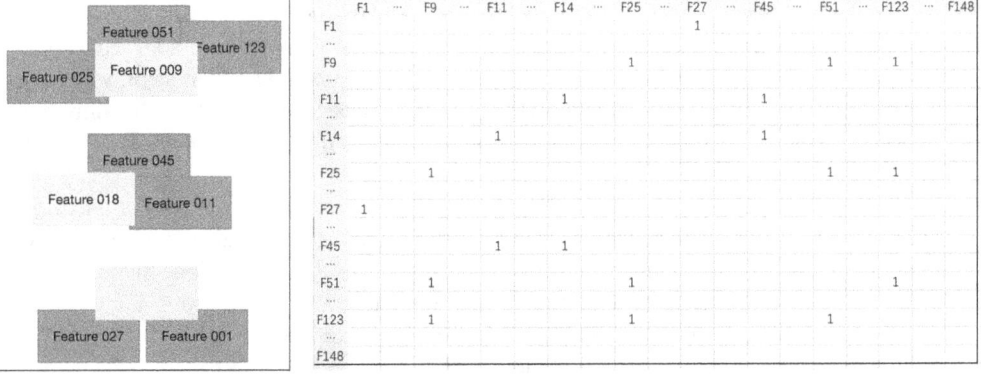

	F1	⋯	F9	⋯	F11	⋯	F14	⋯	F25	⋯	F27	⋯	F45	⋯	F51	⋯	F123	⋯	F148
F1											1								
F9									1						1		1		
F11							1								1				
F14					1										1				
F25			1												1		1		
F27	1																		
F45			1		1														
F51			1								1						1		
F123			1								1				1				
F148																			

FIGURE 5.2 Card sorting results and adjacency matrix.

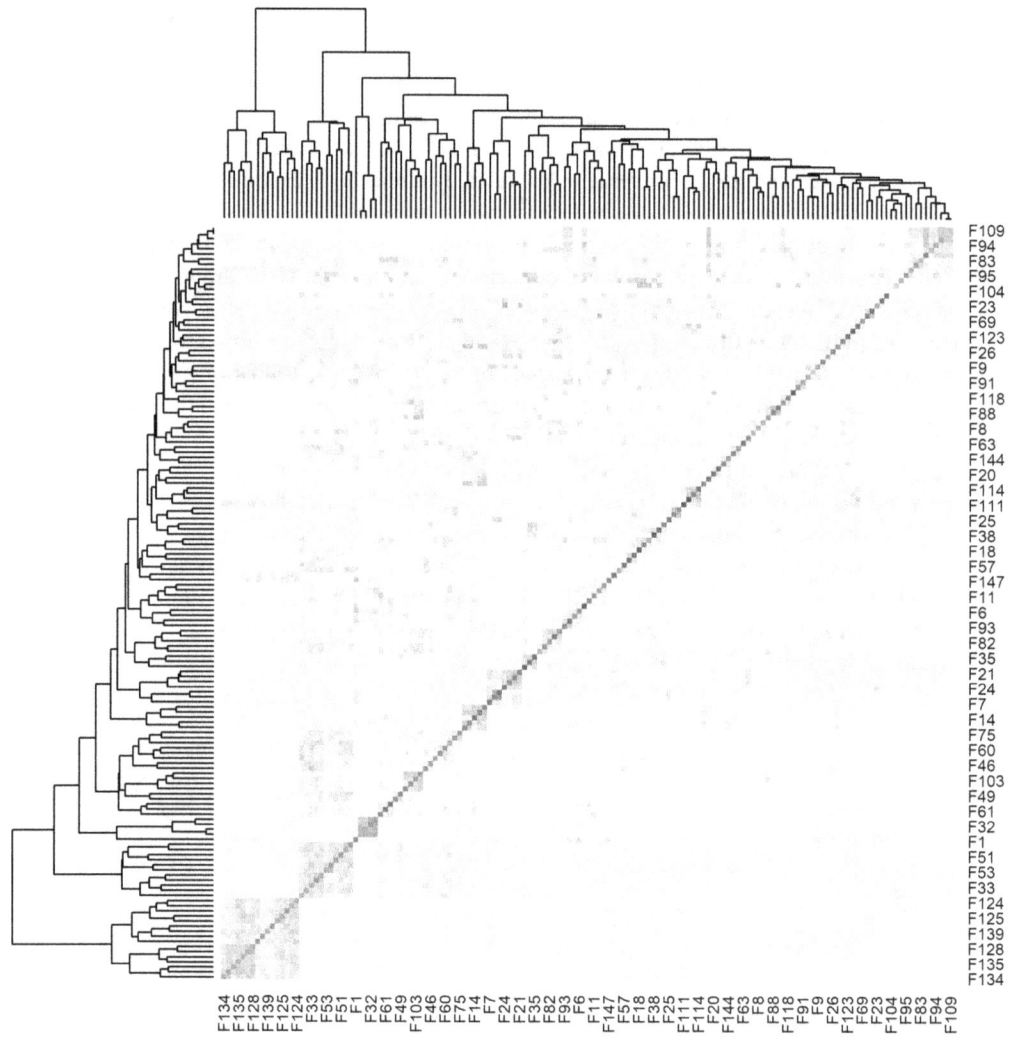

FIGURE 5.3   Dendrograms and correlations among smartphone features.

5. *Visualize the mental models.* This is a 1 vs. *N* comparison between a matrix from the designers and a group of matrices from the users. Given that the designers' conceptual model is reflected in the system design, the current design could be seen as only one correct answer. This answer can be converted into an adjacency matrix using the aforementioned procedure. Thus, the current and perceived designs in the users' mental models can be quantitatively compared.

6. In addition to comparison in terms of adjacency matrices, qualitative analysis of the differences between users and designers is also important. This project analyzed whether users' mental models of information architecture are the same as those of designers and how the differences in mental models were reflected in their interaction behaviors. For example, this study used a clickstream to indicate how users navigate through an information architecture to complement their mental models.

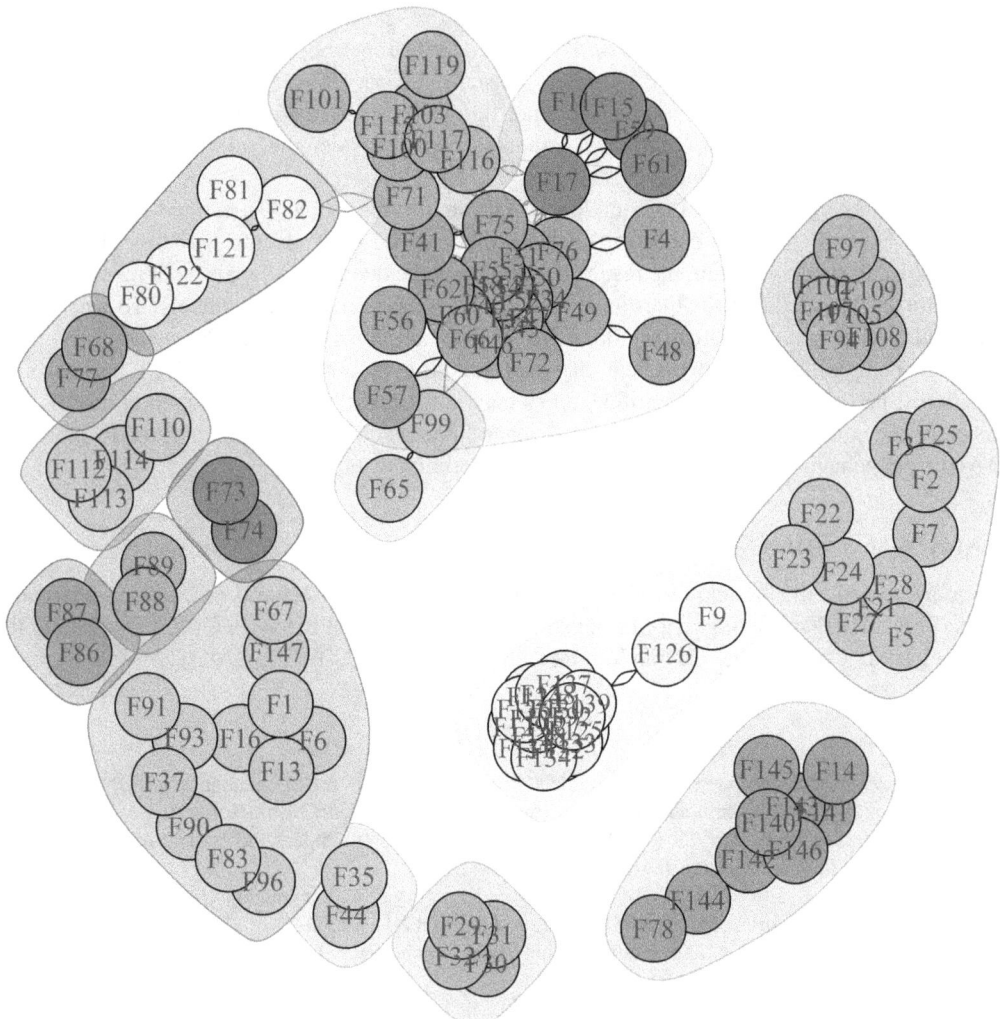

FIGURE 5.4 Clustering of smartphone features with network betweenness more than 0.6.

5.6 CONCLUSION

Mental models are used to understand users' mental representations of how systems work in the real or imagined world. When applied to HCI, the users' mental models usually differ from their conceptual models. The main differences lie in how the information system works and how it interacts with the system, which are the root causes of many interaction problems.

The boundaries of the mental models are not sufficiently clear. There is an ongoing debate on the relationships between mental models and mental representations, schemas and scripts, semantic networks, cognitive maps, predictive cognition, and conceptual models. Specifically, the location of mental models and the structure of mental models have been extensively discussed. Despite these disagreements, there is a consensus that mental models are dynamic, flexible, and incomplete.

Eliciting users' mental models is useful for resolving the differences between their mental models and designers' conceptual models. Given that mental models are within the head, methods based on interviews, card sorting, and concept maps with theoretical frameworks can be applied to indirectly infer mental models through user self-reports or researcher observations. The data were then analyzed through semantic analysis, K-means cluster analysis, HCA, MDS, LPA, 3CM, and network analysis. The results included but were not limited to, semantic maps, dendrograms, cluster maps, and cognitive maps. In addition, it is necessary to consider the validity and dynamics of mental models during interactions.

Future studies should clarify the boundaries of these mental models. Understanding what mental models are not would help to check the validity of elicitation methods and shape the direction of qualitative analysis. In addition, the results derived from the quantitative analysis of mental models still do not answer certain basic questions: How accurate are users' mental models? To what extent do mental models notably increase or decrease the interaction performance? How can users' mental models be modified naturally, or can IT adapt to their evolving mental models? Finally, from a reductionist perspective, neuroscientific studies may not reveal the findings of mental models as a whole. Future studies may better link neuroscience findings on subsets of mental models to those derived from behavioral and cognitive psychology methods.

ACKNOWLEDGMENTS

The authors would like to acknowledge the support from National Natural Science Foundation of China (72171030). The Chongqing Municipal Federation of Social Sciences (2022YC049), and the 2022 Reform in College Elite Curriculum Research Project (CQU-EIE-2022011).

REFERENCES

Allen, R. B. (1997). Mental models and user models. In M. G. Helander, T. K. Landauer, & P. V. Prabhu (Eds.), *Handbook of Human-Computer Interaction* (pp. 49–63). Amsterdam: North-Holland.

Atkinson, C., Thomas, G., & Parry, S. (2020). Using concept mapping to understand motivational interviewing practice. *Qualitative Research Journal*, 20(2), 165–174.

Biemann, T., Ellwart, T., & Rack, O. (2014). Quantifying similarity of team mental models: An introduction of the rRG index. *Group Processes & Intergroup Relations*, 17(1), 125–140.

Braun, V., & Clarke, V. (2006). Using thematic analysis in psychology. *Qualitative Research in Psychology*, 3(2), 77–101.

Brewer, W. F. (1987). Schemas versus mental models in human memory. In P. Morris (Ed.), *Modelling Cognition* (pp. 187–197). Chichester, UK: John Wiley & Sons.

Brooke, J. (1996). Sus: A "quick and dirty" usability. *Usability Evaluation in Industry*, 189(3), 189–194.

Carney, C., Gaspar, J. G., & Horrey, W. J. (2022). Longer-term exposure vs training: Their effect on drivers' mental models of ADAS technology. *Transportation Research Part F: Traffic Psychology and Behaviour*, 91, 329–345.

Carroll, J. M., & Olson, J. R. (1987). Mental models in human-computer interaction. Research issues about what the user of software knows. *Workshop on Software Human Factors: Users' Mental Models*, Washington, DC, May 15–16, 1984.

Clark, A. (2013). Whatever next? Predictive brains, situated agents, and the future of cognitive science. *Behavioral and Brain Sciences*, 36(3), 181–204.

Craik, K. J. W. (1943). *The Nature of Explanation*. Cambridge: Cambridge University Press.

Dray, A., Perez, P., Jones, N., Le Page, C., D'Aquino, P., White, I., & Auatabu, T. (2006). The AtollGame experience: From knowledge engineering to a computer-assisted role playing game. *Journal of Artificial Societies and Social Simulation*, 9(1), 1–11.

Endsley, M. R. (2012). Situation awareness. In G. Salvendy (Ed.), *Handbook of human factors and ergonomics* (pp. 553–568). Hoboken NJ: John Wiley & Sons.

Etienne, M., Du Toit, D. R., & Pollard, S. (2011). ARDI: A co-construction method for participatory modeling in natural resources management. *Ecology and Society*, 16(1), 44.

Gaspar, J. G., Carney, C., Shull, E., & Horrey, W. J. (2020). The impact of driver's mental models of advanced vehicle technologies on safety and performance. Retrieved from https://doi.org/10.7910/DVN/0CWLOZ.

Hartmeyer, R., Stevenson, M. P., & Bentsen, P. (2018). A systematic review of concept mapping-based formative assessment processes in primary and secondary science education. *Assessment in Education: Principles, Policy & Practice*, 25(6), 598–619.

Held, C., Knauff, M., & Vosgerau, G. (2006). General introduction: Current developments in cognitive psychology, neuroscience, and the philosophy of mind. In C. Held, M. Knauff, & G. Vosgerau (Eds.), *Mental Models and the Mind: Current Developments in Cognitive Psychology, Neuroscience, and Philosophy of Mind* (pp. 5–22). Amsterdam, Netherlands: Elsevier.

Hintze, N. R. (2008). First responder problem solving and decision making in today's asymmetrical environment. *Unpublished master's thesis, Naval Postgraduate School*, Monterey, California.

Holland, J. H., Holyoak, K. J., Nisbett, R. E., & Thagard, P. R. (1986). *Induction: Processes of Inference, Learning, and Discovery*. Cambridge, MA: MIT Press.

Iñiguez, G., Battiston, F., & Karsai, M. (2020). Bridging the gap between graphs and networks. *Communications Physics*, 3(1), 88.

Ishikawa, T. (2020). *Human Spatial Cognition and Experience: Mind in the World, World in the Mind*. London: Routledge. Johnson-Laird, P. N. (1975). Models of deduction. In *Reasoning: Representation and Process in Children and Adults* (1st edition). London: Psychology Press. https://www.taylorfrancis.com/chapters/edit/10.4324/9781315682747-2/models-deduction-johnson-laird

Johnson-Laird, P. N. (1980). Mental models in cognitive science. *Cognitive Science*, 4, 71–115.

Johnson-Laird, P. N. (2010). Mental models and human reasoning. *Proceedings of the National Academy of Sciences*, 107(43), 18243–18250.

Jones, N. A., Ross, H., Lynam, T., Perez, P., & Leitch, A. (2011). Mental models: An interdisciplinary synthesis of theory and methods. *Ecology and Society*, 16(1), 46.

Kearney, A. R., & Kaplan, S. (1997). Toward a methodology for the measurement of knowledge structures of ordinary people: The conceptual content cognitive map (3CM). *Environment and Behavior*, 29(5), 579–617.

Kinchin, I. M., Hay, D. B., & Adams, A. (2000). How a qualitative approach to concept map analysis can be used to aid learning by illustrating patterns of conceptual development. *Educational Research*, 42(1), 43–57.

Klein, G., & Baxter, H. C. (2009). Cognitive transformation theory: Contrasting cognitive and behavioral learning. In D. Schmorrow, J. Cohn, & D. Nicholson (Eds.), *The PSI Handbook of Virtual Environments for Training and Education: Developments for the Military and Beyond. Volume I: Learning, Requirements and Metrics* (pp. 50–65). Westport, CT: Praeger Security International.

Klein, G., & Borders, J. (2016). The ShadowBox approach to cognitive skills training: An empirical evaluation. *Journal of Cognitive Engineering and Decision Making*, 10(3), 268–280.

LaMere, K., Mäntyniemi, S., Vanhatalo, J., & Haapasaari, P. (2020). Making the most of mental models: Advancing the methodology for mental model elicitation and documentation with expert stakeholders. *Environmental Modelling & Software*, 124, 104589.

Nersessian, N. J. (1992). In the theoretician's laboratory: Thought experimenting as mental modeling. In *PSA: Proceedings of the Biennial Meeting of the Philosophy of Science Association* (Vol. 1992, No. 2, pp. 291–301). Philosophy of Science Association.

Nersessian, N. J. (2002). The cognitive basis of model-based reasoning in science. In *The Cognitive Basis of Science* (pp. 133–153). Cambridge: Cambridge University Press. https://doi.org/10.1017/CBO9780511613517.008

Newell, A., & Simon, H. (1976). Computer science as empirical inquiry: Symbols and search. *Communications of the Association for Computing Machinery*, 19, 113–126.

Newman, M. E., & Girvan, M. (2004). Finding and evaluating community structure in networks. *Physical Review E*, 69(2), 026113.

Norman, D. A. (1983). Some observations on mental models. In D. Gentner, & A. Stevens (Eds.), *Mental Models* (pp. 7–14). Hillsdale, NJ: Erlbaum.

Norman, D. A. (1988). *The Psychology of Everyday Things*. New York: Basic Books.

Ouyang, X., & Zhou, J. (2018). Smart TV for older adults: A comparative study of the mega menu and tiled menu. In *Human Aspects of IT for the Aged Population. Applications in Health, Assistance, and Entertainment: 4th International Conference, ITAP 2018, Held as Part of HCI International 2018*, Las Vegas, NV, July 15–20, 2018, *Proceedings, Part II* 4 (pp. 362–376). Springer International Publishing.

Özesmi, U., & Özesmi, S. L. (2004). Ecological models based on people's knowledge: A multi-step fuzzy cognitive mapping approach. *Ecological Modelling*, 176(1–2), 43–64.

Payne, S. J. (2007). Mental models in human-computer interaction. In A. Sears, & J. A. Jacko (Eds.), *The Human-Computer Interaction Handbook* (pp. 89–102). Boca Raton, FL: CRC Press.

Rieh, S. Y., Yang, J. Y., Yakel, E., & Markey, K. (2010, August). Conceptualizing institutional repositories: Using co-discovery to uncover mental models. In *Proceedings of the Third Symposium on Information Interaction in Context* (pp. 165–174).

Rutherford, A., & Wilson, J. R. (2004). Models of mental models: An ergonomist-psychologist dialogue. In N. Moray (Ed.), *Ergonomics Major Writings: Psychological Mechanisms and Models in Ergonomics* (pp. 309–323). London: Taylor & Francis.

Sharit, J. (2012). Human error and human reliability analysis. In G. Salvendy (Ed.), *Handbook of Human Factors and Ergonomics* (pp. 734–800). Hoboken NJ: John Wiley & Sons.

Spencer, D. (2009). *Card Sorting: Designing Usable Categories.* Brooklyn, NJ: Rosenfeld Media.

Stanney, K. M., & Salvendy, G. (1995). Information visualization; assisting low spatial individuals with information access tasks through the use of visual mediators. *Ergonomics*, 38(6), 1184–1198.

Stelmach, G. E. (2006). *Mental Models and the Mind: Current Developments in Cognitive Psychology, Neuroscience, and Philosophy of Mind* (Vol. 138). Amsterdam: Elsevier. https://doi.org/10.1016/S0166-4115(06)X8022-6

Tauber, M. J., & Ackermann, D. (Eds.). (1991). *Mental Models and Human-Computer Interaction 2.* North Holland, Amsterdam.

Tolman, E. C. (1948). Cognitive maps in rats and men. *Psychological Review*, 55, 189–208.

Wang, P., Hawk, W. B., & Tenopir, C. (2000). Users' interaction with World Wide Web resources: An exploratory study using a holistic approach. *Information Processing & Management*, 36(2), 229–251.

Ware, C. (2022). Building mental models: Why we present with visualizations. In *Visual Thinking for Information Design* (pp. 143–162). Cambridge: Elsevier. https://doi.org/10.1016/B978-0-12-823567-6.00008-2

Werner, N. E., Rutkowski, R. A., Krause, S., Barton, H. J., Wust, K., Hoonakker, P., & Carayon, P. (2021). Disparate perspectives: Exploring healthcare professionals' misaligned mental models of older adults' transitions of care between the emergency department and skilled nursing facility. *Applied Ergonomics*, 96, 103509.

Wickens, C. D., Lee, J., Liu, Y. D., & Gordon-Becker, S. (2014). *An Introduction to Human Factors Engineering* (pearson new international edition). London: Pearson Education.

Wiley, D. E. (1967). Latent partition analysis. *Psychometrika*, 32(2), 183–193.

Wilson, J. R., & Rutherford, A. (1989). Mental models: Theory and application in human factors. *Human Factors*, 31(6), 617–634.

Xie, B., Zhou, J., & Wang, H. (2017). How influential are mental models on interaction performance? Exploring the gap between users' and designers' mental models through a new quantitative method. *Advances in Human-Computer Interaction*, 2017, 5193258.

Zhang, Y. (2008). Undergraduate students' mental models of the web as an information retrieval system. *Journal of the American Society for Information Science and Technology*, 59(13), 2087–2098. https://doi.org/10.1002/asi.20915

6 Human Error

Manrong She, Zijian Yin, and Zhizhong Li

6.1 INTRODUCTION

Many reports indicated that human error contributed considerably to accidents and incidents in various safety-critical domains, such as 65% of nuclear system failures (Trager, 1985), approximately 70% of aircraft accidents and incidents (Feggetter, 1982; Wiegmann & Shappell, 1999; Shappell et al., 2006), and more than 80% high-consequence marine accidents (Moore, 1993; Hee et al., 1999). The prevalence of human error can also be found in healthcare (Peters & Peters, 2007; Makary & Daniel, 2016) and other service domains. With computerized systems increasingly adopted, avoiding human error when interacting with these systems becomes a focus of design and operation.

For consumer products, although human errors may not cause any safety issues, failure in use could be very disappointing, annoying, and defeat the users. Human error is a cause of user complaints and dissatisfaction with consumer products.

It is not surprising to see human errors being a popular topic in both research and professional communities. However, in different domains and contexts, it may have different meanings and is often not explicitly defined. The definition issue has been recognized and discussed by researchers for decades (e.g. Reason, 1990; Hollnagel, 1998; Woods et al., 2010; Harris & Li, 2011; Liu et al., 2021).

In human-related incident analysis, even experts hold different opinions on the identification of human errors. Below are some hypothetical examples for discussion of the definition of human error:

1. A nuclear power plant (NPP) operator went through a long navigation path (five steps) to reach an intended interface. If he followed an optimal path, only two steps were needed.
2. An NPP operator keyed in a wrong number of control rods to insert, but before the command was released, he found his error and corrected it.
3. In the previous example, it was not the operator himself, but his teammate found the error and corrected it.
4. An NPP operator adjusted the sequence of two steps in a procedure because he knew that it would not cause any unexpected consequences, although the company required operators to follow standard operating procedures strictly.
5. When performing a step to test lamps in a regular inspection procedure, an NPP operator pressed the two REACTOR TRIP buttons (with covers) near the LAMP TEST buttons and caused an unplanned reactor trip. The operator recalled that when he approached the panel, he thought of the criticism by an instructor in the training one day ago on his nonstandard operation on the REACTOR TRIP buttons, and then he performed it as taught. An investigator of this incident questioned why these buttons were arranged together.
6. According to the design philosophy, any two of three condensate water pumps should be set to running while the third one is set to standby. A supervisor decided to stop one of the two running pumps, thinking that one pump working was enough for the functioning of the system and stopping one pump would save the operation cost of more than $1,000 per day. One day, when the running pump failed, an operator tried to switch on a standby pump but failed, finally causing a reactor trip. The control logic of these pumps supposes that

DOI: 10.1201/9781003495109-6

since two pumps are required to be working at the same time, if one of them fails, there should still be one working pump; when both pumps fail, switching on the standby pump does not help, since the operation requirement cannot be met. That is why the standby pump could not be switched on. Such control logic is different from other systems. The operators did not know about the control logic – nobody told them, and they did not read the equipment manual carefully.

7. An emergency response team realized that the situation was different from design assumptions and decided to adjust the emergency operating procedure according to their knowledge. Their response was proved to be correct finally.

8. In the previous example, the adjusted response procedure was finally found to be incorrect. Investigation indicated that it was a complicated situation.

Hagen and Mays (1981) defined human error as deviations from requirements. Similarly, Hollnagel (1998) proposed to identify human error according to a specified performance criterion or standard. Following this idea, whether the above examples are human errors depends on the requirements and performance criteria. Different companies would set up their specifications. For different investigation purposes, different criteria would be adopted. If it is to improve interface design and training, any deviations from "normal operation" (examples 1, 2, 3, and 5) may be considered as human errors; but if it is to promote safety, example (1) will be ignored. If strict procedure is emphasized, examples (4), (6), (7), and (8) would be considered errors, no matter what the results were. The requirements could be different for normal operations and emergency operations.

Reason (1990) defined human error according to the consequence or unintended outcome of human activity. Following this definition, examples (1), (2), (5), and (7) would not be considered human errors, while examples (4) and (8) would be, but (3) would be confusing. The comparison between examples (7) and (8) strongly supports the argument that human error is an after-fact attribution (Woods et al., 2010, p. 2). The use of any performance criteria in identifying human errors also supports this argument. The comparison between examples (2) and (3) demonstrates a confusion that whether an operator commits an error may depend on what the other operators do.

Another classic definition was given by Sanders and McCormick (1993) as "an inappropriate or undesirable human decision or behavior that reduces, or has the potential for reducing effectiveness, safety, or system performance." This definition is broader since potential consequences are included instead of just accounting for actual ones. Human activities may set up unsafe conditions, but after some uncertain time, they will finally result in actual unexpected consequences. Latent errors are quite common and regarded as critical threats to safety. Harris and Li (2011) suggested pilot error as if pilots do things that they should not do, or do not do things that they should do, thus degrading the system and causing accidents. This statement implies some performance requirements and unsafe conditions (system degradation) caused by pilot activities.

Most experts do not consider the intention of performance or nonperformance when defining human error. Neither do they consider whether a human has the ability or possibility to avoid the "error" except Hollnagel (1998). It is fair to consider the ability and possibility aspects when clarifying the responsibility of an operator. A good safety culture is blame-free. It is always the highest priority to identify safety issues, thus a broader definition of human error would be preferred.

For general human-computer interaction studies, human error can be defined as deviation from intended or foreseen usage. Such a definition would help identify interaction design problems or indicate how to improve the design. For safety-critical systems, it would be more meaningful. As pointed out by Sheridan (1997), computer-based systems are growing more complex, and human error could precipitate a disaster. Finding out potential interaction deviations and then analyzing their effects on safety would be one good way to identify hazards in a system.

Deviation from procedures, requirements, or expectations only sometimes leads to unexpected or unsatisfactory consequences. Such an example is (7). Supporting correct deviations or adaption should be considered in the design of safety-critical systems since conditions beyond design considerations would happen occasionally. Unlike procedures for "normal operations," procedures for unpredicted conditions often require operators to make judgments and decisions based on their knowledge. Information provided in system interfaces would be critically important.

In practice, human errors are first attributed to front operators and then to management system issues. Actually, most human errors can also be traced back to interaction design defects. Controlling the source is always the first priority, so interaction design to prevent, catch, and tolerate human errors has received much attention for computerized systems.

In this chapter, we first discuss how humans commit errors (Section 6.2), why they commit errors (Section 6.3), what types of errors they could commit (Section 6.4), and how we analyze human errors in both proactive and retrospective approaches (Section 6.5). Based on this foundation, we then summarize special issues pertinent to the interaction between humans and computers (Section 6.6). This is followed by a discussion on how to prevent and control human errors through design, particularly in the HCI context (Section 6.7). We conclude with a discussion on the challenges and opportunities that new technologies bring to human error (Section 6.8).

6.2 HUMAN ERROR MECHANISMS

To start this chapter, we first discuss the underlying mechanisms of human errors. Studies on why humans go wrong can date back to the 1900s when the followers of Freud and psychodynamics, the behaviorists, and the researchers from Gestalt psychology began to investigate various errors (Read et al., 2021). Following those pioneer studies, more and more relevant theories and models have been developed to explain human errors and human cognition later, especially after several severe accidents such as the Three Mile Island nuclear disaster. In this section, to provide readers with a broad horizon of human error mechanisms, studies from four different but interrelated disciplines will be introduced, namely cognitive psychology and cognitive ergonomics (Section 6.2.1), human reliability analysis (Section 6.2.2), computational cognitive science (Section 6.2.3), and neuroergonomics (Section 6.2.4).

6.2.1 Human Performance Models

Human performance models interpret the process of human cognition and can therefore explain the error mechanisms. Read et al. (2021) summarized four perspectives for understanding human errors, including the mechanistic, individual, interactionist, and systems perspectives. The mechanistic perspective is underpinned by engineering principles and Newtonian science and tends to predict human behavior deterministically. As a step forward, the individual perspective is more focused on human cognition. Nonetheless, the influences of contextual and organizational factors still need to be considered, and this weakness is partly addressed from the interactionist perspective. At last, underpinned by the philosophy of systems thinking, the systems perspective takes the system as the unit of analysis and recognizes the important properties of complex systems, such as dynamicity and non-linear interaction. Following this distinction, Read et al. (2021) provided an overview of the evolution of mainstream models, methods, and underpinning theories, as shown in Figure 6.1. Next, several widely accepted human performance models in the human factors and ergonomics (HF/E) community are described based on Read et al.'s review (2021).

6.2.1.1 Human Information Processing Models

The basic idea of human information processing (HIP) approaches is that human performance is a function of several processing stages (Proctor & Vu, 2012). Among different models, the HIP model proposed by Wickens (1984) depicts a general psychological process of human activities. As presented in Figure 6.2, Wickens's HIP model identifies a series of processing stages through which humans interact with systems to accomplish their tasks. For example, consider the task in which an NPP operator responds to alarms. The process begins with sensation, through which visual and

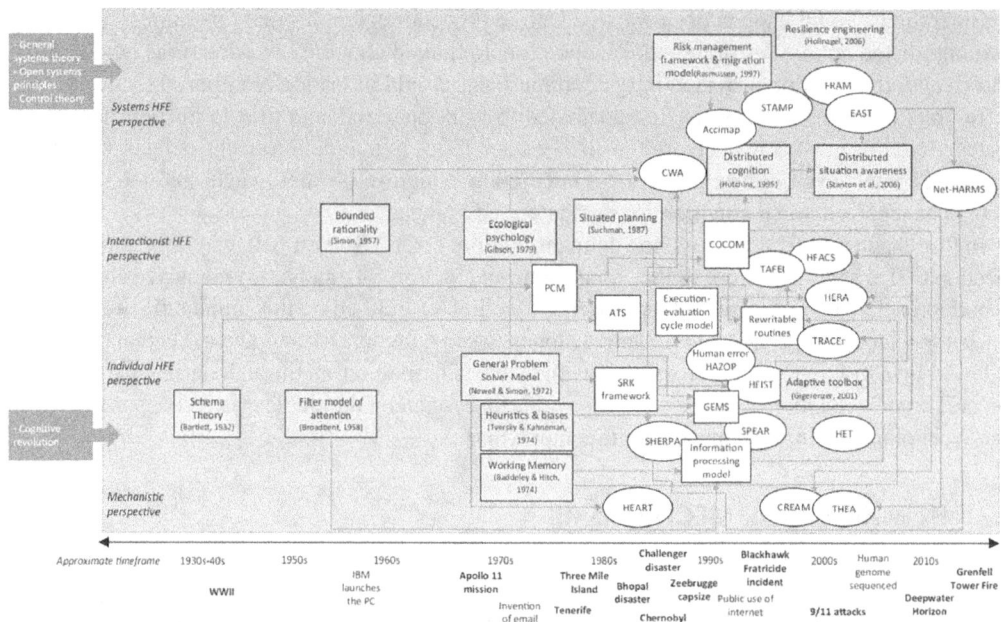

FIGURE 6.1 Overview of the evolution of models (white boxes), methods (white ovals), and underpinning theories (gray boxes). (Read et al. (2021); Reproduced with permission of Taylor & Francis Group.)

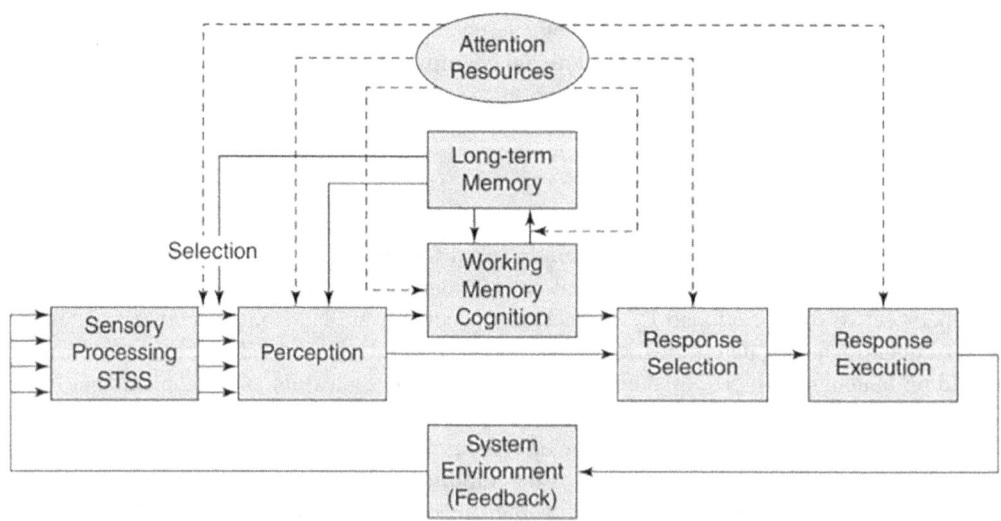

FIGURE 6.2 Human information processing model. (Wickens et al. (2013); Reproduced with permission of Taylor & Francis Group.)

auditory stimuli in the physical world are initially detected. Thus, the operator will see the sudden colorful light during his monitoring of panels or screens. Shortly after, the sensory information will go through the perception stage, which enables the operator to interpret sensory information based on their past experiences stored in the long-term memory (LTM). Hence, the operator can recognize what alarm is triggered. After the perception stage, there usually exist two pathways of further processing. One is the bottom pathway, where response selection is directly triggered.

Explicit rules often direct such a process, for example, a procedural step that requires the operator to perform certain actions after detecting an alarm. Following the response selection, the response is finally executed, i.e. the goal-directed motor interaction with the system. However, sometimes a higher level of cognition is required before response selection, forming the upper pathway. In this case, all information will be temporarily stored and manipulated in the working memory, where perceived information will be integrated with the knowledge stored in LTM. This process is usually known as understanding and sensemaking, or situation assessment, diagnosis, etc. The above processes enable operators to respond to an external stimulus, while there are two more vital elements in the HIP model, i.e. attention and feedback. Attention acts as a filter of perceived information so that only partial information will be selected and processed, as well as the fuel that provides essential but limited resources for every information processing stage, as indicated by the dashed lines in Figure 6.2. As for the feedback loop, response execution may induce some changes in the environment, which are new information to be detected by operators. Wickens's HIP model implies that human errors can occur due to any stage of information processing failure. Such failures, for example, can arise from the threshold of sensory systems, the illusion in perception, the restriction of memory and attention, the heuristics and biases in decision-making, etc. For more discussion about these failures, please refer to Liu et al. (2021).

The HIP model provides an inclusive structure that is linked with numerous theories and models in cognitive psychology, such as the SEEV model of visual attention (Wickens et al., 2001), the working memory model (Baddeley, 2000), and the RPD model of decision-making (Klein, 1993). This information processing view also serves as the basis of the well-known situation awareness model proposed by Endsley (1995b). Having said this, it should be noted that potential problems exist with such stage-based approaches, including the misleading for error prevention and the lack of validity (Sträter, 2005). Furthermore, the traditional linear information processing view has been argued in later studies.

6.2.1.2 Skill-Rule-Knowledge Framework

The Skill-Rule-Knowledge (SRK) framework proposed by Rasmussen (1983) identifies three behavior levels depending on the task requirements of human cognition. According to this framework, skill-based (SB) behaviors are triggered when operators carry out highly-familiar and well-trained activities such as assembly tasks. Such behaviors require only automated sensory-motor performance with little conscious attention or control. At the rule-based (RB) level, behaviors are controlled by stored rules or procedures, which are usually derived from past experience. In such scenarios, operators do not need to formulate a functional understanding but recognize specific signs and select corresponding rules or know-how for control. When it comes to unfamiliar situations where no know-how or rule is available, operators have to formulate explicit goals, analyze the situation, and develop a plan to accomplish the task, all of which constitute knowledge-based (KB) behaviors. Note that the boundaries between these performance levels are not quite distinct but more likely to be smooth transitions. Based on this framework, errors at the SB level relate to variability of force, space, or time coordination; errors at the RB level involve incorrect classification or recognition of situations, erroneous associations to tasks, or memory slips in recall of procedures; errors at KB level relate to goal selection and others (Read et al., 2021).

The SRK framework, alongside Rasmussen's other work in the 1980s, has been applied and adapted in many studies (Waterson et al., 2017). One remarkable application is the development of categorization schemes for human errors. For example, as a classical extension, Reason (1990) proposed the generic error modeling system (GEMS) to represent an integrated picture of error mechanisms based on the SRK framework and provided a detailed taxonomy of error modes at each performance level. Soon to be discussed in Section 6.4, the SRK-based error taxonomy and the GEMS have been applied in many studies later (e.g., Targoutzidis, 2010).

6.2.1.3   Schema-Based Models

A group of models is underpinned by the schema theory, which was popularized by Bartlett (1932). Researchers have proposed diverse definitions for the term *schema*, while a schema can be generally considered as "an organized mental pattern of thoughts or behaviors to help organize world knowledge," according to Neisser's explanation (Plant & Stanton, 2017b). Neisser further compared the schema to a format that "specifies that information must be of a certain sort if it is to be interpreted coherently," as well as a plan for "finding out about objects and events" and "obtaining more information to fill in the format" (Neisser, 1976). Therefore, a schema is a pattern of action as well as a pattern for action. There exist both genotype and phenotype schemas (Neisser, 1976), the former referring to the "wider systemic factors that influence the development of individual cognitive phenomena and behavior" and the latter referring to "local, individual-specific manifestations of the genotype schemata" (Plant & Stanton, 2017b). Readers interested in the schema theory are directed to Plant and Stanton's introduction (2013, 2017b).

6.2.1.3.1   Perceptual Cycle Model

Neisser created a model to explain how schemata are utilized when humans interact with the world, known as the perceptual cycle model (PCM; Neisser, 1976). PCM reveals the cyclic relationships between the world, schema, and action. Environment information in the physical world can modify one's understanding of the situation and update their schema; after that, the schema can direct their anticipation of certain information and guide the formation of mental or physical actions which enable him/her to seek out more information from the environment and further interact with the environment. Neisser's ecological view is in juxtaposition to the information processing view and emphasizes several important characteristics of cognition: (1) It is active rather than passive, as one's schema can direct their behaviors to explore the world; (2) it is cyclical rather than linear as illustrated in the cyclic structure; (3) it is parallel rather than serial, as the information modalities are multiple. PCM demonstrates the top-down and bottom-up processes through which humans interact with the world. Note that the original PCM only provides a high-level explanation of the cognition process. To improve the usability, Plant and Stanton (2017a) developed a schema-action-world (SAW) taxonomy with 28 detailed items to specify the decision-making process, as well as a schema-world-action research method (SWARM) for eliciting perceptual cycle details (Plant & Stanton, 2016). PCM, alongside relevant methods, can help to analyze human errors in decision-making, as inappropriately activated or incomplete schemas would impair the decision-making process. Aligning with Dekker's proposition (2014) that studies on human errors are supposed to answer why the human actions and assessments made sense in that scenario, PCM provides a process-oriented way that allows us to explain the so-called *local rationality* (Dekker, 2011). It has therefore been used to explain human errors involved in many accidents, such as the 1989 Kegworth disaster (Plant & Stanton, 2012), the 2007 Kerang rail crash (Salmon et al., 2013), the 2016 Tesla crash (Banks et al., 2018), and the 2018 WK050 crash (Lynch et al., 2022).

6.2.1.3.2   Activation-Trigger-Schema Model and CS/SAS Model

Unlike PCM, which mainly explains the interaction process between humans and the world, the activation-trigger-schema (ATS) model proposed by Norman (1981) focuses on the process of carrying out action sequences. According to the ATS model, any action sequence is specified by an ensemble of schemas, which is organized in a hierarchical structure. Once an intention is set up, the corresponding high-level schema will be activated, along with its child schemas. All these schemas possess specific triggering conditions and will be triggered only if the triggering conditions are met and that schema is meanwhile at a sufficiently high activation level. During the action sequence, there can be numerous schemas being activated, but only triggered schemas can invoke actions. Norman (1981) applied this model to interpret action slips, indicating that action slips could arise from error formation of the intention, faulty activation of schemas, and faulty triggering of active schemas. This error taxonomy will be further discussed in Section 6.4.1.

As a step forward, Norman and Shallice (1986) distinguished two types of action control, i.e., automatic action, which requires minimal processing resources in well-learned routine tasks, and willed action, which requires active and directed attention in non-routine tasks. In well-learned routine tasks, as mentioned above, the setup of an intention will activate a host of schemas, each with a specific activation value. Once the activation value of one schema exceeds a threshold, the schema is selected and continues to operate until the goal is reached or the schema is blocked. It is then essential for humans to schedule numerous activated schemas and manage the conflicts. This process is achieved, as Norman and Shallice (1986) proposed, through the *contention scheduling* (CS) system, which "resolves competition for selection, preventing competitive use of common or related structures, and negotiating cooperative, shared use of common structures or operations when that is possible." When it comes to novel or complex tasks, a schema might be unavailable to control the desired actions, so an additional system termed *supervisory attentional system* (SAS) is in need to control actions "through the application of extra activation and inhibition to schemas in order to bias their selection by the contention-scheduling mechanisms" (Norman & Shallice, 1986), as illustrated by the vertical threads in Figure 6.3 which show how attention operates on schemas through altering the activation values. In a word, the CS/SAS model indicates that schemas operate only when being selected; the selection of schemas depends solely on the activation values exceeding the threshold; the activation values of schemas can be influenced by multiple factors such as mutual inhibition between schemas with overlapping resource requirements, triggering from the environment, triggering from higher-level schemas, and excitation or inhibition from SAS (Cooper, 2002).

The CS/SAS model has inspired many subsequent studies on the control of action within the discipline of neuropsychology, especially for patients with neurological disorders and diseases as well as healthy adults. Patients with action disorganization syndrome (ADS) or apraxia often make many egregious cognitive errors during everyday tasks, including both omission and commission errors. Many studies have been conducted to explore the disorder of everyday action using the CS/SAS model.

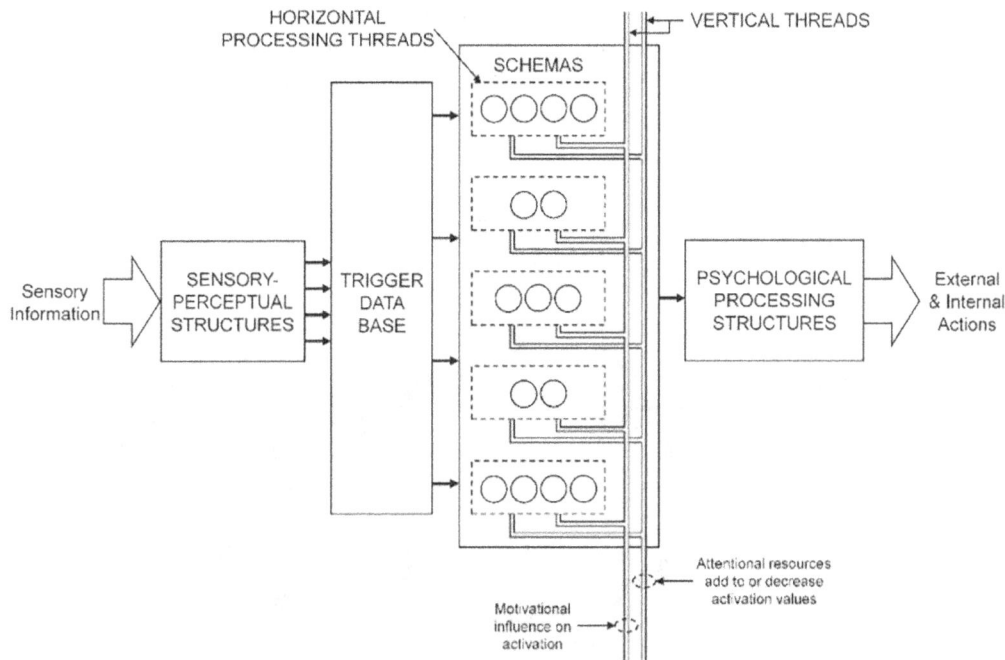

FIGURE 6.3 The overall CS/SAS system: Vertical and horizontal threads. (Norman and Shallice (1986); Reproduced with permission of Springer Nature.)

6.2.1.4 Contextual Control Model

The background discipline of the contextual control model (COCOM; Hollnagel, 1993a) is Cognitive Systems Engineering (CSE), which provides a different perspective to analyze how humans and technology work together by emphasizing the human-technology coagency rather than the interaction of different parts (Hollnagel & Woods, 2005). Hence, the key point of modeling is to understand **what** the joint system does and **why** it does that rather than to explain **how** it does that. Although physically separated, humans and machines now should not be treated as functionally separated. Therefore, both living organisms and machines or artifacts can be *cognitive systems*, which are defined as "a system that can modify its behavior on the basis of experience so as to achieve specific antientropic ends" (Hollnagel & Woods, 2005). The focus on joint system performance, from this viewpoint, should be around how the joint cognitive system maintains control of what it does, where *control* refers to "the ability to direct and manage the development of events, and especially to compensate for disturbances and disruptions in a timely and effective manner" (Hollnagel & Woods, 2005). This cyclic control process can be demonstrated in Figure 6.4, where both feedback control and feedforward control are involved. Moreover, it is emphasized to attend to the "cognition in the wild," which focuses on the situated cognition in the natural environment compared to the "cognition in the mind." Hence, the actions of cognitive systems, through which cognitive systems implement the control, are supposed to be determined by the current context as well as the competence of cognitive systems. Based on the aforementioned philosophies, COCOM is constituted of three main components (Hollnagel & Woods, 2005): (1) *Competence*, which is the set of all possible actions that cognitive systems can carry out, (2) *control*, which is the orderliness of performance and the way competence is applied, and (3) *constructs*, which are cognitive systems' descriptions of the situation and are similar to schemas. COCOM further distinguishes four different control modes that correspond to different characteristics in the orderliness performance, namely (1) scrambled control mode, in which cognitive systems are nearly unable to deal with the situation and the choice of next action is random, (2) opportunistic control mode, in which the planning and anticipation are limited, and the determination of next action mainly depends on the salient features in the current context rather than more stable intentions, (3) tactical control mode, in which performance depends on planning though of limited scope, and often follows available procedures or rules, and (4) strategic control mode, in which cognitive systems take a longer time horizon and consider higher-level goals

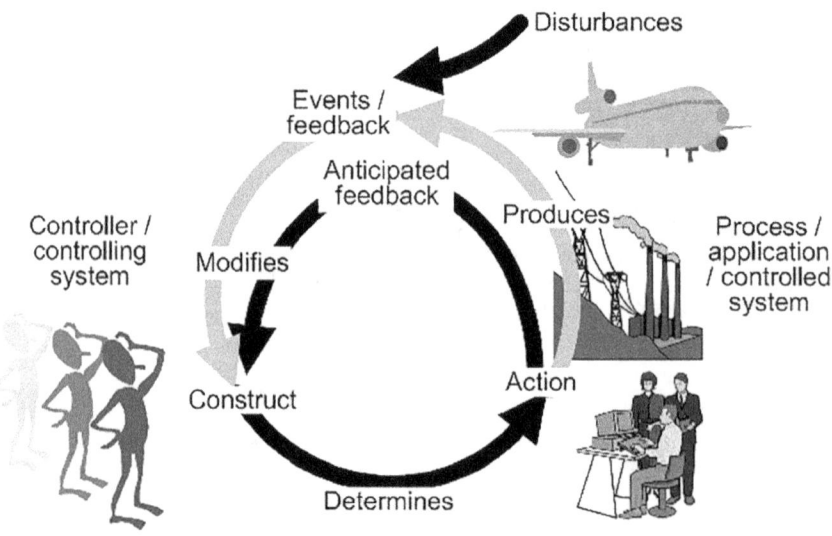

FIGURE 6.4 Contextual control model. (Hollnagel and Woods (2005); Reproduced with permission of Taylor & Francis Group.)

to best deal with the global situation. Based on COCOM, Hollnagel and Woods (2005) further proposed an extended control model (ECOM) to describe concurrent control loops in different layers.

As a functional model with a macro-perspective to interpret human behavior, COCOM has been broadly applied to understand human performance. In addition, COCOM serves as the theoretical model for CREAM (Cognitive Reliability and Error Analysis Method; Hollnagel, 1998), a well-known human reliability analysis method which has been broadly applied to estimate the probability of human error (e.g., Zhang & Tan, 2018) and analyze human-related accidents (e.g., Moura et al., 2016).

6.2.2 Cognitive Models in Human Reliability Analysis

Human reliability analysis (HRA) is an interdisciplinary research area aiming at quantitatively assessing the probability that humans correctly perform system-required activities and do not perform extraneous activities (Swain & Guttmann, 1983). In other words, quantifying the human error probability (HEP) is the primary objective of most HRA studies. Since the 1960s, when military HRA was transferred to civil applications (Pyy, 2000), dozens of HRA methods have been developed (Boring, 2012). Early pioneer methods, such as THERP (Technique for Human Error Rate Prediction; Swain & Guttmann, 1983), HCR (Human Cognitive Reliability; Hannaman et al., 1984), SLIM (Success Likelihood Index Methodology; Embrey, 1986b), and HEART (Human Error Assessment and Reduction Technique; Williams, 1988), were influenced by traditional reliability analysis and treated human as a system component. Such methods, usually referred to as the first-generation methods (Dougherty Jr, 1990), have come under several criticisms, such as the less-than-adequate psychological realism and questionable assumptions about human behaviors (Swain, 1990). Such criticisms motivated researchers to incorporate broader insights from various disciplines, like behavioral science and cognitive psychology, to consolidate the theoretical basis of HRA methods, leading to the development of second-generation methods. One such example is CREAM, as mentioned in the previous section. Several cognitive models in HRA methods have been summarized in Pan et al.'s review (2017). Here, we briefly introduce the cognitive basis model (hereafter referred to as IDHEAS macro-cognition model) of the IDHEAS-G method (General Methodology of an Integrated Human Event Analysis System; Xing et al., 2021), which is a recently developed HRA method by U.S. Nuclear Regulatory Commission (NRC).

The IDHEAS macro-cognition model is a synthesized model that has integrated a range of research on human cognition, with the literature review results documented in the report NUREG-2114 (Whaley et al., 2016). As demonstrated in Figure 6.5, the model starts with the macro-cognitive function (MCF). The term macro-cognition, juxtaposed with the micro-cognition, was originally coined by Cacciabue and Hollnagel (1995) to emphasize cognition in real-world settings rather than in well-controlled experiments, wherein humans were often faced with complicated situations, high time pressure, and risky choices. Five MCFs are identified in this model to characterize the cognitive activities of NPP operators: (1) Detection, which is the process of noticing cues or gathering information from the environment; (2) understanding, which is the integration of obtained information

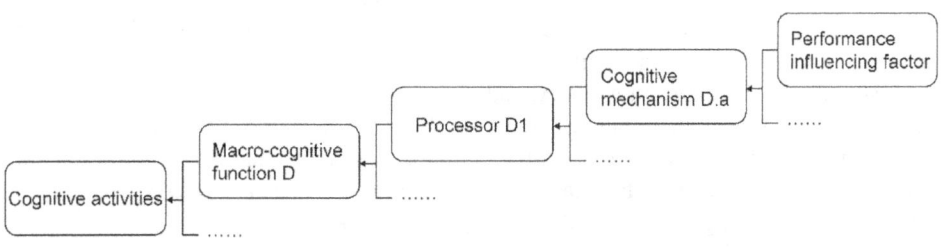

FIGURE 6.5   IDHEAS macro-cognition model. (Adapted from Xing et al. (2021).)

with one's mental model to form the interpretation of current situation; (3) decision-making, which includes selecting strategies, planning, adapting plans, evaluating options, and making judgments; (4) action execution, which is the physical implementation of planned actions; and (5) inter-team coordination, which refers to the interaction and collaboration between different teams. The structured framework is shown in Figure 6.5. Take the MCF understanding, for example, to demonstrate the model details. The processors of understanding include (1) assessing/selecting data, (2) selecting/adapting/developing the mental model, (3) integrating data with the mental model to generate the outcome of understanding, (4) verifying and revising the outcome through iteration of the above processors, and (5) exporting the outcome. Crucial cognitive mechanisms to enable these processors include data, selection of data, mental model, integration of data with the mental model, working memory, and shared cognition within a team. Furthermore, the functioning of each mechanism can be affected by various contextual factors (i.e., performance influencing factors in Figure 6.5); for instance, working memory can be influenced by system transparency, procedures/guidance/instructions, training, scenario familiarity, multi-tasking, etc.

The IDHEAS macro-cognition model serves as the cognitive basis model for the IDHEAS suite (Chang et al., 2016), including the general guidance (IDHEAS-G; Xing et al., 2021) and application-specific methods such as IDHEAS-AtPower (Xing et al., 2017) and IDHEAS-ECA (Xing et al., 2020). In addition, researchers in the HRA community have tried to apply this model in different sectors. For example, Pandya et al. (2018) analyzed generic tasks in radiotherapy based on this model; Liu et al. (2017) applied the model to analyze human error incidents/accidents in petrochemical plants; Yin et al. (2020) applied the model to analyze human error incidents/accidents in NPP commissioning tasks. The results of these applications imply that the IDHEAS macro-cognition model can explain the underlying mechanisms of human errors in diverse sectors.

6.2.3 ERROR MECHANISMS REFLECTED IN COGNITIVE ARCHITECTURES

A cognitive architecture is the overall structure and process of a domain-generic computational cognitive model (Sun, 2004), aiming at modeling the human mind and ultimately building human-level artificial intelligence (AI) (Kotseruba & Tsotsos, 2016). It was estimated that, since the 1950s, around 300 cognitive architectures have been developed, of which at least one-third are still active (Kotseruba & Tsotsos, 2016). There exist three major paradigms of cognitive architectures, namely symbolic architectures, emergent architectures, and hybrid architectures (Duch et al., 2008; Kotseruba & Tsotsos, 2016; Ye et al., 2018). Since different cognitive architectures are based on radically different assumptions and specific world views (Sun, 2004), it is difficult to provide a complete overview in this section. As a compromise, we next briefly introduce a typical hybrid architecture ACT-R, not only because it is a representative model cited by numerous studies but also considering its numerous successful applications as well as the agreement with experimental data.

ACT-R (Adaptive Control of Thought-Rational, also referred to as Atomic Components of Thought) was proposed by Anderson (1993). The underlying assumption is that human cognition should be implemented in terms of neural-like computations (Leiden et al., 2001). Based on that, ACT-R is made up of four main modules along with corresponding buffers and one production system; modules are mechanisms of cognitive activities, and buffers store specific contents (Ritter et al., 2019), as depicted in Figure 6.6. The four main modules are: a visual module for identifying objects, a goal module for managing current goals, a declarative module that mimics the process of retrieving stored information from memory, and a manual module for controlling hands to execute physical actions after receiving command requests through the motor buffer (Ritter et al., 2019). The central production system coordinates the communication and functioning of all modules by means of production rules, which take the form of condition-action pairs. Once a particular condition is met, the production rule triggers the corresponding action.

ACT-R can simulate human errors by appropriately configuring the key components. Lebière et al. (1994) have demonstrated how to model omission errors, by applying the latency threshold to

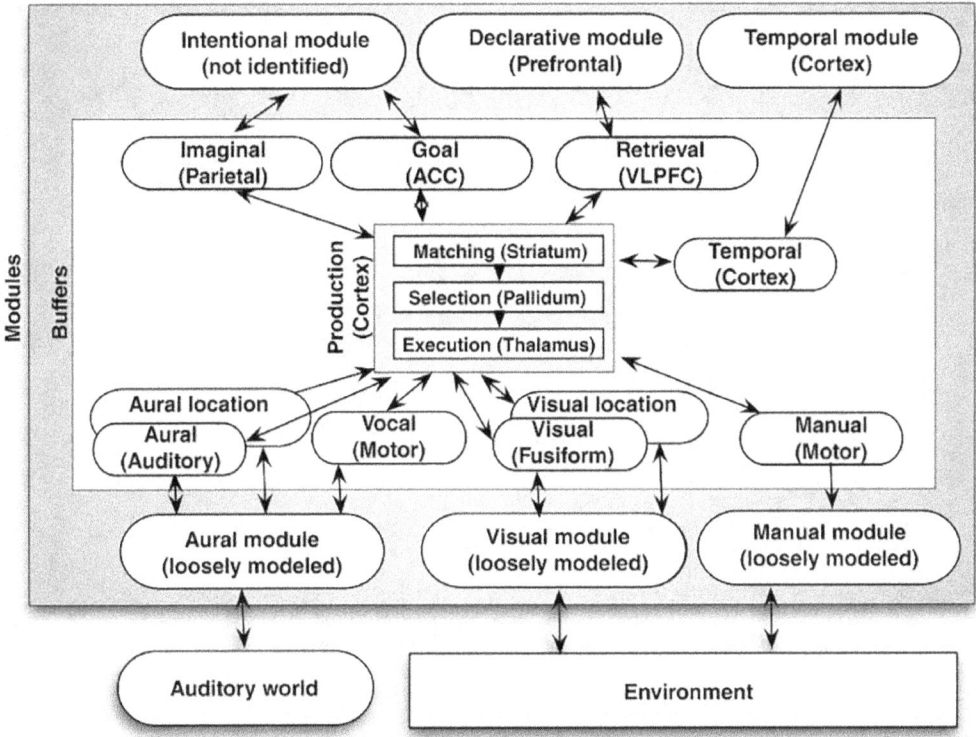

FIGURE 6.6 Schematic diagram of the ACT-R. (Ritter et al. (2019); Reproduced with permission of John Wiley & Sons.)

produce failures of retrieving a chunk with a low activation level, and commission errors, by allowing the partial matching of chunks to be retrieved which could cause a misretrieval. More general error mechanisms which are essential for error modeling in ACT-R, including plan developer, compare actions, monitor, attention, bias mechanism, rule match, schema match, time constraint mechanism, decay, poor learning, retrieval mechanism, plan controller, perceptual mechanism, association developer, and motor mechanism, were briefly assessed in Fotta et al. (2005). In an assessment of several cognitive models, Alvarenga and Frutuoso e Melo (2019) assessed that ACT-R was capable of modeling various errors such as errors related to working memory, monitoring errors, expectation errors, attention errors, and errors in similarity matching and frequency gambling.

6.2.4 Neuro Mechanisms of Human Errors

Preceding sections have talked about human error mechanisms ranging from cognitive functions or information processing stages at the macro-level (e.g., detection, response planning) to cognitive mechanisms or processors at the micro-level (e.g., schema, memory). Then what if we dive deeper? Neuroscience could answer this question, and the more related discipline here is neuroergonomics. Neuroergonomics is an emerging interdisciplinary field that focuses on the human brain in relation to performance at work and other everyday settings (Parasuraman, 2011). Despite certain challenges (Gramann et al., 2021), findings in neuroergonomics research can bring us useful insights into the brain activities underlying cognition in not only laboratory conditions but also real-world settings thanks to the development of portable and mobile neuroimaging techniques (Wascher et al., 2021). For more details about this interdisciplinary, readers are directed to relevant reviews (Ayaz & Dehais, 2021; Mehta & Parasuraman, 2013).

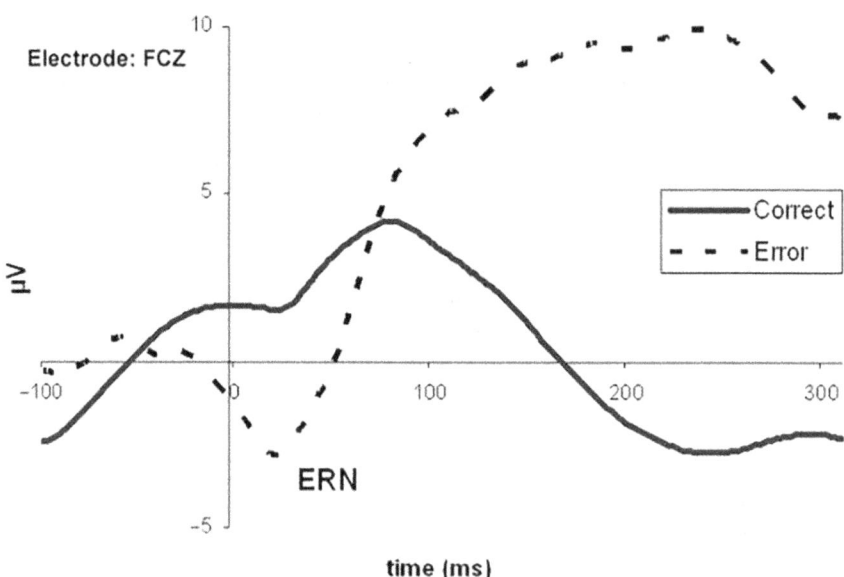

FIGURE 6.7   Response-locked error-related potential waveform from an Eriksen Flanker Task. (Fedota and Parasuraman (2010); Reproduced with permission of Taylor & Francis Group.)

Neuroergonomics can enhance our understanding of human errors, revealing how human performance is constrained by the human brain (Fedota & Parasuraman, 2010). Fedota and Parasuraman (2010) have illustrated how error-related negativity (ERN), a specific type of event-related potential (ERP), can explain the neuro mechanisms of action slips. As demonstrated in Figure 6.7, ERN is characterized by a large negative voltage deflection peaking within 100 ms of an erroneous response (Fedota & Parasuraman, 2010). The ERN dipole was estimated to lie within the anterior cingulate cortex (ACC) (Dehaene et al., 1994; Mathalon et al., 2003), which has been recognized as a region involved in processing behavioral error (Alexander & Brown, 2019). Fedota and Parasuraman (2010) summarized that ERN had relationships with abstract cognitive representations of response execution, and was associated with the error detection process when slips occurred. Yet, there are considerable debates around the functional significance of ERN. Gehring et al. (2012) have summarized several major theories about ERN, namely (1) error detection/comparator theory, which posits that ERN reflects the process that compares the output of the motor system to the best estimate of the correct response, (2) conflict-monitoring theory, which explains ERN as a product of response conflict, i.e., concurrent activation of multiple competing responses, (3) reinforcement learning theory, which proposes that the error signal produced by the monitoring mechanism when events are worse than expected will be conveyed to ACC, to improve task performance by changing control allocation over motor systems, and (4) affective/motivational theory, which hypothesizes that the ERN just represents an affective response to errors. In spite of these debates, ERN has been evoked and detected in visual search of complex stimuli, implying the value of this promising approach in real-world settings (Sawyer et al., 2017).

Besides the aforementioned ERN, numerous research efforts are trying to clarify the neural basis of human mental states (Ayaz & Dehais, 2021). One example is the mental workload, a multidimensional construct that can be explained in terms of demand/resource balance and has a close relationship with human errors (Young et al., 2015). A range of neurophysiological measurements has been used regarding the assessment of the mental workload (Babiloni, 2019), such as EEG, electrooculography (EOG), and heart rate (HR) (Borghini et al., 2014). Previous studies have revealed that the prefrontal cortex (PFC) could be the brain area of interest in probing the mental workload (Ayaz & Dehais, 2021). Some researchers have highlighted the importance of the neurophysiological mechanisms of mental workload and proposed that the investigation into energy mobilization of neurovascular coupling helped understand the mental resources (Mandrick et al., 2016). Yet, some researchers argued that impaired performance due to high mental workload actually resulted from neurological mechanisms to prioritize goals rather than limited resources *per se* (Dehais et al., 2020). In Dehais et al.'s review (2020), several suboptimal neurocognitive states, such as perseveration and effort withdrawal, were considered, which were relevant to performance degradation. For example, perseveration refers to a tendency for humans to persist in a certain action even if the action is no longer relevant to the goal. The inhibitory mechanisms could explain this phenomenon: in complex and novel scenarios, excessive noradrenaline and dopamine released for neuromodulation could suppress the activity of PFC, which is responsible for routine cognitive operations, and activate subcortical areas, which trigger automated schemes and initiate automatic responses (Dehais et al., 2020). These suboptimal neurocognitive states could be mapped along with task engagement and arousal, as demonstrated in Figure 6.8.

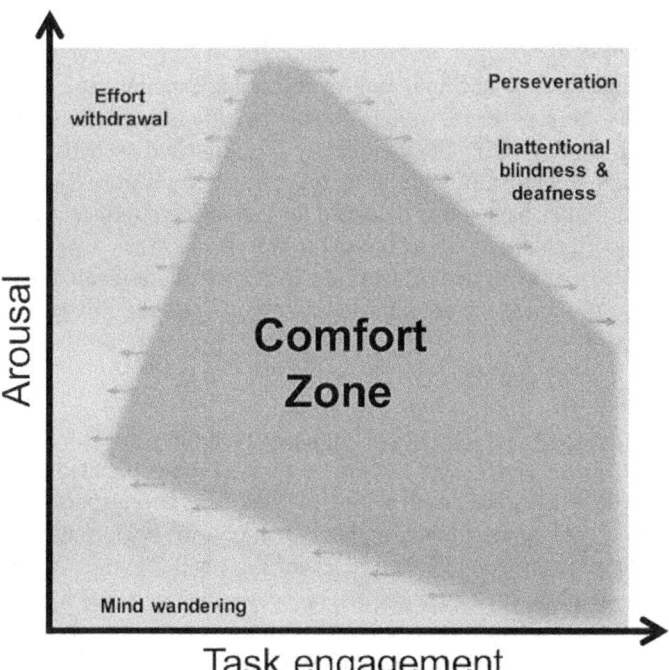

FIGURE 6.8 Performance, arousal, and task engagement. (Dehais et al. (2020); Copyright© 2020 Dehais, Lafont, Roy, and Fairclough.)

6.3 CONTRIBUTING FACTORS

Human performance can be influenced either positively or negatively by various factors, and those negative factors will lead to performance degradation and human errors. Such factors are also known as performance shaping factors (PSFs) or similar concepts in HRA (e.g., Gertman et al., 2005; Swain & Guttmann, 1983; Xing et al., 2021), where PSFs are used to modify HEPs. This section will discuss some salient contributing factors in different components of a general sociotechnical system. The complete set of potential factors can be found in relevant PSF studies (e.g., Groth & Mosleh, 2012; Kim & Jung, 2003). Before the discussion, it is noteworthy that the constructs emerging during the human-computer interaction (HCI) process, such as situation awareness, mental workload, and trust in automation, to name a few, are not deemed as contributing factors here. Nonetheless, such emerging constructs serve as important mediators to explain the forming of human errors. Besides, the underlying mechanisms of the effects of these factors are not discussed in great detail here, some of which are still controversial in extant studies. Readers should keep in mind that all these factors jointly influence human performance with complicated inter-relationships, while the effects also depend heavily on the task type.

6.3.1 Technological Factors

Technological factors are properties of the technology systems which humans interact with, and such technology systems range from traditional industrial systems to automation and autonomy. Below are some common technological factors that could contribute to human performance degradation and human errors.

6.3.1.1 System Complexity

Past decades have witnessed the accumulation of technology increments, which has brought us many benefits, such as more powerful system functions, alongside the cost of increased system complexity (Woods & Cook, 1991). The complexity of automated systems can challenge operators' mental models and significantly undermine their situation awareness (Endsley, 1996, 2016). Previous experimental studies have found evidence for the negative influences of increased system complexity on human performance with regard to the workload (Ntuen & Watson, 1996), task success rate (Lee & Ji, 2019), and error rate (Yang et al., 2021). Hence, rationalizing the internal system complexity is essential to reduce the complicated nature of systems, finally achieving design for safety (Boy & Schmitt, 2013).

6.3.1.2 System Reliability

The effect of system reliability on human performance is not as straightforward as the effect of complexity. Among different systems, one common topic is the automated auxiliary system that could aid operators in their tasks, such as detecting malfunctions or unexpected objects. Some studies reported a degraded performance under higher detection threshold of alarm systems (resulting in higher missed alarm probability and lower false alarm probability) even if the overall reliability (considering both missed alarms and false alarms) of the alarm system was higher and the researchers explained that the cost of increased missed alarm rates was more significant than the benefits of increased overall reliability (Bustamante et al., 2007). This explanation aligns with the findings of Yin et al. (2019) that higher miss rates of participants were reported when the automation system made misses errors than false alarms. Some studies reported an insignificant main effect of reliability on human performance, but the information accuracy of system reliability was found to have a significant effect (Avril, 2023). In fact, many variables could mediate the effect of system reliability, such as trust and trust calibration, under and over-reliance, and complacency, finally leading to different strategies adopted by operators at different reliability levels (Avril et al., 2022). Though operators are sensitive to automation reliability, the concrete effects may manifest themselves in different ways (Chavaillaz et al., 2016).

6.3.1.3 System Transparency

System or automation transparency reflects how much a system is understandable for humans to "see into" it. van de Merwe et al. (2022) reviewed relevant studies in the HF/E community and found that most data revealed the beneficial effect of transparency (e.g., Skraaning & Jamieson, 2019; Stowers et al., 2020) on human performance and situation awareness. Appropriate transparency, based on what we "should see" and what is "most beneficial to see" (Chen et al., 2020), would promisingly benefit human performance and reduce human errors.

6.3.1.4 Human-System Interface (HSI)

HSI is the medium through which humans exchange information with systems. For HSI designers, "what" to present and "how" to present are two important issues to be considered, as poor HSI can be a critical risk jeopardizing safety (Naderpour et al., 2015). For the question of "what" to present, appropriate transparency does not mean an overwhelming barrage of information (Chen et al., 2020), but rather the essential information to support operators' tasks. Just as Lyu et al. (2023) argued, what is essential to human diagnosis is not information quantity/amount but adequacy and relevance. As for "how" to present, there have been numerous studies on the ergonomic design of HSI (e.g., O'Hara & Fleger, 2020). For example, the idea of ecological interface design (EID) was proposed to support the human perception of the work environment (Burns & Hajdukiewicz, 2004).

6.3.2 TASK FACTORS

A task is an activity to be performed by humans to achieve a goal. Tasks in different sectors possess different characteristics and requirements for humans. Some tasks, such as visual search tasks (e.g., Ho et al., 2001), have high requirements on human perception, while other tasks, such as diagnosis tasks (e.g., Lyu et al., 2022), have high requirements on human sensemaking and decision-making. Despite these differences, some common factors could contribute to human errors.

6.3.2.1 Task Complexity

Many methods have been proposed to measure task complexity (e.g., Liu et al., 2012; Park, 2009; Park et al., 2001; Zhao, 1992). Significant correlations between the complexity scores predicted by these methods and error rates have been found in experimental data (Liu & Li, 2014). Therefore, restraining task complexity in practice is meaningful for error reduction. Liu and Li (2012) proposed a task complexity model, which is comprised of ten dimensions, including size, variety, ambiguity, relationship, variability, unreliability, novelty, incongruity, action complexity, and temporal demand. Yet, whether some dimensions are suitable to be deemed as a component of complexity deserves more discussion, such as novelty and temporal demand.

6.3.2.2 Task Familiarity

Task familiarity has been identified as one PSF in HRA studies (e.g., Kang & Seong, 2020; Yin et al., 2021), which would influence HEPs. The effect of this factor on human performance is also salient in experimental studies, known as the practice effect (Kantowitz et al., 2009). This factor includes subjective properties, which are related to humans' experience and training, and objective properties, that the task is either routine or novel for a specific population (e.g., operators, pilots, etc.). Furthermore, as familiarity increases, human behaviors can shift from KB behaviors to RB behaviors and even to SB behaviors. Although intuitively, humans tend to make fewer errors in more familiar tasks, the negative impacts in some situations deserve to be noticed, such as conducting familiar but inappropriate actions in certain situations due to the motor skill automaticity (Poldrack et al., 2005).

6.3.2.3 Time Pressure

The negative effects of high time pressure on human performance (especially correctness) have been observed in many studies (e.g., Rieger & Manzey, 2020, 2022), while some studies have reported an insignificant result (Topi et al., 2005) or contrary result (Trapsilawati et al., 2015). In a meta-analysis study, Szalma et al. (2008) found an overall small but detrimental effect of time pressure on performance, i.e., facilitating speed but impairing accuracy for both perceptual and cognitive tasks, while various moderating variables could influence such effect. Maule and Hockey (1993) summarized two approaches to conceptualize the effects of time constraints on decision-making, namely (1) considering the imposition of a deadline as a stressor and (2) assuming the time constraints to be a factor included in a cost/benefit determination of cognitive strategy. Lyu et al. (2023) proposed that time pressure influenced diagnosis in two aspects, namely (1) during hypothesis generation, altering the number of hypotheses generated, and (2) during information acquisition and hypothesis testing, inducing the use of different strategies to seek information. These effect pathways should be further explored by measuring more mediating variables such as eye movement (Kim & Nembhard, 2019). Last but not least, besides the negative effects of high time pressure encountered mostly in safety-critical domains such as NPP and aviation, the potential byproducts of low time pressure should also be noticed, such as the out-of-the-loop-unfamiliarity (Trapsilawati et al., 2015).

6.3.3   ENVIRONMENTAL FACTORS

Environmental factors have received much attention for a long time. Sanders and McCormick (1993) introduced four common environmental factors: illumination, climate, noise, and motion (e.g., vibration, acceleration, deceleration, and weightlessness). Take noise as an example. It is a kind of stress known to influence human performance in the auditory aspects and is also related to cognitive processing (Szalma & Hancock, 2011). In a meta-analytic study, Szalma and Hancock (2011) compared three theoretical perspectives: (1) Arousal theory (Broadbent, 1978), which explains the noise effect as the increasing arousal just as other stresses, and the arousal level has an inverse-U effect on performance, (2) composite theory (Poulton, 1979), which argues that the noise cannot only cause arousal but also mask inner speech, so the deleterious effects of masking would first be compensated by the increasing arousal and then dominate as the arousal decreases, and (3) maximal adaptability theory (Hancock, 1989), which proposes a tripartite framework to describe stress, including input (consisting of all environmental and task factors), adaptation (representing human capacity to cope with demands), and output (referring to the subsequent behavioral response patterns). The meta-analytic results obtained mixed evidence (both in line with and contrary to) for the arousal and composite theories and were most consistent with the maximal adaptability theory. Despite these different theoretical explanations, some conclusions could be reached from the existing studies considering diverse characteristics of noise (e.g., speech vs. non-speech noise, intermittent vs. continuous noise, noise duration, etc.), including (1) that the noise effects are related to internal speech, but cannot be explained as simple masking, and (2) that the noise effects on performance depend not only on the task type but also on the task parameters and other situational features (Smith, 1989).

6.3.4   INDIVIDUAL AND TEAM FACTORS

To err is human. Front-line personnel (e.g., operators, physicians, pilots, etc.) are the subjects who directly commit errors at the sharp end of a sociotechnical system (Hollnagel, 2004). The aim of the investigation into individual and team factors is not to blame for the errors but rather to understand the vulnerabilities that make humans more prone to errors. Below are some common individual and team factors that could contribute to human errors.

6.3.4.1   Fatigue

Fatigue includes both motor and cognitive fatigue (Behrens et al., 2023), and the latter is our focus here. Cognitive or mental fatigue has been recognized as one critical contributor to performance degradation and consequential accidents (Bendak & Rashid, 2020; Griffith & Mahadevan, 2011).

Ackerman (2011) reviewed four theories of cognitive fatigue, namely (1) Grandjean's theory (1968), in which fatigue is compared to the liquid level in a container that is filled with different sources and could be recovered by sleep and rest, (2) Kahneman's model of attention (1973), which proposes that one's level of arousal is related to the available attentional resources, and the allocation policy would determine the distribution of attentional resources to the criterion task, (3) Hockey's theory (1997), which draws on Kahneman's model of attention, but argues that the individual evaluates discrepancies in the context of a budget for effort expenditure and might revise task goals accordingly, (4) Kanfer and Ackerman's resource allocation model (1989), which extends Kahneman's model to explain the role of individual differences and motivational processes. Ackerman (2011) summarized several fatigue-related task characteristics, including cumulative intellectual demands, penalties for attentional blinks, arousal-related factors, arousal-motivational factors, and motivational factors, and some significant state variables, including sleep deprivation, time of day, recency of last meal, and drugs. These variables should be paid attention to for controlling human fatigue during the course of tasks.

6.3.4.2   Expertise

The differences in behaviors between experts and novices have been identified in various scenarios (e.g., McPherson, 2000; Priest & Lindsay, 1992; Wiedenbeck, 1985). As for life-critical or safety-critical domains, since the professionals are usually well trained and should not be deemed as novices, the construct *mental model* seems more appropriate to characterize individuals' knowledge and expertise. The mental model, a concept similar to the schema (for further discrimination, please refer to Richardson & Ball, 2009), is a cognitive representation of external reality (Jones et al., 2011). The mental model plays a vital role in the formation of situation awareness (Endsley, 1995b), enabling individuals to understand situations and make decisions comprehensively. Such critical effects have been found in many experimental studies (e.g., Lyu & Li, 2019; Sanderson & Murtagh, 1989). Regarding teams, the construct *team mental model* was proposed to describe the mental representation shared across teammates for the key elements within the environment (Mohammed et al., 2010). Yet, it should also be emphasized that possessing a large body of domain knowledge could also put experts at a disadvantage, known as the mental set effect (Wiley, 1998).

6.3.4.3   Motivation

Work motivation is a vibrant research topic in multiple disciplines, such as work and organizational psychology and economics. It is well recognized that motivation is closely related to task/job performance, as performance is often used as a measure to index motivation (Kanfer, 2012), and experiments also confirmed the effect of motivation on HEPs (Wu et al., 2017). There have been abundant studies focused on goal choice and goal pursuit behaviors, covering both intrinsic motivation (performing the task for enjoyment and interest) and extrinsic motivation (performing the task for external incentives) (Kanfer, 2012). Relevant theories include the expectancy-value theory (Vroom, 1964), cognitive evaluation theory (Deci, 1975), and self-determination theory (Deci & Ryan, 2000), to name a few. The details of these theories will not be discussed here, and readers are directed to relevant reviews (e.g., Kanfer, 2012; Kanfer & Chen, 2016).

6.3.4.4   Non-Technical Skills

Non-technical skills (NTSs) are receiving growing attention. Flin et al. (2008) summarized three basic classes of NTSs, namely NTSs related to individual capability (i.e., situation awareness and decision-making), NTSs related to teams (i.e., communication, teamwork, and leadership), and NTSs related to self-regulation (i.e., managing stress and coping with fatigue). Each NTS includes several essential component elements, for example, gathering information, interpreting information, and anticipating future states for the NTS situation awareness. Assessing these NTSs of individuals and providing appropriate training are key issues to reducing human errors and improving system safety.

6.3.5 Organizational Factors

Preceding sections have discussed various factors in a sociotechnical system that could impair human performance and finally lead to errors. In contrast, the latent organizational factors behind human errors at the sharp end should not be overlooked. The effects of organizational factors can be explained by Reason's organizational accident causation model (Reason, 1995b), which identifies two pathways of the effects of organizational errors. The first pathway is the active failure pathway, meaning that organizational errors could create error-forcing conditions at the sharp end to increase error likelihood. For example, an inappropriate work shift scheduling could induce the circadian desynchronization of individuals, eventually leading to the deterioration of human performance (Winget et al., 1984). The second pathway is the latent failure pathway, meaning that organizational errors act directly on the effectiveness of the defenses and barriers of the system. This pathway would not increase the error likelihood directly but rather decrease the likelihood that barriers inhibit errors. The set of organizational factors in specific domains depends on the responsibility and function of the organization. For example, for organizations in NPPs, Jacobs and Haber (1994) proposed 20 safety-related organizational factors, which fell into five categories, including administrative knowledge (consisting of four factors, namely coordination of work, formalization, organizational knowledge, and roles and responsibilities), communications (consisting of three factors, namely external communications, interdepartmental communications, and intradepartmental communications), culture (consisting of four factors, namely organizational culture, ownership, safety culture, and time urgency), decision-making (consisting of five factors, namely centralization, goal prioritization, organizational learning, problem identification, and resource allocation), and human resource allocation (consisting of four factors, namely performance evaluation, personnel selection, technical knowledge, and training).

6.4 HUMAN ERROR CLASSIFICATION

6.4.1 General Classifications

Some typical error taxonomies that categorize general rather than domain-specific errors[1] are demonstrated in Table 6.1. These taxonomies fall into two main perspectives of classification, i.e., behavioral classification and cognitive classification, also distinguished as classification by consequences and by psychological origins (Reason, 1995a), external error mode and internal error mode (Shorrock & Kirwan, 2002), and phenotype and genotype (Hollnagel, 1993b). Yet, it should be emphasized that such discriminations in Table 6.1 tend to be approximate rather than strict since many taxonomies fail to make a clear distinction between phenotype and genotype (Hollnagel, 1993b). Behavioral classification provides a directly observable way to categorize the manifestation of errors, enabling practitioners to elaborate on what can go wrong exactly and the consequences of errors at the sharp end. Cognitive classification digs into the cognitive failures underlying the manifestations and explains how humans go wrong in an implicit manner. Both perspectives are meaningful in enriching our understanding of human errors and are complementary rather than contradictory to each other. Unfortunately, there still lacks a unifying error taxonomy in the HF/E community, nor a unifying theory as discussed in Section 6.2.

6.4.2 Error Taxonomies in Different Domains

In this section, error taxonomies proposed in different domains are briefly discussed. Among these taxonomies, those in a cognitive perspective tend to adapt or extend existing general classifications, some of which have been summarized in Table 6.1, according to the characteristics of specific

TABLE 6.1
Typical Human Error Taxonomies

Error Taxonomies	Brief Summaries	Perspectives of Classification
Classification system for reporting events involving human malfunctions (Rasmussen et al., 1981)	The external modes of malfunction include: (1) The specified or intended task not performed due to omission of a task, omission of an act, inappropriate and inaccurate performance, inappropriate timing, or actions in the wrong sequence, (2) specific, erroneous acts on the system under treatment, including wrong act executed on correct component/equipment, wrong component/equipment, and the wrong time, (3) extraneous act, and (4) coincidence	Behavioral
Error taxonomy in THERP (Technique for Human Error Rate Prediction; Swain & Guttmann, 1983)	Two main groups of errors are distinguished: Error of commission (EOC, actions are conducted incorrectly) and error of omission (EOO, actions are not carried out)	Behavioral
Rouse and Rouse's error taxonomy (1983)	Operators' task is divided into several stages (i.e., the general category of errors), including observation of system state, choice of hypothesis, testing of hypothesis, choice of goal, choice of procedure, and execution of the procedure. Each general category consists of several specific categories of errors. For example, observation of system state includes (1) excessive, (2) misinterpreted, (3) incorrect, (4) incomplete, (5) inappropriate, and (6) lack	Behavioral
Error taxonomy in SHERPA (Systematic Human Error Reduction and Prediction Approach; Embrey, 1986a) and PHEA (Predictive Human Error Analysis; Embrey, 2004)	SHERPA considers errors in five types of human activities: Action, checking, communication, information retrieval, and selection, while PHEA considers one more activity of plan. Each type of activity can fail in different forms. For example, action errors in SHERPA include A1 operation too long/short, A2 operation mistimed, A3 operation in wrong direction, A4 operation too little/much, A5 operation too fast/slow, A6 misalign, A7 right operation on wrong object, A8 wrong operation on right object, A9 operation omitted (Sujan et al., 2020)[a]	Behavioral
Decortis and Cacciabue's temporal error taxonomy (1988)	Five temporal errors are identified: (1) Mis-estimation of sequences of actions and events, (2) mis-estimation of time duration of events, (3) failure in estimating the right moment to take action, (4) failure in anticipating some event, and (5) failure in the synchronization of collective actions	Behavioral
Hollnagel's phenotype classification (1993b)	Three levels of detection are applied: (1) 0-order detection, meaning that the phenotype can be identified by only comparing the current action with the expected action, (2) 1-order detection, meaning that the phenotype can be identified based on the outcomes of several 0-order detections, and (3) 2-order detection, meaning that the phenotype can be identified based on several 1-order detections. The phenotypes based on 0-order detection thus include those classified by time (i.e., premature start of action, delayed start of action, premature finishing of action, delayed finishing of action, and omission) and those classified by sequence (i.e., jump forward, omission, jump backward, repetition, and intrusion). Through combing 0-order detections, eight phenotypes based on 1-order detections are identified, including spurious intrusion, jump/skip, place losing, recovery, side-tracking, capture, reversal, and time compression	Behavioral

(Continued)

TABLE 6.1 (*Continued*)
Typical Human Error Taxonomies

Error Taxonomies	Brief Summaries	Perspectives of Classification
Error taxonomy in CREAM (Cognitive Reliability and Error Analysis Method; Hollnagel, 1998)	Thirteen failure modes are identified for four cognitive functions, namely observation (including O1 observation of wrong object, O2 wrong identification made, and O3 observation not made), interpretation (including I1 faulty diagnosis, I2 decision error, and I3 delayed interpretation), planning (including P1 priority error and P2 inadequate plan formulated), and execution (including E1 execution of wrong type performed, E2 action performed at wrong time, E3 action on wrong object, E4 action performed out of sequence, and E5 action missed)	Behavioral
Norman's categorization of action slips (1981)	Based on the schema theory (see Section 6.2.1), action slips are categorized into three main groups: Slips that result from errors in the formation of the intention (further including mode errors and description errors), slips that result from faulty activation of schemas (further including unintentional activation and loss of activation), and slips that result from faulty triggering of active schemas (further including false triggering and failure to trigger)	Cognitive
Reason's GEMS (Generic Error Modelling System; 1990)	Errors within SB behaviors are slips (actions not as planned) and lapses (memory failures), and within RB and KB behaviors are mistakes (wrong intention). Detailed error modes under each behavioral level are also provided. For example, error modes under SB level include inattention (further including double-capture slips, omissions following interruptions, reduced intentionality, perceptual confusions, and interference errors), and over-attention (further including omissions, repetitions, and reversals)	Cognitive
Endsley's SA (situation awareness) error taxonomy (1995a)	Errors are categorized according to the three level of situation awareness: Level 1 – failure to correctly perceive situation (including data not available, data difficult to detect/perceive, failure to scan or observe data, misperception of data, and memory failure), Level 2 – failure to comprehend situation (including lack of/poor mental model, use of incorrect mental model, over-reliance on default values in model, and memory failure), and Level 3 – failure to project situation into the future (including lack of/poor mental model)	Cognitive
Dörner and Güss's error taxonomy for complex problem solving and dynamic decision-making (2022)	Twenty-four errors are categorized into six steps of problem solving, namely (1) problem identification, (2) goal definition, (3) information gathering, (4) elaboration and prediction, (5) planning, decision-making, and action, and (6) evaluation of outcome and self-reflection	Cognitive
Rasmussen's error taxonomy (1982)	Based on the SRK framework (see Section 6.2.1), the taxonomy distinguishes three facets: External mode of malfunction, internal human malfunction, and mechanisms of human malfunction. External mode of malfunction describes the observable effects of human errors on task performance, including specified task not performed, commission of erroneous act, commission of extraneous act, and sneak-path and accidental timing of several events or faults. Internal human malfunction describes the failure of mental function, including detection, identification, decision, and action, and different behavioral levels (SB, RB, KB) require different mental functions. Mechanisms of human malfunction describe the failure of internal human mechanisms	Integrated

(*Continued*)

TABLE 6.1 (*Continued*)
Typical Human Error Taxonomies

Error Taxonomies	Brief Summaries	Perspectives of Classification
Sutcliffe and Rugg's error taxonomy (1998)	Based on previous work by Reason and Hollnagel, a multi-layer schema is proposed which comprises operational description (i.e., phenotype), cognitive causal categories, social and organizational causes, and design errors. Operational description describes errors according to six dimensions: Timing, action/sequence, force, duration, direction, and object. As for cognitive causal categories, 23 errors are identified under the SRK framework, with SB level including (1) action slips, (2) interruptions, (3) perceptual confusions, (4) plan failures, and (5) mistimed checks, RB level including (6) context error, (7) malformed rule, and (8) rigidity, and KB level including (9) bounded rationality, (10) confirmation bias, (11) halo effect, (12) poor causal-reasoning, (13) representativeness and availability, (14) belief in small samples, (15) hindsight bias, (16) poor future-projection, (17) biased reviewing, (18) thematic vagabonding, (19) encysting, (20) anchoring/ adjustment, (21) inadequate retrieval/matching, (22) tunnel vision, and (23) weak search for evidence	Integrated
Sasou and Reason's team error taxonomy (1999)	Four types of errors are identified: Independent individual errors, dependent individual errors, independent shared errors, and dependent shared errors. Besides, three types of recovery failures are identified: Failure to detect, failure to indicate, and failure to correct	Integrated
Error taxonomy in IDHEAS-G (General Methodology of an Integrated Human Event Analysis System; Xing et al., 2021)	In this three-level cognitive failure mode (CFM) taxonomy, the high-level CFMs represent the failures of macro-cognitive functions, the middle-level CFMs represent the failures of cognitive processors, and the detailed-level CFMs represent behaviorally observable patterns of middle-level CFMs. The five high-level CFMs include failures of detection, understanding, decision-making, action execution, and inter-team coordination. Middle-level CFMs, for example in the failure of detection, include D1 fail to initiate detection, D2 fail to select, identify, or attend to sources of information, D3 fail to perceive, recognize, or classify information, D4 fail to verify the perceived information, and D5 fail to communicate the acquired information. Detailed-level CFMs, for example in D1, include D1-1 detection is not initiated, D1-2 wrong mental model for detection, and D1-3 failure to prioritize information to be detected	Integrated

[a] The detailed error modes in SHERPA reported in different publications might have slight differences.

domains. Taxonomies from a behavioral perspective are often closely related to the task in different domains and exhibit distinctive forms. Note that this section does not aim to conduct a systematic literature review for error taxonomies in each domain but rather to provide readers with an overview of the classification of human errors in diverse sectors. Based on the studies discussed in this section, it appears that human cognitive processes in different domains are technology-neutral and share some commonalities, thus raising the possibility of studying human errors across sectors in a unified manner. However, cross-sectoral comparisons should be treated cautiously, as considerable differences exist not only in the behavioral patterns of errors but also in the task context regarding all levels in a sociotechnical system.

6.4.2.1 Aviation

In the aviation sector, multiple human roles have been studied, such as pilots in the cockpit, air traffic controllers, and maintenance personnel. Nagel (1988) proposed a simple error model for aviation operations in which three information processing stages were considered: information, decision, and action. In an unsafe operation taxonomy, Shappell and Wiegmann (1997) identified three distinct layers, including unsafe supervision, unsafe conditions, and unsafe acts. Where the basic error forms lay, the categorization of unsafe acts directly incorporated Reason's (1990) classification. As for air traffic control (ATC), a well-known human error identification method is the TRACEr (Technique for the Retrospective and Predictive Analysis of Cognitive Errors in air traffic control; Shorrock & Kirwan, 2002). This method distinguished three kinds of errors: External error modes (EEMs, the observable manifestation of actual errors), internal error modes (IEMs, the failures of cognitive functions), and psychological error mechanisms (PEMs, the psychological nature of IEMs). Details of this method will not be further discussed here. Another topic is maintenance errors. Drury (1991) proposed an error taxonomy for aviation maintenance. This taxonomy first identified seven maintenance functions: initiate, access, search, decision, respond, repair, and buy-back. Each function was accomplished through a range of tasks, for example, "initiate," including correct instructions written, correct equipment procured, etc. Each step then consisted of various errors, such as the task "correct instructions written," including incorrect instructions, incomplete instructions, and no instructions available. In Hobbs and Williamson's taxonomy of errors in aircraft maintenance (2003), the authors adapted Reason's taxonomy and added two categories, namely perceptual error in the visual inspection and mischance, to represent unsafe actions nevertheless constitute "correct" behaviors.

6.4.2.2 Healthcare

Healthcare is another sector where human errors have been put much emphasis on. Taib et al. (2011) reviewed 26 medical error taxonomies and found that about two-thirds classified systemic factors, while only a third classified causal mechanisms. The authors advocated using theoretical error concepts, which could help to understand errors thoroughly. Mitchell et al. (2014) reviewed the human factors classification frameworks used in 38 studies in the hospital setting. They found that almost all frameworks described what went wrong from the perspective of either job/task errors or cognitive errors rather than including both perspectives. The authors argued that only capturing job/task errors failed to identify the cause of errors, while only capturing cognitive errors failed to provide information about the error context. Only some studies are introduced here as examples. Shah et al. (2004) proposed an error taxonomy for otolaryngology according to the care flow of patients, namely history and physical, differential or final diagnosis, testing, surgical planning, wrong-site surgery, anesthesia-related, wrong drug/dilution on the surgical field, technical, retained foreign body, equipment-related, postoperative care, medical management, nursing/ancillary, administrative, communication, and miscellaneous. Zhang et al. (2004) proposed a cognitive taxonomy of medical errors based on Reason's classification (1990) and Norman's action theory (1988). In this taxonomy, errors were classified into slips, including execution slips and evaluation slips, and mistakes, including execution mistakes and evaluation mistakes. Each sub-category consisted of several errors in different stages in the execution-evaluation cycle. For instance, execution slips included goal, intention, action specification, and action execution slips. In the Human Factors Classification Framework (HFCF) for patient safety (Mitchell et al., 2016), the human errors involved in precursor events and contributing factors were classified using the SRK framework, as well as the violation classification proposed by Reason (1990).

6.4.2.3 Nuclear Industry

The nuclear industry is a typical safety-critical domain where human errors have been heavily studied. Abundant HRA methods have been developed in the nuclear industry to identify and predict human errors, including THERP (Swain & Guttmann, 1983), ATNEANA (A Technique for Human

Error Analysis; Cooper et al., 1996), SPAR-H (Standardized Plant Analysis Risk-Human Reliability Analysis; Gertman et al., 2005), and IDHEAS-G (Xing et al., 2021) along with its application-specific version IDHEAS-Atpower (Xing et al., 2017) and IDHEAS-ECA (Xing et al., 2020), to name a few. Besides, many databases have been designed to collect human performance data to inform HRA, such as SACADA (Scenario Authoring, Characterization, and Debriefing Application; Chang et al., 2014) and HuREX (Human Reliability data EXtraction; Jung et al., 2020). Most of these methods or databases provide a corresponding error taxonomy, which will not be discussed here. Similar to the aforementioned sectors, human roles in the nuclear industry are diverse and include MCR operators, maintenance personnel, commissioning personnel, etc. Kim et al. (2017) proposed a classification scheme for erroneous behaviors of NPP operators. Four major cognitive activities of operators were identified: Information gathering and reporting, response planning and instruction, situation interpreting, and execution. For each cognitive activity, several task types were summarized, for example, "execution" activity including manipulation - simple (discrete) control, manipulation - simple (continuous) control, manipulation - dynamic manipulation, and notifying/requesting to main control room outside. Based on that, a dichotomy of EOO and EOC was applied for each task type. Lee et al. (2011) investigated potential error modes for operations using soft controls and classified six types of errors: operation omission, wrong object, wrong operation, mode confusion, inadequate operation, and delayed operation. Studies also attended activities outside the MCR. For example, in a study on human errors in NPP commissioning, Yin et al. (2023) did not propose a new error taxonomy but simply adapted the cognitive failure mode (CFM) taxonomy provided in IDHEAS-G.

6.4.2.4   Driving

Driving errors are critical for road safety. Reason et al. (1990) investigated 50 aberrant driver behaviors through a questionnaire survey. Examples of these aberrant behaviors included unknowingly speeding, disregarding speed at night, failing to give way to buses, getting in the wrong lane at roundabouts, etc. These behaviors fell into five classes: slips, lapses, mistakes, unintended and deliberate violations. Later, Stanton and Salmon (2009) developed a generic driver error taxonomy based on an exhaustive literature review, with 24 driver errors identified, including (1) failure to act, (2) wrong action, (3) action mistimed, (4) action too much, (5) action too little, (6) action incomplete, (7) right action on the wrong object, (8) inappropriate action, (9) perceptual failure, (10) wrong assumption, (11) inattention, (12) distraction, (13) misjudgment, (14) looked but failed to see, (15) failed to observe, (16) observation incomplete, (17) right observation on the wrong object, (18) observation mistimed, (19) misread information, (20) misunderstood information, (21) information retrieval incomplete, (22) wrong information retrieved, (23) intentional violation, and (24) unintentional violation. The underlying psychological mechanisms, as well as causal factors, were also reported in their work.

6.4.2.5   Railway and Metro

Railways and the metro are two safety-critical industries in transportation. Gibson et al. (2006) proposed a communication error taxonomy and applied it to railway track maintenance. The taxonomy comprised three dimensions: Communication error criteria (including communication failures, deviations from the grammar, and task communication errors), level of the grammar (including phonology, syntax, semantics, and pragmatics), and EEMs (including omit, wrong action on the right object, right action on the wrong object, wrong action on the wrong object, repeated, mis-ordered, too much/too little, too long/too short, too early/late, and extraneous act). Wang and Fang (2014) proposed an error behavior classification for metro traffic dispatchers. Based on the VACP (visual, auditory, cognitive, and psychomotor) theory, errors were classified according to these four components of information processing resources. For example, visual errors further included detection failure, discrimination failure, inspection failure, location failure, tracking failure, reading failure, and scan failure.

6.4.2.6   Computer Technology and Information and Communications Technology

Computer technology and information and communications technology (ICT) are emerging sectors where human errors are gaining growing attention. Kraemer and Carayon (2007) proposed a conceptual framework of security in computer and information systems, which captured the observed security vulnerabilities, human errors, and contributing factors in the work system. This framework categorized human errors into unintentional errors (including 22 subcategories) and intentional errors (including 23 subcategories). Anu et al. (2018) extended Reason's classification into a taxonomy for software engineering activities. Errors in this taxonomy were categorized into slips, lapses, and mistakes, each consisting of several subcategories. For example, slips included clerical errors and term substitution. In a recent study on network operation and maintenance, Yin et al. (2023) proposed an erroneous behavior taxonomy for the network operation and maintenance technicians. This taxonomy identified a range of detailed erroneous behavior, such as inputting an incorrect parameter, inputting an incorrect template, and omitting to input a character. The failures of MCFs were also considered to distinguish the same behavioral pattern caused by different cognitive failures.

6.4.2.7   Military

O'Connor et al. (2007) proposed a dive team human error taxonomy for the US Navy. This taxonomy identified six categories and 21 subcategories, namely the situation awareness category (including anticipation, problem definition/diagnosis, risk and time assessment, dive status awareness, task awareness, and concentration/avoiding distraction), the decision-making category (including procedural adherence and outcome review), the communication category (including assertiveness/speaking up and information exchange), the team cohesion category (including team climate and conflict solving), the personal resources category (including identifying and managing stress, identifying and managing fatigue, physical and mental fitness, and experience/training), and the supervision/leadership category (including appropriate use of authority, maintaining standards, planning and coordination, workload management, and choice of leadership style).

6.5   HUMAN ERROR ANALYSIS METHODS

Preceding sections have discussed how humans commit errors (Section 6.2), why humans commit errors (Section 6.3), and what kinds of errors humans could commit (Section 6.4). Armed with these theoretical and empirical findings, it is time to introduce human error analysis in practice. Generally speaking, human error analysis methods can be categorized into proactive and retrospective methods, while some can act as both. Proactive methods are often used to predictively identify potential human errors in a given scenario, with some focusing on qualitative error manifestations and others focusing on quantitative likelihood. On the contrary, retrospective methods are used to investigate the causes of an error after it occurred. Though divergent in application, the ultimate goal of both kinds of methods is to reduce human errors and improve system safety.

6.5.1   Proactive Analysis

Proactive analysis methods are also known as human error identification (HEI) methods. Commonly used HEI methods include SHERPA (Systematic Human Error Reduction and Prediction Approach; Embrey, 1986a), PHECA (Potential Human Error Cause Analysis; Whalley, 1988), TAFEI (Task Analysis For Error Identification; Baber & Stanton, 1994), and PHEA (Predictive Human Error Analysis; Embrey, 2004), to name a few. Reviews of these HEI methods can be found in Kirwan (1992, 1998). Stanton (2009) enumerated two representative HEI methods, i.e., SHERPA and TAFEI, and commented that these two methods worked in inherently different ways. The analysis processes of the two methods are briefly introduced here. Originally designed for process industries,

SHERPA requires an eight-step analysis process, including (1) hierarchical task analysis (HTA), (2) task classification, (3) HEI (using a given error mode list as presented in Table 6.1), (4) consequence analysis, (5) recovery analysis, (6) ordinal probability analysis, (7) criticality analysis, and (8) remedy analysis (Stanton, 2005). Aiming at predicting errors in the interaction between user and device, TAFEI comprises three stages, including (1) modeling human activities using HTA, (2) constructing state-space diagrams, and mapping HTA results on state-space diagrams to form the TAFEI diagram, and (3) devising the transition matrix to demonstrate the state transitions during the task (Stanton & Baber, 2005). As commented by Stanton (2009), SHERPA is a divergent error-prediction method and represents the error-list approach, while this could lead to the increments of inter-analyst variability; TAFEI is a convergent error-prediction method and represents the schema-based approach, and is more user-friendly for novices.

HRA methods can be regarded as another family of proactive analysis methods. Except for some methods such as ATNEANA, most HRA methods put emphasis on the quantitative estimation of HEPs. Relevant reviews can be found in Adhikari et al. (2009), Chandler et al. (2006), Pyy (2000), etc. The HRA process generally includes problem definition, task analysis, HEI, representation, human error quantification, impact assessment, error reduction analysis, and documentation and quality assurance (Kirwan, 1994). Concerning HEI, an error-list approach is often adopted to provide analysts with a set of potential error modes (e.g., in CREAM and IDHEAS-G). The quantitative estimation of HEPs can help identify risky errors so that high-risk errors can receive more priority in design and management.

6.5.2 RETROSPECTIVE ANALYSIS

Retrospective human error analysis is conducted to investigate human errors after they occur. It is essential to first talk about a broader topic, i.e., accident analysis since human errors in complex systems and other failures often lead to accidents at last [2]. There have been numerous models proposed for accident analysis. According to Lehto and Salvendy (1991), these models could be classified into general models of the accident process, models of human error and unsafe behavior, and models of the mechanisms of human injury. General models of the accident process can be further divided into sequential models, epidemiological models, energy transfer models, and systems models. These models analyze the causation of accidents in different ways. Sequential models describe an accident as the result of event sequences; epidemiological models describe an accident as the outcome of a combination of factors; systems models view accidents as emergent phenomena on the level of the whole system (Hollnagel, 2004). Correspondingly, lots of methods have been developed based on these different models or perspectives, for example, sequential methods such as AEB (Accident Evolution and Barrier function; Svenson, 2001) and ECFC (Events and Causal Factors Charting; DOE, 1999), epidemiological methods such as HFACS (Human Factors Analysis and Classification System; Shappell & Wiegmann, 2000) and SOAM (Systemic Occurrence Analysis Methodology; Licu et al., 2007), and systems methods such as FRAM (Functional Resonance Accident Model; Hollnagel, 2004), STAMP (Systems-Theoretic Accident Model and Processes; Leveson, 2004), and AcciMap (Rasmussen, 1997).

Within the scope of accident analysis methods, retrospective human error analysis methods specifically concentrate on the failures of human components, with typical methods including HFACS (Shappell & Wiegmann, 2000), TRACEr (Shorrock & Kirwan, 2002), HERA (Human Event Repository and Analysis; Hallbert et al., 2006), and HuRAM+ (Human-related event Root cause Analysis Method plus; Choi & Jung, 2014). Similar to the general accident analysis methods, these methods also focus on identifying the "causes" of human errors, while the causes can be interpreted in different ways. For example, HFACS aims at identifying the organizational factors or latent failures behind human errors, including organizational influences, unsafe supervision, and preconditions of unsafe acts. TRACEr identifies the PEM (psychological mechanisms of errors) as well as the status of PSFs (contextual factors). HuRAM+ records three categories of causes, namely task/system-related factors (e.g., procedure, workload, training, etc.), organizational/safety culture-related factors (e.g., management, administration, etc.), and other factors (Lee, 2022). In a word, the "causes" of human errors can be the failures of human cognition (e.g., working memory overload), the adverse contextual factors (e.g., poor illumination), or the latent failures in organization (e.g., inadequate scheduling). To date, various retrospective analysis methods investigate and record disparate segments of causal information without a unified interpretation of error causes. Therefore, it is suggested that all these causal factors should be considered in the retrospective analysis to reach a comprehensive understanding of human error.

6.6 HUMAN ERROR IN HCI

6.6.1 Significance of Human Error in HCI

The development of information technology has seen a remarkable transformation in safety-critical systems, bringing great benefits to human performance and system safety (Woods & Dekker, 2000). The computer has advantages in collecting, transmitting, and interpreting data, which may alleviate the operator's workload. Embedded systems or functions, such as error detection and alert, diagnosis aid, decision support, and automatic execution, may help to resolve some safety issues (Liu et al., 2021). It should be noted that the introduction of computerized systems does not merely substitute old mediums or channels but rather drastically alter the ways of working. In a conventional physical environment, operators walk along panels, read gauges, and adjust knobs and levers. However, operators no longer need to do this in a computerized environment. They can obtain information from computer screens and control the system through keyboards, mice, and touch screens. This lack of physical restrictions allows massive data, features, functions, and sub-systems to be added to a single platform. Unfortunately, this can overwhelm operators with the sheer amount of information presented to them and can be especially detrimental in high-demanding scenarios (e.g., emergencies) (Woods et al., 2010). The complexity of the computerized system can also make it difficult for operators to understand what they are doing and how they work, necessitating the need for learning and tracing (Roth & O'Hara, 2002). Moreover, computers change the way operators interact with each other. In a conventional environment, they obtain information from the artifacts and interact with other teammates through conversations, gestures, and other body language. In a computerized environment, team members communicate less because nearly all the information is accessible from the computer. This shift in communication can lead to different and incompatible views on the situation, making team coordination more difficult (Chung et al., 2009; Gao et al., 2015; Lin et al., 2016).

Consequently, the introduction of computerized systems has created a range of new challenges in interface management, attention control, knowledge requirements, coordination, and communication (O'Hara & Brown, 2002; Woods et al., 2010). These challenges could create new paths for human errors. Thus, more research into human errors in the field of HCI is necessary, such as identifying new error types, underlying mechanisms, contributing factors, and developing and validating countermeasures.

6.6.2 What Is Special When Interacting with Computerized Systems

Norman (1990) noted that the behavior of a system is not a result of design specifications but that of the interaction between the operator and the system. Similarly, Woods et al. (2010) argued that it is not the technology or the operator alone that creates the problems but rather how the operator utilizes the technology. Many human errors could be attributed to the clumsy use of technology, which is characterized by the mis-coordination between the human and the system (Klein et al., 2004). The clumsy use of technology is shaped by various factors, as presented in Table 6.2. Several factors,

TABLE 6.2

Factors That Influence Task Performance and System Safety in HCI

Category	Description	Factor
System variables	Characteristics of the computerized system	System reliability System complexity Nature of feedback Level of automation
Individual variables	Individual characteristics	Complacency potential Knowledge Training
HSI variables	Interface design features	Virtuality Keyhole Interactivity Agency
Emergent variables	Arising from the interactions between the operator and the system	Workload Trust Situation awareness
Task variables	Context in which the operator and the system interact	Criticality Consequence

such as system reliability, system complexity, and HSI, have been addressed in Section 6.3.1. The factors are classified into five categories: (1) system variables; (2) individual variables; (3) HSI variables; (4) emergent variables; and (5) task variables. System variables refer to the characteristics of the computerized system, such as system reliability (McBride et al., 2014; Yin et al., 2023), system complexity (Liu et al., 2016; Yin et al., 2023), the nature of the feedback provided to the operator (Woods et al., 2010; McBride et al., 2014), and the level of automation (in a broader sense, automation is a special type of computerization, Parasuraman et al., 2000). Individual variables include individual characteristics such as knowledge (McBride et al., 2014), training (Skitka et al., 2000), and complacency potential (Wilson et al., 2020). HSI variables involve interface design features, including virtuality (Woods et al., 2010), keyhole (Woods et al., 2010), interactivity (Woods et al., 2010), and agency (the degree to which an entity has control over the system, Mueller et al., 2020; Woods et al., 2010). Emergent variables are the factors that arise from the interactions between the operator and the system, for example, mental workload (McBride et al., 2014), situation awareness (Endsley, 1995b), and trust in the system (Wilson et al., 2020). Task variables describe the context in which the operator and the system interact. Typical factors are the criticality of the task (Woods et al., 2010) and the consequence of the failure (Mosier et al., 1998). It is important to note that the factors in Table 6.2 are not exhaustive.

The clumsy use of computerized systems may undermine the following cognitive activities of operators. This summary mainly refers to Woods's Impact Flow Diagram, which illustrates how the clumsy use of technology can lead to human error (Woods et al., 2010, p. 159, Figure 10.1). Other relevant literature was also supplemented (e.g., Rasmussen & Vicente, 1989; O'Hara & Brown, 2002).

- *Increased cognitive demands*: Arbitrarily integrating all sources of data, capabilities, and functions into the computerized system would proliferate modes and displays. Operators have to perform interface management tasks, for example, navigation, configuration, arrangement, and interrogation. These extra tasks would increase the operator's cognitive demands and operational complexities.

- *Impaired attention control*: Attention shifts may become more frequent as operators need to perform interface control and primary tasks simultaneously. This challenges the operator's ability to control attention. With limited attentional resources, the primary task performance may be reduced.
- *Information overload, especially in emergencies*: Without effective repetitions of information priority or data trends, operators may not monitor the process well, and they may miss important events, perform for a longer time or with lower accuracy. In situations where massive data flood in a short period, operators may search for fewer possibilities or access the information they are more familiar with, potentially leading to a biased decision.
- *Complicated situation assessment*: Operators take time to collect and integrate data across a set of displays, which complicates the process of situation assessment and increases their mental workload. This may lead to an incomplete or erroneous assessment of the situation.
- *Decreased situation awareness*: Without effective feedback, operators may find it hard to track the system states and performance. They may not know if the system is working normally, leading to poor situation awareness. Even worse, operators may not be able to detect the anomalies or recover from the failures immediately.
- *Impaired mental models*: If the system is not designed in a way that is consistent with the operator's mental models, they may not understand what the system is doing, what it intends to do, and how it works.
- *Decreased knowledge calibration*: Improper system design may mislead users into thinking that they know something when actually they do not. New functions and new modes may create new knowledge requirements for operators to learn about. For example, a lack of knowledge of various mode settings and mode transitions may lead to mode errors.
- *Breakdowns in communication and coordination among the operators*: The operators focusing on the screen narrow the perception range and obstruct the awareness of other teammates. As operators obtain most information from the computer, they tend to communicate less. All these bring difficulties in coordinating work toward the common goal.
- *Over-reliance on the automated system:* Operator complacency is a potential concern when utilizing automated systems. Operators may become over-reliant on the system, resulting in decreased vigilance, lack of situation awareness, and a failure to detect errors or intervene in a timely manner.

Through impacting the operator's cognitive activities, the clumsy use of the computerized systems ultimately influences the operator's behaviors, increasing the potential for various types of errors (e.g., misperception of the information, erroneous assessment of the situation, and erroneous actions). The general classifications of human error, such as Hollnagel's taxonomy of phenotypes and genotypes (Hollnagel, 1993b), Rasmussen's SRK model (Rasmussen, 1983), Reason's taxonomy of slips, lapses, and mistakes (Reason, 1990), and the classifications based on HIP model, are confirmed to be applicable in the field of HCI (Hollnagel, 1991; Stanton, 2009). Rasmussen and Vicente (1989) provided a taxonomy of human errors to identify certain system design improvements for each type of error. This classification divided human errors into four categories: errors related to learning and adaptation, interference among competing cognitive control structures, lack of resources, and stochastic variability. Each category was further subdivided into skill-, rule- and KB errors. Zhang et al. (2004) proposed a taxonomy of human errors in medical contexts, focusing on the level of individuals and their interactions with technology.

Among the various types of errors, mode error receives special attention in the design of computerized systems and is regarded as one of the most vexing errors in HCI (Norman, 1981). A mode error occurs when the operator executes an intention in a way appropriate for one mode when the system is in another (Sarter, 2008). In this sense, mode error can be classified as a phenotype of commission error (Woods et al., 2010). Complicated system modes, a high degree of coupling of the systems, poor feedback toward mode states, and mode transitions are identified as contributors to mode errors (Sarter, 2008). They burden operators with extra memory demands, new knowledge requirements, and complicated situation assessments (Woods et al., 2010). To mitigate the problem of mode error, researchers have proposed countermeasures, including system design (e.g., Hutchins, 1997; Boorman & Mumaw, 2004) and training (e.g., Mumaw et al., 2000).

6.6.3 STUDYING HUMAN ERROR IN HCI

Researchers have devoted themselves to human error classification, analysis, and management emerging from HCI. Studies to classify and analyze human errors related to HCI have been included in the review of Sections 6.4 and 6.5 and thus are skipped in this section. With respect to error management, according to Martinie et al. (2015), human errors can be dealt with at three levels: (1) the system level, for example, to monitor the operator's status and activities within an acceptable range; (2) the interaction level, for example, to prevent the operator's wrong inputs by only allowing the selection of valid values; and (3) the operator level, for example, to provide operators with adequate and complete training. Honig and Oron-Gilad (2018) developed a model to describe how to manage failures in human-robot interaction (HRI), which is called Robot Failure Human Information Processing (RF-HIP) Model. Although intended for HRI, it is believed that the basic idea of the model applies in a broader field of HCI. The model consists of three main parts to manage failures: (1) communicating failures, that is, transmitting the failure-relevant information through visual, audio, speech, or mixed channels; (2) perceiving and comprehending failures, that is, ensuring the perception of relevant information (e.g., by requiring the operator's response to alarms) and facilitating the comprehension of relevant information (e.g., by selecting cues and providing information patterns); and (3) solving failures, that is, motivating the operator to solve the problem, supporting the decision-making (e.g., by optimizing the elicitation and presentation of choices), and acting the decision. A comprehensive summary of human error control strategies and measures is presented in Section 6.7.

Despite the current research progress on human error in HCI, there are still gaps in the following aspects. First, a more systematic and elaborated taxonomy of HCI-relevant human errors, the corresponding cognitive mechanisms, and contributing factors is needed. Second, although various countermeasures have been proposed to prevent or mitigate human errors, the validations of these countermeasures are insufficient. To ensure their effectiveness, more rigorous evaluations are needed. Finally, with the rapid development of automation, the research on human-automation/autonomy/AI interaction (HAI) is flourishing yet full of challenges. A group of HF/E experts have put forth a variety of research needs related to HAI, including models and metrics for HAI, support for situation awareness through automation/intelligent systems, transparency, approaches to HAI, trust in automation/autonomy/AI, and human-system integration (Oswald et al., 2022).

6.7 DESIGN FOR HUMAN ERROR PREVENTION AND CONTROL

The preceding sections have summarized human error mechanisms, contributing factors, and issues particularly pertinent to the interaction between humans and computers as well as new technologies. This section will discuss how to prevent and control human errors through design, with a particular focus on the HCI context.

6.7.1 GENERAL STRATEGIES

Following the risk concept, human error should be controlled first by reducing the occurrence probability and then mitigating the severity of its consequences. Error tolerance ("resilience") is too much emphasized in the literature on human-computer interactions while preventing the occurrence of human errors is not always firstly considered. It is well recognized that humans make errors because they are put in unfavorable conditions. Thus, a basic thinking for human error control is to identify and control the factors (as discussed in Section 6.3) that constitute unfavorable conditions. If unfavorable conditions remain, enhance humans' ability to work in such conditions (Liu et al., 2021). After that, error tolerance is considered as the last measure to mitigate human error consequences.

The general hazard control hierarchy is also applicable to human error control in HCI design. The general control strategies from the highest to the lowest priority are to remove such unfavorable conditions (e.g., automation, providing only safe options), to substitute the conditions with less risky conditions (e.g., using option selection instead of textbox input, allowing smaller change), to control the conditions so that they would less likely lead to human errors (e.g., reducing complexity), to apply administrative measures (e.g., warnings and procedures), and to tolerate human error.

The strategies and control measures have been summarized by Liu et al. (2021) from a general perspective. This part will focus more on design aspects in the HCI context but ignore administrative measures.

6.7.2 DESIGN PHILOSOPHIES

This section presents an overview of three typical design philosophies: human-centered design, EID, and human-computer integrated design. For each design philosophy, the fundamental ideas, design processes and activities, design methods and techniques, applications and practices, and possible impacts on human performance and system safety are briefly discussed.

6.7.2.1 Human-Centered Design

Human-centered design (HCD) can be considered as a philosophy, approach, framework, or practice of system design by involving humans in the whole design process (Williams, 2009). First introduced by Norman and Draper (Norman & Draper, 1986), HCD aims to provide users with usable and useful systems. A typical HCD process encompasses analyzing human needs, requirements, capabilities, and behaviors, implementing design solutions, and evaluating design through usability knowledge and techniques (Norman, 2013). This approach is expected to "enhance effectiveness and efficiency, improve human well-being, user satisfaction, accessibility and sustainability, and counteract possible adverse effects of use on human health, safety and performance" (International Organization for Standardization, 2019).

Several studies have reported slight variations in HCD activities/processes (in brief, referred to as activities); however, the nature of these activities remains consistent. Three classifications of key activities and their correspondence are summarized in Table 6.3. All these activities are iterated with continuous refinement and enhancement until the solution gets closer to the desired one. Refer to ISO 9241-210:2019 for a comprehensive list of requirements and recommendations for HCD principles and activities (International Organization for Standardization, 2019). Maguire (2001) provided a set of methods for each activity, with the evaluation/testing methods being detailed in ISO/TR 16982:2002 (ISO, 2002). Furthermore, Williams (2009) summarized sample deliverables produced during each phase of the UCD process.

TABLE 6.3
Three Classifications of HCD Activities

Source	ISO (2019)	Norman (2013)	Williams (2009)
HCD activities	Plan the human-centered design process	Observation	Design research
	Understand and specify the context of use		
	Specify the user requirements		
	Produce design solutions	Idea generation	Design
		Prototyping	
	Evaluate the design	Testing	Design evaluation

The HCD approach has been widely used for the design of tools, equipment, and systems, such as aircrafts (e.g., Billings, 1991; Mouloua et al., 2003), vehicles (e.g., Fancher et al., 2001; Fridman, 2018), medical devices (e.g., Matheson et al., 2015), mining equipment (e.g., Horberry et al., 2016), and virtual reality devices (e.g., Jerald, 2015). Quantitative and qualitative studies have demonstrated that the HCD can lead to increased effectiveness, improved efficiency, reduced errors, decreased workload, better user acceptance, and/or higher satisfaction (e.g., Errington et al., 2005; Totter et al., 2011; Son et al., 2013; Top et al., 2021).

6.7.2.2   Ecological Interface Design

EID is a theoretical design framework to support problem solving and decision-making for complex sociotechnical systems (Vicente, 2002). EID was first introduced by Rasmussen and Vicente in the late 1980s (Rasmussen & Vicente, 1989). Shortly after, substantial research was published on its principles, designs, evaluations, and practices. The primary goal of EID is to facilitate problem solving and decision-making in novel situations by transforming the process from a cognitive work to a perceptual one (Bennett & Flach, 2019). In other words, the affordance of a work domain (i.e., something that the operator acts on) can be directly perceived, eliminating the need for mental inference, deduction, and calculation (Flach & Vicente, 1989). EID is expected to mitigate the cognitive load associated with navigating, memorizing, and integrating the required information, thereby reducing human errors. The key processes of EID are understanding the abstract, complex, and semantic structure of the work domain and designing graphical interfaces for a direct perception (e.g., configurable displays) (Bennett & Flach, 2019). The former process is usually determined through work domain analysis, particularly through abstraction hierarchy. Abstraction hierarchy typically contains five levels: function purpose, abstract function, generalized function, physical function, and physical form (Rasmussen, 1985). Lower levels of an abstraction hierarchy encompass physical information (i.e., the states of objects in the work domain), while higher levels of abstraction hierarchy encompass functional information (i.e., the functions and purposes of the objects) (Vicente, 2002).

EID has been applied in a variety of areas, including military (e.g., Hall et al., 2012), aviation (e.g., Dinadis & Vicente, 1999), nuclear industry (e.g., Burns, 2000; Chen et al., 2018), chemical industry (e.g., Jamieson, 2007), healthcare (e.g., Sharp & Helmicki, 1998) and so on. Many empirical evaluations of EID have proved it to be remarkably effective in improving performance in terms of higher performance accuracy, shorter completion time, fewer human errors, and/or lower workload (e.g., Vicente, 1996; Sharp & Helmicki, 1998; Hall et al., 2012). Although the utility of EID has been widely recognized, a fundamental challenge remains on how to design proper visual forms or other perceptual modalities to represent the semantic structure of the work domain (Vicente, 2002; Bennett & Flach, 2019).

6.7.2.3   Human-Computer Integrated Design

Human-computer integration describes the partnership or symbiotic relationship between humans and computers (Farooq & Grudin, 2016). In this relationship, humans and computers are integrated either physically or conceptually, allowing them to act as partners to collaborate and cooperate with one another (Rodrigues Barbosa et al., 2023). A similar concept is human-computer/technology symbiosis, which is considered interchangeable with human-computer integration (Farooq & Grudin, 2016). Human-computer/technology symbiosis is identified as one of the seven grand challenges for future HCI (Stephanidis et al., 2019) and is one of the research frontiers in human-AI teams (Oswald et al., 2022). Within the philosophy of human-computer integrated design, the operator/crew's goals, requirements, obligations, responsibilities, the intelligent system and algorithms, the stakeholders, and the contexts of use are taken into account throughout the system lifecycle (Calhoun et al., 2016). Effective human-computer integration is expected to fulfill mission goals and user requirements, maintain the operators in the loop, and adjust to changing circumstances, thereby enhancing human–computer interaction and collaboration (Oswald et al., 2022).

Stephanidis et al. (2019) raised several issues concerning human-computer integration, including designing meaningful human control, ensuring system transparency and accountability, providing adaptive and personalized support for human needs, supporting human skills, detecting and responding to human emotions, considering social and ethical aspects of the design, and ultimately fostering human safety. Calhoun et al. (2016) proposed a set of integrated interfaces to support human-automation collaboration and coordination for multiple remotely piloted aircraft missions. Roth and colleagues put forward seven questions to guide human-automation interaction design (Roth et al., 2018). Calhoun et al. (2021) grouped the seven questions into three categories: situation drivers, visualizations and control mechanisms, and solution generation and presentation. Applying the three categories of guidance, they developed integrated human-automation interfaces for multi-unmanned vehicle control. Human-computer integration can be regarded as an extension of interaction between humans and technology (Rodrigues Barbosa et al., 2023). As a relatively new philosophy, there is a need for further research into comprehensive integrated design principles, elaborated designs in a variety of contexts, and rigorous assessments of their utility.

6.7.3 Design Guidelines to Prevent Human Errors

Error prevention is one of Nielsen's ten usability heuristics for user interface design (Nielsen, 1994). Most, if not all, other heuristics are also related to error prevention. For example, following the consistency heuristic would help avoid confusion. Besides, plenty of experience has been collected from the practice of industrial systems (see Kletz, 2001). There are various interface design guidelines available in the industry and HSI design review guidelines issued by regulatory bodies (such as NUREG-0700, O'Hara & Fleger, 2020). Following these guidelines is generally helpful for human error prevention and control. In this section, some selected guidelines are discussed along with their proper usage.

6.7.3.1 Automation

With automation, the possibilities of human error are removed and thus highly preferred. However, automation has side effects – it may cause undesired and unanticipated consequences, as pointed out by Sarter et al. (1997), Reason (2013), and other scholars. Putting automation as the first consideration in human error prevention does not mean that we should automate everything. However, caution should be taken on the risk of what it may bring, especially for a complex system in which humans are tightly coupled with the system. Automation has been found to contribute to over-reliance, impaired skills, reduced situation awareness, mode errors, and inability to detect or respond to automation failure. Another critical issue related to automation is the authority between humans and automation. In example (6) in the Introduction section, the operator could not override the automation logic. A famous example of such a dilemma is the supremacy of the Maneuvering Characteristics Augmentation System (MCAS) Boeing 737 Max 8 over the pilots. The authority should be reserved for a human operator to make the final decision when appropriate. It is not always possible to automate an operation due to either technology feasibility or implementation cost.

6.7.3.2 Reducing Complexity

Task complexity has a significant influence on human error, as reflected in various human reliability analysis methods. Many task complexity models, dimensions, and factors have been proposed in the literature, as reviewed by Liu and Li (2012), and can be used as guidance to reduce complexity. As an example, the ten dimensions proposed by Liu and Li (2012) can be commonly found in tasks with human–computer interactions and can be considered as a list of complexity sources to control.

6.7.3.3 Navigation Menu, Panel, and Links

Large interactive systems are often designed with multiple-layer interfaces to arrange the contents logically with acceptable layout complexity within a limited display size. Within such interface space, a navigation menu, panel, or set of links reflecting the structure or path of the interface is provided to help users navigate from one place to another as needed and keep awareness of their position inside the interface. However, if not properly designed, deep navigation or links would make users lost in an interface space.

6.7.3.4   Validation and Confirmation

As a common practice, validation and confirmation are used to confirm the correctness and intention of critical inputs (parameters or commands) to a system. Although this measure has prevented many human errors, it is not rare that operators do not make careful validation before they press the [enter] key or click the [OK] button to confirm their inputs because, most times, there are no errors, and thus operators think that this is just a routine operation and gradually form such a quick response pattern. This implies that validation and confirmation should not be abused for non-critical operations, while the validation and confirmation should be emphasized with proper alerting.

6.7.3.5   Feedback

Timely feedback allows a user to be aware of system status change after his/her interaction so that the achievement of the interaction goal can be verified and potential errors can be found and recovered. Unawareness of interaction results would make the user uneasy and feel the system is not in his/her control. Because of this, the user may conduct unnecessary and possibly erroneous operations.

6.7.3.6   Memory Aid

Designers of industrial systems tend to think that they need to have all variables displayed somewhere in the system, making the users switch among displays and keep the data needed for the task in their memory when returning to the task interface. Providing the relevant information right at the task interface or avoiding the need to memorize is highly desired to prevent memory failure, a typical human error mode.

6.7.3.7   Avoiding Overload

As the complexity of systems increases, operators are inundated with too much information, most of which is not relevant to the task they are performing. They have to identify what is needed, locate the needed among many other items, read and memorize the needed items and integrate them into higher-level information, and make judgments. A high cognitive workload can be commonly encountered in computerized systems. This becomes worse under an emergent situation when too many alarms are triggered.

The increase of accessible information amount was purported to result in more seeking and shallower thinking, yet this may be attributed to task strategy but not just overloading (Lyu et al., 2022). For successful teamwork in a computerized environment, information sharing is necessary to maintain good mutual awareness among team members. Unfortunately, many systems rigidly push information. If these systems are not adequately "smart," the pushed information could be irrelevant and overload team members. For this, user-defined information sharing was proposed and found to be effective at improving the operator's diagnosis performance (She et al., 2019).

6.7.3.8   Alarm

Providing alarms is a kind of automation in the detection of abnormal conditions. Without alarming functions, complex system operators have to be burdened with a high workload in observing many variables and detecting their risky deviations. The reliability of such human work is obviously low. An alarm system is thus a must for complex safety-critical systems. As summarized by Wu and Li (2018), a well-designed alarm system serves to: (1) inform the operator that there is a system/parameter deviation; (2) inform the priority and root cause of the deviation; (3) guide the operator's initial response to the deviation; (4) confirm whether the operator's response corrected the deviation;

and (5) aid post-analysis of an accident. Ideally, such a system would significantly reduce operators' workload and increase their operational reliability. However, if not properly designed, operators would be flooded by alarms in an emergency. Methods to overcome or mitigate alarm floods include alarm filtering and suppression, alarm prioritization, and group alarm. The interfaces should present the essential alarms to operators more intuitively, help operators quickly and easily judge the important and urgent alarms, and avoid them being information overloaded.

6.7.3.9 Task Support

In general, the fundamental purpose of interface design is to support users' task performance on the system. Thus, task support verification is included in the human factors review model of NUREG-0711 (O'Hara et al., 2012). Providing task support conforms to the philosophy of HCD, which is defined in ISO 9241-210 (ISO, 2019) as "an approach to interactive systems development that aims to make systems usable and useful by focusing on the users, their needs and requirements, and by applying human factors/ergonomics, usability knowledge, and techniques." Displays are often designed for "shared use" in different tasks in industrial systems. For critical tasks, dedicated displays should be provided to ensure their performance.

6.7.3.10 Team Support

Teamwork is a common form of work organization for complex systems. Team members have different roles and work collaboratively to keep a system safe. HSI design should support teamwork, such as information sharing among team members, to maintain mutual awareness to avoid team-related errors.

6.7.3.11 Providing Information Rather Than Raw Data

As mentioned before, operators are overloaded in information seeking and integration. Thinking that the interfaces of many industrial systems simply provide the raw data from various sensors, information integration is thus especially challenging for operators. What is needed by an operator is often information at high levels rather than the instantaneous sensor values. The operator may want to know whether a variable is going up/down, fluctuating, or stable. He/she cannot get this information directly from the interface but has to observe the variable values for a certain time to reach a judgment. As more complicated examples, the operator needs to judge the relation between two or even more variables. Based on task analysis, information requirements should be identified. Then in interface design, the relevant information items should be provided directly in the interface and lessen the need for information integration by operators as possible. Bars, curves, and other configural figures (Burns & Hajdukiewicz, 2004) can be adopted to support a direct perception of information. The derivation of some high-level information would certainly require the development of algorithms to integrate several variables or predict future situations.

6.7.3.12 Improving Situation Awareness

Situation awareness (SA) is defined as "the perception of the elements in the environment within a volume of time and space, the comprehension of their meaning, and the projection of their status in the near future" (Endsley, 1988). It is found to be the cause of many human errors and is thus critical for human performance in high-consequence domains. Endsley (2021) provides a list of SA-oriented design principles, including general guidelines for supporting situation awareness and guidelines for coping with automation and complexity, design of alarm systems, presentation of information uncertainty, and supporting situation awareness in team operations.

6.7.3.13 Providing Complete Information in a Proper Format

Even after more than half a century of evolution, bad examples of interface design can be found in safety-critical systems nowadays. Below could be an extreme example, as shown in Figure 6.9: the designer intended to put so many contents in a display that he arranged 22 variables at the bottom,

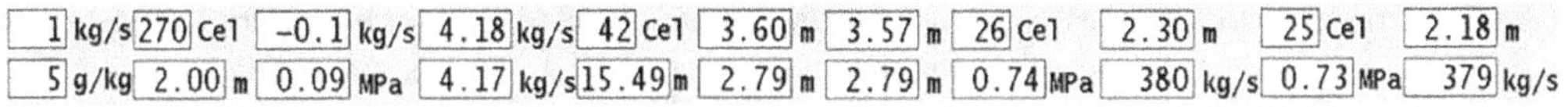

1 kg/s	270 Cel	−0.1 kg/s	4.18 kg/s	42 Cel	3.60 m	3.57 m	26 Cel	2.30 m	25 Cel	2.18 m
5 g/kg	2.00 m	0.09 MPa	4.17 kg/s	15.49 m	2.79 m	2.79 m	0.74 MPa	380 kg/s	0.73 MPa	379 kg/s

FIGURE 6.9   A bad example of interface design.

only showing the variable values and units but dropping the variable names. He thought that operators would know the corresponding variable of each value after training.

It can also be seen from the above that all variable values are displayed with numerical digits. This is not a rare case. Many industrial systems are designed with such an interface style. The advantages of computerization, allowing great graphical flexibility in the information display, have not been fully taken to develop innovative interfaces. Although various configural figures have been proposed in academia, engineers do not know about them – many, if not most, have very limited knowledge of human factors engineering, or even do not know about human factors.

6.7.3.14   Providing Adequate Transparency

Accidents related to the application of automation, AI, and autonomous technologies (referred to as automation in brief) reveal that human interaction with these technologies should be carefully considered in the design of safety-critical systems. These systems should be designed with adequate transparency to support human operators in their awareness of the actions and status of automation, understanding of the logic and mechanisms within automation, and the ability to anticipate future actions and the status of automation. Only with such transparency of automation can human operators have good SA, correctly accept or reject decisions made by automation, timely take over automation, and thus reasonably cooperate with automation – in a word, play their active and positive roles and avoid human-automation conflicts in the system. Thus, transparency is the precondition of human-automation teaming.

6.7.4   Design Guidelines for Error Tolerance

To err is human, but it does not mean that humans only play negative roles in a system. Besides preventing human errors, the design of a safety-critical system should also support human operators to play a positive role, such as detecting and recovering human errors and hardware/software failures. The guidelines mentioned above can also be adapted for this purpose, such as timely feedback and increasing transparency. Important human inputs should be validated. Invalid inputs that may cause unacceptable effects should be prohibited, while those not so critical but suspected should be warned. In the cases where automatic validation is not feasible, timely feedback on the immediate or future effects of human interactions should be provided to support operators' judgment. If there are performance criteria, they should be presented for instant reference.

With contemporary camera monitoring systems and video processing algorithms, behavior deviations during a task process can be detected. These systems are considered external addons but not included in the design of an interactive system.

Reversibility is often discussed in the literature, but it can be feasible for pure computer systems; for systems with actuators, once a command is issued to an actuator, it is often impossible to reverse it. This emphasizes the priority of error prevention and detection over error recovery.

6.8   CONCLUSION

Human errors will continue to be a constant concern for the design and use of interactive systems, especially in safety-critical domains. Despite the disparate definitions, the design of interactive systems should be meticulous in accounting for all kinds of deviations from expected interaction. New classifications of human errors have been developed in various domains; however, a standard classification is needed for data collection, sharing and analysis to support human reliability analysis, training improvement, and other management purposes. It is great to see that human error analysis has become a common practice in many industries for learning from operation experience. Unfortunately, no revolutionary theory of human error mechanisms emerges in the new century.

Many design guidelines are available for avoiding and tolerating human errors. Challenges and opportunities come together with the application of new technologies. In addition, system complexity is increasing. Designers should keep in mind that interfaces are to support users' task performance. Providing information at need and avoiding overload seem to conflict with each other. However, the actual situation is that an excessive amount of irrelevant information is presented, confusing the users or making them strive to seek out the relevant one; even worse, raw data are simply thrown to the users, leading to heavy cognitive work.

The concepts of transparency and SA are easy to understand, but in practice, designers often feel torn between providing information and avoiding overload. Multimodal interaction technologies and design concepts such as configural figures may help, but reducing complexity would be the best solution.

NOTES

1 Some taxonomies, for example the classification of EOO and EOC in THERP, are originally proposed in a certain domain, such as nuclear industry. However, they are evaluated as also applicable to other domains here.
2 Note that the final result could also be an incident, near miss, etc.

REFERENCES

Ackerman, P. L. (2011). 100 years without resting. In P. L. Ackerman (Ed.), *Cognitive Fatigue: Multidisciplinary Perspectives on Current Research and Future Applications* (pp. 11–43). Washington, DC: American Psychological Association.

Adhikari, S., Bayley, C., Bedford, T., Busby, J., Cliffe, A., Devgun, G., Eid, M., French, S., Keshvala, R., Pollard, S., Soane, E., Tracy, D., & Wu, S. (2009). *Human Reliability Analysis: A Review and Critique*, Manchester Business School Working Paper No 589. Manchester: The University of Manchester.

Alexander, W. H., & Brown, J. W. (2019). The role of the anterior cingulate cortex in prediction error and signaling surprise. *Topics in Cognitive Science*, *11*(1), 119–135. https://doi.org/10.1111/tops.12307.

Alvarenga, M. A. B., & Frutuoso e Melo, P. F. (2019). A review of the cognitive basis for human reliability analysis. *Progress in Nuclear Energy*, *117*, 103050. https://doi.org/10.1016/j.pnucene.2019.103050.

Anderson, J. R. (1993). *Rules of the Mind*. Mahwah, NJ: Lawrence Erlbaum Associates, Inc.

Anu, V., Hu, W., Carver, J. C., Walia, G. S., & Bradshaw, G. (2018). Development of a human error taxonomy for software requirements: A systematic literature review. *Information and Software Technology*, *103*, 112–124. https://doi.org/10.1016/j.infsof.2018.06.011.

Avril, E. (2023). Providing different levels of accuracy about the reliability of automation to a human operator: Impact on human performance. *Ergonomics*, *66*(2), 217–226. https://doi.org/10.1080/00140139.2022.2069870.

Avril, E., Cegarra, J., Wioland, L., & Navarro, J. (2022). Automation type and reliability impact on visual automation monitoring and human performance. *International Journal of Human-Computer Interaction*, *38*(1), 64–77. https://doi.org/10.1080/10447318.2021.1925435.

Ayaz, H., & Dehais, F. (2021). Neuroergonomics. In G. Salvendy, & W. Karwowski (Eds.), *Handbook of Human Factors and Ergonomics* (5th ed., pp. 816–841). Hoboken, NJ: John Wiley & Sons, Inc.

Baber, C., & Stanton, N. A. (1994). Task analysis for error identification: A methodology for designing error-tolerant consumer products. *Ergonomics*, *37*(11), 1923–1941. https://doi.org/10.1080/00140139408964958.

Babiloni, F. (2019). Mental workload monitoring: New perspectives from neuroscience. In L. Longo, & M. C. Leva (Eds.), *Human Mental Workload: Models and Applications (Proceedings of 3rd International Symposium, H-WORKLOAD 2019)* (pp. 3–19). Berlin, Heidelberg: Springer International Publishing.

Baddeley, A. (2000). The episodic buffer: A new component of working memory? *Trends in Cognitive Sciences*, *4*(11), 417–423. https://doi.org/10.1016/S1364-6613(00)01538-2.

Banks, V. A., Plant, K. L., & Stanton, N. A. (2018). Driver error or designer error: Using the Perceptual Cycle Model to explore the circumstances surrounding the fatal Tesla crash on 7th May 2016. *Safety Science*, *108*, 278–285. https://doi.org/10.1016/j.ssci.2017.12.023.

Bartlett, F. C. (1932). *Remembering: A Study of Experimental and Social Psychology*. Cambridge: Cambridge University Press.

Behrens, M., Gube, M., Chaabene, H., Prieske, O., Zenon, A., Broscheid, K.-C., Schega, L., Husmann, F., & Weippert, M. (2023). Fatigue and human performance: An updated framework. *Sports Medicine*, *53*(1), 7–31. https://doi.org/10.1007/s40279-022-01748-2.

Bendak, S., & Rashid, H. S. J. (2020). Fatigue in aviation: A systematic review of the literature. *International Journal of Industrial Ergonomics*, *76*, 102928. https://doi.org/10.1016/j.ergon.2020.102928.

Bennett, K. B., & Flach, J. (2019). Ecological interface design: Thirty-plus years of refinement, progress, and potential. *Human Factors*, *61*(4), 513–525.

Billings, C. E. (1991). *Human-Centered Aircraft Automation: A Concept and Guidelines* (Vol. 103885). Washington, DC: National Aeronautics and Space Administration, Ames Research Center.

Boorman, D. J., & Mumaw, R. J. (2004). A new autoflight/FMS interface: Guiding design principles. In A. Pritchett, & A. Jackson (Eds.), *Proceedings of the International Conference on Human-Computer Interaction in Aeronautics [CD-ROM] (pp. 303–321)*. Toulouse, France: EURISCO International.

Borghini, G., Astolfi, L., Vecchiato, G., Mattia, D., & Babiloni, F. (2014). Measuring neurophysiological signals in aircraft pilots and car drivers for the assessment of mental workload, fatigue and drowsiness. *Neuroscience & Biobehavioral Reviews*, *44*, 58–75. https://doi.org/10.1016/j.neubiorev.2012.10.003.

Boring, R. L. (2012). Fifty years of THERP and human reliability analysis. In *Proceedings of the 11th International Probabilistic Safety Assessment and Management Conference (PSAM11) and the Annual European Safety and Reliability Conference (ESREL 2012)* (Vol. 5, pp. 3523–3532), Helsinki, Finland.

Boy, G. A., & Schmitt, K. A. (2013). Design for safety: A cognitive engineering approach to the control and management of nuclear power plants. *Annals of Nuclear Energy*, *52*, 125–136. https://doi.org/10.1016/j.anucene.2012.08.027.

Broadbent, D. E. (1978). The current state of noise research: Reply to poulton. *Psychological Bulletin*, *85*, 1052–1067. https://doi.org/10.1037/0033-2909.85.5.1052.

Burns, C. M. (2000). Putting it all together: Improving display integration in ecological displays. *Human Factors, 42*(2), 226–241.

Burns, C. M., & Hajdukiewicz, J. (2004). *Ecological Interface Design* (1st ed.). Boca Raton, FL: CRC Press. https://doi.org/10.1201/9781315272665.

Bustamante, E. A., Bliss, J. P., & Anderson, B. L. (2007). Effects of varying the threshold of alarm systems and workload on human performance. *Ergonomics*, *50*(7), 1127–1147. https://doi.org/10.1080/00140130701237345.

Cacciabue, P. C., & Hollnagel, E. (1995). Simulation of cognition: Applications. In J.-M. Hoc, & P. C. Cacciabue (Eds.), *Expertise and Technology: Cognition & Human-Computer Cooperation* (1st ed., pp. 55–73). London: Psychology Press.

Calhoun, G., Bartik, J., Ruff, H., Behymer, K., & Frost, E. (2021). Enabling human-autonomy teaming with multi-unmanned vehicle control interfaces. *Human-Intelligent Systems Integration*, *3*(1), 1–20.

Calhoun, G. L., Goodrich, M. A., Dougherty, J. R., & Adams, J. A. (2016). Human-autonomy collaboration and coordination toward multi-RPA missions. In L. J. Rowe, N. J. Cooke, W. Bennett Jr, & D. Q. Joralmon (Eds.). *Remotely Piloted Aircraft Systems: A Human Systems Integration Perspective* (pp. 109–136). Hoboken, NJ: John Wiley & Sons.

Chandler, F. T., Chang, Y. H. J., Mosleh, A., Marble, J. L., Boring, R. L., & Gertman, D. I. (2006). *Human Reliability Analysis Methods: Selection Guidance for NASA*. Washington, DC: National Aeronautics and Space Administration.

Chang, Y. J., Bley, D., Criscione, L., Kirwan, B., Mosleh, A., Madary, T., Nowell, R., Richards, R., Roth, E. M., & Sieben, S. (2014). The SACADA database for human reliability and human performance. *Reliability Engineering & System Safety*, *125*, 117–133.

Chang, Y. J., Xing, J., & Peters, S. (2016, October 2–7). Human reliability analysis method development in the US nuclear regulatory commission. *13th International Conference on Probabilistic Safety Assessment and Management (PSAM 13)*, Seoul, Korea.

Chavaillaz, A., Wastell, D., & Sauer, J. (2016). System reliability, performance and trust in adaptable automation. *Applied Ergonomics*, *52*, 333–342. https://doi.org/10.1016/j.apergo.2015.07.012.

Chen, J. Y. C., Flemisch, F. O., Lyons, J. B., & Neerincx, M. A. (2020). Guest editorial: Agent and system transparency. *IEEE Transactions on Human-Machine Systems*, *50*(3), 189–193. https://doi.org/10.1109/THMS.2020.2988835.

Chen, K., Li, Z., & Jamieson, G. A. (2018). Influence of information layout on diagnosis performance. *IEEE Transactions on Human-Machine Systems, 48*(3), 316–323.

Choi, S. Y., & Jung, W. (2014). Qualitative human event analysis with simulator data by using HuRAM+ and HERA. In R. D. J. M. Steenbergen, P. H. A. J. M. van Gelder, S. Miraglia, & A. C. W. M. T. Vrouwenvelder (Eds.), *Safety, Reliability and Risk Analysis: Beyond the Horizon (Proceedings of the European Safety and Reliability Conference, ESREL 2013)* (pp. 517–522). Boca Raton, FL: CRC Press.

Chung, Y. H., Yoon, W. C., & Min, D. (2009). A model-based framework for the analysis of team communication in nuclear power plants. *Reliability Engineering & System Safety, 94*(6), 1030–1040.

Cooper, R. (2002). Order and disorder in everyday action: The roles of contention scheduling and supervisory attention. *Neurocase, 8*(1–2), 61–79. https://doi.org/10.1093/neucas/8.1.61.

Cooper, S. E., Ramey-Smith, A. M., Wreathall, J., Parry, G. W., Bley, D. C., Luckas, W. J., Taylor, J. H., & Barriere, M. T. (1996). *A Technique for Human Error Analysis (ATHEANA): Technical Basis and Methodology Description*, NUREG/CR-6350. Washington, DC: U.S. Nuclear Regulatory Commission.

Deci, E. L. (1975). *Intrinsic Motivation*. New York: Plenum Publishing.

Deci, E. L., & Ryan, R. M. (2000). The "what" and "why" of goal pursuits: Human needs and the self-determination of behavior. *Psychological Inquiry, 11*(4), 227–268. https://doi.org/10.1207/S15327965PLI1104_01.

Decortis, F., & Cacciabue, P. C. (1988). Temporal dimension in cognitive models. In *Conference Record for 1988 IEEE Fourth Conference on Human Factors and Power Plants* (pp. 279–284). IEEE. https://doi.org/10.1109/HFPP.1988.27514.

Dehaene, S., Posner, M. I., & Tucker, D. M. (1994). Localization of a neural system for error detection and compensation. *Psychological Science, 5*(5), 303–305. https://doi.org/10.1111/j.1467-9280.1994.tb00630.x.

Dehais, F., Lafont, A., Roy, R., & Fairclough, S. (2020). A neuroergonomics approach to mental workload, engagement and human performance. *Frontiers in Neuroscience, 14*. https://www.frontiersin.org/articles/10.3389/fnins.2020.00268.

Dekker, S. (2014). *The Field Guide to Understanding 'Human Error'* (3rd ed.). Boca Raton, FL: CRC Press.

Dekker, S. W. A. (2011). What is rational about killing a patient with an overdose? Enlightenment, continental philosophy and the role of the human subject in system failure. *Ergonomics, 54*(8), 679–683. https://doi.org/10.1080/00140139.2011.592607.

Dinadis, N., & Vicente, K. J. (1999). Designing functional visualizations for aircraft systems status displays. *The International Journal of Aviation Psychology, 9*(3), 241–269.

DOE. (1999). *DOE Workbook: Conducting Accident Investigations*. U.S. Department of Energy.

Dörner, D., & Güss, C. D. (2022). Human error in complex problem solving and dynamic decision making: A taxonomy of 24 errors and a theory. *Computers in Human Behavior Reports, 7*, 100222. https://doi.org/10.1016/j.chbr.2022.100222.

Dougherty Jr, E. M. (1990). Human reliability analysis-where shouldst thou turn? *Reliability Engineering & System Safety, 29*(3), 283–299.

Drury, C. G. (1991). Errors in aviation maintenance: Taxonomy and control. *Proceedings of the Human Factors Society Annual Meeting, 35*(2), 42–46. https://doi.org/10.1518/107118191786755850.

Duch, W., Oentaryo, R. J., & Pasquier, M. (2008). Cognitive architectures: Where do we go from here? In *Proceedings of the 2008 Conference on Artificial General Intelligence 2008: Proceedings of the First AGI Conference* (pp. 122–136), Memphis, TN. IOS Press.

Embrey, D. (2004). Qualitative and quantitative evaluation of human error in risk assessment. In C. Sandom, & R. S. Harvey (Eds.), *Human Factors for Engineers* (pp. 151–202). London: The Institution of Engineering and Technology.

Embrey, D. E. (1986a). SHERPA: A systematic human error reduction and prediction approach. In *Proceedings of the International Topical Meeting on Advances in Human Factors in Nuclear Power Systems* (pp. 184–193), Knoxwill, TN.

Embrey, D. E. (1986b). SLIM-MAUD: A computer-based technique for human reliability assessment. *International Journal of Quality & Reliability Management, 3*(1), 5–12.

Endsley, M. R. (1988). Design and evaluation for situation awareness enhancement. In *Proceedings of the Human Factors Society Annual Meeting* (Vol. 32, No. 2, pp. 97–101). Thousand Oaks, CA: SAGE Publications, Inc.

Endsley, M. R. (1995a). A taxonomy of situation awareness errors. In R. Fuller, N. Johnston, & N. McDonald (Eds.), *Human Factors in Aviation Operations* (pp. 287–292). Aldershot, England: Avebury Aviation, Ashgate Publishing Ltd.

Endsley, M. R. (1995b). Toward a theory of situation awareness in dynamic systems. *Human Factors, 37*(1), 32–64. https://doi.org/10.1518/001872095779049543.

Endsley, M. R. (1996). Automation and situation awareness. In R. Parasuraman, & M. Mouloua (Eds.), *Automation and Human Performance: Theory and Applications* (pp. 163–181). Mahwah, NJ: Lawrence Erlbaum Associates, Inc.

Endsley, M. R. (2016). From here to autonomy: Lessons learned from human-automation research. *Human Factors*, *59*(1), 5–27. https://doi.org/10.1177/0018720816681350.

Endsley, M. R. (2021). Situation awareness. In G. Salvendy, & W. Karwowski (Eds.), *Handbook of Human Factors and Ergonomics* (5th ed., pp. 434–455). Hoboken, NJ: John Wiley & Sons, Inc.

Errington, J., Reising, D. V. C., Bullemer, P., DeMaere, T., Coppard, D., Doe, K., & Bloom, C. (2005). Establishing human performance improvements and economic benefit for a human-centered operator interface: An industrial evaluation. *Proceedings of the Human Factors and Ergonomics Society Annual Meeting*, *49*(23), 2036–2040.

Fancher, P., Bareket, Z., & Ervin, R. (2001). Human-centered design of an ACC-with-braking and forward-crash-warning system. *Vehicle System Dynamics*, *36*(2–3), 203–223.

Farooq, U., & Grudin, J. (2016). Human-computer integration. *Interactions*, *23*(6), 26–32.

Fedota, J. R., & Parasuraman, R. (2010). Neuroergonomics and human error. *Theoretical Issues in Ergonomics Science*, *11*(5), 402–421. https://doi.org/10.1080/14639220902853104.

Feggetter, A. J. (1982). A method for investigating human factor aspects of aircraft accidents and incidents. *Ergonomics*, *25*(11), 1065–1075.

Flach, J. M., & Vicente, K. J. (1989). Complexity, difficulty, direct manipulation and direct perception. Technical Report EPRL-89-03. Urbana-Champaign, IL: Engineering Psychology Research Laboratory, University of Illinois.

Flin, R., O'Connor, P., & Crichton, M. (2008). *Safety at the Sharp End: A Guide to Non-Technical Skills*. Boca Raton, FL: CRC Press.

Fotta, M. E., Byrne, M. D., & Luther, M. S. (2005). Developing a human error modeling architecture (HEMA). In D. D. Schmorrow (Ed.), *Foundations of Augmented Cognition: Volume 11* (pp. 1025–1034). Mahwah, NJ: Lawrence Erlbaum Associates, Inc.

Fridman, L. (2018). Human-centered autonomous vehicle systems: Principles of effective shared autonomy. arXiv preprint arXiv:1810.01835.

Gao, Q., Yu, W., Jiang, X., Song, F., Pan, J., & Li, Z. (2015). An integrated computer-based procedure for teamwork in digital nuclear power plants. *Ergonomics*, *58*(8), 1303–1313.

Gehring, W. J., Liu, Y., Orr, J. M., & Carp, J. (2012). The error-related negativity (ERN/Ne). In E. S. Kappenman, & S. J. Luck (Eds.), *The Oxford Handbook of Event-Related Potential Components*. Oxford: Oxford University Press. https://doi.org/10.1093/oxfordhb/9780195374148.013.0120.

Gertman, D., Blackman, H., Marble, J., Byers, J., & Smith, C. (2005). *The SPAR-H Human Reliability Analysis Method*, NUREG/CR-6883. Washington, DC: U.S. Nuclear Regulatory Commission.

Gibson, W. H., Megaw, E. D., Young, M. S., & Lowe, E. (2006). A taxonomy of human communication errors and application to railway track maintenance. *Cognition, Technology & Work*, *8*(1), 57–66. https://doi.org/10.1007/s10111-005-0020-x.

Gramann, K., McKendrick, R., Baldwin, C., Roy, R. N., Jeunet, C., Mehta, R. K., & Vecchiato, G. (2021). Grand field challenges for cognitive neuroergonomics in the coming decade. *Frontiers in Neuroergonomics*, *2*. https://www.frontiersin.org/articles/10.3389/fnrgo.2021.643969.

Grandjean, E. (1968). Fatigue: Its physiological and psychological significance. *Ergonomics*, *11*(5), 427–436. https://doi.org/10.1080/00140136808930992.

Griffith, C. D., & Mahadevan, S. (2011). Inclusion of fatigue effects in human reliability analysis. *Reliability Engineering & System Safety*, *96*(11), 1437–1447. https://doi.org/10.1016/j.ress.2011.06.005.

Groth, K. M., & Mosleh, A. (2012). A data-informed PIF hierarchy for model-based human reliability analysis. *Reliability Engineering & System Safety*, *108*, 154–174. https://doi.org/10.1016/j.ress.2012.08.006.

Hagen, E. W., & Mays, G. T. (1981). Human factors engineering in the US nuclear arena. *Nuclear Safety*, *22*(3), 337–346.

Hall, D. S., Shattuck, L. G., & Bennett, K. B. (2012). Evaluation of an ecological interface design for military command and control. *Journal of Cognitive Engineering and Decision Making*, *6*(2), 165–193.

Hallbert, B., Boring, R., Gertman, D., Dudenhoeffer, D., Whaley, A., Marble, J., Joe, J., & Lois, E. (2006). *Human Event Repository and Analysis (HERA) System, Volume 1: Overview*, NUREG/CR-6903. Washington, DC: U.S. Nuclear Regulatory Commission.

Hancock, P. A. (1989). A dynamic model of stress and sustained attention. *Human Factors*, *31*(5), 519–537. https://doi.org/10.1177/001872088903100503.

Hannaman, G. W., Spurgin, A. J., & Lucki, Y. (1984). *Human Cognitive Reliability Model for PRA Analysis*, NUS-4531. Electric Power Research Institute.

Harris, D., & Li, W. C. (2011). Error on the flight deck: Interfaces, organizations, and culture. In G. A. Boy (Ed.), *The Handbook of Human-Machine Interaction* (pp. 399–415). Farnham, England: Ashgate Publishing Ltd.

Hee, D. D., Pickrell, B. D., Bea, R. G., Roberts, K. H., & Williamsone, R. B. (1999). Safety Management Assessment System (SMAS): A process for identifying and evaluating human and organization factors in marine system operations with field test results. *Reliability Engineering and System Safety*, 65(2), 125–140.

Ho, G., Scialfa, C. T., Caird, J. K., & Graw, T. (2001). Visual search for traffic signs: The effects of clutter, luminance, and aging. *Human Factors*, 43(2), 194–207. https://doi.org/10.1518/001872001775900922.

Hobbs, A., & Williamson, A. (2003). Associations between errors and contributing factors in aircraft maintenance. *Human Factors*, 45(2), 186–201. https://doi.org/10.1518/hfes.45.2.186.27244.

Hockey, G. R. J. (1997). Compensatory control in the regulation of human performance under stress and high workload: A cognitive-energetical framework. *Biological Psychology*, 45(1), 73–93. https://doi.org/10.1016/S0301-0511(96)05223-4.

Hollnagel, E. (1991). The phenotype of erroneous actions: Implications for HCI design. In G. R. S. Weir, & J. L. Alty (Eds.), *Human-Computer Interaction and Complex Systems* (pp. 73–121). London: Academic Press.

Hollnagel, E. (1993a). *Human Reliability Analysis: Context and Control*. London: Academic Press.

Hollnagel, E. (1993b). The phenotype of erroneous actions. *International Journal of Man-Machine Studies*, 39(1), 1–32. https://doi.org/10.1006/imms.1993.1051.

Hollnagel, E. (1998). *Cognitive Reliability and Error Analysis Method*. Amsterdam, The Netherlands: Elsevier Science.

Hollnagel, E. (2004). *Barriers and Accident Prevention*. New York: Routledge.

Hollnagel, E., & Woods, D. D. (2005). *Joint Cognitive Systems: Foundations of Cognitive Systems Engineering* (1st ed.). Boca Raton, FL: CRC Press.

Honig, S., & Oron-Gilad, T. (2018). Understanding and resolving failures in human-robot interaction: Literature review and model development. *Frontiers in Psychology, 9*, 861.

Horberry, T., Burgess-Limerick, R., Cooke, T., & Steiner, L. (2016). Improving mining equipment safety through human-centered design. *Ergonomics in Design, 24*(3), 29–34.

Hutchins, E. (1997). *The Integrated Mode Management Interface*. NASA Contractor Report NCC 2-591. Moffett Field, CA: NASA-Ames Research Center.

International Organization for Standardization (ISO). (2002). Ergonomics of human-system interaction: Usability methods supporting human-centered design, ISO/TR 16982:2002.

International Organization for Standardization (ISO). (2019). Ergonomics of human-system interaction - Part 210: Human-centered design for interactive systems. ISO 9241-210:2019.

Jacobs, R., & Haber, S. (1994). Organizational processes and nuclear power plant safety. *Reliability Engineering & System Safety*, 45(1), 75–83. https://doi.org/10.1016/0951-8320(94)90078-7.

Jamieson, G. A. (2007). Ecological interface design for petrochemical process control: An empirical assessment. *IEEE Transactions on Systems, Man, and Cybernetics-Part A: Systems and Humans, 37*(6), 906–920.

Jerald, J. (2015). *The VR Book: Human-Centered Design for Virtual Reality*. San Rafael, CA: Morgan & Claypool.

Jones, N. A., Ross, H., Lynam, T., Perez, P., & Leitch, A. (2011). Mental models: An interdisciplinary synthesis of theory and methods. *Ecology and Society*, 16(1). https://www.jstor.org/stable/26268859.

Jung, W., Park, J., Kim, Y., Choi, S. Y., & Kim, S. (2020). HuREX-A framework of HRA data collection from simulators in nuclear power plants. *Reliability Engineering & System Safety*, 194, 106235.

Kahneman, D. (1973). *Attention and Effort*. Upper Saddle River, NJ: Prentice-Hall.

Kanfer, R. (2012). Work motivation: Theory, practice, and future directions. In S. W. J. Kozlowski (Ed.), *The Oxford Handbook of Organizational Psychology*, (Vol. 1, pp. 455–495). Oxford: Oxford University Press.

Kanfer, R., & Ackerman, P. L. (1989). Motivation and cognitive abilities: An integrative/aptitude-treatment interaction approach to skill acquisition. *Journal of Applied Psychology*, 74(4), 657–690. https://doi.org/10.1037/0021-9010.74.4.657.

Kanfer, R., & Chen, G. (2016). Motivation in organizational behavior: History, advances and prospects. *Organizational Behavior and Human Decision Processes*, 136, 6–19. https://doi.org/10.1016/j.obhdp.2016.06.002.

Kang, S., & Seong, P. H. (2020). Performance shaping factor taxonomy for human reliability analysis on mitigating nuclear power plant accidents caused by extreme external hazards. *Annals of Nuclear Energy*, 145, 107533. https://doi.org/10.1016/j.anucene.2020.107533.

Kantowitz, B. H., Roediger III, H. L., & Elmes, D. G. (2009). *Experimental Psychology* (9th ed.). Belmont, CA: Wadsworth Publishing.

Kim, J.-E., & Nembhard, D. A. (2019). Eye movement as a mediator of the relationships among time pressure, feedback, and learning performance. *International Journal of Industrial Ergonomics*, *70*, 116–123. https://doi.org/10.1016/j.ergon.2018.12.006.

Kim, J. W., & Jung, W. (2003). A taxonomy of performance influencing factors for human reliability analysis of emergency tasks. *Journal of Loss Prevention in the Process Industries*, *16*(6), 479–495. https://doi.org/10.1016/S0950-4230(03)00075-5.

Kim, Y., Park, J., & Jung, W. (2017). A classification scheme of erroneous behaviors for human error probability estimations based on simulator data. *Reliability Engineering & System Safety*, *163*, 1–13.

Kirwan, B. (1992). Human error identification in human reliability assessment. Part 1: Overview of approaches. *Applied Ergonomics*, *23*(5), 299–318.

Kirwan, B. (1994). *A Guide to Practical Human Reliability Assessment*. London: Taylor & Francis.

Kirwan, B. (1998). Human error identification techniques for risk assessment of high risk systems-Part 1: Review and evaluation of techniques. *Applied Ergonomics*, *29*(3), 157–177.

Klein, G. A. (1993). A recognition-primed decision (RPD) model of rapid decision making. In G. A. Klein, J. Orasanu, R. Calderwood, & C. E. Zsambok (Eds.), *Decision Making in Action: Models and Methods* (pp. 138–147). New York: Ablex Publishing.

Klein, G., Woods, D. D., Bradshaw, J., Hoffman, R. R., & Feltovich, P. J. (2004). Ten challenges for making automation a "Team Player" in joint human-agent activity. *IEEE Intelligent Systems, 19*(6), 91–95.

Kletz, T. (2001). *An Engineer's View of Human Error*. New York: Routledge.

Kotseruba, I., & Tsotsos, J. K. (2016). A review of 40 years of cognitive architecture research: Core cognitive abilities and practical applications. arXiv e-prints, arXiv:1610.08602v08603. https://doi.org/10.48550/arXiv.1610.08602.

Kraemer, S., & Carayon, P. (2007). Human errors and violations in computer and information security: The viewpoint of network administrators and security specialists. *Applied Ergonomics*, *38*(2), 143–154. https://doi.org/10.1016/j.apergo.2006.03.010.

Lebière, C., Anderson, J. R., & Reder, L. M. (1994). Error modeling in the ACT-R production system. In A. Ram, & K. Eiselt (Eds.), *Proceedings of the Sixteenth Annual Conference of the Cognitive Science Society* (pp. 555–559). New York: Routledge.

Lee, S. (2022, June 26-July 1). Insights for human reliability analysis method gained from root cause analysis of human error events occurred in Korea. *16th Probabilistic Safety Assessment and Management Conference (PSAM 16)*, Honolulu, HI. https://www.iapsam.org/PSAM16/paper.php?ID=K7325.

Lee, S. C., & Ji, Y. G. (2019). Complexity of in-vehicle controllers and their effect on task performance. *International Journal of Human-Computer Interaction*, *35*(1), 65–74. https://doi.org/10.1080/10447318.2018.1428263.

Lee, S. J., Kim, J., & Jang, S.-C. (2011). Human error mode identification for NPP main control room operations using soft controls. *Journal of Nuclear Science and Technology*, *48*(6), 902–910. https://doi.org/10.1080/18811248.2011.9711776.

Lehto, M., & Salvendy, G. (1991). Models of accident causation and their application: Review and reappraisal. *Journal of Engineering and Technology Management*, *8*(2), 173–205.

Leiden, K., Laughery, K. R., Keller, J., French, J., Warwick, W., & Wood, S. D. (2001). A Review of human performance models for the prediction of human error. NASA system-wide accident prevention program, Ames Research Center. https://hsi.arc.nasa.gov/groups/HCSL/publications/HumanErrorModels.pdf.

Leveson, N. (2004). A new accident model for engineering safer systems. *Safety Science*, *42*(4), 237–270. https://doi.org/10.1016/S0925-7535(03)00047-X.

Licu, T., Cioran, F., Hayward, B., & Lowe, A. (2007). EUROCONTROL-Systemic Occurrence Analysis Methodology (SOAM)-A "Reason"-based organisational methodology for analysing incidents and accidents. *Reliability Engineering & System Safety*, *92*, 1162–1169.

Lin, C. J., Hsieh, T.-L., Yang, C.-W., & Huang, R.-J. (2016). The impact of computer-based procedures on team performance, communication, and situation awareness. *International Journal of Industrial Ergonomics, 51*, 21–29.

Liu, P., & Li, Z. (2012). Task complexity: A review and conceptualization framework. *International Journal of Industrial Ergonomics*, *42*(6), 553–568. https://doi.org/10.1016/j.ergon.2012.09.001.

Liu, P., & Li, Z. (2014). Comparison of task complexity measures for emergency operating procedures: Convergent validity and predictive validity. *Reliability Engineering & System Safety*, *121*, 289–293.

Liu, P., Li, Z., & Wang, Z. (2012, June 25–29). Task complexity measure for emergency operating procedures based on resource requirements in human information processing. *11th International Probabilistic Safety Assessment and Management Conference and the Annual European Safety and Reliability Conference*, Helsinki, Finland.

Liu, P., Lv, X., Li, Z., Qiu, Y., Hu, J., & He, J. (2016). Conceptualizing performance shaping factors in main control rooms of nuclear power plants: A preliminary study. In *International Conference on Engineering Psychology and Cognitive Ergonomics* (pp. 322–333). Cham: Springer. https://doi.org/10.1007/978-3-3 19-40030-3_32.

Liu, P., Lyu, X., Qiu, Y., Hu, J., Tong, J., & Li, Z. (2017). Identifying macrocognitive function failures from accident reports: A case study. In S. M. Cetiner, P. Fechtelkotter, & M. Legatt (Eds.), *Advances in Human Factors in Energy: Oil, Gas, Nuclear and Electric Power Industries (Proceedings of the AHFE 2016 International Conference on Human Factors in Energy: Oil, Gas, Nuclear and Electric Power Industries)* (pp. 29–40). Berlin, Heidelberg: Springer.

Liu, P., Zhang, R., Yin, Z., & Li, Z. (2021). Human errors and human reliability. In G. Salvendy, & W. Karwowski (Eds.), *Handbook of Human Factors and Ergonomics* (5th ed., pp. 514–572). Hoboken, NJ: John Wiley & Sons, Inc.

Lynch, K. M., Banks, V. A., Roberts, A. P. J., Radcliffe, S., & Plant, K. L. (2022). Maritime autonomous surface ships: Can we learn from unmanned aerial vehicle incidents using the perceptual cycle model? *Ergonomics*, 1–19. https://doi.org/10.1080/00140139.2022.2126896.

Lyu, X., & Li, Z. (2019). Predictors for human performance in information seeking, information integration, and overall process in diagnostic tasks. *International Journal of Human-Computer Interaction*, 35(19), 1831–1841. https://doi.org/10.1080/10447318.2019.1574097.

Lyu, X., Li, Z., Ma, Q., & She, M. (2022). Effects of accessible information amount and judgment times on human diagnostic performance of nuclear power plant faults. *Ergonomics*, 1–12. https://doi.org/10.1080/ 00140139.2022.2118836.

Lyu, X., She, M., Pan, D., Wu, X., Chen, K., & Li, Z. (2023). Fault diagnosis: Human performance in the digital and automation context. In V. G. Duffy, M. Lehto, Y. Yih, & R. W. Proctor (Eds.), *Human-Automation Interaction: Manufacturing, Services and User Experience* (pp. 265–288). Berlin, Heidelberg: Springer International Publishing. https://doi.org/10.1007/978-3-031-10780-1_14.

Maguire, M. (2001). Methods to support human-centred design. *International Journal of Human-Computer Studies*, 55(4), 587–634.

Makary, M., & Daniel, M. (2016). Medical error-the third leading cause of death in the US. *British Medical Journal*, 353, i2139.

Mandrick, K., Chua, Z., Causse, M., Perrey, S., & Dehais, F. (2016). Why a comprehensive understanding of mental workload through the measurement of neurovascular coupling is a key issue for neuroergonomics? *Frontiers in Human Neuroscience*, 10. https://www.frontiersin.org/articles/10.3389/fnhum.2016.00250.

Martinie, C., Palanque, P., Fahssi, R., Blanquart, J. P., Fayollas, C., & Seguin, C. (2015). Task model-based systematic analysis of both system failures and human errors. *IEEE Transactions on Human-Machine Systems*, 46(2), 243–254.

Mathalon, D. H., Whitfield, S. L., & Ford, J. M. (2003). Anatomy of an error: ERP and fMRI. *Biological Psychology*, 64(1), 119–141. https://doi.org/10.1016/S0301-0511(03)00105-4.

Matheson, G. O., Pacione, C., Shultz, R. K., & Klügl, M. (2015). Leveraging human-centered design in chronic disease prevention. *American Journal of Preventive Medicine*, 48(4), 472–479.

Maule, A. J., & Hockey, G. R. J. (1993). State, stress, and time pressure. In O. Svenson, & A. J. Maule (Eds.), *Time Pressure and Stress in Human Judgment and Decision Making* (pp. 83–101). Berlin, Heidelberg: Springer US. https://doi.org/10.1007/978-1-4757-6846-6_6.

McBride, S. E., Rogers, W. A., & Fisk, A. D. (2014). Understanding human management of automation errors. *Theoretical Issues in Ergonomics Science*, 15(6), 545–577.

McPherson, S. L. (2000). Expert-novice differences in planning strategies during collegiate singles tennis competition. *Journal of Sport and Exercise Psychology*, 22(1), 39–62. https://doi.org/10.1123/jsep.22.1.39.

Mehta, R., & Parasuraman, R. (2013). Neuroergonomics: A review of applications to physical and cognitive work. *Frontiers in Human Neuroscience*, 7. https://www.frontiersin.org/articles/10.3389/fnhum.2013.00889.

Mitchell, R. J., Williamson, A., & Molesworth, B. (2016). Application of a human factors classification framework for patient safety to identify precursor and contributing factors to adverse clinical incidents in hospital. *Applied Ergonomics*, 52, 185–195.

Mitchell, R. J., Williamson, A. M., Molesworth, B., & Chung, A. Z. Q. (2014). A review of the use of human factors classification frameworks that identify causal factors for adverse events in the hospital setting. *Ergonomics*, 57(10), 1443–1472. https://doi.org/10.1080/00140139.2014.933886.

Mohammed, S., Ferzandi, L., & Hamilton, K. (2010). Metaphor no more: A 15-year review of the team mental model construct. *Journal of Management*, *36*(4), 876–910. https://doi.org/10.1177/0149206309356804.

Moore, W. H. (1993). Management of human error in operations of marine systems. Doctoral dissertation, University of California, Berkeley. University Microfilms International.

Mosier, K. L., Skitka, L. J., Heers, S., & Burdick, M. (1998). Automation bias: Decision making and performance in high-tech cockpits. *The International Journal of Aviation Psychology, 8*(1), 47–63.

Mouloua, M., Gilson, R., & Hancock, P. (2003). Human-centered design of unmanned aerial vehicles. *Ergonomics in Design, 11*(1), 6–11.

Moura, R., Beer, M., Patelli, E., Lewis, J., & Knoll, F. (2016). Learning from major accidents to improve system design. *Safety Science*, *84*, 37–45. https://doi.org/10.1016/j.ssci.2015.11.022.

Mueller, F. F., Lopes, P., Strohmeier, P., Ju, W., Seim, C., Weigel, M., ... & Maes, P. (2020). Next steps for human-computer integration. In *Proceedings of the 2020 CHI Conference on Human Factors in Computing Systems* (pp. 1–15), New York. Association for Computing Machinery.

Mumaw, R. J., Boorman, D., Griffin, J., Moodi, M., & Xu, W. (2000). *Training and Design Approaches for Enhancing Automation Awareness (Boeing Document D6-82577)*. Seattle, WA: Boeing Commercial Aviation.

Naderpour, M., Lu, J., & Zhang, G. (2015). A human-system interface risk assessment method based on mental models. *Safety Science*, *79*, 286–297. https://doi.org/10.1016/j.ssci.2015.07.001.

Nagel, D. C. (1988). Human error in aviation operations. In E. L. Wiener, & D. C. Nagel (Eds.), *Human Factors in Aviation* (pp. 263-303). Cambridge, MA: Academic Press.

Neisser, U. (1976). *Cognition and Reality*. New York: W. H. Freeman and Company.

Nielsen, J. (1994). Heuristic evaluation. In J. Nielsen, & R. L. Mack (Eds.), *Usability Inspection Methods* (pp. 25–62). New York: John Wiley & Sons.

Norman, D. (2013). *The Design of Everyday Things: Revised and Expanded Edition*. New York: Basic Books.

Norman, D. A. (1981). Categorization of action slips. *Psychological Review*, *88*(1), 1–15.

Norman, D. A. (1988). *The Psychology of Everyday Things*. New York: Basic Books.

Norman, D. A. (1990). Commentary: Human error and the design of computer systems. *Communications of the ACM, 33*(1), 4–7.

Norman, D., & Draper, S. (1986). *User Centered System Design: New Perspectives on Human-Computer Interaction*. Mahwah, NJ: Lawrence Erlbaum Associates.

Norman, D. A., & Shallice, T. (1986). Attention to action: Willed and automatic control of behavior. In R. J. Davidson, G. E. Schwartz, & D. Shapiro (Eds.), *Consciousness and Self-Regulation: Advances in Research and Theory Volume 4* (1st ed., pp. 1–18). Berlin, Heidelberg: Springer.

Ntuen, C. A., & Watson, A. R. (1996). Workload prediction as a function of system complexity. In *Proceedings Third Annual Symposium on Human Interaction with Complex Systems. HICS'96* (pp. 96–100). IEEE. https://doi.org/10.1109/HUICS.1996.549498.

O'Connor, P., O'Dea, A., & Melton, J. (2007). A methodology for identifying human error in U.S. Navy diving accidents. *Human Factors*, *49*(2), 214–226. https://doi.org/10.1518/001872007X312450.

O'Hara, J. M., & Brown, W. (2002). *The Effects of Interface Management Tasks on Crew Performance and Safety in Complex, Computer-Based Systems: Overview and Main Findings*, NUREG/CR-6690. Washington, DC: United States Nuclear Regulatory Commission.

O'Hara, J. M., & Fleger, S. (2020). *Human-System Interface Design Review Guidelines*, NUREG-0700, Rev. 3. Washington, DC: United States Nuclear Regulatory Commission.

O'Hara, J.M., Higgins, J. C., Fleger, S. A., & Pieringer, P. A. (2012). *Human Factors Engineering Program Review Model*, NUREG-0711, Rev. 3. Washington, DC: United States Nuclear Regulatory Commission.

Oswald, F. L., Endsley, M. R., Chen, J., Chiou, E. K., Draper, M. H., McNeese, N. J., & Roth, E. M. (2022). The National Academies Board on Human-Systems Integration (BOHSI) panel: Human-AI teaming: Research frontiers. *Proceedings of the Human Factors and Ergonomics Society Annual Meeting*, *66*(1), 130–134.

Pan, X., Lin, Y., & He, C. (2017). A review of cognitive models in human reliability analysis. *Quality and Reliability Engineering International*, *33*(7), 1299–1316. https://doi.org/10.1002/qre.2111.

Pandya, D., Podofillini, L., Emert, F., Lomax, A. J., & Dang, V. N. (2018). Developing the foundations of a cognition-based human reliability analysis model via mapping task types and performance-influencing factors: Application to radiotherapy. *Proceedings of the Institution of Mechanical Engineers, Part O: Journal of Risk and Reliability*, *232*(1), 3–37.

Parasuraman, R. (2011). Neuroergonomics: Brain, cognition, and performance at work. *Current Directions in Psychological Science*, *20*(3), 181–186. https://doi.org/10.1177/0963721411409176.

Parasuraman, R., Sheridan, T. B., & Wickens, C. D. (2000). A model for types and levels of human interaction with automation. *IEEE Transactions on Systems, Man, and Cybernetics-Part A: Systems and Humans*, *30*(3), 286–297.

Park, J. (2009). *The Complexity of Proceduralized Tasks*. London: Springer. https://doi.org/10.1007/978-1-84882-791-2.

Park, J., Jung, W., & Ha, J. (2001). Development of the step complexity measure for emergency operating procedures using entropy concepts. *Reliability Engineering & System Safety*, *71*(2), 115–130. https://doi.org/10.1016/S0951-8320(00)00087-9.

Peters, G. A., & Peters, B. J. (2007). *Medical Error and Patient Safety: Human Factors in Medicine*. Boca Raton, FL: CRC Press.

Plant, K. L., & Stanton, N. A. (2012). Why did the pilots shut down the wrong engine? Explaining errors in context using schema theory and the perceptual cycle model. *Safety Science*, *50*(2), 300–315. https://doi.org/10.1016/j.ssci.2011.09.005.

Plant, K. L., & Stanton, N. A. (2013). The explanatory power of schema theory: Theoretical foundations and future applications in Ergonomics. *Ergonomics*, *56*(1), 1–15. https://doi.org/10.1080/00140139.2012.736542.

Plant, K. L., & Stanton, N. A. (2016). The development of the Schema World Action Research Method (SWARM) for the elicitation of perceptual cycle data. *Theoretical Issues in Ergonomics Science*, *17*(4), 376–401. https://doi.org/10.1080/1463922X.2015.1126867.

Plant, K. L., & Stanton, N. A. (2017a). The development of the Schema-Action-World (SAW) taxonomy for understanding decision making in aeronautical critical incidents. *Safety Science*, *99*, 23–35. https://doi.org/10.1016/j.ssci.2016.08.014.

Plant, K. L., & Stanton, N. A. (2017b). *Distributed Cognition and Reality: How Pilots and Crews Make Decisions* (1st ed.). Boca Raton, FL: CRC Press.

Poldrack, R. A., Sabb, F. W., Foerde, K., Tom, S. M., Asarnow, R. F., Bookheimer, S. Y., & Knowlton, B. J. (2005). The neural correlates of motor skill automaticity. *The Journal of Neuroscience*, *25*(22), 5356. https://doi.org/10.1523/JNEUROSCI.3880-04.2005.

Poulton, E. C. (1979). Composite model for human performance in continuous noise. *Psychological Review*, *86*, 361–375. https://doi.org/10.1037/0033-295X.86.4.361.

Priest, A. G., & Lindsay, R. O. (1992). New light on novice-expert differences in physics problem solving. *British Journal of Psychology*, *83*(3), 389–405. https://doi.org/10.1111/j.2044-8295.1992.tb02449.x.

Proctor, R. W., & Vu, K.-P. L. (2012). Human information processing: An overview for human-computer interaction. In J. A. Jacko (Ed.), *The Human-Computer Interaction Handbook: Fundamentals, Evolving Technologies, and Emerging Applications* (3rd ed., pp. 21–40). Boca Raton, FL: CRC Press.

Pyy, P. (2000). Human reliability analysis methods for probabilistic safety assessment, VTT Publications 422. Technical Research Centre of Finland.

Rasmussen, J. (1982). Human errors. A taxonomy for describing human malfunction in industrial installations. *Journal of Occupational Accidents*, *4*(2), 311–333. https://doi.org/10.1016/0376-6349(82)90041-4.

Rasmussen, J. (1983). Skills, rules, and knowledge; signals, signs, and symbols, and other distinctions in human performance models. *IEEE Transactions on Systems, Man, and Cybernetics*, *SMC-13*(3), 257–266.

Rasmussen, J. (1985). The role of hierarchical knowledge representation in decisionmaking and system management. *IEEE Transactions on Systems, Man, and Cybernetics*, *15*(2), 234–243.

Rasmussen, J. (1997). Risk management in a dynamic society: A modelling problem. *Safety Science*, *27*(2), 183–213. https://doi.org/10.1016/S0925-7535(97)00052-0.

Rasmussen, J., & Vicente, K. J. (1989). Coping with human errors through system design: Implications for ecological interface design. *International Journal of Man-Machine Studies*, *31*(5), 517–534.

Rasmussen, J., Pedersen, O. M., Mancini, G., Carnino, A., Griffon, M., & Gagnolet, P. (1981). *Classification System for Reporting Events Involving Human Malfunctions*, RISØ-M-2240. Roskilde, Denmark: Risø National Laboratory.

Read, G. J. M., Shorrock, S., Walker, G. H., & Salmon, P. M. (2021). State of science: Evolving perspectives on 'human error'. *Ergonomics*, *64*(9), 1091–1114. https://doi.org/10.1080/00140139.2021.1953615.

Reason, J. (1990). *Human Error*. Cambridge: Cambridge University Press.

Reason, J. (1995a). Safety in the operating theatre - Part 2: Human error and organisational failure. *Current Anaesthesia & Critical Care*, *6*(2), 121–126. https://doi.org/10.1016/S0953-7112(05)80010-9.

Reason, J. (1995b). A systems approach to organizational error. *Ergonomics*, *38*(8), 1708–1721. https://doi.org/10.1080/00140139508925221.

Reason, J. (2013). *A Life in Error: From Little Slips to Big Disasters*. Farnham, England: Ashgate Publishing Ltd.

Reason, J., Manstead, A., Stradling, S., Baxter, J., & Campbell, K. (1990). Errors and violations on the roads: A real distinction? *Ergonomics*, *33*(10–11), 1315–1332. https://doi.org/10.1080/00140139008925335.

Richardson, M., & Ball, L. J. (2009). Internal representations, external representations and ergonomics: Towards a theoretical integration. *Theoretical Issues in Ergonomics Science*, *10*(4), 335–376. https://doi.org/10.1080/14639220802368872.

Rieger, T., & Manzey, D. (2020). Human performance consequences of automated decision aids: The impact of time pressure. *Human Factors*, *64*(4), 617–634. https://doi.org/10.1177/0018720820965019.

Rieger, T., & Manzey, D. (2022). Understanding the impact of time pressure and automation support in a visual search task. *Human Factors*. https://doi.org/10.1177/00187208221111236.

Ritter, F. E., Tehranchi, F., & Oury, J. D. (2019). ACT-R: A cognitive architecture for modeling cognition. *Wiley Interdisciplinary Reviews: Cognitive Science*, *10*(3), e1488.

Rodrigues Barbosa, G. A., da Silva Fernandes, U., Sales Santos, N., & Oliveira Prates, R. (2023). Human-computer integration as an extension of interaction: Understanding its state-of-the-art and the next challenges. *International Journal of Human-Computer Interaction*, 1–20.

Roth, E., Depass, B., Harter, J., Scott, R., & Wampler, J. (2018). Beyond levels of automation: Developing more detailed guidance for human automation interaction design. *Proceedings of the Human Factors and Ergonomics Society Annual Meeting*, *62*(1), 150–154.

Roth, E., & O'Hara, J. (2002). *Integrating Digital and Conventional Human System Interface Technology: Lessons Learned from a Control Room Modernization Program*, No. NUREG/CR-6749. Washington, DC: U.S. Nuclear Regulatory Commission.

Rouse, W. B., & Rouse, S. H. (1983). Analysis and classification of human error. *IEEE Transactions on Systems, Man, and Cybernetics*, *SMC-13*(4), 539–549. https://doi.org/10.1109/TSMC.1983.6313142.

Salmon, P. M., Read, G. J. M., Stanton, N. A., & Lenné, M. G. (2013). The crash at Kerang: Investigating systemic and psychological factors leading to unintentional non-compliance at rail level crossings. *Accident Analysis & Prevention*, *50*, 1278–1288. https://doi.org/10.1016/j.aap.2012.09.029.

Sanders, M. S., & McCormick, E. J. (1993). *Human Factors in Engineering and Design* (7th ed.). New York: Mcgraw-Hill Book Company.

Sanderson, P. M., & Murtagh, J. M. (1989). Troubleshooting with an inaccurate mental model. In *Proceedings of IEEE International Conference on Systems, Man and Cybernetics* (*Vol.* 1233, pp. 1238–1243*). IEEE*. https://doi.org/10.1109/ICSMC.1989.71501.

Sarter, N. (2008). Investigating mode errors on automated flight decks: Illustrating the problem-driven, cumulative, and interdisciplinary nature of human factors research. *Human Factors, 50*(3), 506–510.

Sarter, N. B., Woods, D. D., & Billings, C. E. (1997). Automation surprises, In G. Salvendy (Ed.), *Handbook of Human Factors and Ergonomics* (2nd ed., pp. 1296–1943). Hoboken, NJ: Wiley.

Sasou, K., & Reason, J. (1999). Team errors: Definition and taxonomy. *Reliability Engineering & System Safety*, *65*(1), 1–9. https://doi.org/10.1016/S0951-8320(98)00074-X.

Sawyer, B. D., Karwowski, W., Xanthopoulos, P., & Hancock, P. A. (2017). Detection of error-related negativity in complex visual stimuli: A new neuroergonomic arrow in the practitioner's quiver. *Ergonomics*, *60*(2), 234–240. https://doi.org/10.1080/00140139.2015.1124928.

Shah, R. K., Kentala, E., Healy, G. B., & Roberson, D. W. (2004). Classification and consequences of errors in otolaryngology. *The Laryngoscope*, *114*(8), 1322–1335. https://doi.org/10.1097/00005537-200408000-00003.

Shappell, S. A., & Wiegmann, D. A. (1997). A human error approach to accident investigation: The taxonomy of unsafe operations. *The International Journal of Aviation Psychology*, *7*(4), 269–291. https://doi.org/10.1207/s15327108ijap0704_2.

Shappell, S. A., & Wiegmann, D. A. (2000). The human factors analysis and classification system-HFACS, DOT/FAA/AM-00/7. Washington, DC: Federal Aviation Administration.

Shappell, S. A., Detwiler, C., Holcomb, K., Hackworth, C. A., & Boquet, A. J. (2006). Human error and commercial aviation accidents: A comprehensive, fine-grained analysis using HFACS, Report No. DOT/FAA/AM-06/18. Washington, DC: US Department of Transportation, Federal Aviation Administration.

Sharp, T. D., & Helmicki, A. J. (1998). The application of the ecological interface design approach to neonatal intensive care medicine. *Proceedings of the Human Factors and Ergonomics Society Annual Meeting*, *42*(3), 350–354.

She, M. R., Li, Z. Z., & Ma, L. (2019). User-defined information sharing for team situation awareness and teamwork. *Ergonomics*, *62*(8), 1098–1112.

Sheridan, T. B. (1997). Task analysis, task allocation and supervisory control. In M. Helander, T. K. Landauer, & P. Prabhu (Eds.), *Handbook of Human-Computer Interaction* (2nd ed., pp. 159–173). Amsterdam, The Netherlands: Elsevier Science B.V.

Shorrock, S. T., & Kirwan, B. (2002). Development and application of a human error identification tool for air traffic control. *Applied Ergonomics*, *33*(4), 319–336. https://doi.org/10.1016/S0003-6870(02)00010-8.

Skitka, L. J., Mosier, K. L., Burdick, M., & Rosenblatt, B. (2000). Automation bias and errors: Are crews better than individuals? *The International Journal of Aviation Psychology, 10*(1), 85–97.

Skraaning, G., & Jamieson, G. A. (2019). Human performance benefits of the automation transparency design principle: Validation and variation. *Human Factors*, *63*(3), 379–401. https://doi.org/10.1177/0018720819887252.

Smith, A. (1989). A review of the effects of noise on human performance. *Scandinavian Journal of Psychology*, *30*(3), 185–206. https://doi.org/10.1111/j.1467-9450.1989.tb01082.x.

Son, H. I., Franchi, A., Chuang, L. L., Kim, J., Bulthoff, H. H., & Giordano, P. R. (2013). Human-centered design and evaluation of haptic cueing for teleoperation of multiple mobile robots. *IEEE Transactions on Cybernetics, 43*(2), 597–609.

Stanton, N. A. (2005). Systematic human error reduction and prediction approach (SHERPA). In N. Stanton, A. Hedge, K. Brookhuis, E. Salas, & H. Hendrick (Eds.), *Handbook of Human Factors and Ergonomics Methods* (pp. 37-1–37-8). Boca Raton, FL: CRC Press.

Stanton, N. A. (2009). Human-error identification in human-computer interaction. In A. Sears, & J. A. Jacko (Eds.), *Human Computer Interaction: Fundamentals* (pp. 123–134). Boca Raton, FL: CRC Press.

Stanton, N. A., & Baber, C. (2005). Task analysis for error identification. In N. Stanton, A. Hedge, K. Brookhuis, E. Salas, & H. Hendrick (Eds.), *Handbook of Human Factors and Ergonomics Methods* (pp. 378–389). Boca Raton, FL: CRC Press.

Stanton, N. A., & Salmon, P. M. (2009). Human error taxonomies applied to driving: A generic driver error taxonomy and its implications for intelligent transport systems. *Safety Science*, *47*(2), 227–237. https://doi.org/10.1016/j.ssci.2008.03.006.

Stephanidis, C., Salvendy, G., Antona, M., Chen, J. Y., Dong, J., Duffy, V. G., … & Zhou, J. (2019). Seven HCI grand challenges. *International Journal of Human-Computer Interaction, 35*(14), 1229–1269.

Stowers, K., Kasdaglis, N., Rupp, M. A., Newton, O. B., Chen, J. Y. C., & Barnes, M. J. (2020). The IMPACT of agent transparency on human performance. *IEEE Transactions on Human-Machine Systems*, *50*(3), 245–253. https://doi.org/10.1109/THMS.2020.2978041.

Sträter, O. (2005). *Cognition and Safety: An Integrated Approach to Systems Design and Assessment* (1st ed.). New York: Routledge. https://doi.org/10.4324/9781315260006.

Sujan, M. A., Embrey, D., & Huang, H. (2020). On the application of human reliability analysis in healthcare: Opportunities and challenges. *Reliability Engineering & System Safety*, *194*, 106189. https://doi.org/10.1016/j.ress.2018.06.017.

Sun, R. (2004). Desiderata for cognitive architectures. *Philosophical Psychology*, *17*(3), 341–373. https://doi.org/10.1080/0951508042000286721.

Sutcliffe, A., & Rugg, G. (1998). A taxonomy of error types for failure analysis and risk assessment. *International Journal of Human-Computer Interaction*, *10*(4), 381–405. https://doi.org/10.1207/s15327590ijhc1004_5.

Svenson, O. (2001). Accident and incident analysis based on the accident evolution and barrier function (AEB) model. *Cognition, Technology & Work*, *3*, 42–52.

Swain, A. D. (1990). Human reliability analysis: Need, status, trends and limitations. *Reliability Engineering & System Safety*, *29*(3), 301–313.

Swain, A. D., & Guttmann, H. E. (1983). *Handbook of Human Reliability Analysis with Emphasis on Nuclear Power Plant Applications*, NUREG/CR-1278. Washington, DC: U.S. Nuclear Regulatory Commission.

Szalma, J. A., Hancock, P. A., & Quinn, S. (2008). A meta-analysis of the effect of time pressure on human performance. *Proceedings of the Human Factors and Ergonomics Society Annual Meeting*, *52*(19), 1513–1516. https://doi.org/10.1177/154193120805201944.

Szalma, J. L., & Hancock, P. A. (2011). Noise effects on human performance: A meta-analytic synthesis. *Psychological Bulletin*, *137*(4), 682–707.

Taib, I. A., McIntosh, A. S., Caponecchia, C., & Baysari, M. T. (2011). A review of medical error taxonomies: A human factors perspective. *Safety Science*, *49*(5), 607–615. https://doi.org/10.1016/j.ssci.2010.12.014.

Targoutzidis, A. (2010). Incorporating human factors into a simplified "bow-tie" approach for workplace risk assessment. *Safety Science*, *48*(2), 145–156. https://doi.org/10.1016/j.ssci.2009.07.005.

Top, F., Pütz, S., & Fottner, J. (2021). Human-centered HMI for crane teleoperation: Intuitive concepts based on mental models, compatibility and mental workload. In *Engineering Psychology and Cognitive Ergonomics: 18th International Conference, EPCE 2021, Held as Part of the 23rd HCI International Conference, HCII 2021, Virtual Event, July 24–29, 2021, Proceedings* (pp. 438–456). Cham: Springer International Publishing.

Topi, H., Valacich, J. S., & Hoffer, J. A. (2005). The effects of task complexity and time availability limitations on human performance in database query tasks. *International Journal of Human-Computer Studies*, *62*(3), 349–379. https://doi.org/10.1016/j.ijhcs.2004.10.003.

Totter, A., Bonaldi, D., & Majoe, D. (2011). A human-centered approach to the design and evaluation of wearable sensors-Framework and case study. In *2011 6th International Conference on Pervasive Computing and Applications*, Port Elizabeth, South Africa (pp. 233–241). IEEE.

Trager Jr, T. A. (1985). *Case study report on loss of safety system function events*. No. AEOD C, 504. Washington, DC: US Nuclear Regulatory Commission.

Trapsilawati, F., Qu, X., Wickens, C. D., & Chen, C.-H. (2015). Human factors assessment of conflict resolution aid reliability and time pressure in future air traffic control. *Ergonomics*, *58*(6), 897–908. https://doi.org/10.1080/00140139.2014.997301.

van de Merwe, K., Mallam, S., & Nazir, S. (2022). Agent transparency, situation awareness, mental workload, and operator performance: A systematic literature review. *Human Factors*. https://doi.org/10.1177/00187208221077804.

Vicente, K. J. (1996). Improving dynamic decision making in complex systems through ecological interface design: A research overview. *System Dynamics Review*, *12*(4), 251–279.

Vicente, K. J. (2002). Ecological interface design: Progress and challenges. *Human Factors, 44*(1), 62–78.

Vroom, V. (1964). *Work and Motivation*. Hoboken, NJ: Wiley.

Wang, J., & Fang, W. (2014). A structured method for the traffic dispatcher error behavior analysis in metro accident investigation. *Safety Science*, *70*, 339–347.

Wascher, E., Reiser, J., Rinkenauer, G., Larrá, M., Dreger, F. A., Schneider, D., Karthaus, M., Getzmann, S., Gutberlet, M., & Arnau, S. (2021). Neuroergonomics on the Go: An evaluation of the potential of mobile EEG for workplace assessment and design. *Human Factors*, *65*(1), 86–106. https://doi.org/10.1177/00187208211007707.

Waterson, P., Le Coze, J.-C., & Andersen, H. B. (2017). Recurring themes in the legacy of Jens Rasmussen. *Applied Ergonomics*, *59*, 471–482. https://doi.org/10.1016/j.apergo.2016.10.002.

Whaley, A. M., Xing, J., Boring, R. L., Hendrickson, S. M. L., Joe, J. C., Le Blanc, K. L., & Morrow, S. L. (2016). *Cognitive Basis for Human Reliability Analysis*, NUREG-2114. Washington, DC: U.S. Nuclear Regulatory Commission.

Whalley, S. P. (1988). Minimising the cause of human error. In G. P. Libberton (Ed.), *Proceedings of 10th Advances in Reliability Technology Symposium* (pp. 114–128). Springer Netherlands. https://doi.org/10.1007/978-94-009-1355-4_11.

Wickens, C. D. (1984). *Engineering Psychology and Human Performance* (1st ed.). Merrill, WI: Merrill.

Wickens, C. D., Helleberg, J., Goh, J., Xu, X., & Horrey, W. J. (2001). *Pilot Task Management: Testing an Attentional Expected Value Model of Visual Scanning*, ARL-01-14/NASA-01-7. Mountain View, CA: NASA Ames Research Center.

Wickens, C. D., Hollands, J. G., Banbury, S., & Parasuraman, R. (2013). *Engineering Psychology and Human Performance* (4th ed.). London: Psychology Press. https://doi.org/10.4324/9781315665177.

Wiedenbeck, S. (1985). Novice/expert differences in programming skills. *International Journal of Man-Machine Studies*, *23*(4), 383–390. https://doi.org/10.1016/S0020-7373(85)80041-9.

Wiegmann, D., & Shappell, S. (1999). Human error and crew resource management failures in Naval aviation mishaps: A review of US Naval Safety Center data, 1990-96. *Aviation, Space, and Environmental Medicine*, *70*(12), 1147–1151.

Wiley, J. (1998). Expertise as mental set: The effects of domain knowledge in creative problem solving. *Memory & Cognition*, *26*(4), 716–730. https://doi.org/10.3758/BF03211392.

Williams, A. (2009). User-centered design, activity-centered design, and goal-directed design: A review of three methods for designing web applications. In *Proceedings of the 27th ACM International Conference on Design of Communication* (pp. 1–8). Bloomington, IN: Association for Computing Machinery.

Williams, J. C. (1988). A data-based method for assessing and reducing human error to improve operational performance. In *Conference Record for 1988 IEEE Fourth Conference on Human Factors and Power Plants* (pp. 436–450). Monterey, CA.

Wilson, K. M., Yang, S., Roady, T., Kuo, J., & Lenné, M. G. (2020). Driver trust & mode confusion in an on-road study of level-2 automated vehicle technology. *Safety Science, 130*, 104845.

Winget, C. M., DeRoshia, C. W., Markley, C. L., & Holley, D. C. (1984). A review of human physiological and performance changes associated with desynchronosis of biological rhythms. *Aviation, Space, and Environmental Medicine*, *55*(12), 1085–1096.

Woods, D. D., & Cook, R. I. (1991). Nosocomial automation: Technology-induced complexity and human performance. In *Conference Proceedings 1991 IEEE International Conference on Systems, Man, and Cybernetics (Vol.1272*, pp. 1279–1282*). IEEE.* https://doi.org/10.1109/ICSMC.1991.169863.

Woods, D., & Dekker, S. (2000). Anticipating the effects of technological change: A new era of dynamics for human factors. *Theoretical Issues in Ergonomics Science, 1*(3), 272–282.

Woods, D. D., Dekker, S., Cook, R., Johannesen, L., & Sarter, N. (2010). *Behind Human Error* (2nd ed.). Farnham, England: Ashgate Publishing Ltd.

Wu, X. J., & Li, Z. Z. (2018). A review of alarm system design for advanced control rooms of nuclear power plants. *International Journal of Human-Computer Interaction, 34*(6), 477–490.

Wu, Z., Pan, X., & Chen, X. (2017). Relation of motivation intensity, stress levels and human performance: A human reliability experiment. In *Proceedings of 2017 11th Asian Control Conference (ASCC)* (pp. 2209–2214). IEEE. https://doi.org/10.1109/ASCC.2017.8287518.

Xing, J., Chang, J., & DeJesus, J. (2020). *Integrated Human Event Analysis System for Event and Condition Assessment (IDHEAS-ECA)*, RIL-2020-02. Washington, DC: U.S. Nuclear Regulatory Commission.

Xing, J., Chang, Y. J., & DeJesus Segarra, J. (2021). *The General Methodology of An Integrated Human Event Analysis System (IDHEAS-G)*, NUREG-2198. Washington, DC: U.S. Nuclear Regulatory Commission.

Xing, J., Parry, G., Presley, M., Forester, J., Hendrickson, S., & Dang, V. (2017). *An Integrated Human Event Analysis System (IDHEAS) for Nuclear Power Plant Internal Events At-Power Application*, NUREG-2199, Volume 1. Washington, DC: U.S. Nuclear Regulatory Commission.

Yang, T., Park, B., Lee, S., Choi, J. H., Park, J., Boring, R. L., & Kim, J. (2021). Experimental analysis of the effects of simulator complexity on human performance. In *Proceedings of 2021 5th International Conference on System Reliability and Safety (ICSRS)* (pp. 85–91). IEEE. https://doi.org/10.1109/ICSRS53853.2021.9660761.

Ye, P., Wang, T., & Wang, F. Y. (2018). A survey of cognitive architectures in the past 20 years. *IEEE Transactions on Cybernetics, 48*(12), 3280–3290. https://doi.org/10.1109/TCYB.2018.2857704.

Yin, Z., Li, Z., Liu, Z., Yang, D., Zhang, J., Long, L., Zhang, Y., & Gong, B. (2023). Collection of IDHEAS-based human error probability data for nuclear power plant commissioning through expert elicitation. *Annals of Nuclear Energy, 181*, 109544. https://doi.org/10.1016/j.anucene.2022.109544.

Yin, Z., Liu, Z., & Li, Z. (2021). Identifying and clustering performance shaping factors for nuclear power plant commissioning tasks. *Human Factors and Ergonomics in Manufacturing & Service Industries, 31*(1), 42–65.

Yin, Z., Liu, Z., Yang, D., & Li, Z. (2020). Using IDHEAS to analyze incident reports in nuclear power plant commissioning: A case study. In D. Harris, & W. C. Li (Eds.), *Engineering Psychology and Cognitive Ergonomics: Cognition and Design (Proceedings of 22nd HCI International Conference, HCII 2020, Part II)* (pp. 90–103). Berlin, Heidelberg: Springer.

Yin, Z., Long, L., Yu, J., Zhang, Y., & Li, Z. (2023, July 23–28). An erroneous behavior taxonomy for operation and maintenance of network systems. *25th International Conference on Human-Computer Interaction (HCII 2023)*, Copenhagen, Denmark.

Yin, Z., Rau, P.-L. P., & Li, Z. (2019). Impacts of automation reliability and failure modes on operators' performance in security screening. In D. Harris (Ed.), *Engineering Psychology and Cognitive Ergonomics (Proceedings of 16th International Conference, EPCE 2019, Held as Part of the 21st HCI International Conference, HCII 2019)* (pp. 137–149). Berlin, Heidelberg: Springer International Publishing.

Young, M. S., Brookhuis, K. A., Wickens, C. D., & Hancock, P. A. (2015). State of science: Mental workload in ergonomics. *Ergonomics, 58*(1), 1–17. https://doi.org/10.1080/00140139.2014.956151.

Zhang, J., Patel, V. L., Johnson, T. R., & Shortliffe, E. H. (2004). A cognitive taxonomy of medical errors. *Journal of Biomedical Informatics, 37*(3), 193–204. https://doi.org/10.1016/j.jbi.2004.04.004.

Zhang, R., & Tan, H. (2018). An integrated human reliability based decision pool generating and decision making method for power supply system in LNG terminal. *Safety Science, 101*, 86–97.

Zhao, B. (1992). A structured analysis and quantitative measurement of task complexity in human-computer interaction (Publication Number 9314110), Ph.D., Purdue University. ProQuest Dissertations & Theses Global. United States – Indiana.

7 Human Actions

Victor Kaptelinin and Marco C. Rozendaal

7.1 INTRODUCTION

Understanding and supporting human actions is a central concern of human-computer interaction (HCI) research and practice. Human actions serve as a meaningful context for specific interactions between people and technology, so getting an insight into human actions is crucial for the analysis, design, and evaluation of interactive technologies. To properly interpret and assess a particular case of human interaction with a technology, one may need to establish whether or not the technology in question helps people to achieve what they want, how safe it is, what people feel about it, etc. (e.g., Sharp et al., 2019). Usually, such assessments need to be made based on the overall context of a meaningful human action, rather than the interaction *per se*. Depending on the context, even a difficult-to-use technology can be considered effective and enjoyable, for instance, if it provides a nice learning or gaming experience.

The aim of this chapter is to present a systematic account of the perspectives on human actions, adopted in HCI, and explore the dynamic relationship between, on the one hand, the particular perspectives adopted in HCI studies and, on the other hand, the specific foci, research agendas, and methodologies of the studies. To do so, this chapter discusses the historical evolution of the views on human actions in HCI and how these developments contributed to the continuous self-reinvention of HCI as a discipline.

Since its emergence as a distinct field of research several decades ago, HCI has undergone a series of transformations, resulting in corresponding revisions of its aims and object of study. The transformations have been variously described as "levels" (Grudin, 1990), "foci" (Grudin, 2005), "waves" (Bødker, 2006, 2015; Cooper & Bowers, 1995), "generations" (Bødker, 2006, 2015), or "paradigms" (Harrison et al., 2007, 2011). An early work by Grudin (1990) differentiates between several levels in the historical trajectory of the foci of interface research and development – from hardware to work settings – which took place between the 1950s and the 1990s. In a later paper, Grudin (2005) identified computer operation, information systems management, and discretionary use as three research foci in the evolution of HCI. Kaptelinin and Nardi (2006) described the developing scope of HCI and interaction design as comprising interrelated trends: an extending view of the human (from *user* to *worker* to *human being*) and an extending view of technology (from *user interface* to *tool* to *artifact*), see Figure 7.1. The notion of "waves" of HCI, reflecting a shift of the main focus of the field from human information processing to actors in meaningful real-life contexts, was introduced by Cooper and Bowers (1995) and further elaborated in subsequent research, most notably by Bødker (2006, 2015). Harrison et al. (2007, 2011) discussed "three paradigms of HCI" as an alternative account of the evolution of HCI, differentiating between three basic sets of underlying metaphors, aims, and questions of interest.[1] The idea of several historical periods in the evolution of HCI is also reflected in Rogers' (2012) taxonomy of HCI theories as classical, modern, and contemporary (by analogy with art history). This chapter generally follows the overall historical progression of HCI research according to the notion of "waves".

The remainder of this chapter is organized as follows. First, we describe how the early focus of HCI on human information processing and technology-defined tasks (the "first-wave HCI") had expanded to include human actors and their purposeful actions (the "second-wave HCI"). Then we present an overview of key theoretical approaches to human actions within the second-wave HCI as well as HCI research informed by such approaches. After that, the analysis proceeds to

DOI: 10.1201/9781003495109-7

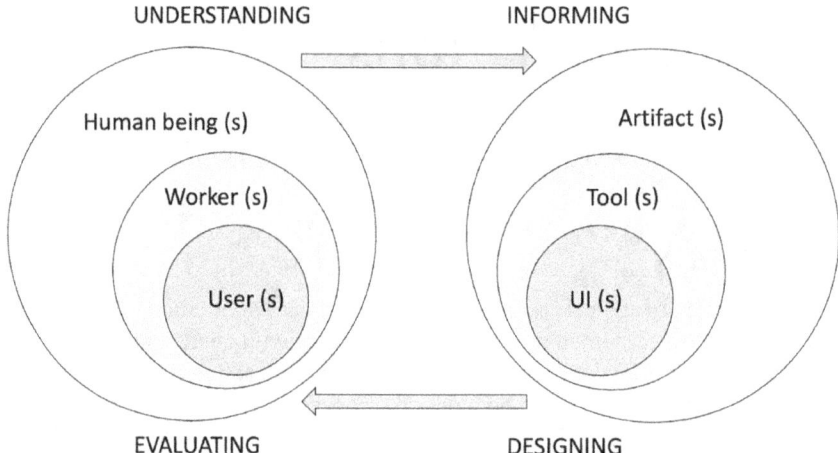

FIGURE 7.1   The expanding scope of interaction design. (Adapted from Kaptelinin and Nardi (2006).)

the "experience turn" in HCI, which foregrounded the centrality of human experience as a whole (generally corresponding to the "third-wave HCI"). The discussion of the "waves of HCI" brings in a range of conceptual perspectives on human actions, informing HCI research, from cognitive psychology (Card et al., 1983; Norman, 1988) to "post-cognitivist" approaches, including activity theory (e.g., Bødker, 1991; Kaptelinin & Nardi, 2006), distributed cognition (Halverson, 2002; Hollan et al., 2000; Hutchins, 1995), phenomenology (e.g., Dourish, 2004; Svanæs, 2013), postphenomenology (e.g., Verbeek, 2015), and philosophy of technology (e.g., Fallman, 2011). The analysis continues with reflections on current and emerging directions of research that aim to shed light on how human agency is being fundamentally transformed by technology (e.g., Frauenberger, 2020; Bardzell, 2010; Coskun et al., 2022). This chapter concludes with a discussion of potential challenges for future HCI research and practice, such as those related to developments in AI-based systems, trans- and posthumanism (including "more than human" centered design), and ethics, which require further exploration of the dynamic relationship between technology and human actions.

7.2   BEYOND HUMAN INFORMATION PROCESSING: FROM THE FIRST-WAVE TO THE SECOND-WAVE HCI

7.2.1   First-Wave HCI: The Emergence and Key Topics of Concern

The beginnings of HCI can be traced to the early 1980s when several developments indicated that a new field of research and practice was emerging. Especially important was the foundational book by Card et al. (1983), entitled "*The Psychology of Human-Computer Interaction*", which gave a name to the new field. The book argued that considering humans as "operators" of machines, typical of earlier studies of human engagements with technology, was not sufficient for understanding how people interact with computers. Card et al. (1983) observed:

> But the user is not an operator. He does not operate the computer, he *communicates with it to accomplish a task*. Thus we are creating a *new arena of human action*: communicating with machines rather than operation of machines.
>
> *Card et al. (1983, p. 7, emphasis added)*

To properly address the challenges brought about by the new reality of human-computer interaction, Card et al. (1983) called for establishing an applied, design-oriented, field of inquiry, informed by information-processing psychology. To make key insights of their contemporary cognitive psychology more accessible and suitable for analyzing human-computer interaction, they distilled them in the form of the *Model Human Processor*, an integrated (and admittedly simplified) representation of a person's perceptual, motor, and cognitive subsystems.

Building on the Model Human Processor, the authors introduced GOMS, a practical approach to modeling human-computer interaction. The names of the approach reflect its key constitutive elements, namely, Goals, Operators, Methods, and Selection rules. A *Goal* is what the user wants to achieve (e.g., to make a phone call). A *Method* describes a procedure for achieving a Goal (e.g., selecting a contact from the contact list), which procedure can be broken down into elementary *Operations* (e.g., open "Phone" app, tap "Contacts", scroll contact list to see the needed contact, tap the contact, tap "Make phone call" button). If there are several methods for achieving the same goal (e.g., in addition to using "Contacts" one can also use "Recent calls" or type in the appropriate number directly), then *Selection Rules* are applied to decide which method should be used when.

GOMS made a major impact on research and practice in the general area of human–technology interaction, and together with some other work conducted around that time (e.g., Nichols, 1982), marked the advent of what Cooper and Bowers (1995) call the "first-wave HCI". In sum, the disciplinary rhetoric behind the first-wave HCI emphasized the importance of addressing the needs of an emerging category of people engaged in direct interactions with technology ("users", as opposed to "operators") by applying information-processing psychology to design better system interfaces helping the users to accomplish their tasks (Cooper & Bowers, 1995).

Card et al. (1983), along with other seminal work marking the emergence of the new field, stimulated a wide range of research and practice efforts, which can be broadly described as "applied cognitive-psychological studies focusing on user interfaces". The research can be broadly divided into two lines of investigation. One of them was using a powerful methodology of experimental studies, adopted from cognitive psychology, to investigate general phenomena and mechanisms of human cognition, immediately relevant to understanding and supporting human interaction with computers. Examples of such studies are analyses of optimal menu structures, command names, or advantages and disadvantages of metaphors in design (Nichols, 1982). Another line of research was using insights from cognitive psychology to develop applied design and evaluation methods, intended for practitioners. In particular, some of the most important developments in usability testing (Roberts & Moran, 1982), as well as modeling of users and tasks (such as GOMS, discussed above) can be directly linked to work in the first-wave HCI.

7.2.2 PERSPECTIVES ON HUMAN ACTIONS IN THE FIRST-WAVE HCI

As mentioned, considering human interaction with computers "a new arena of human action" was key to the rhetoric of the first-wave HCI (Cooper and Bowers, 1995). However, relatively little effort was made to present a systematic account of "human action" in general. Instead, the first-wave HCI research shared, even if not always explicitly, a view on human actions as users' *tasks*, understood in a rather narrow sense. Users were viewed as entities comprising perceptual, motor, and cognitive subsystems (similar to Card et al.'s, 1983, Model Human Processor), and tasks were usually understood as low-level activities, which involved interaction with a system and using the system's functionality to achieve concrete, narrowly defined goals, such as moving a text fragment in a document.

The focus on tasks and user cognition was directly related to the framing of the first-wave HCI as a sub-field of applied cognitive psychology, and to the design of efficient (and generally usable) user interfaces being the dominant concern of the HCI at that time. Low-level, technology-specific tasks are immediately affected by the particular design of an interface, so analyzing them when exploring the space of interface design solutions was apparently well-justified. In other words, the centrality of user interface design in the first-wave HCI essentially determined the analytical focus on low-level user tasks, characterized by a direct interaction with the computer.

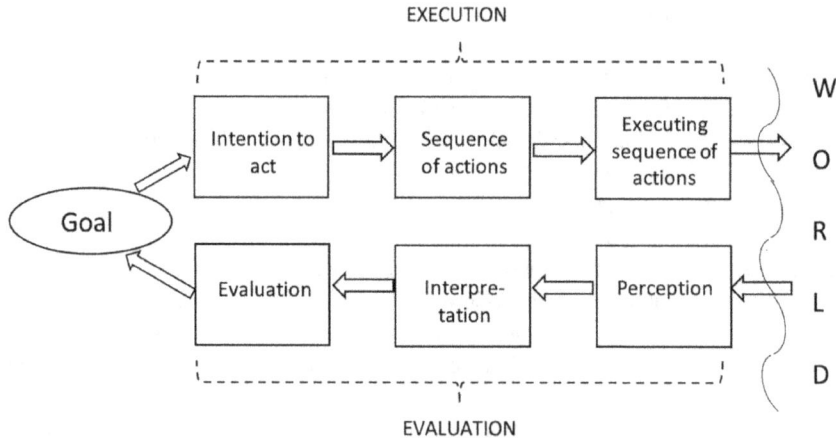

FIGURE 7.2 Norman's model of action. (Adapted from Norman (1988).)

Accordingly, in the first wave, HCI tasks have been often analyzed without making *general* assumptions about human actions. In principle, a general idea of what human actions are is not necessary when the focus is on low-level technology-specific tasks. Such tasks can be analyzed by applying a modeling approach (e.g., GOMS) and simply following its procedures and notational conventions. For instance, an analyst may break down a task into smaller units, describe its overall structure, and assess whether the task can be optimized. In principle, it all can be done by only taking into account such concrete attributes of the task in question as its specific goal, user's sensorimotor and cognitive parameters, and the design of the particular interface encountered by the user, without the need to make any general assumptions.

A notable example of a first-wave HCI model, which does offer a general view on human actions, is Norman's (1988) seven-stage Action Cycle model of action (Figure 7.2). The model, intended to be applied to any goal-oriented interaction with the world and describe a wide range of tasks, made a significant impact on HCI research and practice, serving both as a design aide and as a framework for conceptual analyses (e.g., Vermeulen et al., 2013).

The Action Cycle model starts with the initial step of forming a *goal* and describes the potentially iterative process of achieving the goal by interacting with the *world*. There are two general aspects of the model, which complement each other: execution (making an impact on the world to achieve a goal) and evaluation (assessing whether the impact has actually resulted in achieving the goal). Execution comprises three phases: translating the goal into a concrete intention to act, planning a sequence of actions to implement the intention, and executing the sequence of actions. Evaluation is also performed in three phases, namely: perceiving the state of the world, interpreting the perception, and evaluating the interpretation.

For instance, to share a picture on their phone (goal), a person may decide to cast the picture to a nearby TV set (intention to act), recall what needs to be done in that case (sequence of actions), and actually perform the actions (execution of the action sequence). When the action sequence is executed, the person detects whether the action results in any changes on the TV screen (perceiving the state of the world), see what picture is being displayed (interpreting the perception), and ascertain that this is actually the picture, which is intended to be shared (evaluation of interpretation). If the goal is accomplished, the action is completed. If not (e.g., if the person realizes that a wrong picture is displayed), the cycle is repeated over again until the goal is achieved or abandoned.

7.2.3   TOWARD THE SECOND-WAVE HCI

In many respects, the first-wave HCI has been remarkably successful. By adopting an information-processing framework, it supported collaboration between psychologists and computer
scientists, which helped to establish HCI as a new, and rapidly developing, multi-disciplinary
field of research and practice that addresses important and topical real-life needs in a specific
and rigorous way. The work in the first-wave HCI produced a number of concepts, methods, and
approaches, such as task analysis, usability testing, expert evaluation methodologies, conceptual and empirical studies of users' perception and cognition, etc., which represent some of the
most significant contributions of HCI as a field. It should also be noted that the first-wave-style
HCI still represents a substantial part of contemporary HCI research and practice, and it can be
expected to remain relevant in the future, as cognition will, arguably, always be a central aspect
of human interaction with computers.

Despite all these successes, researchers eventually realized that the potential of cognitive psychology as a theoretical foundation of HCI was rather limited. However central, human cognition
is not the only important factor affecting human interaction with computers. The need to take
into account real-life contexts of technology use – which were not a key focus of the first-wave
HCI – was becoming increasingly obvious, especially in light of the rapid development of computers and their applications. Eventually, it resulted in a transition from the first-wave HCI to the next
phase in the evolution of the field, the second-wave HCI.

According to Cooper and Bowers (1995), the transition took place in the early 1990s as the focus
of the field shifted from the original concern about technology-defined tasks, user interface, and
human information processing to the meaningful actions of human actors in their real-life contexts.
Cooper and Bowers (1995) pointed to Bannon and Bødker (1991) paper, entitled "*From human factors to human actors: The role of psychology and human-computer interaction studies in system
design*", as an especially well-developed analysis of the perceived limitations of the first-wave HCI
(including, e.g., a neglect of human values and motivation).

Probably, the most compelling arguments signaling a conceptual shift in the HCI research community in general can be found in the edited collection, entitled "*Designing Interaction: Psychology
at the User Interface*" (Carroll, 1991), featuring contributions by some of the most prominent HCI
researchers. Many of these contributions reflected a general realization that, to be relevant to the
emerging practical challenges of designing usable and useful interactive systems, the scope of HCI
had to be extended beyond the original focus on information processing and low-level user tasks.
For instance, Bannon and Bødker (1991) presented a critical analysis of cognitive psychology as a
theoretical foundation for HCI and suggested a conceptual reframing of HCI with an emphasis on
mediation and praxis. A key idea of the book, expressed by several contributors, was that the focus
of HCI should be on how people use technology to achieve their meaningful goals in real world
contexts, rather than on just information-processing aspects of the interaction between people and
technology.

While the shift to the second-wave HCI is usually considered to have occurred in the early
1990s (Bødker, 2006; Cooper & Bowers, 1995; Kaptelinin et al., 2003), it was facilitated by influential work taking place several years prior to that time. For instance, Suchman (1987) presented
compelling arguments that human actions should be viewed as *situated*, that is, dynamically and
often unpredictably unfolding in a particular situation, rather than following predefined plans.
Winograd and Flores (1986) and Norman (1986), from their respective conceptual standpoints,
emphasized the need for HCI to understand the use of technology in a wider context of human
relation to the world. In particular, Cypher (1986) showed that real-life contexts are characterized
by managing multiple activity threads, overlapping in time, rather than a linear progression from
one task to another.

7.3   ACTORS, TOOLS, AND PURPOSEFUL ACTIONS: THE SECOND-WAVE HCI

7.3.1   SECOND-WAVE THEORIES

Arguably, at the heart of the transition to the second-wave HCI was a changing perspective on human actions, namely, a shift of the focus from technology-specific tasks to purposeful acting in the world. The notion of "task" implies that a certain action *should* be performed, and the goal of the action has been already decided upon. In the context of the first-wave HCI, it is not really important whether a particular task is initiated by the users themselves or is assigned to them by someone else. Therefore, human agency, as opposed to human cognition, was not, in general, considered a central concern of the research and practice in the field. In contrast, in the second-wave HCI, the analytical focus moved to purposeful human interactions with the world, which had direct implications for the scope and research agenda of the field. The issues of agency, meaning, and context became important concerns and required adopting an appropriate contextual framework for dealing with them.

It was obvious that information-processing psychology could not provide a sufficient foundation for analyses of meaningful human actions in real-life contexts, so HCI research turned to other theories as potential alternative theoretical bases for the second-wave HCI. A number of post-cognitivist, "second wave" (Kaptelinin et al., 2003), conceptual approaches were introduced and employed in HCI. Some of these approaches, namely, activity theory and distributed cognition, are discussed below. These approaches were selected from a wider set of post-cognitivist HCI frameworks (also including, for instance, the language/action perspective (Winograd, 1987) and instrumental genesis (Rabardel, 2003) as the most relevant ones to the discussion in this chapter.

7.3.2   ACTIVITY THEORY

Activity theory is a theoretical approach in social sciences, which originates from the Soviet psychology of the 20th century (Vygotsky, 1981; Leontiev, 1978). Since the 1980s, activity theory has been predominantly developed in the West (e.g., Wertsch, 1981; Engeström, 1987; Nardi, 1993, 1996), and in recent decades it became an international, and also a transdisciplinary approach, employed in various areas of research and practice (e.g., Engeström et al., 1999).

Activity theory was explicitly adopted in some of the most influential analyses marking the shift to the second-wave HCI, either as the main framework (e.g., Bødker, 1989, 1991; Bannon and Bødker, 1991), or as key additional influence (e.g., Norman, 1991). The book by Bødker, entitled "*Through the Interface: A Human Activity Approach to User Interface Design*" (Bødker, 1991) was especially important for introducing activity theory to HCI. In the last decades, a substantial volume of various types of HCI research, including conceptual analyses, empirical studies, method development, and design explorations has been informed by activity theory (Kaptelinin & Nardi, 2006; Clemmensen et al., 2016). In this section, we present the basic concepts of activity theory and give an overview of selected key work in HCI, which was informed by activity-theoretical perspectives on human actions and technology.

7.3.2.1   Leontiev's Activity Theory: Basic Ideas and Principles

Early roots of activity theory can be found in Vygotsky's cultural-historical psychology, which was primarily concerned with the specific mechanisms and phenomena underlying the social origins of the human mind (Vygotsky, 1981). Vygotsky's most fundamental idea was that human mental processes, or "mental functions", are shaped by culturally developed physical and symbolic mediational means, which are appropriated through learning and development. The mechanism underlying the appropriation is internalization, that is, the transformation of externally mediated and socially distributed actions into individual's own mediated "mental functions". The theoretical approach, known as "activity theory" as such, was developed, mostly from the 1950s to the 1970s, by Alexey Leontiev[2] (1978, 1981), a student of Vygotsky's. Leontiev understood human activity as a purposeful, social, and mediated "subject-object" interaction taking place at three levels: motive-oriented activities, goal-oriented actions, and conditions-specific operations (the levels are discussed in more detail below).

The basic underlying ideas of activity theory are (1) the unity and inseparability of human mind and activity, and (2) the social nature of human activity. The former states that the human mind can only be understood as embedded in the overall interaction between humans (subjects) and the word (objects). According to the latter, human interaction with the world is inherently social, and therefore so is the human mind. These general ideas can be specified through the following set of principles of activity theory (Wertsch, 1981; Kaptelinin and Nardi, 2006):

7.3.2.1.1   Object-Orientedness[3]

This principle states that all human activities, including their subjective components, have objective references. Human needs, goals, dreams, emotions, etc., are all related to something objectively existing in the world. Therefore, any analysis of human activities should include an understanding of what these objects are and how people are engaged with them. It should be emphasized that the meaning of "objective" in activity theory is not limited to physically existing entities, and that people interact with *environments* comprising various objects, rather than with each of the objects separately.

7.3.2.1.2   The Hierarchical Structure of Activity

According to Leontiev, humans' interaction with the world can be analyzed at several hierarchical levels (Figure 7.3). The levels generally correspond to the questions of Why, What, and How. At the top level (Why) there are *activities*, directed at their *motives*. In activity theory motives are understood as objects that people should achieve in order to meet certain needs. A simple example would be getting food when we are hungry. To achieve a motive, we need to formulate and attain specific goals (What), of which we are always consciously aware, and which do not necessarily *directly* lead us to the motive (for instance, instead of directly grabbing a piece of food we may grab a phone and order a pizza delivery). To implement an action (How), we need to perform elementary, automatic, and often unconscious, *operations* (such as unlocking a phone screen). The levels are not fixed: for instance, with practice, actions may transform into operations, and, conversely, if there is a breakdown when performing an operation, the operation may suddenly become a conscious action.

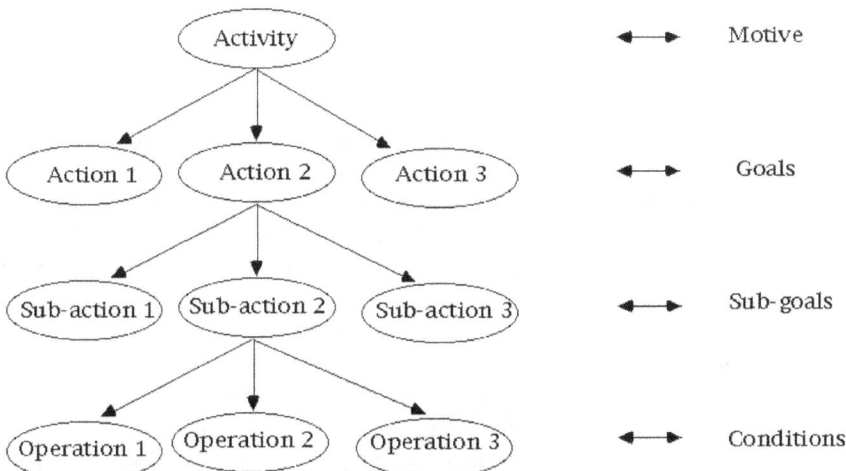

FIGURE 7.3   Leontiev's hierarchical structure of activity.

7.3.2.1.3 Mediation

A fundamentally important aspect of human activities is that they are mediated. We do not interact with the world directly. Instead, a myriad of artifacts are "placed" between us and the world, enabling us to achieve things we would not achieve without them. The design and ways of using the artifacts reflect the experience of other people, contributing to the evolution of mediational means from generation to generation. Some personal artifacts, such as clothes, eyeglasses, and, more recently, laptops and smartphones, may become so integrated into everyday activities that they can be perceived by a person as a part of the self. Such artifacts extend person's physical or mental capabilities to form what in activity theory is called "functional organs". Analyses of mediation in Leontiev's activity theory mostly focused on instrumental mediation, that is, the use of socially developed mediational means as tools for achieving meaningful goals in learning, work, and other purposeful activities.

7.3.2.1.4 Internalization/Externalization

There is a dynamic distribution of activities between individuals and their environments. Building on pioneering work by Vygotsky (1981), these processes are conceptualized in activity theory as internalization and externalization. The processes are taking place along two dimensions, physical and social, corresponding to the distribution of activities between an individual human being and, respectively, their physical and social environments. Internalization and externalization are fundamental mechanisms underlying the shaping of the human mind by culture and society.

7.3.2.1.5 Development

None of the aspects of human interaction with the world, described above, is fixed. There are constant transformations regarding the objects of activities, their structures, mediational means, and the distribution of activities between humans and their environments. Accordingly, a key principle of activity theory is development, which is understood as both an object of analysis and an analytical perspective to be adopted in studies of human activities.

Key aspects of the view on human actions, based on the above principles, can be illustrated by a comparison of that view with Norman's Action Cycle model, discussed in Section 7.2.2 (Figure 7.2). Arguably, the model is generally, even if implicitly, consistent with the principle of *object-orientedness*, as it conceptualizes human actions as goal-oriented interactions with the world. However, the model does not describe actions' motivation (*the hierarchical structure of activity*) and how actions transform over time (*externalization/internalization* and *development*). Finally, it does not differentiate between human interaction with the world with and without technology, so it does not specifically deal with *mediation*.

7.3.2.2 Engeström's Activity System Model

While Leontiev's activity theory emphasizes the social nature of any human activity, it predominantly aims to provide an account of individual human beings rather than explore the social context of human actions. An activity-theoretical framework, which offers concepts and methods for such an exploration, was proposed by Engeström (1987) and Engeström et al. (1999). The framework extends Leontiev's *subject–object* interaction by including *community* as an additional element. The resulting activity system model describes a three-way interaction between these elements, comprising three two-way interactions – *subject-object*, *subject-community*, and *community-object*. Each of these two-way interactions is mediated by a respective mediational means, namely: *tools* (the instruments, which mediate the *subject–object* interaction, *rules* (the conventions and procedures, which mediate the *subject-community* interaction), and *division of labor* (the coordinating mechanisms, which mediate the *community–object* interaction). The model also includes *outcome*, produced by the activity system as a result of transforming the *object* (Figure 7.4). When studying complex real-life cases, a suitable analytical strategy may be to model *networks of activity systems*.

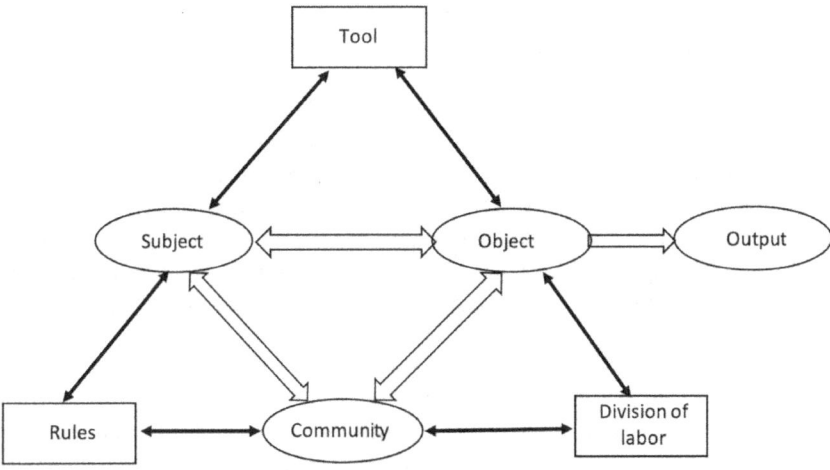

FIGURE 7.4 Engeström's activity system model. (Adapted from Engeström et al. (1999).)

Such networks may indicate, for instance, how outcomes of different activity systems are combined with one another, or how the outcome of a certain activity system may provide an input to an element (e.g., tool) of another activity system.

The activity system model has been widely used in a variety of disciplines, including HCI, as an analytical tool for studying human actions in concrete social contexts (e.g., Law & Sun, 2012). A key advantage of the model is that it supports breaking down an object of study into more manageable components, and suggests potential ways of integrating the components by providing an overall structure of their interrelations.

7.3.2.3   Activity Theory in HCI: General Uses

Clemmensen et al. (2016) reported a literature analysis of applications of activity theory in HCI, which identifies several ways in which the theory has been used in the field. In particular, activity theory has been used to support empirical analyses of HCI phenomena, support design explorations, and develop new conceptual accounts of HCI issues. In all of these cases the most fundamental insight offered by activity theory was that analysis, design, and use of a technology require an understanding of the overall context, in which people use the technology to achieve their meaningful goals.

7.3.2.4   Supporting Empirical Analyses of Human–Technology Interaction

User studies are very central to human–technology interaction research and practice. To assess an existing technology or find out whether there is a need for a new technology, it is crucial to get an insight into real life contexts of actual or potential users. The centrality of such empirical analyses is reflected, for instance, in the ISO standard "Human-centered design for interactive systems", which describes "understanding and specifying the context of use" as a key step in human-centered design life cycle (ISO 9241 - 210:2010). However, many real-life contexts are inherently complex, and, in addition, it might be a non-trivial task to single out the aspects of such contexts, which are relevant from the perspective of human–technology interaction. To deal with this issue, a number of studies turn to activity theory as an analytical framework, helping to structure and focus an inquiry. A common rationale for employing activity-theoretical concepts, such as the concept of mediation, the hierarchical structure of activity, and development, has been looking for guidance in coping with the complexity of everyday contexts when identifying challenges and problem areas in existing practices, as well as opportunities for providing more advanced technological support.

The conceptual framework of activity theory has been used in empirical studies in two ways. The first way was to "internalize" an activity-theoretical perspective and apply the entire framework to guide an empirical analysis. For instance, Bødker (1991) presents a retrospective analysis of a previous participatory design project, UTOPIA, in which the general notion of computers as mediational artifacts in "subject-object" interaction allowed her to identify three aspects of the use of computing technologies: physical aspects (interaction with a computer as a physical thing), handling aspects (interaction with computer applications), and subject/object-directed aspects (interaction with other people and things through technology). Honold (2000) reports an insightful account, generally informed by Engeström's approach, of how interactive technological products (washing machines), designed in one country, were being appropriated as mediational means in a different cultural context.

7.3.2.5 Analytical Tools Informed by Activity Theory

A problem with leaving it to the researchers to adopt activity theory and then make all decisions on how to use it, is that it may require an extensive learning on the part of the researchers, which is not always feasible. A potential solution to this problem is to use the theory to develop analytical tools, which would translate general theoretical insights into concrete and actionable guidelines that can be understood and followed without the need for an extensive study of the theory itself. Several such analytical tools, informed by activity theory, have been employed in HCI research.

The Activity Checklist, proposed by Kaptelinin et al. (1999), explicitly builds on the key principles of Leontiev's approach, namely mediation, object-orientedness, the hierarchical structure of activity, internalization/externalization, and development. It is constructed by systematically combining the principle of mediation with the other four principles, which has produced four subsets of issues and questions, covering a wide range of aspects of the use of technology. The Human-Artifact model (HAM), developed by Bødker and Klokmose (2011), summarizes the aspects of a technological artifact as a mediational means, which are central to the artifact's design and use, in a 2×3 matrix with two columns ("human" and "artifact") and three rows ("why?", "what?", and "how?", corresponding to Leontiev's levels of activity).

Several analytical tools for supporting the design and evaluation of interactive systems have been based on Engeström's activity system model. In particular, Jonassen and Rohrer-Murphy (1999) use the model as a framework for identifying specific components and aspects of constructivist learning environments (CLEs). The Activity-Oriented Design Method (AODM), proposed by Mwanza-Simwami (2011) offers a toolkit, based on Engeström's activity system model, for understanding and supporting learners' interaction with technology. Finally, in a number of studies Engeström's triangular representation activity system model itself is used as a "spatially organized checklist". The nodes of the model, including three types of mediational means – tools, rules, and division of labor – serve as a set of items, structuring analyses of human–technology interaction phenomena (e.g., Law & Sun, 2012).

7.3.2.6 Supporting Design Explorations

Adopting an activity-theoretical view on technology, that is, understanding technology as an instrument used by people to achieve their meaningful goals, has direct implications for design-oriented HCI. In particular, it reveals limitations of the traditional application-centric approach, mostly concerned with designing individual computer applications, and emphasizes the importance of supporting people with integrated configurations of goal-related resources.

A number of concrete systems falling into the general category of "activity-centric computing", were developed in the last decades, with many designs being explicitly informed by activity theory (Bardram et al., 2019). The systems include, for instance, UMEA (Kaptelinin, 2003), Activity-Based Computing (ABC) (Bardram, 2009), Giornata (Voida & Mynatt, 2009), and Laevo (Jeuris et al., 2014). The overarching aim of all these systems was to more efficiently support users' practices, both individual and collaborative ones, by enabling an integrated management of various types of resources, such as tools, documents, contacts, etc., related to a higher-level meaningful goal of an individual user, a group of users, or an organization (often referred to as a "project"). In addition, activity theory was a central theoretical influence for the ABC approach, proposed by Donald Norman and his group at Apple Computer in the mid-1990s (Norman, 1998).

7.3.2.7 Reframing HCI Concepts

Activity theory also informed a range of theoretical explorations in HCI. In particular, activity theory's notion of mediated action was applied to revisit some of the most fundamental HCI concepts, challenging their traditional interpretations and suggesting alternative or further elaborated interpretations.

Beaudouin-Lafon (2000) scrutinizes the notion of direct manipulation (Shneiderman, 1983), commonly implemented in graphical user interfaces (GUIs) to allow the user to act upon user interface objects (e.g., scrolling down a document or dragging and dropping an icon) instead of issuing text commands. Beadouin-Lafon observes that in many cases of presumably "direct" interaction, we do not in fact directly interact with our "objects of interest", such as a text or a 3D object in a digital workspace. Instead, our interaction with GUIs is typically mediated: we use "instruments", such as scrollbars, which, in turn, engage with a respective object of interest. The "instrumental interaction" framework, proposed by Beadouin-Lafon, defines key attributes of interaction instruments and outlines a design space for creating such instruments.

Kaptelinin and Nardi (2012) present an activity-theoretical analysis of another fundamental HCI concept, namely, affordances. They argue that the original notion of affordances, proposed by Gibson (1979), may not be particularly suitable for HCI and interaction design, because it does not specifically describe *technological* affordances. Instead, they offer a *mediated action* view on affordances and identify different kinds of affordances provided by a technological artifact. The most important distinction, according to the mediated action view, is between "handling affordances", that is action possibilities for interacting with the tool in question (e.g., a graspable knife handle), and "effecter affordances", that is, action possibilities for employing the technology to make an effect on an object of interest (e.g., a sharp, cut-with-able, blade).

7.3.3 Distributed Cognition

Distributed cognition (Hutchins, 1995) is another prominent second-wave theory (Bødker, 2006, 2015; Kaptelinin et al., 2003). Like activity theory, it was brought to HCI to provide an account of real-life contexts and extend the scope of the field beyond a narrow focus on interactions within a system comprising two information-processing units, a human and a computer. The conceptual foundations of distributed cognition, are, however, quite different from those of activity theory. The context, in which distributed cognition places humans and technologies, are broadly understood cognitive processes within a larger-scale information-processing system, comprising multiple people and objects in the environment.

The most fundamental assumption of distributed cognition is that cognitive processes are not taking place in humans' brains, or inside human bodies in general. It is claimed that in real-life contexts cognition is distributed, both socially (between people) and physically (between people and objects in the environment). In line with this claim, distributed cognition informed a range of thorough analyses of how information processing and problem solving are performed in complex sociotechnical systems, such as ship bridges or airplane cockpits, through a coordinated interaction between people, tools, and materials in the setting (Hutchins, 1995).

Hollan et al. (2000) argued that distributed cognition's insights are highly relevant for HCI, especially given that technologies are becoming increasingly networked. They call for employing distributed cognition as a theoretical foundation for HCI research, which maintains that cognition should be understood as socially distributed, embodied, and embedded in culture. The theory was also found to provide useful insights for studies in the area of Computer Support for Cooperative Work (CSCW) (Halverson, 2002).

While distributed cognition, as opposed to activity theory, focuses on cognitive systems rather than human purposeful actions, there are some similarities between the approaches. The roots of distributed cognition partly overlap with those of activity theory. In particular, Vygotsky's analysis of internalization has been an inspiration to both of them, and both theories consider internalization a key mechanism underlying the impact of culture on the human mind. Accordingly, both theories view the external and the internal not as two separate worlds with an impregnable border between them, but rather as different components of the human interaction with the world as a whole, which components can dynamically transform into each other. As observed by Halverson (2002), both theories emphasize the centrality of the social and cultural context and share a commitment to ethnography as a research method. Therefore, despite being concerned with cognition rather than action, the analysis informed by distributed cognition may, arguably, provide important insights into human agency as being socially and physically distributed and embedded into the functioning of larger-scale systems.

7.3.4 ETHNOMETHODOLOGY

In terms of methodology, the move from the first-wave to the second-wave HCI was associated with a radical shift from laboratory studies to fieldwork. The need to get an insight into how people actually act and use technology in real-life settings (Suchman, 1987), and the centrality of the notion of context in conceptual analyses within second-wave HCI, determined the adoption of ethnography as a key research methodology in the field (Button & Dourish, 1996). While, in general, "ethnography" was understood rather broadly, and could be applied to any kind of studies conducted in a natural setting, one particular variety of ethnography, ethnomethodology, stood out in terms of both methodological rigor and impact on human–technology interaction research, especially in the field of CSCW.

The basic tenets of ethnomethodology can be summarized as the requirement that researchers (1) avoid theoretical presuppositions and (2) pay close attention to how the social order in a setting is actually produced by the participants in the setting.[4] Because of its clear stand against theoretical presuppositions, ethnomethodology was especially appropriate for revealing unexpected subtle aspects of real-life work contexts, which are crucial for getting the work done but are easy to overlook, especially if researchers already know what they are going to find. In the 1990s, ethnomethodology was successfully applied to provide insightful analyses of work in various organizations, such as London Underground Control rooms (Heath and Luff, 1991), and formulate requirements for technological systems that would successfully support the work. One of the key insights from ethnomethodological analyses was that collaborative work cannot be explained as just following predefined rules and procedures. Instead, it needs to be understood as a collective accomplishment of the participants, based on their continuous awareness of each other's activities, as well as flexible re-arrangement and redistribution of work depending on a particular situation.

7.4 THIRD-WAVE HCI: EXPERIENCE, EMBODIMENT, AND HUMAN–TECHNOLOGY RELATIONS

7.4.1 THE NOTION OF "THIRD-WAVE HCI"

The term "third-wave HCI" was initially introduced by Bødker (2006), who pointed to a number of developments in HCI practice and research, which presented challenges to second-way HCI's predominant focus on the use of tools in the context of work and well-established communities of practice. These developments, which, according to Bødker, were indicators of an emerging "third-wave HCI", included: (1) dealing with broadened and intermixed contexts of use, as technologies spread to private and public spheres; (2) foregrounding culture, emotion, and experience; and (3) moving away from user-centered design and placing more emphasis on designers' own visions. While Bødker herself expressed a rather cautious view, and, in particular, warned against abandoning second-wave theories in the transition to a new wave, the paper has been widely perceived as a harbinger of a new era of HCI.

The exact meaning of the term "third-wave HCI" in current research is somewhat elusive. The term is often explained by pointing out differences from the second-wave HCI, rather than defining what it actually is. For instance, according to Bødker, "the focus of the third wave, to some extent, seems to be defined in terms of what the second wave is not: non-work, non-purposeful, non-rational, etc." (Bødker, 2006). The uncertainty about the exact meaning of the term apparently reflects the actual ambiguity and diversity in HCI research. The state of affairs in the third-wave HCI has been described, for instance, as "a chaos of multiplicity in terms of technologies, use situations, methods, and concepts" (Bødker, 2015) and "fragmentation" (Filimowicz & Tzankova, 2018). Despite the uncertainty regarding the exact meaning of the term "third-wave HCI", there seems to be a general agreement about phenomenology being a key theoretical influence, and experience and embodiment being central research themes, in the third-wave HCI.

7.4.2 Turn to Phenomenology

Phenomenology is a movement in philosophy, represented by such thinkers as Husserl, Heidegger, and Merleau-Ponty, which foregrounds human experience and how the world presents to us in and through experience (Dourish, 2004). Phenomenology assumes that our embodied presence in the world, revealed to us through experience, precedes, and forms a basis for how we make meaning of the world.

Phenomenology was one of the sources of inspiration for the early seminal book "*Understanding Computers and Cognition*" by Winograd and Flores (1986), and a systematic exposition of phenomenology as a theoretical foundation of HCI was offered by Dourish in his influential book "*Where the Action Is: Foundations of Embodied Interaction*" (Dourish, 2004). In this book, Dourish is particularly concerned with *connecting action with meaning*, which, he argues, requires moving computation and interaction outside of the abstract cognitive processes and bringing them into the same physical world in which we engage in a variety of interactions.

Phenomenology's focus on holistic and subjective aspects of human interaction with the world was (along with some other approaches, such as pragmatism[5]) well in line with the overall orientation of studies within the third-wave HCI, which made it a suitable theoretical foundation of the research.

7.4.3 Experience and Embodiment

Focus on experience is probably the most distinct feature of the third-wave HCI. When introducing the notion of third-wave HCI, Bødker (2006) specifically pointed out that emotions and experiences were "keywords in the third wave", and the shift to the third wave was indeed marked by a sharp increase in the number and diversity of studies focusing on various aspects of experience. Some of the key areas of the studies included research methods for analyzing experience (some of the outcomes of this work are presented, for instance, at allaboutux.org; AttrakDiff.de), analytical frameworks for conceptualizing experience (e.g., Forlizzi & Battarbee, 2004; McCarthy & Wright, 2004), and empirical studies of experience in different contexts (e.g., Hassenzahl, 2010, 2018). Two issues, related to analyses of experience in HCI, are especially relevant in the context of the discussion in this chapter: embodiment and pragmatic vs. non-pragmatic engagements with technology.

7.4.3.1 Embodiment

Among the multitude of concepts, employed to understand and support experience, embodiment stands out as being especially strongly connected to phenomenological roots of the third-wave HCI

(Dourish, 2004; Svanæs, 2013). The centrality of embodiment in the third-wave HCI reflects not only an interest in supporting a richer interaction with technology, which would involve entire users' bodies but also an acknowledgment of the fundamental, physical, and social, embeddedness of human beings in the word. Accordingly, an exploration of embodiment in HCI research opens up a possibility for a deeper understanding of how people connect to the world through technology (e.g., Björnfot, 2022).

7.4.3.2 Pragmatic vs. Non-Pragmatic Engagements with Technology

Most analyses of experience in HCI do not contrast a focus on experience to a focus on purposeful actions. On the contrary, it is commonly assumed that purposeful actions are closely linked to experience, being one of its most fundamentally important sources. In other words, it is assumed that experience and purposeful/pragmatic engagements with technology are complementary facets of the human connection to the world. This general position has been advocated by many researchers, including, Bødker (2006), and Hassenzahl (2010).[6] At the same time, as discussed in the section below, some work within the third wave adopts a critical stance toward pragmatically oriented HCI, and explicitly aims to produce designs, which do not aim to have pragmatic value (Hauser et al., 2018).

7.4.4 POSTPHENOMENOLOGY: HUMAN-TECHNOLOGY RELATIONS

As computing technologies increasingly pervade many aspects of our daily lives, they change the way we exist in the world and shape our understanding of it. Let's consider an example of a health technology. Wearable glucose monitoring devices measure glucose-levels in the blood and present this as a value on a display. With such a device, patients may gain a new outlook on life; one that is more secure because health information is readily available but at the same time may also lead to anxiety. For instance, when doubting if the right glucose levels are shown and deciding if, or when, insulin should be administered.

These and other similar technologies affect humans and human actions in more than just pragmatic way. In addition to serving as a tool for achieving a specific goal, they can transform a person's experience of daily life (in potentially both positive and negative ways), afford new behaviors, and affect making judgments and choices. Understanding these effects of technology is currently an important research area in HCI, in which postphenomenology (Ihde, 1990; Rosenberger, 2014; Verbeek, 2015) and philosophy of technology (Fallman, 2011) proved to be valuable sources of theoretical insights.

In his article "*Beyond Interaction*", Verbeek (2015), building on Ihde (1990), outlined a post-phenomenological perspective on human-technology relations. To use the previous example, the monitoring device, serves as an extension of the human (embodiment-relation) and, by translating blood glucose-levels into numbers shown on a display, represents the world to the person (hermeneutic-relation). The device, especially if it automatically delivers insulin without human intervention, can be experienced as a complex piece of technology having its own existence separate from us (alterity-relation). When such a device is worn every day, it can move the background of our experience like clothing (background-relation), until it draws attention to itself, for instance when wearing it becomes uncomfortable (switching back to alterity) or when it needs to be used again (switching back to hermeneutic). Glucose monitoring devices may also appeal to what Verbeek refers to as 'cyborg relations' when in the future, more innovative versions of these devices may become implanted into the human body and begin to function like artificial organs. In this case, Verbeek mentions how technology "merges with the human body into a new hybrid being" (p. 29).

In general, postphenomenology has been used in HCI in a range of studies exploring an alternative to the understanding of technology in a pragmatic sense only, and emphasizing living with rather than using technology. For instance, Hauser et al. (2018) discuss a set of research-through-design studies informed by postphenomenology, which analyzed artifacts as counterfactual material speculations. Jensen and Aagaard (2018) proposed a phenomenological research method to gain insights into artifacts' embodiment and context of use and show how technology can be read in multiple ways and thus affording different purposes and ways of use (Rosenberger, 2014).

7.5 EMERGING CRITICAL AND "MORE THAN HUMAN" PERSPECTIVES IN HCI RESEARCH

7.5.1 NEW CHALLENGES AND EMERGING TRENDS IN HCI RESEARCH

In this section, we discuss some of the current developments in HCI that critically examine human action and agency in a technologically permeated world that is increasingly challenged with onto-logical uncertainties and ecological and humanitarian crises (Light et al., 2017). We outline three strands in contemporary HCI discourse that discuss human action and agency from a relational and entangled perspective: *more-than-human design*, which decentralizes human action and involves participation with, and care for, nonhumans; *performativity*, referring to how human agency is co-produced in sociotechnical configurations; and *design futuring*, an approach to critically exam-ining human action in contemporary culture and speculating about responsible futures.

An assumption, shared by much of the research discussed below, is that technology makes it difficult, or even impossible, to draw a clear line between the social, the artificial, and the natu-ral, as they are becoming increasingly entangled. Reflecting on these developments, Frauenberger (2020) argues that "entanglement theories" that transcend socio-natural and cultural-material dual-isms, including actor network theory (Latour, 2007), postphenomenology (Verbeek, 2015), object-oriented ontology (Bogost, 2012) and agential realism (Barad, 2007), can be expected to become a conceptual foundation for the next wave in HCI and provide a way forward to understanding and dealing with contemporary problems "in the face of rapid technological progress, coupled with a profound social change" (Frauenberger, 2019, p. 2:3).

7.5.2 MORE-THAN-HUMAN DESIGN

Humans are embedded in vast ecologies and systems at multiple scales, from the molecular to the planetary, in which they are interconnected with, and interdependent on, other species. According to Haraway (2016), attaining more livable futures in an ecologically damaged world involves striv-ing for multispecies flourishing and emancipate all entities without privileging any of them, which can be achieved by decentralizing a predominant human orientation on the world and embracing action and agency as part of our coexistence with other species (Haraway, 2016). More-Than-Human Design (MTHD) is a design approach, inspired by this and other similar visions, which focuses on the role and involvement of, and care for, nonhumans, understood as both living organisms and arti-ficial entities like intelligent agents and robots (Giaccardi et al., 2016; Clarke et al., 2019; Wakkary, 2021; Coskun et al., 2022).

In HCI, MTHD approaches have been applied in a variety of studies, exploring the possibilities for involving nonhumans, both natural or artificial entities, in design and use. The topics range from embracing an ecological worldview when designing and implementing media architectures in urban environments (Foth & Caldwell, 2018), and supporting the engagement of trees and animals in the crafting of smart city infrastructures (Clarke et al., 2019), to working with living materials in biode-sign to ensure "a reciprocal and evolving relationship between humans and living artifacts" (Karana et al., 2020, p. 46). MTHD work dealing with the participation of artificial entities includes, for instance, adopting a thing-centered perspective to reveal roles that artifacts play in everyday prac-tices that might otherwise go unnoticed or learning how human and artificial capabilities can foster more-than-human partnerships (Giaccardi, 2020). In general, these and other works on MTHD, call for an awareness, sensitivity, and care to include non-human voices and actions on which human actions are interdependent (Tsing, 2015; de La Bellacasa, 2017).

7.5.3   PERFORMATIVITY

The notion of "performativity" expresses the idea that interaction with technology is not an isolated and prescriptive exchange between humans and artifacts but rather a relational, emergent, and generative process. Performativity according to Barad (2007) is a process in which entities are not pre-given but come into being through the relationships they form with other entities. Performativity therefore prompts an understanding of both identity and agency as emerging from the coming together of human and non-human entities. The embracing of Barad's radical relational stance in HCI is motivated by the question of how humans relate to, and are entangled with, networked and intelligent technologies that display autonomy and have the ability to learn and change.

The general idea of performativity is expressed in a variety of concrete concepts, employed in HCI research. Kuijer and Giaccardi (2018) proposed the concept of co-performance to think about smartness of computing technology as something performed during its use, rather than designed into an artifact before its deployment, as both human and artifact learn and co-develop in their ongoing situated engagements. The concept of fluid assemblages (Redström & Wiltse, 2018), is introduced to provide an account of the character and consequences of data-driven networked computational things. Fluid assemblages are understood as entities, which "do not pre-exist, but are rather enacted and become determinately formed and propertied as things available for use within particular contexts." (Redström & Wiltse, 2018, p. 66). A relational-performative approach to the design of human-robot interaction, proposed by Gemeinboeck (2021), calls for a shift of focus from a representationalist to a performative understanding of agency in human-robot interaction, in which meaning, identity, and agency of both human and robot are shaped through the encounter rather than being defined beforehand.

7.5.4   DESIGN FUTURING

Design Futuring, a term originally coined by Tony Fry (2009), refers to design practices that question and critically examine human action and agency from ethical and political standpoints as alternative presents or future propositions. It is motivated by the need to consider the impacts and long-term consequences of increasingly complex and interconnected technologies on society and the planet. Design futuring in HCI comprises a set of speculative approaches allowing people to experience possible futures (Kozubaev et al., 2020). The approaches employ concepts and methods from speculative design (Dunne & Raby, 2013), critical design thinking (Bardzell & Bardzell, 2013), design fiction (Auger, 2014), and experiencing alternative realities through enactments (Elsden et al., 2017). The basic tenet of design futuring is that imagining alternative presents or future worlds that are crafted in critical ways, afford debate and reflection about the ethical implications of technology and positions human action and agency in a broader societal and temporal context.

The general perspective of design futuring has influenced a variety of recent studies in HCI and related fields. Alvarado Garcia et al. (2021) argue that the fact that HCI has been influenced by a primarily western point of view may limit the diversity of social groups, purposes, and practices, represented in the field. Vervoort et al. (2015) discuss how imaginative scenario building, informed by Goodman's (1978) notion of worldmaking, affords a constructive dialogue among different stakeholders and publics by making visible differences in value systems, viewpoints, and cultures. Similarly, Kozubaev et al. (2020) observe that future visions are influenced by personal experiences, cultural backgrounds, and political agendas (collectively referred to as "positionality"), and therefore it should be made explicit from which point of view a future is envisioned. Taken together, these works provide insights into how human action can be understood as culturally and politically diverse, which impacts the kinds of futures imagined.

7.6 DISCUSSION

7.6.1 THE HISTORICAL PROGRESSION OF HCI'S PERSPECTIVES ON HUMAN ACTIONS: AN OVERVIEW

The main points of the analysis in the previous sections are presented in Table 7.1 and can be summarized as follows. The very emergence of HCI was a response to the challenges caused by the advent of computers and, consequently, interactions with computers becoming what Card et al. (1983) referred to as a "new arena of human action". Supporting the design of "user-friendly" or "intuitive" system interfaces was considered the most present and clear need to be addressed by HCI research. That way of defining the mission of HCI as a field determined the view on human actions as *executing "tasks"*, directly linked to the system interface and typically analyzed by employing algorithmic models. The theoretical foundation of the first-wave HCI was cognitive psychology, according to which the user was viewed as an information-processing system, having an input (sensory functions) and output (motor functions). Interaction with computers during that time often took place in separate physical environments with highly structured organizational procedures, which was probably the reason (or one of the reasons) why non-cognitive, subjective, and contextual aspects of interaction were perceived as less central.

With a realization that computer use is a particular case of human-world interaction in general, limitations of the "user - interface - task" view of human actions were becoming increasingly apparent. A broader view, proposed within the cognitive perspective in HCI, was Norman's (1988) Action Cycle model, which was not limited to following an algorithmic method when interacting with a computer interface. Instead, it describes a *feedback-based interaction* between the person and the world.

The widespread use of computers in various real-life, predominantly work, contexts required a more substantial transformation so the field and eventually resulted in the transition from the first-wave to second-wave HCI (Cooper & Bowers, 1995; Bødker, 2006). To properly understand and support people in these contexts, HCI had to extend its scope from user interface-related tasks to meaningful human actions, that is, become concerned with not only "how", but also "why" people use technology. The extension was achieved by bringing to HCI theoretical frameworks, which consider human actions as embedded in larger-scale activities, framing and motivating the use of technology. The distributed cognition framework (Hutchins, 1995; Hollan et al., 2000) was employed to place interaction with computers into real-life sociotechnical contexts, in which both people and artifacts collectively comprise distributed cognitive systems. Leontiev's activity theory (Leontiev, 1978), especially the notion of the hierarchical, three-level structure of activity (Figure 7.3), was introduced to HCI to conceptually frame user actions within object-oriented activities of individual human subjects, and thus provide an account of the interplay between "why", "what", and "how" (Bødker, 1991). Engeström's activity system model (Engeström et al., 1999), was commonly used in HCI, sometimes in combination with Leontiev's framework, to conceptualize mediated human actions as embedded into the context of collective activity, directed as a shared object and producing a shared outcome (e.g., Kuutti, 1996).

The next fundamental development in the role of interactive technologies in society was a trend beyond work environments, and toward discretionary and private uses (Grudin, 2005; Bødker, 2006). When individuals themselves, rather than the organizations where they work, decide what technology to use and how to use it, utility and usability may not be the only criteria for assessing a technology. Subjective criteria, such as excitement, pleasure, and self-expression, may also be important and even decisive factors of choosing and using a technology. Accordingly, there was a need to understand how to support people with technologies, which are not just tools for effectively and efficiently achieving certain clear goals. To face this challenge, HCI turned to views of human actions as being framed in a holistic relation to the world, such as ones offered by phenomenology

TABLE 7.1

An Historical Progression of HCI's Perspectives on Human Actions

Wave	Human Action	Technology	User	Key Contributions	Theory	Concepts
First	Task execution (human-computer)	Interface	Information-processing system	GOMS, usability methods, experimental studies	Information-processing psychology	Task analysis, mental models
	Goal-oriented action (human-physical world)	Interface	Information-processing system (with I/O)	Norman's action cycle model	Information-processing psychology	Gulfs of execution and evaluation, affordances
Second	Cognitive acts in sociotechnical systems	Component of a distributed cognitive system	Component of a distributed cognitive system	Analyses of sociotechnical systems (such as the bridge of a ship)	Cognitive psychology/distributed cognition	Representation propagation
	Purposeful, mediated social action (individual)	Mediating artifact/individual activity	Subject/need-based agency	Analytical tools for user research, design and evaluation, activity-centric computing	Activity theory (Leontiev)	Mediation, tool, levels of activity, internalization
	Purposeful, mediated social action (collective)	Mediating artifact/collective activity	Subject/activity system agency	Analytical tools for user research, design and evaluation	Activity theory (Engeström)	Activity system, community, contradictions
Third	Meaningful embodied relation to the world	Mediating artifact/equipment and expressive medium	Embodied, sentient subject	Experience- and embodiment-centered analytical tools for design and evaluation	Phenomenology, Postphenomenology, philosophy of technology	Experience, embodiment, intentionality, human-technology relation

(e.g., Dourish, 2004). Experience and embodiment became key concepts in HCI research and practice (e.g., Hassenzahl, 2010; Svanæs, 2013).

Finally, HCI research has been increasingly concerned, especially in recent years, with how the field can contribute to dealing with current social and ecological crises, in which technology is playing a key role as a part of the problem, a part of the solution, or both. There is a growing general realization that to achieve justice, inclusiveness, and sustainability it may not be sufficient to simply continue what has been done before, just in a smarter and more efficient way. A range of contemporary studies, some of which are discussed in Section 7.5 above, call for a radical reassessment of the traditional views on human actions and question not only the means of the actions, but also how their goals, and even the subjects, are commonly understood. The theoretical frameworks, guiding these critical and deconstructive efforts, include feminist approaches (Bardzell, 2010) agential realism (Barad, 2003; Frauenberger, 2020) and postphenomenology (Rosenberger & Verbeek, 2015).

7.6.2 Selected Themes for Future Research

The overview of the historical trends in how HCI has viewed human actions, presented in the previous section, brings in the question of the issues and challenges, which can be expected to drive further development of HCI as a field. While predicting the future is always risky and riddled with uncertainty, the analysis in this chapter, arguably, suggests that certain research themes are likely to become especially important in the future. Below we discuss the following, closely related but distinct, themes: (1) transformative technological mediation, (2) emerging non-human actors, and (3) global implications of design.

7.6.2.1 Transformative Technological Mediation

An implicit assumption underlying many, if not most, HCI studies is that human agency is immutable: the aim of analysis and design is often assumed to be *understanding* people in order to *support* them with appropriate technologies. The historical development of HCI, as it is reflected in the notion of the "waves", can be generally described as a progression toward ever more advanced understanding of humans and their actions and providing ever more advanced support. However, it is increasingly obvious, especially when considering such technologies as social media and mobile devices, that people who use technology are also *changed* by it.

As discussed, there has also been HCI research, especially recently, which questions the immutability of human agency and concerns itself with how technology affects and changes the nature of humans, as well as how the nature has been traditionally understood. So far, such analyses have been mostly positioned as alternatives to the mainstream HCI, typically taking the form of speculative and critical design activities, such as design provocations and probes, characterized by deliberate disruptions of everyday experience and transcendence beyond "the normal". While this approach can be perfectly suitable for an initial exploration phase, future HCI studies are expected to move beyond that phase toward a more impact-oriented and systematic research into the transformative effects of technology on humans and their agency.

In our view, future research in that direction can significantly benefit from existing conceptual analyses of transformative technological mediations, informed by activity theory and postphenomenology. Each of these approaches considers technology a mediational means, which connects a human and the world and, by doing so, makes an impact on the human. Their particular perspectives on transformative mediation are, however, different.

In activity theory, the foundational concept is "activity", understood as purposeful, mediated, and transformative interaction between humans ("subjects") and the world (Leontiev, 1978; Kaptelinin & Nardi, 2006). Essentially, it is assumed that by acting on the world people also transform themselves and, therefore, the transformative potential of technology is determined by its role as a mediator of human purposeful actions. Using the terminology, proposed by Stetsenko (2017), this perspective on transformative mediation can be described as an "activist" one. In postphenomenology, the main focus is on the role of technology in shaping human existence, opening up for particular experiences and actions while foreclosing others (Rosenberger and Verbeek, 2015; Verbeek, 2015). Therefore, the transformative potential of technology, according to postphenomenology, is in establishing various types of relations with the human and producing human-technological assemblages. This perspective on transformative mediation can be described as "relational".

Arguably, human–technology interaction can be transformative in both the activist sense (that is, by using technology to transform the world, humans also transform themselves) and the relational sense (people are transformed through forming human-technological assemblages). Integrating the perspectives, in our view, is a way to get deeper insights into human actions in the context of HCI.

7.6.2.2   Emerging Non-Human Actors

A major challenge for HCI research is the need to expand its focus beyond "human" actions, as there are reasons to believe that humans are not the only actors who should be taken into account in the context of HCI. Currently, there are two lines of development in HCI and related areas that point to that conclusion.

First, in the recent decade or so there has been significant progress in creating intelligent agents, which have certain attributes that are usually associated with humans. Modern AI-based systems, including various types of robots, are capable of sensing, and interacting with, their physical environment, as well as of performing advanced information processing and autonomous decision-making. Moreover, intelligent agents increasingly enter everyday social contexts and therefore become *social actors*,[7] who should possess social and interactive skills, as well as follow social rules (or "robotiquette", Dautenhahn, 2007). In the context of HCI, the status of intelligent agents as new types of actors, and their relation to human actors are topical issues, which are being actively explored, but still remain largely open.

Second, as mentioned, MTHD approach posits that technology can be used to make noticeable and "give voice" to not only humans, but to other living creatures, as well. In this regard, a relevant issue, which needs to be further explored in future HCI research, is the relationship between human actions and agency, on the one hand, and the agency of the emerging non-human actors, on the other hand. In particular, an important problem to address is how/if the notion of "human actions" should be reassessed and revised in light of other actors present in our shared ecosystem. Should it be extended to become in line with the position of a guardian, responsible for their entire environment and other species in it? Should it be limited to one particular voice in a multiplicity of voices? These and similar questions remain largely open.

7.6.2.3   Human Actions, at Scale: Global Implications of the Design of Interactive Technologies

The impact of human activities on the entire environment on Earth has dramatically increased over the last centuries. Currently, the influence exerted by humans on the planet's ecosystems and climate is so significant that our time is sometimes described as the *Anthropocene* epoch, that is, the geological period, which is dominated by human activities.[8]

Arguably, the global impact of human actions should be considered a fundamentally important issue defining the research agenda of HCI as a field (Light et al., 2017). Currently, interactive technology is a key force that directs, coordinates, and integrates individual human actions and determines their overall outcome. Concrete design solutions made when creating such technologies can, therefore, have far-reaching consequences and should be considered as a part of a concerted design effort, required to deal with current social and environmental crises.

The need for HCI to look beyond individual human actions and be concerned with how to support their coordination and integration to collectively create a brighter future – a common theme in several of the research directions, discussed in this chapter – reflects the centrality of ethics, responsibility, sustainability, and inclusiveness in current HCI. An indicator of this centrality is the latest book by Donald Norman, one of the most influential figures in HCI and interaction design, which is dedicated to employing design for creating a better, more meaningful, and sustainable, world (Norman, 2023).

7.7 CONCLUSION

HCI as an area of research is characterized by a diversity of agendas, methods, and perspectives. In this chapter, we argue that, historically, different perspectives in HCI are inextricably related to particular views on human actions. The overall historical progression of the field can be described as a result of continuous responses to new challenges, caused by technological innovations and ever expanding uses of technology in various areas of life. Accordingly, the view of human actions evolved from low-level, technology-specific tasks to purposeful human actions to now forms of social action and being.

A key conclusion, which can be made from the analysis in the chapter is that, despite the seeming fragmentation of HCI research, there is a historical continuity in the field's perspectives on human actions. It is important to note that adopting a new perspective on human actions does not make previous ones obsolete. Instead of being abandoned, previous perspectives are typically embraced and extended by the new ones (even if not always explicitly). This type of cumulative development is likely to continue in the future, as our understanding of human actions can be expected to evolve rather than completely change as a result of a paradigm shift. For instance, analysis of mental models (usually associated with the first-wave HCI) can, arguably, provide valuable insights for current research on explainable AI. Now, when to remain relevant HCI is expected to contribute to addressing new, and unprecedented, societal challenges, related to the transformative impact of technological developments on human (and not only human) lives, it is imperative for the field to make use of all the potential accumulated throughout its history.

NOTES

1 The notions of "waves" and "paradigms" are similar in some respects, but their meanings are also somewhat different. In particular, the "third paradigm" according to (Harrison et al., 2007) roughly corresponds to both the second and the third "waves" according to Bødker (2006).
2 The name is sometimes also spelled in English as "Leont'ev" or "Leontjew".
3 The principle is not related in any way to object-oriented programming.
4 A detailed exposition and discussion of ethnomethodology as an approach in human-technology interaction research is beyond the scope of this chapter and can be found elsewhere (e.g., Heath & Luff, 1991).
5 A discussion of Dewey's pragmatism in relation to design thinking can be found in Dalsgaard (2014).
6 Hassenzahl' position is reflected, for instance, in the very title of his 2010 book: *"Experience Design: Technology for All the Right Reasons"* (Hassenzahl, 2010).
7 Nass et al. (1994) argue that the notion of "social actors" can be applied to even conventional computers.
8 https://www.merriam-webster.com/dictionary/Anthropocene.

REFERENCES

Alvarado Garcia, A., Maestre, J. F., Barcham, M., Iriarte, M., Wong-Villacres, M., Lemus, O. A., ... & Cerratto Pargman, T. (2021). Decolonial pathways: Our manifesto for a decolonizing agenda in HCI research and design. In *Extended Abstracts of the 2021 CHI Conference on Human Factors in Computing Systems* (pp. 1–9).

Auger, J. H. (2014). Living with robots: A speculative design approach. *Journal of Human-Robot Interaction*, *3*(1), 20. https://doi.org/10.5898/JHRI.3.1.Auger.

Bannon, L. J., & Bødker, S. (1991). Beyond the interface: encountering artifacts in use. In *Designing interaction: psychology at the human-computer interface*. (pp. 227–253). Cambridge: Cambridge University Press.

Barad, K. (2003). Posthumanist performativity: Toward an understanding of how matter comes to matter. *Signs: Journal of Women in Culture and Society, 28*(3), 801–831. https://doi.org/10.1086/345321.

Barad, K. (2007). *Meeting the Universe Halfway: Quantum Physics and the Entanglement of Matter and Meaning*. Durham, NC: Duke University Press.

Bardram, J. E. (2009). Activity-based computing for medical work in hospitals. *ACM Transactions on Computer-Human Interaction, 16*(2), 1–36. https://doi.org/10.1145/1534903.1534907.

Bardram, J. E., Jeuris, S., Tell, P., Houben, S., & Voida, S. (2019). Activity-centric computing systems. *Communications of the ACM, 62*(8), 72–81. https://doi.org/10.1145/3325901.

Bardzell, J., & Bardzell, S. (2013). What is "critical" about critical design? In *Proceedings of the SIGCHI Conference on Human Factors in Computing Systems* (pp. 3297–3306). Paris.

Bardzell, S. (2010). Feminist HCI: Taking stock and outlining an agenda for design. In *Proceedings of the SIGCHI Conference on Human Factors in Computing Systems* (pp. 1301–1310). Atlanta, GA.

Beaudouin-Lafon, M. (2000). Instrumental interaction: An interaction model for designing post-WIMP user interfaces. *Proceedings of the SIGCHI Conference on Human Factors in Computing Systems* (pp. 446–453). https://doi.org/10.1145/332040.332473.

Björnfot, P. (2022). *Being Connected to the World through a Robot*. Umeå: Umeå University.

Bødker, S. (1989). A Human Activity Approach to User Interfaces. *Human-Computer Interaction, 4*(3), 171–1195. https://doi.org/10.1207/s15327051hci0403_1

Bødker, S., & Klokmose, C. N. (2011). The Human–Artifact Model: An Activity Theoretical Approach to Artifact Ecologies. *Human-Computer Interaction, 26*(4), 315–371. https://doi.org/10.1080/07370024.2011.626709

Bødker, S. (1991). *Through the Interface: A Human Activity Approach to User Interface Design*. Mahwah, NJ: Lawrence Erlbaum Associates.

Bødker, S. (2006). When second wave HCI meets third wave challenges. *Proceedings of the 4th Nordic Conference on Human-Computer Interaction Changing Roles - NordiCHI'06* (pp. 1–8). https://doi.org/10.1145/1182475.1182476.

Bødker, S. (2015). Third-wave HCI, 10 years later-participation and sharing. *Interactions, 22*(5), 24–31. https://doi.org/10.1145/2804405.

Bogost, I. (2012). *Alien Phenomenology, or, What It's Like to Be a Thing*. Minneapolis, MN: Uiniversity of Minnesota Press.

Button, G., & Dourish, P. (1996). Technomethodology: Paradoxes and possibilities. *Proceedings of the SIGCHI Conference on Human Factors in Computing Systems Common Ground - CHI'96* (pp. 19–26). https://doi.org/10.1145/238386.238394.

Card, S. K., Moran, T. P., & Newell, A. (1983). *The Psychology of Human-Computer Interaction*. Mahwah, NJ: Lawrence Erlbaum Associates.

Carroll, J. M. (Ed.). (1991). *Designing Interaction: Psychology at the Human-Computer Interface*. Cambridge: Cambridge University Press.

Clarke, R., Heitlinger, S., Light, A., Forlano, L., Foth, M., & DiSalvo, C. (2019). More-than-human participation: Design for sustainable smart city futures. *Interactions, 26*(3), 60–63. https://doi.org/10.1145/3319075.

Clemmensen, T., Kaptelinin, V., & Nardi, B. (2016). Making HCI theory work: An analysis of the use of activity theory in HCI research. *Behaviour & Information Technology, 35*(8), 608–627. https://doi.org/10.1080/0144929X.2016.1175507.

Cooper, G., & Bowers, J. (1995). Representing the user: Notes on the disciplinary rhetoric of human-computer interaction. In *The Social and Interactional Dimensions of Human-Computer Interfaces.* (pp. 48–66). Cambridge: Cambridge University Press.

Coskun, A., Cila, N., Nicenboim, I., Frauenberger, C., Wakkary, R., Hassenzahl, M., Mancini, C., Giaccardi, E., & Forlano, L. (2022). More-than-human concepts, methodologies, and practices in HCI. *CHI Conference on Human Factors in Computing Systems Extended Abstracts* (pp. 1–5). https://doi.org/10.1145/3491101.3516503.

Cypher, A. (1986). The structure of users' activities. In D. A. Norman, & S. W. Draper, *User Centered System Design* (pp. 243–264). Boca Raton, FL: CRC Press. https://doi.org/10.1201/b15703-12.

Dalsgaard, P. (2014). Pragmatism and design thinking. *International Journal of Design, 8*(1), 143–155.

Dautenhahn, K. (2007). Socially intelligent robots: Dimensions of human-robot interaction. *Philosophical Transactions of the Royal Society* B, *362*, 679–704.

de La Bellacasa, M. P. (2017). *Matters of Care: Speculative Ethics in More Than Human Worlds* (Vol. 41). Minneapolis, MN: Uiniversity of Minnesota Press.

Dourish, P. (2004). *Where the Action Is: The Foundations of Embodied Interaction* (1st. paperback ed). Cambridge, MA: MIT Press.

Dunne, A., & Raby, F. (2013). *Speculative Everything: Design, Fiction, and Social Dreaming*. Cambridge, MA: MIT Press.

Elsden, C., Chatting, D., Durrant, A. C., Garbett, A., Nissen, B., Vines, J., & Kirk, D. S. (2017). On speculative enactments. *Proceedings of the 2017 CHI Conference on Human Factors in Computing Systems* (pp. 5386–5399). https://doi.org/10.1145/3025453.3025503.

Engeström, Y. (1987). Learning by Expanding: An Activity Theoretical Approach to Developmental Research. Helsinki, Finland: Orienta-Konsultit.

Engeström, Y., Miettinen, R., & Punamäki-Gitai, R.-L. (Eds.). (1999). *Perspectives on Activity Theory*. Cambridge: Cambridge University Press.

Fallman, D. (2011). The new good: Exploring the potential of philosophy of technology to contribute to human-computer interaction. *Proceedings of the SIGCHI Conference on Human Factors in Computing Systems* (pp. 1051–1060). https://doi.org/10.1145/1978942.1979099.

Filimowicz, M., & Tzankova, V. (2018). Introduction I new directions in third wave HCI. In M. Filimowicz, & V. Tzankova (Eds.), *New Directions in Third Wave Human-Computer Interaction: Volume 1-Technologies* (pp. 1–10). Springer International Publishing. https://doi.org/10.1007/978-3-319-73356-2_1.

Forlizzi, J., & Battarbee, K. (2004). Understanding experience in interactive systems. *Proceedings of the 5th Conference on Designing Interactive Systems: Processes, Practices, Methods, and Techniques* (pp. 261–268). https://doi.org/10.1145/1013115.1013152.

Foth, M., & Caldwell, G. A. (2018). More-than-human media architecture. *Proceedings of the 4th Media Architecture Biennale Conference* (pp. 66–75). https://doi.org/10.1145/3284389.3284495.

Frauenberger, C. (2019). Entanglement HCI the next wave? *ACM Transactions on Computer-Human Interaction, 27*(1), 1–27. https://doi.org/10.1145/3364998.

Fry, T. (2009). *Design Futuring* (pp. 71–77). Sydney: University of New South Wales Press.

Gemeinboeck, P. (2021). The aesthetics of encounter: A relational-performative design approach to human-robot interaction. *Frontiers in Robotics and AI, 7*, 577900. https://doi.org/10.3389/frobt.2020.577900.

Giaccardi, E. (2020). Casting things as partners in design: Towards a more-than-human design practice. In H. Wiltse (Ed.), *Relating to Things: Design, Technology and the Artificial* (pp. 99–132). London: Bloomsbury.

Giaccardi, E., Speed, C., Cila, N., & Caldwell, M. L. (2016). Things as co-ethnographers: Implications of a thing perspective for design and anthropology. In R. C. Smith, K. T. Vangkilde, M. G. Kjærsgaard, T. Otto, J. Halse & T. Binder (Eds.), *Design Anthropological Futures* (pp. 235–248). London: Routledge.

Gibson, J. J. (1979). *The Ecological Approach to Visual Perception*. Boston, MA: Houghton, Mifflin and Company.

Goodman, N. (1978). *Ways of Worldmaking* (Vol. 51). Indianapolis, IN: Hackett Publishing.

Grudin, J. (1990). The computer reaches out: The historical continuity of interface design. *Proceedings of the SIGCHI Conference on Human Factors in Computing Systems Empowering People - CHI'90* (pp. 261–268). https://doi.org/10.1145/97243.97284.

Grudin, J. (2005). Three faces of human-computer interaction. *IEEE Annals of the History of Computing, 27*(4), 46–62. https://doi.org/10.1109/MAHC.2005.67.

Halverson, C. A. (2002). Activity theory and distributed cognition: Or what does CSCW need to do with theories? *Computer Supported Cooperative Work (CSCW), 11*(1–2), 243–267. https://doi.org/10.1023/A:1015298005381.

Haraway, D. J. (2016). *Staying with the Trouble: Making Kin in the Chthulucene*. Durham, NC: Duke University Press.

Harrison, S., Sengers, P., & Tatar, D. (2011). Making epistemological trouble: Third-paradigm HCI as successor science. *Interacting with Computers, 23*(5), 385–392. https://doi.org/10.1016/j.intcom.2011.03.005.

Harrison, S., Tatar, D., & Sengers, P. (2007). The three paradigms of HCI. *Alt. Chi. Session at the SIGCHI Conference on Human Factors in Computing Systems*, San Jose, CA (pp. 1–18).

Hassenzahl, M. (2010). Experience sesign: Technology for all the right reasons. *Synthesis Lectures on Human-Centered Informatics, 3*(1), 1–95. https://doi.org/10.2200/S00261ED1V01Y201003HCI008.

Hassenzahl, M. (2018). The thing and I: Understanding the relationship between user and product. In M. Blythe, & A. Monk (Eds.), *Funology 2* (pp. 301–313), Human-Computer Interaction Series. Cham: Springer. https://doi.org/10.1007/978-3-319-68213-6_19.

Hauser, S., Oogjes, D., Wakkary, R., & Verbeek, P.-P. (2018). An annotated portfolio on doing postphenomenology through research products. *Proceedings of the 2018 Designing Interactive Systems Conference* (pp. 459–471). https://doi.org/10.1145/3196709.3196745.

Heath, C., & Luff, P. (1991). Collaborative activity and technological design: Task coordination in London underground control rooms. In L. Bannon, M. Robinson, & K. Schmidt (Eds.), *Proceedings of the Second European Conference on Computer-Supported Cooperative Work ECSCW'91* (pp. 65–80). Springer Netherlands. https://doi.org/10.1007/978-94-011-3506-1_5.

Hollan, J., Hutchins, E., & Kirsh, D. (2000). Distributed cognition: Toward a new foundation for human-computer interaction research. *ACM Transactions on Computer-Human Interaction*, 7(2), 174–196. https://doi.org/10.1145/353485.353487.

Honold, P. (2000). Culture and context: An empirical study for the development of a framework for the elicitation of cultural influence in product usage. *International Journal of Human-Computer Interaction*, 12(3–4), 327–345. https://doi.org/10.1080/10447318.2000.9669062.

Hutchins, E. (1995). *Cognition in the Wild*. Cambridge, MA: MIT Press.

Ihde, D. (1990). *Technology and the Lifeworld: From Garden to Earth*. Bloomington, IN: Indiana University Press.

Jensen, M. M., & Aagaard, J. (2018). A postphenomenological method for HCI research. *Proceedings of the 30th Australian Conference on Computer-Human Interaction* (pp. 242–251). https://doi.org/10.1145/3292147.3292170.

Jeuris, S., Houben, S., & Bardram, J. (2014). Laevo: A temporal desktop interface for integrated knowledge work. *Proceedings of the 27th Annual ACM Symposium on User Interface Software and Technology* (pp. 679–688). https://doi.org/10.1145/2642918.2647391.

Jonassen, D. G., & Rohrer-Murphy, L. (1999). Activity theory as a framework for designing constructivist learning environments. *Educational Technology Research and Development*, 47(1), 61–79.

Kaptelinin, V., Nardi, B. A., & Macaulay, C. (1999). The activity checklist: a tool for representing the "space" of context. *interactions*, 6(4), 27–39. https://doi.org/10.1145/306412.306431

Kaptelinin, V. (2003). UMEA: Translating interaction histories into project contexts. *Proceedings of the 2003 CHI Conference on Human Factors in Computing Systems - CHI'03* (pp. 353–360). https://doi.org/10.1145/642611.642673.

Kaptelinin, V., & Nardi, B. (2012). Affordances in HCI: Toward a mediated action perspective. *Proceedings of the SIGCHI Conference on Human Factors in Computing Systems* (pp. 967–976). https://doi.org/10.1145/2207676.2208541.

Kaptelinin, V., & Nardi, B. A. (2006). *Acting with Technology: Activity Theory and Interaction Design*. Cambridge, MA: MIT Press.

Kaptelinin, V., Nardi, B., Bødker, S., Carroll, J., Hollan, J., Hutchins, E., & Winograd, T. (2003). Post-cognitivist HCI: Second-wave theories. *CHI'03 Extended Abstracts on Human Factors in Computing Systems - CHI'03*(p. 692). https://doi.org/10.1145/765891.765933.

Karana, E., Barati, B., & Giaccardi, E. (2020). Living artefacts: Conceptualizing livingness as a material quality in everyday artefacts. *International Journal of Design*, 14(3), 37–53.

Kozubaev, S., Elsden, C., Howell, N., Søndergaard, M. L. J., Merrill, N., Schulte, B., & Wong, R. Y. (2020). Expanding modes of reflection in design futuring. *Proceedings of the 2020 CHI Conference on Human Factors in Computing Systems* (pp. 1–15). https://doi.org/10.1145/3313831.3376526.

Kuijer, L., & Giaccardi, E. (2018). Co-performance: Conceptualizing the role of artificial agency in the design of everyday life. *Proceedings of the 2018 CHI Conference on Human Factors in Computing Systems* (pp. 1–13). https://doi.org/10.1145/3173574.3173699.

Kuutti, K. (1996). Activity theory as a potential framework for human-computer interaction research. In Context and consciousness: activity theory and human-computer interaction (pp. 17–44). Cambridge, MA: MIT Press.

Latour, B. (2007). *Reassembling the Social: An Introduction to Actor-Network-Theory*. Oxford: Oxford University Press.

Law, E. L.-C., & Sun, X. (2012). Evaluating user experience of adaptive digital educational games with activity theory. *International Journal of Human-Computer Studies*, 70(7), 478–497. https://doi.org/10.1016/j.ijhcs.2012.01.007.

Leontiev (Leont'ev), A. N. (1978). Activity, Consciousness, and Personality. Upper Saddle River, NJ: Prentice-Hall.

Leontiev (Leontyev), A. N. (1981). Problems of the Development of the Mind. Moscow: Progress Publishers.

Light, A., Powell, A., & Shklovski, I. (2017). Design for existential crisis in the anthropocene age. *Proceedings of the 8th International Conference on Communities and Technologies* (pp. 270–279). https://doi.org/10.1145/3083671.3083688.

McCarthy, J., & Wright, P. (2004). *Technology as Experience*. Cambridge, MA: MIT Press.

Mwanza-Simwami, D. (2011). AODM as a framework and model for characterising learner experiences with technology. *Journal of E-Learning and Knowledge Society*, 7(3), 75–85.

Nardi, B. A. (1993). A small matter of programming: Perspectives on end user computing. MIT Press.

Nardi, B. A. (Ed.). (1996). Context and consciousness: Activity theory and human computer interaction. MIT Press.

Nass, C., Steuer, J., & Tauber, E. R. (1994). Computers are social actors. *Proceedings of the 1994 CHI Conference on Human Factors in Computing Systems* (pp. 72–78). https://doi.org/10.1145/191666.191703.

Nichols, J. A. (1982). *Proceedings of the 1982 Conference on Human Factors in Computing Systems*, Gaithersburg, MD, ACM.

Norman, D. A. (1988). *The Psychology of Everyday Things*. New York: Basic Books.

Norman, D. A. (1991). Cognitive artifacts. In *Designing interaction: psychology at the human-computer interface* (pp. 17–38). Cambridge: Cambridge University Press.

Norman, D. A. (1998). *The Invisible Computer: Why Good Products Can Fail, the Personal Computer Is so Complex, and Information Appliances Are the Solution*. Cambridge, MA: MIT Press.

Norman, D. A. (2023). *Design for a Better World: Meaningful, Sustainable, Humanity Centered*. Cambridge, MA: MIT Press.

Norman, D. A., & Draper, S. W. (Eds.). (1986). *User Centered System Design: New Perspectives on Human-Computer Interaction*. Mahwah, NJ: Lawrence Erlbaum Associates.

Rabardel, P. (2003). From artefact to instrument. *Interacting with Computers*, 15(5), 641–645. https://doi.org/10.1016/S0953-5438(03)00056-0.

Redström, J., & Wiltse, H. (2018). *Changing Things: The Future of Objects in a Digital World*. London: Bloomsbury Publishing.

Roberts, T. L., & Moran, T. P. (1982). Evaluation of text editors. *Proceedings of the 1982 Conference on Human Factors in Computing Systems - CHI'82* (pp. 136–141). https://doi.org/10.1145/800049.801770.

Rogers, Y. (2012). *HCI Theory: Classical, Modern, and Contemporary*. San Rafael, CA: Morgan & Claypool.

Rosenberger, R. (2014). Multistability and the agency of mundane artifacts: From speed bumps to subway benches. *Human Studies*, 37(3), 369–392. https://doi.org/10.1007/s10746-014-9317-1.

Rosenberger, R., & Verbeek, P.-P. (2015). A field guide to postphenomenology. In P.-P. V. Rosenberger (Ed.), *Postphenomenological Investigations: Essays on Human-Technology Relations* (pp. 9–41). Lanham, MD: Lexington Books.

Sharp, H., Preece, J., & Rogers, Y. (2019). *Interaction Design, 5e*. Hoboken, NJ: John Wiley & Sons.

Shneiderman, B. (1983). Direct manipulation: A step beyond programming languages. *Computer*, 16(8), 57–69. https://doi.org/10.1109/MC.1983.1654471.

Stetsenko, A. (2017). *The Transformative Mind: Expanding Vygotsky's Approach to Development and Education*. Cambridge: Cambridge University Press.

Suchman, L. A. (1987). *Plans and Situated Actions: The Problem of Human-Machine Communication*. Cambridge: Cambridge University Press.

Svanæs, D. (2013). Interaction design for and with *the lived body*: Some implications of Merleau-ponty's phenomenology. *ACM Transactions on Computer-Human Interaction*, 20(1), 1–30. https://doi.org/10.1145/2442106.2442114.

Tsing, A. L. (2015). The mushroom at the end of the world. In *The Mushroom at the End of the World*. Princeton, NJ: Princeton University Press.

Verbeek, P. P. (2015). COVER STORY beyond interaction: A short introduction to mediation theory. *Interactions*, 22(3), 26–31. https://doi.org/10.1145/2751314.

Vermeulen, J., Luyten, K., van den Hoven, E., & Coninx, K. (2013). Crossing the bridge over Norman's Gulf of Execution: Revealing feedforward's true identity. *Proceedings of the SIGCHI Conference on Human Factors in Computing Systems* (pp. 1931–1940). https://doi.org/10.1145/2470654.2466255.

Vervoort, J. M., Bendor, R., Kelliher, A., Strik, O., & Helfgott, A. E. (2015). Scenarios and the art of worldmaking. *Futures*, 74, 62–70.

Voida, S., & Mynatt, E. D. (2009). It feels better than filing: Everyday work experiences in an activity-based computing system. *Proceedings of the SIGCHI Conference on Human Factors in Computing Systems* (pp. 259–268). https://doi.org/10.1145/1518701.1518744.

Vygotsky, L. S. (1981). *Mind in Society: The Development of Higher Psychological Processes*. Cambridge, MA: Harvard University Press.

Wakkary, R. (2021). *Things We Could Design: For More Than Human-Centered Worlds*. Cambridge, MA: MIT Press.

Wertsch, J. V. (1981). The Concept of activity in Soviet psychology: An Introduction. In The Concept of activity in Soviet psychology. M.E. Sharpe.

Winograd, T. (1987). A language/action perspective on the design of cooperative work. *Human-Computer Interaction*, 3(1), 3–30. https://doi.org/10.1207/s15327051hci0301_2.

Winograd, T., & Flores, F. (1986). *Understanding Computers and Cognition: A New Foundation for Design.* New York: Ablex Publishing Corporation.

8 Affect and Emotion

Johanna Löchner and Björn W. Schuller

8.1 INTRODUCTION

Human-computer interaction (HCI) design has evolved beyond functionality to encompass the realm of emotions and affect. This shift acknowledges the profound impact that emotional experiences have on user satisfaction and engagement. Likewise, various HCI design aspects are focusing on emotional design, cuteness engineering, and the broader influence of affect and emotions on User Experience (UX) (Saariluoma & Jokinen, 2014). Emotional design emphasises the creation of products that elicit specific emotional responses from users. This involves elements like colour schemes, visual aesthetics, sound design, and interactive animations. For instance, a warm colour palette and intuitive navigation can evoke feelings of comfort and trust, enhancing the overall user experience. In more advanced interaction settings, this can further include design of, for example, virtual agents along their voices and wordings, etc. In addition, further direction, such as "cuteness engineering", evolved (Marcus & Ma, 2016). The latter is an intriguing subset of emotional design that centres on incorporating adorable or endearing elements into interfaces. Research shows that elements like anthropomorphic characters or playful animations can induce positive emotions and foster user engagement. This approach has found widespread application in apps, games, and websites targeting a diverse user base.

In fact, considering and integrating affect and emotion in an interface's design can have substantial impact on user experience. This can include, e.g., (1) enhanced engagement and retention: when users form emotional connections with a product, they are more likely to stay engaged and return. Memorable experiences are more likely to lead to brand loyalty (Mostafa & Kasamani, 2021); (2) facilitating usability: emotions can influence cognitive processes, affecting how users perceive and interact with a system. A well-designed emotional interface can make complex tasks feel more manageable and enjoyable; (3) influence decision-making: emotions play a substantial role in decision-making as we will argue below. Affective design can guide users towards desired actions, such as making a purchase, subscribing, or sharing content. However, this requires consideration factors such as (1) cultural considerations: different cultures have distinct emotional cues and responses. HCI designers must be cognisant of cultural nuances to create interfaces that resonate with a global audience; (2) ethical considerations: while emotional design can enhance user experiences, it is crucial to consider the ethical implications. Manipulative designs that exploit emotions for ulterior motives can erode trust and harm user well-being.

Overall, incorporating affect and emotion into HCI design is a dynamic and evolving field with far-reaching implications for user experiences. Emotional design and cuteness engineering are just a few facets of this broader paradigm shift. By understanding the impact of emotions on UX, designers can create interfaces that not only function seamlessly but also resonate on a deeply human level, ultimately leading to more satisfying and memorable interactions. As HCI continues to evolve, it appears imperative that designers remain attuned to the emotional landscape of users, shaping experiences that leave a lasting positive impression.

The remainder of this chapter is structured as follows: We will first introduce the concepts of affect and emotion in Section 8.2. We will then briefly visit the field of affective computing including the automatic analysis, synthesis, and response of/to affect and emotion in HCI in Section 8.3. Finally, after naming current challenges and ways out, we will summarise this chapter's content in Section 8.3.

DOI: 10.1201/9781003495109-8

8.2 AFFECT AND EMOTION

Generally, the early detection of affects and emotions holds significant implications for an individual's overall well-being, influencing moment-to-moment performance, health, academic achievements, decision-making, and social relationships. Specifically, the degree of (un)pleasantness experienced during an event directly reflects a person's ability to cope with it, potentially challenging their well-being if coping strategies or emotion regulation skills are maladaptive. However, emotions are influenced by a multitude of factors, including subjective experiences, memories, and context, leading to various theories for capturing emotions (Scherer, 2009). Affects, emotions, and their concepts are diverse, making the objective, reliable, and valid assessment a complex task, hampered by limitations such as memory biases and social desirability, as well as assumptions about the non-dynamics of human change processes.

In the following, we provide a definition of affect and emotion as well as emotional competence and its implications for mental well-being and decision-making. Further, the assessment of emotions will be discussed. In addition, Figure 8.1 provides an overview on the concepts presented.

8.2.1 BASIC THEORIES AND MECHANISMS OF AFFECT RELEVANT TO HCI

Affect and emotions, though closely related, are distinct psychological constructs. Affect represents an individual's broad and enduring emotional disposition, encompassing the overall emotional tone or mood, and is often described in terms of valence (pleasantness or unpleasantness) and arousal (intensity) (Shiota, Sauter, & Desmet, 2021).

Affect can persist over longer periods, influencing an individual's responses to specific events and shaping their overall psychological experience. In contrast, emotions are brief, intense, and specific psychological states triggered by particular events or stimuli, characterised by cognitive, physiological, expressive, and subjective components, and labelled with distinct emotions such as joy or anger (Scherer, 2009).

Both affect and emotions are integral to our understanding of human emotional experiences and reactions. Furthermore, affect and emotions are intertwined and dynamic concepts and represent a significant area of focus in psychology research. There exists a multitude of theories and concepts

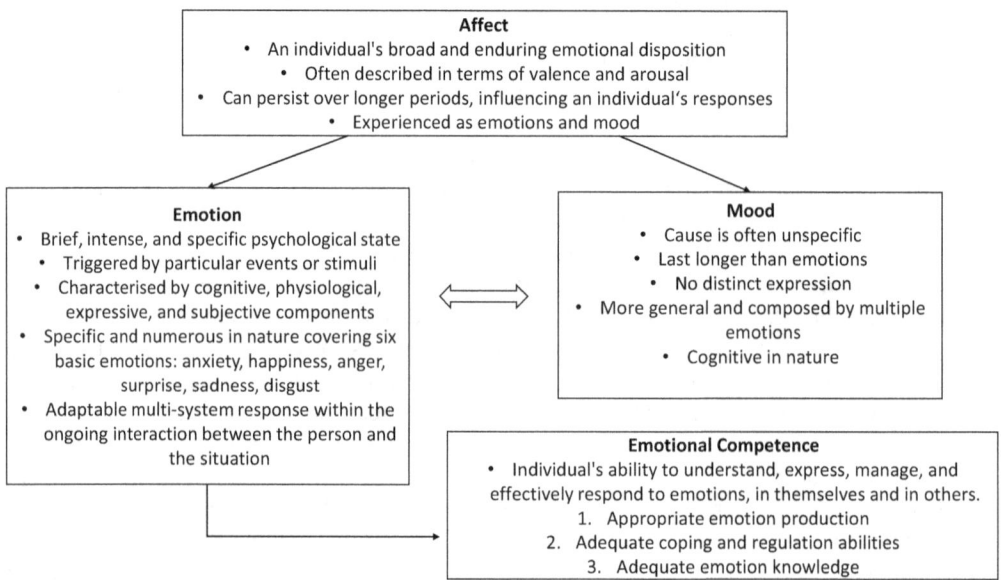

FIGURE 8.1 Affect and emotion – a breakdown.

that attempt to define especially emotions and emotional intelligence, reflecting the intricate interplay among various components (including sensory, cognitive, physiological, expressive, and motivational aspects) over time, that influences affect over time. Over the years, numerous emotion theories have been formulated, with many of them categorising emotions into three primary dimensions: valence, arousal, and dominance (Russel's Circumplex model: Russell, 1980). Further dimensions include novelty or intensity. Ekman distinguishes six basic emotions: happiness, sadness, anger, fear, surprise, and disgust (Ekman, 1992). However, many other classification schemes exist including social emotions or the regulation of emotions among others. Dimensions and classes can thereby be transferred; for example, anger is marked by high arousal, negative valence, and high dominance; fear would be marked accordingly, except for low dominance. Other emotions, such as surprise, would then require additional dimensions such as novelty.

The definition of emotions proposed by Gross and Thompson (2007) amalgamates several fundamental aspects of the emotional process, which can also be found in other emotion theories, like Scherer's Emotional Component Process Model (CPM) (Scherer, 2009). In essence, emotions, according to Gross and Thompson, can be described as a person's interaction with a situation that captures their attention, holds personal significance, and triggers a coordinated, yet adaptable multi-system response within the ongoing interaction between the person and the situation.

From an evolutionary perspective, emotions play a central role in controlling motivation, behaviour, and attention. They prompt us to act, direct our attention to specific stimuli with potentially pleasant or unpleasant consequences, and convey signals that direct our behaviour to obtain rewards or avoid adverse consequences (Petta & Trappl, 2001; Averill, 1982). In today's context, emotions are not only related to survival and primal incentives but also include secondary motivators such as money, status, entertainment, and others. In addition, emotions serve as regulators of the intensity and duration of various behaviours, promote the learning of actions that have been successful under certain circumstances (e.g., pleasure reinforces the repetition of behaviours), and are imprinted in our memory when actions have failed (e.g., through disgust or anger) (Averill, 1982). Similarly, emotions play a central role in guiding our social interactions, which are reinforced by the pleasant or unpleasant effects they elicit (Sanchez-Alvarez, Extremera, & Fernandez-Berrocal, 2016). These dynamics contribute to the formation of bonds and rivalries and provide a framework for us to navigate the social landscape. Some specific emotions may even have unique functions.

In our daily lives, we frequently encounter these emotions and react to them. How individuals respond to various emotional events can vary significantly. The conventional categorisation of emotions into "positive" and "negative" may be debated due to the multifaceted functions of emotions.

Similarly, Scherer's CPM elucidates the emotional process that guides an individual's perception and handling of both negative and positive life experiences (Scherer, 2009). Building upon this model, the Emotional Competence Model postulates that an individual's mental well-being and susceptibility to adverse psychopathology, such as anxiety and depression, significantly hinge on the effective functioning of their emotional processes. This effectiveness relies on several factors, including the individual's subjective emotional response, their perception of the situation, and their ability to appropriately appraise and regulate their emotions (Sanchez-Alvarez, Extremera, & Fernandez-Berrocal, 2016; Mehu & Scherer, 2015). However, these factors undergo substantial variations in the context of development and are contingent on an individual's cognitive and socio-emotional growth (Sanchez-Alvarez, Extremera, & Fernandez-Berrocal, 2016; Zeidner, Matthews, & Roberts, 2012; Saarni, 1984).

A relevant concept to gather influences of human intentions, behaviours, and human decision-making is mirrored in the concept of emotional competence (EC) or emotional intelligence. EC refers to an individual's ability to understand, express, manage, and effectively respond to emotions, both in themselves and in others. It involves a range of skills and capacities related to emotional awareness, regulation, and interpersonal interactions (Saarni, 1984). The Emotional Competence Process (ECP) model, developed by Scherer in 2007, is rooted in an empirically validated theoretical framework of emotion, the CPM (Scherer, 2007, 2001) which draws on extensive research regarding the mechanisms underlying both healthy and maladaptive emotional functioning (Mehu & Scherer, 2015). The ECP model identifies three essential components of EC that have significant implications for mental well-being and mental health: (1) appropriate emotion production: the skill of generating emotions appropriately by evaluating events impartially, including their causes and one's control over outcomes. Shortcomings, like an inappropriate self-image or negative coping assessments, can harm well-being and increase the risk of mental disorders; (2) adequate coping and regulation abilities: this aspect pertains to effective coping and emotion regulation skills, including strategies like reappraisal, constructive responses, and adhering to social norms in managing emotional arousal and expressions. Deficiencies in these skills increase the risk of mental health issues, with rumination, characterised by repetitive dwelling on problems, being particularly harmful; (3) adequate emotion knowledge (awareness, recognition, and understanding skills). This involves understanding personal and situational factors triggering emotions, accurately recognising emotions in oneself and others, and demonstrating empathy and perspective-taking. Deficits can lead to a lack of empathy, difficulty in sharing emotions, and poor socio-emotional skills during interactions. Emotion knowledge correlates with better social competencies, lower internalising and externalising problems in adolescents and adults, empathy, socio-emotional competence, relationship quality, cultural adjustment, and job performance. Training in emotional understanding can reduce problems and enhance social functioning, employability, and well-being.

8.2.2 INFLUENCING HUMAN INTENTIONS, BEHAVIOURS, AND DECISION-MAKING

EC plays a causal role in shaping both mental well-being and mental health. Substantial empirical evidence supports this proposition; studies have consistently demonstrated that a deficiency in emotional intelligence (EI) is correlated with lower levels of mental well-being, impaired (interpersonal) functioning, and an increased likelihood of mental disorders. Conversely, individuals with higher EI tend to exhibit greater resilience and enhanced well-being, as underscored by comprehensive meta-analyses (Sanchez-Alvarez, Extremera, & Fernandez-Berrocal, 2016; Hall, Frank, Holmes, Pfahringer, Reutemann, & Witten, 2009). Investigations by Zeidner, Matthews, and Roberts (2012) have emphasised the need to further develop ability models, particularly in the context of EC, in line with the evolution of emotional competencies outlined by Saarni (1984). Moreover, complementary meta-analyses have affirmed that deficits in EC, including challenges in effectively expressing and comprehending emotions, are linked to heightened anxiety (Mathews, Koehn, Abtahi, & Kerns, 2016) and externalising issues among youth (Trentacosta & Fine, 2010). It is worth noting that compromised emotional regulation skills have been associated with increased psychopathology (Sheppes, Suri, & Gross, 2015), with experimental studies illustrating how emotional regulation strategies can actively influence emotional outcomes (Webb, Miles, & Sheeran, 2012) and hence impair decision-making.

8.2.3 MEASURING AFFECT AND EMOTION IN USABILITY STUDIES

The assessment of an individual's EC and affect plays a pivotal role in decision-making when it comes to user interface design. In psychological studies and usability experiments relating to affect, emotions are typically measured using point estimates derived from questionnaire sum scores and compared between pre- and post-reference point measurement time points. Typically, affect and emotions themselves are not assessed, but rather the appraisal, regulation, or knowledge of emotions. Besides basic research, clinical psychology and psychiatric instruments were merely developed to assess, e.g., anhedonia, anxiety, aggression since such emotions and mood may indicate psychological dysfunction and hence, a psychiatric disorder. E.g., for ensuring a major depressive disorder, a low mood over the majority of time within the last two weeks is one of the core symptoms that must be given to fulfil the diagnostic criteria. Classic well-spread instruments are questionnaires as the Patient Health Questionnaire (PHQ-9) (Kroenke, Spitzer, & Williams, 2001),

Beck's Depression Inventory (BDI) (Beck, Steer, & Brown, 1996) or for assessing well-being with the World Health Organization – Five Well-Being Index (WHO-5) (Topp, Østergaard, Søndergaard, & Bech, 2015). For further understanding of psychological functioning, emotion regulation strategies may be assessed (e.g., with the Difficulties in Emotion Regulation Scale, DERS) (Kaufman, Xia, Fosco, Yaptangco, Skidmore, & Crowell, 2016), or knowledge of emotions by the Geneva Emotion Recognition Test (GERT) (Schlegel, Grandjean, & Scherer, 2014).

However, this approach is beset by several limitations: (1) the incapacity of standardised measures to capture the idiosyncratic nature of individual changes, (2) measurement biases inherent in self-reports, the presence of ceiling and floor effects, and insensitivity to changes at the extremes of the measurement scale, (3) the oversight of dynamic properties in pre- and post-reference point comparisons, and (4) the nonergodic nature of change trajectories (Schiepek, de Felice, Desmet, Aichhorn, & Sammet, 2022; Blome & Augustin, 2015).

Accordingly, the primary approach for assessing outcomes in psychological research and usability studies predominantly relies on the use of standardised questionnaires. Nonetheless, responses collected through these questionnaires only capture a fraction of each individual's unique experiences and can occasionally be quite misleading, i.e., (Desmet, Van Nieuwenhove, De Smet, Meganck, Deeren, Van Huele, et al., 2021) revealed low- to moderate agreement between quanti- and qualitative pre-post measures on Inventory of Interpersonal Problems score, that was due to personal withdrawal from intrapersonal conflicts, that were not captured by the quantitative questionnaire.

Further, assessments of emotions and affect through self-report methods may be susceptible to various forms of bias (Blome & Augustin, 2015). Typically, in questionnaires, individuals retrospectively report on a period preceding the assessment. These recollections over time are prone to biases and gaps in memory experiences, where a discordance exists between the actual experienced states and the subjective evaluation of those experiences (Miron-Shatz, Stone, & Kahneman, 2009). Such deviations can arise, i.e., from an overemphasis on extreme experiences during the recall period and experiences proximate to the assessment time (the peak-end rule), a tendency to rely more on memorable instances than less noticeable ones (the salience memory heuristic), and the influence of personal theories and beliefs shaping the recollection of an event or experience (Miron-Shatz, Stone, & Kahneman, 2009; Stone, Broderick, Shiffman, & Schwartz, 2004).

Also, when assessing emotions and affect, it is further crucial to acknowledge that the current emotional state at the time of assessment can exert an influence on the reconstruction and evaluation of one's overall present experienced state (Meyer, Richter, & Raspe, 2013). Cognition, behaviours, emotions, and moods exhibit dynamic and unstable traits, following diverse patterns over time. Indeed, inflexibility and non-dynamic patterns of these parameters are characteristic of psychopathology (Kashdan & Rottenberg, 2010; Friedman, 2007; Rottenberg, 2005) while mental well-being is characterised by a stable, rhythmic, chaotic, and complex synchronisation among cognition, behaviour, emotions, and moods (Kotsou, Nelis, Gregoire, & Mikolajczak, 2011; Schiepek, Viol, Aas, Kastinger, Kronbichler, Schöller, et al., 2021).

Finally, research of emotions and affect in general and usability studies in more particular aim to draw conclusions and generalisations from a study population to the larger population. This presupposes two conditions: (1) the transfer of the statistical model of an individual to the population as a whole or the existence of a consistent statistical model for all individuals (population homogeneity) and (2) the constancy of statistical temporal parameters over time for all (psychological) change processes (Schiepek, Viol, Aas, Kastinger, Kronbichler, Schöller, et al., 2021). When both conditions are met, ergodicity can be assumed, implying stability and uniform behaviour across the entire population. A stochastic process is considered ergodic if any sample collection accurately represents the overall statistical properties of the entire process. Conversely, a stochastic process becomes nonergodic when its statistics change over time and due to individual responses. Behavioural changes are characterised by non-linearity, transitional phases, and highly individualised change processes. Consequently, the assumption of ergodicity is typically not fulfilled (Schiepek, Viol, Aas, Kastinger, Kronbichler, Schöller, et al., 2021).

These findings underscore the need for evaluation procedures that are attuned to the nuances of individual changes and the intricate nature of outcome measures.

The adoption of ecological momentary assessment (EMA) (synonyms: ambulatory assessment, experience sampling method, and real-time data capture) as an alternative approach for collecting intensive longitudinal data within the natural settings of individuals, facilitates a more precise investigation of individual experiences and states, as well as their interactions with external situational factors. This approach increasingly offers the opportunity to capture the dynamic variability of emotions with a high degree of ecological validity and minimal recall bias of individuals in their natural habitat (Trull & Ebner-Priemer, 2013; Bolger & Laurenceau, 2013). Thereby increasing both ecological validity and generalisability. Aiming to balance participant compliance and the amount of collected data, EMA designs vary greatly regarding their frequency, fixed or flexible response, number of items assessed, and duration. Balancing the EMA interval is key to ensuring participants' adherence (Wrzus & Neubauer, 2023). Meta-analytic results revealed higher compliance rates in studies offering monetary incentives compared with other or no incentives (Wrzus & Neubauer, 2023; Ottenstein & Werner, 2022). The decision on the EMA design as well as the differentiation between event-based and time-based designs should be based on the respective research rationale. Particularly for the assessment of everyday experienced emotions, the Positive and Negative Affect Schedule (PANAS) (McAllister, Vincent, Hassett, Whipple, Oh, Benzo, & Toussaint, 2015) or the multidimensional mood questionnaire (Steyer, Schwenkmezger, Notz, & Eid, 1994) provide a selection of items that can be adapted.

Despite the promises to overcome the before mentioned biases and assumptions, specific guidelines how EMA should be performed to be a reliable and valid measure are still outstanding. In consequence, EMA studies often show high drop-out rates in case of too intense designs or are biased by other factors as unreliable items, unreliable response patterns (e.g., in case a participant just clicks through quickly). In addition, self-reports of emotions depend purely rely on introspection and may lack vital information (Montag & Baumeister, 2019), independent of the nature of the assessment method. Therefore, more 'objective' data on the emotional expression is aimed to gather by capturing the emotional expression component: e.g., physical reaction as skin conductance, heart rate variability, facial expression, or linguistic and paralinguistic parameters, assessed by mobile sensing data collection (Seiferth, Vogel, Aas, Brandhorst, Carlbring, Conzelmann, et al., 2023). This will be discussed next.

8.3   AFFECTIVE COMPUTING

In our rapidly advancing technological landscape, understanding and responding to human emotions has become a pivotal frontier – including in HCI as outlined above. Affective computing, a multidisciplinary field at the intersection of computer science, psychology, and neuroscience, and further supporting disciplines such as linguistics seeks to imbue machines with the ability to interpret, respond to, and even simulate human emotions (Calvo, D'Mello, Gratch, & Kappas, 2015). Affective Computing holds immense potential to revolutionise how we interact with and benefit from technology. The term itself was coined in the 1990s with first patents appearing already in the 1970s and an increasing amount of papers appearing already in the 1980s. Most broadly speaking, affective computing refers to the integration of EI into computing systems. It hence involves the development of algorithms and technologies capable of recognising, interpreting, and responding to human emotions, as well as expressing emotions in a manner that is comprehensible to humans.

8.3.1   Automatic Recognition

Looking at the key components of affective computing, emotion recognition first encompasses the ability of machines to identify and interpret vocal tones, spoken or written words, facial expressions, gestures, physiological signals like heart rate and skin conductance, or touch and interaction data.

Advanced machine learning techniques play a critical role in training algorithms to recognise these emotional cues. While the field has been an early adopter of deep learning, in particular the advent of "end-to-end learning" made the design of automatic emotion recognisers increasingly a matter of designing a suited neural network architecture rather than designing features for the specific modality of analysis (Tzirakis, Zafeiriou, & Schuller, 2018). A further recent breakthrough came with the rise of transformer architectures and self-supervised learning (Wagner, Triantafyllopoulos, Wierstorf, Schmitt, Eyben, Schuller, & Burkhardt, 2023). Ultimately, this led to the advent of large 'foundation' models which seemingly show emergent 'zero-shot learning' abilities out-of-the-box for affective computing tasks (Amin, Cambria, & Schuller, 2023). Nonetheless, we quickly guide through the different input modalities as follows:

Speech(-based) Emotion Recognition is among the most popular modalities. Speech carries a wealth of emotional information through prosody, i.e., intensity, intonation, and rhythm, as well as voice quality aspects and articulatory information (Lee, Chaspari, Provost, & Narayanan, 2023). Advanced machine learning models, including deep neural networks and most recently transformers, have significantly improved the accuracy of emotion recognition from speech (Wagner, Triantafyllopoulos, Wierstorf, Schmitt, Eyben, Schuller, & Burkhardt, 2023). However, challenges persist in handling variations across languages and dialects.

Natural Language Processing (NLP) techniques further enable the analysis of written or spoken text after automatic speech recognition for emotional content. Sentiment analysis, emotion detection, and contextual understanding contribute to interpreting the emotional state of a user from their language. Traditionally, bag-of-word sparse representations of text served as representation for decision-making, but nowadays, pre-trained large language models are forming the state-of-the-art – usually fine-tuned on the target task (Amin, Cambria, Schuller, 2023).

Facial expressions are another cornerstone of human emotion communication. Computer vision algorithms employ techniques like deep learning to track the face, detect and classify facial expressions. For example, based on facial action units (Yang, Hristov, Shen, Lin, & Pantic, 2023) or based on self-learnt representations, providing a further real-time window into the user's emotional state. The lion's share of works uses 2D cameras but also 3D recordings could be successfully employed. In addition to RGB cameras, also thermal imagery can be used to additionally gain insight into blood flow (Filippini, Perpetuini, Cardone, Chiarelli, & Merla, 2020). Further, potentially higher resolution in time can be exploited for the analysis of facial micro-expressions (Zhao, Li, Li, & Pietikäinen, 2023) – expressions only lasting for a few frames, but credited to reveal the genuine emotion.

Recognising emotions through body language is another vision-based option and involves the analysis of gestures, posture, and movement patterns such as gait. Advanced sensors including depth sensors and machine learning models have facilitated significant progress in this domain, enhancing our understanding of non-verbal emotional cues.

Measuring physiological responses, such as skin conductance, heart rate variability, and respiratory patterns, provides valuable insights into emotional states. Affective computing algorithms can process this data to infer emotions, offering a deeper understanding of the user's affective state (Can, Mahesh, & Andre, 2023).

BCIs establish a direct communication channel between the brain and external devices. Through techniques like Electroencephalography (EEG) and functional Magnetic Resonance Imaging (fMRI), models can decode neural activity associated with emotions (Wu, Lu, Hu, & Zeng, 2023). This modality also holds promise for individuals with limited or impaired speech or facial expression capabilities.

Tactile interfaces introduce a dynamic element to emotion recognition, allowing users to interact physically with devices. By analysing touch movement, pressure, temperature, and vibration patterns, systems can infer emotions related to touch and haptic feedback (Olugbade, He, Maiolino, Heylen, & Bianchi-Berthouze, 2023).

Effective emotion recognition often involves combining information from multiple modalities. Fusion techniques, including early fusion (concatenation of features), late fusion (combining classification outputs), and decision-level fusion (weighted integration of classifiers), play a critical role in achieving comprehensive emotional understanding. Likewise, many combinations of modalities are seen in practical usage, with voice acoustics, (spoken) language analysis, and facial expression analysis being among the most popular ones in multimodal fusion.

8.3.2 Automatic Synthesis

The ability to imbue virtual and embodied agents and tactile interfaces with emotionally expressive capabilities represents a significant leap in human-computer interaction. Likewise, we next explore the techniques and methodologies for automatically synthesising emotions across various modalities, including similarly to the analysis speech, language, facial expressions, body posture, and tactile interaction. Physiological data synthesis might be of interest as well, but so far practically not found in HCI applications, but rather for data augmentation in deep learning approaches. By enabling interfaces to emote, one paves the way for potentially more empathetic, responsive, and engaging interactions.

Many of the technical solutions of the past were based on rules, e.g., for emotional speech synthesis (Burkhardt, Reichel, Eyben, & Schuller, 2023) or facial expression generation. However, the rapid progress in generative artificial intelligence (GenAI) recently brought deep learning approaches to the fore, including such as variational autoencoders, generative adversarial networks, or more recently diffusion models. Similar to emergent behaviour found in analysis, one can exploit large 'foundation' models out-of-the-box apparently also for the generation of affective output. While systematic research including human perception evaluation of the generated output is largely absent, an impression of the quality can be seen in Table 8.1 for the generation of affective text directly by the large language model ChatGPT 3: without any specific training or finetuning, the model was prompted to generate the text "I came home" in Ekman's big six emotions. Similarly, Figure 8.2 shows examples of zero-shot generation of emotional faces in these emotions. On purpose, the top row shows 3D renders that could be used in virtual agents, whereas the bottom row shows the corresponding emotion rendered as photography of (non-existent) persons' faces. Obviously, the same avatar or person could have been rendered in different emotions including in different viewing angles or context, making it nowadays very easy to directly produce multimodal content in various target emotions. Other foundation models such as UniAudio (Yang, Tian, Tan, Huang, Liu, Chang, et al., 2023) would allow to do so in audio for speech or music and potentially also further modalities or their combination.

Nonetheless, let us quickly go through the different modalities in synthesis as follows.

Generating emotionally expressive speech involves infusing synthetic voices with appropriate prosody, intonation, and pacing. Text-to-speech (TTS) systems equipped with emotional models can

---

TABLE 8.1

ChatGPT Prompted to "Express in One Short Sentence 'I Came Home' in a [Emotion] Way"

Emotion	Exemplary Response of ChatGPT
Angry	I stormed back into that damned place they call home
Fearful	I cautiously stepped back into the shadowed confines of my so-called sanctuary
Disgusted	I begrudgingly entered that wretched place I'm supposed to call home
Happy	I joyfully returned to the welcoming embrace of my cherished abode
Surprised	I stumbled through the door, startled to find myself back in the familiar confines of my home
Sad	I entered the empty house, the familiar surroundings now suffused with a melancholy air

https://community.openai.com/t/ai-emotions-and-grandiose-claims/15680/1

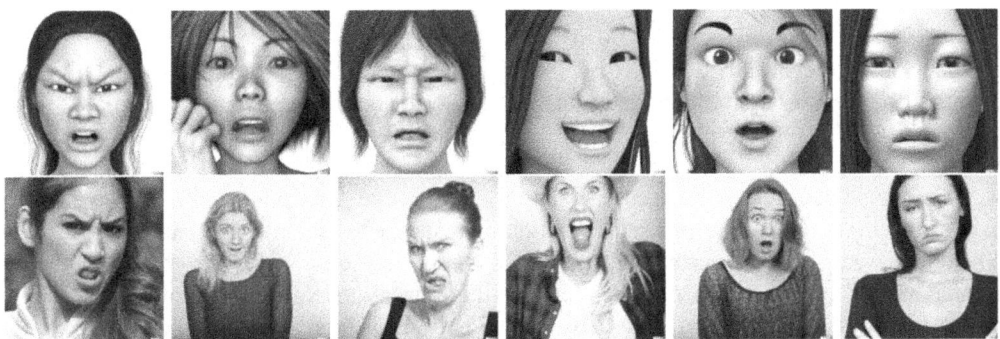

FIGURE 8.2 DALL-E 2 prompted for a "3D render of a [emotion] woman's face with angry, fearful, disgusted, happy, surprised, and sad as emotion (top row). Bottom row: same, but prompted for "[emotion] face" to obtain a photography-style image.

https://dallery.gallery/dall-e-prompts-photography-styles/

dynamically adjust these parameters to convey a range of emotions, from joy and excitement to sadness and concern including subtle nuances and 'mixtures' of emotions such as surprise with a certain percentage of anger or happiness (Zhou, Sisman, Rana, Schuller, & Li, 2022). Further, speech can also be converted in such a manner – usually using deep learning these days (Triantafyllopoulos, Schuller, Iymen, Sezgin, He, Yang, et al., 2023).

Natural Language Generation (NLG) techniques next enable virtual agents or robots to produce text imbued with emotional nuances. By incorporating sentiment analysis and affective computing models, these systems can tailor their responses to match the user's emotional state or desired emotional tone, e.g., to appear empathetic or charismatic (Schuller, Amiriparian, Batliner, Gebhard, Gerzcuk, Karas, et al., 2023).

Creating realistic facial expressions in virtual agents requires sophisticated animation and modelling techniques. Emotion synthesis algorithms partially also leverage facial landmarks and muscle movement simulations to generate expressions that accurately convey emotions, allowing virtual agents to emote in a lifelike manner. Mostly, however, generation is learnt end-to-end without explicit models (de Melo, Gratch, Marsella, & Pelachaud, 2023).

Animating a virtual agent's or robot's body language involves generating movements, postures, and gestures that align with the intended emotional expression. Motion capture technology and procedural animation algorithms enable virtual agents and robots to learn to dynamically adjust their body language to reflect changing emotional states.

Tactile interfaces provide a unique opportunity to convey emotions also through physical interaction. By modulating haptic feedback parameters, such as intensity, frequency, and texture, interfaces can simulate sensations associated with specific emotions, enhancing the user's sensory experience. Also, special crafted hardware devices such as gloves or body-worn suites and smart-clothes equipped with emotional feedback response could be realised successfully (Weda, Mader, van Schaik, Meijer, Kolesnyk, & van Erp, 2023).

Similar to the analysis of emotion, synthesis of it can benefit from a multimodal approach. Integrating emotion synthesis across multiple modalities enables virtual agents to deliver more comprehensive and immersive emotional experiences. Synchronised emotional expression in speech, language, facial expressions, body language, and potentially even tactile feedback creates a cohesive and compelling interaction.

Adapting emotion synthesis to individual users and situational contexts further enhances the authenticity and effectiveness of virtual agent interactions. As shown above, machine learning models can be trained to recognise user-specific emotional cues, and they can accordingly also be trained to tailor the agent's responses accordingly.

8.3.3 Automatic Affective Response

From the above, it seems obvious that machines equipped with affective computing capabilities can adapt their behaviour based on the detected emotions of the user. For instance, a virtual assistant could modify its responses to be more empathetic or supportive when it senses a user is upset or frustrated.

Automatic affective response generation is a cutting-edge field within HCI that focuses on enabling machines to generate emotional reactions in response to user input or contextual cues. This technology leverages sophisticated algorithms and models trained on extensive emotional datasets to imbue virtual agents, robots, or interfaces with the capacity to express empathy, understanding, and even tailored emotional responses. By dynamically adjusting parameters related to speech, language, facial expressions, body language, and even tactile feedback, these systems can convey a wide spectrum of emotions, fostering more engaging and relatable interactions. This capability holds immense promise for applications ranging from virtual assistants providing emotional support to gaming or learning environments that adapt to a player's or student's emotional state (Yannakakis & Melhart, 2023), ultimately enhancing the overall user experience in a deeply personal and meaningful way.

8.4 CURRENT CHALLENGES AND WAYS OUT

While affective computing holds immense promise for revolutionising HCI, it is crucial to acknowledge and address the associated risks and persistent challenges. This section provides a comprehensive examination of the potential pitfalls and ongoing hurdles in the field, emphasising the importance of responsible development and implementation.

8.4.1 Learning from Humans – The Glass Ceiling?

First, at present and likely for oncoming years, Affective Computings systems are learning from humans posing a glass ceiling in obtainable performance. This comes, as humans are usually labelling the data the systems are trained with. While this is usually some (weighted) average opinion from some three to ten or even more human opinions – potentially from experts such as psychologists – it is still subjective in nature. In fact, agreement levels are often only moderate among other humans. As also outlined above, even self-assessment has its limitations, as one may not perfectly remember one's emotion and is not perfectly sure about it. This is added by human annotation mistakes, as some of the models of emotion are continuous in value as the dimensions and potentially also over time. The latter means that values are provided, e.g., on an every 10 ms rate by moving a joystick or alike while observing material to be annotated. Simple models on the other hand such as labelling a whole sequence with labels chosen from few classes only come at the risk of over-simplification, such as the emotion changing within the sequence or the number of classes being insufficient.

The way out would be having the 'ground truth' – the objective actual emotion – as a learning target. This way, affective computing systems could surpass the performance of humans and potentially also recognise the actual emotion rather than the one as judged by other humans. This would, however, require deeper insight into the human brain than current technology allows for, let alone accessible and measureable in real life.

8.4.2 The Data Bottleneck

Coming in line with the immediately above-mentioned uncertainty burden of affective labels is the ever-present bottleneck of limited data availability. While modern large models such as open AI's Whisper are trained with massive data – in Whipser's case roughly 70 years of consecutive speech, and by that more than a human could possibly experience in their life time, speech emotion recognition systems are barely ever trained with several hours. Taking most available databases together still only leads to barely a 100 hours (Gerczuk, Amiriparian, Ottl, & Schuller, 2023). This is similarly true for other modalities and comes, as labelling of data is expensive as several human opinions are usually asked for and sometimes the material has to be sighted multiple times to cater for multiple dimensions – see the last section. In addition, data is often acted or 'elicited' and recorded in lab conditions, which makes it easier to collect emotions rare in real life in sufficient quantity. However, this results often in overly prototypical emotions that do not allow training of systems ready for subtle nuances. Likewise, the amount of 'high quality' real-life samples is only a fraction of today's available emotion data. Rare emotions in these databases include in particular such as social emotions or regulated and faked emotions.

8.4.3 EVALUATION OF ACTUAL APPLICATION

Only very few studies evaluate the actual usage of affective computing systems let alone in real-world application. Rather, most evaluation are performance experiments run in batch mode offline on databases as the ones described in the last section: potentially over-typical and with biased distribution of real-life affect occurrence. Yet, assessment of actual performance in real-life usage would be required, ideally measured by some objective measure, such as performance of a user interface or service that employs affective computing technology.

8.4.4 ROBUSTNESS AND ACCURACY

Affective computing systems heavily rely on accurate emotion recognition and synthesis. However, challenges persist in achieving high levels of accuracy, especially in noisy or ambiguous contexts. Robustness to factors like varying lighting conditions, accents, and facial expressions remains a significant hurdle. A number of additional issues limit the robustness and obtainable accuracy such as the named data bottleneck. For example, emotions are expressed and interpreted differently across cultures, posing a significant challenge for affective computing systems aimed at a global audience. Designing algorithms that can adapt to diverse cultural contexts without bias or misinterpretation is a complex undertaking. Interpreting emotions is also inherently subjective and context-dependent. Affective computing systems must hence grapple with the intricacies of context to provide accurate and meaningful emotional responses. Striking the right balance between context sensitivity and adaptability is a persistent challenge. Despite advances in machine learning and computational power, affective computing still also faces technical limitations, particularly in processing and analysing complex emotional cues. Developing more sophisticated algorithms and leveraging emerging technologies like edge computing will be crucial in overcoming these barriers. At the same time, algorithms that can run efficiently directly on private devices – potentially with low computing power – is required to increase privacy. Similarly, personalisation with only a few examples is needed. Ideally, affective computing systems should have the capability to adapt and learn from user feedback and evolving emotional cues. Implementing effective feedback loops and continual learning mechanisms is a crucial step in creating truly responsive and adaptive systems. Overall, reinforcement learning from real-world deployed affective computing solutions at scale could be the gamechanger to raise robustness and accuracy. In such settings, also approaches to analysis and synthesis could be coupled more tightly to better mutually profit from each other. At present, the two directions are mostly treated independently. However, it has been repeatedly demonstrated that augmentation of training data by synthesis. On the other end, automatic analysis of synthesised affective data can provide feedback on its generation. Generative adversarial models are a first step into this direction, and recent large 'foundation' models in fact often couple these two sides of the same coin more effectively.

8.4.5 AI-RELATED CHALLENGES

A number of challenges in affective computing systems are inherited from the fact that they are based on AI. Modern AI, in particular deep learning, is known for a number of issues, including limited explainability or limited fairness coming amongst others from unevenly distributed training data, which is particularly severe in this field, as data of some age groups as different developmental stages and such of elderly is barely present. Furthermore, sustainability is limited, as there is mostly no centralised archiving of models or often even data used in training. Further concerns include privacy concerns, the consumption of large amounts of energy during training and inference contributing to the ecological crises, and unreliable results lying in the nature of statistical machine learning. The latter is particularly challenging in the case of affective computing given the uncertainty in the learning target as described above. Another challenge in today's AI systems can be safety: AI can increasingly attack AI, such as in adversarial attacks. In fact, it could be shown that affective computing can also be targeted in such ways, e.g., purposefully altering the emotion an AI recognises in data which is changed in ways unnoticeable to humans (Wu, Xu, Fang, Zhang, Yang, Xu, et al., 2023).

Building trust between humans and affective computing systems is, however, paramount for widespread acceptance and adoption – one's emotion is particularly private and sensitive in nature requiring raised trust in comparison to some other HCI-related AI tasks. Ensuring that users understand how emotional data is used and providing transparent explanations for system decisions are hence even more vital steps in fostering trust.

8.4.6 ETHICAL CONSIDERATIONS

A multitude of ethical concerns have been raised in the lifespan of affective computing (Devillers & Cowie, 2023). One of the primary concerns lies in the ethical use of emotional data. There is a risk of infringing on user privacy and autonomy, particularly when collecting and analysing sensitive emotional information. Ensuring transparent data practices and obtaining informed consent are critical safeguards. While emotion synthesis offers tremendous potential for enhancing user experience, it is similarly important to consider ethical implications. Avoiding emotional manipulation are critical aspects of responsible design in particular when providing emotion converting systems.

In fact, affective computing systems, if not carefully designed, have the potential to manipulate users' emotions for various purposes, including marketing or political agendas. Striking a balance between providing personalised experiences and avoiding manipulative practices is therefore a most critical challenge.

8.5 CONCLUSION

In this chapter, we introduced affect and emotion for the usage in human-computer interaction. After arguing how HCI can benefit from artificial EI, we introduced the concepts of affect and emotion from a psychological perspective through the looking glasses of HCI. We further discussed how affect and emotions can influence human intentions, behaviours, and decision-making from the same perspective. We then reviewed how affect and emotion can be measured in usability studies in an active manner, i.e., with the cooperation of users. We further introduced the comparably younger field of affective computing that provides technical solutions and mechanisms to recognise, generate, and respond (to) affect and emotions. While we introduced each of these three aspects for a range of modalities such as audio, video, physiology, or tactile interaction, we argued in particular that a new era came with the advent of large 'foundation' models: general-purpose trained models that show emerging capabilities of affect interpretation and generation. This makes it potentially easier than ever to integrate affective computing into today's and tomorrow's user interfaces. However, numerous challenges persist, which were outlined including points of action to ease on these.

Already, the progress made in collecting emotional data, deep learning, and the integration of multiple modalities in emotion recognition and generation marks a significant advancement in HCI. By leveraging speech, language, facial expressions, body language, physiological data, BCIs, and tactile interaction, we can create systems that perceive and respond to human emotions with unprecedented depth and accuracy. As these technologies continue to advance, and further modalities such as olfactory or gustatory interfaces emerge, we stand on the brink of a new era in human-computer interaction, where our devices and systems are not only intelligent but also emotionally aware and attuned to our individual needs and states of being. Likewise, we approach a future where our interactions with technology are not just intelligent but also more and more deeply human.

Yet, as affective computing continues to advance, it is imperative that we confront and mitigate the associated risks and challenges. Responsible development practices, transparent design principles, and ongoing research are essential in ensuring that affective computing systems enhance HCI without compromising privacy, autonomy, or trust. By acknowledging these risks and striving for responsible innovation, we can harness the full potential of affective computing in a way that benefits society as a whole.

REFERENCES

Amin, M. M., Cambria, E., & Schuller, B. W. (2023). Will affective computing emerge from foundation models and general AI? A first evaluation on ChatGPT. *IEEE Intelligent Systems Magazine*, 38(2) 15–23.

Averill, J. R. (1982). *Anger and Aggression*. New York: Springer.

Beck, A. T., Steer, R. A., & Brown, G. K. (1996). *Beck Depression Inventory-II (BDI-II)*. Agra: Psychological Corporation.

Blome, C., & Augustin, M. (2015). Measuring change in quality of life: Bias in prospective and retrospective evaluation. *Value in Health*, 18(1), 110–115.

Bolger, N., & Laurenceau, J.-P. (2013). *Intensive Longitudinal Methods: An Introduction to Diary and Experience Sampling Research*. New York: Guilford Press.

Burkhardt, F., Reichel, U., Eyben, F., & Schuller, B. (2023). Going retro: Astonishingly simple yet effective rule-based prosody modelling for speech synthesis simulating emotion dimensions. arXiv preprint arXiv:2307.02132.

Calvo, R. A., D'Mello, S., Gratch, J. M., & Kappas, A. (2015). *The Oxford Handbook of Affective Computing*. Oxford: Oxford Library of Psychology.

Can, Y. S., Mahesh, B., & Andre, E. (2023). Approaches, applications, and challenges in physiological emotion recognition: A tutorial overview. *Proceedings of the IEEE*, 111(10), 1287–1313.

de Melo, C. M., Gratch, J., Marsella, S., & Pelachaud, C. (2023). Social functions of machine emotional expressions. *Proceedings of the IEEE*, 111(10), 1382–1397.

Desmet, M., Van Nieuwenhove, K., De Smet, M., Meganck, R., Deeren, B., Van Huele, I., Decock, E., Raemdonck, E., Cornelis, S., Truijens, F., et al. (2021). What too strict a method obscures about the validity of outcome measures. *Psychotherapy Research*, 31(7), 882–894.

Devillers, L., & Cowie, R. (2023). Ethical considerations on affective computing: An overview. *Proceedings of the IEEE*, 111(10), 1445–1458.

Ekman, P. (1992). An argument for basic emotions. *Cognition & Emotion*, 6, 169–200.

Filippini, C., Perpetuini, D., Cardone, D., Chiarelli, A. M., & Merla, A. (2020). Thermal infrared imaging-based affective computing and its application to facilitate human robot interaction: A review. *Applied Sciences*, 10(8), 2924.

Friedman, B. H. (2007). An autonomic flexibility-neurovisceral integration model of anxiety and cardiac vagal tone. *Biological Psychology*, 74, 185–199.

Gerczuk, M., Amiriparian, S., Ottl, S., & Schuller, B. (2023). EmoNet: A transfer learning framework for multi-corpus speech emotion recognition. *IEEE Transactions on Affective Computing*, 14(2), 1472–1487.

Gross, J. J., & Thompson, R. A. (2007). Emotion regulation: Conceptual foundations. In J. J. Gross, (Ed.), *Handbook of Emotion Regulation* (pp. 3–27). New York: The Guilford Press.

Hall, M., Frank, E., Holmes, G., Pfahringer, B., Reutemann, P., & Witten, I. H. (2009). The Weka data mining software: An update. *ACM SIGKDD Explorations Newsletter*, 11, 10–18.

Kashdan, T. B., & Rottenberg, J. (2010). Psychological flexibility as a fundamental aspect of health. *Clinical Psychology Review*, 30, 865–878.

Kaufman, E. A., Xia, M., Fosco, G., Yaptangco, M., Skidmore, C. R., & Crowell, S. E. (2016). The difficulties in emotion regulation scale short form (DERS-SF): Validation and replication in adolescent and adult samples. *Journal of Psychopathology and Behavioral Assessment*, 38, 443–455.

Kotsou, I., Nelis, D., Gregoire, J., & Mikolajczak, M. (2011). Emotional plasticity: Conditions and effects of improving emotional competence in adulthood. *Journal of Applied Psychology*, 96, 827–839.

Kroenke, K., Spitzer, R. L., & Williams, J. B. (2001). The PHQ-9: Validity of a brief depression severity measure. *Journal of General Internal Medicine*, 16, 606–613.

Lee, C.-C., Chaspari, T., Provost, E. M., & Narayanan, S. S. (2023). An engineering view on emotions and speech: From analysis and predictive models to responsible human-centered applications. *Proceedings of the IEEE*, 111(10), 1142–1158.

Marcus, A., & Ma, X. (2016). Cuteness design in the UX: An initial analysis. In: *Design, User Experience, and Usability: Novel User Experiences: 5th International Conference, DUXU 2016, Held as Part of HCI International 2016*, Toronto, Canada, July 17–22, Proceedings, Part II, Springer (pp. 46–56).

Mathews, B. L., Koehn, A. J., Abtahi, M. M., & Kerns, K. A. (2016). Emotional competence and anxiety in childhood and adolescence: A meta-analytic review. *Clinical Child and Family Psychology Review*, 19, 162–184.

McAllister, S. J., Vincent, A., Hassett, A. L., Whipple, M. O., Oh, T. H., Benzo, R. P., & Toussaint, L. L. (2015). Psychological resilience, affective mechanisms and symptom burden in a tertiary-care sample of patients with fibromyalgia. *Stress and Health*, 31, 299–305.

Mehu, M., & Scherer, K. R. (2015). Normal and abnormal emotions-the quandary of diagnosing affective disorder. *Emotion Review*, 7, 201–203.

Meyer, T., Richter, S., & Raspe, H. (2013). Agreement between pre-post measures of change and transition ratings as well as then-tests. *BMC Medical Research Methodology*, 13, 1–10.

Miron-Shatz, T., Stone, A., & Kahneman, D. (2009). Memories of yesterday's emotions: Does the valence of experience affect the memory-experience gap? *Emotion*, 9(6), 885.

Montag, H., & Baumeister, C. (2019). *Digital Phenotyping and Mobile Sensing* (1st ed.). Cham: Springer International Publishing.

Mostafa, R. B., & Kasamani, T. (2021). Brand experience and brand loyalty: Is it a matter of emotions? *Asia Pacific Journal of Marketing and Logistics* 33(4), 1033–1051.

Olugbade, T., He, L., Maiolino, P., Heylen, D., & Bianchi-Berthouze, N. (2023). Touch technology in affective human-, robot-, and virtual-human interactions: A survey. *Proceedings of the IEEE*, 111(10), 1333–1354.

Ottenstein, C., & Werner, L. (2022). Compliance in ambulatory assessment studies: Investigating study and sample characteristics as predictors. *Assessment*, 29, 1765–1776.

Petta, P., & Trappl, R. (2001). *Emotions and Agents*. Cham: Springer.

Rottenberg, J. (2005). Mood and emotion in major depression. *Current Directions in Psychological Science*, 14, 167–170.

Russell, J. A. (1980). A circumplex model of affect. *Journal of Personality and Social Psychology*, 39, 1161–1178. Saariluoma, P., & Jokinen, J. P. (2014). Emotional dimensions of user experience: A user psychological analysis. *International Journal of Human-Computer Interaction*, 30(4), 303–320.

Saarni, C. (1984). An observational study of children's attempts to monitor their expressive behavior. *Child Development*, 55, 1504.

Sanchez-Alvarez, N., Extremera, N., & Fernandez-Berrocal, P. (2016). The relation between emotional intelligence and subjective well-being: A meta-analytic investigation. *Journal of Positive Psychology*, 11, 276–285.

Scherer, K. R. (2001). Appraisal considered as a process of multilevel sequential checking. In A. Schorr, & T. Johnstone (Eds.), *Appraisal Processes in Emotion: Theory, Methods, Research* (pp. 92–120). Oxford: Oxford University Press.

Scherer, K. R. (2007). Component models of emotion can inform the quest for emotional competence. In G. Matthews, M. Zeidner, & R. D. Roberts (Eds.), *The Science of Emotional Intelligence: Knowns and Unknowns* (pp. 101–126). Oxford: Oxford University Press.

Scherer, K. R. (2009). The dynamic architecture of emotion: Evidence for the component process model. *Cognition & Emotion*, 23, 1307–1351.

Schiepek, G., de Felice, G., Desmet, M., Aichhorn, W., & Sammet, I. (2022). How to measure outcome? A perspective from the dynamic complex systems approach. *Counselling and Psychotherapy Research*, 22(4), 937–945.

Schiepek, G., Viol, K., Aas, B., Kastinger, A., Kronbichler, M., Schöller, H., Reiter, E.-M., Said-Yürekli, S., Kronbichler, L., Kravanja-Spannberger, B., et al. (2021). Pathologically reduced neural flexibility recovers during psychotherapy of OCD patients. *NeuroImage: Clinical*, 32, 102844.

Schlegel, K., Grandjean, D., & Scherer, K. R. (2014). Introducing the Geneva emotion recognition test: An example of Rasch-based test development. *Psychological Assessment*, 26, 666–672.

Schuller, B. W., Amiriparian, S., Batliner, A., Gebhard, A., Gerzcuk, M., Karas, V., Kathan, A., Seizer, L., & Löchner, J. (2023). Computational Charisma: A brick by brick blueprint for building charismatic artificial intelligence. *Frontiers in Computer Science, section Human-Media Interaction*, 5(1135201), 1–33.

Seiferth, C., Vogel, L., Aas, B., Brandhorst, I., Carlbring, P., Conzelmann, A., Esfandiari, N., Finkbeiner, M., Hollmann, K., Lautenbacher, H., et al. (2023). How to e-mental health: A guideline for researchers and practitioners using digital technology in the context of mental health. *Nature Mental Health*, 1, 542–554.

Sheppes, G., Suri, G., & Gross, J. J. (2015). Emotion regulation and psychopathology. *Annual Review of Clinical Psychology*, 11, 379–405.

Shiota, M. N., Sauter, D. A., & Desmet, P. M. (2021). What are 'positive' affect and emotion? *Current Opinion in Behavioral Sciences*, 39, 142–146.

Steyer, R., Schwenkmezger, P., Notz, P., & Eid, M. (1994). Testtheoretische analysen des mehrdimensionalen befindlichkeitsfragebogen (MDBF). [Theoretical analysis of a multidimensional mood questionnaire (MDBF). *Diagnostica*, 40, 320–328.

Stone, A. A., Broderick, J. E., Shiffman, S. S., & Schwartz, J. E. (2004). Understanding recall of weekly pain from a momentary assessment perspective: Absolute agreement, between-and within-person consistency, and judged change in weekly pain. *Pain*, 107(1–2), 61–69.

Topp, C. W., Østergaard, S. D., Søndergaard, S., & Bech, P. (2015). The who-5 well-being index: A systematic review of the literature. *Psychotherapy and Psychosomatics*, 84(3), 167–176.

Trentacosta, C. J., & Fine, S. E. (2010). Emotion knowledge, social competence, and behavior problems in childhood and adolescence: A meta-analytic review. *Social Development*, 19, 1–29.

Triantafyllopoulos, A., Schuller, B. W., Iymen, G., Sezgin, M., He, X., Yang, Z., Tzirakis, P., Liu, S., Mertes, S., Andre, E., et al. (2023). An overview of affective speech synthesis and conversion in the deep learning era. *Proceedings of the IEEE*, 111(10), 1355–1381.

Trull, T. J., & Ebner-Priemer, U. (2013). Ambulatory assessment. *Annual Review of Clinical Psychology*, 9, 151–176.

Tzirakis, P., Zafeiriou, S., & Schuller, B. (2018). End2You: The imperial toolkit for multimodal profiling by end-to-end learning. arxiv.org (1802.01115) 1–5.

Wagner, J., Triantafyllopoulos, A., Wierstorf, H., Schmitt, M., Eyben, F., Schuller, B. W., & Burkhardt, F. (2023). Dawn of the transformer era in speech emotion recognition: Closing the valence gap. *IEEE Transactions on Pattern Analysis and Machine Intelligence*, 45(9), 10745–10759.

Webb, T. L., Miles, E., & Sheeran, P. (2012). Dealing with feeling: A meta-analysis of the effectiveness of strategies derived from the process model of emotion regulation. *Psychological Bulletin*, 138, 775–808.

Weda, J., Mader, A. H., van Schaik, M., Meijer, A., Kolesnyk, D., & van Erp, J. B. (2023). Textured materials can enhance tactile actuators for emotional expression and passive touch experiences. In *CHI Conference on Human Factors in Computing Systems, CHI 2023*, New York (pp. 1–4).

Wrzus, C., & Neubauer, A. B. (2023). Ecological momentary assessment: A meta-analysis on designs, samples, and compliance across research fields. *Assessment*, 30, 825–846.

Wu, D., Lu, B.-L., Hu, B., & Zeng, Z. (2023). Affective brain-computer interfaces (ABCIS): A tutorial. *Proceedings of the IEEE*, 111(10), 1314–1332.

Wu, D., Xu, J., Fang, W., Zhang, Y., Yang, L., Xu, X., Luo, H., & Yu, X. (2023). Adversarial attacks and defenses in physiological computing: A systematic review. *National Science Open*, 2(1), 20220023.

Yang, D., Tian, J., Tan, X., Huang, R., Liu, S., Chang, X., Shi, J., Zhao, S., Bian, J., Wu, X., et al. (2023). Uniaudio: An audio foundation model toward universal audio generation. arXiv preprint arXiv:2310.00704.

Yang, J., Hristov, Y., Shen, J., Lin, Y., & Pantic, M. (2023). Toward robust facial action units' detection. *Proceedings of the IEEE*, 111(10), 1198–1214.

Yannakakis, G. N., & Melhart, D. (2023). Affective game computing: A survey. *Proceedings of the IEEE*, 111(10), 1423–1444.

Zeidner, M., Matthews, G., & Roberts, R. D. (2012). The emotional intelligence, health, and wellbeing nexus: What have we learned and what have we missed? *Applied Psychology: Health and Well-Being*, 4, 1–30.

Zhao, G., Li, X., Li, Y., & Pietikäinen, M. (2023). Facial micro-expressions: An overview. *Proceedings of the IEEE*, 111(10), 1215–1235.

Zhou, K., Sisman, B., Rana, R., Schuller, B. W., & Li, H. (2022). Speech synthesis with mixed emotions. *IEEE Transactions on Affective Computing*, 14(4), 16 p.

9 Applying the Science of Social and Organizational Psychology to HCI

Maha Khalid, Gabriela Fernández Castillo, and Eduardo Salas

9.1 INTRODUCTION

Social and organizational psychology plays an important role in the field of Human-Computer Interaction (HCI) by providing insights into how people interact and function within organizations. This understanding can be utilized to design computer systems and interfaces that are tailored to the needs and capabilities of users. This chapter reviews the roots of social and organizational psychology, the historical and theoretical underpinnings necessary to understand micro- and macro-organizational factors at play while providing examples of these factors in the HCI context. Many in the multidisciplinary field of HCI agree that there is a need to design better human-computer interfaces. However, there is disagreement on how to accomplish this task. The aim of this chapter is to bridge the gap between science and practice. Social and organizational psychology contains a wealth of knowledge that can be applied to practical problems within HCI. Understanding social and organizational psychology can influence researchers and practitioners to create systems that are effective, user-friendly, promote collaboration, improve work behavior, and maintain ethical standards.

9.2 THE ROOTS OF SOCIAL AND ORGANIZATIONAL PSYCHOLOGY

9.2.1 ROOTS IN HISTORY

The application of psychology in human behavior and the workplace has roots in the turn of the 20th century. This section provides highlights of individuals and contexts that have played a part in the development of social and organizational psychology and how this development has intersected with HCI. Early literature published in the field highlights the work of five leading figures in social and organizational psychology: Hugo Münsterberg, James McKeen Cattell, Walter Dill Scott, Walter V. Bingham, and Robert Yerkes (Landy, 1997; Katzell & Austin, 1992). Considered to be one of the earliest organizational psychologists, Münsterberg became interested in applying the science of psychology to human problems and conditions in the workplace (Porfeli, 2009). He was considered a revolutionary in applying laboratory experiments to real-world settings. Like Münsterberg, Cattell was an early proponent of the field who gained prominence from developing the first mental test to assess individual differences that influence human behavior (Landy, 1997) (Figure 9.1).

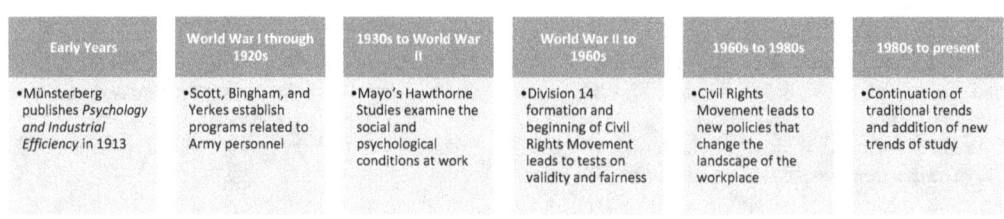

FIGURE 9.1   Brief history of industrial-organizational psychology.

DOI: 10.1201/9781003495109-9

Münsterberg and Cattell were followed by Scott and Bingham who researched personnel selection during World War I. Simultaneously, Yerkes used the application of psychology in war efforts with the development of Army Alpha & Army Beta Intelligence Tests (Yerkes, 1921). These tests examined the mental abilities of military personnel that determined their suitability for various roles. Initially trained as experimental psychologists, the war created opportunities for these scientists to venture from academic to "real world" settings. Katzell and Austin (1992) theorize four forces that accounted for the emergence of industrial and organizational psychology during this period: emergence of empirical studies, influence of Darwinism, capitalism that put an emphasis on efficiency, and growth in industrialization.

From World War I to the Great Depression techniques developed for the Army were improved and used in the private sectors. This helped lay down the groundwork for the discipline to apply the science of psychology in industry. This interplay between science and practice was further demonstrated in Elton Mayo's Hawthorne Studies. For the first time, experimental and observational techniques were used by behavior scientists to study the effect of the environment on employees (the link between improved lighting and higher productivity), supervision, social relations, motivation, and employee satisfaction (Smith, 1998). These advancements solidified roots in the scientist-practitioner model that to this day is a hallmark of industrial and organizational psychology.

The significant contributions made during WWI continued in WWII. Topics related to appraisal procedures, morale, team development strategies, attitudes, and equipment design were studied to address wartime issues (Katzell & Austin, 1992). Lillian Gilbreth, who some consider the founder of the field of modern human factors, obtained her first degree in industrial psychology (Clayton et al., 2015). She designed and studied efficiency and worker experience. After the war, in the shift from a manufacturer-based economy to a service-based economy led to an interest in improving the efficiency of tools and customer satisfaction. This served as an important precursor of HCI.

With the Civil Rights Movement (1964) came an interest in validity and fairness. This had significant effects on employment, hiring, selection, and work discrimination. The importance of diversity and inclusivity shaped HCI in many ways. For instance, user-centered design requires a greater emphasis on the need to design technology that is inclusive and accessible to all people. Incorporating inclusive and accessible technology grants access to a broad range of users, reflecting the real-world context in which the technology will be used.

The need to understand theoretical frameworks of human behavior in organizational psychology led to a cross-fertilization with social psychology. For example, Bandura's (1971) Social Learning Theory was highly influential in understanding organizational behavior. Social Learning Theory posits behaviors are learned through observation, modeling, and imitating others. In the organizational context, an individual's behavior influences, and is influenced by, individual factors and the environment (also known as reciprocal determinism) (Davis & Luthans, 1980). Social psychology provides a deeper understanding of human behavior and social interactions that can inform the design of interfaces and interactions that are more intuitive, natural, and socially responsible. In this chapter, we will focus on social and organizational psychology themes that have evolved over the years and are relevant in the current landscape of HCI.

9.2.2   Theoretical Underpinnings in Social and Organizational Research

The field of psychology has long worked to establish itself as a science. One way it has successfully achieved this is by refining its methods, which started by changing how research is conceptualized. In psychological research, operationalization, the act of taking a concept of interest and specifying precisely how to measure a variable in a concrete manner, is the first step in conducting research that allows for further replication of results to take place (Hoyle, Harris, & Judd, 2002).

Generally, research used in social and organizational psychology falls under two categories: descriptive and experimental. Descriptive research is usually conducted when the phenomena of interest has been previously studied and the researcher desires to contribute to the literature by further describing the relationships that surround the phenomenon (*APA Dictionary of Psychology*, n.d.). In this type of research, no variables are manipulated. On the other hand, experimental research is used when a novel question arises, and an experiment is required to study the variable(s) of interest. This type of testing involves independent and control variable(s).

In HCI, descriptive statistics can provide a detailed understanding of users, their tasks, and their interactions with technology. On the other hand, experimental research can be used to test hypotheses and theories about how people interact with technologies. These research strategies can help inform the design of new technologies and improve the user experience for existing technologies. However, there are some important considerations to take into account when using social and organizational research to inform HCI, namely cross-cultural and international considerations and ethical and legal issues. It is important to ensure that technology is being developed and used in a responsible and ethical manner.

9.2.2.1   Cross-Cultural Research and International Considerations

Most research in psychology focuses on W.E.I.R.D. (Western, Educated, Industrialized, Rich, and Democratic) countries and communities (Henrich, 2020). For this reason, a considerable problem in the field is the application of studies applying research done with W.E.I.R.D characteristics and assuming the findings are equally applicable to people outside those parameters. Therefore, when conducting research, it is important to know the limitations of one's findings and to include diverse participants in the research. Furthermore, in relation to HCI, these considerations should be first priority, as if one is to accomplish efficient technological usability, culture will come into play in everything from communication styles to how the individual interacts with the technology presented. Diversity, equity, and inclusion (DEI) issues have been at the forefront of societal issues, and these should be considered when designing products to ensure accessibility for all.

9.2.2.2   Ethical and Legal Issues

Psychological research demands ethical behavior. According to the American Psychological Association's (APA) ethical guidelines, participants need to be treated fairly and without harm, and mandates researchers to report their results and findings honestly (*APA Dictionary of Psychology*, n.d.). The APA has five principles that guide ethical behavior in the profession: beneficence and nonmaleficence; fidelity and responsibility; integrity, justice, and respect for people's rights and dignity; as exhibited in Figure 9.2 (APA, 2017).

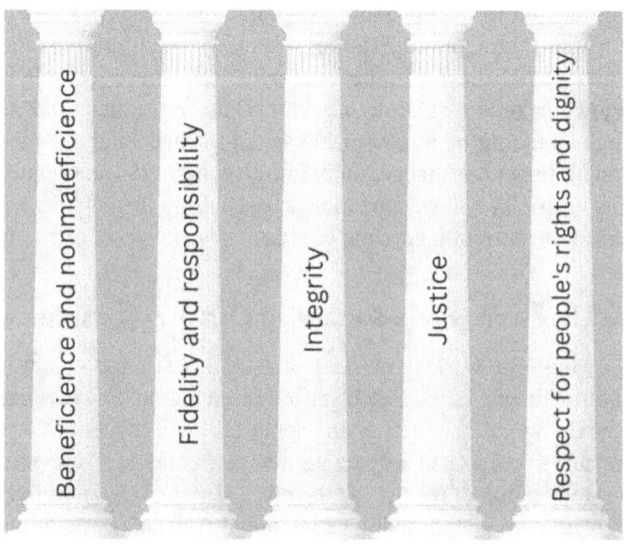

FIGURE 9.2   The five principles that guide ethical behavior.

To conduct ethical research, psychologists and those who work with human participants should aim to uphold these principles and always keep in mind the pillars of consent, confidentiality, and risk of harm. Consent involves having participants be aware of their rights and knowingly participating in the research to follow. Confidentiality means informing participants that their personal information will be protected and their consent will be required for any release of information. Risk of harm involves fully disclosing to participants the risk of participating in the research study and offering the opportunity to drop out at any time. These pillars are most often discussed at the beginning of a research study and, if using research to inform human-computer design, they need to be considered to ensure the protection of participants.

9.3 FUNDAMENTALS OF MICRO-ORGANIZATIONAL PSYCHOLOGY

9.3.1 INDIVIDUAL

9.3.1.1 Attitude

The scientist-practitioner model is rooted in research and scientific practice. It is used in social and organizational psychology and requires specialized attention to the individual in the workplace. Research indicates the need for this specialized attention because there is a bi-directional relationship between an individual and the organization in which they work. In that respect, an individual's affect, attitudes, and behaviors influence the organization in which they work. Correspondingly, the organizations in which individuals work can also influence their affect, attitudes, and behaviors (Brief & Weiss, 2002). The person-environment fit, which is the compatibility between the individual and work environment, can have an impact on many factors including job attraction, selection, and attrition (Kristof-Brown & Guay, 2011).

Attitude is a familiar term to most because it has been studied by different disciplines (e.g., political science, sociology, marketing, industrial and organizational psychology) but attitude was first identified and examined in social psychology (Schleicher, Hansen, & Fox, 2011). In this section, we look at attitude in the organizational context. Attitude refers to an individual's assessment of a person, situation, or other entity, that is characterized by variations in intensity and favorability; this assessment tends to influence an individual's response to that respective object (Schleicher, Hansen, & Fox, 2011; Judge and Kammeyer-Mueller, 2012).

Attitude can be broken down into three components: affect, cognition, and behavior (Schleicher, Hansen, & Fox, 2011; Breckler, 1984; Rosenberg et al., 1960). *Affect* is feelings, *cognition* refers to thoughts, and *behavior* is both the interactions and actions we have. To illustrate the difference between affect, cognition, and behavior, we take the example of an individual who has a positive attitude toward their work colleague. The affective component would be the positive feelings toward the colleague. The cognitive component would be the belief the individual has toward their colleague, believing they are qualified and productive in their role. The behavioral component would be the individual actively seeking opportunities to collaborate on projects with this colleague.

Attitudes are an important factor when considering individual and organizational outcomes. Research on attitude shows its relationship to job satisfaction, organizational commitment, employee performance, retention, and job involvement (Schleicher, Hansen, & Fox, 2011; Judge and Kammeyer-Mueller, 2012). Additionally, researchers have examined attitudes as they relate to work motivation (Locke, 1975). Research demonstrates that motivation is related to job enrichment. Job enrichment can include goal setting, pay raises, increased participation in decision-making, and increased recognition for job accomplishment.

9.3.1.2 Motivation

Motivation, like attitude, has been studied in different fields including sports and educational psychology (Thayer, 1989; Dweck, 1986). The study of motivation has also been influential in social and organizational psychology because it influences job attitudes, group dynamics, training, leadership, and performance appraisals (Diefendorff & Chandler, 2011). Figure 9.3 outlines dominant theories of work motivation.

Despite current theories of work motivation, there is no comprehensive understanding of how different types of motivation theories fit and function together to influence individual behavior in the organizational context (Donovon, 2001). In general, motivation serves to promote the attainment of goals which can lead to an enhancement of work performance and well-being of individuals and the organizations in which they work.

9.3.1.3   Well-Being & Stress

Quality of work life (QWL) has been a point of interest for researchers in the social sciences. Researchers have connected QWL, the relationship between work and non-work to organizational health, worker health, and family health (Hammer & Zimmerman, 2011). Hammer and Zimmerman (2011) define QWL as a personal reaction to the work environment and experience. This includes satisfaction, involvement, commitment, work-life balance, and perceptions of control relating to one's job and organization (p. 651). However, it is important to note that there is no universally accepted definition for QWL. The importance of well-being in relation to an individual's job and organization has been studied over the years. The Society for Industrial and Organizational Psychology (SIOP) (2022) emphasizes the importance of well-being in its mission of transforming science and practice of work that builds effective organizations and promotes employee well-being. Research has shown that work stress can serve as a roadblock to an individual's well-being.

Work stress refers to a process by which an individual responds to and manages demands related to work-related goals over time (Griffin & Clarke, 2011). Brief stress can be motivating and can lead to positive outcomes, but if experienced for an extended period, it can have a negative impact on health and well-being. This is known as chronic stress. To have a comprehensive understanding of the transactional nature between the individual and their environment, it is important to understand how factors of well-being and stress operate in diverse cultural contexts (Erez, 2011). Generating a culturally diverse approach to demystifying well-being requires theoretical and methodological sophistication that go beyond Western theories.

9.3.1.4   Application in HCI

HCI has long sought design strategies that cultivate individual engagement, behavior change, and well-being. Peters, Calvo, and Ryan (2018) developed a model to help understand these aspects of the individual. The METUX model (Motivation, Engagement and Thriving in User Experience) provides a framework that allows scientists and practitioners to improve well-being through increased motivation and sustained engagement with technology. This model encourages the development

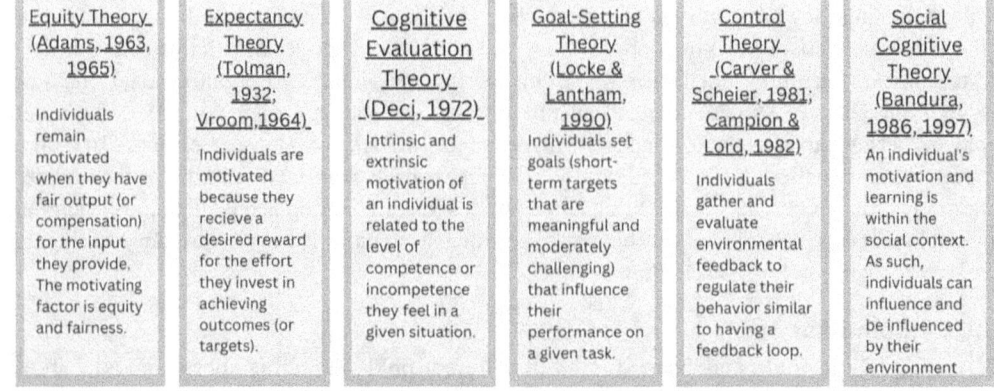

FIGURE 9.3   Dominant theories of work motivation.

of technologies that support psychological well-being and human potential simultaneously. The METUX model builds on the Self-Determination Theory (SDT). SDT outlines three psychological needs in individuals that allow for growth and development: autonomy, competence, and relatedness (Ryan & Deci, 2017). In other words, individuals are motivated when they feel they have a choice and are not being pressured (autonomy), feel competent to complete the task (competence), and feel connected with others (relatedness). METUX can be used to design technologies in which engagement, motivation, and well-being are considered (Peters, Calvo, & Ryan, 2018). As our relationships become increasingly mediated by technology, expertise on individual well-being will become important for future technological design.

9.3.2 PERSONNEL

9.3.2.1 Personnel Recruitment, Selection, and Hiring

Although the issue of personnel recruitment, selection, and hiring might seem to be of interest only to the human resources department of large organizations, understanding how these processes work in an organization is crucial, as these three processes will determine everything in how accessible a technology is to how user-friendly it might be, as the people selected and hired bring their own personal experiences to the table. Therefore, the aim of this section is to inform what organizations look for and how employees are chosen, giving an overview of how organizational psychology perceives these processes. Furthermore, this section will seek to help future workers involved in human-computer design and interactions understand what organizations take into consideration during the recruitment, selection, and hiring process.

Lastly, this section ends with an example of how using seniors to inform human-computer design aids the process of product-design, making them more accessible to the aging population.

Imagine that one of your local restaurants puts out a *help wanted* sign. This restaurant is looking for a chef with experience in Italian cuisine. How does such a restaurant attract people with this experience? It all comes down to a company's recruitment strategies for personnel. Organizations hire personnel because they need human capital resources, defined as the employee's knowledge, skills, abilities, and other characteristics (KSAOs) (Ployhart & Kim, 2013). Personnel recruitment involves seeking out human capital to gain a competitive advantage. For example, your local Italian restaurant trying to gain customers over a franchised chain of Italian restaurants. Therefore, for a recruitment strategy to be successful, it must correctly identify individuals who have the human capital the organization seeks (Barber, 1998). If the only applicants are chefs with experience in Mexican cuisine, then the restaurant's recruitment strategy needs to be reassessed.

There are three main stages in personnel recruitment: (1) identifying and generating a pool of applicants, (2) narrowing the applicant pool into selectees, and (3) the selectee choosing to accept or decline the job offer (Barber, 1998). The local Italian restaurant may post an advertisement on social media stating, "Help Wanted: Chef with experience in Italian cuisine." This ad will aid the restaurant in generating a pool of applicants, but it might experience the issue of too many people applying. The first step of personnel recruitment is now complete, and the restaurant must filter applicants using recruitment activities, usually an interview. In this process, an individual applicant may begin to filter information about the company. The individual takes into consideration information about the organization (e.g., the restaurant's values). This information impacts the individual's decision-making process to either continue with the recruitment process or not (Volpone et al., 2013). This process is simultaneously occurring on the organization's side. Once an applicant pool has been generated and narrowed down, usually in the second step of the recruitment process, the process of personnel selection begins.

The classical view of personnel selection identifies KSAOs that an organization needs for effective job performance (Ployhart & Schneider, 2012). For the restaurant, that would be selecting a chef with experience in Italian cuisine. However, the classical view tends to exclusively focus on the individual's abilities, such as a chef's ability to make pasta, while excluding other aspects of a job, such as working as a team in the kitchen. Therefore, to adequately select personnel, it is imperative to have a holistic approach to this process, evaluating both individual KSAOs but also contextual factors of a job (Ployhart & Schneider, 2012). For the restaurant, this would involve taking into consideration a chef's ability to cook, as well as looking at how they cook under the pressure of a full house.

Once an applicant has been selected, the hiring process begins. The onboarding process allows a new hire to get acquainted with an organization's ins and outs, which is extremely important to that employee's socialization process into the company. Bauer and Erdogan (2011) proposed the model of socialization, as shown in Figure 9.4.

The model proposes three main categories of factors that contribute to an employee's outcomes in an organization. Employee characteristics refer to factors such as openness and a proactive personality. Employee behavior refers to how an individual builds new relationships, how they take feedback, and how they seek out new information. On the other hand, an individual's outcomes in a company are also affected by how they are treated by their organization. If an organization provides insufficient training and a toxic work environment, it is more than likely that outcomes such as commitment and satisfaction will be detrimentally affected (Bauer & Erdogan, 2011).

Another important factor to take into consideration in the process of personnel selection is how biases come into play. As the world leans more toward the automatization of personnel selection processes, a new danger arises. Most consider artificial intelligence (AI) incapable of discrimination, but anything built by humans will have our biases intertwined within it. The dangers of using AI in personnel selection were illustrated by a software built by Amazon engineers to speed up the company's hiring process. The software was meant to filter through thousands of résumés, giving the company only the most qualified of contenders. As reported in 2018, the AI was built by using previous résumés (which happened to be mostly male hires) to filter through potential choices, the software began excluding resumes with the word "woman" and excluded those from women-only colleges (Hamilton, 2018). Therefore, in the modern world, it is more important than ever to understand one's biases and be aware of how these could impact the technology we implement in the hiring process. This is not a warning against the use of AI in personnel selection, but a warning against their use without proper research and implementation.

9.3.2.2 Job Performance & Evaluation

Modernity has pushed the workplace into a lot of automatization processes that involve technology. In order to be able to design such technologies and further evaluate them, one must have a solid understanding of what exactly is taken into consideration when an employee's performance is evaluated in the workplace. For example, as described below, job demands are multifaceted and can involve physical and cognitive demands. If a technology is put into place to help evaluate employee

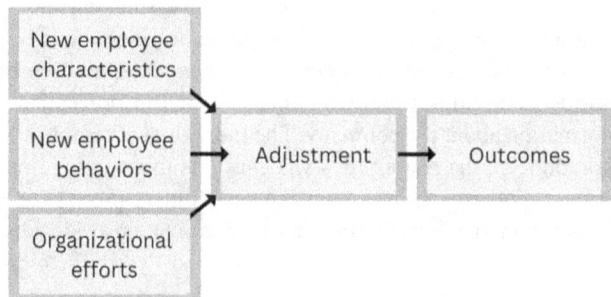

FIGURE 9.4 The model of socialization. (Adapted from Bauer, T. N., & Erdogan, B. (2011). Organizational socialization: The effective onboarding of new employees. In S. Zedeck (Ed.), *APA Handbook of Industrial and Organizational Psychology, Volume 3. Maintaining, Expanding, and Contracting the Organization* (pp. 51–64). https://doi.org/10.1037/12171-002. Copyright 2023 by the American Psychological Association.)

performance but fails to take into consideration the physical demands of the job, employees who are exceptional in these capabilities but perhaps no other aspects of the job will suffer in their performance evaluations. For these reasons, the aim of this section is to familiarize those interested in HCIs with these processes, with the idea being that further designs and applications take this theory into consideration. Finally, the example about aging adults can also illuminate how designing a product that is inclusive of these characteristics can help employees perform better in areas where they might have otherwise fallen short.

Daniela was hired as a Chef at the local Italian restaurant. After Daniela was hired, the restaurant increased in popularity and five-star reviews with comments on Daniela's pizza as being the best in town. These indicators show Daniela as being a high-performing employee. An employee's measurable actions, behaviors, and outcomes that contribute to the organization's values and goals is known as job performance (Pandey, 2019). A popular model for assessing the factors that contribute to job performance is the job demands-resources model (Demerouti et al., 2001). This model defines job demands as psychological, social, and organizational aspects of the job that require effort and skills associated with physical and psychological costs (Demerouti et al., 2001; Pandey, 2019).

Job resources are defined as aspects of the job that aid in completing work goals and reduce job demands (Demerouti et al., 2001; Pandey, 2019). This model is further described in Figure 9.5.

This model would describe Daniela's job demands as being able to stand for prolonged periods of time, knowing how to make Italian food from scratch, and being able to attend to customer needs. Daniela's job resources are her own skills, such as her pizza-making technique, but also the support she receives at her job. If her job provided no training and cultivated a toxic climate, Daniela's performance would likely be negatively affected.

After new employees have been part of an organization for some time, they are evaluated. Job evaluation is defined as a systematic procedure for determining the comparative value of positions within an organization (Armstrong & Baron, 1995). Some popular job evaluation techniques are listed in Figure 9.6.

9.3.2.3   Leadership

All workplaces have some type of hierarchy implemented, further discussed in the organization section. However, in order to truly gauge the field of HCI, having some basic understanding of workers' relationships with their leaders can help further understand the role of these interactions

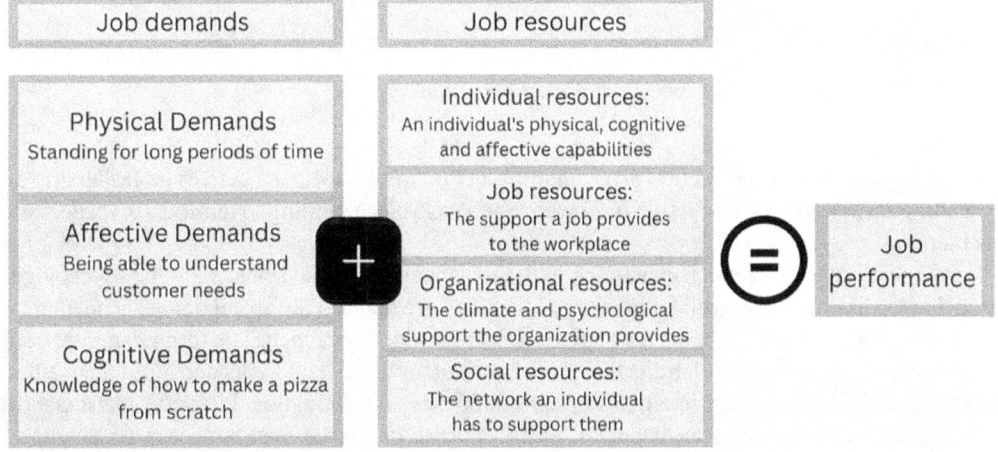

FIGURE 9.5   The job-demands resources (JD-R) model. (Adapted from Demerouti, E., Bakker, A. B., Nachreiner, F., & Schaufeli, W. B. (2001). The job demands-resources model of burnout. *Journal of Applied Psychology*, 86(3), 509. https://doi.org/10.1037/0021-9010.86.3.499. Copyright 2023 by the American Psychological Association.)

Non-analytical techniques	Factor comparison	Point-factor ranking
A job is evaluated as a whole, without breaking it down into its components. An example is job ranking, where jobs are ranked from hardest to easiest.	A job is broken down into its components (such as physical and cognitive factors) and a monetary value is assigned to each factor.	A job is broken down into its components and is given a maximum number of points. An employee is graded based on how many points out of that factor they achieved.

FIGURE 9.6   Popular job evaluation techniques.

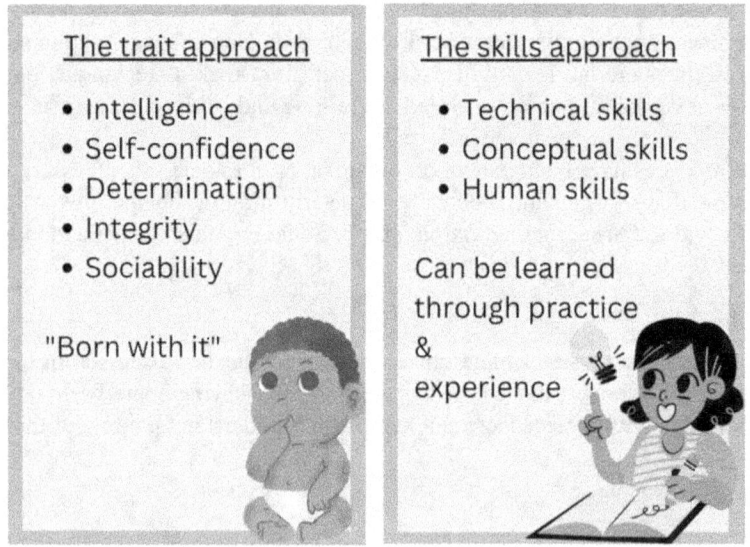

FIGURE 9.7   The psychological debate: are leaders born or made?

in the workplace. This brief section gives an introduction to the two perspectives on leadership and the leading theory on how leadership is viewed in the workplace, aiming to familiarize readers with this literature.

Leadership is a topic of much contention in the social and industrial sciences. Some psychologists argue that leaders are born, while others argue that they are made. This contention gave birth to two main approaches to leadership: the trait approach and the skills approach. The former approach suggests some people are born with traits that give them a natural capacity for leadership. The five traits most cited in the literature are intelligence, self-confidence, determination, integrity, and sociability (Northouse, 2021). On the other hand, the skills approach assumes leadership can be taught and learned, and argues that good leaders have three main abilities: technical skills (such as knowing how to make a pizza), conceptual skills (such as developing a new menu), and human skills (such as positively interacting with customers). However, unlike the trait approach, these can all be acquired through practice and experience (Katz, 1974). Figure 9.7 displays and summarizes these approaches.

Aside from these two approaches to leadership, a dominant theory in the field is the leader-member exchange (LMX) theory. The LMX theory bases itself on the dyadic relationship that takes place between a leader and their followers. According to the theory, the quality of this relationship is predictive of outcomes across individuals, groups, and organizational levels (Gerstner & Day, 1997). Therefore, in LMX theory, the dyad that forms between a leader and their followers is the unit of analysis that allows researchers to predict the outcomes of the relationship. LMX theory has been used to predict everything from stress outcomes to productivity outcomes in the workplace.

9.3.2.4   Application in HCI: Aging

The workplace often views aging workers with a negative lens. The most pervasive of these negative beliefs is that older workers have lower job performance. This view of older workers is known as the "deficit model of aging", stating that as we grow older, our physical and mental capabilities decline (Rau & Adams, 2013). On the other hand, some companies often value older workers, believing that they have accumulated experience that is invaluable to the company. This is known as the "successful aging" approach (Rau & Adams, 2013). Whatever a company's approach to its aging workers might be, the fact remains that most older workers experience age-related disabilities, such as poor vision, limited mobility, or declining memory. While most of these age-related disabilities do not affect the core components of job performance, the application of human factors can aid in the continued success of these employees at the workplace (Rau & Adams, 2013). With the increase of work-from-home policies being adapted by workplaces after COVID-19, it is easy to see how skills and abilities required to interact with computers have become increasingly complex.

9.3.3   TRAINING & DEVELOPMENT

9.3.3.1   Definitions: What Is "Training"?

Although most people are hired for a job based on the experience they have relevant to the job of interest, most workplaces still provide training to familiarize new employees with specific regulations, rules, and their general way of functioning. In the social and organizational sciences, *training* is defined as the conjunction of KSAOs that work together to improve performance in a specific environment (Kozlowski, Salas, & SIOP, 2010). Following the example from our previous section, *Personnel,* we know that Daniela was just hired as a chef at her local Italian restaurant. Therefore, in her training, she needs to acquire knowledge of where all the ingredients are located, how to use a restaurant's oven, and what is on the restaurant's menu. This is all part of Daniela's *knowledge* acquisition that will allow her to perform her job successfully. Additionally, Daniela needs to be told what to do once she arrives, bringing her previous expertise to the table. This is part of her acquisition of *skills*, of which she already has some building blocks but will need more elaborate training. In other words, Daniela might know how to make a pizza sauce but not the restaurant's secret pizza sauce, so her training would elaborate on this skill. Lastly, Daniela will also need training on the restaurant's values and mission, helping construct the necessary *attitude* she needs to have to be able to succeed in her new position. Figure 9.8 breaks down the distinct components of training. Understanding the components that play into training an employee can help further one's understanding of how they will relate themselves to their work environment and the technology involved. The aim of this section is to familiarize the reader with these processes, highlighting what goes into training an employee and how different factors can play into their relationships with technology.

There are many factors that will influence Daniela's training. Most models in the field focus on training transfer and have three main categories of variables: individual difference factors (such as pre-training experience), strategies used during training (such as training activities), and lastly, environmental factors (such as work climate). However, the model of training effectiveness looks at these factors with a longitudinal approach, taking into consideration factors important before, during, and after training (Kozlowski, Salas, & SIOP, 2010).

FIGURE 9.8 Components of training. (Adapted from Salas, E., Wilson, K. A., Priest, H. A., & Guthrie, J. W. (2006). Design, delivery, and evaluation of training systems. In G. Salvendy (Ed.), *Handbook of Human Factors and Ergonomics.* https://doi.org/10.1002/0470048204.ch18. Copyright 2023 by the American Psychological Association.)

The model of training effectiveness relies on the pillar that for a training to be successful, it must first be analyzed at the organizational level, job/task level, and at the individual (person) level. This ensures three key questions are answered: what training is needed, where it is needed, and who needs it (Kozlowski, Salas, & SIOP, 2010). Organizational analysis is needed to evaluate the environmental conditions in which a training will be conducted and what resources are available to conduct it. Job/task analysis is necessary to be able to define comprehensive training objectives that are directly related to the components of a job, and finally, person analysis aids in determining what KSAOs a specific individual might need. In the Italian restaurant's case, this would involve examining how much money the business can dedicate to Daniela's training, breaking down the chef's responsibilities to make sure Daniela is trained and equipped for all of these responsibilities, and catering the training to Daniela's existing knowledge and personality.

Training analysis is vital. An example of its importance can be reflected in diversity training and their bloom in the last couple of years. The majority of Fortune 500 companies offer diversity training (Chang et al., 2019). However, many of these trainings lack proper analysis before their implementation, leading to their effects backfiring (some authors note diversity training can lead to defensiveness in some trainees) (Chang et al., 2019). A recent meta-analysis concluded that diversity training has a small-to-medium size effect, but the majority of the research evaluated did not measure the trainings' long-term impact (Kalinoski et al., 2013). Other research suggests diversity training outcomes as short-lived (Chang et al., 2019).

A major problem with diversity training is their implementation without sufficient evaluation. Therefore, to maximize positive outcomes, researchers recommend gathering data from the implementation of current and future training (Chang et al., 2019). As someone involved in the field of HCI, understanding the need for training and job needs analysis can be vital in order to truly gauge what is to be implemented and how technology will affect other factors and relationships in the workplace. For example, in the case of diversity training, it has been found that simply giving the training once is not enough: knowing this, people involved in the field of HCI could aid in designing training programs that know this information and perhaps even develop modules that invite training over longer periods of time (Cepeda et al., 2006).

9.3.3.2 Feedback & Evaluation

In the educational setting, AI is being used to give students feedback in their work, and even used to evaluate work for plagiarism. These kinds of technology are also reaching the workplace, and therefore, having an understanding of how the feedback process works in the workplace is helpful in order to understand how people come to terms with said feedback. Furthermore, if one is to implement technologies to help workers get better at their job and if the technology is used to evaluate them, understanding how to do this in a way that is designed in the context of a human-centered approach in mind will help improve such technologies. Therefore, the aim of this section is to familiarize the reader with how these processes are perceived in the field of social and organizational psychology, ending with an example of how these technologies are being implemented in the medical field and how their feedback could be improved to maximize medical learning.

According to Salas et al. (2012), feedback is a crucial component of successful training transfer. There are two different types of feedback: process feedback and outcome feedback. The former is preferred, as it breaks down the task into a series of steps where the evaluator is able to determine which parts were performed correctly and incorrectly. This type of feedback tends to be more successful because it allows the trainee to successfully recognize what and where changes can be made. On the other hand, outcome feedback is usually not as successful in changing training outcomes because it focuses solely on whether or not the task at hand was completed successfully. This aside, another important factor to take into consideration when giving feedback is an individual's feedback orientation. Some people take feedback negatively no matter how it is presented, but to improve outcomes, researchers recommend focusing feedback on specific skills and training outcomes, moving away from a person's specific way of doing things and bringing focus to the task at hand. An example of these feedback types can be observed in Figure 9.9.

So far, distinct types of feedback in training have been described. An equally vital component of training is an organization's ability to determine if the training was effective in increasing an individual's KSAOs and if it made the desired impact on the organization's operations. Kirkpatrick's classic typology of evaluation has been used for decades to break down a training's effectiveness (Salas et al., 2012). The typology involves two types of criteria and four levels. The first type of criteria is internal criteria, looking at training outcomes immediately after training. The second type involves external criteria targeting training outcomes after trainees have returned to their jobs.

FIGURE 9.9 Exemplifying different types of feedback.

The first level involves trainee reactions, typically measured via self-report measures, and looks at factors such as the trainee's perception of their instructor and the training material. The second level involves looking at training learning, where an individual's skills need to be evaluated according to training outcomes to observe if the training was effective. In Daniela's case, the first level would involve her completing a survey on how effective she found her training to be and the second level would be her taking a small quiz to demonstrate her new skills. The third level, now located in the external criteria, is looking at behavior on the job. This involves looking at the transfer of learning. In other words, Daniela might know how to make a pizza during training with no evident customer, but she needs to be able to transfer this knowledge into a high-pressure situation. The last level of the typology focuses on results. If the Italian restaurant recently implemented training on how to make pizza dough faster to be able to diminish a customer's waiting time, this part of the typology would entail measuring and evaluating outcomes such as sales, productivity, and overall profits (Salas et al., 2012).

9.3.3.3   Simulation-Based Training

Salas et al. (2009) refer to simulated-based training (SBT) as an instructional setting synthetically generated to convey knowledge, attitudes, concepts, skills, or rules. Simulations provide an environment that allows the learner to be involved actively, allows repeated practice and clinical complexity, offers adjustable levels of difficulty, and can provide feedback during the learning process (Weaver et al., 2010). Social and organizational psychology helps us understand and develop contexts that allow for a more realistic SBT experience.

As mentioned, SBT is utilized to develop professionals' knowledge, skills, and attitudes needed in their work domain (Rosen et al., 2008). This is an important consideration as training in real-life situations can come with safety and ethical concerns – for example, in a health care setting, simulations can provide a controlled environment that reduces the risk of harm to a patient. SBT can be designed, implemented, and evaluated in accordance with the target goals or objectives of a given team. Studies have shown the effectiveness of SBT in various contexts including education and health care (Oh et al., 2015; Chernikova et al., 2020).

In healthcare, for instance, TeamSTEPPS (Team Strategies and Tools to Enhance Performance and Patient Safety) was developed as a customizable SBT (Agency for Healthcare Research and Quality, 2017). TeamSTEPPS is a tool that trains health care professionals to provide quality and safe care in high-risk environments. Team training has been influenced by multiple disciplines including social, cognitive, human factors, and industrial and organization psychology (Bisbey, et al., 2019).

With the rapid growth of technologies, SBT can provide additional benefits to traditional didactic instruction and can increase performance while reducing errors. Social and organizational psychology intersects with HCI as SBT requires theories of human performance that can account for intricate interactions between individuals and their simulated environments. Both fields can inform each other to provide a comprehensive and accurate understanding of human performance in the SBT context.

9.3.3.4   Application in HCI: Simulation-Based Training

While computer and simulation-based trainings were around before COVID-19, the recent pandemic pushed companies and education systems to reimagine how we could learn and virtually train people. SBT has a wide range of possibilities in every field, but it is becoming increasingly relevant in the medical field. A study examining SBT training for laparoscopic surgery found that it is most effective when compared to no intervention, and it is moderately more effective when compared to non-simulation intervention (Zendejas et al., 2013). SBTs, especially those in the medical field, offer the great advantage of endless practice to medical students and residents. Yet, virtual reality (VR) SBTs continue to require improvement, as this same study found box-trainer-type simulators offer better outcomes than VR techniques. Therefore, those with knowledge about HCIs and psychology can collaborate to make the feedback implemented by VR technologies better, maximizing student training. This aside, it is becoming increasingly important to study HCIs and how to improve VR SBTs as medical practices become less interventionist and more reliant on computer techniques.

9.4 FUNDAMENTALS OF MACRO-ORGANIZATIONAL PSYCHOLOGY

9.4.1 TEAMS AND TEAMWORK

The modern workplace is making a strong push toward more teamwork. If one is to be successful in applying HCI principles in the workplace, having an understanding of what a team is and how teams work is crucial to positively apply these principles. The aim of this section is to introduce the reader to what a team is, how an effective team is defined, and how these relationships may come into play when designing products that teams are going to be using (such as a team involved in a highly technical space exploration project).

Many disciplines have made important contributions to the science of teams and teamwork including computer scientists, industrial-organizational psychologists, engineers, and many more (Rosen et al., 2008; Stahl, 2006; Carley, 1997; McComb, 2007). Teamwork is an important factor in HCI as it allows for diverse perspectives and expertise to be brought to the forefront, allowing for collaboration, a more comprehensive end-product, and more efficient problem solving. The diversity of perspectives has helped the field of team sciences evolve to a place where we can better understand the development and management of teams. *Teams* are a group of individuals performing interdependent tasks toward a common mission or objective. Key factors that define a team include: (1) having at least two members, (2) performing tasks related to the work, (3) interdependence among members, (4) having one or more shared goals or objectives, (5) having some form of social interaction among members, (6) maintaining boundaries with other entities, and (7) being situated in a larger context of an organization (Bisbey and Salas, 2019).

There are many moving parts in teams. Team composition is important as every member of a team brings a unique set of characteristics that aggregate and impact performance and task completion. Teams, at their foundation, need individuals who have the skills and abilities needed to do the work. In contrast to individual work, teamwork requires effective performance and coordination of multiple individuals working on a given task (Rosen et al., 2008). It is important to note that teamwork is encapsulated within team performance. *Team performance* is the multilevel process of team members' individual and team-level task work. When talking about team performance, we cannot overlook team effectiveness as they go hand in hand. For example, research has shown a robust relationship between team members' skills and abilities as positively related to team effectiveness.

Team effectiveness refers to the team's outcome relative to a pre-set criterion of evaluation, task performance, team viability, and satisfaction of team members (Rosen et al., 2008; Hackman, 1987). Team composition can interfere with team effectiveness. The positive relationship between team members' skills and abilities to team effectiveness were previously highlighted; however, there are instances where all members of a team are highly skilled yet ineffective. Here other factors such as the preference for teamwork, the mix of personalities within teams, team-level agreeableness, and conscientiousness come into play. Various methods, such as team development and training, can enhance team effectiveness.

Teams exist in dynamic systems. As such, teams need to be effective in diverse environments. Demographic diversity is becoming an increasingly important factor for businesses around the world. However, the simple addition of a diverse member (whether diverse in gender, race, culture, religion, etc.) to a team does not guarantee more effective and positive team outcomes. In order for teams to be effective they need to possess coherent team identity and effective coordination patterns (Li et al., 2018). In fact, research shows that when a diverse member is added to a team, the team can suffer in performance. This is because at the beginning of this addition, team members see themselves as belonging to distinct social categories. Yet, over time, team members can set aside surface-level categorizations (such as gender or race) and bond over psychological attributes (Li et al., 2018).For this change to take place in a positive manner, companies need to make sure they offer the proper support for their teams to thrive.

Decades of team science research have led to the development of eight principles of expert team performance (Bisbey, Traylor, & Salas, 2021). The principles of expert teams, adopted from Bisbey, Traylor, and Salas (2021), are summarized in Figure 9.10.

Expert teams have: (1) shared mental models that allow them to recognize the task, environment, and team and how these components interface, (2) the ability to learn and adapt to situational demands by adapting their approach as needed, (3) clear roles and responsibilities allowing for increased efficacy, (4) a shared vision that provides a clear purpose, direction, and motivation, (5) an effective and dynamic team leadership that encourages each member to leverage their expertise in a given subject matter, (6) the ability to provide a positive and psychologically safe environment that nurtures a safe space for open communication, (7) a cooperative and coordinate effort to perform assigned tasks, and (8) resilience that allows them to thrive under pressure.

9.4.1.1 Application in HCI

Teamwork is an essential component of most, if not all, workplaces. Following the COVID-19 pandemic, many workplaces have shifted to virtual environments. This shift from traditional to virtual teams came with little to no preparation for remote work. Virtual teams are closely related to the HCI discipline in that there is a need to understand how individuals interact in face-to-face teams so

FIGURE 9.10 Eight principles of expert teams. (Adapted from Bisbey, T. M., Traylor, A. M., & Salas, E. (2021). Transforming teams of experts into expert teams: Eight principles of expert team performance. *Journal of Expertise*, 4(2), 200. https://par.nsf.gov/servlets/purl/10357652. In the public domain.)

strategies that are effective in those contexts can be replicated with technology. An important theme of virtual teams is having supportive systems (e.g., team building and feedback) to effectively manage virtual teamwork while providing a psychologically safe environment (Kilcullen, Feitosa, & Salas, 2022). Other factors such as the lack of face-to-face interactions can make it difficult to build relationships and trust in virtual teams. Recognizing the key components of successful teams can be incorporated with HCI to design technologies that facilitate successful virtual teams.

9.4.2 Organizations

Organizations are important to HCI because they provide a context in and for which new technologies and interfaces are designed and developed. Organizations are complex environments that continuously change in response to internal and external factors. The *Social Psychology of Organizations* defined the primary characteristic of an organization as "patterned" human behavior (Katz & Kahn, 2015). Patterns, in this sense, imply shared roles, norms, and values. *Roles* are behaviors that are expected in a functional relationship. *Norms* are expectations in a given role or category. *Values* are ideological beliefs or ideas that serve as aspirational guidelines for behavior and activities. These factors come together to give an organization a distinct organizational *culture*. Organizational culture provides us with a snapshot of how to design, develop, and implement strategies that promote effectiveness.

To further delineate what an organization is, the distinction between formal and informal organizations must be made. *Formal* organizations have an explicitly stated purpose that is often in writing (e.g., your local Italian restaurant), whereas *informal* organizations are usually less explicit in their stated purpose, but members have a shared interest (e.g., pizza lovers club). Organizational psychology focuses on studying formal organizations. Within formal organizations, researchers and practitioners study the structure, function, and processes of an organization. The aim of organizational psychology is to aid organizations in achieving high performance and effectiveness.

The exploration of macro-level issues at the workplace leads us to organizational theory. *Organizational theory* organizes purposeful human action in ideas or models. There are three dominant organizational theories that have been developed over the years. Figure 9.11 outlines some of their contributions to the field.

As seen in Figure 9.11, classical theories have a strong emphasis on order and control. Classical theories assume that to obtain results employees need constant supervision. On the other hand, humanistic theories are oriented toward obtaining results while remaining human-centric. The major weakness of these theories is the notion that only one effective way exists to optimally run an organization. This notion led to the development of contingency organizational theories that accept internal and external factors in driving organizational decisions.

Micro- and macro-level organizational elements impact one another. For instance, the organization's design (macro-level) may have an impact on the behavior of an individual (micro-level). Assessing an organization at the macro-level requires insight into organizational design. Organizational design looks at an organization's strategy, structures, and roles. For example, the multi-national technology corporation IBM has gone through major organizational design changes in recent years. They have continuously looked at the market, have moved from country organizations to global business units, and have been mindful of having increased collaborations within IBM and other organizations. These considerations are important as they can have short-term and long-term implications. Neglecting to address the ways in which an organization can be optimally designed can be costly. There are several determinants of organizational design, including strategy.

An organization's strategy is its long-term goals and ways in which they are reached. Strategy is closely related to the structure of an organization. Strategy is translating the efficiency and/or effectiveness goals of the firm into operational actions, while structure serves as the means to achieving these objectives. The structure of an organization can be assessed through its shape (or structure), which has important implications for effectiveness and productivity. Common organizational

---

CLASSICAL ORGANIZATIONAL THEORIES

Scientific management (Taylor, 1911): work should be broken down into smaller components, those that design work are different from those that perform it, and employees performing specialized functions should be grouped together

Ideal Bureaucracy (Weber, 1947): rewards based on one's contributions in the organization, close supervision of employees, unity of command, information flows top down in an organization.

Administrative Management (Fayol, 1984): employee cohesion, camaraderie, and activities for managers including planning, organizing, commanding, coordinating, and controlling.

HUMANISTIC ORGANIZATIONAL THEORIES

Theory X/Y Leadership Distinction (McGregor, 1960): distinction between two types of managers – Theory X (assumes most people dislike their work so they need to be supervised closely) and Theory Y (assumes work is a natural part of peoples' lives and find work personally rewarding).

The Human Organization (Likert, 1961): System 1 (*exploitative-authoritative leader* imposes decisions on subordinates and uses fear to motivate employees), System 2 (*benevolent authoritative leader* has some trust in employees but there is no teamwork), System 3 (*consultative leader* listens and incorporates some employee ideas), System 4 (*participative leader* values employee and employees are involved in decision making).

CONTINGENCY ORGANIZATIONAL THEORIES

There are no best practices to lead employees or make decisions. Instead, an optimal course of action is dependent (contingent) on the internal and external factors in an organization.

FIGURE 9.11   Contributions of the major organizational theories.

shapes include variations between a flat and narrow organizational structure (Figure 9.12). A flat organizational structure has fewer levels between management and staff-level employees. As such, staff-level employees are more closely involved in decision-making processes, leading to a quicker turnaround response to consumer demands. On the other hand, a narrow organizational structure has more layers and is more hierarchical. This means that management and staff-level employees are more separated by this hierarchy, as can be visualized in Figure 9.12. This structure is more popular in markets that require highly specialized products or where services are consistent.

It is important to note that research has shown there is no "right way" to design an organization; however, successful organizations have clearly defined rules, roles, and responsibilities that fit within the overall system and desired outcomes of that organization. Organizational design is closely linked to culture. According to Ravasi and Schultz (2006), organizational culture consists of a collection of shared mental assumptions that guide interpretation and action within organizations, delineating suitable behavior for diverse situations. As important organizational design is, so is culture. At times, culture can hold tremendous influence over an organization. An organization's culture can be understood by outsiders at times while at other times it may be difficult to grasp how and why individuals within an organization behave. Studying organizational culture is challenging and culture may be difficult to change, but it can be necessary to understand its basic assumptions.

Discussion on culture can invoke what it means to have cultural diversity at the organizational scale as we think of global diversity and the rapidly shifting diversifying domestic population of the US. Various companies have reaped the benefits of diversifying their organizations by conducting business in the international markets. This brings its own set of benefits and challenges. For instance, corporations such as Coca-Cola, IBM, and McDonald's rely on international sales for a substantial portion (if not majority) of their sales and profits. These organizations are constantly investigating creative ways to expand their domestic structure while having an international division. All this requires an aggregate level of analysis that can help us understand how DEI operate in complex organizations nationally and internationally.

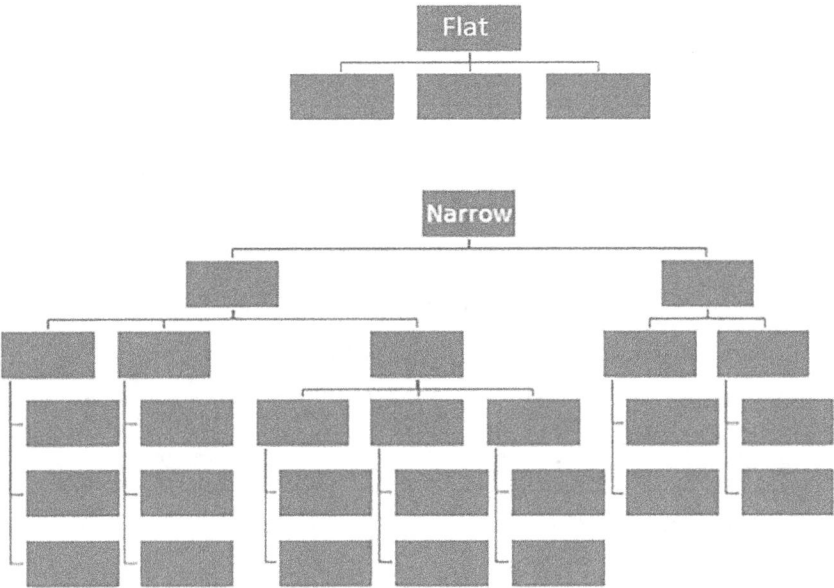

FIGURE 9.12 Comparison between flat and narrow organizational structures.

9.4.2.1 Application in HCI

Recent years have led to the reformulation of organizational design, leading to developments like that of the virtual organization. A virtual organization is characterized by its ability to produce a product or service while collaborating with, and contracting other firms to provide essential components of that product of service. This can be seen in the latest trend of virtual restaurants. Having gained popularity during the COVID-19 pandemic, virtual restaurants exist solely dependent on customers ordering food online. The food is prepared at an existing restaurant, allowing businesses to cut costs by sharing a space and using the same ingredients already present at the restaurant. The primary motivating factor behind creating virtual organizations is cost reduction. In 2019, UberEATS, an online food ordering and delivery platform, helped launch 4,000 virtual restaurants worldwide (Isaac & Yaffe-Bellany, 2019). Having a virtual organizational structure has its own set of challenges that make it difficult to maintain a clear culture, but they also provide us with a fresh outlook on organizational design previously not sought after.

9.5 CONCLUSION AND FUTURE DIRECTIONS

The SIOP has been a key proponent in discussing issues related to industrial and organizational psychology. Since 2014, SIOP has surveyed its members on key issues related to the workplace (Figure 9.13P). SIOP, known as Division 14 of the APA and an organizational affiliate of the Association of Psychological Science (APS), is the largest professional association of industrial and organizational psychologists that promotes research, education, and evidence-based practice in industrial-organizational psychology. Trends in Figure 9.13 provide an insight into the current topics, challenges in the field, and future directions. We can examine how themes from social and organizational psychology have evolved over the years and are relevant in the current landscape of HCI.

In this chapter, we have examined various domains in which social and organizational psychology intersects with HCI. Over the years, social and organizational psychology have provided ways in which individuals and organizations can guide the design of technology that supports productivity, job performance, satisfaction, and organizational culture. An understanding of social and organizational psychology can help inform HCI by providing key insights into user-centered

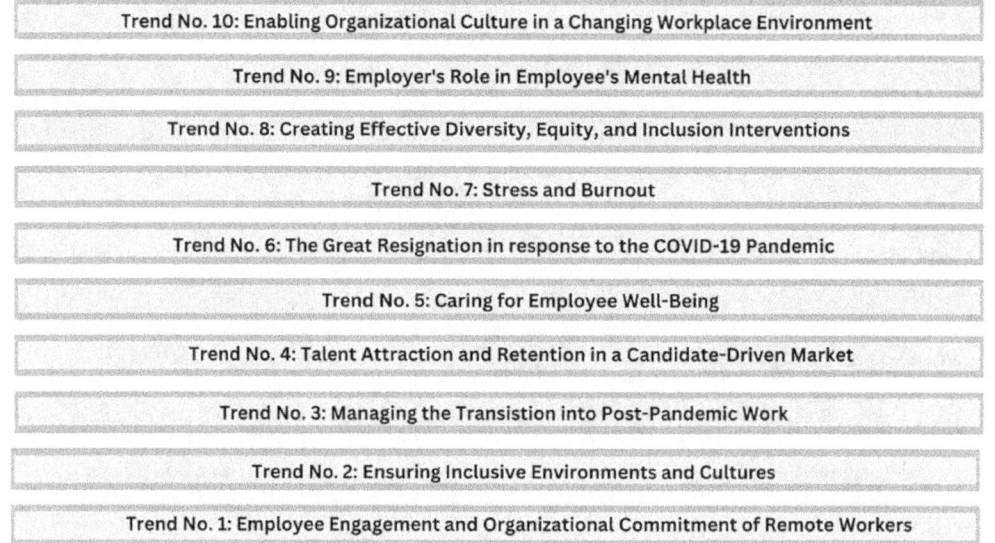

FIGURE 9.13   2022 Top ten work trends from SIOP. (Adapted from SIOP (2022). 2022 Top Ten work trends 2022. https://www.siop.org/Business-Resources/Top-10-Work-Trends. In the public domain.)

design, understanding user experience, and improving accessibility of user interfaces and systems. Cross-disciplinary work can allow for a broader perspective, solutions, and collaborations that may not be possible within a single discipline. By incorporating knowledge from social and organizational psychology, HCI can promote innovation and enhance learning and growth within the discipline.

Over the course of this chapter, we delved into the roots of social and organizational psychology. This section provided an overview of the development of social and organizational psychology and its intersection with HCI. Leading figures in social and organizational psychology, including Hugo Münsterberg and James McKeen Cattell, were discussed along with their contributions to the field. The emergence of industrial and organizational psychology during the World War I period was also explored, along with the advancements made in the discipline to apply the science of psychology in industry. The section also described a basic understanding of psychological research, such as descriptive and experimental research and their application to HCI. The importance of diversity and inclusivity in psychological research was highlighted, as it is necessary in order to be culturally inclusive and have psychological research be more reliable and valid (also necessary for the usability of human-computer designs). The section ended with a consideration of ethical issues in the field.

This chapter then covered the fundamentals of micro-organizational psychology. The section described the individual's attitude, motivation, well-being and stress; finding how it is crucial to take micro-level concepts into account in the development of design in order to engage employees. Then the underlying mechanisms of personnel recruitment, selection, and hiring; job performance and evaluation, alongside leadership, were described. The section highlighted how studying such aspects of the workplace are crucial to determine the accessibility and user-friendliness of technology, ending with an example of design for the aging population. Aside from this, the section highlighted training as an important tool in increasing KSAOs in the workplace. The subsection covered how training should incorporate a training needs analysis, feedback, and how technology can be leveraged to improve training in certain aspects. Another example provided was how surgical outcomes can be improved by HCI design, and how the principles of training can aid in the development of such tools. Overall, this section portrays that understanding the KSAOs of a job, alongside individual characteristics within that job, will aid in the development of HCI design.

Following this, this chapter covered fundamentals of macro-psychology including teamwork and organizations. The modern workplace necessitates the use of teams, and this chapter covered successful team components so that those in the field can incorporate this knowledge and make their teams more effective. Team science is necessary in any field, but the interdisciplinary nature of HCI necessitates some basic understanding of team concepts and principles of effective teams, offered in this section. Following this overview, the section provided an introduction to organizations and how organizational theories have evolved throughout time. The section described how organizations are organized into either flat or hierarchical structures. Two examples on HCI were provided, both highlighting how teams and organizations are necessary to understand if one is to be successful in the new world of virtual work and more hybrid workplaces.

Overall, we hope that this chapter has equipped the reader with the basic understanding of social and organizational psychology; and how its principles should be applied and guide the study of HCI.

ACKNOWLEDGMENTS

This work was partially supported by the U.S. Army Research Institute (ARI) for the Behavioral and Social Sciences and was accomplished under Cooperative Agreement Number W911NF-19-2-0173. This work was also partially supported by the National Aeronautics and Space Administration (NASA) Grant NNX16AP96G to Rice University, as well as NASA Grant NNX17AB55G to Rice University via Johns Hopkins University (Michael Rosen, P.I.). We would also like to thank Dr. Philip Kortum, who advised the authors in the examples provided for HCIs.

REFERENCES

Agency for Healthcare Research and Quality. (2017). About TeamSTEPPS. Retrieved January 26, 2023, from www.ahrq.gov/teamstepps/about-teamstepps/index.htm.

American Psychological Association. (2017). *American Psychological Association, Ethical Principles of Psychologists and Code of Conduct*. Washington, DC: American Psychological Association. https://www.apa.org/ethics/code/.

American Psychological Association. (n.d.). APA dictionary of psychology: Army tests. Retrieved October 3, 2022, from https://dictionary.apa.org/army-tests.

McmArmstrong, M., & Baron, A. (1995). *The Job Evaluation Handbook*. London: CIPD Publishing.

Bandura, A. (1971). *Social Learning Theory*. New York: General Learning Press.

Barber, A. E. (1998). Generating applicants. In *Recruiting Employees: Individual and Organizational Perspectives* (pp. 17–51). Thousand Oaks, CA: SAGE Publications, Inc. https://doi.org/10.4135/9781452243351.

Bauer, T. N., & Erdogan, B. (2011). Organizational socialization: The effective onboarding of new employees. In S. Zedeck (Ed.), *APA Handbook of Industrial and Organizational Psychology*, Volume 3. *Maintaining, Expanding, and Contracting the Organization* (pp. 51–64). Washington, DC: American Psychological Association. https://doi.org/10.1037/12171-002.

Bisbey, T. M., Reyes, D. L., Traylor, A. M., & Salas, E. (2019). Teams of psychologists helping teams: The evolution of the science of team training. *American Psychologist*, 74(3), 278–289. https://doi.org/10.1037/amp0000419.

Bisbey, T., & Salas, E. (2019). Team dynamics and processes in the workplace. In *Oxford Research Encyclopedia of Psychology*.

Bisbey, T. M., Traylor, A. M., & Salas, E. (2021). Transforming teams of experts into expert teams: Eight principles of expert team performance. *Journal of Expertise*, 4(2), 190–207.

Breckler, S. J. (1984). Empirical validation of affect, behavior, and cognition as distinct components of attitude. *Journal of Personality and Social Psychology,* 47(6), 1191–1205. https://doi.org/10.1037/0022-3514.47.6.1191

Brief, A. P., & Weiss, H. M. (2002). Organizational behavior: Affect in the workplace. *Annual Review of Psychology*, 53(1), 279–307. https://doi.org/10.1146/annurev.psych.53.100901.135156.

Carley, K. M. (1997). Extracting team mental models through textual analysis. *Journal of Organizational Behavior: The International Journal of Industrial, Occupational and Organizational Psychology and Behavior*, 18(S1), 533–558.

Cepeda, N. J., Pashler, H., Vul, E., Wixted, J. T., & Rohrer, D. (2006). Distributed practice in verbal recall tasks: A review and quantitative synthesis. *Psychological Bulletin*, *132*(3), 354–380. https://doi.org/10.1037/0 033-2909.132.3.354.

Chang, E., Milkman, K. L., Zarrow, L. J., Brabaw, K., Gromet, D. M., Rebele, R., Massey, C., Duckworth, A. L., & Grant, A. (2019, July 9). Does diversity training work the way it's supposed to? *Harvard Business Review*. https://hbr.org/2019/07/does-diversity-training-work-the-way-its-supposed-to.

Chernikova, O., Heitzmann, N., Stadler, M., Holzberger, D., Seidel, T., & Fischer, F. (2020). Simulation-based learning in higher education: A meta-analysis. *Review of Educational Research*, *90*(4), 499–541. https://doi.org/10.3102/0034654320933544.

Civil Rights Act of 1964, Public Law No. 88-352, 78 Stat. 241 (1964).

Davis, T. R. V., & Luthans, F. (1980). A social learning approach to organizational behavior. *The Academy of Management Review*, *5*(2), 281. https://doi.org/10.2307/257438.

Demerouti, E., Bakker, A. B., Nachreiner, F., & Schaufeli, W. B. (2001). The job demands-resources model of burnout. *Journal of Applied Psychology*, *86*(3), 499–512. https://doi.org/10.1037/0021-9010.86.3.499.

GrDonovon, J. J. (2001). Work motivation. In N. Anderson, D. S. Ones, H. K. Sinangil, & C. Viswesvaran (Eds.), *Handbook of Industrial, Work & Organizational Psychology - Volume 2: Organizational Psychology* (p. 53). Thousand Oaks, CA: SAGE Publications Ltd. https://doi.org/10.4135/9781848608368.n4.

Dweck, C. S. (1986). Motivational processes affecting learning. *The American Psychologist*, *41*(10), 1040–1048. https://doi.org/10.1037/0003-066X.41.10.1040.

Erez, M. (2011). Cross-cultural and global issues in organizational psychology. In S. Zedeck (Ed.), *APA Handbook of Industrial and Organizational Psychology, Volume 3: Maintaining, Expanding, and Contracting the Organization*. (pp. 807–854). Washington, DC: American Psychological Association. https://doi.org/10.1037/12171-023.

Gerstner, C. R., & Day, D. V. (1997). Meta-analytic review of leader-member exchange theory: Correlates and construct issues. *Journal of Applied Psychology*, *82*(6), 827–844. https://doi.org/10.1037/0021-9010.82.6.827.

Griffin, M. A., & Clarke, S. (2011). Stress and well-being at work. In S. Zedeck (Ed.), *APA Handbook of Industrial and Organizational Psychology, Volume 3: Maintaining, Expanding, and Contracting the Organization*. (pp. 359–397). Washington, DC: American Psychological Association. https://doi.org/10.1037/12171-010.

Hamilton, I. A. (2018). Amazon built an AI tool to hire people but had to shut it down because it *was discriminating against women. Business Insider*. Retrieved September 27, 2022, from https://www.businessinsider.com/amazon-built-ai-to-hire-people-discriminated-against-women-2018-10.

Hammer, L. B., & Zimmerman, K. L. (2011). Quality of work life. In S. Zedeck (Ed.), *APA Handbook of Industrial and Organizational Psychology, Volume 3: Maintaining, Expanding, and Contracting the Organization* (pp. 399–431). Washington, DC: American Psychological Association. https://doi.org/10.1037/12171-011.

Henrich, J. P. (2020). *The WEIRDest People in the World: How the West Became Psychologically Peculiar and Particularly Prosperous* (1st ed.). New York: Farrar, Straus and Giroux.

Hoyle, R. H., Harris, M. J., & Judd, C. M. (2002). Research methods in social relations. In S. Zedeck (Ed.), *APA Handbook of Industrial and Organizational Psychology, Volume 3: Maintaining, Expanding, and Contracting the Organization* (pp. 399–431). Washington, DC: American Psychological Association. https://doi.org/10.1037/12171-011.

Isaac, M., & Yaffe-Bellany, D. (2019, August 14). The rise of the virtual restaurant. *The New York Times*. Retrieved October 3, 2022, from https://www.nytimes.com/2019/08/14/technology/uber-eats-ghost-kitchens.html.

Judge, T. A., & Kammeyer-Mueller, J. D. (2012). Job attitudes. *Annual Review of Psychology*, *63*(1), 341–367. https://doi.org/10.1146/annurev-psych-120710-100511.

Kalinoski, Z., Steele-Johnson, D., Peyton, E., Leas, K., Steinke, J., & Bowling, N. (2013). A meta-analytic evaluation of diversity training outcomes. *Journal of Organizational Behavior*, *34*. https://doi.org/10.1002/job.1839.

Katz, D., & Kahn, R. (2015). The social psychology of organizations. In *Organizational Behavior 2* (pp. 152–168). Routledge. Katz, R. L. (1974). *Skills of an Effective Administrator*. Boston, MA: Harvard Business Press.

Katzell, R. A., & Austin, J. T. (1992). From then to now: The development of industrial-organizational psychology in the United States. *Journal of Applied Psychology*, *77*(6), 803–835. https://doi.org/10.1037/0021-9010.77.6.803.

Kilcullen, M., Feitosa, J., & Salas, E. (2022). Insights from the virtual team science: Rapid deployment during COVID-19. *Human Factors*, *64*(8), 1429–1440.

Kozlowski, S. W. J., Salas, E., & Society for Industrial and Organizational Psychology (SIOP). (2010). *Learning, Training, and Development in Organizations.* London: Routledge.

Kristof-Brown, A., & Guay, R. P. (2011). Person-environment fit. In S. Zedeck (Ed.), *APA Handbook of Industrial and Organizational Psychology, Volume 3: Maintaining, Expanding, and Contracting the Organization.* (pp. 3–50). Washington, DC: American Psychological Association. https://doi.org/10.1037/12171-001.

Landy, F. J. (1997). Early influences on the development of industrial and organizational psychology. *Journal of Applied Psychology, 82*(4), 467–477. https://doi.org/10.1037/0021-9010.82.4.467.

Li, J., Meyer, B., Shemla, M., & Wegge, J. (2018). From being diverse to becoming diverse: A dynamic team diversity theory. *Journal of Organizational Behavior, 39*(8), 956–970. https://doi.org/10.1002/job.2272.

Locke, E. A. (1975). Personnel attitudes and motivation. *Annual Review of Psychology, 26*(1), 457–480. https://doi.org/10.1146/annurev.ps.26.020175.002325.

McComb, S. A. (2007). Mental model convergence: The shift from being an individual to being a team member. In *Multi-level Issues in Organizations and Time* (Vol. 6, pp. 95–147). Emerald Group Publishing Limited.

Northouse, P. G. (2021). *Leadership: Theory and Practice.* Thousand Oaks, CA: SAGE Publications.

Oh, P.-J., Jeon, K. D., & Koh, M. S. (2015). The effects of simulation-based learning using standardized patients in nursing students: A meta-analysis. *Nurse Education Today. 35*(5), 6–15.

Pandey, J. (2019). Factors affecting job performance: An integrative review of literature. *Management Research Review: MRN, 42*(2), 263–289. https://doi.org/10.1108/MRR-022018-0051.

Peters, D., Calvo, R. A., & Ryan, R. M. (2018). Designing for motivation, engagement and wellbeing in digital experience. *Frontiers in Psychology, 9*, 797–797. https://doi.org/10.3389/fpsyg.2018.00797.

Ployhart, R. E., & Kim, Y. (2013). Strategic recruiting. In D. M. Cable & K. Y. T. Yu (Eds.), *The Oxford Handbook of Recruitment.* Oxford: Oxford University Press. https://doi.org/10.1093/oxfordhb/9780199756094.013.0002.

Ployhart, R. E., & Schneider, B. (2012). The social and organizational context of personnel selection. In N. Schmitt (Ed.), *The Oxford Handbook of Personnel Assessment and Selection.* Oxford: Oxford University Press. https://doi.org/10.1093/oxfordhb/9780199732579.013.0004.

Porfeli, E. J. (2009). Hugo Münsterberg and the origins of vocational guidance. *The Career Development Quarterly, 57*(3), 225–236. https://doi.org/10.1002/j.2161-0045.2009.tb00108.x.

Rau, B. L., & Adams, G. A. (2013). Recruiting older workers: Realities and needs of the future workforce. In D. M. Cable, & K. Y. T. Yu (Eds.), *The Oxford Handbook of Recruitment.* Oxford: Oxford University Press. https://doi.org/10.1093/oxfordhb/9780199756094.013.0007.

Ravasi, D., & Schultz, M. (2006). Responding to organizational identity threats: Exploring the role of organizational culture. *The Academy of Management Journal, 49*(3), 433–458. https://www.jstor.org/stable/20159775.

Rosen, M. A., Salas, E., Wu, T. S., Silvestri, S., Lazzara, E. H., Lyons, R., Weaver, S. J., & King, H. B. (2008). Promoting teamwork: an event-based approach to simulation-based teamwork training for emergency medicine residents. *Academic Emergency Medicine: Official Journal of the Society for Academic Emergency Medicine, 15*(11), 1190–1198. https://doi-org.ezproxy.rice.edu/10.1111/j.1553-2712.2008.00180.x

Rosenberg, M. J., Hovland, C. I., McGuire, W. J., Abelson, R. P., & Brehm, J. W. (1960). *Attitude Organization and Change: An Analysis of Consistency Among Attitude Components. (Yales Studies in Attitude and Communication.).* Yale University Press.

Ryan, R. M., & Deci, E. L. (2017). *Self-Determination Theory: Basic Psychological Needs in Motivation, Development, and Wellness.* New York: Guilford Publications.

Salas, E., Weaver, S. J., & Shuffler, M. L. (2012). Learning, training, and development in organizations. In S. W. J. Kozlowski (Ed.), *The Oxford Handbook of Organizational Psychology*, Volume 1. (pp. 330–372). Oxford: Oxford University Press. https://doi.org/10.1093/oxfordhb/9780199928309.001.0001.

Salas, E., Wildman, J. L, & Piccolo, R. F. (2009). Using simulation-based training to enhance management education. *Academy of Management Learning & Education, 8*(4), 559–573. https://doi.org/10.5465/AMLE.2009.47785474.

Salas, E., Wilson, K. A., Lazzara, E., King, H. B., & Augenstein, J. S. (2008). Simulation-based training for patient safety: 10 principles that matter. *Journal of Patient Safety, 4*(1). Retrieved from https://commons.erau.edu/publication/974.

Salas, E., Wilson, K. A., Priest, H. A., & Guthrie, J. W. (2006). Design, delivery, and evaluation of training systems. In G. Salvendy (Ed.), *Handbook of Human Factors and Ergonomics* (3rd ed., pp. 472–512). Hoboken, NJL John Wiley & Sons, Inc. https://doi.org/10.1002/0470048204.ch18.

Schleicher, D. J., Hansen, S. D., & Fox, K. E. (2011). Job attitudes and work values. In S. Zedeck (Ed.), *APA Handbook of Industrial and Organizational Psychology, Volume 3: Maintaining, Expanding, and Contracting the Organization.* (pp. 137–189). Washington, DC: American Psychological Association. https://doi.org/10.1037/12171-004.

Smith, J. H. (1998). The Enduring Legacy of Elton Mayo. *Human Relations, 51*(3), 20S–38S. https://doi.org/10.1177/001872679805100302 katzel

Society for Industrial and Organizational Psychology (SIOP). (n.d.). SIOP vision, mission, values, and goals. Retrieved October 3, 2022, from https://www.siop.org/AboutSIOP/Mission#:~:text=Our%20 Mission%20Statement%3A,teaching%20of%20industrial%2Dorganizational%20psychology.

Society for Industrial and Organizational Psychology (SIOP). (2022). Top ten work trends 2022. Retrieved October 3, 2022, from https://www.siop.org/Business-Resources/Top-10WorkTrends.

Stahl, G. (2006). *Group Cognition: Computer Support for Building Collaborative Knowledge (Acting with Technology).* The MIT Press.

Thayer, R. E. (1989). *The Biopsychology of Mood and Arousal.* Oxford: Oxford University Press.

Volpone, S. D., Thomas, K. M., Sinisterra, P., & Johnson, L. (2013). Targeted recruiting: Identifying future employees. In D. M. Cable, & K. Y. T. Yu (Eds.), *The Oxford Handbook of Recruitment.* Oxford: Oxford University Press. https://doi.org/10.1093/oxfordhb/9780199756094.013.018.

Weaver, S., Salas, E., Lyons, R., Lazzara, E., Rosen, M., & DiazGranados, D. (2010). Simulation-based team training at the sharp end: A qualitative study of simulation-based team training design, implementation, and evaluation in healthcare. *Journal of Emergencies, Trauma, and Shock, 3*(4), 369–379.

Yerkes, R. M. (1921). *Psychological Examining in the United States Army: Edited by Robert M. Yerkes* (Vol. 15). US Government Printing Office.

Zendejas, B., Brydges, R., Hamstra, S. J., & Cook, D. A. (2013). State of the evidence on simulation-based training for laparoscopic surgery: A systematic review. *Annals of Surgery, 257*(4), 586–593. https://doi.org/10.1097/SLA.0b013e318288c40b.

10 Decision-Making and Problem-Solving

Don Harris

10.1 INTRODUCTION

Diehl (1991) estimated that decision errors contributed to 56% of accidents in airlines and 53% of accidents in military aviation. Jensen and Benel (1977) reported that 51% of fatal general aviation accidents between 1970 and 1974 were associated with decision errors. Studies of military accidents have found that such errors contributed to 45% of accidents in the US Air Force and 55% in US Naval aviation (Shappell & Wiegmann, 2004); Li and Harris (2005b) observed 53% of accidents in the Republic of China (Taiwan) Air Force had an instance of decision-making error within them. Approximately 71% of accidents in Taiwanese-registered commercial aircraft contained some aspect of a decision error (Li et al., 2008). Put more succinctly, papers reviewing underlying contributory factors suggest that poor decisions are implicated in at least half of all aviation mishaps.

In the US National Motor Vehicle Crash Causation Survey undertaken between 2005 and 2007 which collected on-scene information about factors leading up to the accident, 94% of accidents were attributed to the driver (Singh, 2015). Decision errors (e.g. driving too fast for the conditions, making a false assumption about other drivers' actions; undertaking an illegal manoeuvre) accounted for 33% of road traffic accidents. Yıldırım et al. (2019) reported that 75% of maritime collisions and 73% of ship groundings resulted from decision errors. These figures correspond with those of Chauvin et al. (2013) who in a similar analysis reported that 82% of collisions at sea were attributable to this type of error. In the rail industry, the number of incidents attributable to decision-making varies greatly with respect to the nature of the event. Forty per cent of incidents where the driver failed to pull up correctly at a station were attributable to decision-making, but less than 5% of 'Signals Passes At Danger (SPAD)' events were attributed to such errors (Madigan et al., 2016). Thirty-eight per cent of derailments involved a decision error (Ebrahimi et al., 2021). In non-transport settings, Lenné et al. (2012) reported that 34% of mining accidents in Australia involved a decision error: Berner and Graber (2008) suggested that 15% of diagnostic errors in medicine were decision errors.

At first, these figures may appear alarming, but these statistics need closer consideration. Half of human behaviour involves decision-making. Many decisions are quite trivial and have neither a right nor a wrong answer, so they can never be judged against a criterion to determine if there has been an error. Shall I have tea or coffee? What colour shall I paint the kitchen? Many *post hoc* attributions of error in accident reports have been criticized as being judgements (on the part of the investigator) made with hindsight (Woods et al., 1994). Defining what is and isn't a decision error is not straightforward, hence deciding if one has occurred is equally problematic. Sometimes decisions made that seemed to have no consequence at the time subsequently have unforeseen ramifications many years later.

Whenever discussing decision-making from a scientific perspective, it is implicit that there is a criterion against which quality can be judged. Furthermore, many decisions are made in a complex, uncertain, constantly changing environment, often with incomplete information. In such environments, even good decisions can result in unfavourable outcomes. Retrospectively evaluating decision outcomes is easy: evaluating the quality of the decision and the adequacy of the process at the time it is made is more difficult.

DOI: 10.1201/9781003495109-10

Defining precisely what is meant by decision-making is neither simple nor straightforward, but there is an intimate relationship between decision-making and error. Furthermore, behaviour is rarely predicated upon a single decision but upon a series of related decisions. Where there is a definable, desirable outcome to work towards, some would argue that this is problem-solving. This distinction will be considered in a little more detail later but is probably a moot point.

Initial approaches to the study of decision-making were based on an experimental, rationalist approach centred around a single decision with a single, optimal outcome. More recently, emphasis has shifted to a more real-life approach, aimed at understanding how people make decisions in complex real-world settings, rather than identifying the factors and processes that underlie 'successful' decisions. In complex, dynamic, high jeopardy-risk environments, such as the military, medicine, and transport, 'good' decision-making is essential. In such situations, there are often multiple criteria to satisfy, and it is sometimes difficult to define what is 'success' or the 'best' decision (in contrast to 'failure', which is easy to assess but is often a conclusion reached in retrospect). As a result of the stakes, efforts have been made to improve the decision-making process to help ensure 'good' outcomes. Efforts to train decision-making have often met with mixed success, however, this has often been a result of the poor specification and targeting of the training regime. Modern decision-support systems have provided an alternative approach to training but only in limited circumstances.

Decision-making is pervasive and as a result, the approaches to its study have been complex, diverse, and difficult to draw a boundary around. The key to the academic study of decision-making is predicated upon a desirable or specifiable criterion against which behaviour can be judged. Problem-solving bounds outcomes in such a manner. Decision-making (in general) may not always satisfy this criterion. For example, design decisions judged against product requirements may be judged against explicit criteria, but aesthetic decisions would be difficult to assess in any meaningful manner.

10.2   DECISION-MAKING VS. PROBLEM-SOLVING

Problem-solving and decision-making are two related, but some would argue distinct, activities. There are various opinions concerning the nature of the relationship between the two activities. In both cases, these undertakings involve a number of separate cognitive processes; however, neither should be thought of as a single activity. Taylor (1965) argued that the principal distinction between decision-making and problem-solving is largely related to the ultimate product, as both activities share a great many underlying processes. Taylor suggested that decision-making involved the choice between several alternative potential courses of action. It often had a longer-term focus and may involve a less-specifically defined objective and outcome (e.g. how shall I develop my company? what will be the design for my new house?). In decision-making, objectives must first be established against which alternative courses of action can be developed and evaluated. On the other hand, the result of problem-solving is the solution to a specific issue – there is usually a particular focus to the activity.

Identification and diagnosis are essential steps in problem-solving: to know that you need to solve a problem you first have to realize that you have a problem to solve (cf. decision-making, where in effect you can almost generate your own problem). You then have to diagnose the nature of the problem as a precursor to generating potential courses of action. Hence, a problem is better-defined, time-bound, and outcomes are easier to observe and evaluate (e.g., 'why will my car not start' and 'how do I fix it'?). Causes of problems can be deduced from analysis. Mayer (1983) suggested that problems are characterized by having an initial state and a goal state. Problem-solving is about navigating the path between the two.

In contrast, Klein and Crandall (1995) suggested that decision-making is a component of the problem-solving process, with several different decisions being required during the sub-processes of problem-solving, such as defining goals, generating alternatives, evaluating options, and finally implementing the alternative most likely to be successful. Similar perspectives were shared by Orasanu and Connolly (1993) and Jenkins et al. (2008) who also adopted the standpoint that decision-making was predicated upon the requirement to address a defined problem. If these definitions are applied, much of what is written about decision-making is actually about problem-solving;

however, they are probably best thought of as complimentary activities, rather than discrete categories. Furthermore, problem-solving may require several decisions to be made to effect an outcome.

However, not all decisions require a problem. Decision-making is all pervasive: we make hundreds of decisions every day, mostly trivial in nature, and in many cases, there are no 'poor' outcomes as defined by any objective metric. But such a measure is the key to the scientific study of decision-making or problem-solving. There is assumed to be an intimate association between decision-making and performance from which the effectiveness of the process can be judged.

There is also another key component within problem-solving and decision-making. Judgements are an important part of the process and are characterized by having a value component associated with them. For example, in the clinical decision-making process, judgements may involve assessing the potential consequences (risk vs. benefit) of a particular course of action (Standing, 2020). Risk is frequently cited as a critical judgement in the processes (Janis & Mann, 1977; Jensen et al., 1997) however it is not the only judgement to be made. Ajzen (1996) proposed that underlying any judgements made is the person's belief system, which may also include other issues such as moral evaluations of various potential courses of action. The 'value' of potential outcomes is critical when considering decision-making from the perspective of certain models especially those which adopt a normative/classical decision-making paradigm. However, the notion of 'value' is in itself, often a difficult concept. Judgements are an issue that will be revisited several times in this chapter.

Lehto and Nanda (2021) suggested that the theoretical distinction between decision-making and problem-solving is not useful and that many authors use the terms interchangeably. For the purposes of this discussion, decision-making will be regarded as a component in a problem-solving process: it will be assumed that there will be a well-defined purpose (this allows criteria for evaluation). However, implicit in this perspective is the recognition that a problem exists that requires action to be taken. Situation assessment is a fundamental precursor for all aspects of decision-making (Noble, 1993; Prince & Salas, 1997). As will be seen in the following sections, this is where decision-making in real life and decision-making in the laboratory diverge. When a participant arrives to undertake an experiment, they know that they will be faced with a situation (a problem) that will require action. In real life, especially in complex, dynamic environments, it is not always clear if there is an issue to be addressed and/or the nature of the problem. Hence, all field-based studies of decision-making are initially predicated on understanding the situation.

10.3   ERROR AND ITS RELATIONSHIP TO DECISION-MAKING

Numerous decisions are made on a daily basis, and in most instances, there are no potential 'wrong' outcomes as they are not evaluated against a required level of performance. Hollnagel (1998) suggested three criteria that must be fulfilled before any action can really be described as 'erroneous'.

- *Criterion* principle requires a specified criterion or standard against which performance can be compared. If the action falls short of what is required from an external verifiable viewpoint (what is required by 'the system') then it maybe an 'error'. If the action executed is not what the person intended, there is a failure to perform compared against some internal, cognitive (but less verifiable) criterion. However, what may be an 'error' from an external perspective may not be an 'error' from an internal perspective (see Reason's 'violations' category of unsafe acts: Reason, 1987: 1990).
- *Performance* shortfall criterion demands that the action that leads to a measurable performance shortfall.
- *Volition* criterion necessitates that there must be the opportunity for the person committing the 'error' to act in such a way that would not be considered 'erroneous'.

With reference to Hollnagel's (1998) first criterion, it is impossible for many decisions to ever be ultimately labelled as being 'erroneous'. On reflection, 'bad' decisions can be made. These are defined in retrospect by the context of the activity and the required outcome. There is an intimate link between the scientific study of decision-making and that of human error but at the time when a decision is taken, there is often no standard against which the behaviour can be assessed.

Orasanu and Fischer (1997) described a process decision-making model containing two basic components: situation assessment and selecting a subsequent course of action. Decision errors may arise as a result of failing to recognize a problem or misinterpreting it, or if the issue is identified correctly, adopting an inappropriate course of action. However, when a poor decision becomes an error, it is a moot point. Woods et al. (1994) suggested that 'error' is always a judgement made in hindsight. The same is true in decision-making and problem-solving. Furthermore, in many cases, the time between the 'poor' decision and the unwanted outcome may stretch to years or decades, and the relationship between the decision and outcome may be probabilistic, rather than causative. Design decisions, such as the shape of the roadway section of the Tacoma Narrows Bridge which induced aeroelastic flutter causing its subsequent collapse, or the structural failures of the De Havilland Comet airliners induced by metal fatigue related to the design of the square passenger windows, can only be designated as design errors in retrospect. However, given the state of knowledge at the time, these (poor) design decisions would fail to be categorized as 'erroneous' as they do not fulfil the criterion in Hollnagel's third category: volition. The knowledge relating to the mechanisms of their failure did not exist at the time they were conceived.

Rasmussen (1983, 1986) described three levels of cognitive processing: skill-based; rule-based, and knowledge-based. In the skill-based mode of operation (the least cognitively complex) the person is employing highly learnt physical actions; well-developed skilled behaviours which are sub-conscious (e.g., driving, forming letters and words when writing, or swinging a golf club). There is virtually no conscious monitoring of these behaviours once initiated. In Reason's (1987, 1990) Generic Error Modelling System, based on Rasmussen's model of cognitive processing, these result in errors categorized as 'slips' and/or 'lapses'. In this discussion of decision-making, these skill-based behaviours and related errors need not concern us.

The next highest level of cognitive processing in Rasmussen's hierarchy is rule-based behaviour. These are well-learned 'IF...THEN' rules based upon previous experience (e.g., 'IF the traffic lights are on red THEN stop'). These rules can become quite complex, with many conditional criteria. They are based on rules stored in Long-Term Memory (LTM) and come from many sources (such as training; procedures that have been learnt, or principles related to solutions encountered in previous problems). 'IF... THEN' rules are commonplace in problem-solving, especially when undertaking structured diagnostic procedures ('IF the house lights suddenly go off, THEN first check the circuit breakers'). Such rules will often be self-evident in the following sections discussing the naturalistic decision-making paradigm and training. In Reason's Generic Error Modelling System, the misapplication of these rules is categorized as 'mistakes'. In the HFACS (Human Factors Analysis and Classification System – Shappell & Wiegmann, 2001; Wiegmann & Shappell, 2003) an error classification system developed from Reason (1987, 1990), the misapplication of IF...THEN rules is categorized as a decision error. Many of the previously quoted statistics concerning the prevalence of decision errors are derived from studies employing HFACS.

Rasmussen's highest level of cognitive activity is knowledge-based behaviour. Knowledge-based behaviour is required in novel or unanticipated situations where there are no available rule-based behaviours to draw upon. In these instances, it is required to recognize and categorize the situation, and explicitly formulate goals, actively engaging in problem-solving. As there are no previously learnt rules to apply, solutions must be derived from first principles using processes and knowledge from LTM. This is a conscious activity. Such knowledge-based behaviours are essentially problems in higher-order decision-making. Complex decision-making activities at the knowledge-based level may involve multiple problems to be solved, either sequentially or in parallel (consider the attempts to control the Chernobyl nuclear power station) or have competing, potentially contradictory criteria (e.g., a doctor/surgeon making decisions about the quantity vs. quality of life; long term vs. sort term gains, etc.). Unfortunately, this is also the level of cognitive activity that is most likely to result in a significant error (when assessed in retrospect...) as a result of combining pieces of information using inappropriate principles; not knowing all the relevant facts; or using inappropriate data/information, etc. In Reason's error model, these are also categorized as 'mistakes' but all of these mistakes can be categorized as 'poor' decisions.

Problem-solving can be regarded as a process involving several different types of decisions (see Klein & Crandall, 1995). Using Reason's terminology, at its root a poor outcome from the problem-solving process may be the result of one (or more) mistakes during the course of action. The definition of what is 'correct' and what is 'incorrect' (an error) when considering the outcomes of complex activities is often a verdict based on the final result, not the process (see following discussions concerning naturalistic decision-making and training). At the time of engaging in the design decision-making processes for the de Havilland Comet there were obvious criteria to meet defining success (hold fuselage pressurization; see out of the passenger windows) but less information available concerning potential pitfalls to avoid (see Hollnagel's third criterion: volition). In 1953, no one understood the role of metal fatigue and certainly had not understood this issue during its design.

The obvious exception to these instances are decisions that involve the deliberate breaking of rules ('violations' in the nomenclature of Reason) even if these 'poor' decisions are made for what may initially seem to be ostensibly sensible (or practical) reasons. Such decisions, which deliberately violate norms and procedures, are exemplified by accidents such as the Stresa–Mottarone cable car crash (23 May 2021), where the emergency brake was deliberately disabled as it had been malfunctioning repeatedly resulting in the frequent suspension of the service. When the haulage cable failed, the cabin, carrying 15 passengers, slid back down the cable car way until it collided with a pylon and then fell 54 m, before then further tumbling down the side of the mountain. Only one person survived. Instances such as this accident are interesting from a decision-making perspective, as the utility associated with disabling the safety function outweighed the safety considerations of operating the cable car, an issue visited when considering normative decision-making models.

The study of decision-making and error is inextricably interlinked. Although slightly arbitrary, in the following, emphasis will be placed upon processes and influences in decision-making and problem-solving, rather than upon outcomes, which may ultimately be described as the subsequent 'error'.

10.4   DECISION-MAKING MODELS

Hastie (2001) suggested decisions comprise three fundamental components:

- A course of action to be decided upon (which includes taking no action)
- Belief about states, processes, and events in the world, including the potential outcome stages and the methods by which they will be achieved.
- The values (utilities) associated with potential outcomes.

However, identification and diagnosis are also essential steps preceding the three aspects described above. To know that you need to make a decision you first have to realize that you have a decision to make. This is an issue in situation assessment/situation awareness.

Hastie also described the decision process in terms of three sequential stages: alternative courses of action, consequences, and uncertainty of events. Consequences are the evaluation of potential outcomes (e.g., desirable/undesirable; return on the amount staked). Uncertainty estimates reflect the decision-maker's judgements of the likelihood of occurrence of certain critical events. Preferences refer to the choice of one course of action over another. He drew a clear distinction between 'judgement' and 'decision-making':

> Decision-making refers to the entire process of choosing a course of action. Judgment refers to the components of the larger decision-making process that are concerned with assessing, estimating, and inferring what events will occur and what the decision-maker's evaluative reactions to those outcomes will be.

Hastie (2001, p. 657)

The topic of judgement has already been touched upon and will be revisited again. Goldstein and Hogarth (1997) argued that behavioural research has considered judgement and decision-making as separate research endeavours. Judgement concerned the manner by which people assessed and integrated multiple, potentially incomplete and conflicting cues from the environment using their experience, knowledge, and beliefs. The quality of such judgements was inferred from the correspondence between the judgement and a target criterion condition (Hastie & Rasinski, 1987). In contrast, decision-making was concerned with understanding the processes and outputs underlying preferential choice and action. How do people select a course of action to achieve a goal?

Lehto et al. (2012) defined three basic types of decision-making models. Normative decision models, derived from the economic model of man; naturalistic decision models, which provide a description of the decision-making of experts in real life situations, and behavioural (process) approaches. Many investigations of decision-making and problem-solving do not, however, fall neatly into one of these categories. Behavioural and naturalistic decision models focus on describing human decision-making. The emphasis in naturalistic models tends to be on describing what is actually done, in contrast to behavioural models, which are inclined to illustrate a more idealized process. Normative models adopt a more experimental approach where the emphasis is very much on what *should* be done to make a 'good' decision. Naturalistic and behavioural models also err towards describing decision-making in a dynamic situation, where many related decisions may be required to address a situation. Normative approaches tend to focus on a single decision event in a static context. These distinctions and contrasts, however, are by no means perfect. All three basic types of decision model encompass an element of values and judgement. These approaches are described in the following sub-sections.

10.4.1 NORMATIVE DECISION-MAKING MODELS

For the purposes of this discussion, all decision-making must automatically assume a choice between alternatives and a degree of risk (in the broadest sense of the term). Even electing to do nothing is a choice; in choosing a certain course of action there may be the possibility of rejecting a better outcome (on some *measurable* criterion) one which may be less likely but provide a better return. When gambling or in business, risk may be framed simply in terms of monetary return. In military combat, it may be in terms of the probability of a kill or the amount of damage done; in medicine, it may be the quantity of life when choosing between procedures. The key thing is that the outcome must be *knowable* and *quantifiable* in a meaningful manner.

'Normative' decision-making models are also sometimes termed 'Classical' Decision-Making (CDM) models. Baron (2004) argued that normative models are the product of 'reflection and analysis', rather than being based on data and analysis. They describe what people should do (not what they do, do) thereby providing a basis for characterizing the biases and shortcomings in human decision-making but not describing the actual human cognitive processes *per se*.

Subjective Expected Utility (SEU) is one such model. SEU is an extension of Expected Utility (EU) models, rational models of decision-making that originated from the economic model of man. The EU of a decision is a product of its payoff and probability of occurrence. These can be quantified and the outcome with the highest utility should be the one pursued by any rational decision-maker. A rational decision-maker should also follow the four Von Neumann–Morgenstern axioms for choice of action, preferring the option that maximizes the EU (Von Neumann & Morgenstern, 1953). The four axioms are:

- *Completeness* means a decision-maker has consistent, well-defined preferences and can decide between alternatives.
- *Transitivity* means a decision-maker is consistent in their order of preference of options.

- *Continuity* requires that in the case of a middle outcome preference, there is some probability that the decision-maker is indifferent between the most preferred and least preferred outcome.
- *Independence* requires that a rational decision-maker's preference between two probabilistic outcomes is not influenced when considering these outcomes with respect to a third, independent choice of a fixed probability. It is simply an axiom requiring consistency among preferences.

However, while these axioms may reflect the basis for rational choice when making decisions such as which lottery to play, when applied to other types of decision (such as whether to undergo a potentially risky medical procedure) they are less applicable (see Cohen, 1996). The rationality for such a decision depends on the decision-maker's perspective. The surgeon may make the assessment based on the EU considered over a large sample of patients. The patient considering the procedure will make a one-time decision, where utility is predicated upon longer-term benefits to their health, but also dis-benefits which include the non-trivial risk of their own death.

Under SEU the value of any commodity differs from person to person; this adds a human element. SEU is an approach to describing how people make decisions where there is an element of risk involved. It has two components that are common in all definitions of risk (a probability estimate relating to the likelihood of success and a utility function), but these are now subjective (personal). Utility now reflects the expected value of an outcome but can be very different for different people. For example, a $20 bill found on the street by a homeless person has much higher utility (value) than if the same $20 bill were found by Elon Musk or Jeff Bezos. The EU of an outcome, though, still remains a product of its utility weighted by its probability of occurrence.

EU and SEU suggest that a decision-maker assesses the potential outcomes and then evaluates their utility. Trade-offs between outcomes may be evaluated, for example choosing between a medical procedure that will *certainly* improve a patient's condition for a month and another procedure that *might* improve a patient's condition for a year. On this basis, a rational choice can be made. Furthermore, people making decisions may be characterized on a continuum from highly risk-inclined to risk-averse. A risk-neutral decision-maker will estimate the EU of winning a bet to be equal to the utility of the value of the stake; a risky decision-maker regards the utility of the value of the winning bet to be greater than that of the stake (and *vice versa* for the risk-averse gambler, if there is such a person)!

However, there are some basic issues with SEU theories. From even the most rudimentary analysis of human behaviour it can be observed that in general, people do not behave in a consistent and rational manner. Kahneman and Tversky (1979) noted that the conditions for risk acceptance (gains) and risk aversion (losses) were different when purchasing lottery tickets compared to buying insurance policies. Levy (1992) described how these observations were a problem for SEU theory. The rational assumption should be that there is little utility where stakes are low, but the probability of winning is also very, very low. Winning small amounts where the likelihood of success is high is a rational decision. However, millions of people regularly purchase tickets for national lotteries. The probability of correctly selecting the seven winning numbers on the EuroMillions lottery is 1 in 139,838,160, but it is estimated that 80–100 million people each week purchase a ticket for the EuroMillions draw. SEU can account for such gambles by assuming convex utility functions, but this contradicts other behaviours where exactly the same people seem to be risk averse, for example when buying insurance which involves a relatively small investment to avoid the small probability of a very large loss. SEU can explain one or the other in a given individual, but not both. The objective rationality of most decision-makers is put into question as such behaviours contravene the completeness and transitivity axioms from Von Neumann and Morgenstern (1953). To explain both, it needs to be assumed that people have different utility functions for different domains.

As an alternative, Kahneman and Tversky (1979) proposed Prospect Theory, which postulates that outcomes are evaluated with reference to some pre-determined point rather than being appraised in absolute terms. To illustrate, the perceived value of earning $10 is far smaller when the prospect is of winning $1,010 against a stake of $1,000 than when betting $20 against a stake of $10. Furthermore, people's decision-making becomes more conservative (risk averse) when potential decision outcomes are framed in terms of losses, rather than gains.

While SEU theory makes sense when expressed in quantifiable gains, such as money (and a great deal of normative decision theory is predicated upon a rational, economic model of man), it becomes more difficult to employ in other contexts. For instance, in some circumstances, a pilot may be tempted to push on into bad weather and try to land at their intended destination: they may have almost arrived, and their nominated alternates are some distance away (see Wiggins et al., 2014). The utility of the decision to attempt a landing in bad weather could be high if successful, when measured in terms of time saved, flight pay bonuses, or simply passenger and crew convenience. However, the metrics by which a failure of this gamble paying off will be quite different. They will probably be measured in death, injury, and/or damage.

Multi-Attribute Utility Theory (MAUT) is an alternative normative model that addresses some of the shortcomings of SEU (Keeney & Raiffa, 1976). MAUT incorporates an evaluation of the potential EU of a decision on multiple dimensions. Furthermore, some of these metrics may be deemed to be more important than others (e.g., the aforementioned safety-related measures compared to the expediency measures). It is considered to be a compensatory decision-making strategy as varying values of different aspects can be balanced again with each other in some way; hence all information is considered in the process. Higher values on less important measures can be balanced by lower values on more important criteria.

MAUT or MCDM/MAUA (Multiple Criteria Decision Making/Multi-Attribute Utility Analysis) has gained in popularity as a basis for software-based decision aids (e.g., Logical Decisions™, freeware made available under the MIT license: http://www.logicaldecisions.com). This approach has been used to aid decision-making in a number of complex instances, for example, deciding on the location for a new airport or identifying which nuclear power plant to decommission (Wallenius et al., 2008). Sophisticated MAUT/MCDM/MAUA software allows for many variables to be considered in the decision-making process, measured using various metrics, and allows a variety of outcomes to be considered and compared. For example, when considering which car to purchase next, using a MAUT/MCDM/MAUA-based approach the first step would be to identify the most important variables that will drive the choice (price, performance, fuel consumption, NCAP safety rating, etc.). Each factor is then weighted according to personal preference (e.g., fuel economy may be more important than acceleration). Various formulae are then applied to produce the optimal choice based on the expressed preferences. In this way, preferences and personal biases may be entered into the software, but it is unlikely that the program reflects in any way the human decision-making processes. Indeed, it is designed specifically to aid human decision-making, i.e. augment the human cognitive system and avoid biases, *not* to mimic the psychology of decision-making.

Implicit in all normative approaches is the notion that there is an optimal decision. The decision-maker is characterized as being totally informed, rational, and infinitely sensitive (Edwards, 1954). They recognize and describe what is known about the problem space (knowing not only all potential courses of action open to them but also what the outcomes of such actions will be) and will collect more data if required. They then formally analyse options, applying the principles of utility, probability, and/or risk to arrive at an optimal solution from the range of alternatives derived and compared (Beach & Lipshitz, 1993). All possible outcomes are then evaluated rationally. Models of economic man assume the options available exist on a continuum, rather than being discrete choices. It is presumed that the decision-maker is capable of evaluating these potential outcomes by applying some kind of logical weighting before coming to a definitive conclusion. There is also an implication that the decision-maker is neither time-constrained nor under any degree of stress or organizational influence.

Approaches to the study of decision-making that adopt a perspective based on the normative paradigm have tended to use a quasi-laboratory approach, where problems are characterized using just a few variables and the range of prospective outcomes is pre-defined. Focus is very much upon identifying the input variables that affect the final decision – the output (Lipshitz et al., 2001). From an ecological validity perspective, this is demonstrably not true. Only rarely is any single decision event an end in itself: one decision usually precedes many others, and later decisions are predicated upon earlier decisions. Furthermore, when time is limited, decision-makers tend to adopt a non-compensatory approach. Not all factors are evaluated, and cognitive shortcuts are adopted (Payne et al., 1988). This is very much the case in a dynamic decision-making situation, as commonly faced by the emergency services, the military, and in other critical situations. Both the SEU and MAUT approaches are only applicable when non-real-time decisions are required. They cannot accommodate dynamic decision-making situations. A brief re-examination of the statistics concerning decision errors described in the opening paragraphs reveals that all of these were highly dynamic circumstances.

The general upshot of normative studies has generally been to demonstrate the human biases and irrationality when making decisions, for instance, how framing exactly the same problem in terms of losses rather than gains produces different outcomes (Kahneman & Tversky, 1979; Tversky & Kahneman, 1981). Research has largely demonstrated that people do not adopt an SEU type of approach when making decisions (and hence, by extension also do not pursue a MAUT method), even when the decisions are relatively simple and would not exceed the capacity of the human cognitive system (Kahneman et al., 1982). Azjen (1996) has argued that SEU merely acts as a normative model against which the results from experimental judgements and decisions can be compared. It was never intended as an accurate model of decision-making. Indeed, if Hollnagel's (1998) criteria for an error are applied (in particular the criterion that there must be a standard against which behaviour can be compared) all shortcomings where participants demonstrate biases and non-rational decision-making in experiments can be considered as errors.

10.4.2 Naturalistic Decision-Making Models

Simon (1955) argued that the information processing requirements of the rational decision-making processes involved in normative (CDM) compensatory-type approaches exceeded the limits of the human information processing capacity. Furthermore, the decision-maker often also has limited time to come to a conclusion and limited knowledge about the situation. They do not have complete knowledge of the potential consequences of the alternative approaches that may be employed. Human beings have 'bounded rationality'. Compensatory approaches may be useful when the context is limited, the environment isn't dynamic, and time is available. However, under time pressure people adopt non-compensatory decision-making strategies using cognitive shortcuts (Payne et al., 1988).

To reduce cognitive load and speed up the decision-making process necessitates the use of heuristics. The result is that decision-makers often seek *satisfactory* solutions to problems rather than trying to make optimal decisions, as prescribed by the normative approach. However, the utilisation of heuristics can be a double-edged sword: it can introduce efficiencies in the decision-making process, but their use can also be a source of bias and error. When pursuing a satisfactory decision, a full review of alternatives is not made. The review of options continues until a 'good enough' solution is identified (one that is satisfactory or sufficient depending upon some kind of stopping rule, dependent upon the situation). Options are considered sequentially, not in parallel. However, the downside of this approach is that the best option may never be considered. Naturalistic Decision Making (NDM) examines how subject matter experts use their experience to make decisions in an operational context (Zsambok & Klein, 1997). The NDM paradigm has its origins in the requirement to understand decision-making in a command-and-control context, however it was quickly applied in a wide range of other situations as a useful way of characterising the decision-making process.

In contrast to normative/CDM models, NDM approaches aim to describe the decision-making process, rather than predict its outcome. The research approach is usually qualitative and descriptive, rather than experimental/empirical. Normative decision-making models focus on the quality of the final decision whereas NDM stresses the processes and influences underlying the decision (Klein, 1993; Klein et al., 1989). Studies of NDM take place (usually retrospectively) in an applied setting. The effects of issues such as organizational constraints, social conventions, time, and money are also considered. All decisions are undertaken in a social and environmental context.

Orasanu and Martin (1998) argued that it is often impossible to ascertain the ultimate quality of a decision; outcomes cannot be used as reliable indicators. When operating in an uncertain environment with incomplete information, good decisions can sometimes result in bad outcomes (and *vice versa* – you can 'get lucky'). The effect of chance should not be underestimated. When evaluating a decision (which is always done in retrospect) any assessment should be based on the stakes and the process, *not* on the ultimate outcome. For example, a satisfactory result (one meeting the minimum required standard) when faced with an in-flight emergency is not crashing the aircraft: diverting to the nearest suitable airfield and making a safe landing is a perfectly acceptable outcome. In such a scenario the stakes are high. This outcome is 'good enough' (satisficing – the NDM approach). Maybe in retrospect, the situation *could* have been managed better during the flight (after due consideration of further alternatives and their likelihood – a normative/CDM approach). The aircraft *could* perhaps have continued to its ultimate destination, rather than having to divert and make a precautionary landing, with all its associated expense and inconvenience. In certain contexts, for example disaster response, the failure to decide quickly can have far worse consequences than making a timely sub-optimal, but satisfactory decision (Baumgart et al., 2008). From an error perspective, NDM moves the retrospective analysis of decision errors away from the decision-maker to the systematic causes underlying the outcome (Orasanu et al., 2001).

Orasanu and Connolly (1993) described eight facets of NDM. They encompass:

- *Ill-structured problems.* In complex systems, problems rarely present themselves in a 'neat, complete form' (Orasanu & Connolly, 1993, p. 7) hence decision-makers need to diagnose the nature of a problem before generating any response options. In contrast, normative decision-making studies tend to present all information required and describe a set of potential outcomes from which to select.
- *Uncertain dynamic environments.* Problems are often characterized by incomplete, imperfect, unreliable, and/or ambiguous information and situated in a dynamic, rapidly changing environment (as often found when managing emergencies). Information may rapidly become obsolete.
- *Shifting, ill-defined, or competing goals.* Problems requiring decisions leading to a single, well-understood goal are uncommon. Decision-makers are frequently faced with multiple goals, some of which may not be clear, and which may also be in competition (e.g., cost vs. performance; safety vs. speed). Furthermore, trade-offs may not be simple as a result of operating in a complex, dynamic environment, and it may not always be clear how they may be resolved (*cf.* MAUT where such issues are defined more specifically as a result of having complete knowledge of all the variables represented in the problem space).
- *Action/feedback loops.* Normative approaches focus on a single decision event. In real life, a number of decisions are usually required to address any problem. These may be iterative in nature, with feedback loops verifying progress (or not).
- *Time stress.* Decisions are frequently made under significant time pressure which may result in opting for a less sophisticated strategy, considering only a limited number of options and involving increased use of heuristics.
- *High stakes.* The study of NDM has often been undertaken in high-stakes environments (e.g., disaster response (Jayawardene et al., 2021), military engagements (Alison et al., 2019) even space exploration (Orasanu, 2005)). These are usually retrospective studies of real events, relying on experts and their recall of events (and associated reasoning).

- *Multiple players.* Many problems involve more than a single decision-maker. Such teams may comprise several individuals acting co-operatively or competitively to achieve their goals. Each person may have a different perspective on the situation and what constitutes a satisfactory outcome.
- *Organizational goals and norms.* NDM often takes place in an organizational setting. As a result, the goals can reflect the objectives of the organization as well as those of the decision-maker.

Early NDM studies suggested that people first tried to understand the situation and establish if, from their experience, they had encountered something similar before. They then used their previous experience to frame the current situation and pursue a course of action (Recognition Primed Decision Making – RPD, see Klein et al., 1989). Decision-makers rarely considered all potential options but if the situation was not immediately recognized, they worked sequentially from the most plausible option generated until a potentially satisfactory outcome was produced. Furthermore, Klein et al. (2010) observed that expert decision-makers often pattern-matched the current situation with their previous experience and proposed a course of action without even any conscious awareness of having made a decision.

There are, however, several different categories of naturalistic decision types, of which RPD is just one. Orasanu and Fischer (1997), in a development of Orasanu's (1993) model, described three categories of response: rule-based; choice-based and creative (see Figure 10.1). The type of decision depended upon the assessment of the situation and the speed and type of response required by the context. The situation might be familiar, in which case a pre-learned strategy may be employed; it may require the selection of the most appropriate strategy (from several alternatives), or it may be a completely novel situation, requiring reasoning from fundamental principles. Choice-based responses can also encompass a further form of uncertainty for the decision-maker: undifferentiated alternatives, where it is difficult to anticipate the effects of the potential courses of action (Lipshitz & Strauss, 1997). The latter two options described by Orasanu and Fisher (1997) reflect the higher two levels of cognitive processing proposed by Rasmussen (1983, 1986), rule-based and knowledge-based reasoning, which also form the basis of Reason's Generic Error Modelling System (Reason, 1987, 1990), further emphasising the intimate relationship between the study of decision-making and error.

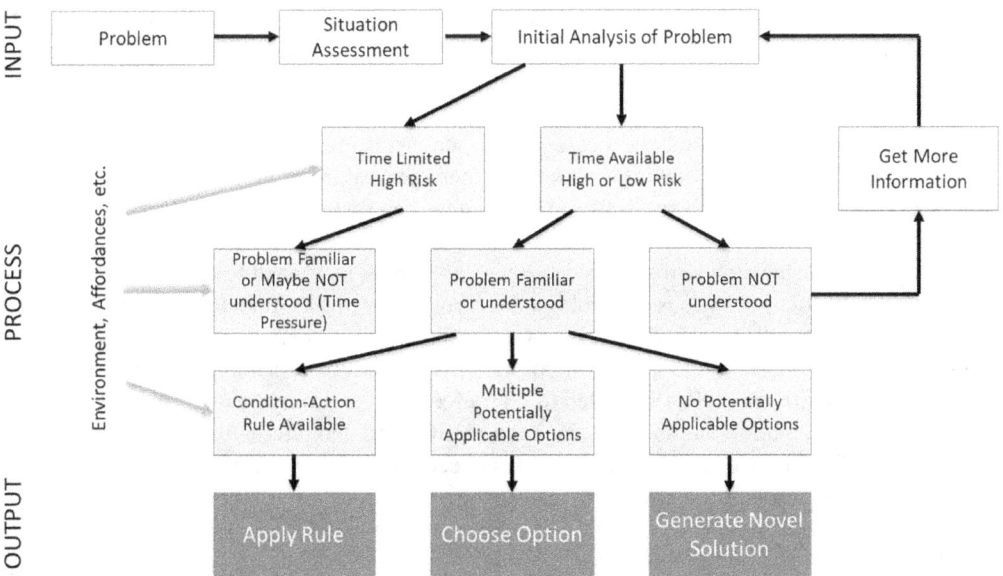

FIGURE 10.1   Categories of NDM decision types and rules for selection. Adapted from Orasanu and Fisher (1997).

All these problem-solving approaches require situation assessment/situation awareness followed by the choice of a course of action. Situation Assessment requires the identification of the problem to be addressed; an assessment of risk; and an estimate of the time available. It is a pre-cursor to Situation Awareness (Noble, 1993; Prince & Salas, 1997). Endsley (1995, 1997) argued that RPD required that the expert decision-maker had a mental model of the system based on current events but developed from past experience, that was dynamically updated, and decisions were based on this representation. Endsley described three levels of Situation Awareness:

- *Level One – Perception*: Recognizing the status, dynamics and attributes of relevant individual elements in the situation.
- *Level Two – Comprehension*: Understand what is perceived: combine data to provide a 'bigger picture' of the situation: Data become information.
- *Level Three – Projection*: Uses that information to anticipate events, predicting what is likely to happen in the near future.

Higher levels of Situation Awareness enable decision-makers to act in a more timely and effective way.

Jensen et al. (1997) and O'Hare (1992) both proposed that risk management is a vital aspect of decision-making as it is essential for the subsequent evaluation of the potential courses of action generated. If the situation is not recognized or understood, diagnostic actions may subsequently proceed but only if there is time available. If the situation is both high risk and time-pressured then action may be taken without completely understanding the problem (Orasanu & Fisher, 1997).

If time is limited (or *very* limited such as the go/no-go decision to continue a take-off or not) the cues available will generate a response where the cue pattern is determined to be of a particular type, and is then matched with an action from memory, which may be in the form of an immediate response or the application of a (hopefully) appropriate standard procedure (see Figure 10.1). When there are several potentially valid options, the decision-maker may need to evaluate these in terms of the risk versus the reward (*cf.* SEU or MAUT), but the outcome of this process is biased towards identifying a satisfactory outcome, not necessarily the best (see the central column in Figure 10.1). Unsatisfactory outcomes are eliminated. In some circumstances, the decision-maker may need to co-ordinate and prioritize several tasks to address the problem.

The previous issues have described circumstances where the problem is recognized. However, this may not always be the case. In complex systems, a decision-maker may require an initial diagnosis of the problem before they can address any potential course of action. Cues may be ambiguous or unfamiliar. In this case, the situation assessment aspect is of vital importance. Noble (1993) suggested that the situation assessment process commences with information from the environment, which is combined with contextual information and more general information from LTM to form an initial representation of the problem. Neisser (1976) suggested that this directs the search for further information. Once there has been an initial appraisal of the nature of the problem, the decision-maker can then either act using standardized procedures or they may then need to create a novel plan of action since no specific guidance is available as this is an entirely new situation (in their experience).

Situation Assessment and the selection and development of potential courses of action reflect the input and output stages of the decision-making process, both of which may be subject to uncertainty. Lipshitz and Strauss (1997) identified three sources of uncertainty: inadequate understanding and incomplete information on the input side, and undifferentiated alternatives on the output side. They suggested five methods used to manage such uncertainty, described in the RAWFS acronym: Reduction; Assumption-based reasoning; Weighing pros and cons; Forestalling and Suppression.

- *Reduction*: Collect more information.
- *Assumption-based reasoning*: Filling gaps in knowledge by making informed assumptions in the absence of information.

- *Weighing pros and cons*: Choose among the alternative courses of action in terms of their potential gains and losses.
- *Forestalling*: Deferring a decision, waiting for further information or options to emerge.
- *Suppression*: Denial (ignoring uncertainty) or rationalization of uncertainty (acknowledging uncertainty but not actually reducing it). A maladaptive strategy.

As their primary research methodology, most NDM studies use structured observation, walk-though/talk-through, real-time questioning, and/or structured interview-based retrospective analysis (e.g., Critical Decision Method or Cognitive Task Analysis). Where multiple decision-makers are involved, Delphi-type techniques have also been employed (e.g. Clayton, 1997). These approaches have been subject to criticism, though. LeBoeuf and Shafir (2001) have criticized such methods on the basis that in highly stressful situations, decision-makers retrospectively reporting on their decision-making processes are unlikely to have accurate introspective access to their memories and the factors influencing their decisions. Harris (2011) has also criticized the NDM approach as it is based on a description of decision-making processes, however, it does not predict behaviour, which is one of the ultimate goals of science, hence it limits its utility. Furthermore, experts are not necessarily available in every situation, but NDM is predicated upon the analysis of SMEs decision-making.

NDM may provide a good description of how expert decision-makers arrive at decisions, but as it does not require a full analysis of the situation and all available options, decision-makers may exhibit a number of shortcomings. Situation Assessment, pattern matching (RPD), situation diagnosis and the development of potential courses of action are all subject to biases (Killion, 2000). RPD may suggest that a situation matches one from previous experience, but this could be incorrect and hence decisions based on such an erroneous initial assumption could be misguided.

Tversky and Kahneman (1974) identified three heuristics to reduce the cognitively demanding task of assessing the probabilities underpinning decision events: these were 'representativeness'; 'availability' and 'anchoring'. These were primarily associated with making judgements, which they argued were associated with the (implicit) estimation of probabilities underpinning decision-making. Rehak et al. (2010) added further to this list by describing other common heuristics in the NDM decision-making processes which may introduce biases:

- *Availability* bias is observed when a situation is classified as being familiar as a result of its frequency of occurrence or recency of occurrence.
- *Representativeness* bias is a product of classifying a situation or event because of its perceived similarity to a prototypical case.
- *Anchoring* (and adjustment) bias is a product of starting from an initial point when formulating a problem and then making adjustments relative to this starting position. The final decision is biased in the direction of the initial anchoring point.
- *Confirmation* bias is defined as the tendency to seek or interpret information to confirm an initial hypothesis (rather than to seek all relevant information, some of which may be disconfirmatory).
- *Hindsight* bias refers to the tendency people have to overestimate the likelihood of an outcome in retrospect.
- *Overconfidence* bias simply reflects mis-placed confidence in the correctness of their diagnosis and decision-making ability.
- *Affect* bias concerns the influence that a person's emotional state has when making a decision.
- *Statistical* biases describe decisions based on inappropriate statistical reasoning concerning the probability or frequency of events.

These biases may affect decision-making at several stages in the process, from recognition of the situation to formulation of the course of action. These are by no means unique to NDM and may also be recognisable in CDM studies and in process decision-making models (see the following section). If these biases are best characterized solely as a component of NDM is arguable. However, these factors are also recognized as being sources of error when considering higher-level reasoning at the knowledge-based level (Reason, 1990), although Lipshitz et al. (2001) have suggested that *'NDM lacks analytical criteria that serve as signpost for error'* (p. 339). Describing such judgements as decision-biases or sources of decision error depends simply upon the context in which the question is asked.

There are many NDM models of decision-making, and a frequent criticism of the paradigm is that each model derived tends to be context-specific (time, location and people involved – Jenkins et al., 2010) which leads to a lack of generalization. CDM/normative approaches make predictions about behaviour but are limited in scope and their investigation context is artificial. Applying Rasmussen's (1997) taxonomy to decision-making research, it is also argued that the NDM approach is of limited utility in the development of decision-support tools. More sophisticated CDM approaches (e.g., MAUT, which forms the basis of several software tools) can cope with more complex problems but are unlikely to actually reflect human cognitive decision-making processes. In contrast, NDM approaches describe these processes, influences, and biases in expert problem-solving but lack generalisability and make no predictions. NDM is domain and context-bound (Lipshitz et al., 2001). 'Ordinary' decisions in 'normal' situations are not considered.

10.4.3 Process Decision Models

Lehto et al. (2012) described a third category of decision model that they labelled 'behavioural' models. These models describe the processes a decision-maker *should* undertake (hence the title of this sub-section) however unlike normative approaches, they also consider human abilities, biases, and limitations. As such, not only do they take into account decision-making activities, but they can also accommodate human error, something omitted from most NDM models (with which they have a great deal in common). Indeed, the descriptions of heuristics and biases in NDM models (see Rehak et al., 2010) have their original roots in these behavioural/process models.

Sjøgren, (2022) cites North Atlantic Treaty Organization (NATO) doctrine which describes two basic approaches to decision-making: the structured (process) approach and NDM. When explaining the structured approach, it describes how *'commanders consider, analyse, and evaluate all relevant factors'* (NATO, 2016, pp. 2–3) but it also stresses that this process-based methodology is not inherently better than NDM. The approaches are not mutually exclusive and choice between the two methods is dependent upon the time and information available, and the experience of the commander.

The 'ideal' process is exemplified by Janis and Mann's (1977) seven-stage method in which they describe what is expected of a skilled decision-maker. They should:

- Generate a range of possible courses of action.
- Survey the range of objectives to be fulfilled.
- Assess the probability and potential costs of both the positive and negative consequences of each course of action.
- Search for new information to evaluate further the alternatives.
- Consider new information, especially when it contradicts the initially chosen course of action.
- Re-assess the probability and potential costs of each course of action considered, including those originally discarded as being unacceptable, before making a final choice.
- Make detailed plans for the implementation of the chosen course of action, including any contingency plans.

This is a highly idealized approach especially when the conditions under which decision-making often takes place are considered (as described by Orasanu & Connolly, 1993). Kahneman and

Tversky (1979) and later Tversky and Kahneman (1992) described some of the shortcomings and biases that may adversely affect this process. Prospect Theory has already been outlined in an earlier section where it was used to identify the shortcomings of normative/classical decision models, but the same criticisms also apply here. The NDM paradigm would suggest that the first two stages in Janis & Mann's model and the last stage are somewhat idealized. Kahneman and Tversky (1979) and Tversky and Kahneman (1992) would be critical of the third and sixth stages in as much as the human decision-maker is poor at estimating the probability of events, may be risk-taking or risk-averse (by nature) and is also prone to biases.

The greatest problem with much behavioural research is not the quality of the work *per se*, but the fact that it tends to focus on what the human decision-maker is not very good at (see Lehto et al., 2012; Lehto & Nanda, 2021 for an overview). A similar criticism was made earlier concerning normative (CDM) approaches. A great deal of underpinning work in this area has concerned the (in) ability to make accurate judgements informing the decision-making process.

Judgements come in two basic guises: those associated with frequencies or probabilities, and those associated with choice (usually concerned with the likely outcome of a course of action). Many authors have demonstrated humans cannot estimate the probability or frequency of events with any accuracy (especially rare events) and hence by extension, they cannot make a meaningful estimate of risk (which also takes into account the consequences of a deleterious occurrence). Typically, decision-makers tend to overestimate rare events in judgement tasks (Erev et al., 1994) but are insensitive to the frequency of rare events when making decisions based on their experience (Hertwig et al., 2004).

In a similar vein to Janis and Mann (1977), Rasmussen developed his taxonomy of cognitive reasoning (skill-based, rule-based, or knowledge-based) into an idealized model of the decision-making processes (Rasmussen & Lind, 1982; Rasmussen, 1986). The decision ladder (see Figure 10.2) can describe both simple and complex decision-making processes within a single model and accommodates the data processing activities required and resultant states of knowledge. Situation Assessment/Awareness activities are described on the left-hand leg of the ladder, and planning and execution on the right-hand side. Both Jenkins et al. (2008) and Lintern (2010) suggested that the NDM paradigm can also be accommodated within the structure provided by Rasmussen's decision ladder.

Lintern (2010) argued that RPD was compatible with those activities described in the processes at the lower levels of Rasmussen's decision ladder. He contended that shortcuts from the input to output side of the decision ladder ('shunts' which are explicit, conscious process that transform one cognitive state into another, or 'leaps' which are a direct connection between two cognitive states with no intervening cognitive processes) were analogous to RPD. However, it can also be suggested that the practices further up the decision ladder reflect the other NDM processes described by Orasanu and Fisher (1997). Naikar (2010) argued that Rasmussen's decision ladder was concerned with describing the decision steps that must be taken independent of typicality, who was involved, or how it was done: in contrast, RPD focussed on the expert decision-maker in a familiar situation. Some other NDM processes, though, do examine the decision-making process in unfamiliar situations (see Orasanu & Fischer, 1997).

Distinct from the NDM approach, as a result of describing the steps and processes required, behavioural models explicitly suggest that training can improve decision-making. For example, STEP (create a Story; Test for conflict; Evaluate the story; develop contingency Plans: Cohen et al., 1997) claims to have been established within the NDM paradigm and has formed the basis for military decision-making training programmes. The initial component in the method (create a Story) has clear foundations in the NDM approach (Recognition/Metacognition model; Cohen et al., 1993, 1996) but the subsequent steps concerned with evaluating the initial evaluation of the situation and proposed courses of action share a great deal in common with the processes described by Janis and Mann (1977) and in Rasmussen's decision ladder (Rasmussen & Lind, 1982; Rasmussen, 1986; Figure 10.2). Training decision-making skills is considered in more depth in the following section.

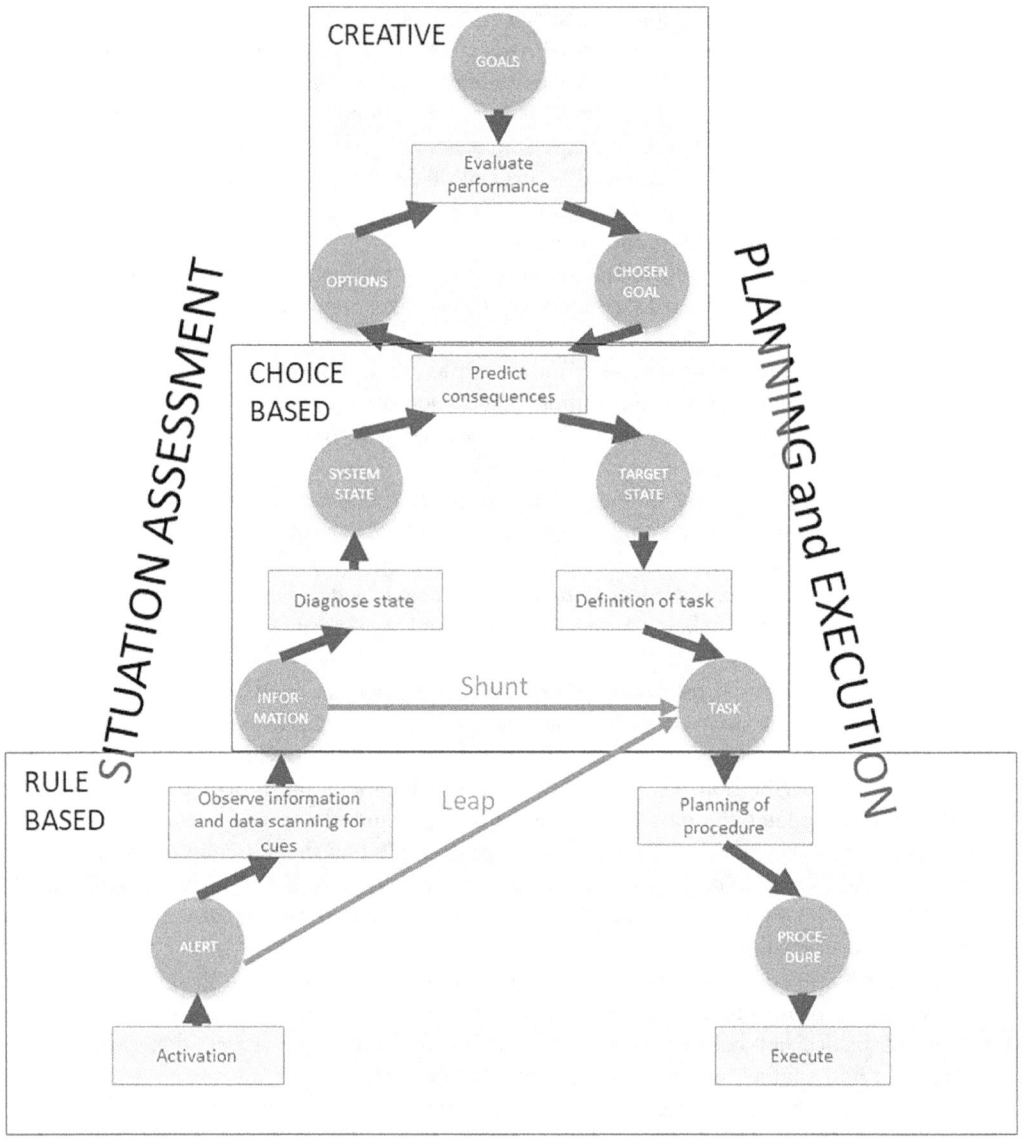

FIGURE 10.2 Rasmussen's decision ladder including 'shunts and 'leaps'. Cognitive states are depicted as circles and cognitive processes as rectangles. Overlaid are Orasanu and Fisher's (1997) categories of NDM. Adapted from Rasmussen et al. (1994) and Lintern (2010).

10.5 TRAINING DECISION-MAKING

Most human behaviour can in some way be described as decision-making and, in most instances, does not require training. Decision-making training is usually only required for personnel involved in complex, dynamic situations where it is not possible to provide decision-aiding either as a result of the nature of the operation or the speed of decision required. This is not to say that decision-making training and decision-aiding are mutually exclusive. They are often complementary. However, Cohen and Freeman (1996) suggest that normative decision theories such as SEU and MAUT are not cognitively compatible with the way experienced decision-makers work and hence cannot form the basis for effective training. Furthermore, such normative approaches require the problem to be fully specified, a luxury not available to decision-makers operating in dynamic, complex, ill-defined, and often high-risk environments (e.g., military engagements, firefighting or disaster relief). As a result, the foundations of decision-making training can generally be found within the process (behavioural) and naturalistic decision-making paradigms.

A training strategy comprises tools (e.g., feedback and simulation) and methods (instruction, demonstration, and practice) to convey training content (Salas & Cannon-Bowers, 1997). There are two main aspects of decision-making training. One aspect is the application of processes to ensure that all relevant facts are considered before determining a course of action. Process-based training comprises generic practices for situation assessment, generating hypotheses, developing options for courses of action, and implementing a decision. This is principally undertaken via explicit instruction with simulated content playing an illustrative role.

The second aspect concerns speeding up the development of heuristics (from experience) and of strategies that can then be applied quickly. Decision training derived from the naturalistic perspective places emphasis on recognition of patterns characteristic of domain-specific situations (RPD). Klein (1998) described four requirements in this respect:

- Engage in deliberate practice.
- Develop a wide range of experience.
- Obtain accurate, diagnostic, and timely feedback.
- Review prior experiences to derive new insights and learn from mistakes.

The methods used tend to be simulations involving demonstration and practice, with less emphasis on explicit instruction (Cohen et al., 1998). However, pattern recognition-based approaches are of limited utility when faced with a decision in a novel or ambiguous situation. Both approaches, though, can also include making decision-makers aware of their potential biases (see Tversky and Kahneman, 1974; Rehak et al., 2010). Buch and Diehl (1984) and Connolly et al. (1989) both found that specific judgement training produced significantly better quality decisions in pilots.

Decision-structuring approaches such as STEP, mentioned previously (Cohen et al., 1997) have characteristics from both schools of thought. STEP was developed as a military training aid. It commences with a recognition-primed assessment of the situation (based on experience) but is then supplemented with subsequent process-based stages to test the initial assumptions and/or acquire further data before coming to a final decision. The authors report that in trials there were significant increases in the amount of conflicting evidence collected; the number of assumptions underlying the evidence identified, and a rise in the number of alternatives generated. The accuracy of assessments also improved.

Orasanu (1993) commented that there was no evidence to suggest that the development of generic training methods to improve all-purpose decision-making was effective, as different types of decision (categorized from an NDM perspective) required different component skills. Caird-Daley et al. (2007) identified the cognitive work requirements within each of Orasanu's (1993) decision types which were used to provide the basis for a Serious Games application to train military decision-making skills. The training foci for each decision type are described in Table 10.1.

The time available is perhaps the single most important factor driving the approach to decision-making training (see Figure 10.1). Zakay and Tsal (1993) found that practice without stress and time pressure did not enhance decision-making if real-life decisions were to be made under the same conditions. Some decisions require an almost immediate response, for example in an aircraft in the advent of an engine failure, should I abort the take-off? The decision to abort is quite simple: Below V1 (the critical go/no-go speed on take-off after which there is not enough runway left to safely stop the aircraft) the rule is to abort. Above V1 continue the take-off and fly away with an engine out. Referring back to Figure 10.1 and Table 10.1, this clearly falls into the category of a time-limited, high-risk situation with a clear rule to apply. The biggest issue for the decision-maker (the pilot) is recognizing that they have a problem. Some engine failures are clear cut: others are more subtle,

TABLE 10.1

Cognitive Work Requirements and Training Foci

Decision Type (Orasanu & Fisher, 1997 Revision in Parentheses)	Training Foci
Go/No-Go Decisions (Time limited, high risk)	• Develop the perceptual patterns in memory that constitute the conditions for aborting an action. • Conduct training under realistic time pressure and including borderline cases.
Recognition-Primed Decisions (Time limited, high risk; condition-action rule available)	• Develop recognition of situational patterns constituting the condition side of a condition-action rule. • Learn the response/action side of the rule and the link between condition and action. • Develop evaluation skills ('what will happen if I take/don't take this action', or 'is there a reason not to take this action')?
Option/Response Selection Decisions (Time available, problem understood, multiple potential responses)	• Train personnel to use heuristics, for example, satisficing (Simon, 1955), elimination by aspects (Tversky, 1972) and dominance-structuring (Montgomery, 1993).
Resource Management Decisions (Problem understood but requires coordination of multiple tasks/resources)	• Acquire knowledge of the time required to complete various tasks, and the interdependencies among them. • Develop scheduling strategies.
Procedural Management (Problem not initially understood, but multiple potential responses after diagnosis)	• Develop situation assessment and risk assessment skills. • Learn the response/action side of potential rules to be applied after diagnosis and the link between their conditions and actions.
Creative Problem-Solving (Problem not understood, no potential available options)	• Develop situation assessment and risk assessment skills. • Develop skills in goal setting, planning, strategizing and evaluation (e.g., case-based reasoning involving presenting many examples of other's experiences).

Source: From Caird-Daley et al. (2007). Based on Orasanu (1993).

involving several parameters becoming 'off normal' before finally the engine fails (borderline cases). This is still a time-pressured situation, though. There is little time to apply diagnostic or decision-making techniques such as those implicit in Janis & Mann's or Rasmussen's structured decision-making approaches. What is required is context-specific expertise derived from previous experience. In these cases, such proficiency must be obtained in a flight simulator (as engine failures are fortunately very rare) so the decision-maker must develop expertise in some other context. Experience derived from this type of training allows for the rapid pattern-matching of cues to prototypes held in LTM.

In the case of a pilot in a flight simulator, the ability to review performance, obtain feedback and learn from mistakes is relatively easy. Similarly in sport, decision-making is also highly dynamic (although less risky) but also requires swift pattern-matching to apply an appropriate solution. However, in this case immersion in real-life situations is readily available and simulation can utilize different, less expensive, interactive technology. Video feedback and de-briefing have been found to improve decision-making performance in a number of sports (e.g., volleyball, basketball, and Australian rules football) enhancing pattern recognition and development of condition-action rules (Gil-Arias et al., 2016; Panchuk et al., 2018; Lorains et al., 2013).

Serious Games are technology-based interactive media with intricate, branching responses to user inputs. Such games take a first-person perspective. They are complex, simulated micro-worlds that allow users to explore the implications of decisions under real-time pressure to assess the consequences of a range of potential responses to situational patterns. Serious Games technology has been used to train decision-making in medical students (Kaczmarczyk et al., 2016). Serious games can provide a high-quality learning environment, presenting complex, sometimes deliberately indeterminate cues to the sources of a problem and also generate feedback concerning the actions taken. However, for a surgeon faced with an emergency the nature of the problem and its solution may be less prescriptive and so more generalized decision-making skills may be required.

There are a number of strategies embodied in mnemonics that have been developed to guide and structure decision-making in practice. Their common aim is to systematize an approach to decision-making that is less adversely affected by biases and workload (O'Hare, 2003). Mnemonic methods have been developed by the military to improve dynamic-decision-making; they have been employed on the aircraft flight deck and are also frequently used by paramedics and surgeons. For example, the SHOR mnemonic (Wohl, 1981) was developed for use by US Air Force to aid command-and-control decision-making under high pressure under severe time constraints. It comprises four steps: Stimuli (gather data; filter it; aggregate it and store or recall it); Hypotheses (create hypotheses about the situation; evaluate them; select one); Options (create response options; evaluate them; select an option); and Response (plan, organize and execute the response). The FORDEC mnemonic was a product of an airline (Lufthansa) Crew Resource Management programme. It comprises six phases: Facts; Options; Risks and Benefits; Decision, Execution; Check (Hörmann, 1995). FORDEC incorporates components addressing situation assessment; option generation, risk analysis and assessing the actual outcome versus intended outcome. It assumes a decision cycle, not a one-off decision. CURVES (Choose and Communicate; Understand; Reason; Value; Emergency; Surrogate) Chow, Czarny, Hughes and Carrese (2010) is a decision-making mnemonic to aid medical professionals in assessing if patients themselves have the capability to make rational decisions about their treatment.

However, mnemonics alone are not enough. Li and Harris (2005a) concluded that mnemonic methods must be matched to the situation if they are to be useful, and they can only be applied successfully in conjunction with training. For training military pilots' decision-making SHOR (Wohl, 1981) was found to be the best method in time-limited and critical, urgent situations but DESIDE (Detect; Estimate; Set safety objectives; Identify; Do; Evaluate: Murray, 1997) was superior for creative problem-solving, knowledge-based decisions which required comprehensive considerations. However, this was only useful when the pilots had time available. Li and Harris (2008) developed these mnemonics into a short decision-making training programme which was formally evaluated using a series of in-flight emergencies implemented in a full-flight simulator. The results showed significant improvements in the quality of pilots' situation assessment and risk management, but this was usually at the expense of the speed of decision-making.

The evaluation of decision-making training is not straightforward, especially in a complex situation with rapidly changing parameters. It is not possible to know or predict everything necessary upon which to base a decision. Orasanu and Martin (1998) pointed out that as NDM research is undertaken in an ecologically valid context *'there is often no clear standard of 'correctness' [. . . and] there is a loose coupling of event outcome and decision process so that outcomes cannot be used as reliable indicators of the quality of the decision'* (Orasanu & Martin, 1998, p. 100). The same applies to evaluating the outcome of decision-making training. The 'correctness' of a decision cannot be based on the outcome: good decisions can result in bad outcomes (and *vice versa*). For example, military and in-flight decisions are often made under uncertainty. The decision-maker can apply an appropriate process, but this may still result in a bad outcome as a result of factors such as the vagaries of the weather or the unanticipated actions of the enemy, hence the quality of decision-making cannot be assessed by its results (Brown et al., 1974). Training success should be judged with respect to the stakes and the processes employed, not the outcome (Mansikka et al, 2024). The effectiveness of decision-making training should be evaluated on the underpinning processes, such as situation assessment and risk management (Li & Harris, 2006, 2008).

10.6   DECISION SUPPORT

The limitations of human decision-making have been outlined, and it has been explained how some generic decision-making models tend to describe what *should* be done, rather than make clear how the human decision-maker actually makes decisions. However, a better understanding of human decision-making also provides insight into the limitations of the information processing system and judgemental biases. For these reasons, decision-support systems can provide essential assistance when faced with complex decisions, often when there are multiple criteria to satisfy. However, emphasis is very much upon *supporting* the decision-maker. These systems are not decision-*making* systems.

MAUT-based software, such as the Logical Decisions™ freeware referred to previously, is a form of flexible software shell to aid decision-makers. It enables many variables to be considered using various metrics and allows several potential outcomes to be considered and compared. It does not, however, model the human decision-making process (which is its strength in this instance) as it enables many more interactions and variables to be considered than could a human (see Simon, 1955): it augments the human decision-making process. This software, however, is a generic decision-making aid: many decision-support systems are application-specific and MAUT-based software (and its like) is also not suitable for dynamic applications.

Decision-support systems usually comprise three basic components: a database; an underpinning system model and a user interface. Intelligent decision-support systems may also include AI (Artificial Intelligence) for data mining, interrogating multiple web-based data sources to identify patterns, associations, and trends. Outputs will depend upon the nature of the application and the underlying system model but may include specific decision-recommendations; providing a number of options (often prioritized) or performance predictions.

There are various generic types of decision-support system (see Table 10.2: Power, 2008). The majority of these are aimed at less dynamic, organizational applications (e.g., making investment decisions); however, they are also finding increasing use as diagnostic tools. Medical decision-support systems may be knowledge-based for diagnostic purposes; data-driven to identify statistical

---

TABLE 10.2
Five Basic Types of Decision-Support System

Category of Decision-Support System	Application	Method of Operation
Data-driven system	Analysis of databases	Query-based system. May use data-mining techniques to elicit underlying trends
Knowledge driven system	Analysis of knowledge bases (including web resources)	Driven by an Expert/Intelligent System using rule-based or case-based reasoning
Model driven system	Generic analysis tool for situation analysis to compare between potential solutions	Uses data and preference parameters provided by users in a decision-making software model (e.g. MAUT) to assist decision-makers analysing a situation
Document driven system	Manages and retrieves electronic documentation using specific keywords or search terms	Web-based search engines
Communication driven system	Promotes computer-based collaborative working allowing several people to work on a shared task	May employ instant messaging software, and internet-based meeting systems and document sharing systems

Source: Based on Power (2008).

patterns of symptoms suggesting a common, underlying cause (Berner & La Lande, 2007: Moreira et al., 2019) or model-driven, for example in the prediction of the propagation of the spread of infectious diseases (Laskowski et al., 2011). Trials of a diagnostic decision-support system by General Practitioners showed a 9% increase in diagnostic accuracy (Kostopoulou et al., 2017). Unlike human decision-making, in this case, the quality of decision has a clear criterion to compare against. A review of clinical decision-support systems by Sutton et al. (2020) also concluded that among other things they had served to reduce the incidence of medication/prescribing errors and other adverse events; promoted adherence to clinical guidelines; reduced costs by the better management of tests and suggesting cheaper medication or treatment options; and provided diagnostic support (based on symptoms, interpretation of medical images and laboratory test results). However, they also concluded that many alerts provided by systems were of little or no consequence; there were concerns about their detrimental effects on doctor's diagnostic skills; and system costs and software maintenance may also be prohibitive. Patient acceptance was also questionable in some cases. Diagnostic decision-support systems have also been used successfully in other application areas, such as aircraft maintenance and scheduling (Zhang et al., 2011: Deng et al., 2021).

These illustrative uses of decision-support systems are applications involving less dynamic or time constrained tasks. Decision-support is also being developed for more dynamic applications. Airport weather decision-support systems have been developed to help provide medium/short-term forecasts around the airport to support decisions concerning scheduling and routing (Shaw et al., 2008). Decision-support systems have also been developed to optimize the flow of air traffic, suggesting prioritisation policies and schedules (Weigang et al., 2008). Shorter-term air traffic control conflict detection and resolution tools have also been created, supporting controllers' decision-making in such circumstances (Özgür & Cavcar, 2008). The system produces resolution advisories based on safety requirements and the relative performance of the types of aircraft involved. Military applications of decision-support systems range from logistics planning to almost real-time decision-making for planning close air support operations (Frame, et al, 2023).

Cummings (2004) suggested that in complex, time-critical environments, a high degree of prescriptive decision-support was not advisable as it was likely the system would not be perfectly reliable. Parasuraman et al. (2000) identified four generic support functions of increasing autonomy, each function building upon the previous functions:

- Data/Information acquisition.
- Data/Information analysis.
- Decision and action selection.
- Action implementation.

In aviation and military applications, the environment is not predictable. In such circumstances, it is essential to avoid simple acceptance of the 'optimal' option presented by a decision-support system. Operators must remain in-the-loop, maintaining their Situation Awareness. Cummings suggested that decision-support systems employed in such circumstances should offer a selection of possible courses of action (Level of Automation 3: see Sheridan & Verplank, 1978) rather than suggesting a single 'optimal' solution. In complex, dynamic situations, decision-aiding should concentrate on the earlier stages in the hierarchy described by Parasuraman et al., 2000). The greater the degree of decision-support, the lower the operator's workload however, as a result, they are concomitantly less aware of system function and pertinent environmental parameters. This can be problematic if required to intervene. However, one of the ironies is that decision-support systems can accommodate and analyse many more parameters than the human decision-maker and are not so prone to bias. The problem then becomes how to make the complex reasoning in such systems available to the operator in order for them to be able to evaluate the quality of the decision options with which they are presented.

10.7 CONCLUSION

Decision-making is all pervasive in life; however, relatively few decisions ultimately need to be evaluated against some standard of correctness. Unsurprisingly, decision-making research has tended to concentrate on such decisions. The decisions that result in an adverse event tend to be those subject to the greatest scrutiny. These are often labelled as 'errors'.

As an inexperienced accident investigator, it was pointed out to me that no one comes to work on a Monday morning with the intention of crashing a helicopter. Once the unsafe act(s) has been established in the investigation process, the question should not be 'why did the pilot make this error'. It should be 'why did the pilot think what they were doing was right'? This leads to an investigation into their decision-making processes. Decisions with a negative outcome can only be understood if it is known what an acceptable outcome looks like *and* the circumstances underpinning the decision when it was made. The product drives the investigation, but the focus should be on the process.

Early research concentrated on the negative aspects of human decision-making (what can't we do; what are we poor at)? This was followed by suggesting what the good decision-maker *should* do (rather than what they *do*, do). Studies of experts in real-life decision situations observed that decision-makers do not always operate in such an idealistic, structured manner. Good-enough decisions are made rapidly on the basis of experience. When this is considered, it is not surprising that decision errors occur but what should be more surprising is that decision-makers get so many decisions right in such circumstances. A sense of perspective is required.

Decision-making research needs to focus on both the strengths and shortcomings of the human decision-maker, learning from the processes by which 'good' decisions are made, incorporating it into training programmes and supplementing human decision-making frailties with targeted decision support. A systemic, distributed cognition approach is required. In the modern technology-rich organization a description of human decision-making behaviour alone is no longer enough.

REFERENCES

Alison, L.J., Shortland, N.D., & Moran, J.M. (2019*). Conflict: How Soldiers Make Impossible Decisions.* Oxford: Oxford University Press. https://doi.org/10.1093/oso/9780190623449.001.0001

Ajzen, I. (1996). The social psychology of decision making. In: E.T. Higgins & A.W. Kruglansk (Eds.), *Social Psychology: Handbook of Basic Principles* (pp. 297–325). New York: Guilford Press.

Baron, J. (2004). Normative models of judgment and decision making. In: D.J. Koehler & N. Harvey (Eds.), *Blackwell Handbook of Judgment and Decision Making* (pp. 19–36). Oxford: Blackwell.

Baumgart, L.A., Bass, E.J., Philips, B., & Kloesel, K. (2008). Emergency Management Decision Making during Severe Weather. *Weather and Forecasting, 23*(6), 1268–1279. https://doi.org/10.1175/2008WAF 2007092.1

Beach, L.R. & Lipshitz, R. (1993). Why classical decision theory is an inappropriate standard for evaluating and aiding most human decision making. In: G. Klein, J. Orasanu, R. Calderwood, & C.E. Zsambok (Eds.), *Decision Making in Action: Models and Methods* (pp. 21–35). Norwood, NJ: Ablex. Berner, E.S., & Graber, M.L. (2008). Overconfidence as a Cause of Diagnostic Error in Medicine. *The American Journal of Medicine, 121*(5), S2–S23. https://doi.org/10.1016/j.amjmed.2008.01.001

Berner, E.S. & La Lande, T.J. (2007). Overview of clinical decision support systems. In: E.S. Berner & T.J. La Lande (Eds.), *Clinical Decision Support Systems* (pp. 3–22). New York: Springer, Health Informatics. https://citations.springernature.com/item?doi=10.1007/978-0-387-38319-4_1

Brown, R.V., Kahr, A.S. & Peterson, C (1974). *Decision Analysis for the Manager.* New York: Holt, Rinehart and Winston.

Buch, G. & Diehl, A. (1984). An Investigation of the Effectiveness of Pilot Judgment Training. *Human Factors, 26*(5), 557–564. https://doi.org/10.1177/001872088402600507

Caird-Daley, A., Harris, D., Bessell, K., & Lowe, M. (2007). *Training Decision Making Using Serious Games.* Human Factors Integration Defence Technology Centre Report (HFIDTC/WP 4.6.1). Yeovil: Aerosystems International/HFI-DTC.

Chauvin, C., Lardjane, S., Morel, G., Clostermann, J-P., & Langard, B. (2013). Human and Organisational Factors in Maritime Accidents: Analysis of Collisions at Sea Using the HFACS. *Accident Analysis & Prevention, 59*(1), 26–37. https://doi.org/10.1016/j.aap.2013.05.006

Chow, G.V., Czarny, M.J., Hughes, M.T., & Carrese, J.A. (2010). CURVES: A Mnemonic for Determining Medical Decision-Making Capacity and Providing Emergency Treatment in the Acute Setting. *Chest, 137*(2), 421–427. https://doi.org/10.1378/chest.09-1133

Clayton, M.J. (1997). Delphi: A Technique to Harness Expert Opinion for Critical Decision-Making Tasks in Education. *Educational Psychology, 17*(4), 373–386. https://doi.org/10.1080/0144341970170401

Cohen, B.J. (1996). Is Expected Utility Theory Normative for Medical Decision Making? *Medical Decision Making, 16*(1), 1–6. https://doi.org/10.1177/0272989X9601600101

Cohen, M.S., Adelman, L., Tolcott, M.A., Bresnick, T.A., & Marvin, F.F. (1993). A Cognitive Framework for Battlefield Commanders' Situation Assessment (Tech. Rep. No. 93-1). Arlington, VA: Cognitive Technologies, Inc.

Cohen, M.S. & Freeman, J.T. (1996). Thinking Naturally About Uncertainty. In: *Proceedings of the Human Factors & Ergonomics Society*, 40th Annual Meeting. Santa Monica, CA: HF&ES.

Cohen, M.S., Freeman, J.T., & Thompson, B.B. (1997, June). Integrated Critical Thinking Training and Decision Support for Tactical Anti-Air Warfare. In: *Proceedings of the 1997 Command and Control Research and Technology Symposium, Washington*, DC.

Cohen, M.S., Freeman, J.T., & Thompson, B.B. (1998). Critical thinking skills in tactical decision making: A model and a training method. In: J. Canon-Bowers & E. Salas (Eds.), *Decision-Making under Stress: Implications for Training & Simulation*. Washington, DC: American Psychological Association Publications.

Cohen, M.S., Freeman, J.T., & Wolf, S. (1996). Metarecognition in Time-Stressed Decision Making: Recognizing, Critiquing, and Correcting. *Human Factors, 38*(2), 206–219. https://doi.org/10.1177/001872089606380203

Connolly, T.J., Blackwell, B.B., & Lester, L.F. (1989). A Simulator-Based Approach to Training in Aeronautical Decision Making. *Aviation Space and Environmental Medicine, 60*(1), 50–2.

Cummings, M.L. (2004). Automation Bias in Intelligent Time Critical Decision Support Systems. In: *Proceedings of AIAA 1st Intelligent Systems Conference*, 20–22 September, Chicago, IL. (pp. 1–6). Reston, VA: American Institute of Aeronautics and Astronautics.

Deng, Q., Santos, B.F. & Verhagen, W.J. (2021). A Novel Decision Support System for Optimizing Aircraft Maintenance Check Schedule and Task Allocation. *Decision Support Systems, 146*, 113545. https://doi.org/10.1016/j.dss.2021.113545

Diehl, A.E. (1991). Human Performance and Systems Safety Considerations in Aviation Mishaps. *International Journal of Aviation Psychology, 1*, 97–106. https://doi.org/10.1207/s15327108ijap0102_1

Ebrahimi, H., Sattari, F., Lefsrud, L., & Macciotta, R. (2021). Analysis of Train Derailments and Collisions to Identify Leading Causes of Loss Incidents in Rail Transport of Dangerous Goods in Canada. *Journal of Loss Prevention in the Process Industries, 72*, 104517. https://doi.org/10.1016/j.jlp.2021.104517

Edwards, W. (1954). The Theory of Decision Making. *Psychological Bulletin, 51*(4), 380–417. https://psycnet.apa.org/doi/10.1037/h0053870

Endsley, M.R. (1995). Toward a Theory of Situation Awareness in Dynamic Systems. *Human Factors, 37*(1), 32–64. https://doi.org/10.1518/001872095779049543

Endsley, M.R. (1997). The role of situation awareness in naturalistic decision making. In: C. Zsambok & G. Klein (Eds.), *Naturalistic Decision Making* (pp. 269–284). Mahwah, NJ: Lawrence Erlbaum.

Erev, I., Wallsten, T.S., & Budescu, D.V. (1994). Simultaneous Over- and Underconfidence: The Role of Error in Judgment Processes. *Psychological Review, 101*, 519–527. https://psycnet.apa.org/doi/10.1037/0033-295X.101.3.519

Frame, M.E., Kaiser, J., Kegley, J., Armstrong, J. & Schlessman, B. (2023). Impacts of Decision Support Systems on Cognition and Performance for Intelligence-Gathering Path Planning. *Military Psychology*. https://doi.org/10.1080/08995605.2023.2178210

Gil-Arias, A., Moreno, M., García-Mas, A., Moreno, A., García-González, L., & Del Villar, F. (2016). Reasoning and Action: Implementation of a Decision-Making Program in Sport. *The Spanish Journal of Psychology, 19*, E60. https://doi.org/10.1017/sjp.2016.58

Goldstein, W.M., & Hogarth, R.M. (1997). Judgment and decision research: Some historical context. In: W.M. Goldstein & R.M. Hogarth (Eds.), *Research on Judgment and Decision Making: Currents, Connections, and Controversies* (pp. 3–65). Cambridge: Cambridge University Press.

Harris, D. (2011). *Human Performance on the Flight* Deck. Aldershot, UK: Ashgate. https://doi.org/10.1201/9781315252988

Hastie, R. (2001). Problems for Judgment and Decision Making. *Annual Review of Psychology, 52*(1), 653–683. https://doi.org/10.1146/annurev.psych.52.1.653

Hastie, R. & Rasinski, K. (1987). The concept of accuracy in social judgment. In: D. Bar-Tal & A. Kruglanski (Eds.), *The Social Psychology of Knowledge* (pp. 193–208). Cambridge: Cambridge University Press.

Hertwig, R., Barron, G., Weber, E., & Erev, I. (2004). Decisions from Experience and the Effect of Rare Events in Risky Choices. *Psychological Science, 15*, 534–539. https://doi.org/10.1111/j.0956-7976.2004.00715.x

Hollnagel, E. (1998). *Cognitive Reliability and Error Analysis Method (CREAM).* Oxford, Elsevier Science.

Hörmann, H.J. (1995). FORDEC: A perspective model for aeronautical decision making. In: R. Fuller, N. Johnston, & N. McDonald (Eds.), *Human Factors in Aviation Operations* (pp. 17–23). Aldershot, UK: Ashgate

Janis, I.L., & Mann, L. (1977). *Decision Making: A Psychological Analysis of Conflict, Choice, and Commitment.* New York: Free Press.

Jayawardene, V., Huggins, T.J., Prasanna, R., & Fakhruddin, B. (2021). The Role of Data and Information Quality during Disaster Response Decision-Making. *Progress in Disaster Science, 12*, 100202. https://doi.org/10.1016/j.pdisas.2021.100202

Jenkins, D.P., Stanton, N.A., Salmon, P.M., & Walker, G.H. (2008). *Decision Making Training for Synthetic Environments: Using the Decision Ladder to Extract Specifications for Synthetic Environments Design and Evaluation* (HFIDTC/2/WP4.6.2/2). Yeovil: Aerosystems International/HFI-DTC.

Jenkins, D.P., Stanton, N.A., Salmon, P.M., Walker, G.H. & Rafferty, L. (2010). Using the Decision-Ladder to Add a Formative Element to Naturalistic Decision-Making Research. *International Journal of Human-Computer Interaction, 26*(2–3), 132–146. https://doi.org/10.1080/10447310903498700

Jensen, R.S. & Benel, R. (1977). *Judgment Evaluation and Instruction in Civil Pilot Training.* Washington, DC: Federal Aviation Administration.

Jensen, R.S., Guilke, J., & Tigner, R. (1997). Understanding expert aviator judgment. In: R. Flin, E. Salas, M. Strub, & L. Martin (Eds.), *Decision Making Under Stress: Emerging Themes and Applications* (pp. 233–242). Aldershot, UK: Ashgate

Kaczmarczyk, J., Davidson, R., Bryden, D., Haselden, S., & Vivekananda-Schmidt, P. (2016). Learning Decision Making through Serious Games. *Clinical Teaching, 13*, 277–282. https://doi.org/10.1111/tct.12426

Kahneman, D., Slovic, S., & Tversky, A. (Eds.). (1982). *Judgment under Uncertainty: Heuristics and Biases.* Cambridge: Cambridge University Press.

Kahneman, D. & Tversky, A. (1979). Prospect theory: An Analysis of Decisions under Risk. *Econometrica, 47*(2), 263–291. Keeney, R.L. & Raiffa, H. (1976). *Decisions with Multiple Objectives: Preferences and Value Trade-offs.* New York: Wiley.

Killion, T. (2000). Decision making and the levels of war. *Military Review, 80*(6), 66–70.

Klein, G.A. (1989). Recognition-primed decisions. In: W.B. Rouse (Ed.), *Advances in Man Machine Systems Research* (Vol. 5, pp. 47–92). Greenwich, CT: JAI Press.

Klein, G.A. (1993). A recognition-primed decision (RPD) model of rapid decision making. In: G. Klein, J. Orasanu, R. Calderwood, & C. Zsambok (Eds.), *Decision Making in Action: Models and Methods* (pp. 138–147). Norwood, NJ: Ablex Publishing Corporation.

Klein, G.A. (1998). *Sources of Power: How People Make Decisions.* Cambridge, MA: MIT Press.

Klein, G.A., Calderwood, R., & Clinton-Cirocco, A. (2010). Rapid on the Fire Ground: The Postscript. *Journal of Cognitive Engineering and Decision Making, 4*(3), 186–209. https://doi.org/10.1518/155534310X12844000801203

Klein, G.A., Calderwood, R., & Macgregor, D. (1989). Critical Decision Method for Eliciting Knowledge. *IEEE Transactions on Systems, Man, and Cybernetics, 19*(3), 462–472. https://doi.org/10.1109/21.31053

Klein, G.A. & Crandall, B.W. (1995). The role of mental simulation in problem solving and decision making. In: P. Hancock, J. Flach, J. Caird, & K. Vicente (Eds.), *Local Applications of the Ecological Approach to Human-Machine Systems* (pp. 324–358). Boca Raton, FL: CRC Press.

Kostopoulou, O., Porat, T., Corrigan, D., Mahmoud, S. & Delaney, B.C. (2017). Diagnostic Accuracy of GPs When Using an Early-Intervention Decision Support System: A High-Fidelity Simulation. *British Journal of General Practice, 67*(656), e201–e208. https://doi.org/10.3399/bjgp16X688417

Laskowski, M., Demianyk, B.C., Witt, J., Mukhi, S.N., Friesen, M.R., & McLeod, R.D. (2011). Agent-Based Modeling of the Spread of Influenza-Like Illness in an Emergency Department: A Simulation Study. *IEEE Transactions on Information Technology in Biomedicine, 15*(6), 877–889. https://doi.org/10.1109/TITB.2011.2163414

LeBoeuf, R.A., & Shafir, E. (2001). Problems and Methods in Naturalistic Decision-Making Research. *Journal of Behavioral Decision Making, 14*(5), 373–375. https://doi.org/10.1002/bdm.392

Lehto, M.R., Fui-Hoon Nah, F., & Yi, J-S. (2012). Decision-making models, decision support, and problem solving. In: G. Salvendy (Ed.), *Handbook of Human Factors and Ergonomics* (4th Edition, pp. 192–242). Hoboken, NJ: John Wiley & Sons.

Lehto, M.R. & Nanda, G. (2021). Decision-making models, decision support, and problem solving. In: G. Salvendy & W. Karwowski (Eds.), *Handbook of Human Factors and Ergonomics* (5th Edition, pp. 159–202). Hoboken, NJ: John Wiley & Sons. https://doi.org/10.1002/9781119636113.ch6

Lenné, M.G., Salmon, P.M., Charles, R., Liu, C., & Trotter, M. (2012). A Systems Approach to Accident Causation in Mining: An Application of the HFACS Method. *Accident Analysis & Prevention, 48*, 111–117. https://doi.org/10.1016/j.aap.2011.05.026

Levy, J.S. (1992). An Introduction to Prospect Theory. *Political Psychology, 13*(2), 171–186. https://www.jstor.org/stable/3791677

Li, W.C. & Harris, D. (2005a). Aeronautical Decision Making: Instructor-Pilot Evaluation of Five Mnemonic Methods. *Aviation, Space and Environmental Medicine, 76*, 1156–1161.

Li, W-C. & Harris, D. (2005b). HFACS Analysis of ROC Air Force Aviation Accidents: Reliability Analysis and Cross-Cultural Comparison. *International Journal of Applied Aviation Studies, 5*, 65–81.

Li, W.C. & Harris, D. (2006). The Evaluation of the Decision Making Processes Employed by Cadet Pilots Following a Short Aeronautical Decision-Making Training Program. *International Journal of Applied Aviation Studies, 6*, 315–333.

Li, W.C. & Harris, D. (2008). The Evaluation of the Effect of a Short Aeronautical Decision-Making Training Program for Military Pilots. *International Journal of Aviation Psychology, 18*, 135–152. https://doi.org/10.1080/10508410801926715

Li, W-C., Harris, D., & Yu, C.S. (2008). Routes to Failure: Analysis of 41 Civil Aviation Accidents from the Republic of China Using the Human Factors Analysis and Classification System. *Accident Analysis and Prevention, 40*, 426–434. https://doi.org/10.1016/j.aap.2007.07.011

Lintern, G. (2010). A Comparison of the Decision Ladder and the Recognition-Primed Decision Model. *Journal of Cognitive Engineering and Decision Making, 4*(4), 304–327. https://doi.org/10.1177/155534341000400404

Lipshitz, R., Klein, G., Orasanu, J., & Salas, E. (2001). Taking Stock of Naturalistic Decision Making. *Journal of Behavioral Decision Making, 14*(5), 331–352. https://doi.org/10.1002/bdm.381

Lipshitz, R. & Strauss, O. (1997). Coping with Uncertainty: A Naturalistic Decision-Making Analysis. *Organizational Behavior and Human Decision Processes, 69*(2), 149–163. https://doi.org/10.1006/obhd.1997.2679

Lorains, M., Ball, K., & MacMahon, C. (2013). An Above Real Time Training Intervention for Sport Decision Making. *Psychology of Sport and Exercise, 14*(5), 670–674. https://doi.org/10.1016/j.psychsport.2013.05.005

Madigan, M., Golightly, D., & Madders, R. (2016). Application of Human Factors Analysis and Classification System (HFACS) to UK Rail Safety of the Line Incidents. *Accident Analysis & Prevention, 97*, 122–131. https://doi.org/10.1016/j.aap.2016.08.023

Mansikka, H., Virtanen, K., Lipponen, T., & Harris, D. (2024). Improving pilots' tactical decisions in air combat training using the critical decision method. *The Aeronautical Journal, 1–14.* doi:10.1017/aer.2024.3

Mayer, R.E. (1983). *Thinking, Problem Solving, Cognition.* New York: W. H. Freeman.

Montgomery, H. (1993). The search for a dominance structure in decision making: Examining the evidence. In: G. Klein, J. Orasanu, R. Calderwood, & C.E. Zsambok (Eds.), *Decision Making in Action: Models and Methods* (pp. 182–187). Norwood, NJ: Ablex.

Moreira, M.W.L., Rodrigues, J.J.P.C, Korotaev, V., Al-Muhtadi, J., & Kumar, N. (2019). A Comprehensive Review on Smart Decision Support Systems for Health Care. *IEEE Systems Journal, 13*(3), 3536–3545. https://doi.org/10.1109/JSYST.2018.2890121

Murray, S.R. (1997). Deliberate Decision Making by Aircraft Pilots: A Simple Reminder to Avoid Decision Making under Panic. *International Journal of Aviation Psychology, 7*, 83–100. https://doi.org/10.1207/s15327108ijap0701_5

Naikar, N. (2010). *A Comparison of the Decision Ladder Template and the Recognition-Primed Decision Model* (DSTO Tech. Rep. DSTO-TR-2397). Fisherman's Bend, Victoria, Australia: Air Operations Division.

Neisser, U. (1976). *Cognition and Reality.* San Francisco, CA: W.H. Freemand and Co.

Noble, D. (1993). A model to support development of situation assessment aids. In: G.A. Klein, J. Orasanu, R. Calderwood, & C.E. Zsambok (Eds.), *Decision Making in Action: Models and Methods* (pp. 287–305). Norwood, NJ: Ablex.

North Atlantic Treaty Organization. (2016). ATP-3.2.2 Command and Control of Allied Land Forces. *Edition B, Version 1*. NATO Standardization Office.

O'Hare, D. (1992). The Artful Decision Maker: A Framework Model for Aeronautical Decision Making. *International Journal of Aviation Psychology, 2*(3), 175–192. https://doi.org/10.1207/s15327108ijap0203_2

O'Hare, D. (2003). Aeronautical decision making: Metaphors, models, and methods. In P.S. Tsang & M.A. Vidulich (Eds.) *Principles and practice of aviation psychology: Human factors in transportation* (pp. 201–237). Mahwah, NJ: Lawrence Erlbaum Associates, Inc.

Orasanu, J. (1993). Decision making in the cockpit. In: E.L. Wiener, B.G. Kanki & R.L. Helmreich (Eds.), *Cockpit Resource Management* (pp. 137–172). San Diego, CA: Academic Press.

Orasanu, J. (2005). Crew Collaboration in Space: A Naturalistic Decision-Making Perspective. *Aviation, Space, and Environmental Medicine, 76*(6), B154–B163.

Orasanu, J. & Connolly, T. (1993). The reinvention of decision making. In: G.A. Klein, J.Orasanu, R. Calderwood, & C.E. Zsambok (Eds.), *Decision Making in Action: Models and Methods* (pp. 3–20). Norwood, NJ: Ablex.

Orasanu, J. & Fischer, U. (1997). Finding decisions in natural environments: The view from the cockpit. In: C.E. Zsambok & G.A. Klein (Eds.), *Naturalistic Decision Making* (pp. 343–357). Mahwah, NJ: Lawrence Erlbaum Associates.

Orasanu, J. & Martin, L. (1998, April). Errors in Aviation Decision Making: A Factor in Accidents and Incidents. In: *Proceedings of the Workshop on Human Error, Safety and Systems Development* (pp. 100–107). NASA-Ames Research Center.

Orasanu, J., Martin, L., & Davison, J. (2001). Cognitive and contextual factors in aviation accidents: Decision errors. In: E. Salas & G. Klein (Eds.), *Linking Expertise and Naturalistic Decision Making* (pp. 209–225). Mahwah, NJ: Lawrence Erlbaum Associates.

Özgür, M. & Cavcar, A. (2008). A Knowledge-Based Conflict Resolution Tool for En-Route Air Traffic Controllers. *Aircraft Engineering and Aerospace Technology, 80*(6), 649–656. https://doi.org/10.1108/00022660810911590

Panchuk, D., Klusemann, M.J., & Hadlow, S.M. (2018). Exploring the Effectiveness of Immersive Video for Training Decision-Making Capability in Elite, Youth Basketball Players. *Frontiers in Psychology, 9*, 2315. https://doi.org/10.3389/fpsyg.2018.02315

Parasuraman, R., Sheridan, T.B., & Wickens, C.D. (2000). A Model for Types and Levels of Human Interaction with Automation. *IEEE Transactions on Systems, Man, and Cybernetics, 30*, 286–297. https://doi.org/10.1109/3468.844354

Payne, J.W., Bettman, J.R., & Johnson, E.J. (1988). Adaptive Strategy Selection in Decision Making. *Journal of Experimental Psychology: Learning, Memory, and Cognition, 14*(3), 534–552. https://psycnet.apa.org/doi/10.1037/0278-7393.14.3.534

Power, D. (2008). Decision support systems: A historical overview. In F. Burstein & C. Holsapple (Eds.), *Handbook on Decision Support Systems 1* (pp. 121–140). Berlin, Heidelberg: Springer. https://doi.org/10.1007/978-3-540-48713-5_7

Prince, C. & Salas, E. (1997). Situation Assessment for Routine Flight and Decision Making. *International Journal of Cognitive Ergonomics, 1*(4), 315–324.

Rasmussen, J. (1983). Skill, Rules and Knowledge: Signals, Signs and Symbols, and Other Distinctions in Human Performance Models. *IEEE Transactions on Systems, Man and Cybernetics, 13*, 257–266. https://doi.org/10.1109/TSMC.1983.6313160

Rasmussen, J. (1986). *Information Processing and Human-Machine Interaction: An Approach to Cognitive Engineering*. Amsterdam: Elsevier.

Rasmussen, J. (1997). Merging paradigms: Decision-Making, management, and cognitive control. In: R. Flin, E. Salas, & L. Martin (Eds.), *Decision-Making Under Stress, Emerging Themes and Applications* (pp. 67–81). Aldershot, UK: Ashgate.

Rasmussen, J. & Lind, M. (1982, June). A model of human decision making in complex systems and its use for design of system control strategies. In: *Proceedings of 1982 American Control Conference* (pp. 270–276). Arlington, VA: IEEE.

Rasmussen, J., Petjersen, A.M., & Goodstein, L.P. (1994). *Cognitive Systems Engineering*. New York: Wiley

Reason, J.T. (1987). Generic error-modelling system (GEMS): A cognitive framework for locating human error forms. In: J. Rasmussen, K. Duncan, & J. Leplat (Eds.), *New Technology and Human Error* (pp. 63–83). London: Wiley.

Reason, J.T. (1990). *Human Error*. Cambridge: Cambridge University Press.

Rehak, L.A., Adams, B., & Belanger, M. (2010). Mapping Biases to the Components of Rationalistic and Naturalistic Decision Making. *Proceedings of the Human Factors and Ergonomics Society Annual Meeting, 54*(4), 324–328. https://doi.org/10.1177/154193121005400412

Salas, E. & Cannon-Bowers, J.A. (1997). Methods, tools, and strategies for team training. In: M.A. Quinones & A. Ehrenstein (Eds.), *Training for a Rapidly Changing Workplace: Applications of Psychological Research* (pp. 249–279). Washington, DC: APA Press.

Shappell, S.A. & Wiegmann, D.A. (2001). Applying Reason: The Human Factors Analysis and Classification System (HFACS). *Human Factors and Aerospace Safety, 1*(1), 59–86.

Shappell, S.A. & Wiegmann, D.A. (2004). HFACS analysis of military and civilian aviation accidents: A North American comparison. In: *Proceedings of International Society of Air Safety Investigators Conference*, Queensland, Australia, 2–8 November, 2004. Sterling, VA: International Society of Air Safety Investigators.

Shaw, B.L., Spencer, P.L., Carpenter, R.L., & Barrere, C.A. (2008). Implementation of the WRF Model for the Dubai International Airport Aviation Weather Decision Support System. In: *Proceedings of 13th Conference on Aviation, Range, and Aerospace Meteorology*. New Orleans, LA: American Meteorological Society.

Sheridan, T.B. & Verplank, W.L. (1978). *Human and Computer Control of Undersea Teleoperators* (Technical Report, Engineering Psychology Program). Cambridge MA: Department of Mechanical Engineering, MIT.

Simon, H.A. (1955). A Behavioral Model of Rational Choice. *Quarterly Journal of Economics, 69*(1), 87–103. https://doi.org/10.2307/1884852

Singh, S. (2015, February). *Critical Reasons for Crashes Investigated in the National Motor Vehicle Crash Causation Survey* (Traffic Safety Facts Crash Stats. Report No. DOT HS 812 115). Washington, DC: National Highway Traffic Safety Administration.

Sjøgren, S. (2022). What Military Commanders Do and How They Do It: Executive Decision-Making in the Context of Standardised Planning Processes and Doctrine. *Scandinavian Journal of Military Studies, 5*(1), 379–397. https://doi.org/10.31374/sjms.146

Standing, M. (2020). *Clinical Judgement and Decision Making in Nursing* (4th Edition). London: SAGE. https://digital.casalini.it/9781526478375

Sutton, R.T., Pincock, D., Baumgart, D.C., Sadowski, D.C., Fedorak, R.N., & Kroeker, K.I. (2020). An Overview of Clinical Decision Support Systems: Benefits, Risks, and Strategies for Success. *NPJ Digital Medicine, 3*(1), 17. https://doi.org/10.1038/s41746-020-0221-y

Taylor, D.W. (1965). Decision making and problem solving. In: J.G. Marsh (Ed.), *Handbook of Organizations* (pp. 48–86). Chicago, IL: Rand McNally.

Tversky, A. (1972). Elimination by Aspects: A Theory of Choice. *Psychological Review, 79*(4), 281. https://psycnet.apa.org/doi/10.1037/h0032955

Tversky, A. & Kahneman, D. (1974). Judgment under uncertainty: Heuristics and biases: Biases in judgments reveal some heuristics of thinking under uncertainty. *Science, 185*(4157), 1124–1131.

Tversky, A. & Kahneman, D. (1981). The Framing of Decisions and the Psychology of Choice. *Science, 211*(4481), 453–458. https://doi.org/10.1126/science.7455683

Tversky, A. & Kahneman, D. (1992). Advances in Prospect Theory: Cumulative Representation of Uncertainty. *Journal of Risk and Uncertainty, 5*(3), 297–323. https://doi.org/10.1007/BF00122574

von Neumann, J. & Morgenstern, O. (1953). *Theory of Games and Economic Behavior*. Princeton, NJ: Princeton University Press. https://doi.org/10.1515/9781400829460

Wallenius, J., Dyer, J.S., Fishburn, P.C., Steuer, R.E., Zionts, S., & Deb, K. (2008). Multiple Criteria Decision Making, Multiattribute Utility Theory: Recent Accomplishments and What Lies Ahead. *Management Science, 54*(7), 1336–1349. https://doi.org/10.1287/mnsc.1070.0838

Weigang, L., de Souza, B.B., Crespo, A.M.F., & Alves, D.P. (2008). Decision Support System in Tactical Air Traffic Flow Management for Air Traffic Flow Controllers. *Journal of Air Transport Management, 14*(6), 329–336. https://doi.org/10.1016/j.jairtraman.2008.08.007

Wiegmann, D.A. & Shappell, S.A. (2003). *A Human Error Approach to Aviation Accident Analysis: The Human Factors Analysis and Classification System*. Aldershot, UK: Ashgate. https://doi.org/10.4324/9781315263878

Wiggins, M.W., Azar, D., Hawken, J., Loveday, T., & Newman, D. (2014). Cue-Utilisation Typologies and Pilots' Pre-Flight and In-Flight Weather Decision-Making. *Safety Science, 65*(1), 118–124. https://doi.org/10.1016/j.ssci.2014.01.006

Wohl, J.G. (1981). Force Management Decision Requirements for Air Force Tactical Command and Control. *IEEE Transactions on Systems, Man, and Cybernetics, 11*(9), 618–639. https://doi.org/10.1109/TSMC.1981.4308760

Woods, D.D., Johannesen, L.J., Cook, R.I., & Sarter, N.B. (1994). Behind Human Error: Cognitive Systems, Computers and Hindsight. CSERIAC Gateway, 5. Columbus, OH: CSERIAC

Yıldırım, U., Başar, E., & Uğurlu, Ö. (2019). Assessment of Collisions and Grounding Accidents with Human Factors Analysis and Classification System (HFACS) and Statistical Methods. *Safety Science, 119*, 412–425. https://doi.org/10.1016/j.ssci.2017.09.022

Zakay, D. & Tsal, Y. (1993). The Impact of Using Forced Decision-Making Strategies on Post-Decisional Confidence. *Journal of Behavioral Decision Making, 6*(1), 53–68. https://doi.org/10.1002/bdm.3960060104

Zhang, P., Zhao, S.W., Tan, B., Yu, L.M., & Hua, K.Q. (2011). Applications of decision support system in aviation maintenance. In: C.S. Jao (Ed.), *Efficient Decision Support Systems-Practice and Challenges in Multidisciplinary Domains* (pp. 397–412). Rijeka, Croatia: InTech.

Zsambok, C.E. & Klein, G.A. (1997). *Naturalistic Decision Making*. Mahwah, NJ: Lawrence Erlbaum Associates.

11 HCI Design Perspectives, Trends and Approaches

David Lamas and Vladimir Tomberg

11.1 INTRODUCTION

Human-computer interaction (HCI) emerged as a distinct discipline in the late 1970s and early 1980s (Carroll, 2016, p. 21). While the study and design of interactions between humans and computers have been ongoing since the advent of computers (Dix, 2010; Grudin, 2011), the establishment of HCI as a discipline gained momentum during this period, marked by several landmarks (Grudin, 1990, 2005, 2011).

One such landmark is the pioneering work of Douglas Engelbart carried out in the 1960s (Engelbart,1962), notably illustrated at the "Mother of All Demos" in 1968 when Douglas Engelbart demonstrated concepts such as the mouse, hypertext, video conferencing and collaborative workspaces (Englebart, 1968). Engelbart's research emphasised the importance of considering the interplay between humans and computers and the potential for computers to enhance human capabilities.

Another landmark was the cognitive revolution in psychology, which gained momentum in the 1960s and 1970s (Chignell, 2023; Grudin, 2005; Bannon, 1986, 1992). It influenced the understanding of human information processing and mental models, bringing attention to the cognitive aspects of HCI and highlighting the need to design interfaces that align with users' mental models and cognitive processes (Carroll and Thomas, 1982; Norman, 1983a; Borgman, 1985).

The establishment of prominent HCI research centres, such as Xerox PARC (Palo Alto Research Center), the University of California, San Diego's Cognitive Science Laboratory, and the University of Maryland's Human-Computer Interaction Lab, during the late 1970s and early 1980s, where foundational research on user interface design, interaction techniques and usability evaluation methods was conducted, is yet another landmark. As is the introduction of the Graphical User Interface (GUI) popularised by Xerox PARC (Palo Alto Research Center) in the 1970s (Grudin, 2005), which marked a significant shift in how we interact with computers. With their visual icons, windows and pointing devices (such as the mouse), GUIs allowed for more intuitive and interactive interactions with computers. This development highlighted the need to study and design user-friendly and efficient interfaces and triggered HCI's turn to design (Rogers, 2012, pp. 68–71).

A landmark is also the Formation of the Special Interest Group on Computer-Human Interaction within the Association for Computing Machinery (ACM SIGCHI) in 1982 (Chignell, 2023; Grudin, 2005). ACM SIGCHI has since played a pivotal role in bringing together researchers and practitioners from various disciplines to exchange knowledge, share research findings and promote the development of HCI (Kumar et al., 2022).

And finally, essential landmarks are the seminal works of Stuart Card, Thomas Moran and Allen Newell, "The Psychology of Human-Computer Interaction," published in 1983 as well as "Designing the User Interface: Strategies for Effective Human-Computer Interaction" by Ben Shneiderman, published in 1986 (first edition).

Strongly driven by technology (Dix, 2010), it's no wonder that HCI has since broadened its horizons even more, considering the advancements in technology, the proliferation of digital devices and the shift from the early non-discretionary use of computers to the massively discretionary use of contemporary digital solutions.

DOI: 10.1201/9781003495109-11

Still, HCI design focuses on designing, developing and improving interactive systems that facilitate effective and satisfying interactions between humans and computers. It encompasses the design of user interfaces, interaction techniques and overall user experiences across a wide range of digital devices and platforms (Long and Dowell, 1989; Long, 1996; Rogers, 2012).

HCI design considers users' needs, capabilities and preferences to create intuitive, usable and engaging interfaces. It involves understanding users' goals, tasks and contexts of use and translating that understanding into design decisions shaping the interaction between humans and computers.

Lately, HCI's considerations have expanded to include the effect and role of technology in society and the world to address topics such as justice and social critique (Constanza-Chock, 2020; Bellini et al., 2022; Bardzell et al., 2018), sustainability (Dourish, 2010; Kumar et al., 2020), development (Dell and Kumar, 2016; Anokwa et al., 2009), fairness (Smith et al., 2023; Lee and Singh, 2021; Richardson et al., 2021; Elahi et al., 2021; decolonial and postcolonialism (Kambunga et al., 2023;; Dourish and Mainwaring, 2012; Irani et al., 2010).

HCI design is an interdisciplinary field that draws upon the bodies of knowledge of areas (Zhang and Galletta, 2015, p. 5; Rogers, 2012, p. 86; Oulasvirta and Hornbæk, 2016; Chignell, 2023). such as Human Factors & Ergonomics, Psychology, Design Studies, Computer Science and Information Sciences, Social Sciences, Humanities, Arts and Neurophysiology and Neuroscience. Informed by this broad knowledge base, HCI designers create meaningful interactive systems that are usable, accessible and engaging and enhance our experiences when interacting with computers and digital devices (Rogers et al., 2023).

11.2   THE ROLE OF THEORY

Theory in HCI design enables a systematic understanding of users, informs design guidelines, promotes user-centred design practices, facilitates evaluation and testing and drives future innovations (Grudin, 2011; Bødker, 2006; Long, 1996). While the extent of its role has been critically evaluated since the inception of the field (Oulasvirta and Hornbæk, 2016; Bødker, 2006; Long, 1996) it remains a foundational aspect upon which HCI design has developed and continues to be fundamental in our understanding of design research and practice, as it is conceptually illustrated in Figure 11.1.

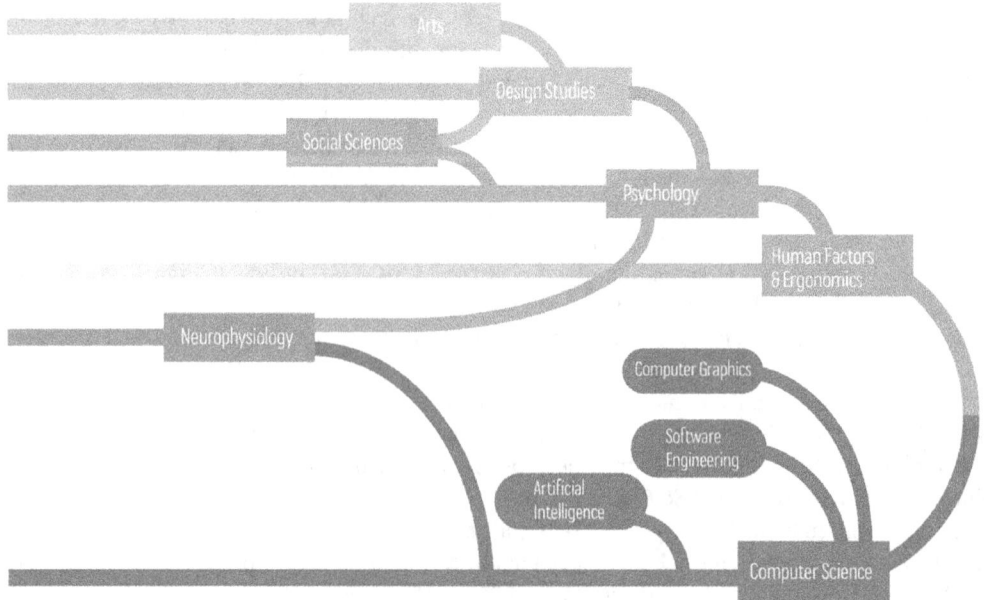

FIGURE 11.1   Disciplinary expansion of HCI. (From the initial disciplines of Human Factors and ergonomics and Computer Science to expand to incorporate Design Studies, Social Sciences, Humanities, Arts and Neurophysiology.)

Although by necessity we offer a partially chronological presentation of the development of the concept of design in HCI, the goal of this chapter is not to explore how HCI has evolved through time, a topic that has been covered extensively by others before (Grudin, 2011; Rogers, 2012; Carroll, 1993, 2016; Chignell, 2023). This chapter owes a debt of gratitude to these works, especially Carroll's Creating a design science of HCI (1993) but it takes its own approach to offer an inclusive understanding of the concept of design in HCI and the role of different theoretical perspectives in the construction of it. For the purposes of clarity our analytical approach has been an integrative one rather than an exhaustive one, looking to identify the key disciplinary currents/continuums that contribute to HCI's own definition of itself and of design. This is with the aim to show both the expansive aspect of HCI's transdisciplinarity while retaining a clear perspective. Our perspective is of an expansive discipline with a common core that informs its culture and vision while remaining flexible and diverse.

11.2.1 HUMAN FACTORS AND ERGONOMICS

The historical origin of HCI, Human Factors and Ergonomics is mostly concerned with "the design and engineering of human-machine systems for the purpose of enhancing human performance." (Dempsey et al., 2000). It has its origins in Taylor's Scientific Management who in the early 1900s applied the contemporary methods and techniques to study worker's performance with the goal of improving efficiency (Grudin, 2011). While his initial approach proved to be not effectively implemented, his initial insight of defining tasks and systematic measurement would remain influential and still mould current practices.

Tasked with the challenge of exploring why Taylor's ideas for improvement were not being effective, F.B. Gilbreth and Lillian M. Gilbreth would expand his work first by developing time-and-movement studies (Meister, 2018, p. 148) and later by combining psychology with scientific management to explore the role of mental aspects and motivation on performance (Grudin, 2011). This ambition of reaching out to other disciplines, and a pragmatic approach to incorporate new tools, theories, frameworks and ideas, coupled with key focus on identifying relevant metrics to assess improvement and to enhance our understanding of the issues of Human-Machine interaction will remain present until today.

It is important to notice that Taylor's ideas have been connected to previous work from Charles Babbage, the father of modern computing, and the larger goals of capitalistic exploitation and management (Whittaker, 2023), a relationship that must be considered critically.

So it is no surprise that the early days of human factors and ergonomics (HF&E) in computing can be traced back to the emergence of computers and the recognition of the need to design systems that accommodate human capabilities and limitations. Its origins can be traced back to research conducted during World War II, particularly in the field of aviation, when researchers studied the interaction between pilots and aircraft controls, aiming to improve usability, safety and performance (Meister, 2018, p. 152). This research laid the foundation for understanding the importance of human-centered design in complex systems.

The start of World War II would prove to be the big push behind the establishment of HF&E (Meister, 2018, p. 184). The demands of the military machine for rapid implementation would see a push for an expansive pragmatic approach as new techniques and ideas were tested and developed in the field. Paul Fitts seminal work on cockpit layout at the Wright Patterson Air Force base would provide the basis for his famous Fitt's law (Fitts, 1954; Meister, 2018, p. 178), which has proven to be robust enough, with considerations, throughout time and devices (Sharif et al., 2020; McGuffin and Balakrishnan, 2005; MacKenzie and Soukoreff, 2003; Card et al., 1978).

In the 1950s and 1960s, researchers began exploring the interaction between humans and early computer systems. The focus was on improving the usability of the interfaces and making the systems more accessible to non-experts (Grudin, 2011, 2005). This era saw the development of early command-line interfaces and efforts to simplify interaction through the use of programming languages and higher-level abstractions (Chignell, 2023).

For instance, the development of ENIAC presented a challenge for ergonomics, and the focus was placed on reducing the Operator's burden, which would become an initial goal for the field (Grudin, 2011).

On the software side of the system, Grace Hopper's pioneering work sought to empower programmers and showed an initial goal of improving the usability of systems and the value of considering the needs of final users for designing the systems (Grudin and Williams, 2013; Hopper, 1952). Much of this initial work formed the basis for proving the value of the future discipline of HCI by enabling more efficient use of costly and complex systems.

As computer usage expanded, researchers started to address the physical comfort and safety aspects of HCI. Ergonomic principles were applied to the design of computer workstations, considering factors such as seating, display placement, keyboard design and lighting to reduce the risk of physical discomfort and musculoskeletal disorders.

Motivated by the high cost of computer usage, initial work during the early days of computing was characterised by non-discretionary use of the systems and focused on selecting and training expert users, improving efficiency and reducing errors and reducing operator's burden (Grudin, 2011; Chignell, 2023) to improve efficiency. Efforts during this time were concerned mostly with procedural issues and optimisation of layout, information presentation, command naming, system response delays.

The challenges of a continuously evolving technology and the promising initial efforts of this era saw also the establishment of the first HCI research centres, starting by Schackel's Human Sciences and Advanced Technology (HUSAT) centre at Loughborough University in and the prolific Xerox PARC, both in 1970 (Grudin, 2011) marking a big impact for HCI as a field.

The formalisation of the field continued with the appearance of the first books in the field, like The Psychology of Computer Programming by Gerald Weinberg (1998) which anticipates the shift of HCI towards psychology or James Martin's "Design of Man-Computer Dialogues" which provided a comprehensive review of the universe of potential input and output applications that characterised the different types and levels of interaction, i.e.: operators, programmers, customers, etc. (Martin, 1973).

In summary, the field of Human Factors and Ergonomics combines tools from psychology with engineering goals and focuses on developing methods to optimise the design of systems for peak performance, and triggered HCI's dive into psychology, addressed in the next section. While HF&E and HCI would part ways in the 1980s, with the arrival of the Personal Computer and the coming of the discretionary user (Grudin, 2005), HF&E also formed a basis for the field of HCI and pioneered the development of metrics, methodologies taken and design improvements based on empirical findings that would become trademarks of the discipline's approach to design.

This is complemented with a very pragmatic and expansive transdisciplinary approach, inherited from HF&E, to incorporate relevant elements, concepts and theories from other disciplines, like Information Theory (Fitt, 1954), Computer Graphics (Sutherland, 1964), Management Studies (Grudin, 2005), Engineering (Meister, 2018, pp. 195–197) and Psychology (Grudin, 2005), an ethos which would later characterise HCI as being a "broad church of theory" (Rogers, 2012, p. 15), but that also brings together a diverse and vibrant community of practitioners (Table 11.1).

11.2.2 PSYCHOLOGY

The limits of HF&E for explaining the complexity of HCI, and the difficulties of developing consistent design principles for solving the design problem of HCI, required expanding the toolset with more powerful conceptual tools, and thus following its pragmatic approach to theory and practice building, saw a big turn to psychology, helped along with advances in Cognitive Psychology and changes in computing that saw an explosion of ideas, tools, methods and theories for HCI that didn't fit anymore in the Human Factors & Engineering umbrella.

TABLE 11.1

Some Key Lessons from Human Factors & Ergonomics for Design in HCI

Contributions	Use Cases
Task analysis and observation of users. Importance of Empirical data for design Time-and-movement studies Number of errors Number of steps	Designers can use empirical data from users to evaluate and refine their designs (Diaper et al., 2003; Ebling and Bonnie, 2000)
Fitt's Law and the importance of layout	Graphical layout impacts greatly the user experience. Designers can consider the ergonomics aspects of their designs, and use tools like Fitt's law for calculating the relationship between size, position and time to interact with it for guiding and evaluating design, setting a right balance between graphic elements size and time, reducing errors and facilitating the user experience (Drewes, 2023; MacKenzie, 1992)
Vinculation between psychology and design Input & output studies from a psychological perspective	Designers can use empirical data from users to evaluate and refine their designs (Meister, 2018)
Importance of motivation	Originally derived from the so-called Hawthorne Effect, designers can take into account motivation and be careful when comparing the performance of motivated testers with what can be expected with the target users (Macefield, 2007)
Usability	Usability is a key goal for designers, which can be achieved by an iterative design approach where testing and validation of design ideas enable further refinements (Grudin, 2011)
Error handling	Designers account for possible errors made by users and incorporate not only strategies to catch them and avoiding or allowing for recovery, i.e. auto-correct, dialog warnings, suggestions, check-ups before allowing to proceed, useful documentation (Ramsey and Atwood, 1979)
Studies about the informational properties of displays	Design can consider the impact of informational density, modulating the amount of items and their complexity in any display to facilitate user's understanding and minimising errors (Ramsey and Atwood, 1979)

Also, in the 1980s, the arrival of Personal Computers oversaw a seismic change in who and how interacted with computers (Grudin, 2011; Chignell, 2023; Carroll, 1993). Humans went from being seen as operators, specialised professionals working with discretionary purpose machines to a more diverse and complex field of users that interacted with general machines for non-discretionary use. This change has been characterised as a change from operators to users (Zhang and Galletta, 2015, p. 49; Grudin, 2011; Bannon, 1986), and with it came a change in focus for the field from learnability to usability (Bannon, 1986). This new focus will forever change the discipline and remain a core goal for HCI design practice.

Cognitive Psychology proved to be an inspiring source for HCI researchers and practitioners of those days (Bannon, 1986; Moran and Card, 1982), with many psychologists joining the field and bringing their theories and concepts into it, enabling insights into the inner mind workings of users (Carroll, 1997; Newell and Card, 1985; Norman, 1988, 1986, 1983b).

HCI expands from ergonomics into mental models, as a way to incorporate not only the behavioural and physical aspects of interaction, but also the internal, mental ones, i.e. motivation, cognitive styles, etc. proposing models such as the Model Human Processor, the GOMS family and the Keystroke-Level Model (Newell and Card, 1985).

These models helped HCI designers understand how users conceptualise and interpret interactive systems and enabled HCI to align the structure and behaviour of interfaces with the users' mental models, thus making systems more intuitive and easier to learn and contributing to the development of user interfaces that are consistent with users' expectations and cognitive processes.

HCI practitioners would also look to other areas of psychology as a source of information for concepts that would help them understand what hinders or enhances the usability of systems. An example of this is the introduction of the concept of Cognitive Load in 1988 from Educational Psychology (Hollender, 2010; Sweller, 1988), as a way to help explain and evaluate the role of information and stimuli on interface design (Duran et al., 2022; Bafna et al., 2020).

Another contribution, in this case from Social Psychology, is the study of technology adoption. Also in 1988, Davis, Bagozzi, Warshaw introduced their Technology Acceptance Model, tackling the key concern of adoption of interactive systems and offering a model for understanding why some novel designs fail to gain acceptance. This model would remain influential in the community and adapted and evolved by HCI practitioners.

Further contributions from psychology include theories, such as bounded rationality (Grimm et al., 1995; Pu et al., 2011) and the recognition-primed decision model, which inform the design of decision support systems and interfaces that present information in a way that facilitates effective decision-making. Also, cognitive psychology research on error recovery processes can inform the design of interfaces (Johnson, 2020) that facilitate users' error correction and provide clear feedback, helping designers understand the causes of user errors and develop strategies for error prevention and recovery, such as confirmation dialogs, undo functions and informative error messages. And HCI designers now routinely apply principles such as chunking (Norman, 2013, p. 83; Vicente and Rasmussen, 1992), progressive disclosure (Sowa, 2009; Shoemaker, 1999; Nakatani and Rohrlich, 1983) and recognition over recall (Nielsen, 1994a; Johnson et al., 1989) to minimise users' cognitive load and enhance their ability to remember and find information (Clarke et al., 2020; Parsons et al., 2020), and use strategies such as scaffolding and cognitive apprenticeship, to guide the design of interfaces that promote effective learning and skill acquisition (Eriksson et al., 2022; Hui et al., 2023; 2018).

It is relevant to mention that a key figure in the field, Don Norman, comes from a cognitive psychology background, and his work reflects the turn from psychology to design in the discipline (Grudin, 2011). Norman's initial work was concerned with Mental Models and was solidly grounded in cognitive psychology, to the extent he would describe his approach as "Cognitive Engineering" (Norman, 1982). The next year he would introduce a first approach to Usability Metrics with his "User Satisfaction Functions (Norman, 1983b). And while as Grudin (2011) noted, "these functions would not hold for long" the idea of usability metrics would be refined and be added to the basic toolkit of HCI design.

Later, in 1986 Norman along with Stephen W. Draper, introduced User-Centred Design expanding on Gould and Lewis (1985). The book argues for placing the user, their goals and needs as the main concern for designers and offers a set of possible dimensions and methods for approaching design. It is relevant that it presented a call for considering the subjective experience of users as a relevant factor for design (Norman and Draper, 1986, p. 3). The introduction of user-centred design (UCD) empowered a new generation of HCI practitioners and fostered the development of new methods and adaptation of transdisciplinary methods to expand the HCI design toolkit.).

This book is also where Norman presented his initial Theory of Action and the concepts for the Gulf of Execution, where the psychological goals of the user need to correspond to the physical elements of the design, and the Gulf of Evaluation where the user interprets the system and defines his next goals (Norman, 1986).

User-Centred Design would become one of the dominant paradigms of Design in HCI and would pioneer many usability testing and inspection methods which show a strong influence from psychology research (Wharton and Lewis, 1994) like thinking-aloud studies, first introduced by Clayton Lewis (1982), cognitive walkthroughs (Polson et al., 1992) or formalisations from psychology experimental results like heuristic evaluation (Nielsen, 1994b) and task analysis (Holzinger, 2005; Diaper, 2003, p. 31; Draper, 1993)

Norman's turn to design continued with his next book, *The Psychology of Everyday Things* (Norman, 1988) which would be later re-edited as *The Design of Everyday Things* (Norman, 2013). In this book he introduced the concept of Affordances, taking the idea from psychology from Gibson (1977) but adapting it for HCI purposes and thus operationalising it for design. The idea of affordances, as the possible actions readily perceivable by the user would remain a core idea in the discipline.

Following this trajectory, Norman's work engaged next with the role of hedonics and affect in design, reflecting on his experiences now as a designer as much as a researcher. This was the basis for his introduction in 1993 of the concept of User Experience as an extension of usability that now incorporated the affective dimension (Norman et al., 1995) and would culminate with Norman's 2004 book, Emotional Design where he expanded his ideas (Norman, 2002, 2004) about hedonics with a call for considering the beauty or attractiveness of designs. His proposal is still grounded in his Psychology background, and influenced by the work of Milton Rosenberg (1960), but it was also grounded in his decades of work as a designer. This evolution represents the trajectory of the HCI from a purely ergonomics perspective, to a psychological one and finally, to a more holistic approach that incorporated behavioural, cognitive and affective requirements.

To conclude, the incorporation of selected contributions from psychology into the body of knowledge of HCI design, enabled designers to conceive interactive systems that are intuitive, efficient and supportive of users' cognitive abilities informed by the understanding of how humans perceive, process and interact with information. As Wharton and Lewis (1994) pointed out, psychology has established several important facts and principles regarding cognition, perception, attention, reaction times and more that can guide designers. It has also provided relevant research methods that have been incorporated into the design practice of HCI. While it was not the skeleton key that was once hoped to be, it provides a strong foundation for the field of HCI design.

The academic psychology formation of these practitioners would inspire a new focus for the discipline on developing and validating theories to inform the understanding and design of interactions and bring a renewed push for giving HCI a scientific approach (Long and Dowell, 1989), that would produce generalisable results (this, combined with the engineering side of HCI would lead to an interest in design process from design studies), and predictive modelling. While this effort didn't provide the expected results, the aim to produce generalisable knowledge, with scientific value would remain and push the discipline to explore other fields like design studies (Carroll, 1993; Gould and Lewis, 1985) or to balance the influence of psychology with the other pillar, computer science (Chignell, 2023; Grudin, 2011; Dix, 2010) to avoid falling into a limiting scope.

From this point onwards, key aspects of the process of laboratory studies from psychology as evaluation and validation of the hypothesis and involving users in the design process would be explicitly included in HCI Design and become a trademark of the house. At the same time as this brings a more rigorous approach to design, this would strengthen the iterative design approach as validation of design ideas with users is essential for iterative design. HCI would make the validation and iteration key elements of its design approach from this point onwards (Table 11.2).

11.2.3 DESIGN STUDIES

As developments from HCI continue, they sparked self-reflection from the community looking into the design process and designer's education for HCI. The challenges arising from a continuously expanding design problem space coupled with the limitations to apply psychological laboratory findings (Grudin, 2011; Gould and Lewis, 1985) to the field saw a "turn to Design" (Rogers, 2012, pp. 68–71) for the discipline. This was not so much a rupture but a movement consistent with the existing HCI ethos inherited from HF&E and Psychology, i.e. a pragmatic approach to incorporate other disciplines and their methods and tools to the HCI corpus and looking for sources and

TABLE 11.2
Some Key Contributions and Lessons from Psychology for Design in HCI

Contributions	Use Cases
Cognitive models Keystroke-Level Model (KLM) Model human processor GOMS	While cognitive models were not as powerful as expected, they are still useful. Designers can use cognitive models for task analysis and design evaluation, comparing between different systems and identify potential improvements in task design (Card et al., 1983)
User-centred design/human-centred design	An influential design paradigm proposed by Norman and others, combining insights from Psychology, HF&E and Design. Designers can use this paradigm to guide their design process by centering the process around the user's needs, goals and experiences. While flexible and wide, it provides a strong ethos for designers today (Norman, 1983a, 2013; Norman et al., 1986)
User experience	Designers need to consider not only the usability and effectiveness of their design, but the affective and hedonic aspects of their designs that have a powerful effect in how user's perceive and use interactive systems (Norman et al., 1995)
Affordances	Affordances as proposed by Norman can help designers to conceptualise the key interactive elements of systems by aligning the user's perceptions with the possibilities the system offers them, and develop design strategies to communicate the interactiveness effectively, manage expectations and enhance user experience (Norman, 2013; Gaver, 1991)
Usability evaluation methods	Informed by the scientific background of many of the HCI pioneers, designers today can use a diverse set of methods to evaluate the usability of their prototypes and inform their design process (Cockton, 2008; Scapin et al. 2009; Lewis, 2014)
Thinking-aloud studies	Designers can use Thinking-Aloud Studies to gain insights about user's minds and perceptions while using systems, as a source of inspiration and validation (Lewis, 1982)
Heuristic evaluation	Heuristic Evaluation remains a popular Usability Evaluation method that offers key insights and guidelines for designers, enabling them to find design problems and evaluate designs in an accessible way that combines well with an iterative design process and reduces costs for the team (Nielsen, 1994a; Nielsen et al. 1990)
Card sorting	Card Sorting is an accessible method that designers use to gain insights into the conceptual models of users and can be used to design the architecture information and navigation flows of interactive systems, for example, it has been used to refine and redesign websites
User personas	Composed based on the design of users, personas enable designers to model and understand the different types of users a system may have and how to design for their needs, goals, motivations and to avoid pain points frictions (Cooper, 2014)
Wizard-of-Oz studies	A method for testing not yet existing systems, testing design ideas and early interface prototypes thus anticipating issues and problems and helping to ground the design process and to define requirements. For example, it has been used extensively to test Smart Assistants and AI systems (Kelley, 2018; Ueno et al., 2022)
SUS	Tools like SUS allow designers to evaluate the usability of their designs with robust instruments that are easy and affordable to implement, thus empowering the validation phase of the design process (Brooke, 2013, 1996)
Cognitive load assessment	Designers can assess cognitive load using methods like the NASA Task Load Index (TLX) or eye-tracking technologies to evaluate their designs and detect potential bottlenecks or issues. Then can apply design and ergonomics principles to solve these issues (Clarke et al., 2020; Zagermann et al., 2016; Sweller, 1988)

(Continued)

TABLE 11.2 (*Continued*)
Some Key Contributions and Lessons from Psychology for Design in HCI

Contributions	Use Cases
Technology acceptance model (adoption)	When dealing with novel systems that change the common practice in some field, designers can use TAM to consider potential obstacles and issues for adoption and how to solve them to ensure adoption. For example, designers can use it to analyse user's data to identify key design requirements or evaluate existing designs for applications in healthcare where future users may be reluctant to new systems or e-learning where users have issues being consistent in their use (Portz et al., 2019; Holden et al., 2010; Camilleri, 2022)

methods for developing generalisable knowledge. Given the focus of HCI on the problem of designing usable interactive systems (Dix, 2010), it was natural to look into the Design Studies and the Design Practice (Gould and Lewis, 1985) as sources for knowledge and disciplinary expansion.

A key inspiration came from the seminal paper, Designerly Ways of Knowing (Cross, 1982) where the design process, and designer's experiences and ideas are seen as a unique source of knowledge relevant to the field, and a necessary complement and answer to the problems of implementing the design ideas derived from lab studies (Long and Dowell, 1989). Already in 1985 Gould and Lewis explored the gap between design principles that came from research and design practice and interrogated designers about their use of such principles or lack thereof.

For HCI, designerly ways of knowing refer, on one hand, not only to the unique knowledge, skills and approaches that designers bring to problem-solving and decision-making processes but also to design practice as a source for knowledge generation (Lowgreen, 2001) and one more tool in the HCI discipline. This is also reflected in the incorporation and further development of Christopher Frayling's Research Through Design methodology into the common practice of HCI research (Zimmerman et al., 2014, 2007; Gaver, 2012). Design Process artefacts, like annotated portfolios (Gaver and Bowers, 2012), design probes (Gaver et al., 1999; Luusua et al., 2015) and prototypes (Sefelin et al., 2003; Nielsen, 1990) or documentation (Zimmerman et al., 2007) and the process itself would be interrogated as sources of data.

From these design experiences, new techniques and methods for generating design ideas, understanding users and refining the design process were developed and became standard practice in the field, like Nielsen's Heuristic Evaluation (1990) or Card Sorting (1995), Cooper's Personas (1988) or Contextual Inquiry (Beyer and Holtzblatt, 1999).

This gap between practice and research would remain a relevant topic in the field of HCI, but the experience of designers would remain a valuable source of data. Seeking to close this gap would also inspire a generation of researchers to look into Design Studies and Design Practice, and even Arts and Artistic Practice as sources of relevant knowledge for HCI also oversaw a new focus on studying the Design Process and how to learn from it and incorporate new ideas (Bobbe et al., 2016; Sanders, 2008; McInerney and Sobiesiak, 1999; Anderson and Olson, 1985).

Reflecting on the Design Process also promoted a more critical mindset to the discipline, going beyond the traditional division of design labour to explore more democratic approaches like Participatory Design, an approach taken from the Scandinavian tradition and that would see to close the gap between users, designers and engineers during the design process (Bødker and Kyng, 2018; Bødker et al., 1995; Muller and Druin, 2012; Sanders et al., 2010; Sanders, 2002). The approach would continue to be critically assessed and refined with new proposals arising from it like Co-Design (Halloran et al., 2009) or Generative Design (Sanders, 2008). With its focus on democratising the power relationship between designers and users, the participatory design family remains an influential approach in HCI design today.

Further, through this contribution to its body of knowledge, HCI recognises the importance of collaboration and interdisciplinary perspectives in the design process. Researchers often engage in collaborations with practitioners, developers, users and other stakeholders to co-create solutions and understand the practical implications of design decisions, and collaborative design activities now foster knowledge exchange, enhance the relevance of design outcomes and ensure the alignment of design solutions with real-world needs and constraints.

The Turn to Design also saw a renewed interest in expanding theory in HCI by incorporating other disciplines and grounding them through reflection in the design process (Rogers, 2012, p. 68; Winograd, 1997). For example, Pattern Languages would be incorporated from Architecture (Wania and Atwood, 2009; Dearden and Finlay, 2006; Alexander, 1977). More ideas of the connections between Architecture and HCI were presented for example in Ruth Dalton's edited book, *Architecture and Interaction*; or between Environmental Design and HCI in Issa and Isaias' 2015 book. Norman's seminal 1988 book, *"The Design of Everyday Things"* would take inspiration from different design sources and combine the psychological basis of 1980s HCI with the natural trajectory to design in the User-Centred Design approach, later expanded to include hedonic and affective aspects, and going beyond the User to become Human-Centred Design (Norman, 2018, 2013).

This cross pollination brought also a push for more critical and socially aware approaches like Critical Design (Bardzell et al., 2018, 2012a,b; Bardzell and Bardzell, 2013) or Value-Sensitive (Friedman, 1996) or Value-Centred Design (Cockton, 2005), or the previously mentioned Sustainable HCI (Knowles et al., 2018; DiSalvo et al., 2010).

This richness of design approaches and visions reflects the centrality of the design practice in HCI and the diversity of the field.

Design Studies expanded HCI as a discipline but also saw an adaptation process to incorporate the disciplinary rigour from Psychology and Science into the design-led processes. With the influence of design studies, HCI embraces the notion of design as a valid research method. Researchers employ design activities not only as a means to create practical solutions but also as a way to explore research questions and generate new knowledge. Design experiments, design probes and design interventions are examples of research approaches within design studies that leverage the design process to gain insights and generate theoretical contributions.

As such, design studies in HCI involve conducting design-focused inquiries that explore various aspects of the design process and investigate and contribute to the development of design methods and techniques that support the creation of effective interactive systems. HCI design researchers explore different approaches to ideation, prototyping, user research, evaluation and other design activities while engaging in reflective and critical analysis of design decisions, methods and artefacts. By examining the strengths and limitations of existing methods and proposing new approaches, design studies contribute to advancing the HCI design toolkit.

The encounter of the Design Studies discipline with the HCI practitioners saw a fruitful exchange of ideas and ontological commitments. HCI remained highly pragmatic and problem-solving focused and with the goal of generating generalisable and applicable knowledge for researchers and practitioners. But it also saw an expansion of the scope of the problems and data sources that the discipline would consider and brought a renewed impulse for expanding the tools and methods available to HCI.

Evaluating and critiquing design artefacts and solutions became a common practice in HCI design. It remains in our core, evaluation and refinement of design ideas is an essential element in any HCI Design Process. Researchers assess the usability and user experience of interactive systems through various evaluation methods to inform iterative improvements and contribute to the refinement of design guidelines and to the establishment of best practices.

This combination of craft, science and discipline (Dix, 2010; Long and Dowell, 1989) and the valorisation of Design Practice as knowledge would reignite the debate of how to characterise HCI; but regardless of the postures, design will remain a central concern in HCI (Long and Dowell, 1989; Chignell, 2023; Dix, 2010; Oulasvirta and Hornbæk, 2016) (Table 11.3).

TABLE 11.3

Some Key Contributions and Lessons from Design Studies for Design in HCI

Contributions	Lessons and Insights
Design a knowledge generation process	Designers can take a more self-reflexive approach to their design process and identify new solutions and challenges for their day-to-day practice. It also fosters a stronger collaboration showcasing the value of learning from each other and previous experiences. HCI has developed a set of practices and design process models based on designer's knowledge. These paradigms and methods continue being refined through this process (Cross, 1982)
Design, science and HCI	HCI design practices are central to HCI and they inform and are informed by scientific practices. This double nature creates a virtuous though complex circle for refining Design
General problem of HCI	Design is the central concern of HCI and it is the unifying aspect for all efforts. Designers participate in generating potential solutions and improvements both to existing or potential applications and sharing knowledge for others. This unifying concern provides a common ground for a diverse community (Long, 1996)
Design process approaches	Since design is a complex and diverse activity, designers can choose between a set of different approaches for guiding their design process taking into account all different kinds of variables like user's needs, designers possibilities, design constraints, etc. The discipline offers a wide set of tools and philosophies which are at the same time, flexible and powerful (Gould and Lewis, 1985; Bongers et al., 2009)
Pattern languages	When faced with common design or usability problems, patterns can offer previous validated solutions that can be reapplied in a new context by designers. Pattern languages offer a corpus of solutions that can be easily adapted and used by designers (Wania et al., 2009)
Design heuristics	The other side of heuristic evaluation, a set of guidelines or principles derived from usability testing and inspection that can guide design preemptively addressing known issues in interaction design. They offer an initial basis for potential design ideas (De Queiroz Pierre, 2015; Bongers et al., 2009)
Prototyping	Prototyping combines the engineering background with design studies, it is a powerful technique that enables designers to test and validate design ideas. HCI has developed a set of prototyping methods from low fidelity, pen and paper prototypes, to interactive mockups and wizard of oz studies that enrich the designer's toolset (Sefelin et al., 2003; Rogers et al., 2023)
Participatory design	Designers can involve users actively in the design process to work together to find solutions to complex problems, improving the acquisition of relevant knowledge and perspectives from the design (Bødker et al., 1995; Rogers et al., 2023)
Co-design	Designers can not only involve but work with users and stakeholders as equals in the design process, using their experience to guide and support the process but sharing the design role and developing technology that is owned by the users (Halloran et al., 2009)
Generative design	Designers can research users through generative methods to create a shared language they can use to communicate with the different stakeholders, thus generating solutions and alternatives to current issues (Sanders, 2008)
Contextual design	Designers can research users in their context through using this method to better understand their needs, goals, drives and desires combining observation, interview and reconstruction of the past (Wixon et al., 1990)
Contextual inquiry	A method originated in Contextual Design for understanding the user, the designer takes the role of an "apprentice" to the user and follows the day to day of the user allowing designers to better ground the design in the real context (Rogers et al., 2023)

(Continued)

TABLE 11.3 (*Continued*)

Some Key Contributions and Lessons from Design Studies for Design in HCI

Contributions	Lessons and Insights
Value-centred design	Designers can use values as the focus of their designs, ensuring the relevance by uncovering not only the needs of the users but the values they share and the value they expect to obtain from the systems we design (Cockton, 2004, 2005)
Research through design	By using a more systematic approach to the design process, designers can generate knowledge that impacts and benefits their design practice and contribute to scholarly research. It enables designers to deepen the reflective aspects of the design process and use them as a source for more generalisable knowledge (Zimmerman, 2007)
Critical design	Designers can use critical design approaches to bring ethics and values to the foreground of design practice, questioning hidden power differentials and agendas and developing socially responsible designs (Bardzell and Bardzell, 2013)

11.2.4 COMPUTER SCIENCE AND INFORMATION SCIENCES

While HCI is a big umbrella term, encompassing an expansive discipline with several influences, it is clearly grounded within Computer Science as its parent discipline (Denning, 2000). Computer science provides the technical foundation and expertise that enables the design, development and optimisation of interactive systems. Its influence spans various aspects, including system architecture, programming, user interfaces, data handling and system performance.

Computer Science (CS) is an extensive domain, with many sub-disciplines like Artificial Intelligence, Theory of Computation, Programming Languages, Architecture, Bioinformatics, Computational Science, HCI and others (Denning, 2000). The seminal ideas in CS were informed by Information Sciences, Cognitive Science and others, form the guiding imaginary of what can be HCI. Key publications in both fields have moulded the ideals and aspirations of what interacting with computers can be and still impulse the design ideas and research agendas of the community.

As Grudin (2005) points out, initial visions that formed our aspirations and goals as designers come from the influential Vannevar Bush essay "As We May Think" (1945) that makes the case for a system that supports and extends our cognition and abilities to understand the world. It will also illustrate a world based around intellectual work instead of physical.

By the 1960s Licklider would further this idea with his proposal for a "man-machine symbiosis" (1960), essentially suggesting a model for HCI analysing different modalities and possibilities for developing and designing such systems. Next, he along with Clark would sketch the requirements and possibilities for "on-line man-computer communication" (1963). In parallel, Engelbart would make his proposal for "Augmenting Human Intellect" suggesting the potential of computer and information systems for such a task and opening the question of how to build and design such systems, and making the point that "[systems] performance can best be improved by considering the whole as a set of interacting components rather than by considering the components in isolation." (Engelbart, 1962). Interestingly, Engelbart's project had a very complex interface designed without much concern about usability. This flaw would lead to the project losing funding (Grudin, 2011).

A much more accessible implementation of similar ideas came from Pauline Atherton's team in Syracuse University. While motivated by Library Sciences needs, Atherton's team developed a series of Information Retrieval Systems for Syracuse University between 1966 and 1970. Their systems, AUDACIOUS, MOLDS, LEEP and SUPARS were iteratively developed with testing with actual users and were the first to be made accessible for the whole campus (Bourne et al., 2003, pp. 69–77).

Another influential development from CS was the development of the GUI, with Sutherland's seminal Sketchpad (Sutherland, 1980, 1964) presented in his thesis for MIT in 1963 that would prove the concept of direct manipulation (Shneiderman, 1982) graphically interacting with computer objects and elements.

Ted Nelson would later imagine a network of digital objects and concepts anticipating our current Internet world (Grudin, 2011; Nelson, 1983). All these initial visions and programmes would form the basis of the whole CS discipline and be highly influential, in different ways and intensities in HCI.

Influential laboratories like Xerox PARC, founded in 1970 precisely to expand the realm of possibilities within Computer Science and Technology (Grudin, 2005), would take many of these ideas and develop designs of what would later become our contemporary computer experience. By 1973 the laboratory had developed the Xerox Alto computer, with a GUI based around the Desktop Metaphor as envisioned by Alan Kay (Smith et al., 1982). This metaphor would become the basis for modern Desktop GUI and influence HCI designers to explore different visual metaphors for interface design. The desktop metaphor would not be contested until the new generation of mobile computing devices where it will prove to be limited (Ailisto et al., 2006).

Later, Mark Weiser proposed the concept of Ubiquitous Computing (Weiser, 1991) where he collected many of the lessons and insights from the work of Xerox PARC and posited the challenge to design for a multimodal, multi-device world.

Other areas of CS will prove to be also influential with HCI and at times aligned. Lessons from software engineering would also influence the approach of HCI practitioners to design, as in the embracement of iterative design approaches proposed by authors like Brooks in his legendary "The mythical man-month" (Brooks, 1974) where he showed the inadequacy of waterfall design process approaches for developing computer systems and the importance of empirical approaches (Carroll, 1993). Concerns regarding command naming (Grudin, 2011; Rosenberg, 1982; Rumelhart and Norman, 1973), related to Programming Languages and Operative Systems, remained a topic of discussion in HCI until the arrival of the GUI, developed in the intersection of HCI and Computer Graphics, to the mass market which caused a change in the discipline.

Another close example is Artificial Intelligence. While the AI and HCI community have not always aligned in times and funding, ideas such as the goal for Natural User Interface, where computers can communicate naturally with humans have long been a goal for both communities (Fu et al., 2018). Topics like context and human behaviour understanding, gesture recognition and voice interaction will be deeply influenced and modulated by the advances and constraints in AI.

Developments like Weizenbaum's conversational program, ELIZA, would be at the same time inspire and challenge HCI and AI, signalling the importance of context and the affective impact of natural language interfaces (Weizenbaum, 1966, 1967) while opening the path for our current iterations of Voiced Assistants Siri or Alexa and Conversational Robots (Murad et al., 2018; Porcheron et al., 2018). Weizenbaum's concerns with such powerful affective impacts would later change his career direction towards a criticism of technology (Weizenbaum, 1986, 1972) and pioneer many of the current topics of concern.

Computer vision, gesture and speech recognition and synthesis, text and image generation and other advances from AI require HCI's involvement in implementing them and offering frameworks for analysis and improvement to address the concerns with trust, explainability (Kim et al., 2023; Mucha et al., 2021), Human-AI collaboration (Fan et al., 2022; Li et al., 2022; Amershi et al., 2019), fairness (Birhane et al., 2022), decision-making (Saxena et al., 2023) safety (Goyal et al., 2023) and usability (Liang et al., 2023; Wang et al., 2022).

Nowadays, Chignell (2023) proposes looking into AI as a way of empowering the community to pursue the original visions of Bush and Engelbart of Human Augmentation that gave birth to our discipline and to move HCI forward.

In summary, it is the collaboration of HCI with the whole of Computer Science that fosters the design and creation of innovative interactive systems. Whereas the rest of CS provide the technical and scientific basis for them, HCI provides the design knowledge to aim for effectiveness, and enhancing the overall usability and user experience, augmenting our abilities and capabilities, thus developing the socio-technical systems that in turn shape individuals and societies (Table 11.4).

TABLE 11.4

Some Key Contributions and Lessons from Computer Science for Design in HCI

Contributions	Lessons and Insights
Augmenting human intellect	Focusing the design on the way to extend our cognition by technology, designers can empower users to accomplish their goals and objectives (Chignell et al., 2023; Engelbart, 1962; Licklider, 1960)
Graphic user interfaces	Designers use visual metaphors and concepts to communicate the status of the system and enable ways of interaction (Sutherland, 1980)
Hypertext	Designers can build informational architectures through developing links between virtual objects (Nielsen, 1990)
Ubiquitous computing	Computing can be embedded and incorporated to a variety of spaces and devices enabling designers to craft continuous interactive experiences for users (Weiser, 1993)
Natural interfaces	By considering the way we communicate between humans, designers can bring a more natural and intuitive interaction experience, reducing the distance between users and computers in their collaboration. A good example of this are the Conversational Interfaces and Gesture Recognition (Glonek et al., 2012)
Iterative design + prototyping from engineering	A lesson learned from Software Engineering, designers can use an iterative design process to refine their solutions and ensure their validity and usability (Brooks, 1974)
Augmented reality	New devices challenge designers to consider ways to build for an enriched, augmented reality considering spatiality both visually and aurally and novel interfaces like haptics. Topics like immersion, transition, presence and usefulness are at the centre of these experiences and require design expertise to implement them successfully (Sutcliffe et al., 2019; Colley et al., 2020)
Artificial intelligence	Designers can use advances in AI both as a method to generate and explore design ideas in human-AI collaboration; as a focus for design bringing their skills into novel interactive systems or to incorporate it to power novel interaction possibilities. It requires a critical perspective to consider not only the benefits but potential issues that may arise from using these systems (Chignell et al. 2023; Engelbart, 1962)

11.2.5 SOCIAL SCIENCES AND HUMANITIES

The fields of social sciences and humanities make significant contributions to design HCI by providing insights into the social, cultural and ethical dimensions of technology design and use. While the role of these fields is not as central to HCI as the previous bodies of knowledge, they nevertheless have had a deep influence in concepts and methods of HCI Design.

While HCI evolved considerably during the years, the pragmatic ethos inherited from Human Factors & Ergonomics and continued by the generation of Psychologists, inspired HCI practitioners to look into other disciplines and fields to complement its knowledge to overcome the detected limitations and expand the discipline toolkit.

Concerns about how to understand the effect and impact of HCI designs and how to better understand users, lead researchers naturally to look for other sources of theories and knowledge when their core disciplines didn't offer a solution. Issues of context, social impact, work practices, and more were added to the original questions of usability. From there, the HCI community began looking into social sciences for methods and concepts that could be useful tools and constructs to adapt and incorporate into HCI design research and practice, and after testing and validation, many of these theories and methods have been adopted as part of the discipline itself.

As stated before, this influence originally came from the generation of psychologists turned HCI Designers that incorporated qualitative methods pioneered in social sciences like in-depth interviews, focus groups and similar to understand users' needs, expectations and requirements.

As previously HF&E had looked into psychology, then psychology looked into social sciences to address the growing interest in considering HCI design within the context of use and the acknowledgement of the social implications of the increasing digitalisation of our professional and personal lives. This, paired with the design goals of ideas like Ubiquitous Computing, brought a need to understand users and even non-users and their environment and brought into our community practices from Anthropology like Ethnography (Millen, 2000; Forsythe, 1999), Ethnomethodology (Martin and Sommerville, 2004; Dourish and Button, 1998) and Participant Observation (Halabi et al., 2013; Rode, 2011).

Social sciences-infused HCI researchers employ qualitative and quantitative methods to conduct contextual inquiry and field studies, and to further understand users' needs, behaviours and experiences to study how social and cultural factors shape and are shaped by interactions with and through technology, providing valuable insights that inform the design process.

Qualitative methods for analysis of data enabled designers and researchers to tap to a vast pool of knowledge and insights from users. Methods like thematic analysis (Braun and Clarke, 2006) have been employed to analyse data from questionnaires, interviews, evaluations and more to generate design guidelines (Bowman et al., 2023). Grounded theory is also widely used in HCI because it enables researchers and designers to take a flexible theoretical approach while analysing data, which resonates with a pragmatic community (Furniss et al., 2011) and enables designers and researchers to explore complex aspects of the user experience as immersion (Brown and Cairns, 2004). Although mostly used in research environments, the method permeates into the overall community. Affinity diagrams, or the KJ method, is a method for visual clustering and sense making of large amounts of dissimilar data (Jokela and Lucero, 2014). It has been used to evaluate interfaces (Lucero, 2015; McQuaid and Bishop, 2001), categorise user types and strategies (Janssen and de Poot, 2006) and has become a standard tool for UX in the industry (Schicker, 2020). Other methods like Narrative Analysis offer an alternative to thematic analysis (Tøndel et al., 2020), aiming to understand the user experience through stories to explore and undercover how users understand the experience.

Other methods are focused on recollecting complex or indirect data. For example, Gaver, along with Dunne and Pacetti developed Cultural Probes as a method to face the challenge of designing for "unfamiliar groups" taking inspiration from the experiences of Situationists (Gaver et al., 1999). Similar approaches have been developed to work with different audiences and participants like Diary Studies (Brown et al., 2000) enabling HCI designers and researchers to explore complex topics and work with sensitive populations (Le Dantec and Edwards, 2008; Badillo-Urquiola et al., 2021).

Nowadays, novel issues regarding power, intersectionality, societal implications and ethics of HCI are being addressed (Fiesler et al., 2022; Pillai et al., 2021) by exploring more potential bridges and ideas from Social Sciences and Humanities and operationalising such ideas into tools and methods for the design research and practice in HCI. Researchers are engaging with issues of colonialism (Alvarado-Garcia et al., 2021; Dourish and Mainwaring, 2012; Philip et al., 2012; Birhane, 2020), intersectionality and power differentials (Erete et al., 2023), racism (Rankin and Henderson, 2021; Ogbonnaya-Ogburu et al., 2020), feminism (Bardzell and Bardzell, 2011; Lindtner et al., 2016; Ahmed and Irani, 2020) and developing critiques and proposals for the community.

The ethical dimensions of technology design and use are emphasised by HCI researchers who investigate ethical issues related to privacy, consent, data protection (Wong and Mulligan, 2019), and algorithmic bias and advocate for responsible design practices (Adomavicius and Yang, 2022; Benabdallah et al., 2022) that prioritise user well-being and social impact, helping to shape guidelines and policies in the field (Spaa et al., 2019; Cockton, 2013; Bødker and Kyng, 2018).

With contributions from philosophy and ethics, HCI researchers also explore the relationship between human values and technology while critically examining the ethical implications of technology design, considering questions of power, autonomy, justice and equity (Constanza-Chock, 2020; Ghoshal et al., 2019; Strohmayer et al., 2019; Fox et al., 2016), and this further contributing to discussions on the societal impact of HCI design and the need to promote the responsible design and use of technology.

Furher, equipped with these contributions, HCI designers are now able to analyse the broader social, cultural and political implications of technology design and use and challenge assumptions and biases in HCI research and design, fostering a more inclusive and socially aware approach.

And finally, it is the convergence between Social Sciences, the Humanities and HCI that enables HCI researcher to delve into topics such as the philosophy of technology, digital culture, media studies and human-computer ethics, providing interdisciplinary insights that enrich the understanding of the complex relationship between humans and technology, a paramount concern now that we are the dawn of our cyborg future (Table 11.5).

TABLE 11.5

Some Key Contributions and Lessons from Social Sciences and Humanities for Design in HCI

Contributions	Lessons and Insights
Ethnography	Originally an anthropological method, ethnography has been taken by the HCI community as a way to conduct field work away from the laboratory space. For designers, it offers a richer understanding of the practice and context of technologies, a way to gain insights about requirements or pain points, and explore the different interrelationships between actors in technology. HCI focused ethnography has developed rapid approaches which are suitable for designers and less demanding of time and resources while still offering powerful insights (Millen, 2000; Forsythe, 1999; Hughes et al., 1995)
Ethnomethodology	Imported from Sociology, ethnomethodology is an analytical orientation (Dourish and Button, 1998) that offers a set of methods to observe and describe the interactions that make a social order. Designers can employ this method to look for patterns or practices in social interaction to uncover insights or problems that can be addressed by a design intervention, or to critique and improve existing design solutions (Dourish and Button, 1998; Martin and Sommerville, 2004)
Qualitative methods	While originally concerned with how to measure and operationalise performance, HCI has expanded its toolkit to add several qualitative methods into our common practice. Qualitative methods support a more nuanced and detailed understanding of users, and empower designers to not only study users and evaluate designs, but to generate design solutions through participatory, co-creation, or generative approaches grounded in qualitative methods (Brown, 2001)
Participant observation	By immersing themselves in the community of users and taking part of it, designers enjoy a more powerful position to observe their users and gain a clearer understanding of the users experiences and perceptions of their world and technology. Participant Observation is also used in Participatory Design approaches (Hayashi and Baranauskas, 2014; Halabi et al., 2013)
Focus groups	A tool originated in Market Research but if carefully applied by selecting the right participants they allow designers to generate and validate design ideas with deep perspective in an efficient way. They are best when used in combination with other methods so each complements each other (Rosenbaum et al., 2002; Rietze et al., 2021; Colin Gibson et al., 2020)
In-depth interviews	One of the most well-known qualitative methods, interviews enable designers to delve deeply into the user's experience, expectations, attitudes, ideas, requirements, goals and worries. They can help designers to explore the design possibilities and generate new ideas, to validate specific features, explore the potential outcomes and impacts of designs or understand the context and interrelationship between different actors in a complex system (Rogers et al., 2023; Spaa et al., 2019; Cho et al., 2019)

(Continued)

TABLE 11.5 (*Continued*)

Some Key Contributions and Lessons from Social Sciences and Humanities for Design in HCI

Contributions	Lessons and Insights
Diary studies	Diary studies are a great way for designers to uncover how technology is used through a period of time by users. They are also very useful for studying sensitive contexts or communities where the presence of a design researcher may not be welcome nor convenient. Some kinds of diaries are photographic ones, which have also been used to gain a better insight of contextual information or generative diaries that have been the starting point for co-designing approaches (Brown et al., 2000; Cho et al., 2019; Badillo-Urquiola et al., 2021)
Thematic analysis	A powerful qualitative data analysis method to identify and structure patterns in data in rich detail. Designers can use thematic analysis to identify concerns, needs, roles, challenges, opportunities and styles when designing interactive systems, to organise the results from validations to better address them and to get a better understanding of users and their experiences (Braun and Clarke, 2006; Bowman et al., 2023)
Grounded theory	Grounded theory is a powerful framework from Anthropology that offers a set of tools and techniques to form a theory that explains a qualitative data set. HCI has used to develop a better understanding of the complex realities of interactive systems. For example, it can be used when designers need to form working understandings or theories of how a system will be experienced, or why some elements of it are used while others not, or to develop better understandings of qualitative data from design experiences, to inform future designs or refine existing ones (Adams et al., 2008; Cole and Gillies, 2022; Furniss et al., 2011)
Affinity diagrams	Affinity Diagrams are a technique that allow designers and HCI practitioners to analyse mixed types of data by identifying a structure of clusters or themes that are identified because they share some common element or value. It is an inductive and iterative process that is very attuned with design practice while offering a systematic approach to analyse data (Rogers et al., 2023; Janssen and de Poot, 2006; McQuaid and Bishop, 2001)
Narrative analysis	By using personal narratives as the unit for analysis, designers can gain a better understanding of the context and lived experience of users, either to understand how a system is experienced and understood by the users, or to gain insights into their context and lives to gain a better understanding of the users' needs, goals, expectations and challenges (Garcia Rodriguez, 2016; Tøndel et al., 2020)
Cultural probes	A design tool developed to develop a deeper conversation beyond the traditional requirements of HCI design. They allow designers to elicit insights, novel ideas and fresh perspectives from users and start design conversations. Probes take many shapes and contents, i.e. postcards, maps, diaries, albums, conceptual art; and they can serve as a springboard for reflection and understanding as an ideation tool, as a generative tool, or to develop novel methods and processes (Gaver et al., 1999; Luusua et al., 2015)
Situated action	An anthropological approach mostly informed by ethnographic methods, Situated Action offers designers a framework of analysis to understand how the changes in the context can affect or modulate the user's goals, needs and overall experience. It focuses on the user-in-a-situation as the unit of analysis and it empowers designers to develop more flexible systems that are more useful and valid. Designers can use situated action to explore issues around changing environments for design like urban environments and wearables or domestic cooking (Rogers, 2012; Van Hove et al., 2020)
Embodied interaction	An approach proposed by Paul Dourish, designers can use embodied interaction as a lens to understand interaction as embodied and explore topics of presence, agency, awareness and control. It can be used to make sense of complex tangible interactions like cockpit design; cognitive multimodal experiences like learning; or novel interaction spaces like Haptic or Sonic Interaction or Augmented and Virtual Reality environments (Dourish, 2001; Benford et al., 2021; Letondal et al., 2018)

(Continued)

TABLE 11.5 (*Continued*)

Some Key Contributions and Lessons from Social Sciences and Humanities for Design in HCI

Contributions	Lessons and Insights
Decolonial and postcolonial approaches	By engaging in decolonial or postcolonial thinking, designers can question the existing ideologies that inform much of our contemporary technology design and deal with issues of power, engagement, inclusion, respect, dignity and justice. The design process can be expanded and strengthened to include more responsible practices and methods by engaging in prototyping for a dialogic engagement foregrounding marginalised voices and knowledge, establishing safe spaces with communities and creating newer experiences of technology that move away from colonial inheritances (Irani et al., 2010; Das and Semaan, 2022; Kambunga et al., 2023)
Critical theory	By taking a critical mindset and foregrounding the ethical aspects of the design practice, HCI designers can uncover and consider the effect of dominant ideologies, flawed understandings and approaches thus exploring alternative design solutions that are more socially responsible and aware and move beyond the traditional narrow goal of usability to consider wider goals of justice, fairness and values (Bardzell and Bardzell, 2013)
Feminist approaches	Contributions from feminist approaches can be incorporated into the design practice informing the design process and design methods, enabling designers to pay attention to marginalised voices and communities, offering critical tools and concepts like intersectionality, ethics of care, reflexivity and safe spaces that enrich designer's analytical tools and deepen the fitness of their design solutions (Bardzell and Bardzell, 2011; Erete et al., 2023)
Activity theory	A conceptual framework that designers can use to analyse complex systems through actions and interactions of users with artefacts considering cultural and historical contexts (Rogers, 2012, p. 55). It can inform finer understandings of the design problem when considering dynamic systems where different actors have complex interactions like working environments or video games (Rogers, 2012; Clemmensen, 2021; Marsh et al., 2021)
Actor-network theory	Another anthropological approach, derived from Latour and others, it provides a framework for understanding Socio-Technical Systems. HCI designers can use ANT to understand their practice as building a network of interactions between different actors in a socio-technical system. It can help designers consider the perceptions, values and attributes that users assign to systems, the strategies they use with complex systems like platforms and gain insights for design and critical action (Das et al., 2023; Wong-Villacres et al., 2019)
Semiotics	Popular in Media And Cultural Studies, Semiotics approaches have been used in HCI to analyse aspects relating to interactive systems. For designers, semiotics can offer a conceptual framework to extract insights from other design experiences and media, explore interaction in a broader context considering cultural and communicative factors and suggest design approaches, strategies and heuristics (de Souza et al., 2001; Andersen, 2001; Bolchini et al., 2009)

11.2.6 NEUROSCIENCE

The goal of understanding human behaviour and inner states has remained in HCI. While psychology provided a strong foundation for HCI, other bodies of knowledge like neuroscience have been incorporated into HCI as well to expand our capabilities to integrate brain-sensing to our practices (Putze et al., 2022). Advances and developments in bioinformatics allowed the expansion of sensor technology thus enabling HCI to begin to use psychophysiological signals as sources of information and materials for design (Cowley et al., 2016).

Interest in the role of emotions in the user experience rose during the 1990s, inspired by the publication of Goleman's book, *Emotional Intelligence* (1995). Two years later, in 1997, Rosalind Picard published her book Affective Computing where she makes a call to creating systems that can detect and express human emotions in their design. This opens a new set of challenges and opportunities to answer Picard's call. Less than a decade later, Stephen Fairclough would conduct a study to evaluate the potential of different physiological signals for designing biocybernetic loops (2004). He will continue developing these ideas and proposing a direction he would call physiological computing.

Neurophysiology, a subfield of neuroscience concerned with the study of the functioning of the nervous system, has an important twofold role in the design of HCI: (1) Creating novel modes of interaction (Riedl et al., 2010; Fairclough, 2004; Cowley et al., 2016), i.e. implicit interaction systems via biocybernetic loops; extending input modalities like in the recent Apple Vision Pro demo where Eye-Tracking and Gesture Detection replace the need for mouse mediated GUI interaction; Brain-Computer Interfaces among others (Choi et al., 2017; Frey et al., 2017); and (2) Evaluating and Validating Interfaces (Benbasat, 2010), i.e. EDA Studies for engagement, Eye-tracking Studies for usability or cognitive load, Emotion Recognition Studies for exploring the hedonic and affective aspects of interfaces. Cowley et al. (2016) offer a comprehensive review of the use of psychophysiological signals in HCI, and while not the focus of the review, identify potential uses for designing or validating interfaces. This general concern with design and validation showcase how central the design problem is for HCI practitioners regardless of their research focus.

The body of knowledge of neurophysiology helps designers understand how the brain processes sensory information and perceives stimuli. With the knowledge of how different sensory modalities (such as vision, hearing and touch) are processed in the brain, designers can optimise the presentation of information and design HCI that are sensorially effective and engaging (Huang et al., 2015). In this case, data collected through neurophysiological techniques, such as, electrodermal activity (EDA) electroencephalography (EEG), eye-tracking and functional magnetic resonance imaging (fMRI), can complement data generated through traditional usability testing methods and provide objective insights into users' cognitive states, emotional responses and attentional focus, helping HCI designers identify usability issues and improve the user experience (Chiossi et al., 2023; Cowley et al., 2016).

The contribution of neurophysiology to the understanding of how emotions are generated and processed in the brain allows HCI designers not only to create emotionally engaging and enjoyable user experiences but also to conceive systems that cater to the mental health and well-being of their users (Terzimehić et al., 2019; Kosunen et al., 2016; Cavazza et al., 2014). For instance, neurophysiology is instrumental in developing neuroadaptive systems, which dynamically adjust their behaviour based on users' cognitive states and intentions assessed through the monitoring of the users' neurophysiological signals in real-time (Fairclough, 2022; Moge et al., 2022; Appriou et al., 2018). Specific examples include Brain-Computer Interfaces, which enable almost direct communication between the brain and a computer system through insights into brain activity patterns, neurofeedback mechanisms and the interpretation of brain signals for control and interaction purposes (Khademi et al., 2022; Choi et al., 2017).

This body of knowledge has enabled HCI practitioners an expanded set of design elements and a new line of inquiry and interaction with users and stakeholders but opened relevant issues regarding privacy and safety (Shackell and Sitbon, 2018). As we move forward, we need to keep these issues and possibilities in mind when designing with psychophysiological signals (Table 11.6).

11.3   DESIGN PARADIGMS

The term "human-computer interaction design" is particularly confusing, as it is well-defined in the literature but not widely used in the industry. Instead, terms such as user experience (UX) design, user interface (UI) design, product or service design and other adjacent terms are often used.

This confusion can lead to misunderstandings and make it difficult to communicate effectively about design. It is therefore important to be clear about the meaning of the terms being used when discussing design.

TABLE 11.6

Some Key Contributions and Lessons from Neurosciences for HCI Design

Contributions	Lessons and Insights
Psychophysiological signals	Taking lessons from psychophysiology and neuroscience, HCI designers have appropriated sensor technologies as a window to understand humans. By sensing different signals produced by the human body, designers can develop systems that are able to sense and perceive humans; or enable novel interaction modalities beyond the keyboard, mouse and controller. Different biosignals are mapped to different affective and cognitive states and can help systems understand users better and adapt to them. They can also support designers to test and validate the effectiveness of their designs by measuring things like affective states, cognitive load, saliency of design elements (Cowley et al. 2016)
Cardiovascular signals (ECG, HR, HRV, BP, PPG)	Cardiovascular activity can be measured through signals like electrocardiograms (ECG) to identify Heart Rate and Heart Rate Variability, blood pressure (BP), Photoplethysmogram (PPG) and even web cameras by measuring changes in the colour of the facial skin. Designers can use these signals to infer the mental and cognitive load of the user, mental stress, or design experiences to help users monitor their health and physical activity (Urrestilla and St-Onge, 2021; Muender et al., 2016; Slovák et al., 2012)
Electrodermal Activity (EDA)	Electrodermal Activity, also sometimes called Galvanic Skin Response, is a measurement of the electrical conductivity of the skin and is correlated with the activity of the nervous system, emotional reactions to stimuli. It is a common tool for testing and validating design, but creative designers can also consider ways of designing implicit interactions through EDA sensors (Albert and Tullis, 2010; Brishtel et al., 2018)
Respiration signals	While less explored than other signals in HCI, alterations in the respiration patterns can be used to gauge human reaction to certain experiences like movies or video games. It has also been used along with other signals for developing interactive mindfulness experiences; or to anticipate certain behaviours like the next speaker in a social context (Sonne et al., 2016; Kuikkaniemi et al., 2010; Prpa et al., 2018)
Electromyography	EMG sensors allow us to detect muscle activation. Although currently other options for detecting facial gestures are more used, facial EMG enables designers and researchers to assess affective states, thus being a powerful tool for testing and validating designs. It also enables designers to deploy novel ways of interaction and control, and has been explored for its use for prosthetics or ergonomics, detecting excessive tension in muscles (i.e. neck and back muscles). A very flexible sensor that has a strong connection with the body, designers can use EMG as a design material for imagining and developing new interactions and systems that combine sensing, anticipating actions and controlling computers through more embodied interactions (Eddy et al., 2023; Perusquía-Hernández et al., 2019; Nakanishi et al., 2022; Zhang et al., 2022)
Electroencephalography	A powerful yet complex compound of signals, EEG signals show the brain activity. While the signals have low spatial accuracy, in contrast with fMRI, they are very time-sensitive, enabling designers and researchers a way to measure reaction time and magnitude of effect from stimuli. For example, through exploring Event-Related Potentials, designers can make sense of design failures related to clear stimuli like sound notifications in complex environments like driving or measuring when a notification can be more effective; or coupled with eye-tracking, to detect sensitive words in a text in an almost immediate manner. EEG signals are also the basis for Brain-Computer Interfaces and a potential design material for designing novel controls (Chuang et al., 2017; Janssen et al., 2019; Li et al., 2021)
Functional Magnetic Resonance Imaging (fMRI)	While less flexible and mobile than other technologies for detecting brain activity, fMRI offers a high spatial accuracy that, in combination with neuroscientific knowledge, offers HCI designers a powerful tol to evaluate design ideas, to explore complex issues related to cognition and perception, like habituation or comprehension among others. For example, Huang et al. (2015) explored how text, icons and pictures are processed differently within the brain in a cross-cultural study, suggesting that each require different considerations when using them in designs (Thanh et al., 2017; Vance et al., 2017; Huang, 2015)

(Continued)

TABLE 11.6 (*Continued*)

Some Key Contributions and Lessons from Neurosciences for HCI Design

Contributions	Lessons and Insights
Near-infrared spectroscopy (fNIRS)	Another source for brain signals, fNIRS uses near infrared light to track hemodynamics which are correlated with brain activity. Much more portable than fMRI it has been used for BCI, workload assessment, emotional regulation interfaces, usability testing applications and validating and complementing traditional inspection methods like thinking-aloud protocols (Howell-Munson et al., 2022; Lukanov et al., 2016; Ashraf et al., 2023; Wang et al., 2021)
Eye-tracking and pupillometry	By tracking changes in gaze direction and pupil dilation, HCI designers can develop a diverse set of applications for their practice. From developing novel input experiences like the ones showcased in Apple Vision Pro device to evaluating interface design and extract insights into the cognitive and affective states of the users like mental workload, attention, engagement, flow and focus or fatigue and sleepiness (Brom et al., 2016; Poole and Linden, 2006; Pai et al., 2023; Liu et al., 2020)
Voice analysis	Analysis of timbre, pitch, cadence, rhythm and other speech components have been used to assess mental and emotional states. Designers can consider adding these elements for developing neuroadaptive Voice-User Interfaces or social interactions like video conference calls. They can also be used for strengthening data analysis or as a tool for testing and validation. While there are limitations to their predictive power (Ma et al., 2023), they are a potential design material for developing emotionally aware interactions (Shu et al., 2018; Nikopoulou et al., 2018; Neubauer et al., 2017; Chang et al., 2011)
Facial recognition	Like many other physiological signals, Facial Recognition is a sensitive topic that concerns aspects of privacy, surveillance, ethics and fairness that go way beyond issues of bias (Buolawmini and Gebru, 2018; Birhane, 2020). Still, facial recognition is used as a replacement for passwords in current devices and when done in a careful setting offer another path for designers to assess the affective impact of designs, or to develop careful interactions around sensitive topics. While designers should be very mindful about issues of power and the potential for abuse and misclassification, facial recognition technologies can offer designers ways to enhance social interactions, foster engagement and support affective computing designs. While there is a large tradition of facial emotion recognition in the field, designers should also be aware that the rationale behind these technologies is contested (Birhane, 2021) and consider how much to trust these models, and when and why to deploy them (Walsh, 2022; Zhang et al., 2021; Peng et al., 2019; Ruan et al., 2023; Lingelbach et al.; 2022)
Thermal signals	While body temperature remains constant in humans, the human body is very sensitive to it. Designers have used thermal interfaces for enriching social interactions, eliciting emotions or supporting self-regulation. Tracking facial temperature changes can support emotional recognition and tracking body temperature can also support elderly people or infants to avoid heat strokes or to enable energy-efficient designs for environments or to support athletic endeavours. While less common than other sources, it offers an exciting opportunity for designing novel interactions that communicate with the body beyond the traditional audio-visual spectrum (Haliburton et al., 2023; Khan et al., 2006; Aryal et al., 2019; Habeeb et al., 2022)
Gesture recognition	Gesture recognition is a wide umbrella that includes different technologies based around different sensor modalities and focusing on different sections of the body. It has the potential to infer cognitive or emotional states of users, enable embodied or touchless interactions, or enable action detection in smart environments. Designers can now look to the whole body as the source for interaction, instead of depending on mediating devices like touchscreens, keyboards or mice (Noroozi et al., 2018; Kamijo et al., 2023; Al-Shamayleh et al., 2018)

(Continued)

TABLE 11.6 (*Continued*)

Some Key Contributions and Lessons from Neurosciences for HCI Design

Contributions	Lessons and Insights
Affective computing	A paradigm originally proposed by Rosalind Picard, Affective Computing challenges HCI to consider how to design systems that can sense and react to the emotional states of users. It requires an interdisciplinary approach to design which is grounded in the understanding of human experience as both cognitive and emotional. It has been explored as a paradigm to design interactive systems for many fields including education, arts, healthcare, video games and more. While Picard's approach is based on the idea of objectively sensing the internal affective states of users, other authors have expanded it to consider other approaches based in phenomenological, enactive or embodied interaction paradigms (Picard, 1999; Wang et al., 2022; Wu et al., 2016; Zheng et al., 2020)
Biocybernetic loop	A concept initially proposed by Pope and taken by Fairclough building upon advances on physiological sensors, a biocybernetic loop is an interaction design where a system interprets human decisions, controls and behaviours via psychophysiological responses. The key aspect of the loop is that the system processes and adapts to the signals. As a conceptual tool it can describe many different designs, but it expands traditional interaction design by foregrounding the role of psychophysiological signals as points of communication between humans and computers. Designers can use this conceptual tool to expand their inventory of types of interaction and develop more flexible systems that can enhance human-computer interaction, i.e. notification management, improving gaming experience (Chanel et al., 2011), detecting fatigue and engagement (Pai et al., 2023), helping manage pain among others (Fairclough and Gilleade, 2012; Pope et al., 1995; Muñoz et al., 2021, 2017)
Physiological computing	Expanding the idea of the biocybernetic loop, Fairclough proposed a design paradigm to expand interaction to include psychophysiological signals, especially the ones related to the central nervous system, offering alternative or complementary modalities for input, control and output of the system. The key implication for design is to think about input, output and control beyond existing paradigms and to consider all the different possible modalities that can be incorporated into the design of human-computer interaction. By considering how systems can be more sensitive and aware of users, designers can expand the current applications and possibilities of technology, developing experiences that are more aware of their users and adaptive and expanding and deepening the communication between humans and computer systems (Fairclough, 2008; Frey et al., 2020; Moge et al., 2022)
Brain-computer interfaces	An application closely related to physiological computing goals, Brain-Computer Interfaces are an interactive system that is directly connected with brain activity via brain signals. They can be invasive, i.e microelectrode arrays or (MEAs) or non-invasive, i.e. EEG, fNIRS. Their main field of application has been medicine and healthcare supporting medical interventions for quality of life improvement and disabilities management and communication and control for individuals with paralysis, rehabilitation or entrainment; but they have also been used in fields like games and entertainment, marketing, emotional regulation, well-being, security or smart environments. Designers can consider how to incorporate BCIs or elements of them into the systems they develop, or novel applications where BCI's can offer still unexplored interaction possibilities (Frey et al., 2017; Choi et al., 2017; Khademi et al., 2022)

When discussing design, it is important to be aware of the different understandings of the term. This can be due to a variety of factors, including cultural differences, the multidisciplinary nature of design, different design and application domains and the different contexts in which design is discussed.

For example, in the context of product design, "design" may refer to the process of creating a physical product. In the context of graphic design, "design" may refer to the process of creating visual communication. In the context of user experience design, "design" may refer to the process of creating interactive experiences. Starting with definitions, it is common to refer to Dan Saffer (2010), who in his well-cited and influential book Designing for Interaction, proposed four major approaches to tackling interaction design projects:

- User-centred design
- Activity-centred design
- Systems design
- Author-Centred Design

The list of the approaches proposed by Saffer is sorted by his preference for the application of these methods. While all the methods are considered eligible for application in specific contexts, UCD has a higher priority because of its obvious orientation to the end users (we will discuss that later) and the Genius design approach is positioned as fully opposite to the UCD approach that doesn't consider the end users at all and therefore is less preferable.

We will start an overview of these methods on the back side order.

In this section, we will explore Saffer's four design approaches trying to shed light on similarities and differences among them.

11.3.1 AUTHOR-CENTRED DESIGN

The "author-centred design" or "genius design" approach is a contradictory one. The name of the approach implies that the designer knows everything that is needed to be known to design a good product. However, the negative examples of products implemented using genius design, such as the Apple Newton and Windows 95, show that relying on the designer's expertise alone may not be enough (Saffer, 2019).

While experienced designers can rely on their experience (which may include past experience with real end users), the choice of genius design for beginners is both common and undesirable. This is because genius design is the simplest way to design. The designer effectively designs for herself, reflecting an understanding of her own end goals, mental models and skills. This understanding of the system model makes it as close as possible to the mental model of the designer. However, even if the product perfectly matches the designer herself, it may not correspond to the mental models and end goals of other users.

In this way, the genius design approach, while a possible way for experienced designers, can be a trap for beginners. It is cheap (no need for user research), quick (for the same reason) and straightforward (not much to evaluate, as the designer knows well what she needs).

In today's design industry, there are many different ways to start a design career. Sometimes, this starts without proper design training. In these cases, we can expect more use of genius design, and as a consequence, more badly designed products.

11.3.2 SYSTEMS-CENTRED DESIGN

System design is an analytical approach to addressing design problems. It views the design process as a system, with a set of entities that act upon each other. According to Saffer (2010), systems design does not discount user goals and needs. In fact, these can be used to set the goal of the system.

System design outlines the four components required for a system (see an example in Figure 11.2):

- *Goal:* The goal of the system is what the system is trying to achieve.
- *Sensor:* The sensor is responsible for gathering information about the environment.
- *Comparator:* The comparator compares the information gathered by the sensor to the goal of the system.
- *Actuator:* The actuator takes action based on the comparison made by the comparator.

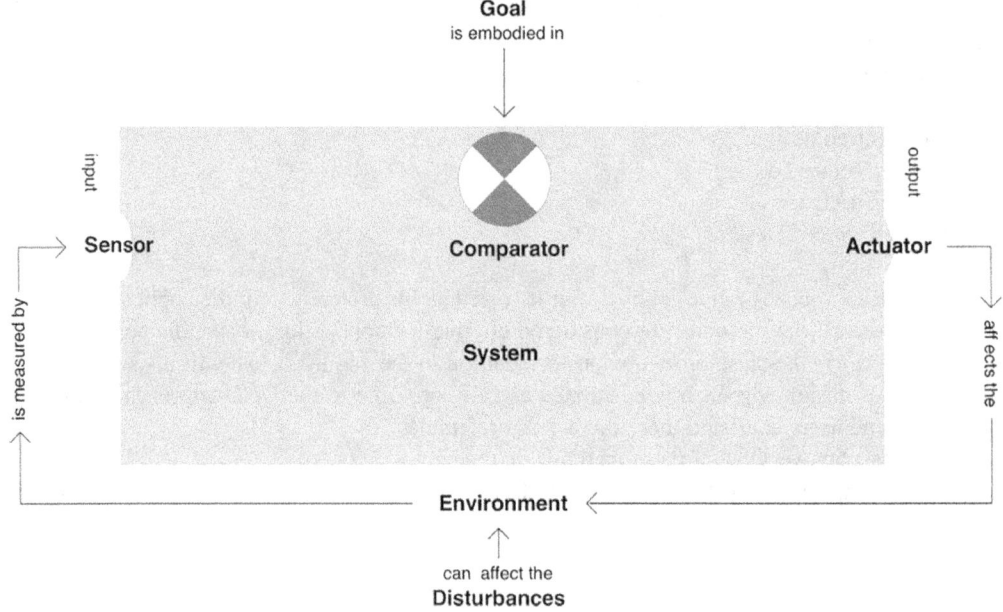

FIGURE 11.2 System design: Feedback: Formal mechanism. (Adapted from Dubberly, H., Pangaro, P. (2010). Introduction to Cybernetics and the Design of Systems.)

System design is a powerful tool for designing user-centred products and services. By understanding the system as a whole, designers can ensure that the product or service meets the needs of the users.

System design is often seen as a process that focuses on the technical aspects of a system, with little or no consideration for the users. However, this is not entirely true. In fact, the first attempts to involve users in system design were made in the 1970s by Boland (1978). This could be considered as the first step in the UCD movement.

In the beginning of the 1980s, Norman (1983a) already discussed system design in the context of cognitive engineering. Cognitive engineering is a discipline that seeks to understand the nature of a person's mental image of a system and the human information processing capabilities of the user. The goal of cognitive engineering is to provide designers with the tools they need to make their products more sensitive and responsive to users' needs.

As part of his discussion of cognitive engineering, Norman proposed four design principles that should contribute to system design:

- *Feedback*: The state of the system should be clearly available to the user, ideally in a form that is unambiguous and that makes the set of options readily available. This helps to avoid mode errors, which are errors that occur when the user is not aware of the current state of the system.
- *The similarity of response sequences*: Different classes of actions should have quite dissimilar command sequences (or menu patterns). This helps to avoid capture and description errors, which are errors that occur when the user accidentally selects the wrong command because the command sequences are too similar.
- *Reversible actions*: Actions should be reversible (as much as possible). This helps to prevent unintentional performance errors, which are errors that occur when the user performs an action that they did not intend to perform.
- *Consistency of the system*: The system should be consistent in its structure and design of commands. This helps to minimise memory problems in retrieving the operations.

These four design principles are still relevant today, and they can be used to create systems that are more user-friendly and effective.

Based on current knowledge, it is evident that the first, third and fourth principles align closely with the Usability Heuristics for User Interface Design proposed by Jacob Nielsen (1994a), who worked closely with Norman. Norman's design principles correspond to Nielsen's heuristics of Visibility of system status, User control and freedom, and Consistency and standards. This demonstrates that the focus on usability has been addressed in system design for quite some time.

However, in system design, the user is primarily viewed as an active component of the system, responsible for executing specific actions, rather than as an individual with psycho-physical capabilities and limitations. The user's role in system design entails carrying out functional tasks such as pressing buttons, executing commands and receiving output messages.

System design plays a crucial role in developing complex systems, encompassing the creation of architectures for various components, interfaces and modules. Here we consider a systematic approach to system design, with a specific focus on the process of architecture creation and the corresponding data necessary for successful implementation. By following this approach, designers can effectively structure and organise the system, facilitating seamless integration and functionality across its constituent elements.

A contemporary example of system design processes can be found in the NASA Systems Engineering Handbook (National Space Administration, 2017). NASA outlines four system design processes: developing stakeholder expectations, defining technical requirements, performing logical decompositions and generating design solutions (Figure 11.3). The framework is based on a comprehensive understanding of and definition of mission objectives and the concept of operations, with a strong emphasis on capturing stakeholder expectations. These expectations are subsequently translated into quality requirements and operational efficiencies throughout the project's life cycle.

NASA's system design process is stakeholder-centred, meaning that the needs and requirements of the stakeholders are the primary drivers of the design process. This is in contrast to user-centered design, which focuses on the needs and requirements of the end users.

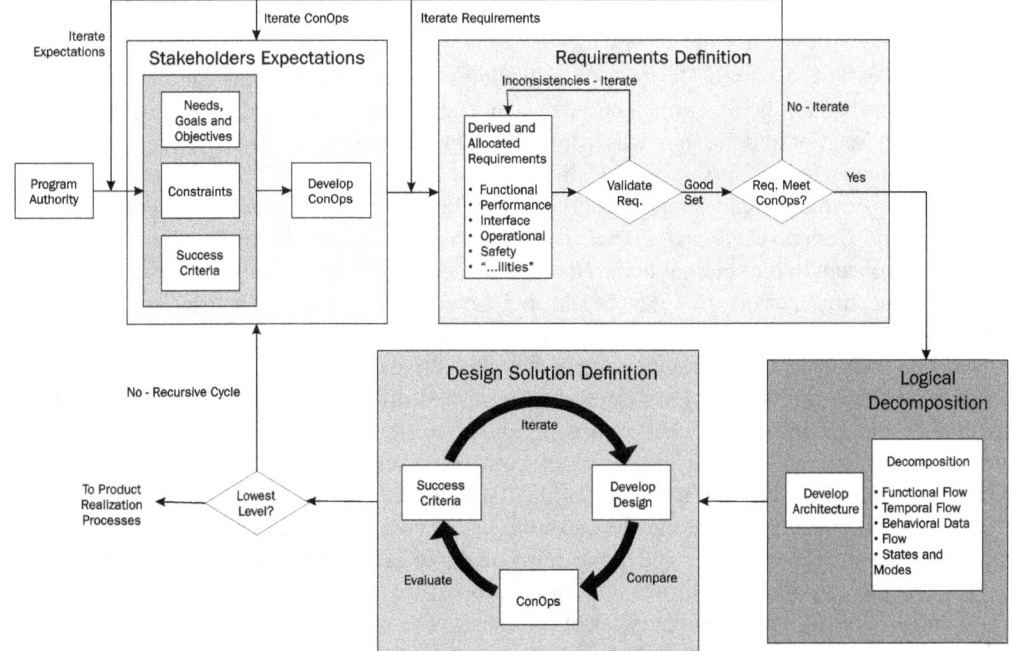

FIGURE 11.3 Interrelationships among the system design processes according to NASA.

In NASA system design, the term "stakeholder" refers to any individual or organisation that has a vested interest in the success of the system. This can include customers, users, developers, maintainers and even government regulators.

The needs and requirements of the stakeholders are captured in a document called the System Requirements Document (SRD). The SRD is a comprehensive document that describes the purpose of the system, its functional requirements, its non-functional requirements and its interfaces.

The SRD is used to guide the system design process. The system designers must ensure that the system meets the needs and requirements of all of the stakeholders.

Reliability is a critical factor in NASA system design because NASA systems are often used in safety-critical applications, where the failure of the system could have catastrophic consequences.

To ensure reliability, NASA system designers use a variety of techniques, including

Redundancy: This involves incorporating multiple copies of critical components into the system. If one component fails, the other components can continue to operate.

Fault tolerance: This involves designing the system so that it can continue to operate even if some of its components fail.

Testing: This involves exhaustively testing the system to identify and fix potential problems.

In some cases, there may be a trade-off between reliability and user experience. For example, a more reliable system may be more complex and difficult to use.

System design is primarily focused on enterprise software companies that develop system-level software, often large-scale distributed systems, with a strong emphasis on reliability because the customers need it. Consumer-oriented companies allocate significant resources to enhance the user experience (UX) in order to cultivate user attraction and engagement. They prioritise creating a delightful UX that entices individuals to actively utilise their platforms. However, these companies may occasionally compromise on reliability to prioritise other aspects of the user experience. This approach will be discussed in the User-Centred Design section.

11.3.3   User-Centred Design

Saffer describes User-Centred Design as one having roots in industrial design and ergonomics, where designers had for the first time focused on a diversity of user types and a variety of user needs.

The term User-Centred Design was coined by Donald Norman and his colleagues from a research laboratory at the University of California San Diego (UCSD) (Chammas et al., 2015). In the mid-80s it became popular after the publication of the book *User-Centered System Design: New Perspectives on Human-Computer Interaction* (Norman and Draper, 1986). Norman built further on the UCD concept in his seminal book *The Psychology of Everyday Things* (Norman, 1988).

In the same time period, in 1985, Gould and Lewis (1985) laid down three principles they believed would lead to a "useful and easy-to-use computer system". They emphasised early focus on users and their tasks, empirical measurement and iterative design.

The newly forming field of User-Centred Design had different methods-, research- and theory-related sources including what was commonly referred to as Usability Engineering and Human-Computer Interaction. These areas of knowledge served as a solid foundation given its focus on the user, the user interface, and, for the first time, the user's goals (Williams, 2009).

Saffer (2009) noticed that the idea to make the life of computer users easier has existed before the appearance of UCD. However, the problem was in the low computing power of computers before the 80s. The increased amount of memory, faster processors and colour displays made the change in possibilities for the new different types of interfaces. The shift from computer system design to design for computer users started and became known as user-centred design.

The first attempts to define design principles for the new user-centred approach were made by Norman in his paper Design Principles for Human-Computer Interfaces (Norman, 1983a).

The usability principles coined by Norman, Gould and Lewis later had been adapted and popularised by Jakob Nielsen as these same basic concepts of heuristics for usability engineering(Abras et al., 2004).

In contrast to other design approaches, the role of a designer in UCD was changed. UCD is the opposite approach to Genius design, the designers are not the users and should not consider their own opinion about the product as the leading opinion.

Designers have to forget about their understanding of the problem and solutions like it is used in Genius design but have to facilitate the achievement of the users' goals. Users are the new stars in the design process. While they have no knowledge of how to design products or interactions, they know quite well what their goals are, what they would like to achieve, and which method or tool would be preferable for them if they had a choice. In such a design approach, the designers consider users as co-creators, and the role of the designers is to listen to users, understand them, convert their needs into design problems, and together with the users, find solutions for these problems together and evaluate them. While the role of the designer is reduced to decision-making, it increases the importance of directing the design process and communicating the project needs with stakeholders. In UCD, end users are considered the main stakeholders, along with businesses, SMEs, and others involved in the process-interested side.

While there are different interpretations of the UCD process, the main workflows correspond in general to the process description of Saffer (2009): Designers conduct thorough upfront research to ascertain the users' goals within the present context. Subsequently, designers initiate the ideation process, involving users to actively contribute to the generation of concepts, a collaborative approach known as participatory design.

According to Williams (2009), the UCD process comprises three phases: design research, design and design evaluation.

The purpose of design research is to understand users and their needs. This involves assessing who the users are, what their needs are, and how they interact with the designed product or service. Users' goals are not easy to define: Saffer compares user goals with Russian dolls, with goals nested inside goals.

The main steps for the design research are:

- *Defining the business stakeholders*: includes identifying who they are, their roles and responsibilities and their needs.
- *Interviewing the business stakeholders*: this should help to understand their needs and requirements in more detail.
- *Conducting background research*: this includes researching the industry, the competition and the target market.
- *Assessment of the work of competitors*: includes understanding their strengths and weaknesses, and how they meet the needs of their customers.
- *Interviewing users*: intended to help to understand the needs and pain points of the people who will be using the product or service.

The typical deliverables for design research are a written report of findings and recommendations, Personas, Process Flow and a Usability Test Plan (Williams, 2009).

Design is the second phase, in user-centred design. This phase represents an iterative process that involves brainstorming, conceptualising and sketching initial drafts of the design. The design phase is based on the findings from the design research and the goal of the design phase is to develop solutions that meet the needs of the users. The design phase should be user-centred, meaning that the team should focus on the needs of the users and ensure that the design meets their needs. For that, the users are actively involved in the co-design process by following the principles of participatory design (Sanders and Rim, 2002).

The specific activities involved in the design phase may vary depending on the project. However, the general principles of brainstorming, conceptualising and sketching will apply in most cases.

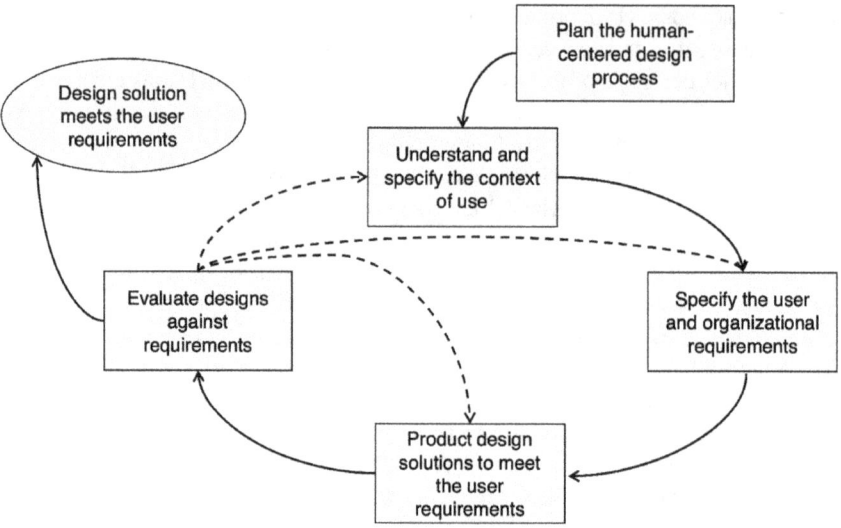

FIGURE 11.4   Human-centred design process according to ISO 9241-210.

The design phase should be iterative, meaning that the team should continuously revisit and refine the design based on feedback from users and design research.

The typical deliverables for the design phase are Sitemaps (for websites), Wireframes, Process Flows (in case, if they are not created during the design research phase), Prototypes and Content Strategy.

The third phase — design evaluation is intended for the evaluation of the design and typically involves testing it for usability.

Different types of usability testing methods can be used in this phase and the two most common deliverables produced during the design evaluation phase of the UCD process are the Usability Test Plan and the Usability Test Report (Williams, 2009).

The technical standards of the User-Centred Design approach are determined by the International Organization of Standardization (Figure 11.4) (Chammas et al., 2015). It includes all mentioned above steps with an additional level of detail.

ISO uses the term "Human-Centred Design" (HCD) instead of "User-Centred Design" by claiming that they are synonymous and by believing that it impacts all the humans involved in the system, not only the end users of the product (Chammas et al., 2015). However, some authors discriminate between the two terms claiming that UCD focuses on individual users, which produces user-friendly designs and outcomes while HCD takes "humans" as its central focus, which lends itself more to "social problem solving" (Digital Adoption Team, 2020).

Another method, very close to UCD, named Goal-Directed Design (GDD) was proposed by Alan Cooper et al. (2014) and developed over a span of years between 1983 and 2014.

The GDD process is a multi-step approach that involves understanding the user's goals, defining the user's tasks, designing the solution and evaluating the solution. Each of these steps is itself a multi-step process, and none of them differs significantly from the multi-step processes and sub-processes used in UCD.

The GDD process and the UCD process are both iterative, meaning that they involve repeated cycles of planning, designing, testing and refining. Both processes also involve a focus on the user, and both use a variety of methods to gather user feedback, such as interviews, surveys and usability testing.

11.3.4 Activity-Centred Design

The next major approach proposed by Saffer is Activity-Centred design (ACD). This approach is based on Activity theory which takes its roots from pedagogical theories coined in the early 20th century by Soviet psychologist Lev Vygotsky and Aleksei N. Leontiev and elaborated by Finnish researcher Yrjö Engeström. Cultural-historical activity theory, as a theoretical framework, provides a lens for comprehending and examining the intricate connection between the human mind (thoughts and emotions) and human activity (actions and behaviours) (Gay and Hembrooke, 2004).

The activity-centred design represents a shift from user-centred design to context-based design, rooted in the principles of Activity theory. The main concepts used in the Activity Theory Model (Engeström, 2001) are a *Subject*, who acts by applying a *Tool* to an *Object* with the goal of achieving a specific *Outcome* (Figure 11.5).

The actions of the subject are applied by using social Rules, can influence or be influenced by communities and typically use Division of labour. ACD appears to bring about a significant shift, in comparison to the User-Centred Design approach. Instead of focusing on the tasks or activities that users need to perform within the application, ACD focuses on identifying the tasks or activities that the application/tool/system should enable (Williams, 2009).

Glanville (2007) stated that design itself is an "activity, leading to an outcome which (in other contexts) is also called design – in this case, used in the form of a noun."

Glanville emphasises a creative side of design expressed in that designers aim to change the world. Creating a new world requires actions. Designers do not just observe the world, they play the role of observers in the world, and hence they are actors. Designers generate knowledge for acting, generate actions, in the world (Glanville, 2007).

The ACD ideas attracted specific interest among education technology scholars who positioned ACD as an approach that emphasises the design of computer-mediated environments to support and structure the interactions and interdependencies of an activity system (Williams, 2009). Eventually, on the basis of ACD, they founded a theoretical framework for computer-supported collaborative learning (CSCL).

Saffer (2010) notices that ACD doesn't focus on the goals and preferences of users, but instead on behaviour surrounding particular tasks. Like the Activity Theory can be applied to different nested environments or systems: micro, meso, exo and macro (Gay and Hembrooke, 2004), the tasks can also have different scales. For example, making a phone call can include several nested tasks like calling a service for assistance, looking up the number on the phone or online, recalling it from memory, and so on. The main difference to the User-Centred Design is focusing not on the user goal

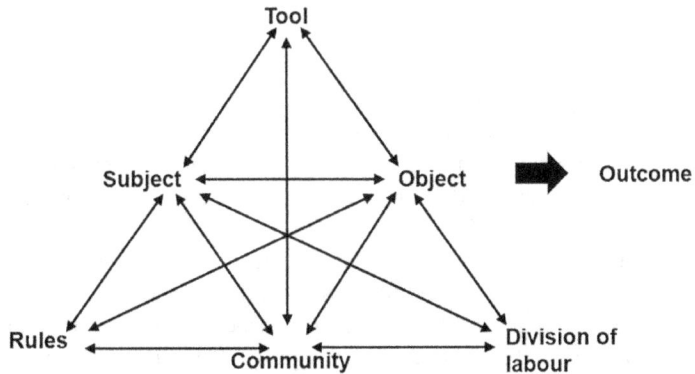

FIGURE 11.5 Engeström's expended activity theory model.

(why the user makes the call and what she wants to achieve by calling) but on the calling activity. The idea of focusing not on the goals but on the activities can be relevant for mass products and services that are hard to adapt to specific groups of users. Examples of such products can be simple as one-screen applications for weather forecasts and complex online systems for personal taxation. Both examples of products are used by wide audiences and designers can shift their focus from the specific user needs to the general behaviour of the product.

Even though the ACD approach has been developing for longer than 20 years, researchers noticed that there are still no definitive guidelines that describe the processes, methods and deliverables that are to be used or produced by the ACD practitioner in business, or the workplace (Williams, 2009).

In some sources, the ACD approach is called Task-Centred Design and is extended to task-oriented design methodologies like Job to be Done (Kalbach, 2020).

11.3.5 CHOICE OF THE DESIGN PARADIGM

Saffer's typology helps to make sense of different approaches to design. By assuming the goals of the system, the ways the system should work and the role of the users, the specific approach or a mix of them can be selected by designers. Still, the choice has to be educated and based on an understanding of a design project's needs from one side and a good understanding of design approaches from another.

Understanding the user's role can be the first step in decision-making. The Genius-based approach may involve users in the phase of testing and deployment. To make it work, a designer is expected to have solid experience and understanding of the design problem. The approach can help when there are not enough resources to implement user research, but the way is risky and fully depends on the level of expertise of the designer.

System design is primarily focused on enterprise software companies that develop system-level software, often large-scale distributed systems, with a strong emphasis on reliability. While the role of the user is not fully excluded, it is not the primary focus. Rather it is addressed as a follow-up after the system design decisions.

In ACD, the activities of the users play the main role and have preferences for the users' goals. While understanding the users is still critically important, the focus is not on the users' needs but rather on the tasks that users have to implement. In ACD the resulting profiles of the users are less specific and more generalised. That allows covering by design the universal tasks for a wider audience.

UCD is primarily concerned with addressing the needs of specific user groups, making it particularly suitable for segmented audiences when developing products intended for use by millions of individuals.

11.4 EMERGING FRONTLINES

Advancements in technology drive research in HCI design. Advances in artificial intelligence fuel interest in computational interaction, which focuses on the design, analysis and evaluation of interactive systems using computational models and methods, combining principles and techniques from HCI and artificial intelligence (AI) to improve the interaction between humans and computers (Chignell, 2023; Oulasvirta et al., 2018).

The goal of computational interaction is to create interactive systems that dynamically respond to users' needs and capabilities and to the characteristics of the context of use (Bi et al. in Oulasvirta et al., 2018, p. 3). It seeks to go beyond traditional approaches that rely on predefined interactions and instead develop systems that can reactively and proactively address preferences, intentions, behaviours and situations. This is achieved through computational models representing and reasoning about various aspects of the interaction process.

Key research areas within computational interaction include statistical modelling (Banovic et al., 2019), optimisation, machine learning (Chung et al., 2020; Kristensson et al., 2019), natural language processing, control theory, signal processing and computer vision. For instance, developing techniques allowing users to actively participate in the training and refinement of machine learning models, making AI systems more transparent and accountable, is a clear frontline addressing pressing concerns with trust and accountability in AI-infused interactive systems.

Other advancements in technology shaping the field of HCI design are Voice User Interfaces (VUIs) (Murad et al., 2022), Augmented Reality (AR) and Virtual Reality (VR) (Kolesnichenko et al., 2019), Tangible User Interfaces (TUI) (Ishii, 2008; Bong et al., 2018), gesture-based Interfaces (Young et al., 2020; Krupka et al., 2017) and multisensory interfaces (Pourjafarian et al., 2019; Wigdor and Morrison, 2010).

VUIs span from voice synthesis and recognition technology to full-fledged conversational agents and require a focus on natural language processing, conversational design and context awareness. While they do offer a natural language interaction, there are still topics on adoption (Cho et al., 2019) and limited use of these systems (Ammari et al., 2019) Current challenges in VUI design are designing VUI-specific design guidelines, a design education not centred on GUI and better prototyping tools (Murad et al., 2022).

AR and VR technologies offer immersive and interactive experiences. Research and design challenges in these cases relate to spatial interactions and understanding user perception in virtual environments.

TUIs bridge the gap between the physical and digital worlds and can be particularly useful in educational, gaming, or collaborative settings. They enable users to directly manipulate physical objects to control and interact with digital information or virtual environments. Tangible User Interfaces use various sensing and tracking technologies to capture user interactions with physical objects, including RFIDs (Radio Frequency Identification), computer vision, touch sensors, accelerometers, or pressure sensors. Ensuring accurate and efficient object recognition and tracking is crucial for seamless interaction and can be challenging.

Gesture-based interfaces leverage depth-sensing cameras or motion sensors to interpret and respond to body movements, allegedly enabling more intuitive and natural interactions. These natural interactions are, however, not deprived of challenges. Open issues include ambiguity in user input, discoverability and learnability, lack of feedback and affordances, environmental factors, fatigue and physical strain, accessibility and inclusivity, not to mention context sensitivity, privacy and security.

Finally, multisensory interfaces involve the integration of multiple sensory modalities to enhance the user experience and interaction with digital systems. Spanning senses such as vision, hearing, touch and sometimes even smell or taste, multisensory interfaces aim to create more immersive, engaging and inclusive interaction experiences. When considering multisensory interfaces, designers must carefully consider which modalities to employ and how to combine them to effectively convey information and create meaningful interactions without overwhelming the user. To do so, designers need to understand the principles of cross-modal perception and consider factors such as the timing, synchronisation and congruence of sensory stimuli to ensure that the integrated experience is coherent and natural. This implies carefully managing the cognitive load and attention demands placed on the user to prevent information overload and ensure that users can effectively process and respond to the stimuli. Multisensory interfaces further have the potential to improve accessibility and inclusivity for users with sensory impairments. Exciting times for HCI design.

11.5 CLOSING REMARKS

While HCI design has made significant progress, researchers and practitioners still have a broad landscape to explore and pursue in addition to the highlighted emerging frontlines. Knowledge gaps in HCI design relate not only to the emerging technologies scoped in the previous section but also to less technology-driven issues such as user experience in long-term use; inclusion and accessibility; ethical and social implications; cross-cultural and global perspectives; designing for specific user groups; designing for complex work environments; and the long-term impact of technology use.

The rapid development of technologies presents new challenges and opportunities for HCI because it is necessary to understand how to design effective and usable interactions with these technologies and assess the potential ethical and societal implications they bring. In this case, the knowledge gap lies in exploring and establishing design principles, interaction techniques and evaluation methods specific to these technologies, in addition to refining the enabling technologies given their predominantly immature state.

However, HCI design challenges go way beyond those driven by technological advances. Although HCI research has made progress in understanding the needs and preferences of specific user groups, such as older adults, children, individuals with disabilities, or novice users, there are still knowledge gaps in designing interfaces that cater to these specific user groups' unique characteristics, abilities and challenges. More research is needed to develop tailored design guidelines and methods for these populations.

Further, despite the progress in designing for accessibility and inclusivity, there are still knowledge gaps in effectively addressing the needs of individuals with diverse abilities, cultural backgrounds and socio-economic contexts. HCI research and practice need to explore more inclusive design approaches, technologies and evaluation methods that can accommodate a broader range of users.

And most HCI studies focus on individual interactions or simple tasks, so research still needs to address the design challenges of complex work environments, such as healthcare settings, emergency response, or collaborative work. Understanding the specific needs and requirements of these environments and designing interfaces that support complex workflows and collaboration is an ongoing area of research.

Furthermore, much of the HCI research has been conducted in Western contexts, and there is a need to broaden the understanding of user experiences and design principles in diverse cultural, social and global contexts. HCI research should strive to incorporate cross-cultural perspectives, considering cultural norms, values and practices to develop more culturally sensitive and globally applicable design guidelines.

Correspondingly, understanding HCI design's ethical and social implications becomes increasingly critical as technology integrates into our lives. There is a need to explore the impact of HCI on privacy, security, trust, algorithmic bias, social interactions and the broader societal implications of emerging technologies. HCI research must address these complex issues and develop responsible design and deployment frameworks.

Finally, as technology increasingly permeates various aspects of our lives, we must explore the long-term impact of technology use on individuals, communities and society. Unfortunately, many HCI studies focus on short-term interactions and initial user experience.

Together with investigating the effects of technology on cognitive abilities, attention spans, social relationships, mental health and well-being, there is a need to understand how user experiences evolve and change over long-term use.

Longitudinal studies and interdisciplinary research are necessary to shed light on the long-term effects of technology use, allowing us to gain insights into users' experiences, satisfaction and adaptation to technology over extended periods can help improve the design and support of interactive systems.

Addressing these knowledge gaps requires further interdisciplinary collaboration and increased contributions to HCI's body of knowledge, as only this way the HCI community can further advance our understanding of human-computer interaction and ensure the design and development of more usable, inclusive and ethically responsible interactive systems.

REFERENCES

Abras, Chadia, Diane Maloney-Krichmar, and Jenny Preece. 2004. "User-Centered Design." In *Berkshire Encyclopedia of Human-Computer Interaction*, edited by William Sims Bainbridge, pp. 445–456. Great Barrington, MA: Berkshire Publishing Group.

Adams, Anne, Peter Lunt, and Paul Cairns. 2008. "A Qualititative Approach to HCI Research." In *Research Methods for Human-Computer Interaction*, edited by Paul Cairns, and Anna Cox, pp. 138–57. Cambridge, UK: Cambridge University Press.

Adomavicius, Gediminas, and Mochen Yang. 2022. "Integrating Behavioral, Economic, and Technical Insights to Understand and Address Algorithmic Bias: A Human-Centric Perspective." *ACM Transactions on Management Information Systems*, 34, 13(3): 1–27.

Ahmed, A., & Irani, L. 2020. "Feminism as a Design Methodology." *Interactions*, 27(6): 42–45.

Ailisto, H., Pohjanheimo, L., Välkkynen, P., Strömmer, E., Tuomisto, T., & Korhonen, I. 2006. "Bridging the Physical and Virtual Worlds by Local Connectivity-based Physical Selection." *Personal and Ubiquitous Computing*, 10: 333–344.

Albert, William, & Tullis, Thomas. 2010. "Chapter 8: Measuring Emotion." In *Measuring the User Experience*, 3rd ed., edited by William Albert and Thomas S. Tullis, pp. 195–216. Burlington, MA: Morgan Kaufmann.

Alexander, Christopher. 1977. *A Pattern Language: Towns, Buildings, Construction.* Oxford: Oxford University Press.

Al-Shamayleh, Ahmad Sami, Rodina Ahmad, Mohammad A. M. Abushariah, Khubaib Amjad Alam, and Nazean Jomhari. 2018. "A Systematic Literature Review on Vision Based Gesture Recognition Techniques." *Multimedia Tools and Applications* 77(21): 28121–84.

Alvarado-Garcia, Adriana, Juan F. Maestre, Manuhuia Barcham, Marilyn Iriarte, Marisol Wong-Villacres, Oscar A. Lemus, Palak Dudani, Pedro Reynolds-Cuéllar, Ruotong Wang, and Teresa Cerratto Pargman. 2021. "Decolonial Pathways: Our Manifesto for a Decolonizing Agenda in HCI Research and Design." In *Extended Abstracts of the 2021 CHI Conference on Human Factors in Computing Systems (CHI EA'21),* pp. 1–9. New York: Association for Computing Machinery.

Amershi, Saleema, Dan Weld, Mihaela Vorvoreanu, Adam Fourney, Besmira Nushi, Penny Collisson, Jina Suh, et al. 2019. "Guidelines for Human-AI Interaction." In *Proceedings of the 2019 CHI Conference on Human Factors in Computing Systems (CHI'19),* pp. 1–13. New York: Association for Computing Machinery.

Ammari, Tawfiq, Jofish Kaye, Janice Y. Tsai, and Frank Bentley. 2019. "Music, Search, and IoT." *ACM Transactions on Computer-Human Interaction: A Publication of the Association for Computing Machinery* 26(3): 1–28.

Andersen, Peter Bøgh. 2001. "What Semiotics Can and Cannot Do for HCI." *Knowledge-Based Systems* 14(8): 419–24.

Nancy S. Anderson and Judith Reitman Olson. 1985. *Methods for Designing Software to Fit Human Needs and Capabilities: Proceedings of the Workshop on Software Human Factors.* Washington, DC: National Academy Press.

Anokwa, Yaw, Thomas N. Smyth, Divya Ramachandran, Jahanzeb Sherwani, Yael Schwartzman, Rowena Luk, Melissa Ho, Neema Moraveji, and Brian DeRenzi. 2009. "Stories from the Field: Reflections on HCI4D Experiences." *Information Technologies & International Development* 5(4): 101.

Appriou, Aurélien, Andrzej Cichocki, and Fabien Lotte. 2018. "Towards Robust Neuroadaptive HCI: Exploring Modern Machine Learning Methods to Estimate Mental Workload From EEG Signals." In *Extended Abstracts of the 2018 CHI Conference on Human Factors in Computing Systems (CHI EA'18),* pp. 1–6. New York: Association for Computing Machinery.

Aryal, Ashrant, and Burcin Becerik-Gerber. 2019. "Skin Temperature Extraction Using Facial Landmark Detection and Thermal Imaging for Comfort Assessment." In *Proceedings of the 6th ACM International Conference on Systems for Energy-Efficient Buildings, Cities, and Transportation (BuildSys'19),* pp. 71–80. New York: Association for Computing Machinery.

Ashraf, Mohsena, Genevieve Patterson, Zachary Kilhoffer, Xu Han, Nolan Brady, Anna Rahn, Nikhitha Atluri, et al. 2023. "Using fNIRS To Understand Adults' Empathy for Children in AI and Cybersecurity Scenarios." In *Proceedings of the 22nd Annual ACM Interaction Design and Children Conference (IDC'23),* pp. 630–34. New York: Association for Computing Machinery.

Badillo-Urquiola, Karla, Zachary Shea, Zainab Agha, Irina Lediaeva, and Pamela Wisniewski. 2021. "Conducting Risky Research with Teens: Co-Designing for the Ethical Treatment and Protection of Adolescents." *Proceedings of the ACM on Human-Computer Interaction*, 231, 4(CSCW3): 1–46.

Bafna, Tanya, John Paulin Paulin Hansen, and Per Baekgaard. 2020. "Cognitive Load during Eye-Typing." In *ACM Symposium on Eye Tracking Research and Applications (ETRA'20 Full Papers 23),* pp. 1–8. New York: Association for Computing Machinery.

Bannon, L. J. 1986. "Computer-Mediated Communication." *User Centered System Design: New Perspectives on Human-Computer Interaction*, pp. 433–456. Hillsdale, NJ: Lawrence Erlbaum.

Bannon, Liam. 1992. "From Human Factors to Human Actors: The Role of Psychology and Human-Computer Interaction Studies in System Design." In *Design at Work: Cooperative Design of Computer Systems*, edited by Joan Greenbaum and Morten Kyng, pp. 25–44. Mahwah, NJ: Lawrence Erlbaum Associates Inc.

Banovic, Nikola, Antti Oulasvirta, and Per Ola Kristensson. 2019. "Computational Modeling in Human-Computer Interaction." In *Extended Abstracts of the 2019 CHI Conference on Human Factors in Computing Systems (CHI EA'19)*, pp. 1–7. New York: Association for Computing Machinery.

Bardzell, Jeffrey, and Shaowen Bardzell. 2013. "What Is 'Critical' about Critical Design?" In *Proceedings of the SIGCHI Conference on Human Factors in Computing Systems*, pp. 3297–3306. New York: ACM. https://doi.org/10.1145/2470654.2466451.

Bardzell, Jeffrey, Shaowen Bardzell, and Mark Blythe. 2018. *Critical Theory and Interaction Design*. Cambridge, MA: MIT Press.

Bardzell, Shaowen, and Jeffrey Bardzell. 2011. "Towards a Feminist HCI Methodology: Social Science, Feminism, and HCI." In *Proceedings of the SIGCHI Conference on Human Factors in Computing Systems (CHI'11)*, pp. 675–84. New York: Association for Computing Machinery.

Bardzell, Shaowen, Daniela K. Rosner, and Jeffrey Bardzell. 2012a. "Crafting Quality in Design: Integrity, Creativity, and Public Sensibility." In *Proceedings of the Designing Interactive Systems Conference (DIS'12)*, pp. 11–20. New York: Association for Computing Machinery.

Bardzell, Shaowen, Jeffrey Bardzell, Jodi Forlizzi, John Zimmerman, and John Antanitis. 2012b. "Critical Design and Critical Theory: The Challenge of Designing for Provocation." In *Proceedings of the Designing Interactive Systems Conference (DIS'12)*, pp. 288–97. New York: Association for Computing Machinery.

Bellini, Rosanna, Debora de Castro Leal, Hazel Anneke Dixon, Sarah E. Fox, and Angelika Strohmayer. 2022. "'There Is No Justice, Just Us': Making Mosaics of Justice in Social Justice Human-Computer Interaction." In *Extended Abstracts of the 2022 CHI Conference on Human Factors in Computing Systems (CHI EA'22)*, pp. 1–6. New York: Association for Computing Machinery.

Benabdallah, Gabrielle, Ashten Alexander, Sourojit Ghosh, Chariell Glogovac-Smith, Lacey Jacoby, Caitlin Lustig, Anh Nguyen, et al. 2022. "Slanted Speculations: Material Encounters with Algorithmic Bias." In *Designing Interactive Systems Conference (DIS'22)*, pp. 85–99. New York: Association for Computing Machinery.

Benbasat, Izak. 2010. "HCI Research: Future Challenges and Directions." *AIS Transactions on Human-Computer Interaction* 2(2): 16–21.

Benford, Steve, Richard Ramchurn, Joe Marshall, Max L. Wilson, Matthew Pike, Sarah Martindale, Adrian Hazzard, et al. 2021. "Contesting Control: Journeys through Surrender, Self-Awareness and Looseness of Control in Embodied Interaction." *Human-Computer Interaction* 36(5–6): 361–89.

Beyer, H., & Holtzblatt, K. 1999. "Contextual design." *Interactions*, 6(1): 32–42.

Birhane, Abeba, Pratyusha Kalluri, Dallas Card, William Agnew, Ravit Dotan, and Michelle Bao. 2022. "The Values Encoded in Machine Learning Research." In *Proceedings of the 2022 ACM Conference on Fairness, Accountability, and Transparency (FAccT'22)*, pp. 173–84. New York: Association for Computing Machinery.

Birhane, Abeba. 2020. "Algorithmic Colonization of Africa." *SCRIPT-Ed* 17(2): 389–409.

Birhane, A. 2021. "The Impossibility of Automating Ambiguity." *Artificial Life*, 27(1): 44–61.

Bobbe, Tina, Jens Krzywinski, Cheryl Woelfel, and Others. 2016. "A Comparison of Design Process Models from Academic Theory and Professional Practice." In *DS 84: Proceedings of the DESIGN 2016 14th International Design Conference*, pp. 1205–14.

Bødker, Susanne. 2006. "When Second Wave HCI Meets Third Wave Challenges." In *Proceedings of the 4th Nordic Conference on Human-Computer Interaction: Changing Roles (NordiCHI'06)*, pp. 1–8. New York: Association for Computing Machinery.

Bødker, Susanne, and Morten Kyng. 2018. "Participatory Design That Matters-Facing the Big Issues." *ACM Transactions on Computer-Human Interaction: A Publication of the Association for Computing Machinery* 25(1): 1–31.

Boland, Richard J. 1978. "The Process and Product of System Design." *Management Science* 24(9): 887–98.

Bolchini, Davide, Rupa Chatterji, and Marco Speroni. 2009. "Developing Heuristics for the Semiotics Inspection of Websites." In *Proceedings of the 27th ACM International Conference on Design of Communication (SIGDOC'09)*, pp. 67–72. New York: Association for Computing Machinery.

Bong, Way Kiat, Weiqin Chen, and Astrid Bergland. 2018. "Tangible User Interface for Social Interactions for the Elderly: A Review of Literature." *Advances in Human-Computer Interaction* 2018(May). https://doi.org/10.1155/2018/7249378.

Bongers, Bert, and Gerrit van der Veer. 2009. "HCI and Design Research Education." In *Creativity and HCI: From Experience to Design in Education*, edited by Paula Kotzé, William Wong, Joaquim Jorge, Alan Dix, and Paula Alexandra Silva, pp. 90–105. Cham: Springer US.

Borgman, Christine L. 1985. "The User's Mental Model of an Information Retrieval System." In *Proceedings of the 8th Annual International ACM SIGIR Conference on Research and Development in Information Retrieval (SIGIR'85)*, pp. 268–73. New York: Association for Computing Machinery.

Bourne, Charles P., and Trudi Bellardo Hahn. 2003. *A History of Online Information Services, 1963-1976*. Cambridge, MA: The MIT Press.

Bowman, Robert, Camille Nadal, Kellie Morrissey, Anja Thieme, and Gavin Doherty. 2023. "Using Thematic Analysis in Healthcare HCI at CHI: A Scoping Review." In *Proceedings of the 2023 CHI Conference on Human Factors in Computing Systems (CHI'23 491)*, pp. 1–18. New York: Association for Computing Machinery.

Braun, Virginia, and Victoria Clarke. 2006. "Using Thematic Analysis in Psychology." *Qualitative Research in Psychology* 3(2): 77–101.

Brishtel, Iuliia, Shoya Ishimaru, Olivier Augereau, Koichi Kise, and Andreas Dengel. 2018. "Assessing Cognitive Workload on Printed and Electronic Media Using Eye-Tracker and EDA Wristband." In *Proceedings of the 23rd International Conference on Intelligent User Interfaces Companion (IUI'18 Companion 45)*, pp. 1–2. New York: Association for Computing Machinery.

Brom, Cyril, Tereza Stárková, Jiří Lukavský, Ondřej Javora, and Edita Bromová. 2016. "Eye Tracking in Emotional Design Research: What Are Its Limitations?" In *Proceedings of the 9th Nordic Conference on Human-Computer Interaction (NordiCHI'16)*, pp. 1–6. New York: Association for Computing Machinery.

Brooke, John. 1996. "SUS: A 'quick and Dirty' Usability Scale. Usability Evaluation in Industry." In *Book Usability Evaluation in Industry*, edited by Patrick W. Jordan, B. Thomas, Ian Lyall McClelland, and Bernard Weerdmeester, pp. 189–94. London: Taylor & Francis.

Brooke, John. 2013. "SUS: A Retrospective." *Journal of Usability Studies* 8(2): 29–40.

Brooks, Frederick P., Jr. 1974. *The Mythical Man Month and Other Essays on Software Engineering*. New York: Addison Wesley Longman Publishing.

Brown, Barry. 2001. "Studying the Internet Experience." HP Laboratories Technical Report HPL 49. https://citeseerx.ist.psu.edu/document?repid=rep1&type=pdf&doi=563a300a287ff45eb897d100f26d59d4d87c62c2.

Brown, Barry A. T., Abigail J. Sellen, and Kenton P. O'hara. 2000. "A Diary Study of Information Capture in Working Life." *CHI'00: Proceedings of the SIGCHI conference on Human Factors in Computing Systems*. https://dl.acm.org/doi/abs/10.1145/332040.332472.

Brown, E., & Cairns, P. 2004, April. A Grounded Investigation of Game Immersion. In *CHI'04 Extended Abstracts on Human Factors in Computing Systems*, pp. 1297–1300.

Bødker, Susanne, Kaj Grønbæk, and Morten Kyng. 1995. "Cooperative Design: Techniques and Experiences from the Scandinavian Scene." In *Readings in Human-Computer Interaction*, edited by Ronald M. Baecker, Jonathan Grudin, William A. S. Buxton, and Saul Greenberg, pp. 215–24. Burlington, MA: Morgan Kaufmann.

Bush, V. 1945. "As We May Think." *The Atlantic Monthly*, 176(1): 101–108.

Camilleri, Mark Anthony, and Adriana Caterina Camilleri. 2022. "Utilitarian and Intrinsic Motivations to Use Mobile Learning Technologies: An Extended Technology Acceptance Model." In *Proceedings of the 8th International Conference on E-Society, E-Learning and E-Technologies (ICSLT'22)*, pp. 76–81. New York: Association for Computing Machinery.

Card, Stuart K., Thomas P. Moran, and Allen Newell. 1983. *The Psychology of Human-Computer Interaction*. London: Taylor & Francis.

Card, Stuart K., William K. English, and Betty J. Burr. 1978. "Evaluation of Mouse, Rate-Controlled Isometric Joystick, Step Keys, and Text Keys for Text Selection on a CRT." *Ergonomics* 21(8): 601–13.

Carroll, John M. 1993. "Creating a Design Science of Human-Computer Interaction." *Interacting with Computers* 5(1): 3–12.

Carroll, John M. 1997. "Human-computer Interaction: Psychology as a Science of Design." *International Journal of Human-Computer Studies* 46(4): 501–22.

Carroll, John M. 2016. "Human-Computer Interaction (brief Intro)." In *The Encyclopedia of Human-Computer Interaction*, edited by Soegaard Mads and Dam Friis, pp. 21–62. Interaction Design Foundation.

Carroll, John M., and John C. Thomas. 1982. "Metaphor and the Cognitive Representation of Computing Systems." *IEEE Transactions on Systems, Man, and Cybernetics* 12(2): 107–16.

Cavazza, Marc, Fred Charles, Gabor Aranyi, Julie Porteous, Stephen W. Gilroy, Gal Raz, Nimrod Jakob Keynan, et al. 2014. "Towards Emotional Regulation through Neurofeedback." In *Proceedings of the 5th Augmented Human International Conference (AH'14)*, pp. 1–8. New York: Association for Computing Machinery.

Chammas, Adriana, Manuela Quaresma, and Cláudia Mont'Alvão. 2015. "A Closer Look on the User Centred Design." *Procedia Manufacturing* 3(January): 5397–5404.

Chanel, G., Rebetez, C., Bétrancourt, M., & Pun, T. 2011. "Emotion Assessment from Physiological Signals for Adaptation of Game Difficulty." *IEEE Transactions on Systems, Man, and Cybernetics-Part A: Systems and Humans*, 41(6): 1052–1063.

Chang, Keng-Hao, Drew Fisher, John Canny, and Björn Hartmann. 2011. "How's My Mood and Stress? An Efficient Speech Analysis Library for Unobtrusive Monitoring on Mobile Phones." In *Proceedings of the 6th International Conference on Body Area Networks (BodyNets'11)*, pp. 71–77. Brussels, BEL: ICST (Institute for Computer Sciences, Social-Informatics and Telecommunications Engineering).

Chignell, Mark, Lu Wang, Atefeh Zare, and Jamy Li. 2023. "The Evolution of HCI and Human Factors: Integrating Human and Artificial Intelligence." *ACM Transactions on Computer-Human Interaction*, 17, 30(2): 1–30.

Chiossi, Francesco, Changkun Ou, and Sven Mayer. 2023. "Exploring Physiological Correlates of Visual Complexity Adaptation: Insights from EDA, ECG, and EEG Data for Adaptation Evaluation in VR Adaptive Systems." In *Extended Abstracts of the 2023 CHI Conference on Human Factors in Computing Systems (CHI EA'23)*, pp. 1–7. New York: Association for Computing Machinery.

Cho, Minji, Sang-Su Lee, and Kun-Pyo Lee. 2019. "Once a Kind Friend Is Now a Thing: Understanding How Conversational Agents at Home Are Forgotten." In *Proceedings of the 2019 on Designing Interactive Systems Conference (DIS'19)*, pp. 1557–69. New York: Association for Computing Machinery.

Choi, Inchul, Ilsun Rhiu, Yushin Lee, Myung Hwan Yun, and Chang S. Nam. 2017. "A Systematic Review of Hybrid Brain-Computer Interfaces: Taxonomy and Usability Perspectives." *PloS One* 12(4): e0176674.

Chuang, Lewis L., Christiane Glatz, and Stas Krupenia. 2017. "Using EEG to Understand Why Behavior to Auditory In-Vehicle Notifications Differs Across Test Environments." In *Proceedings of the 9th International Conference on Automotive User Interfaces and Interactive Vehicular Applications (AutomotiveUI'17)*, pp. 123–33. New York: Association for Computing Machinery.

Chung, Mu-Huan, Mark Chignell, Lu Wang, Alexandra Jovicic, and Abhay Raman. 2020. "Interactive Machine Learning for Data Exfiltration Detection: Active Learning with Human Expertise." In *2020 IEEE International Conference on Systems, Man, and Cybernetics (SMC)*, pp. 280–87.

Clarke, Martina A., Ryan M. Schuetzler, John R. Windle, Emily Pachunka, and Ann Fruhling. 2020. "Usability and Cognitive Load in the Design of a Personal Health Record." *Health Policy and Technology* 9(2): 218–24.

Clemmensen, Torkil. 2021. "Socio-Technical HCI Design in a Wider Context." In *Human Work Interaction Design: A Platform for Theory and Action*, edited by Torkil Clemmensen, pp. 267–80. Cham: Springer International Publishing.

Cockton, Gilbert. 2004. "Value-Centred HCI." In *Proceedings of the Third Nordic Conference on Human-Computer Interaction (NordiCHI'04)*, pp. 149–60. New York: Association for Computing Machinery.

Cockton, Gilbert. 2005. "A Development Framework for Value-Centred Design." In *Extended Abstracts on Human Factors in Computing Systems (CHI EA'05)*, pp. 1292–95. New York: Association for Computing Machinery.

Cockton, Gilbert. 2008. "Putting Value into E-Valu-Ation." In *Maturing Usability: Quality in Software, Interaction and Value*, edited by Effie Lai-Chong Law, Ebba Thora Hvannberg, and Gilbert Cockton, pp. 287–317. London: Springer London.

Cockton, Gilbert. 2013. "Design Isn't a Shape and It Hasn't Got a Centre: Thinking BIG about Post-Centric Interaction Design." In *Proceedings of the International Conference on Multimedia, Interaction, Design and Innovation (MIDI'13)*, pp. 1–16. New York: Association for Computing Machinery.

Cole, Tom, and Marco Gillies. 2022. "More than a Bit of Coding: (un-)Grounded (non-)Theory in HCI." In *Extended Abstracts of the 2022 CHI Conference on Human Factors in Computing Systems (CHI EA'22)*, pp. 1–11. New York: Association for Computing Machinery.

Colin Gibson, Ryan Mark D. Dunlop, and Matt-Mouley Bouamrane. 2020. "Lessons from Expert Focus Groups on How to Better Support Adults with Mild Intellectual Disabilities to Engage in Co-Design." In *Proceedings of the 22nd International ACM SIGACCESS Conference on Computers and Accessibility (ASSETS'20 48)*, pp. 1–12. New York: Association for Computing Machinery.

Colley, Ashley, Mari Suoheimo, and Jonna Häkkilä. 2020. "Exploring VR and AR Tools for Service Design." In *Proceedings of the 19th International Conference on Mobile and Ubiquitous Multimedia (MUM'20)*, pp. 309–11. New York: Association for Computing Machinery.

Constanza-Chock, Sasha. 2020. Design Justice. In *Community-Led Practices to Build the Worlds We Need*, edited by Sandra Braman. Cambridge, MA: MIT Press.

Cooper, H. M. 1988. "Organizing Knowledge Syntheses: A Taxonomy of Literature Reviews." *Knowledge in Society*, 1(1): 104.

Cooper, Alan, Robert Reimann, David Cronin, and Christopher Noessel. 2014. *About Face: The Essentials of Interaction Design*, 4th ed. Nashville, TN: John Wiley & Sons.

Cowley, Benjamin, Marco Filetti, Kristian Lukander, Jari Torniainen, Andreas Henelius, Lauri Ahonen, Oswald Barral, et al. 2016. "The Psychophysiology Primer: A Guide to Methods and a Broad Review with a Focus on Human-Computer Interaction." *Foundations and Trends(r) in Human-Computer Interaction* 9(3–4): 151–308.

Cross, Nigel. 1982. "Designerly Ways of Knowing." *Design Studies* 3(4): 221–27.

Das, Dipto, A. K. M. Najmul Islam, S. M. Taiabul Haque, Jukka Vuorinen, and Syed Ishtiaque Ahmed. 2023. "Understanding the Strategies and Practices of Facebook Microcelebrities for Engaging in Sociopolitical Discourses." In *Proceedings of the 2022 International Conference on Information and Communication Technologies and Development (ICTD'22)*, pp. 1–19. New York: Association for Computing Machinery.

Das, Dipto, and Bryan Semaan. 2022. "Decolonial and Postcolonial Computing Research: A Scientometric Exploration." In *Companion Publication of the 2022 Conference on Computer Supported Cooperative Work and Social Computing (CSCW'22 Companion)*, pp. 168–74. New York: Association for Computing Machinery.

De Queiroz Pierre, Raisa da Silva 2015. Heuristics in Design: A Literature Review. *Procedia Manufacturing*, 3, 6571–6578.

de Souza, Clarisse Sieckenius, Simone Diniz Junqueira Barbosa, and Raquel Oliveira Prates. 2001. "A Semiotic Engineering Approach to HCI." In *Extended Abstracts on Human Factors in Computing Systems (CHI EA'01)*, pp. 55–56. New York: Association for Computing Machinery.

Dearden, Andy, and Janet Finlay. 2006. "Pattern Languages in HCI: A Critical Review." *Human-Computer Interaction* 21(1): 49–102.

Dell, Nicola, and Neha Kumar. 2016. "The Ins and Outs of HCI for Development." In *Proceedings of the 2016 CHI Conference on Human Factors in Computing Systems (CHI'16), pp.* 2220–32. New York: Association for Computing Machinery.

Dempsey, Patrick G., Michael S. Wogalter, and Peter A. Hancock. 2000. "What's in a Name? Using Terms from Definitions to Examine the Fundamental Foundation of Human Factors and Ergonomics Science." *Theoretical Issues in Ergonomics Science* 1(1): 3–10.

Denning, Peter J. 2000. "Computer Science: The Discipline." *Encyclopedia of Computer Science* 32(1): 9–23.

Diaper, Dan, and Neville Stanton. 2003. *The Handbook of Task Analysis for Human-Computer Interaction*. Boca Raton, FL: CRC Press.

Digital Adoption Team. 2020. "User-Centered Design vs. Human-Centered Design: What's the Difference?" *Digital Adoption*. January 26, 2020. https://www.digital-adoption.com/user-centered-design-vs-human-centered-design/.

DiSalvo, Carl, Phoebe Sengers, and Hrönn Brynjarsdóttir. 2010. "Mapping the Landscape of Sustainable HCI." In *Proceedings of the SIGCHI Conference on Human Factors in Computing Systems (CHI'10)*, pp. 1975–84. New York: Association for Computing Machinery.

Dix, Alan. 2010. "Human-Computer Interaction: A Stable Discipline, a Nascent Science, and the Growth of the Long Tail." *Interacting with Computers* 22(1): 13–27.

Dourish, Paul, and Graham Button. 1998. "On 'Technomethodology': Foundational Relationships between Ethnomethodology and System Design." *Human-Computer Interaction* 13(4): 395–432.

Dourish, Paul, and Scott D. Mainwaring. 2012. "Ubicomp's Colonial Impulse." In *Proceedings of the 2012 ACM Conference on Ubiquitous Computing (UbiComp'12), pp.* 133–42. New York: Association for Computing Machinery.

Dourish, Paul. 2001. *Where the Action Is: The Foundations of Embodied Interaction*. Cambridge, MA: MIT Press.

Dourish, Paul. 2010. "HCI and Environmental Sustainability: The Politics of Design and the Design of Politics." In *Proceedings of the 8th ACM Conference on Designing Interactive Systems (DIS'10)*, pp. 1–10. New York: Association for Computing Machinery.

Draper, Stephen W. 1993. "The Notion of Task in HCI." In *INTERACT'93 and CHI'93 Conference Companion on Human Factors in Computing Systems*, pp. 207–8. New York: Association for Computing Machinery.

Drewes, Heiko. 2023. "The Fitts' Law Filter Bubble." In *Extended Abstracts of the 2023 CHI Conference on Human Factors in Computing Systems (CHI EA'23)*, pp. 1–5. New York: Association for Computing Machinery.

Duran, Rodrigo, Albina Zavgorodniaia, and Juha Sorva. 2022. "Cognitive Load Theory in Computing Education Research: A Review." *ACM Transactions on Computing Education*, 40, 22(4): 1–27.

Ebling, Maria R., and Bonnie E. John. 2000. "On the Contributions of Different Empirical Data in Usability Testing." In *Proceedings of the 3rd Conference on Designing Interactive Systems: Processes, Practices, Methods, and Techniques (DIS'00)*, pp. 289–96. New York: Association for Computing Machinery.

Eddy, Ethan, Erik J. Scheme, and Scott Bateman. 2023. "A Framework and Call to Action for the Future Development of EMG-Based Input in HCI." In *Proceedings of the 2023 CHI Conference on Human Factors in Computing Systems (CHI'23)*, pp. 1–23. New York: Association for Computing Machinery.

Elahi, Mehdi, Himan Abdollahpouri, Masoud Mansoury, and Helma Torkamaan. 2021. "Beyond Algorithmic Fairness in Recommender Systems." In *Adjunct Proceedings of the 29th ACM Conference on User Modeling, Adaptation and Personalization (UMAP'21)*, pp. 41–46. New York: Association for Computing Machinery.

Engelbart, Douglas C. 1962. "Augmenting Human Intellect: A Conceptual Framework Available." *SRI AFOSR-3223*. Air Force, Office of Scientific Research.

Engeström, Yrjö. 2001. "Expansive Learning at Work: Toward an Activity Theoretical Reconceptualization." *Journal of Education and Work* 14(1): 133–56.

Englebart, Douglas C. 1968. "The Mother of All Demos." *Presented at the Fall Joint Computer Conference*, San Francisco, CA, December 9. https://www.youtube.com/watch?v=yJDv-zdhzMY.

Erete, Sheena, Yolanda Rankin, and Jakita Thomas. 2023. "A Method to the Madness: Applying an Intersectional Analysis of Structural Oppression and Power in HCI and Design." *ACM Transactions on Computer-Human Interaction*, 24, 30(2): 1–45.

Eriksson, Eva, Gökçe Elif Baykal, and Olof Torgersson. 2022. "The Role of Learning Theory in Child-Computer Interaction - A Semi-Systematic Literature Review." In *Proceedings of the 21st Annual ACM Interaction Design and Children Conference (IDC'22)*, pp. 50–68. New York: Association for Computing Machinery.

Fairclough, Stephen H. 2008. "Fundamentals of Physiological Computing." *Interacting with Computers* 21(1–2): 133–45.

Fairclough, Stephen H. 2022. "Chapter 1: Designing Human-Computer Interaction with Neuroadaptive Technology." In *Current Research in Neuroadaptive Technology*, edited by Stephen H. Fairclough and Thorsten O. Zander, pp. 1–15. Cambridge, MA: Academic Press.

Fairclough, Stephen H., and Louise Venables. 2004. "Effects of Task Demand and Time-on-Task on Psychophysiological Candidates for Biocybernetic Control." *Proceedings of the Human Factors and Ergonomics Society... Annual Meeting Human Factors and Ergonomics Society* 48(1): 85–89.

Fairclough, Stephen, and Kiel Gilleade. 2012. "Construction of the Biocybernetic Loop: A Case Study." In *Proceedings of the 14th ACM International Conference on Multimodal Interaction (ICMI'12)*, pp. 571–78. New York: Association for Computing Machinery.

Fan, Mingming, Xianyou Yang, Tsztung Yu, Qingzi Vera Liao, and Jian Zhao. 2022. "Human-AI Collaboration for UX Evaluation: Effects of Explanation and Synchronization." *Proceedings of the ACM on Human-Computer Interaction*, 96, 6(CSCW1): 1–32.

Fiesler, Casey, Christopher Frauenberger, Michael Muller, Jessica Vitak, and Michael Zimmer. 2022. "Research Ethics in HCI: A SIGCHI Community Discussion." In *Extended Abstracts of the 2022 CHI Conference on Human Factors in Computing Systems (CHI EA'22)*, pp. 1–3. New York: Association for Computing Machinery.

Fitts, P. M. 1954. "The Information Capacity of the Human Motor System in Controlling the Amplitude of Movement." *Journal of Experimental Psychology*, 47(6): 381.

Forsythe, Diana E. 1999. "'it's Just a Matter of Common Sense': Ethnography as Invisible Work." *Computer Supported Cooperative Work: CSCW: An International Journal* 8(1–2): 127–45.

Fox, Sarah, Mariam Asad, Katherine Lo, Jill P. Dimond, Lynn S. Dombrowski, and Shaowen Bardzell. 2016. "Exploring Social Justice, Design, and HCI." In *Proceedings of the 2016 CHI Conference Extended Abstracts on Human Factors in Computing Systems (CHI EA'16)*, pp. 3293–3300. New York: Association for Computing Machinery.

Frey, Jérémy, Gilad Ostrin, May Grabli, and Jessica R. Cauchard. 2020. "Physiologically Driven Storytelling: Concept and Software Tool." In *Proceedings of the 2020 CHI Conference on Human Factors in Computing Systems (CHI'20)*, pp. 1–13. New York: Association for Computing Machinery.

Frey, Jérémy, Jelena Mladenović, Fabien Lotte, Camille Jeunet, and Léa Pillette. 2017. "When HCI Meets Neurotechnologies: What You Should Know about Brain-Computer Interfaces." In *Proceedings of the 2017 CHI Conference Extended Abstracts on Human Factors in Computing Systems (CHI EA'17)*, pp. 1253–56. New York: Association for Computing Machinery.

Friedman, Batya. 1996. "Value-Sensitive Design." *Interactions* 3(6): 16–23.

Furniss, Dominic, Ann Blandford, and Paul Curzon. 2011. "Confessions from a Grounded Theory PhD: Experiences and Lessons Learnt." In *Proceedings of the SIGCHI Conference on Human Factors in Computing Systems (CHI'11)*, pp. 113–22. New York: Association for Computing Machinery.

Garcia Rodriguez, Marisa C. 2016. "Chapter 7: 'The Stories We Tell Each Other': Using Technology for Resistance and Resilience Through Online Narrative Communities." In *Emotions, Technology, and Health*, edited by Sharon Y. Tettegah and Yolanda Evie Garcia, pp. 125–47. San Diego, CA: Academic Press.

Gaver, Bill, and John Bowers. 2012. "Annotated Portfolios." *Interactions* 19(4): 40–49.

Gaver, Bill, Tony Dunne, and Elena Pacenti. 1999. "Design: Cultural Probes." *Interactions* 6(1): 21–29.

Gaver, William W. 1991. "Technology Affordances." In *Proceedings of the SIGCHI Conference on Human Factors in Computing Systems (CHI'91)*, pp. 79–84. New York: Association for Computing Machinery.

Gaver, William. 2012. "What Should We Expect from Research through Design?" In *Proceedings of the SIGCHI Conference on Human Factors in Computing Systems (CHI'12)*, pp. 937–46. New York: Association for Computing Machinery.

Gay, Geraldine, and Helene Hembrooke. 2004. *Activity-Centered Design: An Ecological Approach to Designing Smart Tools and Usable Systems*. Cambridge, MA: MIT Press.

Ghoshal, Sucheta, Andrea Grimes Parker, Christopher A. Le Dantec, Carl Disalvo, Lilly Irani, and Amy Bruckman. 2019. "Design and the Politics of Collaboration: A Grassroots Perspective." In *Conference Companion Publication of the 2019 on Computer Supported Cooperative Work and Social Computing (CSCW'19)*, pp. 468–73. New York: Association for Computing Machinery.

Gibson, J. J. 1977. "The Theory of Affordances." *Hilldale, USA*, 1(2): 67–82.

Glanville, Ranulph. 2007. "Try Again. Fail Again. Fail Better: The Cybernetics in Design and the Design in Cybernetics." *Kybernetes. The International Journal of Cybernetics, Systems and Management Sciences* 36(9/10): 1173–1206.

Glonek, Grzegorz, and Maria Pietruszka. 2012. "Natural User Interfaces (NUI): Review." *Journal of Applied Computer Science & Mathematics* 20: 27–45.

Goleman, Daniel. 1995. *Emotional Intelligence*. New York: Bantam Books.

Gould, John D., and Clayton Lewis. 1985. "Designing for Usability: Key Principles and What Designers Think." *Communications of the ACM* 28(3): 300–311.

Goyal, Nitesh, Sungsoo Ray Hong, Regan L. Mandryk, Toby Jia-Jun Li, Kurt Luther, and Dakuo Wang. 2023. "SHAI 2023: Workshop on Designing for Safety in Human-AI Interactions." In *Companion Proceedings of the 28th International Conference on Intelligent User Interfaces (IUI'23)*, pp. 199–201. New York: Association for Computing Machinery.

Grimm, Thomas, Johann Mitlöhner, and Werner Schönfeldinger. 1995. "Bounded Rationality and Adaptive Agents in Economic Modeling." In *Proceedings of the International Conference on Applied Programming Languages (APL'95)*, pp. 56–62. New York: Association for Computing Machinery.

Grudin, Jonathan. 1990. "The Computer Reaches Out: The Historical Continuity of Interface Design." In *Proceedings of the SIGCHI Conference on Human Factors in Computing Systems (CHI'90)*, pp. 261–68. New York: Association for Computing Machinery.

Grudin, Jonathan. 2005. "Three Faces of Human-Computer Interaction." *IEEE Annals of the History of Computing* 27(4): 46–62.

Grudin, Jonathan. 2011. "A Moving Target: The Evolution of HCI." In *Human-Computer Interaction Handbook*, 3rd ed., edited by Julie Jacko. London: Taylor & Francis.

Grudin, Jonathan, and Gayna Williams. 2013. "Two Women Who Pioneered User-Centered Design." *Interactions* 20(6): 15–20.

Habeeb, Dana, James Clawson, Arash Zakeresfahani, and Zebulon Holtz. 2022. "Investigating and Validating On-Body Temperature Sensors for Personal Heat Exposure Tracking." In *Proceedings of the 2022 CHI Conference on Human Factors in Computing Systems (CHI'22)*, pp. 1–14. New York: Association for Computing Machinery.

Halabi, Ammar, Basile Zimmermann, and Michele Courant. 2013. "Participate, Collaborate, and Decide: Defining Design Problems in a Syrian Community." In *Proceedings of the Sixth International Conference on Information and Communications Technologies and Development (ICTD'13)*, Volume 2, pp. 49–52. New York: Association for Computing Machinery.

Haliburton, Luke, Svenja Yvonne Schött, Linda Hirsch, Robin Welsch, and Albrecht Schmidt. 2023. "Feeling the Temperature of the Room: Unobtrusive Thermal Display of Engagement during Group Communication." *Proceedings of the ACM on Interactive, Mobile, Wearable and Ubiquitous Technologies (IMWUT)*, 14, 7(1): 1–21.

Halloran, John, Eva Hornecker, Mark Stringer, Eric Harris, and Geraldine Fitzpatrick. 2009. "The Value of Values: Resourcing Co-Design of Ubiquitous Computing." *CoDesign* 5(4): 245–73.

Hayashi, Elaine C. S., and Maria Cecília C. Baranauskas. 2014. "Participant Observation and Experiences in the Design for Affectibility." In *HCI International 2014: Posters' Extended Abstracts*, pp. 36–41. Cham: Springer International Publishing.

Holden, Richard J., and Ben-Tzion Karsh. 2010. "The Technology Acceptance Model: Its Past and Its Future in Health Care." *Journal of Biomedical Informatics* 43(1): 159–72.

Hollender, Nina, Cristian Hofmann, Michael Deneke, and Bernhard Schmitz. 2010. "Integrating Cognitive Load Theory and Concepts of Human-computer Interaction." *Computers in Human Behavior* 26(6): 1278–88.

Holzinger, Andreas. 2005. "Usability Engineering Methods for Software Developers." *Communications of the ACM* 48(1): 71–74.

Hopper, Grace Murray. 1952. "The Education of a Computer." In *Proceedings of the 1952 ACM National Meeting (Pittsburgh) on – ACM'52*. New York: ACM Press. https://doi.org/10.1145/609784.609818.

Howell-Munson, Alicia, Christopher Micek, Ziheng Li, Michael Clements, Andrew C. Nolan, Jackson Powell, Erin T. Solovey, and Rodica Neamtu. 2022. "BrainEx: Interactive Visual Exploration and Discovery of Sequence Similarity in Brain Signals." *Proceedings of the ACM on Human-Computer Interaction*, 162, 6(EICS): 1–41.

Huang, Sheng-Cheng, Randolph G. Bias, and David Schnyer. 2015. "How Are Icons Processed by the Brain? Neuroimaging Measures of Four Types of Visual Stimuli Used in Information Systems." *Journal of the Association for Information Science and Technology* 66(4): 702–20.

Hughes, John, Val King, Tom Rodden, and Hans Andersen. 1995. "The Role of Ethnography in Interactive Systems Design." *Interactions* 2(2): 56–65.

Hui, Julie S., Darren Gergle, and Elizabeth M. Gerber. 2018. "IntroAssist: A Tool to Support Writing Introductory Help Requests." In *Proceedings of the 2018 CHI Conference on Human Factors in Computing Systems (CHI'18)*, pp. 1–13. New York: Association for Computing Machinery.

Hui, Julie, and Michelle L. Sprouse. 2023. "Lettersmith: Scaffolding Written Professional Communication Among College Students." In *Proceedings of the 2023 CHI Conference on Human Factors in Computing Systems (CHI'23)*, pp. 1–17. New York: Association for Computing Machinery.

Irani, Lilly, Janet Vertesi, Paul Dourish, Kavita Philip, and Rebecca E. Grinter. 2010. "Postcolonial Computing: A Lens on Design and Development." In *Proceedings of the SIGCHI Conference on Human Factors in Computing Systems (CHI'10)*, pp. 1311–20. New York: Association for Computing Machinery.

Ishii, Hiroshi. 2008. "The Tangible User Interface and Its Evolution." *Communications of the ACM* 51(6): 32–36.

Janssen, Christian P., Remo M. A. van der Heiden, Stella F. Donker, and J. Leon Kenemans. 2019. "Measuring Susceptibility to Alerts While Encountering Mental Workload." In *Proceedings of the 11th International Conference on Automotive User Interfaces and Interactive Vehicular Applications: Adjunct Proceedings (AutomotiveUI'19)*, pp. 415–20. New York: Association for Computing Machinery.

Janssen, Ruud, and Henk de Poot. 2006. "Information Overload: Why Some People Seem to Suffer More than Others." In *Proceedings of the 4th Nordic Conference on Human-Computer Interaction: Changing Roles (NordiCHI'06)*, pp. 397–400. New York: Association for Computing Machinery.

Johnson, Jeff. 2020. *Designing with the Mind in Mind: Simple Guide to Understanding User Interface Design Guidelines*. Amsterdam, The Netherlands: Elsevier Science.

Johnson, Jeff, Teressa L. Roberts, William Verplank, David C. Smith, Charles H. Irby, Marian Beard, and Kevin Mackey. 1989. "The Xerox Star: A Retrospective." *Computer* 22(9): 11–26.

Jokela, Tero, and Andrés Lucero. 2014. "MixedNotes: A Digital Tool to Prepare Physical Notes for Affinity Diagramming." In *Proceedings of the 18th International Academic MindTrek Conference: Media Business, Management, Content & Services (AcademicMindTrek'14)*, pp. 3–6. New York: Association for Computing Machinery.

Kalbach, Jim. 2020. *The Jobs to Be Done Playbook: Align Your Markets, Organizations, and Strategy Around Customer Needs*. Seattle, WA: Two Waves Books.

Kambunga, Asnath Paula, Rachel Charlotte Smith, Heike Winschiers-Theophilus, and Ton Otto. 2023. "Decolonial Design Practices: Creating Safe Spaces for Plural Voices on Contested Pasts, Presents, and Futures." *Design Studies* 86 (May): 101170.

Kamijo, Takeshi, Albert J. J. M. van Breemen, Xiao Ma, Santhosh Shanmugam, Thijs Bel, Gerard de Haas, Bart Peeters, et al. 2023. "A Touchless User Interface Based on a near-Infrared-Sensitive Transparent Optical Imager." *Nature Electronics* 6(6): 451–61.

Kelley, Jeff. 2018. "Wizard of Oz (WoZ)-A Yellow Brick Journey." *Journal of Usability Studies* 13(3): 119–24.

Khademi, Sadaf, Mehrnoosh Neghabi, Morteza Farahi, Mehdi Shirzadi, and Hamid Reza Marateb. 2022. "2- A Comprehensive Review of the Movement Imaginary Brain-Computer Interface Methods: Challenges and Future Directions." In *Artificial Intelligence-Based Brain-Computer Interface*, edited by Varun Bajaj and Ganesh Ram Sinha, pp. 23–74. Cambridge, MA: Academic Press.

Khan, Masood Mehmood, Michael Ingleby, and Robert D. Ward. 2006. "Automated Facial Expression Classification and Affect Interpretation Using Infrared Measurement of Facial Skin Temperature Variations." *ACM Transactions on Autonomous and Adaptive Systems*, 1(1): 91–113.

Kim, Sunnie S. Y., Elizabeth Anne Watkins, Olga Russakovsky, Ruth Fong, and Andrés Monroy-Hernández. 2023. "'Help Me Help the AI': Understanding How Explainability Can Support Human-AI Interaction." In *Proceedings of the 2023 CHI Conference on Human Factors in Computing Systems (CHI'23)*, pp. 1–17. New York: Association for Computing Machinery.

Knowles, Bran, Oliver Bates, and Maria Håkansson. 2018. "This Changes Sustainable HCI." In *Proceedings of the 2018 CHI Conference on Human Factors in Computing Systems (CHI'18)*, pp. 1–12. New York: Association for Computing Machinery.

Kolesnichenko, Anya, Joshua McVeigh-Schultz, and Katherine Isbister. 2019. "Understanding Emerging Design Practices for Avatar Systems in the Commercial Social VR Ecology." In *Proceedings of the 2019 on Designing Interactive Systems Conference (DIS'19)*, pp. 241–52. New York: Association for Computing Machinery.

Kosunen, I., Salminen, M., Järvelä, S., Ruonala, A., Ravaja, N., & Jacucci, G. 2016, March. "RelaWorld: Neuroadaptive and Immersive Virtual Reality Meditation System." In *Proceedings of the 21st International Conference on Intelligent User Interfaces*, pp. 208–217.

Kristensson, Per Ola, Nikola Banovic, Antti Oulasvirta, and John Williamson. 2019. "Computational Interaction with Bayesian Methods." In *Extended Abstracts of the 2019 CHI Conference on Human Factors in Computing Systems (CHI EA'19)*, pp. 1–6. New York: Association for Computing Machinery.

Krupka, Eyal, Kfir Karmon, Noam Bloom, Daniel Freedman, Ilya Gurvich, Aviv Hurvitz, Ido Leichter, et al. 2017. "Toward Realistic Hands Gesture Interface: Keeping It Simple for Developers and Machines." In *Proceedings of the 2017 CHI Conference on Human Factors in Computing Systems (CHI'17)*, pp. 1887–98. New York: Association for Computing Machinery.

Kuikkaniemi, Kai, Toni Laitinen, Marko Turpeinen, Timo Saari, Ilkka Kosunen, and Niklas Ravaja. 2010. "The Influence of Implicit and Explicit Biofeedback in First-Person Shooter Games." In *Proceedings of the SIGCHI Conference on Human Factors in Computing Systems (CHI'10)*, pp. 859–68. New York: Association for Computing Machinery.

Kumar, Neha, Julie A. Adams, Bill Buxton, Linda Candy, Pablo Cesar, Leigh Clark, Benjamin R. Cowan, et al. 2022. "A Chronology of SIGCHI Conferences: 1983 to 2022." *Interactions* 29(6): 34–41.

Kumar, Neha, Vikram Kamath Cannanure, Dilrukshi Gamage, Annu Sible Prabhakar, Christian Sturm, Cuauhtémoc Rivera Loaiza, Dina Sabie, Md Moinuddin Bhuiyan, and Mario A. Moreno Rocha. 2020. "HCI Across Borders and Sustainable Development Goals." In *Extended Abstracts of the 2020 CHI Conference on Human Factors in Computing Systems (CHI EA'20)*, pp. 1–8. New York: Association for Computing Machinery.

Le Dantec, Christopher A., and W. Keith Edwards. 2008. "Designs on Dignity: Perceptions of Technology among the Homeless." In *Proceedings of the SIGCHI Conference on Human Factors in Computing Systems (CHI'08)*, pp. 627–36. New York: Association for Computing Machinery.

Lee, Michelle Seng Ah, and Jat Singh. 2021. "The Landscape and Gaps in Open Source Fairness Toolkits." In *Proceedings of the 2021 CHI Conference on Human Factors in Computing Systems (CHI'21)*, pp. 1–13. New York: Association for Computing Machinery.

Letondal, Catherine, Jean-Luc Vinot, Sylvain Pauchet, Caroline Boussiron, Stéphanie Rey, Valentin Becquet, and Claire Lavenir. 2018. "Being in the Sky: Framing Tangible and Embodied Interaction for Future Airliner Cockpits." In *Proceedings of the Twelfth International Conference on Tangible, Embedded, and Embodied Interaction (TEI'18)*, pp. 656–66. New York: Association for Computing Machinery.

Lewis, Clayton. 1982. "Using the' Thinking-Aloud' Method in Cognitive Interface Design." RC9265 (#40713). IBM Thomas J. Watson Research Center.

Lewis, James R. 2014. "Usability: Lessons Learned … and yet to Be Learned." *International Journal of Human-Computer Interaction* 30(9): 663–84.

Li, Huimin, Ying Zeng, Xiyu Song, Li Tong, Jun Shu, and Bin Yan. 2021. "Sensitive Text Information Detection Based on Single-Trial EEG Signals." In *4th International Conference on Biometric Engineering and Applications (ICBEA'21)*, pp. 13–18. New York: Association for Computing Machinery.

Li, Tianyi, Mihaela Vorvoreanu, Derek DeBellis, and Saleema Amershi. 2022. "Assessing Human-AI Interaction Early through Factorial Surveys: A Study on the Guidelines for Human-AI Interaction." *ACM Transactions on Computer-Human Interaction*. https://doi.org/10.1145/3511605.

Liang, Jenny T., Chenyang Yang, and Brad A. Myers. 2023. "Understanding the Usability of AI Programming Assistants." arXiv [cs.SE]. arXiv.

Licklider, Joseph Carl Robnett. 1960. "Man-Computer Symbiosis." *IRE Transactions on Human Factors in Electronics* HFE-1(1): 4–11.

Lindtner, Silvia, Shaowen Bardzell, and Jeffrey Bardzell. 2016. "Reconstituting the Utopian Vision of Making: HCI after Technosolutionism." In *Proceedings of the 2016 CHI Conference on Human Factors in Computing Systems (CHI'16)*, pp. 1390–1402. New York: Association for Computing Machinery.

Lingelbach, Katharina, Nektaria Tagalidou, Patrick S. Markey, Bettina Föll, Matthias Peissner, and Mathias Vukelić. 2022. "Examining Joy of Use and Usability during Mobile Phone Interactions within a Multimodal Methods Approach." In *Proceedings of Mensch Und Computer 2022 (MuC'22)*, pp. 276–85. New York: Association for Computing Machinery.

Liu, Shengxi, Xiaomei Tao, and Qiong Gui. 2020. "Research on Emotional State in Online Learning by Eye Tracking Technology." In *Proceedings of the 4th International Conference on Intelligent Information Processing (ICIIP'19)*, pp. 471–77. New York: Association for Computing Machinery.

Long, John. 1996. "Specifying Relations between Research and the Design of Human-Computer Interactions." *International Journal of Human-Computer Studies* 44(6): 875–920.

Long, John, and John Dowell. 1989. "Conceptions of the Discipline of HCI: Craft, Applied Science, and Engineering." *Proceedings of the 5th Conference of the BCS-HCI Group on People and Computers V*, edited by A. Sutcliffe and L. Macaulay, pp. 9–32. Cambridge: Cambridge University Press.

Lowgren, J. 2001. "From HCI to Interaction Design." In *Human Computer Interaction: Issues and Challenges*, pp. 29–43. IGI Global.

Lucero, A. 2015. "Using Affinity Diagrams to Evaluate Interactive Prototypes." In *Human-Computer Interaction–INTERACT 2015: 15th IFIP TC 13 International Conference, Bamberg, Germany, September 14-18, 2015, Proceedings, Part II 15*, pp. 231–248. Springer International Publishing.

Lukanov, Kristiyan, Horia A. Maior, and Max L. Wilson. 2016. "Using fNIRS in Usability Testing: Understanding the Effect of Web Form Layout on Mental Workload." In *Proceedings of the 2016 CHI Conference on Human Factors in Computing Systems (CHI'16)*, pp. 4011–16. New York: Association for Computing Machinery.

Luusua, Anna, Johanna Ylipulli, Marko Jurmu, Henrika Pihlajaniemi, Piia Markkanen, and Timo Ojala. 2015. "Evaluation Probes." In *Proceedings of the 33rd Annual ACM Conference on Human Factors in Computing Systems (CHI'15)*, pp. 85–94. New York: Association for Computing Machinery.

Macefield, Ritch. 2007. "Usability Studies and the Hawthorne Effect." *Journal of Usability Studies* 2(3): 145–54.

MacKenzie, I. Scott. 1992. "Fitts' Law as a Research and Design Tool in Human-Computer Interaction." *Human-Computer Interaction* 7(1): 91–139.

MacKenzie, I. Scott, and R. William Soukoreff. 2003. "Card, English, and Burr (1978): 25 Years Later." In *Extended Abstracts on Human Factors in Computing Systems (CHI EA'03)*, pp. 760–61. New York: Association for Computing Machinery.

Marsh, Tim, Ashima Thomas, and Eng Tat Khoo. 2021. "Between Game Mechanics and Immersive Storytelling: Design Using an Extended Activity Theory Framework." In *Serious Games*, edited by Bobbie Fletcher, Minhua Ma, Stefan Göbel, Jannicke Baalsrud Hauge, Tim Marsh, pp. 113–28. Cham: Springer International Publishing.

Martin, David, and Ian Sommerville. 2004. "Patterns of Cooperative Interaction: Linking Ethnomethodology and Design." *ACM Transactions on Computer-Human Interaction* 11(1): 59–89.

Martin, James. 1973. *Design of Man-Computer Dialogues*. Hoboken, NJ: Prentice-Hall.

McGuffin, Michael J., and Ravin Balakrishnan. 2005. "Fitts' Law and Expanding Targets." *ACM Transactions on Computer-Human Interaction: A Publication of the Association for Computing Machinery* 12(4): 388–422.

McInerney, Paul, and Rick Sobiesiak. 1999. "The UI Design Process: Planning, Managing, and Documenting UI Design Work." In *Extended Abstracts on Human Factors in Computing Systems (CHI EA'99)*, p. 177. New York: Association for Computing Machinery.

McQuaid, Heather L., and David Bishop. 2001. "An Integrated Method for Evaluating Interfaces." In *Extended Abstracts on Human Factors in Computing Systems (CHI EA'01)*, pp. 287–88. New York: Association for Computing Machinery.

Meister, David. 2018. *History of Human Factors and Ergonomics*. London: Taylor & Francis Group.

Millen, David R. 2000. "Rapid Ethnography: Time Deepening Strategies for HCI Field Research." In *Proceedings of the 3rd Conference on Designing Interactive Systems: Processes, Practices, Methods, and Techniques (DIS'00)*, pp. 280–86. New York: Association for Computing Machinery.

Moge, Clara, Katherine Wang, and Youngjun Cho. 2022. "Shared User Interfaces of Physiological Data: Systematic Review of Social Biofeedback Systems and Contexts in HCI." In *Proceedings of the 2022 CHI Conference on Human Factors in Computing Systems (CHI'22)*, pp. 1–16. New York: Association for Computing Machinery.

Moran, Thomas P., and Stuart K. Card. 1982. "Applying Cognitive Psychology to Computer Systems: A Graduate Seminar in Psychology." In *Proceedings of the 1982 Conference on Human Factors in Computing Systems (CHI'82)*, pp. 295–98. New York: Association for Computing Machinery.

Mucha, Henrik, Sebastian Robert, Ruediger Breitschwerdt, and Michael Fellmann. 2021. "Interfaces for Explanations in Human-AI Interaction: Proposing a Design Evaluation Approach." In *Extended Abstracts of the 2021 CHI Conference on Human Factors in Computing Systems (CHI EA'21)*, pp. 1–6. New York: Association for Computing Machinery.

Muender, Thomas, Matthew K. Miller, Max V. Birk, and Regan L. Mandryk. 2016. "Extracting Heart Rate from Videos of Online Participants." In *Proceedings of the 2016 CHI Conference on Human Factors in Computing Systems (CHI'16)*, pp. 4562–67. New York: Association for Computing Machinery.

Muller, Michael J., and Allison Druin. 2012. "Participatory Design: The Third Space in Human-computer Interaction." In *The Human-Computer Interaction Handbook: Fundamentals, Evolving Technologies, and Emerging Applications*, edited by Julie A. Jacko, pp. 1125–1154. Boca Raton, FL: CRC Press. https://doi.org/10.1201/b11963-ch-49.

Muñoz, John Esteban, Élvio Rúbio Gouveia, Mónica Cameirão, and Sergi Bermudez I. Badia. 2017. "The Biocybernetic Loop Engine: An Integrated Tool for Creating Physiologically Adaptive Videogames." In *Proceedings of the 4th International Conference on Physiological Computing Systems*. SCITEPRESS - Science and Technology Publications. https://doi.org/10.5220/0006429800450054.

Muñoz, John E., Luis Quintero, Chad L. Stephens, and Alan Pope. 2021. "Taxonomy of Physiologically Adaptive Systems and Design Framework." In *Adaptive Instructional Systems. Design and Evaluation*, edited by Robert A. Sottilare and Jessica Schwarz, pp. 559–76. Cham: Springer International Publishing.

Murad, Christine, Cosmin Munteanu, Leigh Clark, and Benjamin R. Cowan. 2018. "Design Guidelines for Hands-Free Speech Interaction." In *Proceedings of the 20th International Conference on Human-Computer Interaction with Mobile Devices and Services Adjunct (MobileHCI'18)*, pp. 269–76. New York: Association for Computing Machinery.

Murad, Christine, Humaira Tasnim, and Cosmin Munteanu. 2022. "'Voice-First Interfaces in a GUI-First Design World': Barriers and Opportunities to Supporting VUI Designers on-the-Job." In *Proceedings of the 4th Conference on Conversational User Interfaces (CUI'22)*, pp. 1–10. New York: Association for Computing Machinery.

Nakanishi, Yukiya, Masaaki Fukuoka, Shunichi Kasahara, and Maki Sugimoto. 2022. "Synchronous and Asynchronous Manipulation Switching of Multiple Robotic Embodiment Using EMG and Eye Gaze." In *Proceedings of the Augmented Humans International Conference 2022 (Ahs'22)*, pp. 94–103. New York: Association for Computing Machinery.

Nakatani, Lloyd H., and John A. Rohrlich. 1983. "Soft Machines: A Philosophy of User-Computer Interface Design." In *Proceedings of the SIGCHI Conference on Human Factors in Computing Systems (CHI'83)*, pp. 19–23. New York: Association for Computing Machinery.

National Space Administration. 2017. *NASA Systems Engineering Handbook*, NASA SP-2016-6105 Rev2. CreateSpace Independent Publishing Platform.

Nelson, Theodor H. 1983. *Literary Machines: The Report On, and Of, Project Xanadu Concerning Word Processing, Electronic Publishing, Hypertext, Thinkertoys, Tomorrow's Intellectual Revolution, and Certain Other Topics Including Knowledge, Education and Freedom*. Swarthmore, PA: Ted Nelson.

Neubauer, Catherine, Mathieu Chollet, Sharon Mozgai, Mark Dennison, Peter Khooshabeh, and Stefan Scherer. 2017. "The Relationship between Task-Induced Stress, Vocal Changes, and Physiological State during a Dyadic Team Task." In *Proceedings of the 19th ACM International Conference on Multimodal Interaction (ICMI'17)*, pp. 426–32. New York: Association for Computing Machinery.

Newell, Allen, and Stuart K. Card. 1985. "The Prospects for Psychological Science in Human-Computer Interaction." *Human-Computer Interaction* 1(3): 209–42.

Nielsen, Jacob. 1995. "Applying Discount Usability Engineering." *IEEE Software* 12(1): 98–100.

Nielsen, Jakob, and Robert L. Mack, eds. 1994. *Usability Inspection Methods*. Nashville, TN: John Wiley & Sons.

Nielsen, Jakob, and Rolf Molich. 1990. "Heuristic Evaluation of User Interfaces." In *Proceedings of the SIGCHI Conference on Human Factors in Computing Systems (CHI'90)*, pp. 249–56. New York: Association for Computing Machinery.

Nielsen, Jakob. 1990. "The Art of Navigating through Hypertext." *Communications of the ACM* 33(3): 296–310.

Nielsen, Jakob. 1994a. "Heuristic Evaluation." In *Usability Inspection Methods*, edited by Jakob Nielsen and Robert L. Mack, pp. 25–62. New York, John Wiley & Sons.

Nielsen, Jakob. 1994b. "Enhancing the Explanatory Power of Usability Heuristics." In *Proceedings of the SIGCHI Conference on Human Factors in Computing Systems (CHI'94)*, pp. 152–58. New York: Association for Computing Machinery.

Nielsen, Jakob. 1994c. "Usability Inspection Methods." In *Conference Companion on Human Factors in Computing Systems (CHI'94)*, pp. 413–14. New York: Association for Computing Machinery.

Nikopoulou, Rozalia, Ioannis Vernikos, Evaggelos Spyrou, and Phivos Mylonas. 2018. "Emotion Recognition from Speech: A Classroom Experiment." In *Proceedings of the 11th PErvasive Technologies Related to Assistive Environments Conference (PETRA'18)*, pp. 104–5. New York: Association for Computing Machinery.

Norman, Donald A. 1982. "Steps toward a Cognitive Engineering: Design Rules Based on Analyses of Human Error." In *Proceedings of the 1982 Conference on Human Factors in Computing Systems (CHI'82)*, pp. 378–82. New York: Association for Computing Machinery.

Norman, Donald A. 1983a. "Design Principles for Human-Computer Interfaces." In *Proceedings of the SIGCHI Conference on Human Factors in Computing Systems (CHI'83)*, pp. 1–10. New York: Association for Computing Machinery.

Norman, Donald A. 1983b. "Some Observations on Mental Models." In *Mental Models*, edited by Dedre Gentner and Albert L. Stevens, pp. 7–14. New York: Psychology Press. https://doi.org/10.4324/9781315802725-5.

Norman, Donald A. 1988. *The Psychology of Everyday Things*. London, England: Basic Books.

Norman, Donald A. 2002. "Emotion & Design: Attractive Things Work Better." *Interactions* 9(4): 36–42.

Norman, Donald A. 2013. *The Design of Everyday Things: Revised and Expanded Edition*. New York: Basic Books.

Norman, Donald A. 2018. "Emotional Design." *Ubiquity* 2004(January): 1.

Norman, Donald A., Jim Miller, and Austin Henderson. 1995. "What You See, Some of What's in the Future, and How We Go about Doing It." In *Conference Companion on Human Factors in Computing Systems - CHI'95*. New York: ACM Press. https://doi.org/10.1145/223355.223477.

Norman, Donald A., and Stephen W. Draper. 1986. *User Centered System Design; New Perspectives on Human-Computer Interaction*. Mahwah, NJ: Lawrence Erlbaum Associates.

Noroozi, Fatemeh, Ciprian Adrian Corneanu, Dorota Kamińska, Tomasz Sapiński, Sergio Escalera, and Gholamreza Anbarjafari. 2018. "Survey on Emotional Body Gesture Recognition." *IEEE Transactions on Affective Computing*, 12(2): 505–23. https://ieeexplore.ieee.org/abstract/document/8493586/.

Ogbonnaya-Ogburu, Ihudiya Finda, Angela D. R. Smith, Alexandra To, and Kentaro Toyama. 2020. "Critical Race Theory for HCI." In *Proceedings of the 2020 CHI Conference on Human Factors in Computing Systems (CHI'20)*, pp. 1–16. New York: Association for Computing Machinery.

Oulasvirta, Antti, Andrew Howes, Per Ola Kristensson, and Xiaojun Bi. 2018. *Computational Interaction*. Oxford: Oxford University Press.

Oulasvirta, Antti, and Kasper Hornbæk. 2016. "HCI Research as Problem-Solving." In *Proceedings of the 2016 CHI Conference on Human Factors in Computing Systems (CHI'16)*, pp. 4956–67. New York: Association for Computing Machinery.

Pai, Akshay Palimar, Jayasankar Santhosh, and Shoya Ishimaru. 2023. "Real-Time Feedback on Reader's Engagement and Emotion Estimated by Eye-Tracking and Physiological Sensing." In *Adjunct Proceedings of the 2022 ACM International Joint Conference on Pervasive and Ubiquitous Computing and the 2022 ACM International Symposium on Wearable Computers (UbiComp/ISWC'22 Adjunct)*, pp. 97–98. New York: Association for Computing Machinery.

Parsons, Paul, Ali Baigelenov, Ya-Hsin Hung, and Connor Schrank. 2020. "What Design Methods Do DataVis Practitioners Know and Use?" In *Extended Abstracts of the 2020 CHI Conference on Human Factors in Computing Systems (CHI EA'20)*, pp. 1–8. New York: Association for Computing Machinery.

Peng, Shuna, Yang Dong, Weisha Wang, Jieyi Hu, and Weiyang Dong. 2019. "The Affective Facial Recognition Task: The Influence of Cognitive Styles and Exposure Times." *Journal of Visual Communication and Image Representation* 65(December): 102674.

Perusquía-Hernández, Monica, Saho Ayabe-Kanamura, Kenji Suzuki, and Shiro Kumano. 2019. "The Invisible Potential of Facial Electromyography: A Comparison of EMG and Computer Vision When Distinguishing Posed from Spontaneous Smiles." In *Proceedings of the 2019 CHI Conference on Human Factors in Computing Systems (CHI'19)*, pp. 1–9. New York: Association for Computing Machinery.

Philip, Kavita, Lilly Irani, and Paul Dourish. 2012. "Postcolonial Computing: A Tactical Survey." *Science, Technology & Human Values* 37(1): 3–29.

Picard, Rosalind W. 1999. "Affective Computing for HCI." In *Proceedings of the 8th International Conference on Human-Computer Interaction*, edited by Hans-Jorg Bullinger and Jurgen Ziegler, pp. 829–33. HCI International.

Pillai, Ajit G., Ahmet Baki Kocaballi, Tuck Wah Leong, Rafael A. Calvo, Nassim Parvin, Katie Shilton, Jenny Waycott, Casey Fiesler, John C. Havens, and Naseem Ahmadpour. 2021. "Co-Designing Resources for Ethics Education in HCI." In *Extended Abstracts of the 2021 CHI Conference on Human Factors in Computing Systems (CHI EA'21)*, pp. 1–5. New York: Association for Computing Machinery.

Polson, Peter G., Clayton Lewis, John Rieman, and Cathleen Wharton. 1992. "Cognitive Walkthroughs: A Method for Theory-Based Evaluation of User Interfaces." *International Journal of Man-Machine Studies* 36(5): 741–73.

Poole, Alex, and Linden J. Ball. 2006. "Eye Tracking in HCI and Usability Research." In *Encyclopedia of Human Computer Interaction*, edited by Claude Ghaoui, pp. 211–19. Hershey, PA: IGI Global.

Pope, Alan T., Edward H. Bogart, and Debbie S. Bartolome. 1995. "Biocybernetic System Evaluates Indices of Operator Engagement in Automated Task." *Biological Psychology* 40(1): 187–95.

Porcheron, Martin, Joel E. Fischer, Stuart Reeves, and Sarah Sharples. 2018. "Voice Interfaces in Everyday Life." In *Proceedings of the 2018 CHI Conference on Human Factors in Computing Systems (CHI'18)*, pp. 1–12. New York: Association for Computing Machinery.

Portz, Jennifer Dickman, Elizabeth A. Bayliss, Sheana Bull, Rebecca S. Boxer, David B. Bekelman, Kathy Gleason, and Sara Czaja. 2019. "Using the Technology Acceptance Model to Explore User Experience, Intent to Use, and Use Behavior of a Patient Portal Among Older Adults With Multiple Chronic Conditions: Descriptive Qualitative Study." *Journal of Medical Internet Research* 21(4): e11604.

Pourjafarian, Narjes, Anusha Withana, Joseph A. Paradiso, and Jürgen Steimle. 2019. "Multi-Touch Kit: A Do-It-Yourself Technique for Capacitive Multi-Touch Sensing Using a Commodity Microcontroller." In *Proceedings of the 32nd Annual ACM Symposium on User Interface Software and Technology (UIST'19)*, pp. 1071–83. New York: Association for Computing Machinery.

Prpa, Mirjana, Thecla Schiphorst, Kivanç Tatar, and Philippe Pasquier. 2018. "Respire: A Breath Away from the Experience in Virtual Environment." In *Extended Abstracts of the 2018 CHI Conference on Human Factors in Computing Systems (CHI EA'18)*, 1–6. New York: Association for Computing Machinery.

Pu, Pearl, Li Chen, and Rong Hu. 2011. "A User-Centric Evaluation Framework for Recommender Systems." In *Proceedings of the Fifth ACM Conference on Recommender Systems (RecSys'11)*, pp. 157–64. New York: Association for Computing Machinery.

Putze, Felix, Susanne Putze, Merle Sagehorn, Christopher Micek, and Erin T. Solovey. 2022. "Understanding HCI Practices and Challenges of Experiment Reporting with Brain Signals: Towards Reproducibility and Reuse." *ACM Transactions on Computer-Human Interaction*, 31, 29(4): 1–43.

Ramsey, H. Rudy, and Atwood, Michael. 1979. Human Factors in Computer Systems: Review of the Literature and Development of Design Aids. Technical Report No. SAI-79-111-DEN. Englewood, CO: Science Applications Inc.

Rankin, Yolanda A., and Kallayah K. Henderson. 2021. "Resisting Racism in Tech Design: Centering the Experiences of Black Youth." *Proceedings of the ACM on Human-Computer Interaction*, 192, 5(CSCW1): 1–32.

Richardson, Brianna, Jean Garcia-Gathright, Samuel F. Way, Jennifer Thom, and Henriette Cramer. 2021. "Towards Fairness in Practice: A Practitioner-Oriented Rubric for Evaluating Fair ML Toolkits." In *Proceedings of the 2021 CHI Conference on Human Factors in Computing Systems (CHI'21 236)*, pp. 1–13. New York: Association for Computing Machinery.

Riedl, René, Adriane B. Randolph, Jan vom Brocke, Pierre-Majorique Léger, and Angelika Dimoka. 2010. "The Potential of Neuroscience for Human-Computer Interaction Research," *SIGHCI 2010 Proceedings*.

Rietze, Jessica, Isabell Bürkner, Anne Pfister, and Rainer Blum. 2021. "Online Focus Groups with and for the Elderly: Specifics, Challenges, Recommendations: Online-Fokusgruppen Mit Und Für Senior*innen: Besonderheiten, Herausforderungen, Empfehlungen." In *Proceedings of Mensch Und Computer 2021 (MuC'21)*, pp. 194–98. New York: Association for Computing Machinery.

Rode, Jennifer A. 2011. "Reflexivity in Digital Anthropology." In *Proceedings of the SIGCHI Conference on Human Factors in Computing Systems (CHI'11)*, pp. 123–32. New York: Association for Computing Machinery.

Rode, Jennifer, Mark Blythe, and Bonnie Nardi. 2012. "Qualitative Research in HCI." In *Extended Abstracts on Human Factors in Computing Systems (CHI EA'12)*, pp. 2803–6. New York: Association for Computing Machinery.

Rogers, Yvonne. 2012. HCI Theory: Classical, Modern, and Contemporary. In *Synthesis Lectures on Human-Centered Informatics*, edited by John M. Carroll, p. 14. San Rafael, CA: Morgan & Claypool.

Rogers, Yvonne, Helen Sharp, and Jennifer Preece. 2023. *Interaction Design: Beyond Human-Computer Interaction*. Hoboken, NJ: John Wiley & Sons.

Rosenbaum, Stephanie, Gilbert Cockton, Kara Coyne, Michael Muller, and Thyra Rauch. 2002. "Focus Groups in HCI: Wealth of Information or Waste of Resources?" In *Extended Abstracts on Human Factors in Computing Systems (CHI EA'02)*, pp. 702–3. New York: Association for Computing Machinery.

Rosenberg, Jarrett. 1982. "Evaluating the Suggestiveness of Command Names." In *Proceedings of the 1982 Conference on Human Factors in Computing Systems (CHI'82)*, pp. 12–16. New York: Association for Computing Machinery.

Rosenberg, Milton J. 1960. "Cognitive, Affective, and Behavioral Components of Attitudes." In *Attitude Organization and Change: An Analysis of Consistency among Attitude Components*, edited by M. J. Rosenberg and Carl I. Hovland, pp. 1–14. New Haven, CT: Yale University Press.

Ruan, Xingran, Charaka Palansuriya, Aurora Constantin, and Konstantinos Tsiakas. 2023. "Supporting Children's Metacognition with a Facial Emotion Recognition Based Intelligent Tutor System." In *Proceedings of the 22nd Annual ACM Interaction Design and Children Conference (IDC'23)*, pp. 502–6. New York: Association for Computing Machinery.

Rumelhart, David E., and Donald A. Norman. 1973. "Active Semantic Networks as a Model of Human Memory." In *Proceedings of IJCAI*, pp. 450–57.

Saffer, Dan. 2009. *Designing for Interaction: Creating Innovative Applications and Devices*, 2nd ed. Upper Saddle River, NJ: New Riders Publishing.

Sanders, Elizabeth B.-N. 2002. "From User-Centered to Participatory Design Approaches." In *Design and the Social Sciences: Making Connections*, edited by Jorge Frascara, pp. 18–25. Boca Raton, FL: CRC Press.

Sanders, Elizabeth B.-N, Eva Brandt, and Thomas Binder. 2010. "A Framework for Organizing the Tools and Techniques of Participatory Design." In *Proceedings of the 11th Biennial Participatory Design Conference (PDC'10)*, pp. 195–98. New York: Association for Computing Machinery.

Sanders, Liz. 2008. "An Evolving Map of Design Practice and Design Research." *Interactions* 15(6): 13–17.

Saxena, Devansh, Erina Seh-Young Moon, Aryan Chaurasia, Yixin Guan, and Shion Guha. 2023. "Rethinking 'Risk' in Algorithmic Systems through a Computational Narrative Analysis of Casenotes in Child-Welfare." In *Proceedings of the 2023 CHI Conference on Human Factors in Computing Systems (CHI'23)*, pp. 1–19. New York: Association for Computing Machinery.

Scapin, Dominique L., Gilbert Cockton, Alan Woolrich, Mark Springett, Christian Stary, and Marco Winckler. 2009. Maturation of Usability Evaluation Methods: Retrospect and Prospect. In *Final Reports of COST294-MAUSE Working Groups*, edited by Effie L.-C. Law. Columbus, OH: IRIT Press.

Schicker, Eva. 2020. "Affinity Mapping and Why It Is Important in Your UX Strategy." *UsabilityGeek*. March 3, 2020. https://medium.com/usabilitygeek/affinity-mapping-and-why-it-is-important-in-your-ux-strategy-322675234f9e.

Sefelin, Reinhard, Manfred Tscheligi, and Verena Giller. 2003. "Paper Prototyping - What Is It Good for? A Comparison of Paper- and Computer-Based Low-Fidelity Prototyping." In *Extended Abstracts on Human Factors in Computing Systems (CHI EA'03)*, pp. 778–79. New York: Association for Computing Machinery.

Shackell, Cameron, and Laurianne Sitbon. 2018. "Cognitive Externalities and HCI: Towards the Recognition and Protection of Cognitive Rights." In *Extended Abstracts of the 2018 CHI Conference on Human Factors in Computing Systems (CHI EA'18)*, pp. 1–10. New York: Association for Computing Machinery.

Sharif, Ather, Victoria Pao, Katharina Reinecke, and Jacob O. Wobbrock. 2020. "The Reliability of Fitts's Law as a Movement Model for People with and without Limited Fine Motor Function." In *Proceedings of the 22nd International ACM SIGACCESS Conference on Computers and Accessibility (ASSETS'20)*, pp. 1–15. New York: Association for Computing Machinery.

Shneiderman, Ben. 1982. "The Future of Interactive Systems and the Emergence of Direct Manipulation." *Behaviour & Information Technology* 1(3): 237–56.

Shoemaker, Phillip B. 1999. "Designing Interfaces for Handheld Computers." In *Extended Abstracts on Human Factors in Computing Systems (CHI EA'99)*, pp. 126–27. New York: Association for Computing Machinery.

Shu, Lin, Jinyan Xie, Mingyue Yang, Ziyi Li, Zhenqi Li, Dan Liao, Xiangmin Xu, and Xinyi Yang. 2018. "A Review of Emotion Recognition Using Physiological Signals." *Sensors* 18(7). https://doi.org/10.3390/s18072074.

Slovák, Petr, Joris Janssen, and Geraldine Fitzpatrick. 2012. "Understanding Heart Rate Sharing: Towards Unpacking Physiosocial Space." In *Proceedings of the SIGCHI Conference on Human Factors in Computing Systems (CHI'12)*, pp. 859–68. New York: Association for Computing Machinery.

Smith, David Canfield, Charles Irby, Ralph Kimball, and Eric Harslem. 1982. "The Star User Interface: An Overview." In *Proceedings of the* June 7-10, *1982, National Computer Conference (AFIPS'82)*, pp. 515–28. New York: Association for Computing Machinery.

Smith, Jessie J., Lex Beattie, and Henriette Cramer. 2023. "Scoping Fairness Objectives and Identifying Fairness Metrics for Recommender Systems: The Practitioners' Perspective." In *Proceedings of the ACM Web Conference 2023 (WWW'23)*, pp. 3648–59. New York: Association for Computing Machinery.

Smith, Sidney L. 1963. "Man-Computer Information Transfer." In *Electronic Information Display Systems*, edited by James H. Howard, pp. 284–99. Washington, DC: Spartan Books.

Sonne, Tobias, and Mads Møller Jensen. 2016. "ChillFish: A Respiration Game for Children with ADHD." In *Proceedings of the TEI'16: Tenth International Conference on Tangible, Embedded, and Embodied Interaction (TEI'16)*, pp. 271–78. New York: Association for Computing Machinery.

Sowa, G. 2009. "User Interface: Standards and Research Tools." *Journal of Applied Computer Science Methods* 1(1). https://yadda.icm.edu.pl/baztech/element/bwmeta1.element.baztech-e0c3d45d-6d15-45e9-83ca-6953e3d659af.

Spaa, Anne, Abigail Durrant, Chris Elsden, and John Vines. 2019. "Understanding the Boundaries between Policymaking and HCI." In *Proceedings of the 2019 CHI Conference on Human Factors in Computing Systems (CHI'19)*, pp. 1–15. New York: Association for Computing Machinery.

Strohmayer, Angelika, Jenn Clamen, and Mary Laing. 2019. "Technologies for Social Justice: Lessons from Sex Workers on the Front Lines." In *Proceedings of the 2019 CHI Conference on Human Factors in Computing Systems (CHI'19)*, pp. 1–14. New York: Association for Computing Machinery.

Sutcliffe, Alistair G., Charalambos Poullis, Andreas Gregoriades, Irene Katsouri, Aimilia Tzanavari, and Kyriakos Herakleous. 2019. "Reflecting on the Design Process for Virtual Reality Applications." *International Journal of Human-Computer Interaction* 35(2): 168–79.

Sutherland, Ivan Edward. 1964. "Sketch Pad a Man-Machine Graphical Communication System." In Proceedings of the SHARE Design Automation Workshop (DAC'64), pp. 6.329–6.346. New York: Association for Computing Machinery.

Sutherland, Ivan Edward. 1980. *Sketchpad: A Man-Machine Graphical Communication System*. New York: Garland Publishing.

Sweller, John. 1988. "Cognitive Load during Problem Solving: Effects on Learning." *Cognitive Science* 12(2): 257–85.

Terzimehić, Nađa, Renate Häuslschmid, Heinrich Hussmann, and Monica C. Schraefel. 2019. "A Review & Analysis of Mindfulness Research in HCI: Framing Current Lines of Research and Future Opportunities." In *Proceedings of the 2019 CHI Conference on Human Factors in Computing Systems (CHI'19)*, pp. 1–13. New York: Association for Computing Machinery.

Thanh Vi, Chi, Kasper Hornbæk, and Sriram Subramanian. 2017. "Neuroanatomical Correlates of Perceived Usability." In *Proceedings of the 30th Annual ACM Symposium on User Interface Software and Technology (UIST'17)*, pp. 519–32. New York: Association for Computing Machinery.

Tøndel, Inger Anne, Daniela Soares Cruzes, and Martin Gilje Jaatun. 2020. "Using Situational and Narrative Analysis for Investigating the Messiness of Software Security." In *Proceedings of the 14th ACM / IEEE International Symposium on Empirical Software Engineering and Measurement (ESEM) (ESEM'20)*, pp. 1–6. New York: Association for Computing Machinery.

Ueno, Takane, Yuto Sawa, Yeongdae Kim, Jacqueline Urakami, Hiroki Oura, and Katie Seaborn. 2022. "Trust in Human-AI Interaction: Scoping Out Models, Measures, and Methods." In *Extended Abstracts of the 2022 CHI Conference on Human Factors in Computing Systems (CHI EA'22)*, pp. 1–7. New York: Association for Computing Machinery.

Urrestilla, Nerea, and David St-Onge. 2021. "Measuring Cognitive Load: Heart-Rate Variability and Pupillometry Assessment." In *Companion Publication of the 2020 International Conference on Multimodal Interaction (ICMI'20 Companion)*, pp. 405–10. New York: Association for Computing Machinery.

Van Hove, Stephanie, Anissa All, Peter Conradie, and Lieven De Marez. 2020. "Holistic Assessment of Situated Cooking Interactions: Preliminary Results of an Observational Study." In *Design, User Experience, and Usability. Design for Contemporary Interactive Environments*, pp. 158–74. Cham: Springer International Publishing.

Vance, Anthony, Brock Kirwan, Daniel Bjornn, Jeffrey Jenkins, and Bonnie Brinton Anderson. 2017. "What Do We Really Know about How Habituation to Warnings Occurs Over Time? A Longitudinal fMRI Study of Habituation and Polymorphic Warnings." In *Proceedings of the 2017 CHI Conference on Human Factors in Computing Systems (CHI'17)*, pp. 2215–27. New York: Association for Computing Machinery.

Vicente, Kim J., and Jens Rasmussen. 1992. "Ecological Interface Design: Theoretical Foundations." *IEEE Transactions on Systems, Man, and Cybernetics* 22(4): 589–606.

Walsh, Toby. 2022. "The Troubling Future for Facial Recognition Software." *Communications of the ACM* 65(3): 35–36.

Wang, Liang, Zhe Huang, Ziyu Zhou, Devon McKeon, Giles Blaney, Michael C. Hughes, and Robert J. K. Jacob. 2021. "Taming fNIRS-Based BCI Input for Better Calibration and Broader Use." In *The 34th Annual ACM Symposium on User Interface Software and Technology (UIST'21)*, pp. 179–97. New York: Association for Computing Machinery.

Wang, Yan, Wei Song, Wei Tao, Antonio Liotta, Dawei Yang, Xinlei Li, Shuyong Gao, et al. 2022. "A Systematic Review on Affective Computing: Emotion Models, Databases, and Recent Advances." *An International Journal on Information Fusion* 83–84(July): 19–52.

Wania, Christine E., and Michael E. Atwood. 2009. "Pattern Languages in the Wild: Exploring Pattern Languages in the Laboratory and in the Real World." In *Proceedings of the 4th International Conference on Design Science Research in Information Systems and Technology (DESRIST'09)*, pp. 1–15. New York: Association for Computing Machinery.

Weinberg, Gerald M. 1998. *The Psychology of Computer Programming: Silver* Anniversary Edition, 2nd ed. New York: Dorset House Publishing.

Weiser, Mark. 1991. "The Computer for the 21 St Century." *Scientific American* 265(3): 94–105.

Weiser, Mark. 1993. "Hot Topics-Ubiquitous Computing." *Computer* 26(10): 71–72.

Weizenbaum, Joseph. 1966. "ELIZA-a Computer Program for the Study of Natural Language Communication between Man and Machine." *Communications of the ACM* 9(1): 36–45.

Weizenbaum, Joseph. 1967. "Contextual Understanding by Computers." *Communications of the ACM* 10(8): 474–80.

Weizenbaum, Joseph. 1972. "On the Impact of the Computer on Society." *Science* 176(4035): 609–14.

Weizenbaum, Joseph. 1986. "Not without Us." *SIGCAS Computers and Society* 16(2–3): 2–7.

Whang, Michael. 2008. "+++++++ Usability Tests of the WMU Libraries' Web Site." *Journal of Web Librarianship* 2(2–3): 205–18.

Wharton, Cathleen, and Clayton Lewis. 1994. "Role of Psychology in Usability Inspection Methods." CU-ICS-93-06. University of Colorado at Boulder.

Whittaker, Meredith. 2023. "Origin Stories: Plantations, Computers, and Industrial Control." *Logic Magazine*, March.

Wigdor, Daniel, and Gerald Morrison. 2010. "Designing User Interfaces for Multi-Touch and Surface-Gesture Devices." In *CHI'10 Extended Abstracts on Human Factors in Computing Systems (CHI EA'10)*, pp. 3193–96. New York: Association for Computing Machinery.

Williams, Ashley. 2009. "User-Centered Design, Activity-Centered Design, and Goal-Directed Design: A Review of Three Methods for Designing Web Applications." In *Proceedings of the 27th ACM International Conference on Design of Communication (SIGDOC'09)*, pp. 1–8. New York: Association for Computing Machinery.

Winograd, Terry, Peter Denning, and Robert Metcalfe. 1997. "From Computing Machinery to Interaction Design." *Citeseer*. https://citeseerx.ist.psu.edu/document?repid=rep1&type=pdf&doi=7d3d22750e0d9d 4665454ac28ee1f0bcb61f21d4.

Wixon, Dennis, Karen Holtzblatt, and Stephen Knox. 1990. "Contextual Design: An Emergent View of System Design." In *Proceedings of the SIGCHI Conference on Human Factors in Computing Systems (CHI'90)*, pp. 329–36. New York: Association for Computing Machinery.

Wong, Richmond Y., and Deirdre K. Mulligan. 2019. "Bringing Design to the Privacy Table: Broadening 'Design' in 'Privacy by Design' Through the Lens of HCI." In *Proceedings of the 2019 CHI Conference on Human Factors in Computing Systems (CHI'19)*, pp. 1–17. New York: Association for Computing Machinery.

Wong-Villacres, Marisol, Neha Kumar, and Betsy DiSalvo. 2019. "The Parenting Actor-Network of Latino Immigrants in the United States." In *Proceedings of the 2019 CHI Conference on Human Factors in Computing Systems (CHI'19)*, pp. 1–12. New York: Association for Computing Machinery.

Wu, Chih-Hung, Yueh-Min Huang, and Jan-Pan Hwang. 2016. "Review of Affective Computing in Education/learning: Trends and Challenges." *British Journal of Educational Technology: Journal of the Council for Educational Technology* 47(6): 1304–23.

Young, Gareth, Hamish Milne, Daniel Griffiths, Elliot Padfield, Robert Blenkinsopp, and Orestis Georgiou. 2020. "Designing Mid-Air Haptic Gesture Controlled User Interfaces for Cars." *Proceedings of the ACM on Human-Computer Interaction*, 81, 4(EICS): 1–23.

Zagermann, Johannes, Ulrike Pfeil, and Harald Reiterer. 2016. "Measuring Cognitive Load Using Eye Tracking Technology in Visual Computing." In Proceedings of the Sixth Workshop on Beyond Time and Errors on Novel Evaluation Methods for Visualization (BELIV'16), pp. 78–85. New York: Association for Computing Machinery.

Zhang, Ping, and Dennis F. Galletta. 2015. *Human-Computer Interaction and Management Information Systems: Foundations.* London: Routledge.

Zhang, Qiwu, and Junru Zhu. 2022. "The Application of EMG and Machine Learning in Human Machine Interface." In *2022 2nd International Conference on Bioinformatics and Intelligent Computing (BIC 2022)*, pp. 465–69. New York: Association for Computing Machinery.

Zhang, Shikun, Yuanyuan Feng, and Norman Sadeh. 2021. "Facial Recognition: Understanding Privacy Concerns and Attitudes across Increasingly Diverse Deployment Scenarios." In *Proceedings of the Seventeenth USENIX Conference on Usable Privacy and Security* (SOUPS'21), pp. 243–62. USENIX Association.

Zheng, Caroline Yan, Cherie Lacey, and Mark Paterson. 2020. "Affect and Embodiment in HRI." In *Companion of the 2020 ACM/IEEE International Conference on Human-Robot Interaction (HRI'20)*, pp. 667–68. New York: Association for Computing Machinery.

Zimmerman, John, and Jodi Forlizzi. 2014. "Research through Design in HCI." In *Ways of Knowing in HCI*, edited by Judith S. Olson and Wendy A. Kellogg, pp. 167–89. New York: Springer.

Zimmerman, John, Jodi Forlizzi, and Shelley Evenson. 2007. "Research through Design as a Method for Interaction Design Research in HCI." In *Proceedings of the SIGCHI Conference on Human Factors in Computing Systems (CHI'07)*, pp. 493–502. New York: Association for Computing Machinery.

12 User Acceptance of Interactive Technologies

Andrina Granić

12.1 BACKGROUND AND A BRIEF HISTORY OF TECHNOLOGY ADOPTION

More than half a century of research has resulted in a large number of models and theories of technology acceptance and adoption. The proposed theoretical perspectives have been widely used to facilitate the assessment of various information and communication technology (ICT) that relate to different types of technologies, interactive systems, environments, tools, applications, services, and devices that significantly impact our daily lives in various ways. Understanding user acceptance and actual use of innovations and new technologies is one of the most extensive areas of research, which is a high priority for researchers and practitioners in the field. In this context, technology adoption at individual level (TechA@IL) is the focus of this chapter, as it is a mature branch of research aimed at understanding what constitutes the general acceptance of technological innovations.

In general, an *innovation* is "an idea, practice, or object perceived as new by the individual" (Rogers & Shoemaker, 1971, as cited in Zaltman & Lin, 1971, p. 657) and is one of the many things that can be perceived as new from an adopter's point of view. This position is very similar to the one that considers innovation as "any idea, practice, or material artifact perceived to be new by the relevant unit of adoption" (Zaltman & Lin, 1971, p. 656), where the unit of adoption can range from a single person to a "city" or "state". The broad problem of innovation adoption has been considered from different angles in the past, while in the ICT field, when it comes to interactive technology, it has been treated under the term innovation diffusion research (Rogers, 1962, 1995), information technology (IT) implementation (Cooper & Zmud, 1990), technology acceptance research (Davis, 1986, 1989; Venkatesh et al., 2003), personal computer (PC) utilization (Thompson et al., 1991), and more recently, information systems (IS) continuance research (post-adoption) (Bhattacherjee, 2001).

A major theoretical paradigm underlying research on individual adoption of ICT is rooted in the literature on adoption and diffusion of innovations, in particular Innovation Diffusion Theory (IDT) (refer also to Figure 12.1), which assumes that individual perceptions of how to use an innovation have a significant impact on users' adoption (Rogers, 1962, 1995; Moore & Benbasat, 1991). Other influential theoretical approaches that attempt to explain the relationship among users' perceptions, attitudes, and ultimate use of a technology include the Theory of Reasoned Action (TRA) (Fishbein & Ajzen, 1975), the Theory of Planned Behavior, TPB (), the Technology Acceptance Model, TAM (Davis, 1986; Davis et al., 1989), and the Augmented TAM, A-TAM (Taylor & Todd, 1995b), as well as the Unified Theory of Acceptance and Use of Technology (UTAUT) (Venkatesh et al., 2003). In this context, perceptions have been conceptualized differently; while TAM (Davis, 1986; Davis et al., 1989) postulates only two perceptions, Moore and Benbasat (1991) in their Perceived Characteristics of Innovation (PCI) refined the IDT model to include eight perceived characteristics of the use of an innovation. Besides, actual use/behavior, i.e., adoption, has been operationalized in various ways as the outcome that theories and models of technology acceptance and adoption attempt to explain. Some theoretical perspectives (e.g., TRA, TPB, TAM, A-TAM, and UTAUT) use behavioral intention as the dependent variable based on the assumption that usage intentions are predictors of future usage behavior, while other approaches predict individuals' actual behavior without considering intentions (e.g., IDT, PCI, Social Cognitive Theory (SCT) (Bandura, 1986), Model of Personal Computer Utilization (MPCU) (Thompson et al., 1991)).

DOI: 10.1201/9781003495109-12

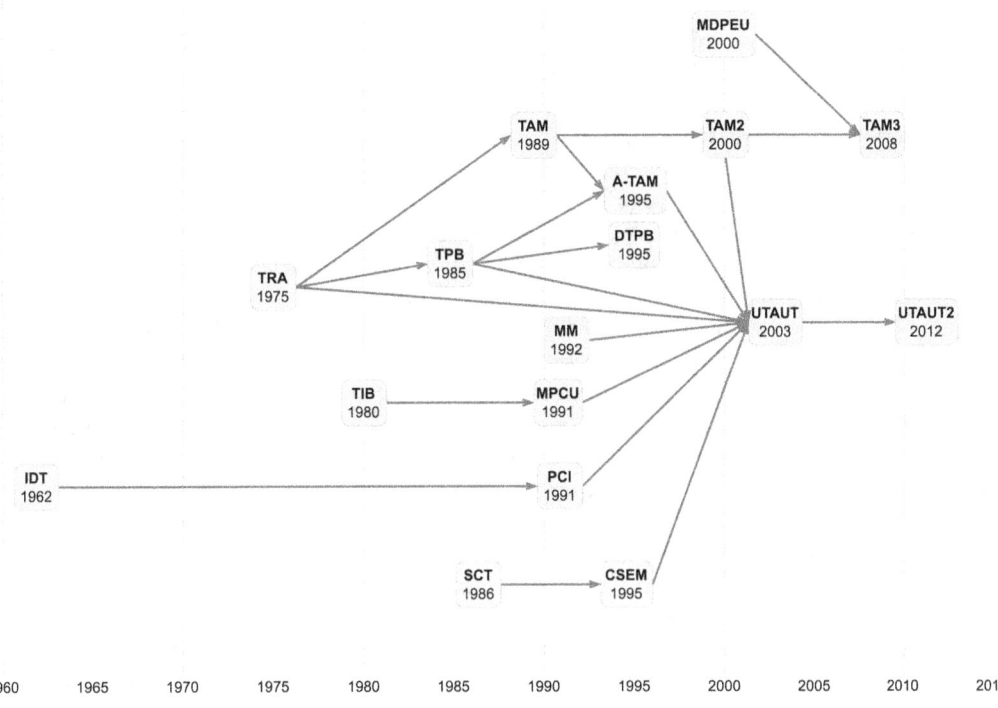

FIGURE 12.1 Chronological overview and relational linkages among influential theories and models of technology acceptance and adoption (Granić, 2023a; Davis & Granić, 2024).

Furthermore, the literature on technology adoption suggests that *initial acceptance behavior* (also referred to as *initial adoption* or *pre-adoption beliefs*) of a technology is primarily based on indirect experiences (an individual's perception and/or affect) with a particular technology, whereas *post-adoption* (also referred to as *continued* or *sustained use*) is based on past experiences (Agarwal & Prasad, 1997; Karahanna et al., 1999). Thus, *long-term engagement* and *sustained use in the future* are primarily driven by rational considerations when users have concrete knowledge about the technology based on their experience and are able to consciously identify and articulate the benefits offered by the innovation (*ibid.*). Despite differences in theoretical formulations and constructs, however, all of these lines of research seek to understand and explain what makes new technology well accepted by adopters. As mentioned earlier, the key to adoption is that the adopter (individual or organization) must perceive the idea, behavior, or product as new and innovative, so technology adoption can refer to either organizational adoption or individual adoption. It is important to emphasize that this chapter focuses on individual adoption, while post-adoption approaches are outside the scope of this chapter.

12.1.1 Advances in Technology Adoption at the Individual Level TechA@IL

12.1.1.1 Leading Theories on Technology Adoption at the Organizational Level

Leading theories describing the adoption of technological innovations at an organization level presented in this section include (in chronological order): *Innovation Diffusion Theory* (IDT) by Rogers (1962, 1995), *Model of the Information Technology (IT) Implementation Process* by Cooper and Zmud (1990), *Technology, Organization and Environment (TOE) Framework* by Tornatzky and Fleischer (1990), *Electronic Data Interchange (EDI) adoption model* by Iacovou, Benbasat, and Dexter (1995), and *Model of Acceptance with Peer Support* (MAPS) by Sykes, Venkatesh, and Gosain (2009).

Innovation Diffusion Theory (IDT) is one of the most popular models for studying the adoption and use of innovations by individuals and organizations, introduced by Rogers (1962, 1995). Diffusion is defined as "the process by which an innovation is communicated through certain channels over time among the members of a social system" (Rogers, 1995, p. 5). In this context, "innovation" is one of the various things that can be perceived as new from the adopter's perspective and can be described by five characteristics: relative advantage, compatibility, complexity, traceability, and observability. IDT assumes that an adopter (individual or organization) first learns about an innovation, then develops a positive or negative attitude toward it, and performs actions that lead to the adoption or rejection of the innovation before being convinced to use the innovation.

In Cooper and Zmud's (1990) *Model of the Information Technology (IT) Implementation Process*, IT implementation is defined as "an organizational effort directed toward diffusing appropriate information technology within a user community" (*ibid.* p. 124). Their model represents a variation of the stage model of IT implementation activities proposed by Kwon and Zmud (1987), a model based on the literature on organizational change, innovation, and technological diffusion. In addition, a well-accepted Cooper and Zmud's model, which also considers some of the post-adoption behaviors, includes six stages, namely initiation, adoption, adaptation, acceptance, routinization, and infusion. Accordingly, the model covers an implementation process ranging from review of organizational requirements to the successful use of the technology in regular practice.

Recognizing that innovation adoption is significantly influenced by the technological, organizational, and environmental context within an organization, Tornatzky and Fleischer (1990) postulated the *Technology, Organization and Environment (TOE) Framework*. The TOE framework describes the process organizations go through in adopting and implementing technological innovations and remains one of the most well-known and widely used theories of organizational adoption. It identifies three aspects of the organizational context that influence this process, specifically technological, organizational, and environmental, which thus represent both constraints and opportunities for technological innovation.

Iacovou, Benbasat, and Dexter (1995) introduced a generic adoption framework called *Electronic Data Interchange (EDI) adoption model* and described the characteristics that influence organizations in adopting IT innovations. According to Iacovou and colleagues, there are three main determinants of innovation adoption: the perceived benefits the organization wants to achieve (technical), organizational readiness (organizational), and external pressures on the organization (reactive).

Sykes et al. (2009) integrated constructs from individual-level technology adoption research and social network constructs to develop the *MAPS*. The authors theorized that "individual's embeddedness in the social network of the organizational unit implementing a new technology can enhance their understanding of technology use" (Sykes et al., 2009, p. 371). Indeed, in such a context, an individual's coworkers can be very helpful in overcoming problems that impede the adoption and use of an organization's technological innovation.

12.1.1.2   Chronological Overview of the Most Influential Approaches of TechA@IL

With regard to technology adoption at the individual level (TechA@IL), many theories and models have been proposed that aim to explain the main dependent constructs of interest, particularly the behavioral intention to use a specific interactive technology and its actual use. *Behavioral intention* refers to an individual's subjective likelihood of performing a particular behavior (Fishbein & Ajzen, 1975), while *system use*, possibly conceptualized as duration, frequency, and intensity (Venkatesh et al., 2008), exists as the ultimate dependent variable in models of technology adoption (Davis et al., 1989; Venkatesh et al., 2003). This subsection provides a brief overview of the most influential theories and models of technology adoption at the individual level, while the following sections present some arbitrarily selected hybrid models in TechA@IL and illustrative research examples of the application of TechA@IL theories and models in a wide range of contexts.

Figure 12.1 provides a chronological representation and an illustrative overview of the relational linkages among the major theoretical perspectives on technology acceptance and adoption, thus shedding light on more than half a century of research in the field of TechA@IL.

As emphasized earlier, *Innovation Diffusion Theory (IDT)* (Rogers, 1962, 1995) is an accepted theory that attempts to explain how and why new technological advances spread through societies and cultures, from introduction to widespread adoption at the individual and organizational levels.

To overcome the lack of a theoretical foundation and the inadequate definition and measurement of constructs, Moore and Benbasat (1991) focused on measuring the perceptions of potential adopters of IT. Perceptions of adoption are initially based on the five characteristics of innovation derived by Rogers in his IDT. While Rogers' definitions are based on the perception of the innovation itself, the proposed refined IDT, called *Perceived Characteristics of Innovating (PCI)*, changes the focus to the perception of the use of the innovation (Moore & Benbasat, 1991). Thus, in PCI, all characteristics are redefined in terms of potential users' use, trial, or observation of the innovation, and eight perceived characteristics emerge: voluntariness, image, relative advantage, compatibility, ease of use, result demonstrability, trialability, and visibility.

Social Cognitive Theory (SCT), Bandura's (1986) widely accepted, empirically validated model of individual behavior, is one of the most powerful theories of human behavior (Venkatesh et al., 2003). Bandura used a social cognitive framework to explore the dynamic interplay of affect, thought, and action in bringing about personal and social change. SCT is an interactive model of causation in which environmental influences (such as social pressures or particular situational characteristics), cognitive and other personal factors (including personality and demographic characteristics), and behavior act as interacting determinants of one another. This relationship is referred to as "triadic reciprocality" (Bandura, 1986).

Compeau and Higgins (1995) applied and extended Bandura's SCT to the context of computer use and introduced the *Computer Self-Efficacy Model (CSEM)*, a model that supports the SCT perspective on computer behavior. The model incorporates elements of all three SCT forces, namely environmental, cognitive, and behavioral. It has been found that computer self-efficacy, i.e., an individual's belief in his or her ability to use computers competently, plays an important role in shaping the individual's feelings and behaviors. Individuals with high self-efficacy beliefs use computers more often, enjoy using them more, and are less fearful of computers. In addition, outcome expectations, particularly related to work performance, have a significant impact on affect and directly influence computer use. Finally, affect and anxiety also have a significant effect on computer use and thus have a direct influence on use. According to Venkatesh et al. (2003), the theory underlying CSEM enables the extension of the model to predict individual adoption and use of IT in general.

Theory of Interpersonal Behavior (TIB) was established by Triandis (1980), who recognized the key role of social and affective factors in the formation of behavioral intentions. It also recognizes the importance of past behaviors (habits) in current behavior. Based on these observations, Triandis postulates that behavior is a function of intention, along with the strength of habit (past behavior) and various facilitating conditions. Behavioral intention is influenced by social factors, affect, and perceived consequences. Social factors include norms, roles, and self-concept, while the construct of affect encompasses all emotional reactions (positive and negative) to a decision and is thus distinct from the rational evaluations of consequences represented by the construct of perceived consequences.

Model of Personal Computer Utilization (MPCU) (Thompson et al., 1991) is largely derived from Triandis' TIB. Thompson and colleagues adapted Triandis' model to the context of IS and used the model to predict individual acceptance and use of PC. However, the model itself is useful for predicting acceptance and adoption of a variety of information technologies (Venkatesh et al., 2003). Because Thompson and colleagues wanted to predict actual behavior (PC utilization) rather than intention, behavioral intention is excluded from the proposed model. In addition, habits are also excluded because, according to the authors, in the context of PC use, past use has a redundant relationship to current use. Factors that influence the use of PC include consistency with the job, complexity of use, long-term consequences of use, affect toward use, social factors that influence use, and facilitating conditions for use of PC.

Assuming that individuals are generally rational and use available information systematically, Fishbein and Ajzen (1975) developed the *Theory of Reasoned Action (TRA)* to predict and understand behavior and attitudes. TRA considers behavioral intentions rather than attitudes as the most important predictors/antecedents of behavior. According to TRA, the principal predictor of an individual's behavior is his or her behavioral intention, while the two determinants of an intention are the individual's attitude toward performing the behavior (attitude) and the perceived social influence of people who matter to the individual (subjective norm). Behavioral intention is usually defined as an individual's subjective likelihood that he or she will perform a particular behavior, attitude refers to an individual's degree of evaluative affect toward the target behavior, while subjective norm refers to the individual's perception that the vast majority of people significant to the individual believe that he or she should or should not perform the behavior in question (Fishbein & Ajzen, 1975).

Ajzen's *Theory of Planned Behavior (TPB)* (Ajzen, 1985, 1991) represents an extension of Fishbein and Ajzen's TRA (1975) to account for situations in which individuals do not have complete voluntary control over their behavior. TPB assumes that behavior is a direct function of intention (which is consistent with other models of behavior) as well as perceived behavioral control. Behavioral intention is formed by one's attitude and subjective norm, constructs that are the same as in TRA, along with perceived behavioral control, a third construct that reflects perceptions of internal and/or external constraints and limitations on behavior (Ajzen, 1985, 1991). In addition, each of the predictors of intention is in turn determined by the underlying belief structures, specifically attitudinal beliefs, normative beliefs, and control beliefs, respectively.

Decomposed Theory of Planned Behavior (DTPB) introduced by Taylor and Todd (1995a) is a version of Ajzen's TPB model, which similarly proposes the same three predictors of behavioral intention. However, in DTPB, the uniform belief structure underlying each antecedent (specifically, attitude, subjective norms, and perceived behavioral control), representing a variety of dimensions, is decomposed into a multidimensional belief construct. Indeed, decomposing beliefs allows for clearer and more understandable relationships among predictors of intention. Thus, DTPB should be able to provide a more comprehensive understanding of technology use.

Technology Acceptance Model (TAM) (Davis, 1986, 1989; Davis et al., 1989), one of the most significant adaptations of TRA, assumes two main beliefs, perceived usefulness and perceived ease of use, determine an individual's attitude toward use as well as actual use of a system. In his dissertation version of TAM, Davis (1986) postulated that an individual's attitude toward use is a key determinant of actual use or rejection of a system/technology. To appreciate the importance of intentions, Davis et al. (1989) version of TAM shapes usage behavior (i.e., actual system usage) as a direct function of behavioral intention. Behavioral intention is in turn influenced by attitude toward use, which reflects positive or negative feelings toward technology use, and perceived usefulness, which reflects the belief that technology use will improve performance. Attitude is directly influenced by perceived usefulness and perceived ease of use. The introduction of intentions as a mediating variable in the model is important for both substantive and pragmatic reasons (Taylor and Todd, 1995a). The formation of an intention to perform a behavior is assumed to be a necessary precursor to behavior (Fishbein & Ajzen, 1975). It has been observed that the inclusion of intention increases the predictive power of models such as TRA and TAM compared to models that do not include intention (Taylor & Todd, 1995a). Overall, TAM parsimony has been shown to be a valid and powerful approach to explaining technology acceptance that is applicable to a variety of technologies and contexts (Davis & Granić, 2024; Al-Emran & Granić, 2021).

TAM does not consider the influence of social and control factors on behavior, variables that have been shown to have a significant impact on usage behavior. These variables are also important determinants of behavior in TPB (Ajzen, 1991), where social influences (subjective norm) are modeled as determinants of behavioral intention, and perceived behavioral control is modeled as a determinant of both behavioral intention and behavior. As a result, Taylor and Todd (1995b) added subjective norm and perceived behavioral control to TAM to provide a more complete test of the important determinants of technology use. The complete model was termed the *Augmented TAM (A-TAM)* (Taylor and Todd, 1995b), which is sometimes referred to in the literature as the Combined TAM and TPB (C-TAM-TPB).

Extended TAM (TAM2) (Venkatesh & Davis, 2000) is a theoretical extension of TAM that includes subjective norm as an additional predictor of intention in the case of mandatory settings. With the goal of improving the predictive power of perceived usefulness, Venkatesh and Davis, using TAM as a starting point, proposed TAM2, an enhanced model that includes additional theoretical constructs linking social influence (subjective norm, voluntariness, and image) and cognitive instrumental processes (job relevance, output quality, and result demonstrability) (Venkatesh & Davis, 2000).

While TAM2 identifies general determinants of perceived usefulness (*ibid.*), general determinants of perceived ease of use were identified by Venkatesh in *Model of the Determinants of Perceived Ease of Use (MDPEU)* (Venkatesh, 2000). Two main sets of antecedents to perceived ease of use are identified: anchors and adjustments. Anchors are considered to be general beliefs about computers and computer use, whereas adjustments are considered to be beliefs formed based on direct experience with the target system. For both groups, some determinants are suggested from previous research identifying predictors of perceived ease of use (Davis et al., 1992).

To understand how different interventions may affect the known determinants of technology adoption and use, Venkatesh and Bala (2008) combined TAM2 and MDPEU and developed an integrated model known as *Technology Acceptance Model 3 (TAM3)*. TAM3 provides a comprehensive network of determinants of perceived usefulness and perceived ease of use (i.e., individual technology adoption and use). It implies that experience plays a crucial role in shaping the relationships between different constructs in the technology adoption process. Specifically, it suggests that experience will moderate the relationships between perceived ease of use and perceived usefulness, computer anxiety and perceived ease of use, as well as perceived ease of use and behavioral intention. Increased experience strengthens the relationship between perceived ease of use and perceived usefulness, reduces computer anxiety, and enhances perceived ease of use, as well as reinforces the influence of perceived ease of use on behavioral intention. These moderating effects highlight the importance of user experience and skill development in shaping perceptions and intentions toward technology adoption and use.

Because the influence of enjoyment on intentions to use has not been previously studied, Davis, Bagozzi, and Warshaw (1992) applied motivational theory to study the adoption and use of new technologies. The *Motivational Model (MM)*, based on the psychological aspects of technology acceptance, assumes that individuals' behavior and system use are influenced by intrinsic and extrinsic motivation. Intrinsic motivation arises from the user's internal drive to perform an activity without obvious reinforcement and is related to feelings of pleasure and satisfaction. In contrast, extrinsic motivation arises when the user wants to perform an activity because he or she perceives it as helpful in achieving goals other than the activity itself. An example of intrinsic motivation is enjoyment, while usefulness can be considered an example of extrinsic motivation.

Venkatesh et al. (2003) reviewed existing theories and models of new technology acceptance and proposed *Unified Theory of Acceptance and Use of Technology (UTAUT)* by reviewing and integrating eight prominent user acceptance models, specifically TRA, TAM2 (extended TAM), MM, TPB, A-TAM, MPCU, PCI (refined IDT), and CSEM (extended SCT). The UTAUT model is intended to capture the essential elements of these eight already established models. Following the UTAUT examination, Venkatesh and colleagues (2003) proposed three main constructs that directly determine behavioral intention, specifically performance expectancy, effort expectancy, and social influences. In addition, behavioral intention and facilitating conditions are considered predictors of actual behavior (use). Although UTAUT has been criticized for the excessive number of independent variables used to predict intentions and behavior (Bagozzi, 2007), it is considered quite robust in assessing and predicting technology acceptance compared to other models (Venkatesh et al., 2003).

Venkatesh et al. (2012) extended the original UTAUT and developed the *Extended UTAUT (UTAUT2)*, paying particular attention to the context of consumer use. To formulate UTAUT2, they added three new constructs that directly influence behavioral intention, namely hedonic motivation, price value, and habit. Moreover, in addition to behavioral intention and facilitating conditions, the model postulated habit as an additional predictor of use. Venkatesh and colleagues hypothesized that individual differences (age, gender, and experience) will moderate the relationships between the constructs and behavioral intention and actual behavior. The hypothesis regarding the moderating effects of age, gender, and experience on the relationships between the constructs and behavioral intention was supported (*ibid.*). Age, gender, and experience were found to influence how the constructs within the UTAUT2 model impact behavioral intention and actual behavior in the context of consumer use. However, the specific nature and magnitude of these moderating effects can vary depending on the particular context and technology being studied.

12.1.2   Hybrid Models in TechA@IL

In an effort to provide a comprehensive overview of the current state of research in TechA@IL, this section presents several theoretical hybrid approaches. To increase the explanatory power of technology adoption in different application domains, single models/theories have been integrated with other models of technology adoption (such as TAM, TPB, UTAUT, SCT; TR) and post-adoption (such as Information Systems Success Model (ISSM), Expectation–Confirmation Theory (ECT)), as well as with a number of theories/models from other disciplines, such as Flow Theory (FT), Protection Motivation Theory (PMT), Push-Pull-Mooring (PPM) model, and Uses and Gratifications Theory (UGT), among many others. This research revealed that TAM (Davis, 1986, 1989), which has already proven to be a powerful model applicable to a variety of technologies and contexts at the individual level (see also some recent reviews in the field of technology adoption (Al-Emran & Granić, 2021; Granić, 2022; Salahshour Rad et al., 2018; Marangunić & Granić, 2015)), has proven to be successful in predicting user acceptance when combined with a variety of other theoretical viewpoints.

A number of indicative hybrid approaches and relevant research examples from a variety of fields are presented below and illustrated in Figure 12.2. To describe the underlying rationale driving the selection of the presented approaches, the exploration, innovation, and goal of generating a comprehensive overview of the field are considered. By drawing from diverse perspectives, the aim is to offer a more holistic and comprehensive understanding of the adoption phenomenon. By including a range of theoretical hybrid approaches, the intention is to capture the diversity and richness of the field. The selection also aimed to reflect current trends in the field, where researchers are increasingly exploring interdisciplinary connections and blending theories from various domains to address the complex nature of technology adoption. This observation highlights the continued relevance and influence of TAM in the TechA@IL field, even when combined with other models or theories. While the presented works reflect an arbitrary selection, it is important to acknowledge that alternative contributions and relevant research exist. The arbitrary approach serves as a starting point for exploration, recognizing that different selections might yield different perspectives and interpretations.

- TAM (Davis, 1986, 1989), a widely used powerful model of individual-level technology adoption, is combined with UTAUT (Venkatesh et al., 2003) as another rich theoretical approach to explaining behavioral intentions and technology use, to explore predictive factors that influence pre-service teachers' intentions to use learning management system (LMS) in developing countries (Buabeng-Andoh & Baah, 2020).
- TPB (Ajzen 1985, 1991), which states that behavior is a direct function of behavioral intention and perceived behavioral control, is used with TAM to identify aspects that affect the adoption of mobile learning at the university level and to explain how perceptions influence the adoption of m-learning among students (Gómez-Ramirez et al., 2019).
- TPB is again used with TAM to examine the relationship between the hybrid model and the adoption of communication platforms (Zoom) in higher education in developing countries (Ly et al., 2023).
- TPB and TAM are integrated into the study that developed an extended TAM with a TPB model to predict and explain customers' behavioral intentions toward online banking adoption (Lee, 2009).

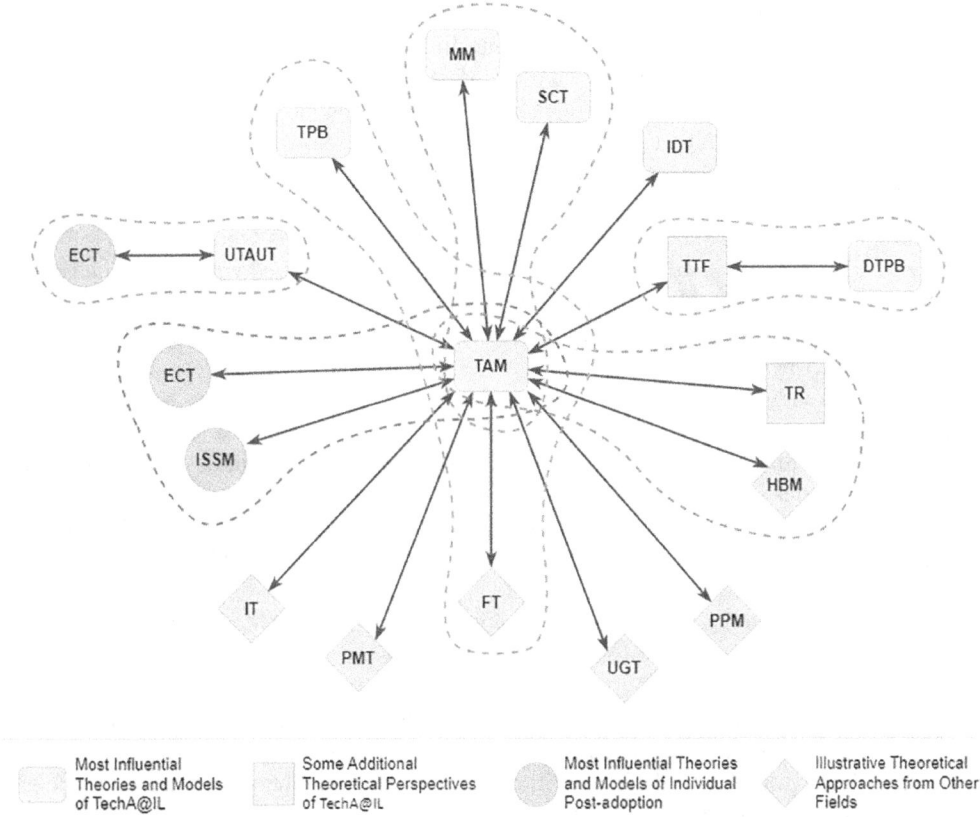

FIGURE 12.2 Hybrid models in TechA@IL for predicting user acceptance of interactive technologies across application domains.

- IDT (Rogers, 1962, 1995), one of the most popular models for studying the adoption and use of innovations at both individual and organizational levels, is integrated with TAM to explore the diffusion and adoption of an open source learning platform (OSLP) among university students, as OSLPs are a dominant component of education today (Huang et al., 2020).
- IDT and TAM are again combined in a research framework that aims to investigate the factors that affect the adoption of products based on haptic technology, such as driving simulation games or home entertainment devices with haptic technology (Jongchul & Sung-Joon, 2014).
- Task-Technology Fit (TTF) (Goodhue & Thompson, 1995), a model that explains the correlation between individuals' performance and information technologies, is integrated with TAM to propose a unified model that provides a more comprehensive understanding of behaviors related to intentions to continue using Massive Open Online Courses (MOOCs) (Wu & Chen, 2017).
- DTPB (Taylor and Todd, 1995a) and the TTF concept are used as a research framework to explain students' adoption intention with respect to brain-computer interface (BCI) games in a learning context (Wang et al., 2022).
- SCT (Bandura, 1986), one of the strongest theories of human behavior, is integrated with TAM, as well as MM (Davis et al., 1992), which posits that individuals' behavior and use of technology are influenced by intrinsic and extrinsic motivation, in a study designed to examine the factors that influence students' intentions to continue using blogs for learning (Ifinedo, 2017).

- Technology Readiness (TR) (Parasuraman, 2000), a multiple-item scale measuring readiness to adopt new technologies (through four dimensions: optimism and innovativeness, as drivers of technology readiness, and discomfort and insecurity as barriers), is combined with TAM to explain the adoption of Industry 4.0 based on various technologies such as big data, cloud, industrial internet, simulation, augmented reality, additive manufacturing, cybersecurity, advanced manufacturing, and the like (Castillo-Vergara et al., 2022).
- UGT, a theory of mass communication that focuses on the needs, motivations, and gratifications of media users and was first introduced in the 1940s when scholars began to study why people choose to consume different forms of media, is applied with TAM to identify key influencers of wearables adoption (Travers, 2015).
- PPM model (Moon, 1995), a prevailing paradigm first used in the study of human migration that describes why people move from one place to another for a period of time was used concurrently with TAM to build a conceptual model of the factors influencing the intention to use a metaverse educational application platform (Wang & Shin, 2022).
- FT (Csikszentmihalyi, 1975), a theory from positive psychology that describes a state in which a person is so involved in an activity that nothing else seems to matter, is combined with TAM in a theoretical framework designed to explain users' attitudes and behavioral intentions in virtual reality (VR) surfing experience and its adoption in leisure and tourism (Huang et al., 2023).
- FT is again integrated with two other theoretical perspectives, TAM and TPB, to examine the intrinsic and extrinsic motivations that influence users' acceptance of instant messaging (Lu et al., 2009).
- Health Belief Model (HBM), one of the best known and most widely used theories in health behavior research developed by social psychologists in the 1950s, is integrated with TR and TAM to understand the determinants of users' intention to continue using food delivery applications (Silva et al., 2022).
- PMT, a theory postulated by Rogers (1975) to better understand fear appeals and considered a special case of a more general category of theories that use expectancy and value constructs, is combined with TAM to study students' behavioral intentions to adopt wearable technologies in learning activities, specifically smartwatch devices (Al-Emran et al., 2021).
- Interactivity Theory (IT), which focuses on the study of feedback as a concept and recognizes that new media tools have changed the way people communicate, is combined with TAM to examine the link between interactive features and personal cognition and to test key factors that affect users' intention to continue using mobile banking applications (Yin & Lin, 2022).
- ISSM (DeLone & McLean, 1992), as an influential and robust theoretical foundation for the study of technology post-adoption, is combined with TAM to examine students' behavioral intentions to use social media, particularly their perceptions of academic performance and satisfaction (Al-Rahmi et al., 2021).
- ISSM and TAM have also joined to understand and study the factors that influence consumer adoption and continued use of sports branded apps (Won et al., 2022).
- ECT (Oliver, 1980), a leading cognitive theory in the field of consumer satisfaction that seeks to explain post-adoption satisfaction as a function of expectations, perceived performance, and disconfirmation of beliefs, is used in a study that extends UTAUT with the goal of investigating students' perspectives on the acceptance of m-learning in higher education (Alowayr & Al-Azawei, 2021).
- ECT is also integrated with two other approaches, specifically TAM along with ISSM, to examine and identify several factors as likely predictors of continuance intention in an e-learning context (Roca et al., 2006).

12.1.3 Applying TechA@IL Theories and Models to a Wide Range of Contexts

To test their applicability, on the one hand, and to increase their predictive power, on the other hand, original models and theories, their extensions and modifications, as well as hybrid theoretical approaches have been widely used to facilitate the assessment of various interactive technologies. Due to the extensive body of work worldwide, this account can only give a representative sample of numerous studies in various domains of application, which undoubtedly cannot be all-inclusive. Research has revealed that the most studied domains are social media, the workplace, business, healthcare, and the extensively studied area of education. Since it would be practically impossible to provide an in-depth discussion of each area in this chapter, the goal of this section is not to carry out a thorough analysis of each field, but to provide a high-level discussion and refer the reader to exemplary research in each field. Therefore, related (systematic) reviews are presented first, followed by illustrative primary studies that have been conducted in the aforementioned contexts.

12.1.3.1 Education

In addition to providing a holistic view of the key factors that affect and reliably predict the adoption of educational technologies in general, Granić's (2022) systematic review also provided insights into the types of technologies and participants, the modes of delivery, and widely used theories and models of technology adoption and acceptance. On the other hand, Sneesl and colleagues (2022) conducted a systematic review that considered only higher education institutions (HEIs) and found that Internet-of-Things (IoT) technologies lead to widespread adoption of a smart campus concept and facilitate the adoption process for HEI administrators and decision-makers. Liu and colleagues (2020), focusing on academics' adoption of learning technologies, systematically examined the literature and expanded the term "technology" to include types of innovative teaching practices that use learning technologies in the context of HEI.

Several reviews summarizing empirical research in education have focused on specific models/theories, mainly TAM and UTAUT. Aiming to present the current state of research on the application of TAM in education, Granić and Marangunić (2019) systematically reviewed the literature on TAM in learning and teaching, considering a variety of learning domains, learning technologies, and user types. More recently, while focusing on higher education, Rosli and colleagues (2022) conducted a systematic review to gain insight into the evolution of TAM throughout the pandemic, such as an understanding of the prevailing external variables for TAM and the technologies studied. Regarding mobile learning (m-learning) as a validated mode for delivery, Aytekin and colleagues (2022) presented a systematic review of m-learning adoption in the context of UTAUT, while Al-Emran and colleagues (2018) presented a systematic review and synthesis of relevant TAM studies related to m-learning from 2006 to 2018.

Considering the primary studies, new research has found that TAM is the most influential model in studying the acceptance of educational technology by students, teachers, and other stakeholders (Granić, 2022, 2023b). For example, while the extended UTAUT (UTAUT2) has been used to evaluate the acceptance of blended learning in executive student education (Dakduk et al., 2018), and to explore pre-service teachers' acceptance of learning management software (Raman & Don, 2013), numerous studies have provided empirical evidence of the predictive validity of TAM, with recent research including electronic learning (Prasetyo et al., 2021), mobile learning (Lai, 2020), Virtual Reality Environments (VLEs) (Fussell & Truong, 2021), MOOCs (Al-Adwan, 2020), and LMSs (Dampson, 2021). In addition, several acceptance studies have explored the applicability of TAM for various assistive technologies in education, from social media platforms (Al-Rahmi et al., 2021) to learning support technologies through teaching assistant robots (Park & Kwon, 2016), simulators (Lemay et al., 2018), VR (Lin & Yeh, 2019), and augmented reality technologies (Jang et al., 2021), among many others.

12.1.3.2 Health

With the advancement of all types of technology, coupled with concerns for patient safety, efficient healing, and well-being, healthcare technology is gaining more and more attention. The systematic review conducted by AlQudah and colleagues (2021) provided insight into the impact of technologies and services in healthcare and examined the studies that empirically assessed the various technologies in healthcare in terms of technology acceptance models and theories. On the other hand, Kavandi and Jaana (2020), in their systematic review focusing only on the elderly, uncovered factors that significantly influence seniors' adoption of various health information technologies. In their review, Nadal and colleagues (2020) examined how technology acceptance is interpreted and measured in the mobile health (m-health) literature. While Cruz and colleagues (2020) in their systematic review examined instruments, approaches, scales, and assessment tools used to evaluate technology acceptance and adoption of ICT for people with dementia, the identification of methodological tools and approaches for measuring acceptance of ambient assisted living (AAL) in rehabilitation research was the focus of the literature review conducted by Choukou and colleagues (2021). Due to the recent development of a variety of sports applications and the resulting need to assess consumer behaviors related to their use, Angosto and colleagues (2020) systematically reviewed the literature on consumer intentions to use mobile applications for fitness and physical activity. Finally, based on the theoretical framework of TAM, Tao and colleagues (2019) summarized existing studies that empirically examined user acceptance of consumer-oriented health information technologies in a systematic review. Considering primary studies that address health technology adoption, they examined users' intentions to accept and use electronic health records (Razmak & Bélanger, 2018), health service delivery (Dhaggara et al., 2020), telehealth technology (Chau & Hu, 2002), clinical IS (Pai & Huang, 2011), and mobile health apps for lifestyle and therapy (Schomakers et al., 2022).

12.1.3.3 Social Media

The social media landscape is rapidly evolving, driven by advances in digital technology and the human need to communicate and interact. However, when it comes to social media adoption and use in general, there is a lack of comprehensive literature reviews. This is rather surprising given the popularity and high user acceptance of social media, which have enabled its spread to all levels of society. Despite targeting also applied educational theories, Al-Qaysi and colleagues (2020a) analyzed IS used to study social media acceptance and adoption in their systematic review. The latest systematic literature review of disciplines, applications, and influential factors regarding social media adoption in education by Al-Qaysi and colleagues (2023) focused on TAM as the basic theoretical model. Since TAM has been the common theoretical approach regarding user adoption of social media, the main objective of two earlier systematic reviews was to identify studies based on TAM that observed what exactly leads to user acceptance of the technology underlying social media (Al-Qaysi et al., 2020b; Wirtz and Gottel, 2016). In addition, the rapid development of new platforms for disseminating and sharing information has influenced primary research looking at social network adoption (Makki et al., 2017; Chang and Chen, 2018) and instant messaging (Lu et al., 2009).

12.1.3.4 Workplace

An overview of the aspects that influence the adoption and use of technology in organizations is provided in the review conducted by Saghafian and colleagues (2021). With the aim of understanding the need for integrated models for technology adoption, an earlier study reviewed studies published between 2000 and 2012 on the adoption of technologies such as the Internet, e-commerce, enterprise resource planning (ERP), radio frequency identification (RFID), EDI, and knowledge management at the enterprise level (Gangwar et al., 2014). Regarding only small and medium enterprises (SMEs), the new systematic review of the literature conducted by Zamani (2022) over the last decade has revealed several categories of influential approaches to technology adoption in SMEs, while Chouki and colleagues (2020) presented a systematic literature review to gain a better understanding of the barriers that hinder the adoption of IT in SMEs. Finally, to understand what factors lead users to reject or adopt different types of intelligent environments, FakhrHosseini and colleagues (2022) reviewed nine prominent theories and models of technology adoption and discussed their core constructs and the extent to which they predict user behavior. A number of primary studies were influenced by the fact that jobs are highly dependent on IT and its ever-expanding capabilities. For example, the general use of IT in the workplace in terms of technology acceptance was examined by McFarland and Hamilton (2006), the acceptance of management software was explored by Hernández and colleagues (2008), a specific ERP technology was analyzed by Bueno and Salmeron (2008), while the customer relationship management information system was the focus of research by Tung and colleagues (2009).

12.1.3.5 Business

As the transmission of funds or data and the buying and selling of goods and services increasingly occur online, there is growing interest in conducting primary research that addresses the ways in which technology regulates the business world. For example, the adoption of online banking was studied by Ahmad and Al-Zubi (2016), while mobile banking was the focus of research conducted by Sharma and Sharma (2019). In addition, consumer behavior in e-commerce was researched by Huseynov and Özkan Yıldırım (2019). While Gu et al. (2019) focused on mobile shopping in their study, Kınış and Tanova (2022) conducted a more recent study to investigate the determinants of mobile e-wallet adoption.

12.1.3.6 Broad Range of Context

Predicting technology adoption and use through replication and application of original theories and models, as well as their extensions and modifications, has informed numerous primary studies conducted in a wide range of contexts, most notably public acceptance of autonomous vehicle technologies (Hewitt et al., 2019), travelers' usage intentions for in-vehicle global positioning system products (GPS) (Chen & Chen, 2011), sensory technologies in the online shopping process (Kim & Forsythe, 2008), end-user response to biometric authentication systems (Kanak & Sogukpinar, 2017), gaming adoption in mobile social networks (Chen et al., 2017), young consumers' smartphone usage intentions (Rigopoulou et al., 2017), the adoption of products based on haptic technologies (Jongchul & Sung-Joon, 2014), the acceptance of augmented reality (AR) and wearable technologies in knowledge-intensive training (Guest et al., 2018), the acceptance of wearable tracking systems for passengers to ensure safety (Kwee-Meier et al., 2016), and the adoption of wearable devices in general, which is a meaningful milestone for predicting the future development of IT devices (Chang et al., 2016). Finally, some recent studies addressed different individual technologies. While Billanes and Enevoldsen (2021) provided an overview of the likelihood of acceptance of various technologies such as smart energy technologies (e.g., smart meters, smart thermostats), mobile payments, smart glasses, autonomous/driverless vehicles, and IT in general, the study by Guner and Acarturk (2020) examined the use and acceptance of ICTs, particularly computers, the Internet, and cell phones, by the elderly.

12.2 FACTORS INFLUENCING USER DECISION ON TECHNOLOGY ADOPTION

12.2.1 Core Predictors for the Behavior and Actual Adoption of TechA@IL

As shown in the previous section, 17 core theoretical perspectives in TechA@IL (visualized in Figure 12.1) emerged from the research and have been very well accepted by the research community and practitioners in TechA@IL over the past decades (Granić, 2023a). Most of them postulate behavioral intention as a dependent construct (except IDT, PCI, SCT, CSEM, and MPCU). Behavioral intention simply refers to what individuals intend to do and is usually defined as an individual's subjective likelihood of performing a particular behavior (Fishbein & Ajzen, 1975). Triandis's (1980) TIB model posits that social factors, affect, and perceived consequences influence behavioral intentions, which in turn influence behavior. As emphasized earlier, few of the models presented do not consider intention to use but rather aim to predict the individual's actual behavior, notably IDT (Rogers, 1962, 1995) and PCI as its refined version (Moore & Benbasat, 1991), SCT (Bandura, 1986) and CSEM as its extension (Compeau & Higgins, 1995), and MPCU (Thompson et al., 1991).

Consistent with the theory underlying all intention models, the remaining models presented model usage behavior as a direct function of behavioral intention. This dependent construct is referred to as *actual system use* (e.g., TAM (Davis, 1989; Davis et al., 1989), MDPEU (Venkatesh, 2000)), *behavior* (e.g., TRA (Fishbein & Ajzen, 1975), A-TAM (Taylor & Todd, 1995b)), and *use/ usage behavior* (e.g., DTPB (Taylor & Todd, 1995a), TAM2 (Venkatesh & Davis, 2000), UTAUT (Venkatesh et al., 2003)).

Elaborating on Eason's three-decade-old classical ergonomic framework of human, machine, task, and environment (Eason, 1991), a *Triad of Predictors* is proposed to influence behavioral intention and cause actual technology adoption (Granić, 2023a). Three dimensions of influence are identified, so the postulated, existing constructs (core predictors) influencing TechA@IL are thematically categorized as follows:

- *User aspects* encompass specific traits or qualities of individual users that can influence their behaviors related to technology acceptance and adoption (see Table 12.1),
- *Task and technological aspects* include predictors related to both technological features and task-related factors within the context of technology acceptance and adoption (see Table 12.2), and
- *Social and environmental aspects* embrace predictors attributed to social and environmental factors within the realm of technology acceptance and adoption (see Table 12.3).

The tables presented provide an understanding of the *Triad of Predictors* of behavioral intention and/or adoption of TechA@IL according to the frequency of their use in the postulated most influential theories and models (visualized in Figure 12.1). It can be observed that *subjective norm(s)* is the most frequently used core predictor postulated as a predictor of social and environmental aspects in six theories/models. Other commonly studied predictors include *perceived usefulness* and *perceived ease of use*, two constructs related to task and technological aspects, and *attitude toward behavior* and *perceived behavioral control*, constructs related to individual aspects of users. Finally, *facilitating conditions (for use)*, another predictor related to social and environmental aspects, are also often thought to directly influence and predict behavioral intention and actual technology use.

12.2.2 External Predictors Used to Modify the Most Influential Perspectives of TechA@IL

Modern interactive technologies are impacting our lives in a variety of ways, drastically affecting and changing the functionality, accessibility, and use of literally every aspect of our daily lives. To address the challenges of the rapidly changing interactive technology landscape, researchers worldwide are striving to develop and validate a variety of approaches to explain technology adoption and use. In this context, the core theoretical perspectives in TechA@IL, along with their extensions and modifications, as well as the hybrid theoretical approaches developed in TechA@IL have been widely used to predict user adoption of various interactive technologies. Again, this section can only provide a glimpse of representative research in TechA@IL and is certainly not all-inclusive. Therefore, technology adoption for specific application domains is presented below, introducing various *external predictors* that influence and predict behavioral intentions and actual technology use in each case. Thus, in addition to the existing core predictors of behavioral intentions and actual adoption presented in the previous section, external predictors are used to modify and extend the influential theoretical perspectives in TechA@IL for the specific context.

TABLE 12.1
TechA@IL: Core Predictors Related to *User Aspects*

Predictor	Theory/Model	Definition of the Construct
Attitude toward behavior/using	TRA, TPB, DTPB, TAM, A-TAM	An individual's positive or negative feelings (evaluative affect) about performing the target behavior (Fishbein & Ajzen, 1975); Reflects feelings of favourableness or unfavorableness toward performing a behavior (Taylor & Todd, 1995a)
Perceived behavioral control	TPB, DTPB, A-TAM	An individual's perception of the ease or difficulty of performing the behavior of interest (Ajzen, 1985, 1991); Reflects perceptions of internal and external constraints on behavior (Taylor & Todd, 1995b)
Habit(s)	TIB, UTAUT2	Situation-behavior sequences that occur without self-instruction, while the individual is usually not conscious of these sequences (Triandis, 1980); The extent to which people tend to perform behaviors automatically because of learning (Venkatesh et al., 2012)
Affect	TIB, CSEM	An individual's liking for computer use and his or her actual behavior (Compeau & Higgins, 1995)
Affect toward use	MPCU	The feelings of joy, elation, or pleasure, or depression, disgust, displeasure, or hate associated by an individual with a particular act (Triandis, 1980)
Anxiety	CSEM	Evoking anxious or emotional reactions when it comes to performing behavior (Compeau & Higgins, 1995)
(Computer) self-efficacy	CSEM	An individual judgment of one's capability to use technology to accomplish a particular task or job (Compeau & Higgins, 1995)
Intrinsic motivation	MM	The performance of an activity for no apparent reinforcement other than the process of performing the activity per se (Davis et al., 1992)
Extrinsic motivation	MM	The performance of an activity because it is perceived to be instrumental in achieving valued outcomes that are distinct from the activity itself (Davis et al., 1992)
Hedonic motivation	UTAUT2	the fun or pleasure derived from using a technology (Venkatesh et al., 2012)
Voluntariness of use	PCI	The degree to which the use of the innovation is perceived as being voluntary, or of free will (Moore & Benbasat, 1991)
Perceived consequences	TIB	Each act is perceived as having potential consequences that have value, together with a probability that the consequence will occur (Triandis, 1980)

Several studies looking at Internet use from the perspective of the individual user have been presented in a series of studies. Pan and Jordan-Marsh (2010) applied an extended TAM model to examine the Internet use behavior of the elderly population in a developing country. Specific external predictors affecting behavioral intention and adoption examined in their study include:

- *subjective norm:* defined as a person's perception that most people who matter to him or her believe that he or she should or should not perform the behavior in question (Fishbein & Ajzen, 1975),
- *facilitating conditions:* refer to the extent to which a person believes that there is an organizational and technical infrastructure in place to support the use of the system (Venkatesh et al., 2003).

TABLE 12.2

TechA@IL: Core Predictors Related to *Task and Technological Aspects*

Predictor	Theory/Model	Definition of the Construct
Perceived usefulness	TAM, A-TAM, TAM2, MDPEU, TAM3	The degree to which an individual believes that using a particular system would enhance his or her job performance (Davis, 1986)
Perceived ease of use	TAM, PCI, TAM2, MDPEU, TAM3	The degree to which an individual believes that using a particular system would be free of physical and mental effort (Davis, 1986)
Complexity (of use)	IDT, MPCU	The degree to which an innovation is perceived as being difficult to use (Rogers, 1962, 1995)
Effort expectancy	UTAUT, UTAUT2	The degree of ease associated with the use of the system (Venkatesh et al., 2003)
Performance expectancy	UTAUT, UTAUT2	The degree to which an individual believes that using the system will help her/him to attain gains in job performance (Venkatesh et al., 2003)
Relative advantage	IDT, PCI	The degree to which using the innovation is perceived as being better than using its precursor (Rogers, 1962, 1995)
Trialability	IDT, PCI	The degree to which an innovation may be experimented with before adoption (Rogers, 1962, 1995)
Compatibility	IDT, PCI	The degree to which an innovation is perceived as being consistent with the existing values, needs, and past experiences of potential adopters (Rogers, 1962, 1995)
Outcome expectations	CSEM	The personal and performance-related consequences of the behavior, in relation to sense of accomplishment and job-related outcomes (Compeau & Higgins, 1995)
Job fit (with use)	MPCU	The extent to which an individual believes that using a technology can enhance the performance of his or her job (Thompson et al., 1991)
Long-term consequences (of use)	MPCU	Outcomes that have a pay-off in the future, such as increasing the flexibility to change jobs or increasing the opportunities for more meaningful work (Thompson et al., 1991)
Price value	UTAUT2	Consumers' cognitive trade-off between the perceived benefits of the applications and the monetary cost for using them (Venkatesh et al., 2012)

The World Wide Web, then seen as an emerging new IT, was the focus of earlier research by Moon and Kim (2001), who examined individual use of the WWW by extending TAM to include the theory of individuals' intrinsic motivation:

- *perceived playfulness:* the strength of one's belief that interaction with the WWW fulfills the user's intrinsic motives (Moon & Kim, 2001).

The evolution of the Web encouraged Alexandrakis et al. (2020) to explore Web 2.0 as a possible extension of TAM and to gain further insight into older adults' motivations for technology use:

- *future time perspective:* refers to the subjective estimation of remaining life time, which consequently affects priorities and decisions (emotion-based or knowledge-based) (Alexandrakis et al. 2020),
- *loneliness:* refers to a negative subjective feeling triggered by a perceived imbalance between desired social relationships and actual relationships (Pinquart & Sorensen, 2001, as cited in Alexandrakis et al. 2020),
- *age:-* refers to a person's age measured from birth to a certain date.

TABLE 12.3

TechA@IL: Core Predictors Related to *Social and Environmental Aspects*

Predictor	Theory/Model	Definition of the Construct
Subjective norm(s)	TRA, TPB, DTPB, A-TAM, TAM2, TAM3	The person's perception that most people who are important to her/him think she/he should or should not perform the behavior in question (Fishbein & Ajzen, 1975; Venkatesh & Davis, 2000)
Facilitating conditions (for use)	TIB, MPCU, UTAUT, UTAUT2	Objective factors, "out there" in the environment, that several judges or observers can agree to make an act easy to do (Triandis, 1980); The degree to which an individual believes that an organizational and technical infrastructure exists to support use of the system (Venkatesh et al., 2003)
Social influences	UTAUT, UTAUT2	The degree to which an individual perceives that important others believe he or she should use the new system (Venkatesh et al., 2003)
Social factors (influencing use)	TIB, MPCU	The individual's internalization of the reference groups' subjective culture, and specific interpersonal agreements that the individual has made with others, in specific social situations (Triandis, 1980)
Image	PCI	The degree to which the use of an innovation is perceived to enhance one's image or status in one's social system (Moore & Benbasat, 1991)
Observability	IDT	The degree to which the results of an innovation are observable to others (Rogers, 1962, 1995)
Result demonstrability	PCI	The tangibility of the results of using the innovation, including their observability and communicability Moore & Benbasat, 1991)
Visibility	PCI	The degree to which one can see others using the system in the organization (Venkatesh et al., 2003, adapted from Moore & Benbasat, 1991)

Based on the Perceived Risk Theory, Featherman and Pavlou (2003) integrated and empirically tested seven risk facets (performance, financial, time, psychological, social, privacy, and overall risk) in the TAM model, proposing a model for e-service adoption in the context of e-bill payment services:

- *perceived risk:* defined as the potential loss in pursuing a desired outcome of using an electronic service (Featherman and Pavlou, 2003).

As Mobility as a Service (MaaS) is evolving as a new type of user-centric transportation solution, Ye and colleagues (2020) built an analysis framework for MaaS acceptance and intent based on the UTAUT model:

- *perceived risk:* refers to the negative outcome expected when a consumer uses the service and the psychological expectation of the severity of the consequences when the consequences occur (Ye et al., 2020),
- *individual innovation:* the ability of an individual to discover and accept new things; it is used to evaluate a person's acceptance of new things (Ye et al., 2020).

Aiming to investigate the adoption intention of Near Field Communication (NFC) mobile payment (m-payment) services among consumers, Aris and colleagues (2022) extended the underlying theoretical perspectives, in particular TAM and the Mobile Technology Acceptance Model (MTAM) proposed by Ooi and Tan (2016):

- *technology availability:* defined as the level of individual perception of the existence of sufficient technological infrastructure to support the use of NFC payments (Aris et al., 2022),
- *consumers' technology readiness:* interpreted as individuals' enthusiasm and comfortability to use new technological innovations to get their jobs done (Parasuraman and Colby, 2015, as cited in Aris et al., 2022).

Consumer behavior in electronic commerce (e-commerce), as one of the extensively researched areas, was the focus of Huseynov and Özkan Yıldırım's (2019) research, in which the extended TAM was used as a theoretical model to study online customers' buying behavior on e-commerce platforms:

- *perceived enjoyment:* motivational factors that influence consumers' attitudes and behavioral intentions; the most important ones are consumers' hedonic and utilitarian online shopping motivations (Huseynov & Özkan Yıldırım, 2019),
- *perceived compatibility:* the extent to which online shopping platforms are perceived to be consistent with the potential adopters' lifestyle, existing values, previous experiences, expectations, and needs of potential users (Huseynov & Özkan Yıldırım, 2019),
- *perceived social pressure:* indicates the extent to which the significant referents (e.g., relatives, friends, and media) influence the individuals' particular behavior (Huseynov & Özkan Yıldırım, 2019).

In particular, the establishment of branded apps in conjunction with the use of online marketplaces has led to one of the most effective ways to succeed in direct-to-consumer e-commerce. Given the importance of sports apps, Won and colleagues (2022) examined the factors that influence consumers' intention to use a sports brand app by integrating the ISSM and the extended TAM:

- *perceived enjoyment:* positive feeling elicited by the hedonic features of technologies that intrinsically motivate users' adoption or continuance behaviors (Cloke, 2014, as cited in Won et al., 2022).

While the technology to access the Internet may be ubiquitous, shopping in an electronic medium is anything but intuitive and relaxed. By extending the TAM model, Vijayasarathy's (2004) research sought to explain consumers' intentions to use online shopping:

- *self-efficacy:* consumers' self-assessment of their capabilities to shop online (Vijayasarathy, 2004),
- *normative beliefs:* the extent to which a consumer believes that people they care about would recommend that they shop online (Vijayasarathy, 2004).

Because online shopping is largely dependent on user satisfaction and other factors that potentially increase customers' loyalty intentions, Chiu et al. (2009) examined customers' loyalty intentions toward online shopping by combining TAM with trust and fairness:

- *loyalty intention:* defined as the subjective probability that a customer will continue to purchase products from an online shop in the future (Chiu et al., 2009); in this study, loyalty intention is used as a surrogate for actual behavior.

In the evolving field of electronic payments and based on the Perceived Risk Theory (as in the study of electronic services), Lee (2009) summarized five specific risk facets (financial, privacy, performance, social, and time risk) and integrated them into the TAM and TPB models to propose a theoretical model to explain customers' behavioral intentions when using online banking:

- *perceived risk:* in online banking defined as the subjectively determined expectation of loss by an online bank user when considering a particular online transaction (Lee, 2009),
- *perceived benefit:* in online banking divided into two main types, referred to as direct and indirect advantages; direct advantages refer to immediate and tangible benefits that customers would enjoy by using online banking, while indirect advantages are those benefits that are less tangible and difficult to measure (Lee, 2008, as cited in Lee, 2009).

Although e-wallet (mobile payment) technology, i.e., using a mobile application to process financial transactions, is a beneficial way to process transactions, research has shown that the adoption rate is contrary to expectations. For example, Kınış and Tanova (2022) recently conducted a study to investigate the determinants of mobile e-wallet system adoption using the extended TAM model:

- *assurance of reimbursement:* providing incentives, such as instant information about the refund, along with various types of guarantees or refund promises to influence consumer confidence in deciding whether or not to use mobile payment systems (Kınış & Tanova, 2022).

The study conducted by Park and Kim (2014) examined cognitive factors that contribute to influencing users' perceptions and attitudes toward mobile cloud computing services by integrating these factors with TAM:

- *satisfaction:* the summary psychological state that results when emotions related to unconfirmed expectations are associated with the consumer's prior feelings about the consumption experience (Oliver, 1981, as cited in Bhattacherjee, 2001),
- *service and system quality:* refers to the perceived level of overall performance of a particular system and its service (DeLone and McLean, 1992, as cited in Park and Kim, 2014).

Dhagarra and colleagues (2020) complemented TAM by empirically examining the influence of behavioral characteristics and cognitive beliefs on patients' behavioral intention to accept technology in health care delivery:

- *trust:* measures the patient's positive attitude toward the health care provider to reduce consumer fear of provider opportunism (Dhagarra et al., 2020),
- *privacy concerns;* measures the patient's perception that he or she cannot rely on the provider, especially if he or she has to disclose personal information (Dhagarra et al., 2020).

The study by Zahid and colleagues (2022) addresses e-government, i.e., the provision of government information and services to citizens via the Internet or other digital means, and aims to extend the DTPB to include multidimensional trust (distinguishing among relationships at three levels: economic, social, and structural bonds) to better predict and explain the adoption of e-government services:

- *trust:* in the context of e-government, trust refers to an individual's subjective belief in the integrity and capabilities of the government agency providing the e-government services (Lallmahomed et al., 2017, as cited in Zahid et al., 2022).

Aiming to identify factors that influence citizens' intention to adopt new e-government systems, Rana et al. (2017) proposed a moderatorless UTAUT model called the Unified Model of E-Government Adoption (UMEGA):

- *attitude:* defined as the extent to which an individual makes a favorable or unfavorable evaluation or appraisal of the behavior in question (Ajzen, 1991).

In terms of improving mobile library services and using TAM as a theoretical background, Yoon's (2016) study used user satisfaction factors as antecedents for intention to use mobile library applications in academic libraries:

- *satisfaction:* the summarized psychological state that results when the emotions associated with unconfirmed expectations are linked to the consumer's prior feelings about the consumption experience (Oliver, 1981, as cited in Bhattacherjee, 2001).

Green IT, a forward-looking and environmentally friendly information technology, is a developing trend in information technology, so social and environmental responsibility are important factors when it comes to its acceptance. In this context, Yoon (2018) proposed a technology acceptance model for Green IT by adding normative variables (descriptive, injunctive, and personal norms) to TAM:

- *personal norms:* moral commitment, which refers to the sense of personal obligation to perform a certain behavior based on internalized values (Schwartz, 1977, as cited in Yoon, 2018),
- *descriptive norms:* people's perceptions of what most people do without making judgments (Yoon, 2018),
- *injunctive norms:* perceptions of what most people approve or disapprove of (Cialdini et al., 1990, as cited in Yoon, 2018); they are roughly equivalent to a subjective norm as operationalized in TRA (Yoon, 2018),
- *environmental beliefs:* a set of beliefs about adverse consequences caused by environmental degradation (Stern and Dietz, 1994, as cited in Yoon, 2018).

Decision support systems (DSS), a class of IS used to support management decisions and actions, were the focus of Djamasbi and colleagues (2010), who, extending the TAM model, examined the effects of positive mood on the acceptance of a DSS supporting a complex planning task:

- *positive mood:* implying access to an abundant quality and quantity of positive thoughts to support cognitive processes (Isen, 2008, as cited in Djamasbi et al., 2010).

In addition, Mir and Padma (2020) proposed an integrated technology acceptance framework using the TAM, Technology-Organization-Environment (TOE) model and the Human-Organization-Technology-fit (HOT -fit) framework for evaluating DSS in the agricultural sector:

- *social norms:* the assumed behavior that an individual is expected to adopt in a particular group, community, or culture (Mir and Padma, 2020).

To address the research gap on the acceptance of wearable tracking systems that must be used ubiquitously to ensure passenger safety, Kwee-Meier and colleagues (2016) extended TAM by incorporating a set of additional context-specific predictors for the dependent variable intention to use:

- *perceived security risk:* defined as the perceived risk of location data being accessed and/ or misused without authorization (Kwee-Meier et al., 2016),
- *privacy concern:* sensitivity to location identification in the event of an emergency and anonymous continuous tracking (Raschke et al., 2014, as cited in Kwee-Meier et al., 2016),
- *trust:* the need for proper functioning (the technical reliability of the location system in the event of a disaster) and efficacy (the location identification that would help them be found) (Kwee-Meier et al., 2016),
- *social influence:* the degree to which an individual perceives that significant others believe he or she should use the new system (Venkatesh et al., 2003).

The study conducted by Acikgoz and Perez Vega (2022) aimed to understand the driving forces behind voice assistant usage habits using TAM as a foundational theory:

- *trust:* as one of the most powerful elements to overcome uncertainty, described as a person's willingness and belief to have confidence in another person (Kumar, 1996, as cited in Acikgoz and Perez Vega, 2022).

The purpose of the study presented by Sagnier and colleagues (2020) was to test an extended TAM developed for investigating user acceptance of VR:

- *presence:* a subjective sensation that allows users to interact with and feel connected to a world outside of themselves (Thornson, et al., 2009, as cited in Sagnier et al., 2020),
- *cybersickness:* a relatively common negative consequence of exposure to virtual environments, with symptoms varying by person, type of equipment, and type of use (Nichols & Patel, 2002, as cited in Sagnier et al., 2020).

Yu and Huang's (2020) study investigated consumers' intention to use smart libraries via TAM, with the goal of helping smart libraries use artificial intelligence (AI) to establish services in an extremely competitive environment:

- *subjective norm:* a person's perception that most people he or she cares about think he or she should or should not perform the behavior in question (Fishbein and Ajzen, 1975),
- *perceived behavior control:* the extent to which a person can control his or her behavioral practice (Yu and Huang, 2020).

Lee and Lehto's (2013) study was conducted using TAM to identify determinants that influence intention to use an online video sharing service (YouTube) for procedural learning:

- *user satisfaction:* the extent to which users are satisfied and content with their previous use of an information system (Szymanski & Hise, 2000, as cited in Lee & Lehto, 2013).

The abundance of personal information shared on social media prompted Öztürk and colleagues' (2022) research to highlight the positive or negative impact of privacy and trust concerns on consumers' intention to use social media applications. In this study, the extended TAM was used:

- *trust:* the attitude that a representative will help another individual achieve his or her goals in a situation characterized by uncertainty and vulnerability (Lee and See, 2004, as cited in Öztürk, et al. 2022),
- *privacy concern:* consisting of two subdimensions: social privacy, defined as "how we manage our personal information, usability, and the accessibility of others to our information," while information privacy as "control of the technologies used by institutions and organizations to access, use, and analyze them for their interests" (Rader, 2014, as cited in Öztürk, et al. 2022).

Hasija and Esper (2022) examined the role of organizational factors in reconciling the differences between the potential benefits of AI for supply chain management (SCM) and its actual acceptance and use. The authors proposed theoretical extensions and added contextual dimensionality for SCM work situations to the "social influence" aspect of the UTAUT model:

- *AI trustworthiness:* involves emphasizing trust in education and training efforts in the early stages of adoption stages (priming) and trust in the workplace and workforce stability post-AI adoption (reinforcing) (Hasija & Esper, 2022).

Human-robot collaboration is progressing, with social robots becoming capable of complex and human-like capabilities such as developing creative outputs. Meeners (2022) hypothesized that functional elements of the TAM2 model, augmented with social-emotional elements, would predict intention to collaborate and that these would differ for different forms of a creative social robot:

- *perceived humanness:* concerns similarity to humans, which can promote anthropomorphism, i.e., the attribution of human or animal-like characteristics to nonliving objects (Breazeal, 2003, as cited in Meeners, 2022),
- *perceived social interactivity:* concerns the user's perception that a social robot has intentions, feelings, and beliefs (Breazeal, 2003, as cited in Meeners, 2022), and thus the expectation that social robots interact in accordance with their social mental models (Wirtz et al., 2018, cited in Meeners, 2022),
- *perceived social presence:* concerns the user's perception of dealing with a social entity, i.e., the feeling of being in someone's company (Heerink et al., 2009, as cited in Meeners, 2022).

While a growing number of studies have addressed ICT acceptance among older adults, most of which have focused on computer and Internet use, Zhu and Cheng (2022) examined smartphone use and acceptance among older adults by extending the TAM model:

- *social support:* refers to the provision of psychological and material resources by social networks to strengthen a person's ability to cope with technological stress and problems (Cohen, 2004, as cited in Zhu & Cheng, 2022).

Naaz and colleagues (2022) proposed body odor, a biometric trait unique to each individual, as one of the alternatives for authentication. Aiming to learn more about user acceptance of this technology, the authors extended the TAM model and proposed a technology acceptance model for biometric techniques based on body odor, which they named OdorTAM:

- *willingness:* generally suggests possible actions of the participant/responder (Shan, 2012, as cited in Naaz et al., 2022).

Hewitt and colleagues' (2019) study of emerging autonomous vehicle technologies combined efforts to measure public acceptance of autonomous vehicles by combining elements of generic technology acceptance models (i.e., UTAUT), car acceptance models (i.e. CTAM (Osswald et al., 2012)), and autonomy levels. As a result, they propose the Autonomous Vehicle Acceptance Model (AVAM), an adaptation of UTAUT and CTAM for autonomous vehicle technologies:

- *anxiety:* defined in the car context as the extent to which a person responds to a situation with apprehension, discomfort, or feelings of arousal (Osswald et al., 2012),
- *self-efficacy:* defined as a person's belief in his or her ability and competence to use a technology (e.g., a radio) to accomplish a specific task (Osswald et al., 2012),
- *attitude toward using technology:* defined as an individual's overall affective response to using a system; this determinant aims to reflect the user's beliefs about system use and its implications; it addresses attributes of in-vehicle IS that go beyond mere functionality and seeks to determine attitude toward system use through affective elements such as enjoyment and likability (Osswald et al., 2012),
- *perceived safety:* the extent to which an individual believes that using a system will affect his or her well-being (Osswald et al., 2012).

12.3 GUIDELINES FOR RESEARCHERS AND PRACTITIONERS

This section explains the procedure of utilizing the vast body of knowledge in the domain of technology acceptance and adoption, encompassing diverse theoretical perspectives and concepts, to tailor existing models for designing an individual study within a specific application domain. Thorough guidance is offered to researchers and practitioners regarding the design and empirical validation of their theoretical approach when applied to a particular interactive technology. The primary aim is to meticulously choose a core (baseline) model and subsequently enhance it by integrating existing constructs (core predictors) and potentially novel ones (external predictors), as well as context-dependent factors that are pertinent and unique to the specific context and technology under examination. The relevant body of research is duly acknowledged, with notable mention given to the guidelines proposed by Hong et al. (2014) for developing context-specific models, as well as the recommendations for future UTAUT-based research on technology acceptance and use put forth by Venkatesh et al. (2016).

The *Technology Acceptance and Adoption Research Approach* (TA2RA, pronounced as "Tay-Two-Ra") emphasizes a systematic approach that adheres to actionable guidelines and delineates a series of steps to facilitate the use of existing research in the field of technology adoption at the individual level (TechA@IL). The TA2RA approach comprises guidelines to assist researchers in developing context-specific research models by guiding them through the various stages of the process (Table 12.4). These stages encompass drawing from existing approaches, identifying pertinent constructs, conducting empirical studies, and iteratively refining the research model. By following this approach, researchers can systematically navigate through the necessary steps, ultimately leading to the creation of a robust and contextually appropriate research model. It is important to recognize that the examples provided and the names of existing models, constructs, and methods serve merely as a starting point. The selection of models and theories, relevant constructs, methods, and techniques should be contingent upon the unique research context and objectives. Consequently, it is crucial to ensure that the chosen theoretical approaches and constructs align with the individual research goals and the distinctive characteristics of the interactive technology and specific application domain.

It is imperative to emphasize that the TA2RA offers a comprehensive framework, wherein the precise procedural stages may exhibit variations contingent upon the specific application domain and research objectives at hand. This approach holds considerable potential to yield advantages for novice researchers and practitioners engaged in the realm of technology acceptance and adoption, owing to multiple reasons enumerated below:

- *Structured approach:* The guidelines provide a structured approach to conducting research in the field of technology adoption. Novice researchers often benefit from a step-by-step framework that helps them navigate the complexities of developing a research model and conducting empirical studies. It provides them with a clear roadmap and reduces the chances of missing important steps or elements in their research process.
- *Comprehensive literature review:* The guidelines emphasize the importance of conducting a comprehensive literature review. Novice researchers may not be familiar with the existing theories, models, constructs, and variables in the field of technology adoption. By conducting a thorough literature review, they can gain a solid understanding of the foundational concepts. It helps them build a strong theoretical foundation for their research and ensures that they are building upon existing knowledge.
- *Integration of existing knowledge:* The guidelines encourage novice researchers to draw from existing research approaches and constructs. This approach allows them to benefit from the extensive knowledge and empirical evidence that has already been accumulated in the field. By integrating existing knowledge into their research, they can leverage the strengths of established theories/models while focusing on the unique aspects of their specific application domain.

TABLE 12.4
Technology Acceptance and Adoption Research Approach (TA2RA)

TA2RA: Technology Acceptance and Adoption Research Approach

Conduct Literature Review:

- Conduct a comprehensive literature review on theories and models of technology acceptance and adoption in your specific application domain.
- Identify key theories and models that have been widely used and validated in previous research.
- Understand the variables and constructs (core predictors) commonly employed in these theories and models.
- Example: Conduct a literature review on technology acceptance and adoption theories/models in educational context and identify core theoretical perspectives in TechA@IL such as the Technology Acceptance Model (TAM), Unified Theory of Acceptance and Use of Technology (UTAUT), Innovation Diffusion Theory (IDT), and Social Cognitive Theory (SCT). Identify determinants of behavioral intention (core predictors) like perceived usefulness, perceived ease of use, social influence, subjective norm, attitude toward using, self-efficacy, affect, and relative advantage.

Identify Relevant Constructs:

- Analyze the detected research models and theories to identify the relevant constructs (external predictors) applicable to your particular application domain.
- Determine the key factors and variables that influence technology acceptance and adoption within your specific context.
- Example: Identify determinants of behavioral intention (external predictors) relevant to your context and application domain, such as compatibility, complexity, satisfaction, trust, facilitating conditions, perceived social presence, and perceived risk. These constructs can be derived from theories like the Innovation Diffusion Theory (IDT), Model of Personal Computer Utilization (MPCU), Perceived Risk Theory, Protection Motivation Theory (PMT), and Uses and Gratifications Theory (UGT).

Select a Core Model:

- Choose a well-established model of technology acceptance and adoption as a baseline for your research in a particular context.
- Examples of the most influential approaches of TechA@IL include the Technology Acceptance Model (TAM), Unified Theory of Acceptance and Use of Technology (UTAUT) or the Innovation Diffusion Theory (IDT).
- Ensure that the chosen core (baseline) model aligns with the key factors and variables relevant to your specific application domain.
- Example: In the widely recognized Technology Acceptance Model (TAM), perceived usefulness and perceived ease of use are considered core predictors. Choose among other core models like UTAUT, which incorporates performance expectancy, effort expectancy, and social influence, or the Theory of Planned Behavior (TPB) which includes attitudes, subjective norms, and perceived behavioral control.

Adapt and Customize the Model:

- Modify the core model to potentially incorporate the specific constructs and variables identified in your literature review.
- Consider adding or modifying external predictors of behavioral intention to better capture the unique aspects of technology adoption in your specific domain.
- Consider including context-dependent factors that act as determinants (antecedents) of core predictors of behavioral intention.
- Consider including moderating factors that have the ability to moderate relationships between the constructs within your model.
- Categorize the new constructs based on the three dimensions of influence: individual, task and technological, and social and environmental aspects (Triad of Predictors).
- Example: Adapt the Technology Acceptance Model (TAM) by adding context-specific constructs like compatibility with existing systems, perceived security, and privacy concerns. Customize Unified Theory of Acceptance and Use of Technology (UTAUT) by incorporating industry-specific variables such as organizational support, industry regulations, or specific user characteristics.

(Continued)

TABLE 12.4 (*Continued*)

Technology Acceptance and Adoption Research Approach (TA2RA)

TA2RA: Technology Acceptance and Adoption Research Approach

Develop Hypotheses:

- Based on your adapted model, formulate research hypotheses that describe the relationships between the identified constructs and variables.
- Ensure that your hypotheses are aligned with the theoretical foundations and empirical evidence from previous research.
- Specify the direction and strength of the relationships.
- Example: Formulate hypotheses based on the adapted model like "Perceived compatibility positively influences the intention to use technology in the healthcare industry" or "Organizational support positively moderates the relationship between perceived ease of use and intention to adopt technology in the education sector."

Design and Conduct Empirical Research:

- Develop a research procedure that aligns with your research objectives, such as surveys, experiments, or case studies.
- Collect data from participants within your specific application domain to test your research hypotheses.
- Ensure that your sample size is sufficient for statistical analysis and represents the target population.
- Example: Choose appropriate research methods like surveys, experiments, or interviews. Collect data from participants in your specific domain. Use validated measurement tools such as the Technology Acceptance Model Questionnaire (TAMQ), UTAUT questionnaire, or the Diffusion of Innovations (DOI) scales to measure constructs like perceived usefulness, compatibility, or resistance to change.

Analyze and Interpret Results:

- Use appropriate statistical techniques to analyze your data, such as regression analysis, Structural Equation Modeling (SEM), or qualitative data analysis.
- Evaluate the significance and strength of the relationships between the identified constructs and variables.
- Interpret the results in the context of your research objectives, contributing to both theoretical and practical implications.
- Example: Analyze collected data using statistical techniques like multiple linear regression analysis. Interpret the results to examine the relationships between constructs, such as the significant positive effect of perceived usefulness on behavioral intention.

Refine and Iterate:

- Based on the findings from your empirical research, refine and iterate your context-specific research model as necessary.
- Consider the limitations and suggestions for future research that arise from your study.
- Continuously engage with the existing literature to refine your research model and contribute to the advancement of knowledge in your specific application domain.
- Example: If the results suggest that perceived security is a crucial determinant in the context of technology adoption, you may include it as a construct in the revised model.

- *Contextual relevance.* The guidelines emphasize the importance of considering context-specific factors. Novice researchers may not be aware of the contextual nuances and complexities that can influence technology acceptance and adoption in different domains. By conducting contextual analysis, they can gain insights into the specific determinants and factors relevant to their research context. This ensures that their research is tailored to the specific needs and challenges of the domain, leading to more relevant and impactful findings.
- *Iterative process and validation.* The guidelines highlight the iterative nature of research and the importance of empirical validation. Novice researchers may not have a clear understanding of the need for iterative refinement of their research models and hypotheses. By conducting empirical research, analyzing data, and refining their models based on the findings, they can ensure the validity and robustness of their research. It also helps them gain practical insights and contributes to the continuous improvement of their work.

- *Practical implications.* The guidelines emphasize the practical implications of research in technology adoption. Novice researchers and practitioners are often interested in how their work can be applied in real-world settings. By considering the practical implications of their research, they can provide actionable insights, recommendations, and guidelines that can inform relevant decision-making and interventions.

In summary, the *Technology Acceptance and Adoption Research Approach* (TA2RA) provides novice researchers and practitioners with a methodical and well-informed approach for conducting research within the realm of technology adoption. It helps them build on existing knowledge, account for contextually contingent factors, validate their findings, and contribute to both theoretical and practical advancements in the field. This framework is expected to facilitate the assessment and prediction of future usage behavior across diverse types of innovative interactive technologies, thereby capturing the attention of researchers and practitioners immersed in their respective research and practical undertakings within specific application domains.

12.4   SUMMARY AND FUTURE DIRECTIONS

With regard to the adoption of interactive technologies in general and at the individual level, numerous theories and models have been put forward to predict and explain human behavior related to the acceptance and use of innovations and new technologies. Despite differences in theoretical formulations and constructs, all of these lines of research seek to understand the factors that drive user acceptance and to explain why certain innovations are well received and adopted by users. Since it would be practically impossible to discuss a number of relevant aspects in detail, the goal of this chapter is to provide a high-level introduction and discussion, and then refer the reader to exemplary research on each aspect.

This chapter first presents the background and brief history of technology adoption, focusing on advances in technology adoption at the individual level (TechA@IL). It provides a chronological overview of the relational linkages among 17 most influential theoretical approaches from TechA@IL, some exemplary and randomly selected hybrid approaches from a variety of fields, and the presentation of a representative sample of numerous studies in different application domains such as education, business, social media, health care, workplace, etc. In addition, the chapter provides insight into the factors that influence users' decisions about technology adoption. Three dimensions of influence are identified so that the key constructs (core predictors) influencing TechA@IL resulting from the leading approaches presented earlier are thematically categorized as individual, task & technology, and/or social & environmental. In addition to the core predictors of behavioral intentions and actual adoption, a whole range of different external predictors are presented that serve to modify and extend the influential theoretical perspectives in TechA@IL for the specific context. Finally, the chapter presents the *Technology Acceptance and Adoption Research Approach* (TA2RA) as a concrete and systematic guidance to aid researchers and practitioners in developing context-specific research models. It is expected that the TA2RA approach will prove valuable in future TechA@IL research, taking into account researchers' unique objectives and the distinctive attributes of interactive technologies and specific application domains.

In summary, the ongoing development of various types of interactive technologies and the growing number and diversity of users are opening up new research directions that will help improve our understanding of technology adoption and sustainable use in the future. Although extensive work has already been done, there is still great potential for further advances, investigations, and practices in this research area. Future work could therefore take new research directions, which are briefly outlined below.

First, to explore user experience (UX) aspects in technology adoption, as the experiential component in Human-Computer Interactions (HCI) field is not yet well understood (Hornbæk & Hertzum, 2017). Therefore, both research aspects should be combined, one focusing on utilitarian aspects (for example, the TAM model) and the other on experiential and hedonic aspects (for example, the UX research) of technology use.

Second, to improve the explanatory power of individual models of technology adoption, it may be useful in future work to investigate their integration with already established theories and models from other disciplines, such as:

- *Theory of Trying (TofT)* (Bagozzi & Warshaw, 1990), a theory from social psychology developed to show that intentions do not always lead to a particular action, and therefore focused on the assessment of the attempt to perform an action.
- *Flow Theory* (Csikszentmihalyi, 1975), a theory from positive psychology that describes a state in which a person is so involved in an activity that nothing else seems to matter; the concept of flow became a key element for optimal experience theory, which places subjective user experience at the center of developmental processes.
- *Expectation-Confirmation Model (ECM)* (Bhattacherjee, 2001), a model based on Oliver's (1980) leading cognitive theory in the field of consumer satisfaction that attempts to explain post-adoption satisfaction (ECT) with a focus on continuance, i.e., post-adoption behavior.
- *Model of Personal Computer Utilization (MPCU)* (Thompson et al., 1991), an adapted model of Triandis (TIB) for predicting individual acceptance and use of PCs, which has also proven useful for predicting acceptance and actual behavior toward a variety of information technologies.
- *Value-based Adoption Model (VAM)* (Kim et al., 2007), which aims to explain technology adoption based on Davis's model (TAM), but argues that perceived value based on benefits (usefulness and enjoyment) and sacrifice (technicality and perceived fee) determines adoption intention.

Third, to explore the predictive power of models when it comes to the expected adoption and actual use of so-called emerging or cutting-edge technologies, such as:

- *Extended Reality (XR)*, an umbrella term for immersive technologies that can merge the physical and virtual worlds and which refers to Augmented Reality (AR), Virtual Reality (VR), Mixed Reality (MR), and related applications in a variety of domains such as entertainment, healthcare, education, marketing, real estate, travel and tourism, and so on.
- *Smart Home* as a specific branch of Internet of Things (IoT) that focuses on household appliances and devices.
- *Internet of Toys (IoToys)* as one of several emerging applications of IoT and related applications in for example healthcare and education.
- *Internet of Musical Things (IoMusT)* as another evolving research area that encompasses the extension of the IoT paradigm into the musical domain.
- *Wearable Technology*, or simply "wearables", as smart IoT devices worn by individuals to track, analyze, and transmit personal data.
- *Artificial Intelligence (AI)* and *AI-based smart products* in a variety of areas including lifestyle (autonomous vehicles, spam filters), education (voice assistants, personalized learning), electronic commerce (personalized shopping), and the like.

REFERENCES

Acikgoz F., & Perez Vega, R. (2022). The Role of Privacy Cynicism in Consumer Habits with Voice Assistants: A Technology Acceptance Model Perspective, *International Journal of Human-Computer Interaction*, 38(12), 1138–1152. https://doi.org/10.1080/10447318.2021.1987677

Agarwal, R., & Prasad, J. (1997). The Role of Innovation Characteristics and Perceived Voluntariness in the Acceptance of Information Technologies. *Decision Sciences*, 28, 557–582. https://doi.org/10.1111/j.1540-5915.1997.tb01322.x

Ahmad, H., & Al-Zubi, A. (2016). Determinants of Internet Banking Adoption among Customers of Commercial Banks: An Empirical Study in the Jordanian Banking Sector. *International Journal of Business and Management*, 11(3), 95. https://doi.org/10.5539/ijbm.v11n3p95

Ajzen, I. (1985). From Intentions to Actions: A Theory of Planned Behavior. In: Kuhl, J., & Beckmann J. (Eds.), *Action Control. SSSP* Springer *Series in Social Psychology*. Berlin, Heidelberg: Springer. https://doi.org/10.1007/978-3-642-69746-3_2

Ajzen, I. (1991). The Theory of Planned Behavior. *Organizational Behavior and Human Decision Processes*, 50(2), 179–211. https://doi.org/10.1016/0749-5978(91)90020-T

Al-Adwan, A.S. (2020). Investigating the drivers and barriers to MOOCs adoption: The perspective of TAM. *Education and Information Technologies Volume, 25*, 5771–5795. doi:10.1007/s10639-020-10250-z

Al-Emran, M., Mezhuyev, V., & Kamaludin A. (2018). Technology Acceptance Model in M-Learning Context: A Systematic Review. *Computers & Education*. https://doi.org/10.1016/j.compedu.2018.06.008

Al-Emran, M., & Granić, A. (2021). Is It Still Valid or Outdated? A Bibliometric Analysis of the Technology Acceptance Model and Its Applications from 2010 to 2020. In Al-Emran, N., & Shaalan, K. (Eds.), *Recent Advances in Technology Acceptance Models and Theories. Studies in Systems, Decision and Control*, Switzerland AG: Springer Nature, pp. 1–12. https://doi.org/10.1007/978-3-030-64987-6_1

Al-Emran, M., Granić, A., Al-Sharafi, M., Nisreen, A., & Sarrab, M. (2021). Examining the Roles of Students' Beliefs and Security Concerns for Using Smartwatches in Higher Education. *Journal of Enterprise Information Management*, 34(4), 1229–1251. https://doi.org/10.1108/JEIM-02-2020-0052

Al-Qaysi, N., Granić, A., Al-Emran, M., Ramayah, T., Garces, E., & Daim, T. U. (2023). Social Media Adoption in Education: A Systematic Review of Disciplines, Applications, and Influential Factors. *Technology in Society*, 73(C), 102249. https://doi.org/10.1016/j.techsoc.2023.102249

Alexandrakis, D., Chorianopoulos, K., & Tselios, N. (2020). Older Adults and Web 2.0 Storytelling Technologies: Probing the Technology Acceptance Model through an Age-Related Perspective. *International Journal of Human-Computer Interaction*, 36(17), 1–13. https://doi.org/10.1080/10447318.2020.1768673

Alowayr, A., & Al-Azawei, A. (2021). Predicting Mobile Learning Acceptance: An Integrated Model and Empirical Study Based on Higher Education Students' Perceptions. *Australasian Journal of Educational Technology*, 37(3), 38–55. https://doi.org/10.14742/ajet.6154

Al-Qaysi, N., Mohamad-Nordin, N., & Al-Emran, M. (2020a). Employing the Technology Acceptance Model in Social Media: A Systematic Review. *Education and Information Technologies*, 25, 4961–5002. https://doi.org/10.1007/s10639-020-10197-1

Al-Qaysi, N., Mohamad-Nordin, N., & Al-Emran, M. (2020b). A Systematic Review of Social Media Acceptance from the Perspective of Educational and Information Systems Theories and Models. *Journal of Educational Computing Research*, 57(8), 2085–2109. https://doi.org/10.1177/0735633118817879

AlQudah, A. A., Al-Emran, M., & Shaalan, K. (2021). Technology Acceptance in Healthcare: A Systematic Review. *Applied Sciences-Basel*, 11(22), 10537–10537. https://doi.org/10.3390/app112210537

Al-Rahmi, A. M., Shamsuddin, A., Alturki, U., Aldraiweesh, A., Yusof, F. M., Al-Rahmi, W. M., & Aljeraiwi, A. A. (2021). The Influence of Information System Success and Technology Acceptance Model on Social Media Factors in Education. *Sustainability*, 13(14), 7770–7770. https://doi.org/10.3390/su13147770

Angosto, S., García-Fernández, J., Valantine, I., & Grimaldi-Puyana, M. (2020). The Intention to Use Fitness and Physical Activity Apps: A Systematic Review. *Sustainability*, 12, 6641. https://doi.org/10.3390/su12166641

Aris, F., Ismail, K., & Mohezar, S. (2022). Fostering Mobile Payment Adoption: A Case of Near Field Communication (NFC). *International Journal of Business and Society*, 23(3), 1535–1553. https://doi.org/10.33736/ijbs.5180.2022

Aytekin, A., Özköse, H., & Ayaz, A. (2022). Unified Theory of Acceptance and Use of Technology (UTAUT) in Mobile Learning Adoption: Systematic Literature Review and Bibliometric Analysis. *COLLNET Journal of Scientometrics and Information Management*, 16(1), 75–116. https://doi.org/10.1080/09737766.2021.2007037

Bagozzi, R. P. (2007). The Lagacy of the Technology Acceptance Model and a Proposal for a Paradigm Shift. *Journal of the Association for Information Systems*, 8(4), 244–254. https://doi.org/10.17705/1jais.00122

Bagozzi, R. P., & Warshaw, P. R. (1990). Trying to Consume. *Journal of Consumer Research*, 17(2), 127–140. https://doi.org/10.1086/208543

Bandura, A. (1986). *Social Foundations of Thought and Action: A Social Cognitive Theory*. Englewood Cliffs, NJ: Prentice Hall, Inc.

Bhattacherjee, A. (2001). Understanding Information Systems Continuance: An Expectation-Confirmation Model. *MIS Quarterly*, 25(3), 351–370. https://doi.org/10.1016/10.2307/3250921

Billanes, J., & Enevoldsen, P. (2021). A Critical Analysis of Ten Influential Factors to Energy Technology Acceptance and Adoption. *Energy Reports*, 7, 6899–6907. https://doi.org/10.1016/j.egyr.2021.09.118

Buabeng-Andoh, C., & Baah, C. (2020). Pre-service Teachers' Intention to Use Learning Management System: An Integration of UTAUT and TAM. *Interactive Technology and Smart Education*, 17(4), 455–474. https://doi.org/10.1108/ITSE-02-2020-0028

Bueno, S., & Salmeron, J. L. (2008). TAM-Based Success Modeling in ERP. *Interacting with Computers*, 20(6), 515–523. https://doi.org/10.1016/j.intcom.2008.08.003

Castillo-Vergara, M., Álvarez-Marín, A., Villavicencio Pinto, E., & Valdez-Juárez, L. E. (2022). Technological Acceptance of Industry 4.0 by Students from Rural Areas. *Electronics*, 11(14), 2109. https://doi.org/10.3390/electronics11142109

Chang, C.-C., & Chen, P.-Y. (2018). Analysis of Critical Factors for Social Games Based on Extended Technology Acceptance Model: A DEMATEL Approach. *Behaviour & Information Technology*. https://doi.org/10.1080/0144929X.2018.1480654

Chang, H. S., Lee, S. C., & Ji, Y. G. (2016). Wearable Device Adoption Model with TAM and TTF. *International Journal of Mobile Communications*, 14(5), 518–537. https://doi.org/10.1504/IJMC.2016.078726

Chau, P. Y. K., & Hu, P. J. (2002). Examining a Model of Information Technology Acceptance by Individual Professionals: An Exploratory Study. *Journal of Management Information Systems*, 18(4), 191–229. https://doi.org/10.1080/07421222.2002.11045699

Chen, C.-F., & Chen, P.-C. (2011). Applying the TAM to Travelers' Usage Intentions of GPS Devices. *Expert Systems with Applications*, 38(5), 6217–6221. https://doi.org/10.1016/j.eswa.2010.11.047

Chen, H., Rong, W., Ma, X., Qu, Y., & Xiong, Z. (2017). An Extended Technology Acceptance Model for Mobile Social Gaming Service Popularity Analysis. *Mobile Information Systems*, 2017, 1–12. https://doi.org/10.1155/2017/3906953

Chiu, C.-M., Lin, H.-Y., Sun, S.-Y., & Hsu, M.-H. (2009). Understanding Customers' Loyalty Intentions towards Online Shopping: An Integration of Technology Acceptance Model and Fairness Theory. *Behaviour & Information Technology*, 28(4), 347–360. https://doi.org/10.1080/01449290801892492

Chouki, M., Talea, M., Okar, C., & Chroqui, R. (2020). Barriers to Information Technology Adoption within Small and Medium Enterprises: A Systematic Literature Review. *International Journal of Innovation and Technology Management*, 17(1). https://doi.org/10.1142/S0219877020500078

Choukou, M. A., Shortly, T., Leclerc, N., Freier, D., Lessard, G., Demers, L., & Auger, C. (2021). Evaluating the Acceptance of Ambient Assisted Living Technology (AALT) in Rehabilitation: A Scoping Review. *International Journal of Medical Informatics*, 150, 4461–4461

Compeau, D. R., & Higgins, C. A. (1995). Computer Self-Efficacy: Development of a Measure and Initial Test. *MIS Quarterly*, 19(2), 189–211. https://doi.org/10.2307/249688

Cooper, R. B., & Zmud, R. W. (1990). Information Technology Implementation Research: A Technological Diffusion Approach. *Management Science*, 36(2), 123–139. https://www.jstor.org/stable/2661451

Cruz, A. M., Daum, C., Comeau, A., uevara Salamanca J. D., McLennan, L., Neubauer, N., & Liu, L. (2020). Acceptance, Adoption, and Usability of Information and Communication Technologies for People Living with Dementia and Their Care Partners: A Systematic Review. *Disability and Rehabilitation: Assistive Technology*. https://doi.org/10.1080/17483107.2020.1864671

Csikszentmihalyi, M. (1975). *Beyond Boredom and Anxiety*. San Francisco, CA: Jossey-Bass.

Dakduk, S., Santalla-Banderali, Z., & van der Woude, D. (2018). Acceptance of Blended Learning in Executive Education. *Online Teaching. SAGE Open*, 1–16. https://doi.org/10.1177/2158244018800647

Dampson, D.G. (2021). Determinants of learning management system adoption in an era of COVID-19: Evidence from a Ghanaian university. *European Journal of Education and Pedagogy,* 2(3), 80–87. doi:10.24018/ejedu.2021.2.3.94

Davis, F. D. (1986). A technology acceptance model for empirically testing new end-user information systems: theory and results. Doctoral dissertation. MIT Sloan School of Management, Cambridge, MA

Davis, F. D. (1989). Perceived Usefulness, Perceived Ease of Use, and User Acceptance of Information Technology. *MIS Quarterly*, 13(3), 319–340. https://doi.org/10.2307/249008

Davis, F. D., Bagozzi, R. P., & Warshaw, P. R. (1989). User Acceptance of Computer Technology: A Comparison of Two Theoretical Models. *Management Science*, 35(8), 982–1003. https://doi.org/10.1287/mnsc.35.8.982

Davis, F. D., Bagozzi, R. P., & Warshaw, P. R. (1992). Extrinsic and Intrinsic Motivation to Use Computers in the Workplace. *Journal of Applied Social Psychology*, 22, 1111–1132. https://doi.org/10.1111/j.1559-1816.1992.tb00945.x

Davis, F. D., & Granić, A. (2024). *The Technology Acceptance Model - 30 Years of TAM*. Cham: Springer. doi: 10.1007/978-3-030-45274-2

DeLone, W. H., & McLean, E. R. (1992). Information Systems Success: The Quest for the Dependent Variable. *Information System Research*, 3(1), 60–95. https://doi.org/10.1287/isre.3.1.60

Dhagarra, D., Goswami, M., & Kumar, G. (2020). Impact of Trust and Privacy Concerns on Technology Acceptance in Healthcare: An Indian Perspective. *International Journal of Medical Informatics*, 141, 104164. https://doi.org/10.1016/j.ijmedinf.2020.104164

Djamasbi, S., Strong, D. M., & Dishaw, M. (2010). Affect and Acceptance: Examining the Effects of Positive Mood on the Technology Acceptance Model. *Decision Support Systems*, 48(2), 383–394. https://doi.org/10.1016/j.dss.2009.10.002

Eason, K. D. (1991). Ergonomic Perspectives on Advances in Human-Computer Interaction. *Ergonomics*, 34(6), 721–741. https://doi.org/10.1080/00140139108967347

FakhrHosseini, S., Chan, K., Lee, C., Jeon, M., Son, H., Rudnik, J., & Coughlin, J. (2022). User Adoption of Intelligent Environments: A Review of Technology Adoption Models, Challenges, and Prospects. *International Journal of Human-Computer Interaction*. https://doi.org/10.1080/10447318.2022.2118851

Featherman, M. S., & Pavlou, P. A. (2003). Predicting E-Services Adoption: A Perceived Risk Facets Perspective. *International Journal of Human-Computer Studies*, 59(4), 451–474. https://doi.org/10.1016/S1071-5819(03)00111-3

Fishbein, M., & Ajzen, I. (1975). *Belief, Attitude, Intention and Behavior: An Introduction to Theory and Research*. Reading, MA: Addison-Wesley.

Fussell, S. G., & Truong, D. (2021). Using virtual reality for dynamic learning: An extended technology acceptance model. *Virtual Reality*. https://doi.org/10.1007/s10055-021-00554-x

Gangwar, H., Date, H., & Raoot, A. D. (2014). Review on IT Adoption: Insights from Recent Technologies. *Journal of Enterprise Information Management*, 27(4), 488. https://doi.org/10.1108/JEIM-08-2012-0047

Gómez-Ramirez, I., Valencia-Arias, A., & Duque, L. (2019). Approach to M-Learning Acceptance among University Students: An Integrated Model of TPB and TAM. *International Review of Research in Open and Distributed Learning*, 20(3), 141–164. https://doi.org/10.19173/irrodl.v20i4.4061

Goodhue, D. L., & Thompson, R. L. (1995). Task-Technology Fit and Individual Performance. *MIS Quarterly*, 19(2), 213–236. https://doi.org/10.2307/249689

Granić, A. (2022). Educational Technology Adoption: A Systematic Review. Education and Information Technologies 27, 9725–9744. https://doi.org/10.1007/s10639-022-10951-7

Granić, A. (2023a). Technology Adoption at Individual Level: Toward an Integrated Overview. *Universal Access in the Information Society*. https://doi.org/10.1007/s10209-023-00974-3

Granić, A. (2023b). Technology Acceptance and Adoption in Education. In: Zawacki-Richter, O., & Jung, I. (Eds.), *Handbook of Open, Distance and Digital Education*. Singapore: Springer. https://doi.org/10.1007/978-981-19-2080-6_11

Granić, A., & Marangunić, N. (2019). Technology Acceptance Model in Educational Context: A Systematic Literature Review. *British Journal of Educational Technology*, 50(5), 2572–2593. https://doi.org/10.1111/bjet.12864

Gu, D., Khan, S., Khan, I. U., & Khan, S. U. (2019). Understanding Mobile Tourism Shopping in Pakistan: An Integrating Framework of Innovation Diffusion Theory and Technology Acceptance Model. *Mobile Information Systems*, 2019, Article ID 1490617. https://doi.org/10.1155/2019/1490617

Guest, W., Wild, F., Vovk, A., Lefrere, P., Klemke, R., Fominykh, M., & Kuula, T. (2018). Technology Acceptance Model for Augmented Reality and Wearable Technologies. *Journal of Universal Computer Science*, 24(2), 192–219.

Guner, H., & Acarturk, C. (2020). The Use and Acceptance of ICT by Senior Citizens: A Comparison of Technology Acceptance Model (TAM) for Elderly and Young Adults. *Universal Access in the Information Society*, 19, 311–330. https://doi.org/10.1007/s10209-018-0642-4

Hasija, A., & Esper, T. L. (2022). In Artificial Intelligence (AI) we trust: A qualitative investigation of AI technology acceptance. *Journal of Business Logistics*, 43, 388–412. https://doi.org/10.1111/jbl.12301

Hernández, B., Jiménez, J., & Martín, M. J. (2008). Extending the Technology Acceptance Model to Include the IT Decision-Maker: A Study of Business Management Software. *Technovation*, 28(3), 112–121. https://doi.org/10.1016/j.technovation.2007.11.002

Hewitt, C., Amanatidis, T., Sarkar, A., & Politis, I. (2019). Assessing Public Perception of Self-Driving Cars: the Autonomous Vehicle Acceptance Model. *IUI'19*, March 17–20, 2019, Marina del Ray, CA.

Hong, W., Chan, F. K. Y., Thong, J. Y. L., Chasalow, L. C., & Dhillon, G. (2014). A Framework and Guidelines for Context-Specific Theorizing in Information Systems Research. *Information Systems Research*, 25(1), 111–136. https://doi.org/10.1287/isre.2013.0501

Hornbæk, K., & Hertzum, M. (2017). Technology Acceptance and User Experience: A Review of the Experiential Component in HCI. *ACM Transactions on Computer-Human Interaction*, 24(5), 33 https://doi.org/10.1145/3127358

Huang, C.-Y., Wang, H.-Y., Yang, C.-L., & Shiau, S. (2020). A Derivation of Factors Influencing the Diffusion and Adoption of an Open Source Learning Platform. *Sustainability*, 12(18), 7532. https://doi.org/10.3390/su12187532

Huang, Y.-C., Li, L.-N., Lee, H.-Y., Browning, M., & Yu, C-P. (2023). Surfing in Virtual Reality: An Application of Extended Technology Acceptance Model with Flow Theory. *Computers in Human Behavior Reports*, 9, 100252. https://doi.org/10.1016/j.chbr.2022.100252

Huseynov, F., & Özkan Yıldırım, S. (2019). Online Consumer Typologies and Their Shopping Behaviors in B2C E-Commerce Platforms. *SAGE Open*, 9(2), 9. https://doi.org/10.1177/215824401985463

Iacovou, C. L., Benbasat, I., & Dexter, A. S. (1995). Electronic Data Interchange and Small Organizations: Adoption and Impact of Technology. *MIS Quarterly*, 19(4), 465–85. https://doi.org/10.2307/249629

Ifinedo, P. (2017). Examining Students' Intention to Continue Using Blogs for Learning: Perspectives from Technology Acceptance, Motivational, and Social-Cognitive Frameworks. *Computers in Human Behavior*, 72, 189–199. https://doi.org/10.1016/j.chb.2016.12.049

Jang, J., Ko, Y., Shin, W.S., & Han, I. (2021). Augmented reality and virtual reality for learning: An examination using an extended Technology Acceptance Model. *IEEE ACCESS*, 9, 6798–6809. doi: 10.1109/ACCESS.2020.3048708

Jongchul, O., & Sung-Joon, Y. (2014). Validation of Haptic Enabling Technology Acceptance Model (HE-TAM): Integration of IDT and TAM. *Telematics and Informatics*, 31(4), 585–596.

Kanak, A., & Sogukpinar, I. (2017). BioTAM: A Technology Acceptance Model for Biometric Authentication Systems. *IET Biometrics*, 6(6), 457–467. https://doi.org/10.1049/iet-bmt.2016.0148

Karahanna, E., Straub, D. W., & Chervany, N. L. (1999). Information Technology Adoption across Time: A Cross-sectional Comparison of Pre-adoption and Post-adoption Beliefs. *MIS Quarterly*, 23(2), 183–213. https://doi.org/10.2307/249751

Kavandi, H., & Jaana, M. (2020). Factors That Affect Health Information Technology Adoption by Seniors: A Systematic Review. Health & Social Care in the Community, 28(6), 1827–1842. https://doi.org/10.1111/hsc.13011

Kim, H. W., Chan, H. C., & Gupta, S. (2007). Value-Based Adoption of Mobile Internet: An Empirical Investigation. *Decision Support Systems*, 43(1), 111–126. https://doi.org/10.1016/j.dss.2005.05.009

Kim, J., & Forsythe, S. (2008). Sensory Enabling Technology Acceptance Model (SE-TAM): A Multiple-Group Structural Model Comparison. *Psychology & Marketing*, 25(9), 901–922. https://doi.org/10.1002/mar.20245

Kınış, F., & Tanova, C. (2022). Can I Trust My Phone to Replace My Wallet? The Determinants of E-Wallet Adoption in North Cyprus. *Journal of Theoretical and Applied Electronic Commerce Research*, 17(4), 1696–1715. https://doi.org/10.3390/jtaer17040086

Kwee-Meier, S. T., Bützler, J. E., & Schlick, C. (2016). Development and Validation of a Technology Acceptance Model for Safety-Enhancing, Wearable Locating Systems. *Behaviour & Information Technology*, 35(5), 394–409. https://doi.org/10.1080/0144929X.2016.1141986

Kwon, T. H., & Zmud, R. W. (1987). Unifying the Fragmented Models of Information Systems Implementation. In Boland, R. J., & Hirschheim, R. A. (Eds.), *Critical Issues in Information Systems Research* (pp. 227–251). New York: John Wiley & Sons, Inc.

Lai, H.J. (2020). Investigating older adults' decisions to use mobile devices for learning, based on the unified theory of acceptance and use of technology. *Interactive Learning Environments*, 28(7), 890–901. doi:10.1080/10494820.2018.1546748

Lee, D. Y., & Lehto, M. R. (2013). User Acceptance of YouTube for Procedural Learning: An Extension of the Technology Acceptance Model. *Computers & Education*, 61, 193–208. https://doi.org/10.1016/j.compedu.2012.10.001

Lee, M.-C. (2009). Factors Influencing the Adoption of Internet Banking: An Integration of TAM and TPB with Perceived Risk and Perceived Benefit. *Electronic Commerce Research and Applications*, 8(3), 130–141. https://doi.org/10.1016/j.elerap.2008.11.006

Lemay, D.J., Morin, M.M., Bazelais, P., & Doleck, T. (2018). Modeling students' perceptions of simulation-based learning using the Technology Acceptance Model. *Clinical Simulation in Nursing*, 20, 28–37. doi:10.1016/j.ecns.2018.04.004

Lin, P.H. &; Yeh, S.C. (2019). How motion-control influences a VR-supported technology for mental rotation learning: from the perspectives of playfulness, gender difference and Technology Acceptance Model. *International Journal of Human-Computer Interaction*, 35(18), 1736–1746. doi:10.1080/10447318.2019.1571784

Liu, Q., Geertshuis, S., & Grainger, R. (2020). Understanding Academics' Adoption of Learning Technologies: A Systematic Review. *Computers & Education*. https://doi.org/10.1016/j.compedu.2020.103857

Lu, Y., Zhou, T., & Wang, B. (2009). Exploring Chinese Users' Acceptance of Instant Messaging Using the Theory of Planned Behavior, the Technology Acceptance Model, and the Flow Theory. *Computers in Human Behavior*, 25(1), 29–39. https://doi.org/10.1016/j.chb.2008.06.002

Ly, B., Ly, R., & Hor, S. (2023). Zoom Classrooms and Adoption Behavior among Cambodian Students. *Computers in Human Behavior Reports*, 9, 100266. https://doi.org/10.1016/j.chbr.2022.100266

Makki, T. W., DeCook, J. R., Kadylak, T., & Lee, O. J. (2017). The Social Value of Snapchat: An Exploration of Affiliation Motivation, the Technology Acceptance Model, and Relational Maintenance in Snapchat Use. *International Journal of Human-Computer Interaction*, 34(1). https://doi.org/10.1080/10447318.2017.1357903

Marangunić, N., & Granić, A. (2015). Technology Acceptance Model: A Literature Review from 1986 to 2013. *Universal Access in the Information Society*, 14(1), 81–95. https://doi.org/10.1007/s10209-014-0348-1

McFarland, D. J., & Hamilton, D. (2006). Adding Contextual Specificity to the Technology Acceptance Model. *Computers in Human Behavior*, 22(3), 427–447. https://doi.org/10.1016/j.chb.2004.09.009

Meeners, M. (2022). Human-robot collaboration in creative innovation processes: The influence of functional, relational and social-emotional elements on the intention to collaborate with a creative social robot in the work environment. Thesis. University of Twente. https://essay.utwente.nl/89482/1/Meeners_MA_BMS.pdf

Mir, S. A., & Padma, T. (2020). Integrated Technology Acceptance Model for the Evaluation of Agricultural Decision Support Systems. *Journal of Global Information Technology Management*, 23(2), 138–164. https://doi.org/10.1080/1097198X.2020.1752083

Moon, B. (1995). Paradigms in Migration Research: Exploring 'Moorings' as a Schema. *Progress in Human Geography*, 19(4), 504–524. https://doi.org/10.1177/030913259501900404

Moon, J. W., & Kim, Y. G. (2001). Extending the TAM for a World-Wide-Web context. *Information & Management*, 38(4), 217–230. https://doi.org/10.1016/s0378-7206(00)00061-6

Moore, G. C., & Benbasat, I. (1991). Development of an Instrument to Measure the Perceptions of Adopting an Information Technology Innovation. *Information Systems Research*, 2(3), 173–191. https://doi.org/10.1287/isre.2.3.192

Naaz, S., Ali Khan, S., Siddiqui, F., Saquib Sohail, S., Øivind Madsen, D., & Ahmad, A. (2022). OdorTAM: Technology Acceptance Model for Biometric Authentication System Using Human Body Odor. *International Journal of Environmental Research and Public Health*, 19(24), 16777. https://doi.org/10.3390/ijerph192416777

Nadal, C., Sas, C., & Doherty, G. (2020). Technology Acceptance in Mobile Health: Scoping Review of Definitions, Models, and Measurement. *Journal of Medical Internet Research*, 22(7), e17256. https://doi.org/10.2196/17256

Oliver, R. L. (1980). A Cognitive Model of the Antecedents and Consequences of Satisfaction Decisions. *Journal of Marketing Research*, 17 (4), 460–469. https://doi.org/10.1177/002224378001700405

Ooi, K. B., & Tan, G. W. H. (2016). Mobile Technology Acceptance Model: An Investigation Using Mobile Users to Explore Smartphone Credit Card. *Expert Systems with Applications*, 59, 33–46. https://doi.org/10.1016/j.eswa.2016.04.015

Osswald, S., Wurhofer, D., Trösterer, S., Beck, E., & Tscheligi, M. (2012). Predicting Information Technology Usage in the Car: Towards a Car Technology Acceptance Model. In *Proceedings of AutoUI. ACM*, New York, 51–58

Öztürk, A., Erkan, C., & Erkuş, U. (2022). The Effect of Privacy Perception on Social Media on Attitude Towards Social Media Usage. *Journal of Yasar University*, 17/65, 79–94

Pai, F.-Y., & Huang, K.-I. (2011). Applying the Technology Acceptance Model to the Introduction of Healthcare Information Systems. *Technological Forecasting and Social Change*, 78(4), 650–660. https://doi.org/10.1016/j.techfore.2010.11.007

Pan, S., & Jordan-Marsh, M. (2010). Internet Use Intention and Adoption among Chinese Older Adults: From the Expanded Technology Acceptance Model Perspective. *Computers in Human Behavior*, 26(5), 1111–1119. https://doi.org/10.1016/j.chb.2010.03.015

Parasuraman, A. (2000). Technology Readiness Index (TRI): A Multiple-Item Scale to Measure Readiness to Embrace New Technologies. *Journal of Service Research*, 2, 307–320. https://doi.org/10.1177/109467050024001

Park, E., & Kim, K. J. (2014). An Integrated Adoption Model of Mobile Cloud Services: Exploration of Key Determinants and Extension of Technology Acceptance Model. *Telematics and Informatics*, 31(3), 376–385. https://doi.org/10.1016/j.tele.2013.11.008

Park, E., & Kwon, S.J. (2016). The adoption of teaching assistant robots: a technology acceptance model approach. *Program-Electronic Library and Information Systems*, 50(4), 354–366. doi:10.1108/PROG-02-2016-0017

Prasetyo, Y.T., Ong, A.K.S., Concepcion, G.K.F., Navata, F.M.B., Robles, R.A.V., Tomagos, I.J.T., Young, M.N., Diaz, J.F.T., Nadlifatin, R., & Redi, A.A.N.P. (2021). Determining factors affecting acceptance of e-learning platforms during the COVID-19 pandemic: Integrating extended Technology Acceptance Model and DeLone & McLean IS Success Model. *Sustainability*, 13(15), 8365–8365. doi:10.3390/su13158365

Raman, A., & Don, Y. (2013). Preservice Teachers' Acceptance of Learning Management Software: An Application of the UTAUT2 Model. *International Education Studies*, 6(7), 157–164. https://doi.org/10.5539/ies.v6n7p157

Rana, N. P., Dwivedi, Y. K., Lal, B., Williams, M. D., & Clement, M. (2017). Citizens' Adoption of an Electronic Government System: Towards a Unified View. *Information Systems Frontiers*, 19(3), 549–568. https://doi.org/10.1007/s10796-015-9613-y

Razmak, J., & Bélanger, C. (2018). Using the Technology Acceptance Model to Predict Patient Attitude toward Personal Health Records in Regional Communities. *Information Technology & People*, 31(2), 306–326. https://doi.org/10.1108/ITP-07-2016-0160

Rigopoulou, I. D., Chaniotakis, I. E., & Kehagias, J. D. (2017). An Extended Technology Acceptance Model for Predicting Smartphone Adoption among Young Consumers in Greece. *International Journal of Mobile Communications*, 15(4), 372. https://doi.org/10.1504/ijmc.2017.084860

Roca, J. C., Chiu, C. M., & Martinez, F. J. (2006). Understanding e-Learning Continuance Intention: An Extension of the Technology Acceptance Model. *International Journal of Human-Computer Studies*, 64(8), 683–696. https://doi.org/10.1016/j.ijhcs.2006.01.003

Rogers, E. (1962). *Diffusion of Innovations*. New York: The Free Press

Rogers, E. (1995). *Diffusion of Innovations*, 4th Ed. New York: The Free Press

Rogers, R. W. (1975). A Protection Motivation Theory of Fear Appeals and Attitude Change. *Journal of Psychology*, 91(1), 93–114. https://doi.org/10.1080/00223980.1975.9915803

Rosli, M. S., Saleh, N. S., Ali, A. Md., Abu Bakar, S., & Tahir, L. M. (2022). A Systematic Review of the Technology Acceptance Model for the Sustainability of Higher Education during the COVID-19 Pandemic and Identified Research Gaps. *Sustainability*, 14(18), 11389–11389 https://doi.org/10.3390/su141811389

Saghafian, M., Laumann, K., & Skogstad, M. R. (2021). Stagewise Overview of Issues Influencing Organizational Technology Adoption and Use. *Frontiers in Psychology*, 12, 30145–30145. https://doi.org/10.3389/fpsyg.2021.630145

Sagnier, C., Loup-Escande, E., Lourdeaux, D., Thouvenin, I., & Valléry, G. (2020). User Acceptance of Virtual Reality: An Extended Technology Acceptance Model. *International Journal of Human-Computer Interaction*, 1–15. https://doi.org/10.1080/10447318.2019.1708612

Salahshour Rad, M., Nilashi, M., & Mohamed Dahlan, H. (2018). Information Technology Adoption: A Review of the Literature and Classification. *Universal Access in the Information Society*, 17, 361–390. https://doi.org/10.1007/s10209-017-0534-z

Schomakers, E. M., Lidynia, C., Vervier, L. S., Calero Valdez, A., & Ziefle, M. (2022). Applying an Extended UTAUT2 Model to Explain User Acceptance of Lifestyle and Therapy Mobile Health Apps: Survey Study. *JMIR Mhealth Uhealth*, 10(1), e27095. https://doi.org/10.2196/27095.

Sharma, S. K., & Sharma, M. (2019). Examining the Role of Trust and Quality Dimensions in the Actual Usage of Mobile Banking Services: An Empirical Investigation. *International Journal of Information Management*, 44, 65–75. https://doi.org/10.1016/j.ijinfomgt.2018.09.013

Silva, G. M., Dias, A., & Rodrigues, M. S. (2022). Continuity of Use of Food Delivery Apps: An Integrated Approach to the Health Belief Model and the Technology Readiness and Acceptance Model. *Journal of Open Innovation: Technology, Market, and Complexity*, 8(3), 114; https://doi.org/10.3390/joitmc8030114

Sneesl, R., Jusoh, Y. Y., Jabar, M. A., & Abdullah, S. (2022). Revising Technology Adoption Factors for IoT-Based Smart Campuses: A Systematic Review. *Sustainability*, 14(8), 4840–4840. https://doi.org/10.3390/su14084840

Sykes, T. A., Venkatesh, V., & Gosain, S. (2009). Model of Acceptance with Peer Support: A Social Network Perspective to Understand Employees' System Use. *MIS Quarterly*, 33(2), 371–393. https://doi.org/10.2307/20650296

Tao, D., Wang, T., Wang, T., Zhang, T., Zhang, X., & Qu, X. (2019). A Systematic Review and Meta-Analysis of User Acceptance of Consumer-Oriented Health Information Technologies. *Computers in Human Behavior*. https://doi.org/10.1016/j.chb.2019.09.023

Taylor, S., & Todd, P. A. (1995a). Understanding Information Technology Usage: A Test of Competing Models. *Information Systems Research*, 6(2), 144–176. https://doi.org/10.1287/isre.6.2.144

Taylor, S., & Todd, P. A. (1995b). Assessing IT Usage: The Role of Prior Experience. *MIS Quarterly*, 19(4), 561–570. https://doi.org/249633

Thompson, R. L., Higgins, C. A., & Howell, J. M. (1991). Personal Computing: Toward a Conceptual Model of Utilization. *MIS Quarterly*, 15(1), 124–143. https://doi.org/10.2307/249443

Tornatzky, L., & Fleischer, M. (1990). *The Process of Technology Innovation*. Lexington: Lexington Books.

Travers, J. (2015). Uses and gratifications of wearable technology adoption. Thesis at University of Missouri. https://mospace.umsystem.edu/xmlui/handle/10355/58517

Triandis, H. C. (1980). Values, Attitudes, and Interpersonal Behavior. *Nebraska Symposium on Motivation, 1979: Beliefs, Attitudes, and Values*, University of Nebraska Press, Lincoln, NE, 195–259.

Tung, F.-C., Lee, M. S., Chen, C.-C., & Hsu, Y.-S. (2009). An Extension of Financial Cost and TAM Model with IDT for Exploring Users' Behavioral Intentions to Use the CRM Information System. *Social Behavior and Personality: An International Journal*, 37(5), 621–626. https://doi.org/10.2224/sbp.2009.37.5.621

Venkatesh V., Thong, J. Y., & Xu, X. (2012). Consumer Acceptance and Use of Information Technology: Extending the Unified Theory of Acceptance and Use of Technology. *MIS Quarterly*, 36, 157–78. https://doi.org/10.2307/41410412

Venkatesh, V. (2000). Determinants of Perceived Ease of Use: Integrating Control, Intrinsic Motivation, and Emotion into the Technology Acceptance Model. *Information Systems Research*, 11(4), 342–365.

Venkatesh, V., & Bala, H. (2008). Technology Acceptance Model 3 and a Research Agenda on Interventions. *Decision Sciences*, 39(2), 273–315. https://doi.org/10.1111/j.1540-5915.2008.00192.x

Venkatesh, V., & Davis, F. D. (2000). A Theoretical Extension of the Technology Acceptance Model: Four Longitudinal Field Studies. *Management Science*, 46(2), 186–204. https://doi.org/10.1287/mnsc.46.2.186.11926

Venkatesh, V., Brown, S. A., Maruping, L. M., & Bala, H. (2008). Predicting Different Conceptualizations of System Use: The Competing Roles of Behavioral Intention, Facilitating Conditions, and Behavioral Expectation. *MIS Quarterly*, 32(3), 483–502.

Venkatesh, V., Morris, M. G., Davis, G. B., & Davis, F. D. (2003). User Acceptance of Information Technology: Toward a Unified View. *MIS Quarterly*, 27(3), 425–478. https://doi.org/10.2307/30036540

Venkatesh, V., Thong, J. Y. L., & Xu, X. (2016). Unified Theory of Acceptance and Use of Technology: A Synthesis and the Road Ahead. *Journal of the Association for Information Systems*, 17(5), 328–376. https://ssrn.com/abstract=2800121

Vijayasarathy, L. R. (2004). Predicting Consumer Intentions to Use On-Line Shopping: The Case for an Augmented Technology Acceptance Model. *Information & Management*, 41(6), 747–762. https://doi.org/10.1016/j.im.2003.08.011

Wang, G., & Shin, C. (2022). Influencing Factors of Usage Intention of Metaverse Education Application Platform: Empirical Evidence Based on PPM and TAM Models. *Sustainability*, 14, 17037. https://doi.org/10.3390/su142417037

Wang, Y.-M., Wei, C.-L., & Wang, M.-W. (2022). Factors Influencing Students' Adoption Intention of Brain-Computer Interfaces in a Game-Learning Context. *Library Hi Tech*. https://doi.org/10.1108/LHT-12-2021-0506

Wirtz, B. W., & Gottel, V. (2016). Technology Acceptance in Social Media: Review, Synthesis and Directions for Future Empirical Research. *Journal of Electronic Commerce Research*, 17(2), 97–115.

Won, D., Chiu, W., & Byun, H. (2022). Factors Influencing Consumer Use of a Sport-Branded App: The Technology Acceptance Model Integrating App Quality and Perceived Enjoyment. *Asia Pacific Journal of Marketing and Logistics*. https://doi.org/10.1108/APJML-09-2021-0709

Wu, B., & Chen, X. (2017). Continuance Intention to Use MOOCs: Integrating the Technology Acceptance Model (TAM) and Task Technology Fit (TTF) Model. *Computers in Human Behavior*, 67, 1–12. https://doi.org/10.1016/j.chb.2016.10.028

Ye, J., Zheng, J., & Yi, F. (2020). A Study on Users' Willingness to Accept Mobility as a Service Based on UTAUT Model. *Technological Forecasting and Social Change*, 157, Article 120066. https://doi.org/10.1016/j.techfore.2020.120066

Yin, L. X., & Lin, H. C. (2022). Predictors of Customers' Continuance Intention of Mobile Banking from the Perspective of the Interactivity Theory. *Economic Research /Ekonomska Istraživanja*, 35(1), 6820–6849. https://doi.org/10.1080/1331677X.2022.2053782

Yoon, C. (2018). Extending the TAM for Green IT: A Normative Perspective. *Computers in Human Behavior*, 83, 129–139. https://doi.org/10.1016/j.chb.2018.01.032

Yoon, H.-Y. (2016). User Acceptance of Mobile Library Applications in Academic Libraries: An Application of the Technology Acceptance Model. *The Journal of Academic Librarianship*, 42(6), 687–693. https://doi.org/10.1016/j.acalib.2016.08.003

Yu, K., & Huang, G. (2020), Exploring Consumers' Intent to Use Smart Libraries with Technology Acceptance Model. *The Electronic Library*, 38(3), 447–461. https://doi.org/10.1108/EL-08-2019-0188

Zahid, H., Ali, S., Abu-Shanab, E., & Javed, H. M. U. (2022). Determinants of Intention to use E-Government Services: An Integrated Marketing Relation View. *Telematics and Informatics*, 68(1), 101778. https://doi.org/10.1016/j.tele.2022.101778

Zaltman, G., & Lin, N. (1971). On the Nature of Innovations. *American Behavioral Scientist*, 14(5), 651–673. https://doi.org/10.1177/000276427101400503

Zamani, S. Z. (2022). Small and Medium Enterprises (SMEs) Facing an Evolving Technological Era: A Systematic Literature Review on the Adoption of Technologies in SMEs. *European Journal of Innovation Management*, 25(6), 735–757

Zhu, X., & Cheng, X. (2022). Staying Connected: Smartphone Acceptance and Use Level Differences of Older Adults in China. *Universal Access in the Information* Society. https://doi.org/10.1007/s10209-022-00933-4

13 Privacy and Trust in HCI

Bart P. Knijnenburg and Nathan J. McNeese

13.1 INTRODUCTION

As computer systems increasingly permeate our daily lives and – both directly and indirectly – affect all aspects of our well-being, studying our trust in these systems becomes crucial to their effective operation. In this chapter, we particularly consider computing technology developed in recent decades, in particular artificial intelligence (AI) systems and social technologies.

AI systems work to leverage data collected about people to make personalized and context-dependent suggestions, adaptations, or decisions. The often non-deterministic and autonomous operation of such systems requires unprecedented levels of trust in their accurate and unbiased operation. Relatedly, important privacy concerns may arise from the data collected and inferences made by these systems.

Social technologies have partially shifted our social interactions (and thereby, the formation and maintenance of social relationships) to the Internet. Given the typically larger scale/lower bandwidth nature of computer-mediated social interactions, it is important to study how interpersonal trust occurs in a computer-mediated context. Furthermore, we have entrusted the platforms that manage these interpersonal interactions with a wealth of personal information about ourselves. Again, important privacy concerns may arise regarding the access that others (e.g., other users, but also advertisers and governmental organizations) have to this data.

This chapter provides a survey and introduction to the scientific study of privacy and trust within the field of human-computer interaction (HCI). The first section covers several theoretical perspectives that HCI researchers and practitioners can employ to conceptualize privacy, trust, and the bi-directional relationship between them (noting how trust may assuage privacy concerns, how privacy concerns can reduce or hamper the formation of trust, and/or how privacy concerns have a weaker impact on disclosure when trust is high). This is then followed by a section discussing the various methods HCI researchers and practitioners can employ to measure privacy and trust. We end this chapter with a summary of the HCI opportunities in trust and privacy and a discussion of the gaps and challenges in existing research.

13.2 THEORETICAL PERSPECTIVES ON PRIVACY AND TRUST

This section discusses the prevailing theoretical perspectives on privacy, trust, and their inter-relatedness. The selected theoretical perspectives are socio-technical in nature and therefore apply beyond the domain of HCI. Therefore, particular HCI opportunities are highlighted throughout this section.

13.2.1 PRIVACY AS INFORMATION DISCLOSURE

The traditional perspective on computing systems is that of a device that can store, process, and disseminate information (Denning et al., 1988). This information has historically always included *personal* information (Westin, 1967), and the amount of personal information stored, processed, and disseminated by computing systems is ever increasing (Kumaraguru & Cranor, 2005). It should thus come as no surprise that, traditionally, the most pervasive theoretical perspective on digital privacy has been an information-centric perspective of privacy (Solove, 2002).

DOI: 10.1201/9781003495109-13

Within this information-centric perspective, computer scientists have enthusiastically adopted information-*theoretic* definitions of privacy. The logical or numerical nature of such privacy "metrics" allows computer scientists to develop computational privacy-preserving solutions (e.g. *k*-anonymity (Sweeney, 2002), differential privacy (Dwork & Naor, 2010), cryptography (Menezes et al., 2018)) that optimize them. However, the exact nature of the connection between these information-theoretic definitions of privacy and actual human experiences remains unclear (making it impossible to answer questions like "what level of k-anonymity is acceptable or appropriate for most end-users?")

Within the context of HCI, the information-centric perspective of privacy revolves around the *disclosure* of users' personal information – voluntary or mandatory, direct or mediated – to (a) computer system(s) (Buck et al., 2022; Smith et al., 2011). This subsection covers two contradictory theories of users' information disclosure behavior – "privacy calculus" (Laufer & Wolfe, 1977) and "privacy heuristics" (Acquisti & Grossklags, 2006) – as well as an attempt to reconcile the two, leveraging the behavioral psychological concept of "dual-route processing models" (Angst & Agarwal, 2009).

13.2.1.1   Privacy Calculus

The main theory of human interaction within the information disclosure perspective on privacy is the "privacy calculus" theory (Laufer & Wolfe, 1977). This theory sees the data subject[1] as a cognitively unbounded information processor who rationally weighs the risks and benefits of disclosing personal information. Privacy calculus is a privacy-specific instance of general human decision-making theories (Li, 2012; Rust et al., 2002; Stone, 1981), which argue that people gather information about various aspects of each choice option, assign a value to each of these aspects, trade off the different aspects, and then choose the option that maximizes their utility (Bettman et al., 1998; Fishbein & Ajzen, 1975; Simon, 1959). In privacy calculus, the two aspects that are traded-off are perceived privacy risk and the perceived benefit of disclosure.

Perceived privacy risk is the fear that "potential loss of control over personal information, such as when information about you is used without your knowledge or permission" (Featherman & Pavlou, 2003; Jacoby & Kaplan, 1972; Li, 2012). This loss of control can lead to unintended uses and distribution of the information (Malhotra et al., 2004; Olivero & Lunt, 2004; Sheehan & Hoy, 2000; Van Slyke et al., 2006); this is particularly relevant in the realm of AI, where systems tend to make inferences about data that may, on the one hand, be incorrect[2] or biased (Knijnenburg & Kobsa, 2013b), or on the other hand, so accurate[3] as to become "creepy" (Phelan et al., 2016; Tene & Polonetsky, 2013). Moreover, privacy risk tends to rise with the accumulation of large amounts of data (Barocas & Nissenbaum, 2014; Malhotra et al., 2004); again, this is particularly prominent in the realm of AI and social networks, where algorithms may be used to re-identify users[4] in large volumes of anonymous marketing data (Narayanan & Shmatikov, 2008, 2009).

Perceived benefit is the expectation that the disclosure of information leads to an advantageous outcome. In some cases, the disclosure of personal information is required for the basic functionality offered by the computing system (e.g. a shopping website requesting shipping information (Preibusch et al., 2012)), while in other cases disclosure is optional, creating an additional benefit. Such additional benefits can include convenience (e.g. the use of a login or cookies to save settings (Degeling et al., 2019)), personalization (e.g. the use of tracked data or explicit feedback to provide personalized recommendations (Friedman et al., 2015)), interpersonal connection (e.g. systems that offer social networking or collaboration (Wisniewski et al., 2015a)) or direct material benefits (e.g. discounts (Woodruff et al., 2014)).

Practically speaking, the idea behind privacy calculus is that people weigh the risks and benefits when deciding whether to disclose the requested information. One critique of this existing privacy research is that it tends to study how perceived risks and benefits influence users' general intention to use a computing system, rather than their concrete behavioral decision to disclose their personal information – the former being only weakly related to the latter (Norberg et al., 2007; Spiekermann et al., 2001). That said, there are several studies that show evidence that people's perceptions of risks and benefits of disclosure influence their disclosure and/or system use behavior:

- Between 58.2% (Metzger, 2007) and 72% (Hoffman et al., 1999) of respondents in consumer surveys cite risk as a reason not to disclose their personal information;
- In scientific studies, perceived risk generally has a negative effect on disclosure intentions (Li et al., 2010, 2011; Norberg et al., 2007);
- Privacy risk may lead users to restrict access to their personal information (Li & Santhanam, 2008; Petronio, 2002), influence their intention to transact in a web shop (Kim et al., 2008; Pavlou, 2003), or their intention to adopt an online service (Featherman & Pavlou, 2003). The effect of privacy risk on users' purchase intentions has been shown to be even stronger than the effect of the economic risk of the transaction (Bhatnagar et al., 2000; Dinev & Hart, 2006), and intentions decrease even further when a service requests information that does not serve the purpose of the request (Phelps et al., 2000);
- While users' privacy concerns can inhibit their use of personalized services and advertising (Awad & Krishnan, 2006; Wilkinson et al., 2021), they seem to be willing to give up privacy for personalization (Hann et al., 2007; Olivero & Lunt, 2004), as long as this gives them benefits (Phelps et al., 2000);
- In terms of benefits, users particularly value content relevance, time savings, enjoyment, and novelty – offering such benefits may override their initial privacy concerns (Hagel & Rayport, 1999; Ho & Tam, 2006; Hui et al., 2006).

HCI Opportunity: In general, the perceived risks and benefits of disclosure are both important determinants of users' willingness to adopt and provide personal information to a computing system, and researchers therefore claim that they should both meet a certain threshold (Treiblmaier & Pollach, 2007) or that they at least should be in balance (Chellappa & Sin, 2005; Xu et al., 2009, 2011). HCI practitioners seeking to increase the adoption of their app or service are therefore recommended to highlight the benefits of their system and to mitigate any perceived risks associated with disclosure (White, 2004).

If, on the other hand, the HCI researcher or practitioner aims to support users' privacy decision-making practices, they could do this by giving users intuitive information (transparency) and interface options (control). A vast body of relevant research on this topic can be found in the proceedings of the Symposium on Usable Privacy and Security (SOUPS) and the Privacy Enhancing Technologies Symposium (PETS). Of particular interest may be the design guidelines and taxonomies that have recently been developed to help HCI researchers and practitioners get a better understanding of this field of research (cf. Feng et al., 2021; NIST, 2019; Schaub et al., 2015; Wilkinson et al., 2021).

13.2.1.2   Privacy Paradoxes and Heuristics

Researchers have had difficulties reconciling the privacy calculus theory with various "paradoxes" that have plagued the HCI privacy research field. Perhaps the most basic paradox (often labeled the "privacy paradox" (Gerber et al., 2018; Kokolakis, 2017; Norberg et al., 2007)) shows a remarkable gap between people's attitudes about privacy (and/or their intentions to disclose their personal information) and their actual information disclosure – where in most cases people disclose more information than their attitudes or intentions would suggest (Barth & de Jong, 2017). Another paradox, related to control, observes that while people generally claim to want control over their privacy, they often do not take action to protect their privacy when given the opportunity to do so (Acquisti & Gross, 2006; Compañó & Lusoli, 2010; Knijnenburg et al., 2013c; Kolter & Pernul, 2009; Wenning & Schunter, 2006). Finally, Nissenbaum (2011) emphasizes a paradox around transparency, noting that privacy notices that provide sufficient detail are "unlikely to be understood, let alone read," but that notices that are simplified for human consumption are equally ineffective because they "drain away important details".

These paradoxes emphasize the criticism raised against the privacy calculus theory (Acquisti et al., 2022; Knijnenburg et al., 2017) for making unrealistic assumptions about decision-makers' agency (Hargittai & Marwick, 2016), the availability of complete information about the risks and benefits associated with disclosure (Karwatzki et al., 2017), users' awareness and understanding of this information (Mirkovic et al., 2015), users' cognitive resources to make a coherent trade-off between these risks and benefits (Sundar et al., 2013; Wilson & Valacich, 2012), and their motivation to make this trade-off (Athey et al., 2017; Choi et al., 2018). These critics of the privacy calculus theory often contend that privacy decisions are subject to bounded rationality (Simon, 1982), in that users apply a series of heuristics when making such decisions (Acquisti et al., 2022; Adjerid et al., 2013; Knijnenburg et al., 2017; Lowry et al., 2012). This heuristic behavior, in turn, leads to certain observable decision externalities – things that should not affect users' privacy decisions, but do, such as the default setting and how the decision is framed (Johnson et al., 2002; Knijnenburg & Kobsa, 2014; Lai & Hui, 2004), the available options to choose from (Knijnenburg et al., 2013b; Tang et al., 2012), the order in which the privacy requests are made (Acquisti et al., 2012; Knijnenburg & Kobsa, 2013b), the professionalism of the user interface that makes the privacy request (John et al., 2011), as well as social cues (Acquisti et al., 2012; Knijnenburg & Kobsa, 2013b).

The existence of these decision externalities might explain the effectiveness of "privacy dark patterns" (Bösch et al., 2016; Forbrukerrådet, 2018) that aim to trick people into submitting to more permissible tracking and data collection practices than they would normally allow – the most prominent example of this being the prevalence of such dark patterns in cookie consent requests that have pervaded the online experience of Europeans (Degeling et al., 2019).

HCI Opportunity: Given the difficulties users face in making privacy decisions, privacy advocates have called for systems to be designed with an inherent adherence to privacy, as opposed to being added *ad hoc* through a myriad of complex privacy settings. This notion of "privacy by design" (Cavoukian, 2013) seeks to invent and promote design patterns and principles that are inherently more private. Others note, though, that the principles of privacy by design often lack concrete specification, making it difficult for practitioners to implement this notion in real-world systems (Shapiro, 2009; Spiekermann, 2012; Stark et al., 2016).

Another HCI opportunity in light of heuristic privacy decision-making is the idea of leveraging the aforementioned decision externalities to nudge people (Thaler & Sunstein, 2008) toward better privacy decision-making practices (Acquisti, 2012; Balebako et al., 2011; Wang et al., 2014), e.g. through justifications (Acquisti et al., 2012; Kobsa & Teltzrow, 2005; Wang & Benbasat, 2007) or privacy seals (Egelman et al., 2009; Rifon et al., 2005; Xu et al., 2009). However, research shows that privacy nudging often does not work (Besmer et al., 2010; Hui et al., 2007; Patil et al., 2011), can have unintended consequences (Anaraky et al., 2020; Bahirat et al., 2021), and may take a too strongly paternalistic approach to guiding users (Knijnenburg et al., 2017; Smith et al., 2013).

A final opportunity for HCI researchers and practitioners is to provide *user-tailored* privacy decision support (Knijnenburg et al., 2022a; Knijnenburg, 2017). The idea behind this approach is that while people differ substantially in their privacy decision-making practices, there are observable behavioral patterns that can be modeled via machine learning (Dong et al., 2016; Knijnenburg et al., 2013a; Sanchez et al., 2019a; Wisniewski et al., 2017; Xie et al., 2014). This would allow a system to measure users' privacy preferences (e.g. by observing their existing privacy decision-making practices) (Abuelgasim & Kayem, 2016; Knijnenburg et al., 2013a; Liu et al., 2016; Sanchez et al., 2019b) and subsequently adapt the system to these preferences (Bahirat et al., 2018; He et al., 2019; Namara et al., 2022; Wilkinson et al., 2017). This adaptive mechanism is not meant to completely automate the privacy decision-making process (Namara et al., 2018a), but rather to actively help users bridge the gulfs of execution and evaluation (cf. Norman, 2013) that exist between their privacy preferences and the available system settings (Liu et al., 2011; Madejski et al., 2012; Strater & Lipford, 2008).

13.2.1.3   Dual Process Theories of Privacy Decision-Making

With ample evidence supporting both the privacy calculus theory and the existence of heuristic approaches to privacy decision-making, researchers have recently argued for an integration of these two modes of decision-making using *dual process theories*. Popularized by Kahneman's book *Thinking, Fast and Slow* (Kahneman, 2011), dual process theories suggest that users engage in more deliberate or more heuristic attitude formation and decision practices depending on the contextual and personal circumstances. For example, the Elaboration Likelihood Model (ELM (Petty & Cacioppo, 1986; Petty & Wegener, 1999) – arguably the most prominent contemporary dual process theory) specifies a "central route" where people make decisions based on a deliberate evaluation of substantive information, such as superior or distinctive aspects of the decision options (Lord et al., 1995; Petty et al., 1983), and a "peripheral route" where decisions are based on heuristic aspects and general feelings (Bansal et al., 2008; Shamdasani et al., 2001; Slovic et al., 2004), the former being aligned with the privacy calculus theory and the latter being aligned with heuristic privacy decision-making practices.

Unlike most other dual process theories, the ELM specifies two antecedents to a user's tendency to use either route of processing: motivation and self-efficacy (Cacioppo et al., 1986; Petty et al., 1983). This aligns with findings in privacy research that suggest that people use shortcuts and heuristics because they are incapable (Liu et al., 2011; Madejski et al., 2012) or not motivated (Compañó & Lusoli, 2010; Knijnenburg & Bulgurcu, 2023) to make a more elaborate privacy decision.

ELM research in the context of privacy has shown that the relative impact of different types of privacy-related cues is indeed determined by the degree of elaboration users engage in (Angst & Agarwal, 2009; Knijnenburg & Bulgurcu, 2023; Kobsa et al., 2016; Lowry et al., 2012; Yang et al., 2006; Zhou, 2012b). For instance, Yang et al. (2006) demonstrated that objective information (a central cue) about privacy has a stronger impact on the perceptions and trust of people with high anxiety and high involvement, while third-party seals (a peripheral cue) have a stronger impact on the perceptions and trust of people with low anxiety and/or involvement. Similarly, studies in the domain of mobile banking (Zhou, 2012b) and online websites (Bansal et al., 2008) show that content-based arguments and information quality are more important for people with high levels of privacy concern, and self-efficacy, design, company information, and reputation are more important for people with low levels of privacy concern or self-efficacy. Kobsa et al. (2016) showed that people with high privacy concerns prefer their data to be stored locally, while people with low concerns tend to trust reputable companies with their data. Finally, Angst and Agarwal's (2009) study about users' attitudes toward electronic health records shows that the strength of the presented arguments is more likely to affect people with high privacy concerns and high involvement than those with low concerns and involvement.

Finally, recent research shows that the mere presence of "privacy nudges" may influence users to opt for a more heuristic decision-making approach: evidence shows that users who encounter information disclosure decisions with a pre-selected default setting or a one-sided framing make less deliberate decisions (Bahirat et al., 2021; Knijnenburg & Bulgurcu, 2023), regardless of whether these nudges are meant to make them disclose more, or less information. Similarly, disclosure justifications seem to exacerbate the effect of default and framing nudges regardless of whether or not these justifications align with the nudge (Anaraky et al., 2020).

HCI Opportunity: The fact that users put varying amounts of deliberation into their privacy decisions opens an opportunity for HCI researchers and practitioners to design privacy-setting interfaces that are attuned to increased deliberation. For instance, Knijnenburg and Bulgurcu (2023) showed that while online form-autocompletion tools tend to reduce the deliberation users put into filling out forms (cf. Preibusch et al., 2012), adding interface elements that make it easier to engage in deliberate decision-making does indeed improve users' purpose-specificity in information disclosure. Similarly, in developing interfaces that effect user-tailored privacy suggestions, Namara et al. (2022)

demonstrated that privacy suggestions (rather than fully automated interventions) increase user engagement with privacy – even in situations where no suggestions are made (Namara, 2022).

13.2.2 PRIVACY AS INTERPERSONAL BOUNDARY REGULATION

As computing systems are increasingly used to mediate human collaboration (i.e., computer-supported collaborative work) and communication (i.e., social media, social networks), privacy scholars have proposed conceptions of digital privacy as a dialectical process of managing interpersonal boundaries with others. This interpersonal perspective on privacy predates the digital era (cf. Altman, 1975), but has been adapted to the specific considerations of computer-mediated interactions.

Considering this interpersonal perspective on privacy, an important note must be made about users' interactions with autonomous AI systems (i.e., *agents*). Research shows that users may interchangeably treat such systems as tools or autonomous entities, depending on the context (Bickmore & Cassell, 2001; Knijnenburg & Willemsen, 2016; Nass et al., 1994). Consequently, their privacy considerations may align more with the information disclosure perspective or the interpersonal boundary regulation perspective, depending on how they perceive the system (Page et al., 2018). At the time of writing, the field of autonomous AI systems is undergoing major advances, paired with increased exposure of "common users" to systems that can simulate believable human-like interactions. Given these continuing advances, it is likely that users will develop new privacy norms regarding their interactions with autonomous systems (Proferes, 2022), and this area of research is understandably still in its infancy. For now, we advise researchers and practitioners to draw upon either of the two privacy perspectives, depending on how users perceive the system under investigation (i.e., as a tool or as an agent).

13.2.2.1 Altman's Theory

In a book predating much of the modern computing era, Irwin Altman outlined a conceptualization of privacy as an "interpersonal boundary process by which a person or group regulates interaction with others" (Altman, 1975). Altman's theory takes a dialectic stance, where privacy is not just a restriction of interactions, but rather a dynamic balance between restricting and seeking interactions with others, depending on one's needs. In the offline world, this process is often regulated or negotiated through implicit behavioral cues, such as physical proximity, body language, and eye contact (Altman, 1975). In the online world, this process requires the use of established (e.g., privacy settings) or *ad hoc* (e.g., workarounds or negotiated norms) privacy mechanisms.

Importantly, in a networked world, these privacy mechanisms extend far beyond disclosure (Palen & Dourish, 2003). For instance, Wisniewski et al. (2017) established a set of 11 coherent privacy practices of Facebook users that include setting privacy settings, withholding or moderating outgoing information, restricting incoming information, blocking people or apps, and sharing information selectively with certain people only. They further demonstrated that different users engage in different subsets of these practices, establishing six cohesive "privacy management strategies." In categorizing the various interpersonal privacy management practices, Lampinen et al. (2011) defined three important dimensions: behavioral vs. mental practices, individual vs. collaborative practices, and preventative vs. corrective strategies. Similarly, Wisniewski et al. (2012) defined a taxonomy of five boundary types, spanning up ten distinct boundaries: relationship boundaries (connection and context), network boundaries (discovery and intersection), territorial boundaries (inward-facing and outward-facing), disclosure boundaries (self-disclosure and confidant-disclosure), and interactional boundaries (disabling and blocking).

HCI Opportunity: In the context of interpersonal interaction, it is important for HCI researchers and practitioners to treat privacy as a multidimensional construct. For example, the privacy provided or afforded by new social computing technologies can be assessed against the dimensions outlined by Lampinen et al. (2011). Similarly, in designing privacy interfaces for social networks, designers could use the taxonomy of boundaries proposed by Wisniewski et al. as a guideline (Wisniewski et al., 2012). Finally, HCI practitioners should carefully consider how the communication principles of newly designed social media platforms align with users' preferred communication strategies (Page et al., 2013a), and study users' audience management practices if the technology enables users to broadcast their content to a broader audience (Litt, 2012).

13.2.2.2   Petronio's Communication Privacy Management Theory

Built upon Altman's Theory, Petronio's Communication Privacy Management Theory (Petronio, 2002) expands the notion of interpersonal boundaries, differentiating between personal boundaries (how one manages the disclosure of private information) and collective boundaries (how one manages information that is already shared with others). Personal boundaries are governed by disclosure privacy principles such as those outlined by Altman. Collective boundaries, on the other hand, are managed through co-ownership relationships that Petronio calls "boundary alliances." These boundary alliances are subject to certain rules and expectations around the privilege and responsibilities of being a confidant. Notably, the more widely certain information is shared, the higher its *permeability.*

Collective boundaries with varying levels of permeability provide a more nuanced distinction than the dichotomy of "private" and "public." This nuance is key to our understanding of privacy on social media, where a user who decides to share information using the "public" setting may still have an expectation of low permeability, depending on their usual audience (Litt, 2012). If a post of such a user gets re-shared by others, this can create boundary turbulence, despite the original post being public.[5]

HCI Opportunity: A key element to Petronio's conception of privacy is the collective management of privacy boundaries. HCI researchers and practitioners are encouraged to study privacy as a collective practice (Cho & Filippova, 2016; Jia & Xu, 2016; Murphy et al., 2014) and to create interaction mechanisms that allow users to form and control the "boundary alliances" that are created when information is collectively managed (Hu et al., 2011; Kolter et al., 2009; Squicciarini et al., 2009). Note that in digital environments, collective management does not only pertain to personal information about an individual that is managed by a group of peers, but also digital assets that in themselves pertain to a group of individuals (e.g., photos, videos, or posts that are tagged with multiple people) (Besmer & Richter Lipford, 2010; Wisniewski et al., 2015b). Furthermore, we note that even more so than other privacy management activities, collective or collaborative privacy management activities are culturally determined (Cho et al., 2018; Dourish & Anderson, 2006). As such, cultural practices must be respected if social media technologies are to meaningfully support collective privacy management.

13.2.2.3   Nissenbaum's Contextual Integrity Theory

The final theoretical perspective on interpersonal privacy we discuss in this chapter is Nissenbaum's Contextual Integrity theory. Nissenbaum's conceptualization of privacy takes a normative perspective, relating privacy activities to the normative expectations embedded in the social context in which they occur (Nissenbaum, 2009). Like the concept of "schemas" and "scripts" in social psychology, the privacy norms embedded in the social context (as well as the type of information being shared) dictate the rights and responsibilities of actors (i.e., senders, receivers, data subjects), which Nissenbaum calls the "transmission principles" of the situation.

According to Nissenbaum's theory, much of the privacy violations – and subsequent protection activities – that occur in social settings stem from misunderstandings or changes regarding these context-relevant transmission principles: interpersonal relationships may change over time (Page et al., 2012), and people may encounter each other in unexpected contexts. Considering the latter issue, social media users struggle with the problem of *context collapse,*[6] where multiple heretofore separated social circles coexist on the same platform. This makes it difficult for people to tailor their self-presentation to a specific social circle (Marwick & Boyd, 2011; Vitak et al., 2015), and may also result in ambiguous transmission principles (e.g., is it appropriate to talk about a colleague's social media post in a business meeting?). Nissenbaum does not give direct suggestions on how to resolve such privacy conflicts, but the contextual normative nature of the theory suggests that the *appropriateness* of a privacy action (e.g., a disclosure) may be derived from the context, the actors, the type of information, and the purpose of the privacy action (Lederer et al., 2003; Patil & Lai, 2005).

HCI Opportunity: Although Contextual Integrity has been a popular framework among HCI researchers, scholars with in-depth knowledge of the theory have recently contended that many HCI researchers do not fully engage with the true meaning of the theory (Badillo-Urquiola et al., 2018; Xia, 2023). Arguably, the difficulty lies in the establishment of the transmission principles: the concept of purpose-specificity has been used in a commercial context to establish accepted transmission principles and study users' adherence to privacy norms (Knijnenburg & Bulgurcu, 2023), but such principles are much more difficult to establish in interpersonal contexts. A user-tailored privacy approach can be applied here to infer transmission principles from similar past interactions, but this would undermine the normative nature of the Contextual Integrity approach (Knijnenburg et al., 2022a; Knijnenburg et al., 2017).

A further complication arises from the fact that norms – including privacy norms – are continually evolving, especially in the context of ever-changing social technologies (Proferes, 2022): What was a clear privacy *faux pas* 5 years ago may be perfectly acceptable today. Similarly, as we highlighted in the introduction to this section on interpersonal privacy theories, it would be impossible to predict the privacy norms around the AI technologies that will exist 5 years from now, which makes it very difficult for HCI practitioners tasked with minimizing privacy violations to design for all possible eventualities. Indeed, scholars tracking the evolution of social networks have shown significant shifts in privacy perceptions and practices (Stutzman et al., 2013), with the platforms themselves usually reluctantly lagging behind public opinion. Importantly, in domains where norms have not yet been established, there exists an opportunity for regulators to guide the future development of norms by applying a normative lens to privacy regulation (Compañó & Lusoli, 2010; Hull, 2015).

13.2.3 TRUST IN HUMAN-TECHNOLOGY RELATIONSHIPS

Trust is a foundational concept and behavior driving human interactions at an individual, team, and societal level (Mayer et al., 1995). Most interactions, both with other humans or with technology, are guided by trust and trustworthiness (Glaeser et al., 2000; Sapienza et al., 2013; Cho et al., 2015), and this concept moderates both the content and flow of interaction. Over the years, there have been many definitions of trust posited (PytlikZillig & Kimbrough, 2016), with one of the most canonical coming from the work of Mayer and colleagues (1995), who define trust as "the willingness of a party to be vulnerable to the actions of another party based on the expectation that the other will perform a particular action important to the trustor, irrespective of the ability to monitor or control that other part" (p. 712). As noted by the authors, this definition is highly dependent on two actors interacting in a manner that could exhibit a violation toward the trustor. Thus, a key part of Mayer and colleagues' definition is that of vulnerability. The critical aspect of vulnerability is that of the importance and potential of losing something that one cares about. As also noted by Mayer, vulnerability is linked to risks (taking a risk that may result in losing something you care about), and thus, trust is the exhibition of being willing to take a risk. Within the past 25 years, research on trust generally has increased. Specifically, research on trust in technology has burgeoned. As technology has become more and more sophisticated in its abilities, its interactions with humans have become more frequent and richer. In addition, as noted earlier in this chapter, the methods and manner in which human personal data is collected and then engrained into numerous technologies has only increased and led to humans' trust-specific relationships with technology being embedded with privacy concerns. The importance of trust and privacy have grown in parallel with each other (Waldman, 2018) and acknowledging both conceptually, but also from a real life-living standpoint, is necessary. Each plays off each other and is a necessity for humans living in a modern computing age.

Thus, in sum, the need for and importance of trust in these specific human-technology-mediated interactions have grown much in part due to privacy implications.

Certainly, this entire chapter could be focused on human-human trust, as there are decades of insights and findings on this (Madhavan & Wiegmann, 2007 for a comparison of human-human and human-automation trust), but the focus here is on human-technology trust. Indeed, within the domain of human-technology trust, there are decades of insights as well, so what is presented here will be a small piece of the large puzzle. This section will be scoped mainly in terms of how technology has advanced through periods of automation and, more recently, autonomy (most notably linked to AI systems). As previously outlined, technology has greatly advanced in the past few decades, and most of that advancement has been related to paradigm shifts resulting in technologies becoming "smarter" and thus not requiring as much human control or interdependence as once needed. It is easy to see that as technology has become less reliant on human intervention and humans lose control over said technology, the relationship between humans' trust in and with technology has become more complex. We are currently seeing this complexity potentially peaking, as AI has been introduced to the population masses in a way that it never has previously. Below, we highlight the literature relating to trust in automation and trust in autonomy.

13.2.3.1 Trust and Automation

The history of trust in automation technology is mainly related back to the community of human factors, which is where HCI branched off from and still maintains an active presence (Meister, 2018). Human factors have long been interested in humans' behaviors, needs, and wants from and with technology and using those insights to better design said technology. The focus in the Human Factors community has long been on the control processes embedded within certain technologies and humans' interactions with them. Thus, there is a strong flavor of human interaction-based research with automation technologies. Automation technologies are defined by Parasurman and Riley (1997) as "the execution by a machine agent (usually a computer) of a function that was previously carried out by a human" (p. 231; see also Parsons, 1985). This is a broad definition, but it is exhibited by technological history that has shown automation technology first taking over many physical-based activities and then moving toward more cognitive-based activities. Parasuraman et al. (2000) later updated his own definition to provide more details: "a device or system that accomplishes (partially or fully) a function that was previously, or conceivably could be, carried out (partially or fully) by a human operator" (p. 286). This definition then led to the levels of automation approach, which, in a simplified manner and explanation, brought forth 10 levels of automation, with the higher levels (closer to 10) exemplifying the need for less human control and intervention. Thus, a great deal of research in automation and trust has been focused on the levels of automation paradigm, seeking to understand further the complexities of trust with lower and higher level automation, ultimately searching for a sweet spot in automation that will lead to the best and most calibrated human trust.

There are numerous findings relating to the relationships among and between trust and automation. Some of these findings are summarized below. Yet, the most consistent and general finding is that as levels of automation increase, humans' trust in technology becomes more complicated, and levels of trust and trustworthiness may decrease. Often, this decrease comes from an overreliance on automation, thus resulting in errors and then a loss of trust (see work related to the perfect automation schema; Lyons & Guznov, 2018). This finding is not always true, as trust is inherently highly individualistic, technologies vary in their capability and interaction method, and context plays an incredibly important role, but in most studies, this finding reigns true at some level. The trust in automation literature also brought forth an additional definition of trust that is viewed foundationally at the same level as Mayer. Work by Lee and See (2004) reviewed trust accounting specifically for context, automation characteristics, and cognitive processes impacted by trust; this work led to the definition of trust: "the attitude that an agent will help achieve an individual's goals in a situation characterized by uncertainty and vulnerability" (p. 33). One can see the focus of the human-technology relationship in this definition, most notably that the agent, in this case, technology, will *help* the human achieve their goals. Yet, Mayer's definition is still apparent in this definition as vulnerability is still at play. Lee and See's definition, in this author's opinion, is basically the foundation of Mayers' definition, just accounting for the introduction of technology. We highly encourage the reader here to look at Lee and See's canonical paper (2004) as there is a much more in-depth description of the multifaceted characteristics that impact human-technology-mediated trust.

It was noted that there have been many findings in the trust and automation literature. Work by Hoff and Bashir provided a comprehensive literature review of the trust in automation work. Some of the findings that they most notably highlight are the importance of different types of trust and their impacts on the setting of human-automation interaction. In addition, this has also been an overall trend and finding in the trust literature: trust needs to both be conceptualized and measured with more specificity and nuance than just the general concept of trust (Couch & Jones, 1997; Ullman & Malle, 2018). Specifically, they note that dispositional trust, situational trust, and learned trust are all at play during these types of interactions. There is value in quickly overviewing what each of these types of trust represents as it provides an additional level of understanding to trust and a more concrete illumination regarding trust and automation. Dispositional trust focuses on the human themselves and their individual makeup, which leads them to either have a propensity to distrust or trust a technology. This trust is very much void of the context or the specific technology. Hoff and Bashir (2014) noted that this type of trust stems from long-term individual behaviors coming from biological and environmental characteristics and that these are most likely to be very stable. Specific characteristics of dispositional trust may be personality (Merritt & Ilgen, 2008), age (Pak et al., 2012), culture (Huerta et al., 2012), and gender (Tung, 2011). Regarding situational trust, this is highly dependent on the context and dependent on both external factors of the actual environment and internal contextualized factors of individual human beings, which may change based on the environment. The type of learned trust depends on a human being's previous experience and how they use that experience to mediate their trust in an automated technology. If a human has more frequent interactions with technology, then they have more data points on how to mediate their own trust in technology. In addition to the frequency of interactions, the quality of interactions directly dictates trust (such that more poor interactions with automated technology lead to less trust).

13.2.3.2 Trust and Autonomy

The next step in technological advancement is autonomy. This is the type of technology that society is increasingly becoming aware of and interacting with. Whereas automation mainly allows for a sense of human intervention and control based on how the technology was originally programmed, autonomy does not provide this same level of control and often requires humans to rely on the technology in a manner where they are not familiar with how it will react. Thus, this type of technology serves the potential to both increase the importance of human-technology relationships but also muddy them in a manner of complexity we have never seen before.

McNeese (the second author of this chapter) has worked extensively with autonomy and, more specifically, with human-autonomy interactions. As technology has advanced, humans are often left wondering how to interact with and trust autonomous technology due to a lack of a mental model guiding them in this interaction. Mental models often guide humans in almost everything they do, as they are a cognitive schema based on previous experience that allows us to predict and then implement human behavior. In the case of technology, we have mental models in place that are typically guiding our interaction with said technology. For instance, based on previous experience, when interacting with an at-home smart assistant, we know to use a turn-taking communication pattern (you ask, and then the technology responds). This is due to a mental model of understanding this type of human-automated technology interaction. Yet, in the case of human-autonomy technology interaction, the human is in a bit of a conundrum in terms of not having a fully developed mental model to guide interaction, and thus this directly harms trust development. For the most part, most of society does not have rich and frequent interactions with autonomy; thus, most of society lacks experience; thus, most of society lacks human-autonomy technology mental models. What actually occurs when most humans are interacting with autonomy is a type of recognition-primed decision-making (Klein, 1993), where the human is searching for the most similar experience to guide them because they do not have an exact experience that fits the current interaction. The second author posits that humans combine both their human-human interaction mental models and human-automated technology mental models to help them respond to human-autonomy technology interaction, and what typically happens is a misaligned interaction and decreased trust (Schelble et al., 2022c).

To better understand the complicated relationship of trust in human-autonomy interactions, one must understand autonomy. In terms of definitions, there are many that have been put forth, but the authors of this work prefer the following: "systems which have a set of intelligence-based capabilities that allow it to respond to situations that were not programmed or anticipated in the design (i.e., decision-based responses). Autonomous systems have a degree of self-government and self-directed behavior (with the human's proxy for decisions)" (p. 3, USAF, 2013). Work by McNeese and colleagues (O'Neill et al., 2022) has outlined the definition of autonomy and its distinction from automation. The levels of automation approach that were previously noted makes differentiating automation and autonomy quite difficult; as you near higher levels, the original level approach notes that the technology is getting more autonomous, yet it is still classified as a level of automation. Very recently, McNeese and colleagues (2022) worked to align the levels of automation approach with autonomy; in this work, they took the levels of automation and redefined it with levels of autonomy to better understand how autonomy is nuanced and can also fall into differing levels.

A great deal of recent work has focused on the relationship and role of trust within autonomous systems. Work ranging from recommender systems (O'Donovan & Smyth, 2005), autonomous vehicles (Abraham et al., 2017), smart home autonomy (Michler et al., 2020), and autonomous teammates (McNeese et al., 2023), otherwise known as human-autonomy or human-AI teaming (McNeese et al., 2018), have all found compelling insights into how humans perceive trust in this relationship and paradigm. In general, science is still organizing itself regarding foundational findings and outcomes relating to trust in autonomous technology. Simply, there are many findings that both align with each other, but there are many that are misaligned. Yet, at this point in time, it is clear that the mixed bag of results relating to trust and autonomy are tied to a few main areas: (1) context matters more than ever; where and what humans are interacting with is highly varied, and this results in differing findings; (2) experience matters; people with more experience with autonomy are generally more trusting; and (3) individual differences are of paramount importance in deciding trust in autonomy. When compared with some of the findings in trust in automation, we can certainly see some overlapping trends with dispositional trust, situational trust, and learned trust. More research is needed as the technology is ever changing. The rapid acceleration of technological growth is part of the problem in identifying firm high-level empirical findings.

Indeed, it is difficult to stabilize and identify high-level trends in trust and autonomy; but it is possible to better understand findings relating to this relationship when a more specific area is investigated. For the purposes of providing specific findings, a quick overview of the work on trust in human-autonomy teams (McNeese et al., 2018; O'Neill et al., 2022; Zhang et al., 2021; McNeese et al., 2023) is presented. McNeese & and colleagues have studied trust in human-autonomy teaming a multitude of times with different focuses in different contexts. In a remotely piloted aircraft system context, it was found that lower levels of trust in the autonomous agent were present in lower-performing teams and that trust in the autonomous agent decreased over time regardless of performance (McNeese et al., 2019; 2021). In a similar context, findings empirically link team interaction dynamics to the development of trust and that trust in the autonomous teammate is specifically linked to failures of the autonomous teammate (Demir et al., 2021). In addition, research has explored the relationship of trust and ethical behavior in human-autonomy teaming, finding that trust in the autonomous teammate is empirically influenced by ethical violations and that attempts to repair trust in the autonomous teammate did not work when the autonomous teammate conducted an ethical violation (Textor et al., 2022). In related research, results indicate that human-autonomy teams with an unethical autonomous teammate had lower team trust and lower trust in autonomy (Schelble et al., 2022a). Qualitative work focused on the same paradigm of ethics and trust shows that humans indicate a parental relationship with autonomous teammates, one in which the autonomy must earn the trust of the human (Lopez et al., 2023). In addition to this empirical work, much has been written about the conceptual nature of trust in human-autonomy teams: see (Huang et al., 2021; Caldwell et al., 2022; Ezer et al., 2019; Schelble et al., 2022b). Albeit this is a small review of the trust in human-autonomy teaming literature, it gives one an idea of how potentially convoluted and expansive the trust in human-autonomy interactions literature is as a whole.

13.2.4 DESIGNING FOR TRUST IN AUTONOMY: REPAIR AND CALIBRATION

Clearly, trust is a critical component to establishing and maintaining safe and optimal human-technology interactions. Thus, it lends credence to the idea that we must be thinking about the concept at the forefront of the design of autonomy that will have forward-facing interactions with humans. There is a strong need to take a human-centered design approach to autonomy, not just to create trustworthy interactions but for many reasons, such as safety, reliability, accuracy, and optimal performance. Yet, designing autonomy for humans to trust is easier said than done, and many have been working on this for years; as noted in the automation section, there is a long history of trust in technology. The work here will continue as the multifaceted nature of the human-technology relationship continues to evolve, thus necessitating the need for continual introspection and refinement of what the key criteria are for designing and building trust. Currently, there is an abundance of work examining how to design and build trust by utilizing methods such as explainability (Weitz et al., 2019), anthropomorphism (Jensen et al., 2021), virtual reality (Morra et al., 2019), and virtual assistants (Galdon et al., 2021) to name a few.

Among the largest areas focused on designing for trust are trust repair and trust calibration. Trust calibration is an endeavor that scientifically attempts to pinpoint what the right amount of human trust a human should have in a technology. One can think about trust as being a spectrum of having too little or too much, with obvious issues stemming from each. Trusting too much is known as overtrust, often resulting in humans becoming over-reliant on technology and resulting in significantly negative outcomes (Parasuraman & Manzey, 2010; Robinette et al., 2016). Trusting too little is known as undertrust, and it can lead to a lack of monitoring and an unbalanced workload (de Visser et al., 2020). Calibration is concerned with identifying what the appropriate amount of trust for the human, the task, and the overall context (not too little and not too much) is. There is incredible complexity and nuance to this, as there are many variables at play that may change the equation resulting in proposer calibration.

In a similar vein, trust repair in the context of human-technology interaction is focused on attempting to repair the loss of trust that a human may have in a technology. It is important to note that trust repair is a general strategy not specific to technology, and recently a robust review of the concept was published by Sharma and colleagues (2023). In said review, trust repair is defined as "any increase in trust above the post-transgression level and complete repair as an increase in trust to the pre-transgression level" (p 363, Sharma et al., 2023). De Visser and colleagues (2018) have led the way in studying trust repair in the human-technology domain, outlining the conceptual importance of the strategy while also providing a robust review of the literature. Notably, there are many trust repair strategies that have been devised and studied, some including denial, apology, explanations, and promises (Esterwood & Robert, 2021). The effectiveness of each strategy is mixed and, much like most things in this area, highly dependent on context. In general, though, trust repair strategies have shown promise to bring back human trust in technology if designed and applied in an appropriate manner and context.

13.2.5   Trust as a Component of Acceptance

Humans' acceptance of technology is paramount to their adoption and use of technology, and often acceptance is explicitly linked to trust. The most pronounced and relevant research focus in this area is the technology acceptance model that seeks to predict human behavior of whether humans will reject or accept technology (see review Marangunić & Granić, 2015). In general, it has been found that the two most critical components allowing for prediction are perceived utility and ease-of-use.

> Recently, in hopes of improving both perceived utility and ease-of-use, autonomy research has shifted its focus toward the idea of designing AI to be human-centered. Specifically, this human-centeredness is often achieved by creating design recommendations for AI systems that researchers and developers can use as guidelines for building AI systems that benefit humans (Amershi et al., 2019). These recommendations include recommendations for AI systems that consider the differences between individuals that may impact how they perceive, accept, and interact with AI (Yi et al., 2005). These recommendations can often be targeted toward increasing the perceived utility of an AI system, such as increases in algorithmic precision and explainability and the creation of educational materials (Arya et al., 2020; Nadarzynski et al., 2019). However, recommendations can also specifically target the ease-of-use of AI tools, such as the use of voice interaction, conversational speech, or visual communication (Vashistha et al., 2019, Dalton et al., 2018).
>
> *(Quoted section taken from Flathmann (2023) with permission).*

There are many works that have linked trust to the technology acceptance model (see Pavlou, 2001, 2003; Belanche et al., 2012; Vorm & Combs, 2022). A meta-analysis on the relationship of trust and the technology acceptance model from 2011 indicates a significant influence of trust on the model's constructs. Moving forward, the role of trust must be accounted for if we are to design technology that is aligned with human acceptance.

13.2.6   Relationship between Privacy and Trust

Although this chapter has addressed a wide variety of theoretical perspectives on privacy and trust, we have yet to carefully delineate any theoretical perspectives on how these two concepts are related to each other. An important interdisciplinary perspective on this relationship is provided by the "APCO model" (Smith et al., 2011), which outlines the antecedents and outcomes of privacy concerns. Interestingly, both the APCO model and its successor (Dinev et al., 2015) present the relationship between privacy and trust as bidirectional: Some see privacy concerns as a mediator between trust and disclosure (or other privacy-related behaviors), where a lack of trust may heighten privacy concerns (Malhotra et al., 2004; Van Slyke et al., 2006; Xu et al., 2005; Zhou, 2012a), while others see the formation (or lack) of trust as a mediating step between privacy concerns and disclosure (Dinev et al., 2006; Dinev & Hart, 2006). Yet other researchers suggest that trust moderates the effects of privacy concerns on information disclosure and other privacy-related behaviors: privacy concerns are less likely to influence disclosure when trust is high (Smith et al., 2011). Taking a different perspective, research on teamwork has found that privacy concerns may create a barrier to the formation of trust in teams (Musick et al., 2023), as the formation of trust within teams is usually predicated on extensive sharing of information between team members.

HCI Opportunity: As trust is an important factor in determining users' system-specific privacy concerns, and subsequently their disclosure behaviors, it can be beneficial for data-intensive applications to build a trust relationship with the user. For instance, trust was shown to be a determining factor in the emergency adoption of collaboration technology during the COVID-19 pandemic (Namara & Knijnenburg, 2021). Kobsa et al. (2016) showed that trust can have a rational influence (rooted in users' perceptions of risk and their system-specific privacy concerns) as well as a heuristic influence (rooted in the affect heuristic). Along these lines, HCI practitioners could aim to build trust rationally by making sure that users privacy concerns are always accounted for (i.e., not miti-gated retroactively in reaction to a privacy incident), or heuristically by leveraging trust-inspiring design, or a strong brand name.

13.3   MEASUREMENT OF PRIVACY AND TRUST

Privacy and trust play a pervasive role in HCI – whether it be toward computing systems, or inter-personally (mediated by computers). As such, it is important in many HCI studies to accurately measure privacy and/or trust.

As mentioned, users' privacy and trust-related perceptions and attitudes do not always align with their behaviors (cf. the "privacy paradox" (Gerber et al., 2018; Kokolakis, 2017; Norberg et al., 2007)). Conversely, behaviors that may appear like expressions of trust or meant to pro-tect one's privacy may sometimes have an unexpected cause (for instance, users may refuse to disclose certain personal information not because of privacy concerns, but because they think the information does not accurately represent them (Knijnenburg & Kobsa, 2013b)). As such, it is important to determine for each study or field trial whether the concepts of privacy and/or trust should be measured subjectively (i.e., as perceptions or attitudes), objectively (i.e., as (an) expressed behavior(s)), or both.

Given the existence of attitude-behavior gaps in both privacy (Kokolakis, 2017) and trust (Dunning et al., 2012), the use of behavioral indicators to measure these concepts must be approached with extreme care. For instance, increased disclosure may be caused by heuristic influ-ences (Acquisti & Grossklags, 2006) rather than reduced privacy concerns: a cookie notice that causes an increase in acceptance rates may *seem* like it reduced users' concerns, but it may simply be misleading users to select "accept", thereby exacerbating the problem in the long run (Degeling et al., 2019). To avoid such situations, researchers are advised to triangulate behavioral data with subjective measures that aim to measure trust and/or privacy in a more unambiguous manner (Knijnenburg et al., 2012).

Privacy and trust are both complex, multifaceted concepts that have context-dependent defini-tions (Knijnenburg et al., 2022b). As such, measuring users' subjective perceptions and/or attitudes of these concepts is a challenging endeavor. Researchers in the field of psychometrics argue that it is best to measure such complex constructs with multi-item scales (Diamantopoulos et al., 2012), and that these scales must be appropriately tailored to the correct context and/or target (DeVellis, 2011). In this section, we cover existing scales (and, where applicable, behavioral indicators) to measure trust and privacy as dispositions, toward a system, and toward other people. In many cases, HCI researchers can (and should) adopt these existing measurement scales in their work, thereby ensur-ing robust measurement and improving comparability between research efforts. In other cases, it may be necessary for HCI researchers to develop their own scales (or carefully adapt existing scales), so as to carefully match a novel context and/or target. Researchers are encouraged to consult DeVellis (2011) for a comprehensive primer on scale development.

13.3.1   INFORMATION PRIVACY AND TRUST IN TECHNOLOGY AS DISPOSITIONS

We first cover information privacy and trust in technology as personal traits or dispositions – in other words, the tendency of a person to be concerned about their privacy, or to be trusting, in their interaction with a computing system (e.g., a hardware device, computer program, Web site, or app). Personal traits tend to have normative roots (Utz & Kramer, 2009; Xu et al., 2008), which has several consequences. For one, they are relatively stable (Bansal et al., 2010), meaning they do not vary in the short term and tend to only be influenced by sustained experiences or life-changing events. Furthermore, their normative basis means that a person's privacy and trust dispositions are informed by their culture.

13.3.1.1   Information Privacy Concerns

The most concerted effort toward scale development for privacy has arguably occurred in the area of *information privacy concerns*. A very substantial contribution in this area has been the recurring privacy surveys by Westin and Harris Interactive, which started in the early '80s (Harris Interactive Inc., 2000; Harris et al., 2003a; Westin et al., 1981). They identified and subsequently refined three types of users in their surveys: privacy fundamentalists, the unconcerned, and a pragmatic majority (Harris et al., 2003a). Westin's categorization is based on users' answers to three questions that measure their general privacy attitudes on a four-point scale (Kumaraguru & Cranor, 2005). While the existence of three user types has found substantial adoption in privacy literature, researchers have recently begun to criticize Westin's categorization. For instance, in a broad 2014 survey of Internet users, Woodruff et al. (2014) found a much larger contingent of privacy fundamentalists (larger than the pragmatic majority). They also found that neither Westin's categorization nor the underlying questions showed a strong correlation with users' behavioral intentions and their reactions to scenarios outlining privacy consequences. They argue that the multidimensionality of privacy attitudes may be one reason for this lack of correlation.

An early attempt at a multidimensional measurement scale was made by Smith et al. (1996). Their Concern For Information Privacy (CFIP) scale consists of 15 items divided over four dimensions that measure individuals' general concerns about information privacy practices around *collection, unauthorized secondary use, improper access*, and *errors*. The psychometric properties of this scale were confirmed in 2002 by Stewart and Segars (2002).

Whereas the CFIP scales were developed before the rise of the Internet, Malhotra et al. (2004) developed the Internet Users' Information Privacy Concerns (IUIPC) scales with the explicit purpose of capturing privacy concerns in an online world. Malhotra et al.'s scale establishes three dimensions – *collection, control,* and *awareness*, with the IUIPC scale itself serving as a higher-order factor. While the IUIPC scales are used extensively to measure users' general privacy concerns, some researchers find low discriminant validity between the scale's dimensions (Groß, 2021; Knijnenburg & Kobsa, 2013b) (suggesting that they are conceptually indistinguishable), while others have found low predictive validity with general risk beliefs and general trusting beliefs (Sipior et al., 2013) (suggesting that higher levels of concern as measured by IUIPC are not predictive of higher levels of risk or lower levels of trust).

More recently, researchers have started to realize that the variety of Internet-based technologies is so high, that measurement instruments for privacy concerns must be tailored to a more specific context than "the Internet." This has sparked a search for the optimal balance of general applicability and context-specificity for privacy measurement scales (Knijnenburg et al., 2013a). Good examples of more specific privacy measurement scales are the Mobile Users' Information Privacy Concerns (MUIPC) scales (Xu et al., 2012), the Perceived Surveillance scale (Segijn et al., 2022), the video surveillance scale (Koshimizu et al., 2006) and the Attitudes Towards Cybervetting (ATC) scale (Cook et al., 2020).

13.3.1.2   Information Privacy Protection Behaviors

Information privacy-related behaviors can generally be divided into two categories: disclosure behaviors and protection behaviors. While several scholars have attempted to define a set of cohesive disclosure behavior scales (e.g. Lusoli et al., 2012; Phelps et al., 2000; 2001), Knijnenburg et al. (2013a) argue that the disclosure of different types of information, and even the dimensionality of the disclosure behaviors themselves, are highly dependent on the recipient of the information.

A more generic set of scales can be developed in the realm of protection behaviors. Buchanan et al. (2007) uncover two dimensions among 12 protection behaviors: a *general caution* dimension and a *technical protection* dimension, both of which are correlated with the IUIPC scale. Likewise, Lusoli et al. (2012) found six dimensions of privacy protection behavior: reactive practices (e.g. spam- and spyware filters), proactive practices (e.g. contacting websites about their privacy practices), withholding information, minimizing disclosure, avoiding the use of technology, and lying.

HCI Opportunity: While personal traits like information privacy concerns and trust dispositions are generally considered stable and thus outside the direct influence of computer systems, it is still useful for HCI researchers and practitioners to measure these traits in their user experiments or field trials. People's privacy and trust dispositions may influence subsequent privacy or trust-related attitudes and behaviors toward a system *alongside* the influence that the system itself may have (Knijnenburg et al., 2012), thus acting as a *precision variable* in statistical analyses. Furthermore, users' information privacy concerns and trust dispositions may *moderate* the effect a system has on their privacy attitudes or disclosure behavior (Ghaiumy Anaraky, 2022; Knijnenburg & Kobsa, 2013a; Kobsa et al., 2016; Namara & Knijnenburg, 2021); concerned/distrusting individuals are generally more sensitive toward a system's privacy practices, whereas unconcerned/trusting individuals are generally more amenable to reap additional benefits offered by the system.

13.3.2 INTERPERSONAL PRIVACY AND TRUST AS DISPOSITIONS

Whereas the previous subsection covered users' privacy and trust dispositions from an interacting-with-technology perspective, we here cover *interpersonal* privacy and trust dispositions, primarily because of the large theoretical discrepancy between the information-centric and interpersonal perspectives on trust and privacy.

Interestingly, despite a prevalence of theories about computer-mediated interpersonal privacy and trust, few comprehensive efforts have been made to carefully measure users' interpersonal privacy and trust dispositions (Page et al., 2013c). Below, we cover existing attitudinal and behavioral scales and call for additional research in this direction.

13.3.2.1 Interpersonal Privacy Attitudes and Concerns

In contrast to the substantial efforts to develop standardized scales to measure information privacy concerns, no widely known scales exist that measure general interpersonal privacy concerns. One *ad hoc* scale development attempt was conducted at the CSCW2013 Networked Privacy workshop (Page et al., 2013c), and the resulting scale got adapted over time and used in several works (Knijnenburg & Kobsa, 2014; Li, 2020; Najafian et al., 2023). This scale has not undergone a rigorous validation process, but it has generally produced robust results.

Perhaps the most comprehensive effort to develop a scale that measures interpersonal privacy as a personal trait comes in the form of the Boundary Preservation (and Enhancement) scales developed by Page et al. (2012, 2013b). The concept of Boundary Preservation Concerns (BPC) originally arose in the context of location-sharing social media, where a single item was used to measure the extent to which someone is worried that their relationship with others will change when using location-sharing social media. Page et al. found BPC to be a root cause of several subsequent privacy concerns (Page et al., 2012, 2013b). In a subsequent survey study (Page et al., 2019) the concept was applied to social media in general, and the scale was expanded to eight items measuring two dimensions: Boundary Preservation Concerns and Boundary Enhancement (BE; i.e., the belief that the use of social media can *improve* interpersonal relationships). Note that BPC and BE are platform-specific (and thus not truly personal traits), with users of different platforms expressing different levels of BPC and BE. Furthermore, BPC and BE have been found to be significantly predictive of social media platform use (Page et al., 2019).

Whereas BPC and BE are platform-specific, a truly platform-independent personal trait related to interpersonal privacy is the "FYI communication style" trait (Page et al., 2013a) which measures whether someone prefers to communicate information to and receive information from others in a direct, one-to-one, synchronous manner (low FYI) or an indirect, one-to-many, asynchronous manner (high FYI). This scale was also initially developed for location-sharing social media but later extended to a broader social media context (Page et al., 2019). Users who score high on the FYI communication style scale tend to score lower on PBC and higher on BE regardless of the platform. Moreover, the latest version of the FYI communication style scale consists of two dimensions: *FYImy* measures one's preferences for communicating information to others, while *FYIothers* measures one's preferences for receiving information from others (Page et al., 2019).

13.3.2.2 Boundary Management Behaviors

Interpersonal privacy behaviors have been studied rather extensively, and their dimensionality tends to be considerably more complex than the information privacy protection behaviors outlined in the previous section. Specifically, Karr-Wisniewski et al. (2011) defined ten dimensions of boundary management behaviors along five boundary types: relationship boundaries can be regulated in terms of connection (who has access to one's social network) and context (what interpersonal interactions are allowed for each relationship); network boundaries are regulated in terms of discovery (who has access to one's connections) and intersection (how interaction between connections or groups of connections are regulated); the regulation of territorial boundaries concerns both inward-facing boundaries (regulating incoming context) and outward-facing boundaries (regulating semi-public content available to others); disclosure boundaries involve self-disclosure (how one discloses information to one's network) and confidant-disclosure (how co-owned information is managed); interactional boundaries are regulated through disabling (turning off certain interactive features) and blocking (disallowing access to oneself for certain individuals). Note that the measurement of these boundary regulation behaviors is platform-specific, as each social media platform has a different set of features to regulate these boundaries (Wisniewski et al., 2012). The varying default settings and prominence of these features make comparisons across platforms rather challenging. Furthermore, the ever-changing nature of social media platforms makes it difficult to study these behaviors over time.

Research shows that while users' privacy boundary management behaviors may be a consequence of privacy concerns (Wisniewski et al., 2012), their implementation does not necessarily reduce the effectiveness of the social media platform to support social connections. In fact, research shows that users who are able to regulate their privacy boundaries to a level that matches their privacy requirements perceive higher levels of social connectedness and social capital than those whose boundary regulation behaviors lag their privacy desires (i.e., those who experience a state of *social crowding*) (Wisniewski et al., 2015a).

Whereas the boundary management behaviors identified by Karr-Wisniewski et al. (2011) focus on individual users' actions to manage their privacy, several other researchers have focused on users' collaborative privacy management behaviors. Cho and Filippova (2016) conducted a focus group study with social network users in Singapore and identified four types of "networked" privacy management behaviors: collaborative behaviors (privacy management through discussion and coordination with confidants), corrective behaviors (changing or undoing confidant disclosures), information control behaviors (managing boundaries to reduce the amount of confidant disclosures) and preventive behaviors (reducing disclosure about oneself or others to reduce the spread of personal information). In a follow-up survey study conducted in both Singapore and the USA, they found that the prevalence of these behaviors was significantly associated with users' privacy concern and to a lesser extent with their self- and/or collective efficacy.

In a parallel effort, Jia and Xu (2016) identified three categories of collaborative privacy management behaviors (based on established measurements and theoretical conceptions of privacy management): collaborative ownership management (collaboratively deciding what collective information to post online), collaborative access management (collaboratively deciding to limit or reduce access to collective information posted online), and collaborative extension management (collaboratively deciding to limit further disclosure of collective information beyond the current audience). They found that these behaviors were significantly motivated by collective disclosures (i.e., more disclosure would lead to more management behaviors) and by the group's propensity to value privacy.

13.3.2.3   Interpersonal Trust and Trust in AI

Interpersonal trust informs our relationships with not only other humans but also social structures that guide our daily behavior. Traditionally, interpersonal trust has seen quite a bit of focus within the social learning community, with one of the canonical works defining it as "a generalized expectancy held by an individual that the word, promise, oral or written statement of another individual or group can be relied on" (Rotter, 1967, 1980). Although there has certainly been an abundance of research on trust, in its general nature, over the years, there is still a great deal not known about how trust is developed and maintained and shapes and interacts with interpersonal processes (Simpson, 2007). As noted by Simpson (2007), the lack of concrete knowledge pertaining to interpersonal trust is most likely due to the complicated nature of the concept to begin with. Indeed, there are so many inputs that go into developing trust, and each is highly contextually dependent, meaning that trust can develop in many ways and manifest through behaviors in an equally complex manner. So, while there are certainly pockets of knowledge pertaining to what trust is, the insights become much more muddied when one seeks to understand how it develops and helps mediate interactions.

One of the most prominent perspectives that guides interpersonal trust is that of the interdependence theory-based approach (Simpson, 2007; Rempel et al., 2001). This approach focuses on the closeness of trust in relationships, classifying them as low, medium, and high, finding that humans involved in relationships that have high trust are more optimistic and benevolent toward their partner's motives and intentions. Not surprisingly, lower trust level leads to more cautious behaviors and perceptions. Indeed, trust acts as an interpersonal moderator for humans, often manifesting itself through more positive behaviors when high trust is present.

As previously discussed, the perceptions of humans' trust in both automation and autonomy are continuing to develop in real time, with experience and individual differences leading the way in molding those perceptions. The role of interpersonal trust within human-autonomy interactions is indeed an area where much more research is needed. Further work in this area will help richen the interactions that will occur between humans and AI agents moving forward.

HCI Opportunity: Despite the existence of several nuanced theories of interpersonal boundary regulation, current efforts toward the measurement of interpersonal privacy take a predominantly behavioral approach. The field of HCI research would benefit substantially from a more concerted effort to develop a universal instrument to measure "computer-mediated interpersonal privacy concerns." Arguably, the existence of such a scale would support the integration of the many research efforts that exist in this area and allow for comparisons of user concerns across platforms and over time. We acknowledge that the continual evolution of social media platforms likely requires further iterations of such a scale over time, but early efforts in this regard (Knijnenburg & Kobsa, 2014; Li, 2020; Najafian et al., 2023) demonstrate that developing a core set of items to measure users' concerns on social media is both plausible and useful.

13.3.3   Privacy and Trust toward a Specific System, Person, or AI

Moving beyond the measurement of privacy and trust as personal traits, we now discuss efforts to measure users' trust and privacy concerns regarding a specific target (i.e., a specific computing system, person, or AI). These measurement efforts are of particular interest to HCI practitioners and researchers, since many HCI studies consider the evaluation of a specific (new or existing) system, AI agent, or social media platform. As such, measurements of privacy and trust toward a system, person or AI can serve as dependent variables in such studies.

13.3.3.1   System-Specific and Person-Specific Privacy Concerns

No comprehensive efforts exist to develop a measurement scale for system-specific privacy concerns: this is arguably because such scales would have to be tailored to the specific system under investigation anyway (DeVellis, 2011). Despite the lack of a validated scale, several studies have attempted to measure system-specific privacy concerns, most prominently in the area of person-alized systems (Knijnenburg et al., 2010, 2012; Kobsa et al., 2016). These studies show that sys-tem-specific privacy concerns are influenced by people's general privacy concerns, their perception of the privacy protection offered by the system, and their trust in the provider of the system. System specific-privacy concerns may in turn reduce users' intention to provide information to the person-alized system, as well as their overall satisfaction with the system (the latter effect is particularly prominent among users with high general privacy concerns and high privacy self-efficacy) (Kobsa et al., 2016). A related concept that has regularly been considered is "perceived over-disclosure threat" which measures to what extent users believe that a system causes them to over-disclose their personal information (Knijnenburg, 2015; Knijnenburg & Cherry, 2016; Knijnenburg & Jin, 2013; Knijnenburg & Kobsa, 2013b, 2014). Studies measuring this concept find that, unsurprisingly, per-ceived over-disclosure threat increases with disclosure, but decreases with sufficient transparency and control. In turn, perceived over-disclosure threat reduces trust and satisfaction.

Another effort to measure system-specific privacy concerns is the Privacy Beliefs and Judgments (PB&J) framework. Hepler and Blasiola (2021) created a scale that measures for 16 "top-of-mind" privacy concerns whether the user thinks that a certain application or site engages in the con-cern-inducing behavior (Belief), and whether the user thinks it would be good or bad (or neither) if the app or site indeed engaged in that behavior (Judgment). The 16 items were grouped into inter-personal concerns, data collection and use concerns, and data access concerns. Note that this scale is specifically developed for social network applications. A nice thing about this scale is that, like, e.g., the Protection Motivation Theory (PMT), it disentangles users' beliefs (how likely is this to happen?) from their judgments (how would you feel about it?).

On the interpersonal side, concerns regarding a specific person are rarely measured in existing research. An exception is a study by Li et al. (2022b), who measured the perceived risk of sharing information with a specific friend/contact using a four-item scale. Their work shows that perceived risk significantly influences demographic information disclosure, social information disclosure, and acceptance of friend requests. Furthermore, perceived risk is influenced by users' disposition to privacy (i.e., their information and interactional privacy concerns) and by the contextual factors of the relationship with the other person (i.e., whether the user knows the person offline, whether they share group membership, whether they live in the same city, whether they have mutual friends, whether they go to the same college, and whether the other person posts about interesting topics).

13.3.3.2   Disclosure Behavior

On the behavioral side, Knijnenburg et al. (2013a) conducted a careful review of existing studies investigating users' patterns of information disclosure toward a system. They show that most stud-ies either consider users' disclosures to a system as independent decisions (e.g. Acquisti et al., 2012; Joinson et al., 2008) or as a single summated "disclosure tendency" score (e.g. De Souza & Dick, 2009; John et al., 2011; Joinson et al., 2010; Knapp & Kirk, 2003; Metzger, 2004, 2006, 2007). They instead advocate for an approach where disclosure behavior is treated as a multidimensional concept, where individual disclosure behaviors are neither completely independent nor constitute a single "disclosure tendency" score, but rather a series of sub-dimensions (e.g. Khalil & Connelly, 2006; Knijnenburg et al., 2013a; Phelps et al., 2000; White, 2004).

Moreover, Knijnenburg et al. (2013a) argue that there is no universal dimensional structure of information disclosure, but that this dimensional structure crucially depends on the context (e.g., the domain, the type of system that is receiving the information). Their work describes an elabo-rate process to identify the dimensional structure of disclosure behavior using Exploratory and Confirmatory Factor Analysis, as well as a means to cluster participants along the disclosure dimen-sions using Mixture Factor Analysis (Knijnenburg et al., 2013a).

In the context of interpersonal disclosure, Olson et al. (2004, 2005) conducted a comprehensive study measuring users' tendency to disclose 40 different items to 19 different types of people. They clustered both the items and the people. On the "item" side, they uncovered six clusters: email content, credit card data, and transgressions; failures, opinions, salary, and SSN; phone numbers, age, marital status, and successes; pregnancy, health information, and affiliations; work-related documents, websites, and availability; work email and phone number. On the "people" side, they identified five clusters: the public/a competitor, coworkers, managers, and trusted coworkers, family, and one's spouse. Knijnenburg and Kobsa (2014) added further granularity to this segmentation of recipients in a study measuring users' disclosure tendencies of 8 items toward 55 different groups of recipients. Their study presents several acceptable segmentations, from very coarse (two groups), to very granular (14 groups).

13.3.3.3   Trust Attitudes: Competence, Benevolence, Integrity, Functionality, Helpfulness, Reliability

There are many attitudes that directly impact trust, yet competence, benevolence, and integrity are among the most impactful. Competency is the ability to perform a task, duty, or role in a manner that is expected (Robert, 2002). Competency is moderated both by technical skill and ability but also one's social ability. In general, a more competent individual or technology will lead to higher levels of trust. Similarly, benevolence addresses the understanding that the person or technology that one is working with will act in a manner that is for the good of another person or the team (Mayer et al., 1995). Like competency, higher levels of benevolence lead to higher levels of trust. Finally, integrity is defined by the trustee confirming to a set of rules/principles that the trustor finds to be aligned with their expectations (Kim et al., 2003). Again, higher levels of integrity lead to higher level of trust.

When these attitudes are applied to trust in technology and acknowledged in HCI, they are often framed through their counterpart concepts: functionality, helpfulness, and reliability (McKnight et al., 2009). Functionality is the counterpart to competence and the technology is referred to in a manner that it can complete its functions given the affordances it is capable of. Helpfulness is the counterpart of benevolence, and technologically we do not align this with helpful emotions (as we do with humans), but there is the expectation that it will do what it is asked of and provide help when needed. As for reliability, its counterpart is integrity. There is an expectation that the technology will operate in a manner that it was designed to.

HCI Opportunity: As information systems increasingly depend on access to users' personal information, HCI researchers and practitioners should engage in efforts to determine the dimensionality of disclosure behaviors in their specific domain. As Knijnenburg et al. (2013a) point out, this is particularly useful in domains where data-intensive personalization efforts are crucial to the operation of the system (e.g. recommender systems, AI-based systems) Specifically, since personalized systems usually request access to a large amount of information that is beneficial but not crucial for the system to work, an in-depth understanding of users' multidimensional disclosure tendencies allows such systems to avoid requesting information the user is unwilling to disclose (or, conversely, to prioritize request for information that the user is likely willing to disclose) (Knijnenburg, 2015). Such adaptive information requests are yet another example of user-tailored privacy (Knijnenburg et al., 2022a).

13.3.4   Issues in Measuring Privacy and Trust

The measurement of users' privacy and trust-related dispositions and attitudes is not without issues. Perhaps the foremost issue is the fact that many HCI researchers and practitioners take a rather *ad hoc* approach to measurement: despite recommendations to deploy careful measurement practices (DeVellis, 2011; Gardner, 1996; Podsakoff et al., 2003) complex attitudes and dispositions are often measured using researcher-generated single item metrics.

Throughout this section, we have recommended that HCI researchers and practitioners adopt or adapt existing measurement scales. However, even this cannot be done without careful consideration of the context in which the scale is intended to be deployed. Next, we discuss issues around the context-dependence – and particularly the cultural dependence – of privacy and trust-related measurement scales.

13.3.4.1 Context-Dependence

At the beginning of this section, we noted how the well-studied attitude-behavior gap in privacy (Norberg et al., 2007) and trust (Dunning et al., 2012) research impacts measurement efforts. Research has shown that the attitude-behavior gap is widest when attitudes and behaviors are considered at a generic level (e.g., as general dispositions), and that this gap narrows as measurements become more contextualized (Manstead, 1996).

As such, privacy researchers can expect the gap between attitudes and disclosure behavior to be much narrower if they aim to measure users' privacy concerns regarding the disclosure of a specific item to a specific recipient (Knijnenburg et al., 2013a; Knijnenburg & Bulgurcu, 2023). Note, though, that even at this context-specific level, there may exist a discrepancy between users' privacy concerns and their actual privacy behaviors: the privacy paradox remains an undying conundrum in privacy research (Norberg et al., 2007; Sun et al., 2020). This emphasizes the importance of measuring both attitudes and behaviors in privacy-related research, and to carefully consider which of these two metrics one should design for or adapt to (Knijnenburg et al., 2017).

Simultaneously, HCI researchers and practitioners who aim to adopt an existing privacy or trust-related measurement scale should be considerate of potential discrepancies between the context in which the scale was developed and the context in which they intend to deploy it. If these two contexts deviate to some extent, it may be fruitful to re-validate the scale, and perhaps even to adapt some of the items to the next context (DeVellis, 2011).

13.3.4.2 Cultural Differences

On a similar note, HCI researchers and practitioners should consider the cultural context in which the privacy and trust-related measurement scales they aim to use are developed. Privacy and trust are culturally created and embedded phenomena, meaning that their antecedents, consequences, and cognitive interpretations may differ across cultures (Berkovsky et al., 2018; Li, 2022).

Several researchers have studied such cross-cultural differences in the context of both privacy (Chen et al., 2008; Cho et al., 2018; Dourish & Anderson, 2006; Harris et al., 2003b; Li et al., 2017, 2022a, b; Namara et al., 2018b; Ozdemir et al., 2016; Rui & Stefanone, 2013) and trust (Berkovsky et al., 2018; Jarvenpaa et al., 1999; Kim, 2008). This research finds both main-effect differences in trust and privacy concerns between countries as well as interaction effects (i.e., the antecedents and/or consequences of privacy differ in strength between cultures) (Berkovsky et al., 2018; Li et al., 2017, 2022a). Researchers can generalize these cross-country differences by explaining them as *cross-cultural* differences (Berkovsky et al., 2018; Li et al., 2017), where culture is usually measured at the country level using Hofstede's dimensions (Hofstede, 2011) or Schwartz's cultural values (Schwartz, 1994). Cultures are of course more granular than the country level, but research has shown that measuring culture at the individual level is particularly difficult (Ghaiumy Anaraky et al., 2021).

Another issue concerns the generalizability of privacy and trust-related measurement scales across cultures. As the cognitive interpretations of privacy and trust are culture-specific, it is possible that the measurement scales we use to measure privacy and trust-related concepts do not apply universally across cultures. Indeed, several studies have demonstrated a certain amount of "measurement non-invariance" in privacy and trust-related scales (Cho et al., 2018; Li et al., 2021; Reeskens & Hooghe, 2007). HCI researchers and practitioners should thus aim to re-validate scales that were developed in a different cultural context and be particularly careful about measurement invariance when conducting cross-cultural research studies.

HCI Opportunity: Due to the context-specific nature of system-specific privacy concerns, measurement scales for system-specific privacy concerns must be tailored to the domain in order to be useful. Therefore, to improve the methodological rigor of domain-specific privacy research, HCI researchers and practitioners in various domains should engage in efforts to develop and test a system-specific privacy concerns scale for their specific domain.

On a similar note, HCI researchers and practitioners should make a concerted effort to determine the cultural (non-)invariance of their privacy and trust-related measurement scales. Where differences exist, efforts can be made to make measurement scales more universal, or to create culture-specific variants of existing measurement scales.

13.4   MOVING FORWARD

This chapter has presented a high-level overview of HCI-relevant research on privacy and trust. As a primer for researchers and practitioners, we aimed to focus this chapter on timeless aspects of privacy and trust that can serve as a basis for contemporary research efforts. As such, we covered several theoretical perspectives that can be employed to conceptualize privacy and trust, followed by an overview of the methods that can be employed to measure these concepts.

Importantly, throughout this chapter, we have outlined a number of opportunities for HCI researchers and practitioners in the space of privacy. These can be summarized as follows:

- To increase the adoption of an app or service, highlight the benefits of the system and mitigate any perceived risks associated with disclosure.
- To support users' privacy decision-making practices, give users intuitive information (transparency) and interface options (control).
- Design systems with an inherent adherence to privacy, as opposed to privacy features being added *ad hoc*.
- Leverage the concept of "privacy nudging" but be aware of its potential unintended consequences.
- Provide user-tailored privacy decision support to actively help users bridge the gulf of execution and evaluation that exist between users' privacy preferences and the available system settings.
- Design privacy-setting interfaces that are attuned to increased deliberation.
- Assess the privacy afforded by new and existing social media platforms against the multidimensional criteria outlined in existing guidelines.
- Develop interaction mechanisms that allow users to form and control the "boundary alliances" that are created when information is collectively managed.
- To meaningfully support collective privacy management, respect culturally specific practices.
- Build trust rationally by making sure that users privacy concerns are always accounted for, or heuristically by leveraging trust-inspiring design, or a strong brand name.
- In researching trust and privacy, account for the potentially moderating effect of privacy concerns and trust dispositions.
- Support the development of a universal instrument to measure "computer-mediated interpersonal privacy concerns."
- Determine the dimensionality of disclosure behaviors in your specific domain of practice/research.
- Develop and test a system-specific privacy concerns scale for your specific domain of practice/research.
- Determine the cultural (non-)invariance of any privacy and trust-related measurement scales used in your studies.

Reflecting upon this chapter, we see a clear distinction between the domain of information privacy and trust in technology (the domain of most commercial and corporate information systems) and the domain of interpersonal privacy and interpersonal trust (the domain of computer-mediated communication and social media). The relation of AI systems to these two domains, however, is somewhat ambiguous, since AI agents can be treated either as a system or as an autonomous entity.

Even within each domain, we find that privacy and trust-related dispositions, attitudes, and behaviors crucially depend on the context. This includes the type of boundary that is being negotiated, the situation in which the privacy or trust decision takes place, and the characteristics of the other party. This context-dependency may make privacy and trust difficult to design for – *user-tailored* solutions may have an advantage here.

This chapter does not cover domain-specific *solutions* that improve privacy and/or trust. Due to the context-dependent nature of privacy and trust and the ever-evolving nature of computing technology, we considered this to be a futile effort. Interested readers in contemporary privacy solutions can consider selected chapters of the Modern Socio-Technical Perspectives on Privacy book (Knijnenburg et al., 2022b) (e.g., the chapters on privacy-enhancing technologies, social media, tracking and personalization, healthcare, the Internet of Things). For solutions regarding trust, it is best to look toward the rich literature in this area. There is an overwhelming amount of research on the general concept of trust, yet a great deal is needed that focuses on human-autonomy-based trust.

Moreover, this chapter does not cover privacy and trust-related regulations – again, because we expect significant advances in this domain to quickly render anything we could include in the chapter outdated. Interested readers can read up on recent advances such as the European General Data Protection Regulation (e.g., Diamantopoulou et al., 2022), the California Consumer Privacy Act (e.g. Stallings, 2020), and the EU Artificial Intelligence Act (e.g., Veale & Borgesius, 2021), keeping in mind that these regulations are still in flux.

Furthermore, while the theoretical perspectives on privacy and trust covered in this chapter provide a descriptive account of privacy and trust in HCI (i.e., how privacy and trust occur in computing systems), we have refrained from covering moral or ethical frameworks of privacy and trust that provide a prescriptive angle (i.e., how privacy and trust ought to be handled by computing systems). However, this omission should not be mistaken for an implied lack of importance. Indeed, we strongly believe that privacy researchers and practitioners should consider the ethical implications of their work, especially when it comes to controversial topics like privacy and trust. Navigating the ethics of privacy and trust is a wicked problem, though, because privacy and trust have external objective (i.e., computational models of privacy and trust), subjective (i.e., people's opinions about privacy and trust) and behavioral (i.e., what can be inferred from their actions) definitions. These definitions often do not align, and it is difficult to determine for which "version" of privacy and trust HCI designers and practitioners should optimize their systems (Knijnenburg et al., 2017). This problem is further exacerbated in situations where the privacy and trust of multiple stakeholders must be reconciled (Ekstrand et al., 2018).

As a careful consideration of ethical perspectives on privacy and trust would likely double the length of this chapter, we instead refer the interested reader to several excellent resources (e.g., (Kisselburgh & Beever, 2022; Proferes, 2022)) in case they wish to further explore and address these issues. We have positioned this chapter as a practical primer that offers HCI researchers and practitioners the knowledge and awareness of theoretical and measurement issues that are indispensable in enabling them to do so.

NOTES

1 Since the persons whose information is in the information system do not have to be its primary users, we prefer to refer to them as "data subjects."

2 A famous example of privacy violation through incorrect predictions revolves around the AI-powered DVR TiVo, which one user described as persistently recording TV shows with gay themes after they watched particular items; its algorithm apparently "overfitting" a previously encountered social information pattern (Zaslow, 2002).

3 A famous example of privacy violation through creepy predictions revolves around a 14-year-old girl whose father found out via a personalized Target advertisement that she was pregnant: Target had predicted the pregnancy based on her shopping patterns and sent her parental home an ad leaflet with baby products in the mail (Duhigg, 2012).

4 A famous example of re-identification revolves around a dataset Netflix released to researchers as part of a contest to improve its algorithm; researchers were able to "de-anonymize" the data by cross-referencing ratings with (public) IMDb profiles. A closeted lesbian sued Netflix in response, alleging that the de-anonymization procedure could "out" her based on her viewing behavior (Singel, 2009).

5 A famous example of privacy violation through unexpected re-sharing revolves around a communications director who shared off-key jokes on her personal Twitter account as she was boarding a trans-atlantic flight: by the time her flight landed, one of her tweets had caused an international outrage (Ronson, 2015).

6 A special example of context collapse revolves around Facebook's "Beacon" advertisement system, which would automatically broadcast users' online purchases made outside the Facebook platform to their Facebook friends. This novel transmission principle was so unexpected that a backlash and class-action lawsuit ensued (Schiffman, 2007), and Facebook eventually had to shut down the system.

REFERENCES

Abraham, H., Lee, C., Brady, S., Fitzgerald, C., Mehler, B., Reimer, B., & Coughlin, J. F. (2017, January). Autonomous vehicles and alternatives to driving: trust, preferences, and effects of age. *Proceedings of the Transportation Research Board 96th Annual Meeting* (pp. 8–12). Washington, DC: Transportation Research Board.

Abuelgasim, A., & Kayem, A. (2016). *An Approach to Personalized Privacy Policy Recommendations on Online Social Networks*. 126–137. https://www.scitepress.org/DigitalLibrary/PublicationsDetail. aspx?ID=HJHeqEO0nxk=&t=1

Acquisti, A. (2012). Nudging privacy: The behavioral economics of personal information. *Digital Enlightenment Yearbook* 2012, 193–197.

Acquisti, A., & Gross, R. (2006). Imagined Communities: Awareness, Information Sharing, and Privacy on the Facebook. In G. Danezis & P. Golle (Eds.), *Privacy Enhancing Technologies* (Vol. 4258, pp. 36–58). Berlin/Heidelberg: Springer. https://www.springerlink.com/content/gx00n8nh88252822/abstract/

Acquisti, A., & Grossklags, J. (2006). Privacy and Rationality. In K. J. Strandburg & D. S. Raicu (Eds.), *Privacy and Technologies of Identity: A Cross-Disciplinary Conversation* (pp. 15–29). Springer US. https://doi. org/10.1007/0-387-28222-X_2

Acquisti, A., Brandimarte, L., & Loewenstein, G. (2022). Privacy and Behavioral Economics. In B. P. Knijnenburg, X. Page, P. Wisniewski, H. R. Lipford, N. Proferes, & J. Romano (Eds.), *Modern Socio-Technical Perspectives on Privacy* (pp. 61–77). Springer International Publishing. https://doi. org/10.1007/978-3-030-82786-1_4

Acquisti, A., John, L. K., & Loewenstein, G. (2012). The Impact of Relative Standards on the Propensity to Disclose. *Journal of Marketing Research*, *49*(2), 160–174. https://doi.org/10.1509/jmr.09.0215

Adjerid, I., Acquisti, A., Brandimarte, L., & Loewenstein, G. (2013). Sleights of Privacy: Framing, Disclosures, and the Limits of Transparency. *Proceedings of the Ninth Symposium on Usable Privacy and Security* (pp. 9:1–9:11). https://doi.org/10.1145/2501604.2501613

Altman, I. (1975). *The Environment and Social Behavior: Privacy, Personal Space, Territory, and Crowding*. Monterey, CA: Brooks/Cole Publishing Company. https://www.eric.ed.gov/ERICWebPortal/ detail?accno=ED131515

Amershi, S., Weld, D., Vorvoreanu, M., Fourney, A., Nushi, B., Collisson, P., ... & Horvitz, E. (2019, May). Guidelines for Human-AI Interaction. *Proceedings of the 2019 CHI Conference on Human Factors in Computing Systems* (pp. 1–13). Glasgow.

Anaraky, R., Knijnenburg, B., & Risius, M. (2020). Exacerbating Mindless Compliance: The Danger of Justifications during Privacy Decision Making in the Context of Facebook Applications. *AIS Transactions on Human-Computer Interaction*, *12*(2), 70–95. https://doi.org/10.17705/1thci.00129

Angst, C. M., & Agarwal, R. (2009). Adoption of Electronic Health Records in the Presence of Privacy Concerns: The Elaboration Likelihood Model and Individual Persuasion. *MIS Quarterly*, *33*(2), 339–370.

Arya, V., Bellamy, R. K., Chen, P. Y., Dhurandhar, A., Hind, M., Hoffman, S. C., ... & Zhang, Y. (2020, January). AI Explainability 360: Hands-on Tutorial. *Proceedings of the 2020 Conference on Fairness, Accountability, and Transparency* (pp. 696–696). Barcelona, Spain.

Athey, S., Catalini, C., & Tucker, C. (2017). *The Digital Privacy Paradox: Small Money, Small Costs, Small Talk* (Working Paper 23488). National Bureau of Economic Research. https://doi.org/10.3386/w23488

Awad, N. F., & Krishnan, M. S. (2006). The Personalization Privacy Paradox: An Empirical Evaluation of Information Transparency and the Willingness to be Profiled Online for Personalization. *MIS Quarterly*, *30*(1), 13–28.

Badillo-Urquiola, K., Page, X., & Wisniewski, P. (2018, September 13). Literature Review: Examining Contextual Integrity within Human-Computer Interaction. *Symposium on Applications of Contextual Integrity*. Princeton, NJ. https://doi.org/10.2139/ssrn.3309331

Bahirat, P., He, Y., Menon, A., & Knijnenburg, B. (2018). A Data-Driven Approach to Developing IoT Privacy-Setting Interfaces. *23rd International Conference on Intelligent User Interfaces* (pp. 165–176). https://doi.org/10.1145/3172944.3172982

Bahirat, P., Willemsen, M., He, Y., Sun, Q., & Knijnenburg, B. (2021). Overlooking Context: How do Defaults and Framing Reduce Deliberation in Smart Home Privacy Decision-Making? *Proceedings of the 2021 CHI Conference on Human Factors in Computing Systems* (pp. 1–18). Association for Computing Machinery. https://doi.org/10.1145/3411764.3445672

Balebako, R., Leon, P. G., Mugan, J., Acquisti, A., Cranor, L. F., & Sadeh, N. (2011). Nudging Users towards Privacy on Mobile Devices. *CHI 2011 Workshop on Persuasion, Influence, Nudge and Coercion through Mobile Devices* (pp. 23–26). https://www.andrew.cmu.edu/user/jmugan/Publications/chiworkshop.pdf

Bansal, G., Zahedi, F. "Mariam," & Gefen, D. (2010). The Impact of Personal Dispositions on Information Sensitivity, Privacy Concern and Trust in Disclosing Health Information Online. *Decision Support Systems*, *49*(2), 138–150. https://doi.org/10.1016/j.dss.2010.01.010

Bansal, G., Zahedi, F., & Gefen, D. (2008). The Moderating Influence of Privacy Concern on the Efficacy of Privacy Assurance Mechanisms for Building Trust: A Multiple-Context Investigation. *ICIS 2008 Proceedings*. https://aisel.aisnet.org/icis2008/7

Barocas, S., & Nissenbaum, H. (2014). Big Data's End Run around Procedural Privacy Protections. *Communications of the ACM*, *57*(11), 31–33. https://doi.org/10.1145/2668897

Barth, S., & de Jong, M. D. T. (2017). The Privacy Paradox - Investigating Discrepancies between Expressed Privacy Concerns and Actual Online Behavior - A Systematic Literature Review. *Telematics and Informatics*, *34*(7), 1038–1058. https://doi.org/10.1016/j.tele.2017.04.013

Belanche, D., Casaló, L. V., & Flavián, C. (2012). Integrating Trust and Personal Values into the Technology Acceptance Model: The Case of E-Government Services Adoption. *Cuadernos de Economía y Dirección de la Empresa*, *15*(4), 192–204.

Berkovsky, S., Taib, R., Hijikata, Y., Braslavsku, P., & Knijnenburg, B. (2018). A Cross-Cultural Analysis of Trust in Recommender Systems. *Proceedings of the 26th Conference on User Modeling, Adaptation and Personalization* (pp. 285–289). https://doi.org/10.1145/3209219.3209251

Besmer, A., & Richter Lipford, H. (2010). Moving beyond Untagging: Photo Privacy in a Tagged World. *Proceedings of the SIGCHI Conference on Human Factors in Computing Systems* (pp. 1563–1572). https://dl.acm.org/citation.cfm?id=1753560

Besmer, A., Watson, J., & Lipford, H. R. (2010). The Impact of Social Navigation on Privacy Policy Configuration. *Proceedings of the Sixth Symposium on Usable Privacy and Security* (pp. 7:1–7:10). https://doi.org/10.1145/1837110.1837120

Bettman, J. R., Luce, M. F., & Payne, J. W. (1998). Constructive Consumer Choice Processes. *Journal of Consumer Research*, *25*(3), 187–217. https://doi.org/10.1086/209535

Bhatnagar, A., Misra, S., & Rao, H. R. (2000). On Risk, Convenience, and Internet Shopping Behavior. *Communications of the ACM*, *43*(11), 98–105. https://doi.org/10.1145/353360.353371

Bickmore, T., & Cassell, J. (2001). Relational Agents: A Model and Implementation of Building User Trust. *Proceedings of the SIGCHI Conference on Human Factors in Computing Systems* (pp. 396–403). https://doi.org/10.1145/365024.365304

Bösch, C., Erb, B., Kargl, F., Kopp, H., & Pfattheicher, S. (2016). Tales from the Dark Side: Privacy Dark Strategies and Privacy Dark Patterns. *Proceedings on Privacy Enhancing Technologies*, *2016*(4), 237–254. https://doi.org/10.1515/popets-2016-0038

Buchanan, T., Paine, C., Joinson, A. N., & Reips, U.-D. (2007). Development of Measures of Online Privacy Concern and Protection for Use on the Internet. *Journal of the American Society for Information Sciences and Technology*, *58*(2), 157–165. https://doi.org/10.1002/asi.20459

Buck, C., Dinev, T., & Anaraky, R. G. (2022). Revisiting APCO. In B. P. Knijnenburg, X. Page, P. Wisniewski, H. R. Lipford, N. Proferes, & J. Romano (Eds.), *Modern Socio-Technical Perspectives on Privacy* (pp. 43–60). Cham: Springer International Publishing. https://doi.org/10.1007/978-3-030-82786-1_3

Cacioppo, J. T., Petty, R. E., Kao, C. F., & Rodriguez, R. (1986). Central and Peripheral Routes to Persuasion: An Individual Difference Perspective. *Journal of Personality and Social Psychology*, *51*(5), 1032.

Caldwell, S., Sweetser, P., O'Donnell, N., Knight, M. J., Aitchison, M., Gedeon, T., ... & Conroy, D. (2022). An Agile New Research Framework for Hybrid Human-AI Teaming: Trust, Transparency, and Transferability. *ACM Transactions on Interactive Intelligent Systems (TiiS)*, *12*(3), 1–36.

Cavoukian, A. (2013). Privacy by Design and the Promise of SmartData. In I. Harvey, A. Cavoukian, G. Tomko, D. Borrett, H. Kwan, & D. Hatzinakos (Eds.), *SmartData* (pp. 1–9). New York: Springer. https://link.springer.com/chapter/10.1007/978-1-4614-6409-9_1

Chellappa, R. K., & Sin, R. G. (2005). Personalization versus Privacy: An Empirical Examination of the Online Consumer's Dilemma. *Information Technology and Management*, *6*(2–3), 181–202.

Chen, H.-G., Chen, C. C., Lo, L., & Yang, S. C. (2008). Online Privacy Control via Anonymity and Pseudonym: Cross-Cultural Implications. *Behaviour & Information Technology*, *27*(3), 229–242. https://doi.org/10.1080/01449290601156817

Cho, H., & Filippova, A. (2016). *Networked Privacy Management in Facebook: A Mixed-Methods and Multinational Study*. CSCW.

Cho, H., Knijnenburg, B., Kobsa, A., & Li, Y. (2018). Collective Privacy Management in Social Media: A Cross-Cultural Validation. *ACM Transactions on Computer-Human Interaction*, *25*(3), 17:1–17:33. https://doi.org/10.1145/3193120

Cho, J. H., Chan, K., & Adali, S. (2015). A Survey on Trust Modeling. *ACM Computing Surveys (CSUR)*, *48*(2), 1–40.

Choi, H., Park, J., & Jung, Y. (2018). The Role of Privacy Fatigue in Online Privacy Behavior. *Computers in Human Behavior*, *81*, 42–51. https://doi.org/10.1016/j.chb.2017.12.001

Compañó, R., & Lusoli, W. (2010). The Policy Maker's Anguish: Regulating Personal Data Behavior between Paradoxes and Dilemmas. In T. Moore, D. Pym, & C. Ioannidis (Eds.), *Economics of Information Security and Privacy* (pp. 169–185). Springer US. https://doi.org/10.1007/978-1-4419-6967-5_9

Cook, R., Jones-Chick, R., Roulin, N., & O'Rourke, K. (2020). Job Seekers' Attitudes toward Cybervetting: Scale Development, Validation, and Platform Comparison. *International Journal of Selection and Assessment*, *28*(4), 383–398. https://doi.org/10.1111/ijsa.12300

Couch, L. L., & Jones, W. H. (1997). Measuring Levels of Trust. *Journal of Research in Personality*, *31*(3), 319–336.

Dalton, J., Ajayi, V., & Main, R. (2018, June). Vote Goat: Conversational Movie Recommendation. *The 41st international ACM SIGIR Conference on Research & Development in Information Retrieval* (pp. 1285–1288). Ann Arbor, MI.

De Souza, Z., & Dick, G. N. (2009). Disclosure of Information by Children in Social Networking-Not Just a Case of "You Show Me Yours and I'll Show You Mine." *International Journal of Information Management*, *29*(4), 255–261. https://doi.org/10.1016/j.ijinfomgt.2009.03.006

De Visser, E. J., Pak, R., & Shaw, T. H. (2018). From 'Automation' to 'Autonomy': The Importance of Trust Repair in Human-Machine Interaction. *Ergonomics*, *61*(10), 1409–1427.

De Visser, E. J., Peeters, M. M., Jung, M. F., Kohn, S., Shaw, T. H., Pak, R., & Neerincx, M. A. (2020). Towards a Theory of Longitudinal Trust Calibration in Human-Robot Teams. *International Journal of Social Robotics*, *12*(2), 459–478.

Degeling, M., Utz, C., Lentzsch, C., Hosseini, H., Schaub, F., & Holz, T. (2019). We Value Your Privacy ... Now Take Some Cookies. *Informatik Spektrum*, *42*(5), 345–346. https://doi.org/10.1007/s00287-019-01201-1

Demir, M., McNeese, N. J., Gorman, J. C., Cooke, N. J., Myers, C. W., & Grimm, D. A. (2021). Exploration of Teammate Trust and Interaction Dynamics in Human-Autonomy Teaming. *IEEE Transactions on Human-Machine Systems*, *51*(6), 696–705. https://doi.org/10.1109/THMS.2021.3115058

Denning, P., Comer, D. E., Gries, D., Mulder, M. C., Tucker, A. B., Turner, A. J., & Young, P. R. (1988). Computing as a Discipline: Preliminary Report of the ACM Task Force on the Core of Computer Science. *Proceedings of the Nineteenth SIGCSE Technical Symposium on Computer Science Education (p. 41)*. https://doi.org/10.1145/52964.52975

DeVellis, R. F. (2011). *Scale Development: Theory and Applications*. Thousand Oaks, CA: SAGE.

Diamantopoulos, A., Sarstedt, M., Fuchs, C., Wilczynski, P., & Kaiser, S. (2012). Guidelines for Choosing between Multi-Item and Single-Item Scales for Construct Measurement: A Predictive Validity Perspective. *Journal of the Academy of Marketing Science*, *40*(3), 434–449. https://doi.org/10.1007/s11747-011-0300-3

Diamantopoulou, V., Lambrinoudakis, C., King, J., & Gritzalis, S. (2022). EU GDPR: Toward a Regulatory Initiative for Deploying a Private Digital Era. In B. P. Knijnenburg, X. Page, P. Wisniewski, H. R. Lipford, N. Proferes, & J. Romano (Eds.), *Modern Socio-Technical Perspectives on Privacy* (pp. 427–448). Springer International Publishing. https://doi.org/10.1007/978-3-030-82786-1_18

Dinev, T., & Hart, P. (2006). An Extended Privacy Calculus Model for E-Commerce Transactions. *Information Systems Research, 17*(1), 61–80. https://doi.org/10.1287/isre.1060.0080

Dinev, T., Bellotto, M., Hart, P., Russo, V., Serra, I., & Colautti, C. (2006). Privacy Calculus Model in E-Commerce-A Study of Italy and the United States. *European Journal of Information Systems, 15*(4), 389–402. https://doi.org.janus.libr.tue.nl/10.1057/palgrave.ejis.3000590

Dinev, T., McConnell, A. R., & Smith, H. J. (2015). Research Commentary-Informing Privacy Research through Information Systems, Psychology, and Behavioral Economics: Thinking Outside the "APCO" Box. *Information Systems Research, 26*(4), 639–655. https://doi.org/10.1287/isre.2015.0600

Dong, C., Jin, H., & Knijnenburg, B. P. (2016). PPM: A Privacy Prediction Model for Online Social Networks. *Proceedings of The International Conference on Social Informatics* (pp. 400–420). https://doi.org/10.10 07/978-3-319-47874-6_28

Dourish, P., & Anderson, K. (2006). Collective Information Practice: Emploring Privacy and Security as Social and Cultural Phenomena. *Human-Computer Interaction, 21*(3), 319–342. https://doi.org/10.1207/ s15327051hci2103_2

Duhigg, C. (2012, February 16). How Companies Learn Your Secrets. *The New York Times*. https://www.nytimes.com/2012/02/19/magazine/shopping-habits.html

Dunning, D., Fetchenhauer, D., & Schlösser, T. M. (2012). Trust as a Social and Emotional Act: Noneconomic Considerations in Trust Behavior. *Journal of Economic Psychology, 33*(3), 686–694.

Dwork, C., & Naor, M. (2010). On the Difficulties of Disclosure Prevention in Statistical Databases or the Case for Differential Privacy. *Journal of Privacy and Confidentiality, 2*(1). https://repository.cmu.edu/ jpc/vol2/iss1/8

Egelman, S., Tsai, J., Cranor, L. F., & Acquisti, A. (2009). Timing Is Everything?: The Effects of Timing and Placement of Online Privacy Indicators. *Proceedings of the 27th International Conference on Human Factors in Computing Systems* (pp. 319–328). https://doi.org/10.1145/1518701.1518752

Ekstrand, M. D., Joshaghani, R., & Mehrpouyan, H. (2018). Privacy for All: Ensuring Fair and Equitable Privacy Protections. *Conference on Fairness, Accountability and Transparency* (pp. 35–47). https://pro-ceedings.mlr.press/v81/ekstrand18a.html

Esterwood, C., & Robert, L. P. (2021, August). Do You Still Trust Me? Human-Robot Trust Repair Strategies. *2021 30th IEEE International Conference on Robot & Human Interactive Communication (RO-MAN)* (pp. 183–188). IEEE.

Ezer, N., Bruni, S., Cai, Y., Hepenstal, S. J., Miller, C. A., & Schmorrow, D. D. (2019, November). Trust Engineering for Human-AI Teams. In *Proceedings of the Human Factors and Ergonomics Society Annual Meeting* (Vol. 63, No. 1, pp. 322–326). Los Angeles, CA: SAGE Publications.

Featherman, M. S., & Pavlou, P. A. (2003). Predicting e-services adoption: A perceived risk facets perspective. *International Journal of Human-Computer Studies, 59*(4), 451–474. https://doi.org/10.1016/ S1071-5819(03)00111-3

Feng, Y., Yao, Y., & Sadeh, N. (2021). A Design Space for Privacy Choices: Towards Meaningful Privacy Control in the Internet of Things. *Proceedings of the 2021 CHI Conference on Human Factors in Computing Systems* (pp. 1–16). https://doi.org/10.1145/3411764.3445148

Fishbein, M., & Ajzen, I. (1975). *Belief, Attitude, Intention, and Behavior: An Introduction to Theory and Research*. Reading, MA: Addison-Wesley Pub. Co.

Flathmann, C. (2023, February). How to Make Agents and Influence Teammates: Understanding the Social Influence AI Teammates Have in Human-AI Teams. Dissertations, 3339.

Forbrukerrådet. (2018). *Deceived by Design: How Tech Companies Use Dark Patterns to Discourage Us from Exercising Our Rights to Privacy*. https://consumerwatchdog.org/sites/default/files/2018-06/2018-06-25%20Deceived%20by%20design%20-%20Final.pdf

Friedman, A., Knijnenburg, B. P., Vanhecke, K., Martens, L., & Berkovsky, S. (2015). Privacy Aspects of Recommender Systems. In F. Ricci, L. Rokach, & B. Shapira (Eds.), *Recommender Systems Handbook* (2nd ed., pp. 649–688). New York: Springer US. https://doi.org/10.1007/978-1-4899-7637-6_19

Galdon, F., Hall, A., & Wang, S. J. (2021). Designing Trust in Highly Automated Virtual Assistants: A Taxonomy of Levels of Autonomy. *Artificial Intelligence in Industry 4.0: A Collection of Innovative Research Case-studies that are Reworking the Way We Look at Industry 4.0 Thanks to Artificial Intelligence* (pp. 199–211).

Gardner, P. L. (1996). The Dimensionality of Attitude Scales: A Widely Misunderstood Idea. *International Journal of Science Education, 18*(8), 913–919. https://doi.org/10.1080/0950069960180804

Gerber, N., Gerber, P., & Volkamer, M. (2018). Explaining the Privacy Paradox: A Systematic Review of Literature Investigating Privacy Attitude and Behavior. *Computers & Security, 77*, 226–261. https://doi.org/10.1016/j.cose.2018.04.002

Ghaiumy Anaraky, R. (2022). *Empowering Older Adults with Their Information Privacy Management* [PhD Thesis, Clemson University]. https://tigerprints.clemson.edu/all_dissertations/3183

Ghaiumy Anaraky, R., Li, Y., & Knijnenburg, B. (2021). Difficulties of Measuring Culture in Privacy Studies. *Proceedings of the ACM on Human-Computer Interaction*, 5(CSCW2), 378:1–378:26. https://doi.org/10.1145/3479522

Glaeser, E. L., Laibson, D. I., Scheinkman, J. A., & Soutter, C. L. (2000). Measuring Trust. *The Quarterly Journal of Economics*, *115*(3), 811–846.

Groß, T. (2021). Validity and Reliability of the Scale Internet Users' *Information Privacy Concerns (IUIPC)*. *Proceedings on Privacy Enhancing Technologies*. https://petsymposium.org/popets/2021/popets-2021-0026.php

Hagel, J., & Rayport, J. F. (1999). The Coming Battle for Customer Information. In *Creating Value in the Network Economy* (pp. 159–171). Boston, MA: Harvard Business School Press. https://hbr.org/1997/01/the-coming-battle-for-customer-information/ar/3

Hann, I.-H., Hui, K.-L., Lee, S.-Y., & Png, I. (2007). Overcoming Online Information Privacy Concerns: An Information-Processing Theory Approach. *Journal of Management Information Systems*, *24*(2), 13–42. https://doi.org/10.2753/MIS0742-1222240202

Hargittai, E., & Marwick, A. (2016). "What Can I Really Do?" Explaining the Privacy Paradox with Online Apathy. *International Journal of Communication*, *10*, 21.

Harris Interactive inc. (2000). *A Survey of Consumer Privacy Attitudes and Behaviors* (Privacy and American Business Newsletter). Harris Interactive, Inc. https://www.bbbonline.org/UnderstandingPrivacy/library/harrissummary.pdf

Harris, L., Westin, A. F., & associates. (2003a). *Most People Are "Privacy Pragmatists" Who, While Concerned about Privacy,* Will Sometimes Trade It Off for Other Benefits. Equifax Inc.

Harris, M. M., Hoye, G. V., & Lievens, F. (2003b). Privacy and Attitudes towards Internet-Based Selection Systems: A Cross-Cultural Comparison. *International Journal of Selection and Assessment*, *11*(2–3), 230–236.

He, Y., Bahirat, P., Knijnenburg, B. P., & Menon, A. (2019). A Data-Driven Approach to Designing for Privacy in Household IoT. *ACM Transactions on Interactive Intelligent Systems*, *10*(1), 10:1–10:47. https://doi.org/10.1145/3241378

Hepler, J., & Blasiola, S. (2021). *Users' Top-of-Mind Privacy Concerns* [TTC Labs research report]. Meta Research. https://www.ttclabs.net/research/users-top-of-mind-privacy-concerns

Ho, S. Y., & Tam, K. (2006). Understanding the Impact of Web Personalization on User Information Processing and Decision Outcomes. *MIS Quarterly*, *30*(4), 865–890.

Hoff, K. A., & Bashir, M. (2014). Trust in Automation. *Human Factors*. https://doi.org/10.1177/0018720814547570

Hoffman, D. L., Novak, T. P., & Peralta, M. (1999). Building consumer trust online. *Communications of the ACM*, *42*(4), 80–85. https://doi.org/10.1145/299157.299175

Hofstede, G. (2011). Dimensionalizing Cultures: The Hofstede Model in Context. *Online Readings in Psychology and Culture*, *2*(1). https://doi.org/10.9707/2307-0919.1014

Hu, H., Ahn, G.-J., & Jorgensen, J. (2011). Detecting and Resolving Privacy Conflicts for Collaborative Data Sharing in Online Social Networks. In *Proceedings of the 27th Annual Computer Security Applications Conference* (pp. 103–112). https://dl.acm.org/citation.cfm?id=2076747

Huang, L., Cooke, N. J., Gutzwiller, R. S., Berman, S., Chiou, E. K., Demir, M., & Zhang, W. (2021). Distributed Dynamic Team Trust in Human, Artificial Intelligence, and Robot Teaming. In C. S. Nam & J. B. Lyons (Eds.), *Trust in human-robot interaction* (pp. 301–319). London: Academic Press.

Huerta, E., Glandon, T., & Petrides, Y. (2012). Framing, Decision-Aid Systems, and Culture: Exploring Influences on Fraud Investigations. *International Journal of Accounting Information Systems*, *13*, 316–333.

Hui, K.-L., Tan, B. C. Y., & Goh, C.-Y. (2006). Online Information Disclosure: Motivators and Measurements. *ACM Transactions on Internet Technology*, *6*(4), 415–441. https://doi.org/10.1145/1183463.1183467

Hui, K.-L., Teo, H. H., & Lee, S.-Y. T. (2007). The Value of Privacy Assurance: An Exploratory Field Experiment. *MIS Quarterly*, *31*(1), 19–33.

Hull, G. (2015). Successful Failure: What Foucault Can Teach Us about Privacy Self-Management in a World of Facebook and Big Data. *Ethics and Information Technology*, *17*(2), 89–101. https://doi.org/10.1007/s10676-015-9363-z

Jacoby, J., & Kaplan, L. B. (1972). The Components of Perceived Risk. In M. Venkatesan (Ed.), *Proceedings of the Third Annual Conference of the Association for Consumer Research* (pp. 382–393). Association for Consumer Research. https://www.acrwebsite.org/search/view-conference-proceedings.aspx?Id=12016

Jarvenpaa, S. L., Tractinsky, N., & Saarinen, L. (1999). Consumer Trust in an Internet Store: A Cross-Cultural Validation. *Journal of Computer-Mediated Communication*, 5(2), JCMC526. https://doi.org/10.1111/j.1083-6101.1999.tb00337.x

Jensen, T., Khan, M. M. H., Fahim, M. A. A., & Albayram, Y. (2021, June). Trust and Anthropomorphism in Tandem: The Interrelated Nature of Automated Agent Appearance and Reliability in Trustworthiness Perceptions. *Designing Interactive Systems Conference* 2021 (pp. 1470–1480).

Jia, H., & Xu, H. (2016). Measuring individuals' concerns over collective privacy on social networking sites. *Cyberpsychology: Journal of Psychosocial Research on Cyberspace*, 10(1). https://cyberpsychology.eu/article/view/6184

John, L. K., Acquisti, A., & Loewenstein, G. (2011). Strangers on a Plane: Context-Dependent Willingness to Divulge Sensitive Information. *Journal of Consumer Research*, 37(5), 858–873. https://doi.org/10.1086/656423

Johnson, E. J., Bellman, S., & Lohse, G. L. (2002). Defaults, Framing and Privacy: Why Opting In≠Opting Out. *Marketing Letters*, 13(1), 5–15. https://doi.org/10.1023/A:1015044207315

Joinson, A. N., Paine, C., Buchanan, T., & Reips, U.-D. (2008). Measuring Self-Disclosure Online: Blurring and Non-Response to Sensitive Items in Web-Based Surveys. *Computers in Human Behavior*, 24(5), 2158–2171. https://doi.org/16/j.chb.2007.10.005

Joinson, A. N., Reips, U.-D., Buchanan, T., & Schofield, C. B. P. (2010). Privacy, Trust, and Self-Disclosure Online. *Human-Computer Interaction*, 25(1), 1–24. https://doi.org/10.1080/07370020903586662

Kahneman, D. (2011). *Thinking, Fast and Slow*. Macmillan.

Karr-Wisniewski, P., Wilson, D., & Richter-Lipford. (2011, August 6). A New Social Order: Mechanisms for Social Network Site Boundary Regulation. *AMCIS 2011 Proceedings - All Submissions*. https://aisel.aisnet.org/amcis2011_submissions/101

Karwatzki, S., Dytynko, O., Trenz, M., & Veit, D. (2017). Beyond the Personalization-Privacy Paradox: Privacy Valuation, Transparency Features, and Service Personalization. *Journal of Management Information Systems*, 34(2), 369–400. https://doi.org/10.1080/07421222.2017.1334467

Khalil, A., & Connelly, K. (2006). Context-aware telephony: Privacy preferences and sharing patterns. *Proceedings of the 2006 20th Anniversary Conference on Computer Supported Cooperative Work (pp. 469–478)*. ACM. https://doi.org/10.1145/1180875.1180947

Kim, D. J. (2008). Self-Perception-Based versus Transference-Based Trust Determinants in Computer-Mediated Transactions: A Cross-Cultural Comparison Study. *Journal of Management Information Systems*, 24(4), 13–45.

Kim, D. J., Ferrin, D. L., & Rao, H. R. (2008). A Trust-Based Consumer Decision-Making Model in Electronic Commerce: The Role of Trust, Perceived Risk, and Their Antecedents. *Decision Support Systems*, 44(2), 544–564. https://doi.org/10.1016/j.dss.2007.07.001

Kim, D. J., Ferrin, D. L., & Rao, H. R., (2003). Antecedents of Consumer Trust in B-to-C Electronic Commerce. *Proceedings of Ninth Americas Conference on Information Systems* (pp. 157–167). Tampa, FL.

Kisselburgh, L., & Beever, J. (2022). The Ethics of Privacy in Research and Design: Principles, Practices, and Potential. In B. P. Knijnenburg, X. Page, P. Wisniewski, H. R. Lipford, N. Proferes, & J. Romano (Eds.), *Modern Socio-Technical Perspectives on Privacy*(pp. 395–426). Springer International Publishing. https://doi.org/10.1007/978-3-030-82786-1_17

Klein, G. A. (1993). A Recognition-Primed Decision (RPD) Model of Rapid Decision Making. *Decision Making in Action: Models and Methods*, 5(4), 138–147.

Knapp, H., & Kirk, S. A. (2003). Using Pencil and Paper, Internet and Touch-Tone Phones for Self-Administered Surveys: Does Methodology Matter? *Computers in Human Behavior*, 19(1), 117–134. https://doi.org/10.1016/S0747-5632(02)00008-0

Knijnenburg, B. P. (2015). *A User-Tailored Approach to Privacy Decision Support* [Ph.D. Thesis, University of California, Irvine]. https://search.proquest.com/docview/1725139739/abstract

Knijnenburg, B. P. (2017). Privacy? I Can't Even! Making a Case for User-Tailored Privacy. *IEEE Security & Privacy*, 15(4), 62–67.

Knijnenburg, B. P., & Bulgurcu, B. (2023). Designing Alternative Form-Autocompletion Tools to Enhance Privacy Decision Making and Prevent Unintended Disclosure. *ACM Transactions on Computer-Human Interaction*, 30, 91.

Knijnenburg, B. P., & Cherry, D. (2016, June 22). Comics as a Medium for Privacy Notices. *SOUPS 2016 Workshop on the Future of Privacy Notices and Indicators*.

Knijnenburg, B. P., & Jin, H. (2013). The Persuasive Effect of Privacy Recommendations. Twelfth Annual Workshop on HCI Research in MIS. https://aisel.aisnet.org/sighci2013/16

Knijnenburg, B. P., & Kobsa, A. (2013a). Helping Users with Information Disclosure Decisions: Potential for Adaptation. *Proceedings of the 2013 ACM International Conference on Intelligent User Interfaces* (pp. 407–416). https://doi.org/10.1145/2449396.2449448

Knijnenburg, B. P., & Kobsa, A. (2013b). Making Decisions about Privacy: Information Disclosure in Context-Aware Recommender Systems. *ACM Transactions on Interactive Intelligent Systems*, *3*(3), 20:1–20:23. https://doi.org/10.1145/2499670

Knijnenburg, B. P., & Kobsa, A. (2014). Increasing Sharing Tendency without Reducing Satisfaction: Finding the Best Privacy-Settings User Interface for Social Networks. *ICIS 2014 Proceedings*. https://aisel.aisnet.org/icis2014/proceedings/ISSecurity/4

Knijnenburg, B. P., & Willemsen, M. C. (2016). Inferring Capabilities of Intelligent Agents from Their External Traits. *ACM Trans. Interact. Intell. Syst.*, *6*(4), 28:1–28:25. https://doi.org/10.1145/2963106

Knijnenburg, B. P., Anaraky, R. G., Wilkinson, D., Namara, M., He, Y., Cherry, D., & Ash, E. (2022a). User-Tailored Privacy. In B. P. Knijnenburg, X. Page, P. Wisniewski, H. R. Lipford, N. Proferes, & J. Romano (Eds.), *Modern Socio-Technical Perspectives on Privacy* (pp. 367–393). Springer International Publishing. https://doi.org/10.1007/978-3-030-82786-1_16

Knijnenburg, B. P., Kobsa, A., & Jin, H. (2013a). Dimensionality of Information Disclosure Behavior. *International Journal of Human-Computer Studies*, *71*(12), 1144–1162. https://doi.org/10.1016/j.ijhcs.2013.06.003

Knijnenburg, B. P., Kobsa, A., & Jin, H. (2013b). Preference-Based Location Sharing: Are More Privacy Options Really Better? *Proceedings of the SIGCHI Conference on Human Factors in Computing Systems* (pp. 2667–2676). https://doi.org/10.1145/2470654.2481369

Knijnenburg, B. P., Kobsa, A., & Jin, H. (2013c). *Counteracting the Negative Effect of Form Auto-completion on the Privacy Calculus. ICIS 2013 Proceedings.* https://aisel.aisnet.org/icis2013/proceedings/SecurityOfIS/2

Knijnenburg, B. P., Kobsa, A., & Jin, H. (2014). *Segmenting the Recipients of Personal Information* [Under review].

Knijnenburg, B. P., Page, X., Wisniewski, P., Lipford, H. R., Proferes, N., & Romano, J. (2022b). Introduction and Overview. In B. P. Knijnenburg, X. Page, P. Wisniewski, H. R. Lipford, N. Proferes, & J. Romano (Eds.), *Modern Socio-Technical Perspectives on Privacy* (pp. 1–11). Springer International Publishing. https://doi.org/10.1007/978-3-030-82786-1_1

Knijnenburg, B. P., Raybourn, E. M., Cherry, D., Wilkinson, D., Sivakumar, S., & Sloan, H. (2017, February 25). Death to the Privacy Calculus? *Proceedings of the 2017 Networked Privacy Workshop at CSCW.* https://papers.ssrn.com/abstract=2923806

Knijnenburg, B. P., Willemsen, M. C., & Hirtbach, S. (2010). Receiving Recommendations and Providing Feedback: The User-Experience of a Recommender System. In F. Buccafurri & G. Semeraro (Eds.), *E-Commerce and Web Technologies* (Vol. 61, pp. 207–216). Springer. https://www.springerlink.com/index/10.1007/978-3-642-15208-5_19

Knijnenburg, B. P., Willemsen, M. C., Gantner, Z., Soncu, H., & Newell, C. (2012). Explaining the User Experience of Recommender Systems. *User Modeling and User-Adapted Interaction*, *22*(4–5), 441–504. https://doi.org/10.1007/s11257-011-9118-4

Kobsa, A., & Teltzrow, M. (2005). Contextualized Communication of Privacy Practices and Personalization Benefits: Impacts on Users' Data Sharing Behavior. In D. Martin & A. Serjantov (Eds.), *Privacy Enhancing Technologies: Fourth International Workshop, PET 2004,* Toronto, Canada (Vol. 3424, pp. 329–343). Springer Verlag. https://doi.org/10.1007/11423409_21

Kobsa, A., Cho, H., & Knijnenburg, B. P. (2016). The Effect of Personalization Provider Characteristics on Privacy Attitudes and Behaviors: An Elaboration Likelihood Model Approach. *Journal of the Association for Information Science and Technology*. https://doi.org/10.1002/asi.23629

Kokolakis, S. (2017). Privacy Attitudes and Privacy Behaviour: A Review of Current Research on the Privacy Paradox Phenomenon. *Computers & Security*, *64*, 122–134. https://doi.org/10.1016/j.cose.2015.07.002

Kolter, J., & Pernul, G. (2009). Generating User-Understandable Privacy Preferences. *Conference on Availability, Reliability and Security* (pp. 299–306). https://doi.org/10.1109/ARES.2009.89

Kolter, J., Kernchen, T., & Pernul, G. (2009). Collaborative Privacy: A Community-Based Privacy Infrastructure. In D. Gritzalis & J. Lopez (Eds.), *Emerging Challenges for Security, Privacy and Trust: 24th IFIP TC 11 International Information Security Conference, SEC 2009* (pp. 226–236). Pafos: Springer Verlag.

Koshimizu, T., Toriyama, T., & Babaguchi, N. (2006). Factors on the Sense of Privacy in Video Surveillance. *Proceedings of the 3rd ACM Workshop on Continuous Archival and Retrieval of Personal Experiences* (pp. 35–44). https://doi.org/10.1145/1178657.1178665

Kumaraguru, P., & Cranor, L. F. (2005). *Privacy Indexes: A Survey of Westin's Studies* (Technical Report CMU-ISRI-5-138). Institute for Software Research International, School of Computer Science, Carnegie Mellon University. https://www.casos.cs.cmu.edu/publications/papers/CMU-ISRI-05-138.pdf

Lai, Y.-L., & Hui, K.-L. (2004). Opting-in or Opting-Out on the Internet: Does It Really Matter? *ICIS 2004: Twenty-Fifth International Conference on Information Systems* (pp. 781–792). Washington, DC. https://aisel.aisnet.org/icis2004/63

Lampinen, A., Lehtinen, V., Lehmuskallio, A., & Tamminen, S. (2011). We're in It Together: Interpersonal Management of Disclosure in Social Network Services. *Proceedings of the SIGCHI Conference on Human Factors in Computing Systems* (pp. 3217–3226). Vancouver. https://dl.acm.org/citation.cfm?id=1979420

Laufer, R. S., & Wolfe, M. (1977). Privacy as a Concept and a Social Issue: A Multidimensional Developmental Theory. *Journal of Social Issues*, *33*(3), 22–42. https://doi.org/10.1111/j.1540-4560.1977.tb01880.x

Lederer, S., Mankoff, J., & Dey, A. K. (2003). Who Wants to Kow What When? Privacy Preference Determinants in Ubiquitous Computing. *Proceedings of the SIGCHI Conference on Human Factors in Computing Systems* (pp. 724–725). https://doi.org/10.1145/765891.765952

Lee, J. D., & See, K. A. (2004). Trust in Automation: Designing for Appropriate Reliance. *Human Factors*, *46*(1), 50–80. https://doi.org/10.1518/hfes.46.1.50_30392

Li, H., Sarathy, R., & Xu, H. (2010). Understanding Situational Online Information Disclosure as a Privacy Calculus. *Journal of Computer Information Systems*, *51*(1), 62–71.

Li, H., Sarathy, R., & Xu, H. (2011). The Role of Affect and Cognition on Online Consumers' Decision to Disclose Personal Information to Unfamiliar Online Vendors. *Decision Support Systems*, *51*(3), 434–445. https://doi.org/10.1016/j.dss.2011.01.017

Li, X., & Santhanam, R. (2008). Will It Be Disclosure or Fabrication of Personal Information? An Examination of Persuasion Strategies on Prospective Employees. *International Journal of Information Security and Privacy*, *2*(4), 91–109. https://doi.org/10.4018/jisp.2008100105

Li, Y. (2012). Theories in Online Information Privacy Research: A Critical Review and an Integrated Framework. *Decision Support Systems*, *54*(1), 471–481. https://doi.org/10.1016/j.dss.2012.06.010

Li, Y. (2020). Investigating Obfuscation as a Tool to Enhance Photo Privacy on Social Networks Sites [PhD Thesis]. https://tigerprints.clemson.edu/all_dissertations/2694

Li, Y. (2022). Cross-Cultural Privacy Differences. In B. P. Knijnenburg, X. Page, P. Wisniewski, H. R. Lipford, N. Proferes, & J. Romano (Eds.), *Modern Socio-Technical Perspectives on Privacy* (pp. 267–292). Springer International Publishing. https://doi.org/10.1007/978-3-030-82786-1_12

Li, Y., Cho, H., Anaraky, R. G., Knijnenburg, B., & Kobsa, A. (2022a). Antecedents of Collective Privacy Management in Social Network Sites: A Cross-Country Analysis. *CCF Transactions on Pervasive Computing and Interaction*, *4*(2), 106–123. https://doi.org/10.1007/s42486-022-00092-8

Li, Y., Ghaiumy Anaraky, R., & Knijnenburg, B. (2021). How Not to Measure Social Network Privacy: A Cross-Country Investigation. *Proceedings of the ACM on Human-Computer Interaction*, *5*(CSCW1), 1–32. https://doi.org/10.1145/3449218

Li, Y., Kobsa, A., Knijnenburg, B. P., & Nguyen, M.-H. C. (2017). Cross-Cultural Privacy Prediction. *Proceedings on Privacy Enhancing Technologies*, *2017*(2), 113–132. https://doi.org/10.1515/popets-2017-0019

Li, Y., Rho, E. H. R., & Kobsa, A. (2022b). Cultural Differences in the Effects of Contextual Factors and Privacy Concerns on Users' Privacy Decision on Social Networking Sites. *Behaviour & Information Technology*, *41*(3), 655–677. https://doi.org/10.1080/0144929X.2020.1831608

Litt, E. (2012). Knock, Knock. Who's There? The Imagined Audience. *Journal of Broadcasting & Electronic Media*, *56*(3), 330–345. https://doi.org/10.1080/08838151.2012.705195

Liu, B., Andersen, M. S., Schaub, F., Almuhimedi, H., Zhang, S. (Aerin), Sadeh, N., Agarwal, Y., & Acquisti, A. (2016). Follow My Recommendations: A Personalized Privacy Assistant for Mobile App Permissions. *Twelfth Symposium on Usable Privacy and Security* (pp. 27–41).

Liu, Y., Gummadi, K. P., Krishnamurthy, B., & Mislove, A. (2011). Analyzing Facebook Privacy Settings: User Expectations vs. Reality. *Proceedings of the 2011 ACM SIGCOMM Conference on Internet Measurement Conference* (pp. 61–70). https://doi.org/10.1145/2068816.2068823

Lopez, J., Textor, C., Lancaster, C., Schelble, B., Freeman, G., Zhang, R., McNeese, N., & Pak, R. (2023). The Complex Relationship of AI Ethics and Trust in Human-AI Teaming: Insights from Advanced Real-World Subject Matter Experts. *AI and Ethics*, 1–21.

Lord, K. R., Lee, M.-S., & Sauer, P. L. (1995). The Combined Influence Hypothesis: Central and Peripheral Antecedents of Attitude toward the Ad. *Journal of Advertising*, *24*(1), 73–85. https://doi.org/10.1093/comjnl/bxs103

Lowry, P. B., Moody, G., Vance, A., Jensen, M., Jenkins, J., & Wells, T. (2012). Using an Elaboration Likelihood Approach to Better Understand the Persuasiveness of Website Privacy Assurance Cues for Online Consumers. *Journal of the American Society for Information Science and Technology*, *63*(4), 755–776. https://doi.org/10.1002/asi.21705

Lusoli, W., Bacigalupo, M., Lupiáñez-Villanueva, F., Andrade, N., Monteleone, S., & Maghiros, I. (2012). *Pan-European Survey of Practices, Attitudes and Policy Preferences as Regards Personal Identity Data Management* (SSRN Scholarly Paper ID 2086579). Social Science Research Network. https://papers.ssrn.com/abstract=2086579

Lyons, J. B., & Guznov, S. Y. (2018). Individual Differences in Human-Machine Trust: A Multi-Study Look at the Perfect Automation Schema. *Theoretical Issues in Ergonomics Science*, *20*(4), 440–458.

Madejski, M., Johnson, M., & Bellovin, S. M. (2012). A Study of Privacy Settings Errors in an Online Social Network. *Fourth International Workshop on SECurity and SOCial Networking* (pp. 340–345). https://doi.org/10.1109/PerComW.2012.6197507

Madhavan, P., & Wiegmann, D. A. (2007). Similarities and Differences between Human-Human and Human-Automation Trust: An Integrative Review. *Theoretical Issues in Ergonomics Science*, *8*(4), 277–301.

Malhotra, N. K., Kim, S. S., & Agarwal, J. (2004). Internet Users' Information Privacy Concerns (IUIPC): The Construct, the Scale, and a Nomological Framework. *Information Systems Research*, *15*(4), 336–355. https://doi.org/10.1287/isre.1040.0032

Manstead, A. S. R. (1996). Attitudes and Behaviour. *Applied Social Psychology* (pp. 3–29). SAGE Publications Ltd. https://doi.org/10.4135/9781446250556

Marangunić, N., & Granić, A. (2015). Technology Acceptance Model: A Literature Review from 1986 to 2013. *Universal Access in the Information Society*, *14*, 81–95.

Marwick, A. E., & Boyd, D. (2011). I Tweet Honestly, I Tweet Passionately: Twitter Users, Context Collapse, and the Imagined Audience. *New Media & Society*, *13*(1), 114–133. https://doi.org/10.1177/1461444810365313

Mayer, R. C., Davis, J. H., & Schoorman, F. D. (1995). An Integrative Model of Organizational Trust. *The Academy of Management Review*, *20*(3), 709–734. https://doi.org/10.2307/258792

McKnight, H., Carter, M., and Clay, P. (2009). Trust in Technology: Development of a Set of Constructs and Measures. *Proceedings of Diffusion Interest Group in Informa- tion Technology*, *10*, pp. 1–12.

McNeese, N. J., Demir, M., Chiou, E. K., & Cooke, N. J. (2021). Trust and Team Performance in Human-Autonomy Teaming. *International Journal of Electronic Commerce*, *25*(1), 51–72.

McNeese, N. J., Demir, M., Cooke, N. J., & Myers, C. (2018). Teaming with a Synthetic Teammate: Insights into Human-Autonomy Teaming. *Human factors*, *60*(2), 262–273.

McNeese, N. J., Flathmann, C., O'Neill, T. A., & Salas, E. (2023). Stepping Out of the Shadow of Human-Human Teaming: Crafting a Unique Identity for Human-Autonomy Teams. *Computers in Human Behavior*, *148*, 107874.

McNeese, N., Demir, M., Chiou, E., Cooke, N., & Yanikian, G. (2019). Understanding the Role of Trust in Human-Autonomy Teaming. *Proceedings of the 52nd Hawaii International Conference on System Sciences*. Wailea, HI.

Meister, D. (2018). *The History of Human Factors and Ergonomics*. Boca Raton, FL: CRC Press.

Menezes, A. J., Oorschot, P. C. van, & Vanstone, S. A. (2018). *Handbook of Applied Cryptography*. Boca Raton, FL: CRC Press.

Merritt, S. M., & Ilgen, D. R. (2008). Not All Trust Is Created Equal: Dispositional and History-Based Trust in Human-Automation Interaction. *Human Factors*, *50*, 194–210.

Metzger, M. J. (2004). Privacy, Trust, and Disclosure: Exploring Barriers to Electronic Commerce. *Journal of Computer-Mediated Communication*, *9*(4). https://doi.org/10.1111/j.1083-6101.2004.tb00292.x

Metzger, M. J. (2006). Effects of Site, Vendor, and Consumer Characteristics on Web Site Trust and Disclosure. *Communication Research*, *33*(3), 155–179.

Metzger, M. J. (2007). Communication Privacy Management in Electronic Commerce. *Journal of Computer-Mediated Communication*, *12*(2), 335–361. https://doi.org/10.1111/j.1083-6101.2007.00328.x

Michler, O., Decker, R., & Stummer, C. (2020). To Trust or Not to Trust Smart Consumer Products: A Literature Review of Trust-Building Factors. *Management Review Quarterly*, *70*, 391–420.

Mirkovic, J., Dark, M., Du, W., Vigna, G., & Denning, T. (2015). Evaluating Cybersecurity Education Interventions: Three Case Studies. *IEEE Security Privacy*, *13*(3), 63–69. https://doi.org/10.1109/MSP.2015.57

Morra, L., Lamberti, F., Prattic, F. G., La Rosa, S., & Montuschi, P. (2019). Building Trust in Autonomous Vehicles: Role of Virtual Reality Driving Simulators in HMI Design. *IEEE Transactions on Vehicular Technology*, *68*(10), 9438–9450.

Murphy, A. R., Reddy, M. C., & Xu, H. (2014). Privacy Practices in Collaborative Environments: A Study of Emergency Department Staff. *Proceedings of the 17th ACM Conference on Computer Supported Cooperative Work & Social Computing* (pp. 269–282). https://dl.acm.org/citation.cfm?id=2531643

Musick, G., Gilman, E. S., Duan, W., McNeese, N. J., Knijnenburg, B. P., & O'Neill, T. (2023). Knowing Unknown Teammates: Exploring Anonymity and Explanations in a Teammate Information-Sharing Recommender System. *Proceedings of the ACM on Human-Computer Interaction*, *7*(CSCW2). https://doi.org/10.1145/3610075

Nadarzynski, T., Miles, O., Cowie, A., & Ridge, D. (2019). Acceptability of Artificial Intelligence (AI)-Led Chatbot Services in Healthcare: A Mixed-Methods Study. *Digital Health*, *5*, 2055207619871808.

Najafian, S., Musick, G., Knijnenburg, B., & Tintarev, N. (2023). How Do People Make Decisions in Disclosing Personal Information in Tourism Group Recommendations in Competitive versus Cooperative Conditions? *User Modeling and User-Adapted Interaction*. https://doi.org/10.1007/s11257-023-09375-w

Namara, M. (2022). *Evaluating Privacy Adaptation Presentation Methods to support Social Media Users in their Privacy-Related Decision-Making Process* [Ph.D. Thesis, Clemson University]. https://tigerprints.clemson.edu/all_dissertations/3087

Namara, M., & Knijnenburg, B. P. (2021). The Differential Effect of Privacy-Related Trust on Groupware Application Adoption and Use during the COVID-19 pandemic. *Proceedings of the ACM on Human-Computer Interaction*, *5*(CSCW2), 405:1–405:34. https://doi.org/10.1145/3479549

Namara, M., Sloan, H., & Knijnenburg, B. P. (2022). The Effectiveness of Adaptation Methods in Improving User Engagement and Privacy Protection on Social Network Sites. *Proceedings on Privacy Enhancing Technologies*. https://petsymposium.org/popets/2022/popets-2022-0031.php

Namara, M., Sloan, H., Jaiswal, P., & Knijnenburg, B. P. (2018a). The Potential for User-Tailored Privacy on Facebook. *IEEE Symposium on Privacy-Aware Computing*.

Namara, M., Wilkinson, D., Lowens, B. M., Knijnenburg, B. P., Orji, R., & Sekou, R. L. (2018b). Cross-cultural Perspectives on eHealth Privacy in Africa. *Proceedings of the Second African Conference for Human Computer Interaction: Thriving Communities* (pp. 7:1–7:11). https://doi.org/10.1145/3283458.3283472

Narayanan, A., & Shmatikov, V. (2008). Robust De-anonymization of Large Sparse Datasets. *2008 IEEE Symposium on Security and Privacy* (pp. 111–125). https://doi.org/10.1109/SP.2008.33

Narayanan, A., & Shmatikov, V. (2009). De-anonymizing Social Networks. *2009 30th IEEE Symposium on Security and Privacy* (pp. 173–187). https://doi.org/10.1109/SP.2009.22

Nass, C., Steuer, J., & Tauber, E. R. (1994). Computers Are Social Actors. *Proceedings of the SIGCHI Conference on Human Factors in Computing Systems* (pp. 72–78). https://doi.org/10.1145/191666.191703

Nissenbaum, H. (2009). *Privacy in Context: Technology, Policy, and the Integrity of Social Life*. Stanford, CA: Stanford University Press.

Nissenbaum, H. (2011). A Contextual Approach to Privacy Online. *Daedalus*, *140*(4), 32–48. https://doi.org/10.1162/DAED_a_00113

NIST. (2019). NIST Privacy Framework: An Enterprise Risk Management Tool [Discussion Draft]. National Institute of Standards and Technology.

Norberg, P. A., Horne, D. R., & Horne, D. A. (2007). The Privacy Paradox: Personal Information Disclosure Intentions versus Behaviors. *Journal of Consumer Affairs*, *41*(1), 100–126. https://doi.org/10.1111/j.1745-6606.2006.00070.x

Norman, D. (2013). *The Design of Everyday Things:* Revised and Expanded Edition. New York: Basic Books.

O'Donovan, J., & Smyth, B. (2005, January). Trust in Recommender Systems. *Proceedings of the 10th International Conference on Intelligent User Interfaces* (pp. 167–174). San Diego, CA.

O'Neill, T., McNeese, N., Barron, A., & Schelble, B. (2022). Human-Autonomy Teaming: A Review and Analysis of the Empirical Literature. *Human Factors*, *64*(5), 904–938.

Olivero, N., & Lunt, P. (2004). Privacy versus Willingness to Disclose in E-commerce Exchanges: The Effect of Risk Awareness on the Relative Role of Trust and Control. *Journal of Economic Psychology*, *25*(2), 243–262. https://doi.org/10.1016/S0167-4870(02)00172-1

Olson, J. S., Grudin, J., & Horvitz, E. (2004). *Toward Understanding Preferences for Sharing and Privacy* (Technical Report 2004-138). Microsoft Research.

Olson, J. S., Grudin, J., & Horvitz, E. (2005). A Study of Preferences for Sharing and Privacy. *CHI'05 Extended Abstracts* (pp. 1985–1988). https://doi.org/10.1145/1056808.1057073

Ozdemir, Z. D., Benamati, J. H., & Smith, H. J. (2016). A Cross-Cultural Comparison of Information Privacy Concerns in Singapore, Sweden and the United States. *Proceedings of the 18th Annual International Conference on Electronic Commerce: E-Commerce in Smart Connected World* (p. 4). https://dl.acm.org/citation.cfm?id=2971607

Page, X., Anaraky, R. G., & Knijnenburg, B. P. (2019). How Communication Style Shapes Relationship Boundary Regulation and Social Media Adoption. *Proceedings of the 10th International Conference on Social Media and Society* (pp. 126–135). https://doi.org/10.1145/3328529.3328553

Page, X., Bahirat, P., Safi, M. I., Knijnenburg, B. P., & Wisniewski, P. (2018). The Internet of What? Understanding Differences in Perceptions and Adoption for the Internet of Things. *Proceedings of the ACM on Interactive, Mobile, Wearable and Ubiquitous Technologies*, 2(4), 183:1–183:22. https://doi.org/10.1145/3287061

Page, X., Knijnenburg, B. P., & Kobsa, A. (2013a). FYI: Communication Style Preferences Underlie Differences in Location-Sharing Adoption and Usage. *Proceedings of the 2013 ACM International Joint Conference on Pervasive and Ubiquitous Computing* (pp. 153–162). https://doi.org/10.1145/2493432.2493487

Page, X., Knijnenburg, B. P., & Kobsa, A. (2013b). What a Tangled Web We Weave: Lying Backfires in Location-sharing Social Media. *Proceedings of the 2013 Conference on Computer Supported Cooperative Work* (pp. 273–284). https://doi.org/10.1145/2441776.2441808

Page, X., Kobsa, A., & Knijnenburg, B. P. (2012). Don't Disturb My Circles! Boundary Preservation Is at the Center of Location-Sharing Concerns. *Proceedings of the Sixth International AAAI Conference on Weblogs and Social Media* (pp. 266–273). https://www.aaai.org/ocs/index.php/ICWSM/ICWSM12/paper/view/4679

Page, X., Tang, K., Stutzman, F., & Lampinen, A. (2013c). Measuring Networked Social Privacy. *Proceedings of the 2013 Conference on Computer Supported Cooperative Work Companion* (pp. 315–320). San Antonio, TX.

Pak, R., Fink, N., Price, M., Bass, B., & Sturre, L. (2012). Decision Support Aids with Anthropomorphic Characteristics Influence Trust and Performance in Younger and Older Adults. *Ergonomics*, 55, 1059–1072.

Palen, L., & Dourish, P. (2003). Unpacking "Privacy" for a Networked World. *Proceedings of* CHI'2003 (pp. 129–136). Fort Lauderdale, FL.

Parasuraman, R., & Manzey, D. H. (2010) Complacency and Bias in Human Use of Automation: An Attentional Integration. *Human Factors*, 52(3), 381–410.

Parasuraman, R., Sheridan, T. B., & Wickens, C. D. (2000). A model for types and levels of human interaction with automation. *IEEE Transactions on Systems, Man, and Cybernetics - Part A: Systems and Humans*, 30(3), 286–297. https://doi.org/10.1109/3468.844354

Parasuraman, R., & Riley, V. (1997). Humans and Automation: Use, Misuse, Disuse, Abuse. *Human Factors*, 39(2), 230–253. https://doi.org/10.1518/001872097778543886

Parsons, H. M. (1985). Automation and the Individual: Comprehensive and Comparative Views. *Human Factors*, 27(1), 99–111. https://doi.org/10.1177/001872088502700109

Patil, S., & Lai, J. (2005). Who Gets to Know What When: Configuring Privacy Permissions in an Awareness Application. *Proceedings of the SIGCHI Conference on Human Factors in Computing Systems* (pp. 101–110). https://doi.org/10.1145/1054972.1054987

Patil, S., Page, X., & Kobsa, A. (2011). With a Little Help from My Friends: Can Social Navigation Inform Interpersonal Privacy Preferences? *Proceedings of the ACM 2011 Conference on Computer Supported Cooperative Work* (pp. 391–394). https://doi.org/10.1145/1958824.1958885

Pavlou, P. (2001). Integrating Trust in Electronic Commerce with the Technology Acceptance Model: Model Development and Validation. *AMCIS 2001 Proceedings* (p. 159).

Pavlou, P. A. (2003). Consumer Acceptance of Electronic Commerce: Integrating Trust and Risk with the Technology Acceptance Model. *International Journal of Electronic Commerce*, 7(3), 101–134.

Petronio, S. (2002). *Boundaries of Privacy: Dialectics of Disclosure*. Albany, NY: State University of New York Press.

Petty, R. E., & Cacioppo, J. T. (1986). The Elaboration Likelihood Model of Persuasion. In L. Berkowitz (Ed.), *Advances in Experimental Social Psychology: Volume 19* (pp. 123–205). Academic Press. https://doi.org/10.1016/S0065-2601(08)60214-2

Petty, R. E., & Wegener, D. T. (1999). The Elaboration Likelihood Model: Current Status and Controversies. In S. Chaiken & Y. Trope (Eds.), *Dual-Process Theories in Social Psychology* (pp. 37–72). New York: Guilford Press.

Petty, R. E., Cacioppo, J. T., & Schumann, D. (1983). Central and Peripheral Routes to Advertising Effectiveness: The Moderating Role of Involvement. *Journal of Consumer Research*, 10(2), 135–146.

Phelan, C., Lampe, C., & Resnick, P. (2016). It's Creepy, But It Doesn't Bother Me. *Proceedings of the 2016 CHI Conference on Human Factors in Computing Systems* (pp. 5240–5251). https://doi.org/10.1145/2858036.2858381

Phelps, J. E., D'Souza, G., & Nowak, G. J. (2001). Antecedents and Consequences of Consumer Privacy Concerns; An Empirical Investigation. *Journal of Interactive Marketing (John Wiley & Sons)*, 15(4), 2–17.

Phelps, J., Nowak, G., & Ferrell, E. (2000). Privacy Concerns and Consumer Willingness to Provide Personal Information. *Journal of Public Policy & Marketing*, 19(1), 27–41. https://doi.org/10.1509/jppm.19.1.27.16941

Podsakoff, P. M., MacKenzie, S. B., Lee, J.-Y., & Podsakoff, N. P. (2003). Common Method Biases in Behavioral Research: A Critical Review of the Literature and Recommended Remedies. *Journal of Applied Psychology*, 88(5), 879–903. https://doi.org/10.1037/0021-9010.88.5.879

Preibusch, S., Krol, K., & Beresford, A. R. (2012). The Privacy Economics of Voluntary Over-disclosure in Web Forms. *10th Annual Workshop on the Economics of Information Security*. https://weis2012.econinfosec.org/papers/Preibusch_WEIS2012.pdf

Proferes, N. (2022). The Development of Privacy Norms. In B. P. Knijnenburg, X. Page, P. Wisniewski, H. R. Lipford, N. Proferes, & J. Romano (Eds.), *Modern Socio-Technical Perspectives on Privacy* (pp. 79–90). Springer International Publishing. https://doi.org/10.1007/978-3-030-82786-1_5

PytlikZillig, L. M., & Kimbrough, C. D. (2016). Consensus on Conceptualizations and Definitions of Trust: Are We There Yet? In E. Shockley, T. M. S. Neal, L. M. PytlikZillig, & B. H. Bornstein (Eds.), *Interdisciplinary Perspectives on Trust: Towards Theoretical and Methodological Integration* (pp. 17–47). Cham: Springer International Publishing. https://doi.org/10.1007/978-3-319-22261-5_2

Reeskens, T., & Hooghe, M. (2007). Cross-Cultural Measurement Equivalence of Generalized Trust. Evidence from the European Social Survey (2002 and 2004). *Social Indicators Research*, 85(3), 515–532. https://doi.org/10.1007/s11205-007-9100-z

Rempel, J. K., Ross, M., & Holmes, J. G. (2001). Trust and Communicated Attributions in Close Relationships. *Journal of personality and social psychology*, 81(1), 57.

Rifon, N. J., LaRose, R., & Choi, S. M. (2005). Your Privacy Is Sealed: Effects of Web Privacy Seals on Trust and Personal Disclosures. *Journal of Consumer Affairs*, 39(2), 339–360. https://doi.org/10.1111/j.1745-6606.2005.00018.x

Robert, A., Roe. (2002). What Makes a Competent Psychologist?. *European Psychologist*, 7(3), 192–202.

Robinette, P., Li, W., Allen, R., Howard, A. M., & Wagner, A. R. (2016). Overtrust of Robots in Emergency Evacuation Scenarios. In: *The Eleventh ACM/IEEE International Conference on Human Robot Interaction* (pp. 101–108). IEEE Press.

Ronson, J. (2015, February 12). How One Stupid Tweet Blew Up Justine Sacco's Life. *The New York Times*. https://www.nytimes.com/2015/02/15/magazine/how-one-stupid-tweet-ruined-justine-saccos-life.html

Rotter, J. B. (1967). A New Scale for the Measurement of Interpersonal Trust. *Journal of Personality*, 35(4), 651–665.

Rotter, J. B. (1980). Interpersonal Trust, Trustworthiness, and Gullibility. *American Psychologist*, 35(1), 1.

Rui, J., & Stefanone, M. A. (2013). Strategic Self-Presentation Online: A Cross-Cultural Study. *Computers in Human Behavior*, 29(1), 110–118. https://doi.org/10.1016/j.chb.2012.07.022

Rust, R. T., Kannan, P. K., & Peng, N. (2002). The Customer Economics of Internet Privacy. *Journal of the Academy of Marketing Science*, 30(4), 455–464. https://doi.org/10.1177/009207002236917

Sanchez, O. R., Torre, I., & Knijnenburg, B. P. (2019a). Semantic-Based Privacy Settings Negotiation and Management. *Future Generation Computer Systems*. https://doi.org/10.1016/j.future.2019.10.024

Sanchez, O. R., Torre, I., He, Y., & Knijnenburg, B. P. (2019b). A Recommendation Approach for User Privacy Preferences in the Fitness Domain. *User Modeling and User-Adapted Interaction*. https://doi.org/10.1007/s11257-019-09246-3

Sapienza, P., Toldra-Simats, A., & Zingales, L. (2013). Understanding Trust. *The Economic Journal*, 123(573), 1313–1332.

Schaub, F., Balebako, R., Durity, A. L., & Cranor, L. F. (2015). A Design Space for Effective Privacy Notices. *Eleventh Symposium On Usable Privacy and Security (SOUPS 2015)* (pp. 1–17). https://www.usenix.org/conference/soups2015/proceedings/presentation/schaub

Schelble, B. G., Flathmann, C., McNeese, N. J., Freeman, G., & Mallick, R. (2022c). Let's Think Together! Assessing Shared Mental Models, Performance, and Trust in Human-Agent Teams. *Proceedings of the ACM on Human-Computer Interaction*, 6(GROUP), 1–29.

Schelble, B. G., Flathmann, C., Scalia, M., Zhou, S., Myers, C., McNeese, N. J., ... & Freeman, G. (2022b). Addressing the Spread of Trust and Distrust in Distributed Human-AI Teaming Constellations. CHI TRAIT Workshop (2022b). New Orleans, LA.

Schelble, B. G., Lopez, J., Textor, C., Zhang, R., McNeese, N. J., Pak, R., & Freeman, G. (2022a). Towards Ethical AI: Empirically Investigating Dimensions of AI Ethics, Trust Repair, and Performance in Human-AI Teaming. *Human Factors*, 66(4), 1037–1055. https://doi.org/10.1177/00187208221116952.

Schiffman, B. (2007, December 5). Facebook CEO Apologizes, Lets Users Turn Off Beacon. *Wired*. https://www.wired.com/2007/12/facebook-ceo-apologizes-lets-users-turn-off-beacon/

Schwartz, S. H. (1994). Beyond Individualism/Collectivism: New Cultural Dimensions of Values. In U. Kim, H. C. Triandis, Ç. Kâğitçibaşi, S.- C, & G. Yoon (Eds.), *Individualism and Collectivism: Theory, Method, and Applications* (pp. 85–119). Thousand Oaks, CA: SAGE Publications, Inc.

Segijn, C. M., Opree, S. J., & Ooijen, I. van. (2022). The Validation of the Perceived Surveillance Scale. *Cyberpsychology: Journal of Psychosocial Research on Cyberspace*, 16(3), Article 3. https://doi.org/10.5817/CP2022-3-9

Shamdasani, P. N., Stanaland, A. J., & Tan, J. (2001). Location, Location, Location: Insights for Advertising Placement on the Web. *Journal of Advertising Research*, 41(4), 7–21.

Shapiro, S. S. (2009). Privacy by Design: Moving from Art to Practice. *Communications of the ACM*, 53, 27–29. https://doi.org/10.1145/1743546.1743559

Sharma, K., Schoorman, F. D., & Ballinger, G. A. (2023). How Can It Be Made Right Again? A Review of Trust Repair Research. *Journal of Management*, 49(1), 363–399.

Sheehan, K. B., & Hoy, M. G. (2000). Dimensions of Privacy Concern among Online Consumers. *Journal of Public Policy & Marketing*, 19(1), 62–73. https://doi.org/10.1509/jppm.19.1.62.16949

Simon, H. A. (1959). Theories of Decision-Making in Economics and Behavioral Science. *The American Economic Review*, 49(3), 253–283. https://doi.org/10.2307/1809901

Simon, H. A. (1982). *Models of Bounded Rationality: Empirically Grounded Economic Reason*. Cambridge, MA: MIT Press.

Simpson, J. A. (2007). Foundations of Interpersonal Trust. In A.W. Kruglanski, & E.T. Higgins (Eds.), *Social Psychology: Handbook of Basic Principles* (Vol. 2, pp. 587–607). London: The Guilford Press.

Singel, R. (2009, December 17). Netflix Spilled Your Brokeback Mountain Secret, Lawsuit Claims. *WIRED*. https://www.wired.com/2009/12/netflix-privacy-lawsuit/?+wired27b+%2528Blog+-+27B+Stroke+6+%2528Threat+Level%2529%2529/

Sipior, J., Ward, B. T., & Connolly, R. (2013). Empirically Assessing the Continued Applicability of the IUIPC Construct. *Journal of Enterprise Information Management*, 26(6), 4–4.

Slovic, P., Finucane, M. L., Peters, E., & MacGregor, D. G. (2004). Risk as Analysis and Risk as Feelings: Some Thoughts about Affect, Reason, Risk, and Rationality. *Risk Analysis*, 24(2), 311–322. https://doi.org/10.1111/j.0272-4332.2004.00433.x

Smith, H. J., Dinev, T., & Xu, H. (2011). Information Privacy Research: An Interdisciplinary Review. *MIS Quarterly*, 35(4), 989–1016.

Smith, H. J., Milberg, S. J., & Burke, S. J. (1996). Information Privacy: Measuring Individuals' Concerns about Organizational Practices. *MIS Quarterly*, 20(2), 167–196. https://doi.org/10.2307/249477

Smith, N. C., Goldstein, D. G., & Johnson, E. J. (2013). Choice without Awareness: Ethical and Policy Implications of Defaults. *Journal of Public Policy & Marketing*, 32(2), 159–172. https://doi.org/10.1509/jppm.10.114

Solove, D. J. (2002). Conceptualizing Privacy. *California Law Review*, 90, 1087–1155.

Spiekermann, S. (2012). The Challenges of Privacy by Design. *Communications of the ACM*, 55(7), 38–40. https://doi.org/10.1145/2209249.2209263

Spiekermann, S., Grossklags, J., & Berendt, B. (2001). E-privacy in 2nd Generation E-Commerce: Privacy Preferences versus Actual Behavior. *Proceedings of the 3rd ACM Conference on Electronic Commerce* (pp. 38–47).

Squicciarini, A. C., Shehab, M., & Paci, F. (2009). Collective Privacy Management in Social Networks. *Proceedings of the 18th International Conference on World Wide Web* (pp. 521–530). https://dl.acm.org/citation.cfm?id=1526780

Stallings, W. (2020). Handling of Personal Information and Deidentified, Aggregated, and Pseudonymized Information under the California Consumer Privacy Act. *IEEE Security & Privacy*, 18(1), 61–64. https://doi.org/10.1109/MSEC.2019.2953324

Stark, L., King, J., Page, X., Lampinen, A., Vitak, J., Wisniewski, P., Whalen, T., & Good, N. (2016). Bridging the Gap between Privacy by Design and Privacy in Practice. *Proceedings of the 2016 CHI Conference Extended Abstracts on Human Factors in Computing Systems* (pp. 3415–3422). https://doi.org/10.1145/2851581.2856503

Stewart, K. A., & Segars, A. H. (2002). An Empirical Examination of the Concern for Information Privacy Instrument. *Information Systems Research*, *13*(1), 36–49. https://doi.org/10.1287/isre.13.1.36.97

Stone, D. L. (1981). The Effects of the Valence of Outcomes for Providing Data and the Perceived Relevance of the Data Requested On Privacy-Related Behaviors, Beliefs, and Attitudes (Thesis 31634 PhD) [Ph.D. Thesis]. Purdue University.

Strater, K., & Lipford, H. R. (2008). Strategies and Struggles with Privacy in an Online Social Networking Community. *Proceedings of the 22nd British HCI Group Annual Conference on People and Computers* (pp. 111–119). Swindon.

Stutzman, F., Gross, R., & Acquisti, A. (2013). Silent Listeners: The Evolution of Privacy and Disclosure on Facebook. *Journal of Privacy and Confidentiality*, *4*(2), 7–41.

Sun, Q., Willemsen, M. C., & Knijnenburg, B. P. (2020). Unpacking the Intention-Behavior Gap in Privacy Decision Making for the Internet of Things (IoT) Using Aspect Listing. *Computers & Security*, *97*, 101924. https://doi.org/10.1016/j.cose.2020.101924

Sundar, S. S., Kang, H., Wu, M., Go, E., & Zhang, B. (2013). Unlocking the Privacy Paradox: Do cognitive Heuristics Hold the Key? *CHI'13 Extended Abstracts on Human Factors in Computing Systems* (pp. 811–816). https://doi.org/10.1145/2468356.2468501

Sweeney, L. (2002). k-Anonymity: A Model for Protecting Privacy. *International Journal on Uncertainty, Fuzziness, and Knowledge-Based Systems*, *10*(5), 557–570. https://doi.org/10.1142/S0218488502001648

Tang, K., Hong, J., & Siewiorek, D. (2012). The Implications of Offering More Disclosure Choices for Social Location Sharing. *Proceedings of the SIGCHI Conference on Human Factors in Computing Systems* (pp. 391–394). https://doi.org/10.1145/2207676.2207730

Tene, O., & Polonetsky, J. (2013). A Theory of Creepy: Technology, Privacy and Shifting Social Norms. *Yale Journal of Law and Technology*, *16*, 59–102.

Textor, C., Zhang, R., Lopez, J., Schelble, B. G., McNeese, N. J., Freeman, G., ... & de Visser, E. J. (2022). Exploring the Relationship between Ethics and Trust in Human-Artificial Intelligence Teaming: A Mixed Methods Approach. *Journal of Cognitive Engineering and Decision Making*, *16*(4), 252–281.

Tung, F.-W. (2011). Influence of Gender and Age on the Attitudes of Children towards Humanoid Robots. In J. A. Jacko (Ed.), *Human-Computer Interaction, Part IV* (pp. 637–646). Berlin, Germany: Springer-Verlag.

Thaler, R. H., & Sunstein, C. (2008). *Nudge: Improving Decisions about Health, Wealth, and Happiness*. New Haven, CT: Yale University Press.

Treiblmaier, H., & Pollach, I. (2007). Users' Perceptions of Benefits and Costs of Personalization. *ICIS 2007 Proceedings*. https://aisel.aisnet.org/icis2007/141

Ullman, D., & Malle, B. F. (2018). What Does It Mean to Trust A Robot? Steps toward a Multidimensional Measure of Trust. *Companion of the 2018 ACM/IEEE international conference on human-robot interaction* (pp. 263–264). Chicago, IL.

USAF. (2019). *Autonomous Horizons: The Way Forward*. Air University Press, Maxwell AFB, AL

Utz, S., & Kramer, N. (2009). The Privacy Paradox on Social Network Sites Revisited: The Role of Individual Characteristics and Group Norms. *Cyberpsychology: Journal of Psychosocial Research on Cyberspace*, *3*(2). https://www.cyberpsychology.eu/view.php?cisloclanku=2009111001&article=1

Van Slyke, C., Shim, J. T., Johnson, R., & Jiang, J. J. (2006). Concern for Information Privacy and Online Consumer Purchasing. *Journal of the Association for Information Systems*, *7*(1). https://aisel.aisnet.org/jais/vol7/iss1/16

Vashistha, P., Singh, J. P., Jain, P., & Kumar, J. (2019, June). Raspberry Pi Based Voice-Operated Personal Assistant (Neobot). *2019 3rd International conference on Electronics, Communication and Aerospace Technology (ICECA)* (pp. 974–978). Coimbatore: IEEE.

Veale, M., & Borgesius, F. Z. (2021). Demystifying the Draft EU Artificial Intelligence Act-Analysing the Good, the Bad, and the Unclear Elements of the Proposed Approach. *Computer Law Review International*, *22*(4), 97–112. https://doi.org/10.9785/cri-2021-220402

Vitak, J., Blasiola, S., Litt, E., & Patil, S. (2015). Balancing Audience and Privacy Tensions on Social Network Sites: Strategies of Highly Engaged Users. *International Journal of Communication*, *9*, 1485–1504.

Vorm, E. S., & Combs, D. J. (2022). Integrating Transparency, Trust, and Acceptance: The Intelligent Systems Technology Acceptance Model (ISTAM). *International Journal of Human-Computer Interaction*, *38*(18–20), 1828–1845.

Waldman, A. E. (2018). *Privacy as Trust: Information Privacy for an Information Age*. Cambridge, MA: Cambridge University Press.

Wang, W., & Benbasat, I. (2007). Recommendation Agents for Electronic Commerce: Effects of Explanation Facilities on Trusting Beliefs. *Journal of Management Information Systems*, *23*(4), 217–246. https://doi.org/10.2753/MIS0742-1222230410

Wang, Y., Leon, P. G., Acquisti, A., Cranor, L. F., Forget, A., & Sadeh, N. (2014). A Field Trial of Privacy Nudges for Facebook. *Proceedings of the 32nd Annual ACM Conference on Human Factors in Computing Systems* (pp. 2367–2376). https://doi.org/10.1145/2556288.2557413

Weitz, K., Schiller, D., Schlagowski, R., Huber, T., & André, E. (2019, July). "Do You Trust Me?" Increasing User-Trust by Integrating Virtual Agents in Explainable AI Interaction Design. *Proceedings of the 19th ACM International Conference on Intelligent Virtual Agents (pp. 7–9)*. Paris.

Wenning, R., & Schunter, M. (2006). *The Platform for Privacy Preferences 1.1 (P3P1.1) Specification*. W3C Working Group Note. https://www.w3.org/TR/P3P11/

Westin, A. F. (1967). Special report: Legal safeguards to insure privacy in a computer society. *Communications of the ACM, 10*(9), 533–537.

Westin, A. F., Harris, L., & Associates. (1981). *The Dimensions of Privacy: A National Opinion Research Survey of Attitudes toward Privacy*. New York: Garland Publishing.

White, T. B. (2004). Consumer Disclosure and Disclosure Avoidance: A Motivational Framework. *Journal of Consumer Psychology, 14*(1–2), 41–51. https://doi.org/10.1207/s15327663jcp1401&2_6

Wilkinson, D., Namara, M., Patil, K., Guo, L., Manda, A., & Knijnenburg, B. (2021). The Pursuit of Transparency and Control: A Classification of Ad Explanations in Social Media. *Proceedings of the 54th Hawaii International Conference on System Sciences* (p. 763). https://doi.org/10.24251/HICSS.2021.093

Wilkinson, D., Sivakumar, S., Cherry, D., Knijnenburg, B. P., Raybourn, E. M., Wisniewski, P., & Sloan, H. (2017). User-Tailored Privacy by Design. *Proceedings of the Usable Security Mini Conference 2017*. https://doi.org/10.14722/usec.2017.23007

Wilson, D., & Valacich, J. (2012, December 14). Unpacking the Privacy Paradox: Irrational Decision-Making within the Privacy Calculus. *ICIS 2012 Proceedings*. https://aisel.aisnet.org/icis2012/proceedings/ResearchInProgress/101

Wisniewski, P. J., Knijnenburg, B. P., & Lipford, H. R. (2017). Making Privacy Personal: Profiling Social Network Users to Inform Privacy Education and Nudging. *International Journal of Human-Computer Studies, 98*, 95–108. https://doi.org/10.1016/j.ijhcs.2016.09.006

Wisniewski, P., Islam, A. K. M. N., Knijnenburg, B. P., & Patil, S. (2015a). Give Social Network Users the Privacy They Want. *Proceedings of the 18th ACM Conference on Computer Supported Cooperative Work & Social Computing* (pp. 1427–1441). https://doi.org/10.1145/2675133.2675256

Wisniewski, P., Lipford, H., & Wilson, D. (2012). Fighting for My Space: Coping Mechanisms for SNS Boundary Regulation. *Proceedings of the SIGCHI Conference on Human Factors in Computing Systems* (pp. 609–618). https://dl.acm.org/citation.cfm?id=2207761

Wisniewski, P., Xu, H., Lipford, H., & Bello-Ogunu, E. (2015b). Facebook Apps and Tagging: The Trade-Off between Personal Privacy and Engaging with Friends. *Journal of the Association for Information Science and Technology, 66*(9), 1883–1896. https://doi.org/10.1002/asi.23299

Woodruff, A., Pihur, V., Consolvo, S., Schmidt, L., Brandimarte, L., & Acquisti, A. (2014). Would a Privacy Fundamentalist Sell Their DNA for $1000... If Nothing Bad Happened as a Result? The Westin Categories, Behavioral Intentions, and Consequences. *Symposium on Usable Privacy and Security (SOUPS)*. https://www.usenix.org/system/files/conference/soups2014/soups14-paper-woodruff.pdf

Xia, H. (2023). A Critique of Using Contextual Integrity to (Re)consider Privacy in HCI. *Information for a Better World: Normality, Virtuality, Physicality, Inclusivity: 18th International Conference, IConference 2023, Virtual Event*, March 13–17, 2023, Proceedings, Part II (pp. 251–256). https://doi.org/10.1007/978-3-031-28032-0_21

Xie, J., Knijnenburg, B. P., & Jin, H. (2014). Location Sharing Privacy Preference: Analysis and Personalized Recommendation. *Proceedings of the 19th International Conference on Intelligent User Interfaces* (pp. 189–198). https://doi.org/10.1145/2557500.2557504

Xu, H., Dinev, T., Smith, H. J., & Hart, P. (2008). Examining the Formation of Individual's Privacy Concerns: Toward an Integrative View. *ICIS 2008 Proceedings*. Paris.

Xu, H., Gupta, S., Rosson, M. B., & Carroll, J. M. (2012). Measuring Mobile Users' Concerns for Information Privacy. *ICIS 2012 Proceedings*. Orlando, FL.

Xu, H., Luo, X. (Robert), Carroll, J. M., & Rosson, M. B. (2011). The Personalization Privacy Paradox: An Exploratory Study of Decision Making Process for Location-Aware Marketing. *Decision Support Systems, 51*(1), 42–52. https://doi.org/10.1016/j.dss.2010.11.017

Xu, H., Teo, H.-H., & Tan, B. C. Y. (2005). Predicting the Adoption of Location-Based Services: The Role of Trust and Perceived Privacy Risk. *Proceedings of the International Conference on Information Systems* (pp. 861–874). https://aisel.aisnet.org/icis2005/71

Xu, H., Teo, H.-H., Tan, B. C. Y., & Agarwal, R. (2009). The Role of Push-Pull Technology in Privacy Calculus: The Case of Location-Based Services. *Journal of Management Information Systems, 26*(3), 135–174. https://doi.org/10.2753/MIS0742-1222260305

Yang, S.-C., Hung, W.-C., Sung, K., & Farn, C.-K. (2006). Investigating Initial Trust toward e-Tailers from the Elaboration Likelihood Model Perspective. *Psychology and Marketing*, *23*(5), 429–445. https://doi. org/10.1002/mar.20120

Yi, Y., Wu, Z., & Tung, L. L. (2005). How Individual Differences Influence Technology Usage Behavior? Toward an Integrated Framework. *Journal of Computer Information Systems*, *46*(2), 52–63.

Zaslow, J. (2002, November 26). If TiVo Thinks You Are Gay, Here's How to Set It Straight. *Wall Street Journal (Eastern Edition)*, A.1. https://www.wsj.com/articles/SB1038261936872356908

Zhang, R., McNeese, N. J., Freeman, G., & Musick, G. (2021). "An Ideal Human" Expectations of AI Teammates in Human-AI Teaming. *Proceedings of the ACM on Human-Computer Interaction*, *4*(CSCW3), 1–25.

Zhou, T. (2012a). Examining Location-based Services Usage from the Perspectives of Unified Theory of Acceptance and Use of Technology and Privacy Risk. *Journal of Electronic Commerce Research*, *13*(2), 135–144.

Zhou, T. (2012b). Understanding users' initial trust in mobile banking: An elaboration likelihood perspective. *Computers in Human Behavior*, *28*(4), 1518–1525. https://doi.org/10.1016/j.chb.2012.03.021

14 Ethics in HCI

Janet C. Read

14.1 INTRODUCTION

This chapter opens with a short history of ethics and with some definitions that may be useful. We then explore the general ethics around research using ethical codes as a prompt for this discussion. This is followed with sections that consider how we can work with participants in HCI research, then a look at design and the challenges around design as they refer to ethics with special sections on persuasive design and on emerging technologies. The narrative is interspersed with small cases that add interest which mainly come from the authors' own work with children who, as edge users in HCI, highlight ethical concerns that help us all to do more ethical work.

For many people, ethics is associated with protocols and procedures that are aligned to research studies. Ethics is thought about in terms of rules and preventions and can be looked on as something that has to be got over and done with before the real fun of research and design can happen. In this chapter, it is hoped that some of the useful tips will make this process easier but also that the provocations will lead the reader to think, somewhat more laterally, about the ethics around their own work. It is hoped that readers may find, in this chapter, a reason to do things differently – as they design, interact with users, or think about new technologies. There are different ways to choose participants, different lenses to evaluate designs, different methods to gather consent, and different questions to be asked about our HCI work.

14.1.1 A BRIEF HISTORY OF ETHICS

In essence, ethics is the branch of philosophy concerned with the evaluation of morals. Morals determine whether something is just or right, and so ethics is concerned with determining if there can be a systematic way to decide if something is right or wrong. Often in ethics, there are dilemmas around whether it can be acceptable for one person to be harmed in order to benefit another. It can be argued that ethics has its roots in the moral tales of ancient times like Homer's Iliad. Socrates is generally considered the father of ethics as he sought to use rational methods to establish what he called moral truths. Plato and Aristotle followed, and ethical ideas, like 'do no harm' were included in religious texts. Other Greeks, notably the stoics (Zeno and Epicurus and others), had much to say about ethics, and it was with their thinking, on what makes a good life, that virtue ethics finds its roots. Ethical debate was not constrained to the Greeks; from China, Confucius promoted the rule of 'do only to others what you want done to yourself' and in other ancient civilizations, attitudes, and laws, around what it was to be good came from many cultures. An interesting example from Buddhism emphasizes that harm can come from unintended actions. Channa reportedly gave food to the Buddha which made him (the Buddha) ill and caused his death but as the intention of Channa was to be generous to the Buddha, so Channa should not be condemned. This is an interesting point on which to ponder – can our ignorance of possible consequences ever be an excuse? In most cases, we would have to say that it is our duty to carefully consider any unintended consequences of our actions.

Around the eighteenth century, ethical dilemmas started to be widely debated. This was when the ideas of universalism (it's only okay if it's good for everyone) and utilitarianism (the right thing to do is that that brings the most benefit to the most people) became a common means to debate ethics.

Indeed, at that time, ethics was largely a topic of philosophical debate; actions shortly after World War II changed that landscape with the discovery of, and then concern about, scientific experiments on humans.

The Nuremberg Code is a set of ethical research principles for human experimentation and while it came about after World War II it is important to stress that informed consent, beneficence (that of doing good) and non-maleficence (that of doing no harm), had already been referred to in earlier codes. The Nuremberg Code is an example of preventative ethics in so far as it is a set of rules that were initially intended to prevent wrong happening. It stipulated that participants' consent was essential, that results should be for the good of society, that unnecessary harm should be avoided, that risk should not exceed benefit, that proper planning should take place, that those doing research should be trained and well qualified, and that a participant should always be able to quit. This code has since been a springboard for many ethical codes that are tailored to different research and practice fields. Parts of the code are incorporated into the International Covenant on Civil and Political Rights (ICCPR), which is upheld by more than 170 countries, including in Article 4.2 – freedom from medical or scientific experimentation without consent and in Articles 24 and 55 – the right to privacy.

14.1.2 PRACTICAL ETHICS – CODES AND VALUES

Ethics codes, and their application and consequences, tend to dominate much of the practical ethics that concern HCI practitioners. Within HCI, ethics codes have tended to emerge from the disciplines that underpin our work: computing, IT, design, engineering. As with the early codes, these typically tell us what we should, or should not, be doing. Many such codes have been discussed in the literature (Anderson et al., 1993; Gotterbarn et al., 2017), and they all have been shown to be of some use. Beyond academia, businesses have also been keen to produce codes of ethics but, across all domains, the value of a code has been criticized with one author saying that *A code in isolation is a veneer for being ethical and in essence is misleading* (Wood & Rimmer, 2003). To be useful, an ethical code typically needs regulatory teeth or a code of conduct. The GDPR(EU) and CCPA(US) are both regulatory codes that set out how data and humans should be treated. These have the power to be able to determine right behaviours on emerging technologies, such as facial recognition and say how these technologies can and should be used. They can also stipulate consequences for those who break their codes. Some ethical codes are associated with professional life to promote good behaviour with the regulation that anyone found breaking the code can be evicted from the profession – examples include codes of medical ethics and codes for legal professionals. The Association of Computing Machinery (ACM) Ethical Code (https://www.acm.org/code-of-ethics) has seven basic principles (beneficence, non-maleficence, honesty, fairness, respect for others work, privacy, and confidentiality) that match those found elsewhere as well as additional principles associated with being a computer professional and being a leader.

*** *QUESTION? What should be the 'punishment' for breaking an ethics code for HCI?*

Whilst a code can exist for all people, values dictate how we, as individuals, behave when faced with ethical dilemmas. As mentioned earlier, an ethical dilemma is a situation in which it is not obvious what the right action is. The Australian Computer Society (ACS) standard of conduct (https://www.acs.org.au) acknowledges that individuals have to make judgement calls when it states, at the end of the code, that *In summary, a member is expected to act at all times in a manner likely to be judged by informed, respected and experienced peers, in possession of all the facts, as the most ethical way to act in the circumstances.* Research has suggested that individual values play a key role in determining and resolving ethical dilemmas (Moser, 1988), and so we cannot easily navigate ethical dilemmas without considering values.

EXAMINING VALUES

There are several ways that we can explore our values. One is to make a bucket list that cata-logues things we would like to do with our time here on Earth. Our choices will reflect what we hold dear to us. In our group, we use the CHECk (Read et al., 2013) toolkit to examine choices we are about to make when designing research or design studies with children – we ask ourselves;

1. What are we aiming to discover?
2. Why (this question)?
3. Why are we using these methods?
4. Why are we using these children?

The process of asking ourselves these questions will lead us to think about how important that paper is to us, we will start to examine our research values as we think about methods and participants, we will examine worth values as we justify why this question should be answered with these children in this way. Our values dictate our choices.

In 1959, the issue of doing research on animals prompted the Three Rs checklist (Russell & Burch, 1959), which asked, of researchers working with animals:

* Does this animal need to be used, is there a REPLACEMENT that is not an animal?
* Do this many animals need to be used, can we REDUCE the number?
* Does this study have to have all these elements – can it be REFINED to reduce stress?

We can apply these same questions to all our research and design work with human participants in HCI, and they can challenge us to think about what we are doing with human participants. In terms of changing our potential study to better serve our participant base, the extent to which we would consider changing our initial plan is typically dictated by what we value the most. When consider-ing the design of any work in HCI, we constantly get an opportunity to examine our values. One provocation from Brown et al. (2016) asks, *"Do those vulnerable people need to be 'researched' on? – and if so - any research with them should benefit them more than us"*. This leads us to the very important topic of research participation.

14.2 HCI AND THE ETHICS OF RESEARCH PARTICIPATION

By design, HCI research typically includes participants. These are the humans who come to our study to carry out a user study or usability test, to inform the design of a new product, to participate in an experimental study or to be observed while using our technology. All too often in HCI work we take the easy option of finding an easy way to recruit set of individuals, we cash in on their will-ingness to participate without really thinking about what they are giving us as well as their time.

14.2.1 Recruitment of Participants

In HCI studies that include participants, the recruitment of those participants, and the extent to which they understand their involvement and have agency to discuss their participation is an area where ethical judgements need to be made. Recruitment can be to a study in a lab setting, to a longitudinal field study, or to an online research or evaluation event. The purpose of participation can vary between being solely testers of an artefact to being co-designers and co-researchers. In some instances, participants may be recruited to meet a demography or may be simply a convenience group. When recruiting participants, we need to ask:

- Who will be chosen?
- Will this be a biased sample?
- Are there special approaches needed for this group?
- What is the value to the individuals and how is that approached?

Ethically, the choice on who gets to participate in research and design work should adhere to the principle of fairness. The ethics of selection are problematic (see the next paragraph) but, where a group of individuals are identified as candidates for participation, we must ask carefully – should anyone from that group be excluded? This is highlighted in work with children where a school class may comprise 35 children and the 'experiment' might only want 32! In these cases, we would always say that all children should participate even if three of them have to have their data randomly removed from the final write up for the paper to give us a balanced design (Frauenberger et al., 2018).

*** QUESTION? When is it okay to remove a participant's data from a study?

Selection bias is where participants are recruited that under-represent certain groups of people. Offenwanger et al. (2021) interviewed 13 HCI researchers and examined over 1,000 papers to provide empirical evidence for the under-representation of women, the invisibility of non-binary participants and a deteriorating representation of women in MTurk studies. The authors interestingly noted that "Studies recruit all women intentionally and all men by coincidence". It is also apparent that most HCI research and design is done with WEIRD (Western, Educated, Industrialized, Rich, and Democratic) participants (Henrich et al., 2010). In a study of five years of HCI papers, it was shown that over 70% were based on western participant samples and that over half the countries in the world had never provided participants for HCI research (Linxen et al., 2021). Additionally, based on the data Linxen et al., (2021) could glean, over 70% of participants were college-educated. These examples clearly demonstrate a need for diversification in the selection of participants but also, to monitor participant trends, it is important to better describe participant groups. There are cases where it is acceptable to recruit a homogenous group of individuals to a study – perhaps there is a good reason to use university students aged between 20 and 22 as they can be split into two groups to consider the effects of, say, a new interaction method. The point here is to ask the question – why this group? In writing up research or design work, the author should always explain the rationale behind selection.

Having chosen a population from which to take sample participants, the practicalities of recruitment then follow. This is where consideration must be made of special approaches that might be needed as well as care taken to ensure that participants are valued. Where recruitment will be in dangerous settings, in care settings, with individuals with additional needs or with minors, special approaches are needed. In our own work with children, we have always sought to build relationships with schools and with children so that participation is well understood by all stakeholders. Explanations of the value and extent of participation are expected as part of the evaluation of research. Most boards that examine ethics (Institutional review boards (IRB), Independent Ethics Committees (IEC), Ethical Review Boards (ERB), or Research Ethics Boards (REB)), expect information sheets to be sent to participants that clearly articulate why participants are being recruited, what the perceived value of the work is and what, if any, compensation, will be given.

In HCI, experimental work and user studies are reasonably easy to explain and justify to participants. In an experimental study, we can begin by highlighting the value of finding out which of two competing situations, products or solutions is more beneficial and in doing the experiment the participant is seldom left unsure or uncertain and can probably understand their contribution to the work. User studies can include observations, logging and surveys and they are also relatively easy to justify and explain. Typically, we present to a group a software solution or a prototype and expect the group to interact with that in some way to find bugs, explore use patterns or gather feedback on their enjoyment and/or success. In these situations, the individual can easily understand what is being asked of them and can easily make an informed judgement as to whether to participate. An exception to this is where studies use deception, for example using protocols like Wizard of Oz where a human 'wizard' mimics the actions of an intelligent system (that has typically not been made to work yet) in order to evaluate user acceptance, explore user behaviour or gather user feedback. In our own work on Wizard of Oz with children, we circumvented some of the potential ethical problems by (1) having the wizard in the room which meant that there was less deception than hiding the wizard away (although one can argue that children would still be in the dark about what the wizard was actually doing while sat at a laptop in the room) and (2) encouraging children, after the experiment, to act as wizards in order that they could better understand what they had experienced (Read et al., 2005).

Participating in design work is an area where value has especially been studied owing to both the nature of that participation – which typically requires more time than a user study or an experiment – but also due to the participatory design philosophy which is around empowerment, democracy, and inclusion. Within participatory design communities, there is ongoing interest in the value of participation to both participants and designers. In exploring the role of users in design in HCI, Ehn (2008) referred to participatory design (designing for use before use, so naturally being relatively speculative and open ended) and meta-design (designing for design after design, which is to say that design that is deferred till after the project is defined). In this paper, Ehn (2008) outlined two values that motivate HCI researchers to include users in design – one being the social and rational idea of democracy, the other being the importance of bringing the participants' 'tacit knowledge' into the design process. For individuals taking part in such design activities, value is associated with a belief that outcomes will be used and this value will to some extent depend on the proximity of that use to the individual. Rashid et al. (2006) captured this proximity in a hierarchical way which is paraphrased for participatory design situations as follows:

- What I design – I will use
- What I design – people close to me will use
- What I design – a small group of people – similar to me, but not known to me – will use
- What I design – a group of people – as yet not really clearly describable – will use

Understanding this value chain is important when recruiting participants into participatory design sessions. In a design session in which the user is contributing designs for a product only he or she will ever use, then the value to that individual will be high – providing he or she perceives that these designs will be used – i.e., will be integrated into a product. However, in most cases, when we include users in design sessions, it is not for a product solely for themselves, and in some cases, there is never a product made. This requires a different consideration of value when explaining to participants why they should contribute. The need for design sessions to have a meaning beyond the session itself is a main theme of the critique by Bødker and Kyng (2018) who implored researchers and designers to focus on participation in design activities that address areas where dramatic, potentially negative, changes are underway; you could consider this to include areas such as climate change, political reform, etc. Their rationale for this is that as a community of HCI practitioners we should be concerned that our research and design work has real impact but also that such motivations for design would form a strong basis for engagement and action by all participating parties. Two practical ways, therefore, to provide additional value to participants when doing design work with us in HCI are a) to ensure that those who participate feel valued and can contribute value and b) to work with UX practitioners and industry so that outputs can 'go to market' and have impact in the real world.

***QUESTION? Is it ever wrong to engage with users in a participatory design session?*
In summary, Vines et al. (2013) provoked HCI professionals to ask:

- who initiates, directs, and benefits from participation?
- what form of participation does the user take?
- how can users actively participate in the processes?

They highlight ethical problems including the need to explicitly state the influencing factors that lead to some individuals participating over others and stress the need to be clear about how individuals who do participate have their voices heard.

CASE: GOOD PRACTICE FROM THE CHILD COMPUTER INTERACTION COMMUNITY

Since 2019 all papers submitted to IDC and to IJCCI must include a section titled "Selection and Participation of Children" in which the authors of the paper should describe how children were selected (if there were no children simply write: no children participated in this work), what consent processes were followed (i.e. did they consent and if so what they were told), how they were treated, how data sharing was communicated, and especially any additional ethical considerations.

CASE: REPORTING BACK IN HCI WORK

In a recent study (Read et al., 2022), we explored the extent to which academic papers reported how participants in studies were informed of the outcomes of the research they had done. Reporting back is when we go back, typically sometime later, to our participants and tell them what we found. This can be challenging but, in our paper, we challenge the HCI community to build this into their research designs and include it in their write ups of their work, so that they can adequately and appropriately explain to participants, where at all possible, what their participation meant.

14.2.2 INFORMED CONSENT

If fairness is the ethical backbone to selection, then volunteerism is the ethical backbone to consent. Simply put, the gathering of consent is the process by which individuals show their willingness, as volunteers, to participate. The primary elements of informed consent are: (Beauchamp & Childress, 2019)

- competence to understand and decide – which refers to a person's ability to make and communicate a decision to consent to participate.
- voluntariness – a choice being made of a person's free will, as opposed to being made as the result of coercion or duress.
- understanding of material information – *facts that would be likely to influence a decision* to participate.

Consent constitutes a legally valid authorization to work with an individual and take away their data/ideas/results and so universities and companies alike are very keen to ensure that consent is gathered, dated, and signed from participants in research and design. Also, as stated earlier, consent is enshrined in human rights law – so, for all these reasons – universities and companies will have clear procedures for gathering consent that will typically include the giving out of clear information and the signing of a form.

While consent can often be easily gained, it is important in HCI to take a user-centred approach to consent and seek to make the gathering of consent as transparent and user friendly as possible while also seeking to get *'the best consent possible'*. When we are looking to gather consent from individuals who are interacting with technology – be that in an online study or as part of a mobile or ubiquitous technology study – it can be relatively easy to gather consent by making the process almost unconscious. An interesting study in 2010 with 80,000 participants, (Böhme & Köpsell, 2010) showed how individuals clicked 'accept' on a dialogue box that popped up on an online study with very little thought. They compared the familiar words found on End User License Agreements (EULAs) with a more meaningful label 'I do take part' and noted that, despite all other aspects being the same, the familiar text drew the participants to choose it significantly more than the more meaningful text. This example serves to remind us that it can be very easy, with the right trickery, to gain consent from individuals. Thankfully, HCI researchers have considered how the design of technology related to consent gathering can be improved to give greater confidence; for example, participants in an online study were initially asked for their consent and then began answering questions. After a few questions – from which answers were only held on the local device – the participants were reminded that their data was being gathered and they were shown that the study had found out some things about them – before being asked if they were still happy to continue (Morrison et al., 2014).

Another, better than basic, approach is where consent is gathered before and after participation. Even beyond this, Luger and Rodden (2014) suggested that we should not just record consent to interact with technology in a moment in time but should facilitate and sustain continued agency throughout use, which would suggest that we constantly remind our participants that they can withdraw/re-consent as they work with us. Some of the debate on rich consent has been informed from feminist and empowerment thinking. The FRIES model of sexual consent, which is used in schools, defines consent as being Freely given, Reversible, Informed, Enthusiastic and Specific. This inspired Strengers et al. (2021) to consider if a similar approach could be applied to technology use. Their acronym 'TEASE' is rather less easy to remember, but the basic ideas are good; when deciding to give consent to interact with technology we might actively ask ourselves is it inherently safe, does it react to us in an appropriate way, is it likely to meet our expectations, will we be able to quit at any time and can we decide to what extent we consent. This latter point – on the extent of consent is very important; if there are layers to our consent, then we need to try hard to articulate them in the consent gathering process. We might ask if we can record a user's mouse clicks, record what they say using an audio file, or record all their expressions using video – these are different layers of consent, and we must be okay with individuals perhaps not consenting to everything we wanted!

Challenges in HCI studies include gathering consent from individuals in large in the wild-type studies and getting consent from children and other vulnerable groups. Considering how we gather consent in HCI work in the wild is tricky and anyone who has tried to describe such studies for ethics boards will understand the difficulties. As Brown et al. (2016) stated in their five provocations for ethical HCI, in some cases consent cannot be gathered; this is not a deal breaker, as explained in the ethics code of the UK research funder ESRC that states: *'Informed consent may be impractical or meaningless in some research, such as research on crowd behaviour, or may be contrary to the research design, as is sometimes the case in psychological experiments where consent would compromise the objectives of the research – covert research may be undertaken when it might provide unique forms of evidence or where overt observation might alter the phenomenon being studied'*. Covert work does happen in HCI work. In Deep Cover HCI, an instance is described where research was carried out without consent. In this case, an installation was placed, and observations were made. The observations were covert – that is, the individuals walking up to the installation were unaware of what was being gathered – and the researchers were outsiders; in other words, the people who were being (remotely) observed were not known to them (Williamson & Sundén, 2015). The authors made the case that this setup (covert observation and interpreted by outsiders) was a suitable arrangement for consent to not be required.

In terms of consent, vulnerable populations are those who are considered to not be able to consent for themselves. Children are a much-studied example of a user group that can be described as vulnerable. In legal terms, in most countries, minors under 18 cannot consent for themselves. Many commentators, and research studies, have rightly pointed out that in most cases children can make decisions for themselves (e.g. Miller et al., 2004), but the legal aspects of consent generally leave HCI practitioners needing to gather consent from adults and then gathering assent from children. This is not without problems. Rode (2009) wrote a fairly scathing attack on the implications of gathering informed consent in the context of a study in which children were going to be asked about their parents' rules for online engagement If they, during the interview, confessed to having broken these rules and potentially put themselves in danger, then the researcher would HAVE to tell the grown-ups. The author highlighted in this work the conflict between the children's privacy – viz. their rights – and the ethical codes. By listening to the children's stories about the rules their parents had put in place, Rode also observed that parents did not always act in their children's best interests; in one situation, a father had attempted to steal his own child's identity to purchase a home, in another, a liberal family who let their child have unfettered internet access eventually had to deal with their child getting into an online relationship and an unplanned pregnancy. These examples beg the question *is it always good that parents are in charge of a consent process?* A similar observation is seen when social media posts show parents putting pictures of their children online – presumably without the child's permission (Kumar et al., 2018).

Whether or not we believe that children should be able to consent fully to participate, say, in a design study, the legalities tend to prevent that approach and in these cases the gathering of, and managing of assent, becomes hugely important. The same must be the approach with individuals who may not have the language to consent such as those with special needs or the very elderly, etc. Assent, as an agreement after a caregivers' consent, only requires a basic comprehension of procedures and purpose and the ability to indicate a preference. Assent avoids the issues of competence that require a participant to clearly understand the right to withdraw, to understand randomization as it applies to selection and allocation to tasks, and to understand rights associated with voluntariness (Grisso & Vierling, 1978).

CASE: HOW WE GATHER CHILDREN'S ASSENT?

In our work with children, we have worked very hard on developing a protocol for gathering assent from children. We begin with telling the children who we are and where we have come from in order to give them clarity about what we are doing with them. We explain what a university is, we explain what our jobs are, and we tell the children that we find things out and write about them so we can get more money while also hopefully making things better for children around the world. We then explain what science is and what research is – we explain that we use 'data' to answer questions and we show examples to help them in this. We tell the children that they might be giving us data while we work with them. This might be numbers or ticks on a survey, it might be inside the technology they are using, it might be ideas. We explain about personal data and how that is different from numbers. We then tell them about what we will do with things they hand in, stressing that they do not need to hand in anything. We tell them that we might make lots of money from their ideas or data and we ask them about what we should do if we do make lots of money – then we generally tell them that we don't expect to make any money at all ☹. We tell the children about who it is that is paying for us to be with them, and we tell the children why that matters. We then state, and the children understand, that by handing in anything, they have effectively assented. Since carrying out this protocol we have always been pleased when children have chosen to do the activity with us and have chosen to not hand anything in. This is evidence, we believe, that we are empowering children to choose.

CASE: ETHICS REVIEW BOARDS AND FILLING OUT THE PAPERWORK

An ethics review board is a panel of reviewers that is used to evaluate the ethics of a research or design study. Common in universities, these boards typically require copies of information sheets, consent forms, risk assessments and research instruments. While not evaluating the research per se, this panel will be looking to ensure that the research complies with the local and national ethical codes. Getting ethical approval can feel like jumping through hoops, but it is a good first step towards doing ethical work. It is important to know the limitations of such boards, their role is protective; good ethics goes much further and, in some cases, doing good ethics can upset the ethics review boards as it may mean thinking outside the box.

14.2.3 PRIVACY AND CONFIDENTIALITY

Anyone who has ever filled out an ethics review form will have come across questions about anonymization, where the prevailing attitude (preventism) is that the more anonymization there is, the easier it will be to 'tick off' the ethics. Anonymization is promoted as a best practice in terms of protecting participants' privacy and rights and it relates primarily to the removal or obscuring of names, and other identifying items, from papers and reports. Critics claim that anonymization is potentially politically situated, referring to how women, in earlier times, were not allowed their names, and so anonymization is to put down an individual's being-ness. There is also a belief that it is dangerous for people to know your name/identity and so anonymization is for protection (Moore, 2012). While there is considerable debate about this, for the time being, the encouragement is always to anonymize if possible and to use pseudonyms, change details like gender and age, or aggregate data, to 'conceal' identities.

While in most cases, anonymization might be very much preferred, one of Brown et al. (2016)'s five provocations suggests that the pursuit of anonymity should be questioned. They proposed that rather than be anonymous, perhaps the participants might want to join into a project as a co-creators. This raises an ethical tension around the need to safeguard participants while also being aware of their desire to be seen and heard (Wiles et al., 2012). For an individual wishing to non-anonymize their participants there can be considerable pressure from peers, publishers and ethics boards who will all see this as risky. In many cases, participants are not anonymous to the research or design team but are anonymous in the subsequent paper or report. In the paper designing PETS by Druin et al. (1999), the KidsTeam (who carried out the design work with adults) are named as authors.

CASE: ANONYMOUS, FREE TO PARTICIPATE, CHILDREN

In a recent evaluation study with children aged 3 and 4, we actively sought to ensure anonymity while measuring how they learned English using an app. Traditionally learning is measured using pre-test and post-tests that require some logging of children to codes in order to match up sets of data. In our study, we were comparing learning with an app to learning with other methods so that would also have required a coding system to place children in different groups. In this case we delegated all the allocation of children to groups to the teachers and we kept no records. In assessing learning, we gave children different coloured cards that they deposited in boxes relating to different answers so gathered 'group' results – this method also allowed children to not participate.

14.2.4   RESEARCH WITH MINORS

The concerns about participation, consent and privacy are highlighted when working with minors or indeed with any 'vulnerable' population. Some of these concerns have already been touched on in the earlier sections, but it is worth here taking account of the special cases of working with children in HCI research. The first important consideration regards where children are found. Research and design with children typically occurs in homes, schools, or public spaces. There can be studies in laboratory settings, but these require a lot of organization, and in general, it is easier to work with children in places that are designed for them.

When children participate in Child Computer Interaction (CCI) work, it is very important to be clear about their inclusion and the meaning of their contributions (Iversen et al., 2017). From an ethical position, it is not appropriate to work with children and then to discard, without justification, their contributions. As an example, a constant tension in design work with children is to ensure that they understand the value of their contributions (Guha et al., 2004). Read et al. (2016) have developed a tool to account for ideas as they pass from the children to the design team which goes some way towards acknowledging the need for probity when working with children in this way.

As explored earlier, gathering consent from children is seldom done as legally their parents have to consent for them. However, that does not exempt researchers from needing to work hard to ensure children understand that their participation is voluntary. As there is generally data gathered from children it is important to be clear why that data is being gathered and why those children are the ones who are giving this data up. The second of the two checklists in (Read et al., 2013) provides a set of questions to help in the setting up of a study with children:

- Why are we doing this research? What do we tell (the children)?
- Who is funding the research? What do we tell (the children)?
- What might happen in the long term? What do we tell (the children)?
- What might we publish? What do we tell (the children)?

While this is not a substitute for an ethics review board application, it puts communication with the children to the centre and steers the research/evaluation team towards an honest conversation with the children. Associated with these approaches the children need, in all cases, to be empowered to not take part, to be able to not submit their data and to withdraw their data if possible after they know what they have agreed to. Privacy and anonymization with children are problematic as they will all want to tell you their names and, given half a chance, most would want their pictures on the papers and their names in the acknowledgements. The usual approaches to dealing with anonymization like using codes can be problematic as children are often likely to forget codes and are also likely to tell you who they are in any case. An approach we have used is to have a tear off strip on data sheets where they can write their name but then have it removed by the teacher before the work comes to us. With coded data, we have always asked the teachers to deal with that as they can maintain a master list which removes the need for us to know anything about the children.

***QUESTION? Once you have talked with children about research and about how you want to do it all right – what do you tell the child who says can I be on the paper then?

CASE: ACCOUNTING FOR IDEAS

Having done many design sessions with class sized groups of children, and gathered in many design ideas, our group were concerned about how we could explain to children the process by which their ideas were considered and analyzed. We had talked to children about representation and about fairness, but we wanted to study how this could be quantified. In two studies we explored systematic ways to ensure large numbers (20+) of children's designs were properly considered. In the first study (Read et al., 2014), we first noted the best ideas from each child's design then bundled children's designs into groups of four and chose (1) the four best ideas from those four and (2) the best idea from each child. We then incorporated the chosen ideas into a design and theorized that one idea per child resulted in a not-so-great design. In a follow-on study (Read et al., 2016), we looked at designs through lenses and then included our own design activity, informed by the lenses, and in this case were able to track back – to the children's ideas – where our own inspirations had come from. These two methods both point to a more explainable participatory design approach that can help children understand what they have contributed.

CASE: RED POST BOXES – WHAT DATA AM I SENDING?

In a recent study with children aged 9 and 10, our first session of ten began with all children filling in a small piece of paper with their name on and a score out of ten about what they knew about the topic we were studying. The rest of the 1-hour session was given over to playing three games, one after the other in teams of four or five. Before each game, each child 'rated' the game using a Smileyometer and after each game they scored the game again. These ratings were gathered on a piece of A4 paper that had no room for a name and simply required them to tick one of five images. When they had played all the games, we then asked them to use the remaining time to design a new game – they sketched this out.

The three outputs from this activity were used to explain about data and about what they might choose to hand in. We used the first to explain that, as it had their names on it, it was 'personal' data and that we would on no account want that – and we quickly collected that data up and gave it directly to the teacher. We then talked about what use we could get from the ticked game ratings and explained how this was not personal. We then discussed the ideas they had and explained that while these were not personal data, they nonetheless were personal ideas and they might be less keen to share these. In the room we situated a big red post box and explained that over the coming weeks – what they posted in there could not be retrieved and would be ours to keep. We then invited children to carefully consider whether they wanted to give us their ideas, their game scores or nothing.

14.3 ETHICS IN HCI DESIGN

A considerable amount of HCI work is concerned with the design of technology and this may not always rely on participant studies. HCI work typically either begins with users and ends with a design, or starts with a design and ends with users. Especially in this latter case, the ethical questions that we want to be asking mainly relate to beneficence – that is, the ethical principle of doing no harm.

14.3.1 ETHICAL DESIGN PRACTICE

A technology that a human will use can give material answers to the ethical question of how to act (Verbeek, 2006). Much of the work in HCI is concerned with designing artefacts that will eventually be used with humans in situations where they (the humans) may change things. The design of artefacts in HCI typically takes on one of two forms. In one form, it is an expert design activity which results in a design that is then evaluated with users or situated in a place as a concept or

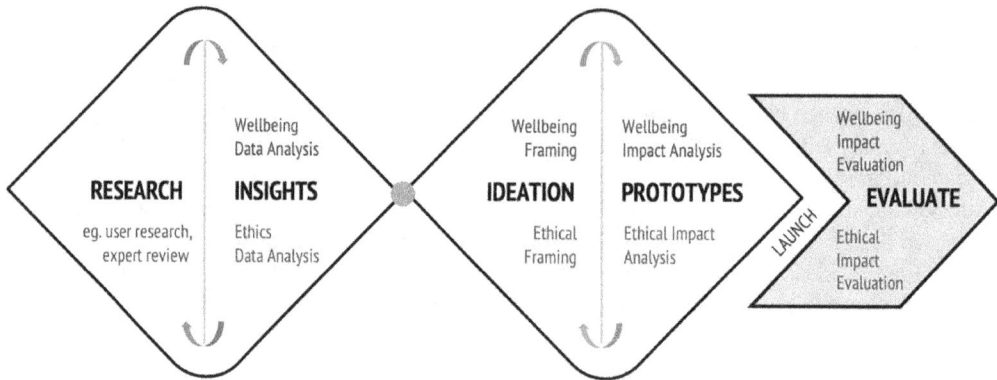

FIGURE 14.1   Responsible design process framework. A process for technology development in which well-being support and ethical impact analysis are incorporated (Peters et al., 2020).

product; in the other it is considered as a participatory process, sometimes ending only with ideas or very low fidelity prototypes, and often used as a means to understand a space rather than to result in a technology.

In the first instance, the consideration of ethics should take place alongside the design process. Peters et al. (2020), in their paper about responsible AI design, described a framework based on the double diamond, which is a well-used design model, in which well-being and ethics align with the diamonds (Figure 14.1).

At the insight stage, an ethical lens is put against the data to check for biases, during ideation ethical reflection is integrated, and when prototypes are built questions as to their use are asked like 'Will this prototype be useable to all people? Will this prototype cause harm to some groups?' and even in evaluation, effort is made to look for unintended use patterns. In his Materializing Morality paper, Verbeek (2006) wrote about how ethical questions regarding the design of technologies have to go beyond the functionality of the technology and the goals for which they are designed. He stresses that designers need to try to figure out if their products will promote behaviours that were unintended and that could be dangerous or unhelpful. The main requirement is that designers need to connect the context of design with the context of use, assessing their design carefully. One way to make products more ethical is to make them explainable. Schneiderman (2020) described how decisions that affect people's lives – like being turned down for a mortgage or a job interview – raise questions from the 'victim' which should be explained to the user. In developing the explanations for systems, it turns out that the designers often then realize that their own design decisions have come with biases so the process of making explainable systems better protects the user but also acts as an ethical design checklist.

14.3.2   Design for Good

There is a growing enthusiasm for design that brings a positive outcome to people, the planet or to societies. Broadly described as 'design for good' or 'HCI for good', this is about design that improves lives thorough human centred design. As the title of his book, Cary (2017) described 'design for good' as a collaborative, dignified, empathic process that results in designs that improve people's lives while not costing the earth. In HCI, design for good is more usually referred to as design for social impact and incorporates much of the work done around Designing with and for Marginalized Communities, ICT4D (Information and Communications Technologies for Development which is a decentralized movement dedicated to making access to digital technologies more equitable), HCI4D (HCI for Development which is a community for everyone who is interested in the role of technology in diverse domains such as, but not limited to: conflict zones; early-grade reading;

infant mortality; rural and urban community development; disenfranchised and marginalized populations) and designing for low carbon and sustainability (Bates et al., 2017).

Design for Good takes much of its ethical steer from virtue ethics, which is that branch of ethics, rooted in ancient stoicism, that gives space for greater discretion and judgement and for inner motivation and commitment. Where decisions are made based on honesty, courage, compassion, and gratitude (four of the virtues) then it is argued that good will prevail. Not only that, but also the enactor of such work will gain fulfilment and will flourish. Returning to Bødker and Kyng, (2018),'s manifesto that our design work should have a real purpose, design for good has to be a space where we can focus our efforts. It brings with it its own ethical dilemmas; one of these is with the potential exploitation of individuals and/or lack of understanding of local contexts; another is the challenge of the possible disruption caused when new technologies are brought to places where they do not easily fit. Mthoko and Pade-Khene (2013) highlighted four themes that need to be considered when doing research in developing countries: collaboration and participation, socio-economic context, cost and benefits and underlying stakeholder interests. In one study in the hills in Uganda, technology taken to children, and donated to the school after the study, was later sold by the school's proprietors; the technology being locally worth too much to leave for children to learn from.

14.3.3 Ethics in Sensitive Contexts

Following on from the potential concerns in some design for good settings, where much good can be done, a sensitive context is any context in which extra care must be taken. In Sultana et al.'s (2020) ethnographic study in rural Bangladesh, the authors sought to better understand the wellbeing practices of a rural community steeped in witchcraft traditions. Access to the participants was facilitated by a local non-profit NGO working in that area who introduced the researchers via a public meeting. Individuals were recruited using a snowball sampling approach, and they agreed to participate orally as they did not have the skills to read and write. Over a 3-year period, the villagers were observed, interviewed, and invited to focus groups. The Global Code of Conduct for Research in Resource Poor Settings provides guidance for such situations with four headings of fairness, care, respect, and honesty (Pennington, 2018). From this code, three relevant articles for HCI researchers are:

- Article 1 – Research has to be locally relevant.
- Article 3 – Understandable and accessible feedback about the findings of the research must be given to local communities and research participants.
- Article 9 – Community assent should be obtained locally.

Another sensitive area for HCI design is where technology relates to personal issues. Gender transition (Haimson et al., 2016) and bereavement (Brubaker et al., 2012) have both been studied in HCI contexts using analysis of social media. In the former, the authors used social media to invite transitioning adults to anonymously reflect using open-response survey questions; they recognized the limitation of this as a method but felt that it was the right choice to protect participants. In the latter case, social media profiles associated with people at least 3 years post-mortem were analyzed to look at the language around grief. Given that this was an example of research for which consent is not obtained, there could be ethical questions around the use of such data, although legally such data is outside traditional data constraints.

14.3.4 Ethics of Persuasive Technologies

A persuasive technology is one that is specifically designed to persuade a user to do something or change something (Fogg, 1998). Such technologies can be very open about their intentions, others can be more subtle. Ethical challenges arise over involvement in the development of some such technologies, to the interaction designed into such technologies to persuade, the norms that promote change, and finally to any deception that might be needed in the design and evaluation of such products.

Persuasive technologies are not inherently good or bad. Many have very noble aims, and some are just promoting well understood good habits – it can be hard to see very much wrong with an App that encourages children to eat broccoli once a week. In most HCI work, it seems that researchers do not tend to get involved in the design of harmful technologies although occasionally there can be questions asked at a review point around 'is that ethical?' where the question is not about the research but is about the product being considered. In their exploration of the ethics of persuasive technologies Berdichevsky and Neuenschwander (1999) explored how, in a traditional act of persuasion, responsibility for any outcome is shared between the persuader and the persuaded but when technology is involved, the designer's motivations feed into a persuasive technology that may have intended and unintended consequences for the persuaded. As with all ethics of design, it is then the responsibility of the designer to work hard to examine as many unintended consequences as possible. The intention outcome matrix of Stibe and Cugelman (2016) demonstrates the different situations around persuasive technology, using the term 'backfiring' to capture those situations where an unintended consequence is detrimental. Note that this diagram (Figure 14.2) also introduces the term dark patterns.

Introduced by Henry Brignull on his website darkpatterns.org, dark patterns are essentially the 'bad' elements of persuasive technology where the user is tricked into doing something that they had not wanted to do. In our own work with teenagers exploring monetization in games as a dark pattern, we have explored how app developers could make money in a more ethical way – a surprising idea from the youths was that the government should simply pay for games for teenagers, in the same way that it pays for healthcare and education (Fitton & Read, 2023). In an article on persuasive technology, Gram-Hansen (2010) framed ethical discussion around the Greek word Kairos which is roughly described as *"to say the right thing at the right time"*. Applying this principle into the design of persuasive technology should, theoretically, reduce the risk of unintended consequences but, to make that a reality we would need to understand the user's state, or take them to a place where we can anticipate their state ready for the persuasion. Gram-Hansen (2010) stressed, in this article, the particular problems with ubiquitous persuasive technology that, because volunteerism is absent and the technology is initiating the interaction, could be changing human behaviour without full disclosure.

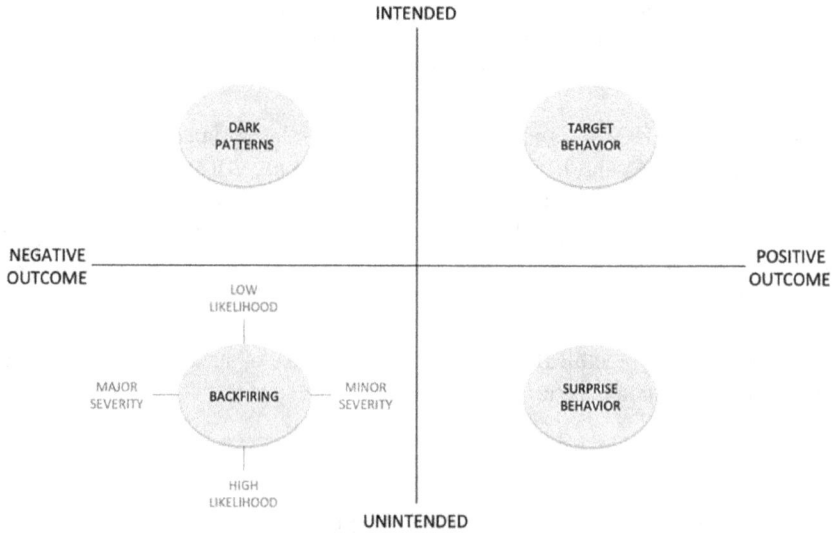

FIGURE 14.2 Intention Outcome Matrix (Stibe & Cugelman, 2016).

When interacting with users to evaluate persuasive technologies, HCI practitioners may have to employ some deceit in order to not influence outcomes. Most such technologies are evaluated without divulging the intended consequences and so care has to be taken to ensure that after use, the participants fully understand what they have been engaging in (Torning & Oinas-Kukkonen, 2009). Whilst less common, some 'covering up' may also be needed at the design stage. In one study, teenagers with behavioural problems were invited to participate in the design of a serious game to 'improve their own emotional intelligence' – in framing the design sessions it was necessary to implement strategies to hide the fact that the intended game was related so closely to them (Mazzone et al., 2008).

14.4   ETHICS AND EMERGING TECHNOLOGIES

In the book, *Klara and the Sun* by Kazuo Ishiguro, a companion robot stands in the companion robot shop hoping to be purchased. When a little girl eventually chooses her and takes her home, Klara the robot finds herself having to be the friend to the child and the friend to the mother. Klara must watch and listen and learn, and the story, told through her perspective, draws you into a world that is close enough to the near future that you can imagine living in it. When considering the ethics of emerging technologies, be they robots, de-personalized AI or surveillance systems, or extended realities, we rely on our understandings of the here and now as well as on the contexts in which such technologies will be used.

Robots are perhaps a good place to start thinking about this as they can be humanized and can become parts of our social families. Coeckelbergh (2021) used virtue ethics (from the stoics of ancient Greece) to consider how we might behave with social robots by considering that, whilst the robot is a machine, it has a moral standing on account of its relationship to a human and can therefore be treated badly. In Klara, worn-out robots are replaced, and old versions thrown out for scrap – this 'feels' wrong and has us wondering if the robot is making us less humane just by virtue of it being replaceable. Taking a virtue ethics approach, we can identify that being mean to a robot is bad – not because of the robot – but because it is not a nice thing to do (Vallor, 2016). Of course, in most conversations about ethics and robots – we paint the opposite picture – of the robots somehow taking charge and developing their own 'bad' behaviours and therefore needing to be legally constrained and potentially even criminalized (Lemley & Casey, 2019).

Frameworks for discussing the ethics of emerging technology generally hinge around the quandary of anticipatory thinking or forecasting. The ethics of emerging devices squarely sits amid uncertainty around their future uses and their future contributions to the wider social constructs of humanity (Sollie, 2007). The ethicist must be able to say something meaningful when asked about the technology and in this case can really only either speak according to that which is known or otherwise has to speculate. A driverless car can still run over a child because we know what cars do and how they situate; it is much harder to think forward to driverless cars behaving not as cars but as activists, maybe self-organizing and deciding to block a motorway just to get themselves a better deal on car parking.

An ethical technical assessment is one approach where the designers of a new technology actively look into the near future, predict what could happen, and attempt to feed that back into the design process in an iterative constantly evolving way (Palm & Hansson, 2006). Techno – ethical scenarios begin not with the technology but with a moral landscape and then see how the new technology fits into that, highlighting promises and expectations of the new item, formulating objections that may arise from its adoption, presenting arguments for and against such objections before then coming back with an ethical judgement presented as a future scenario (Boenink et al., 2010). For many of us, these imaginings might be difficult and the approach of Stahl et al. (2010), who used ethical projections and ideas from literature and other places, might be easier to apply – in this case, we could conclude that maybe one day our robot companion would seek to kill us as we thought about upgrading him!!

THE KILLER BALL BEARING

In my own first experience of an ethics board, I sat on a panel of engineering ethics where a proposal had come in for a PhD study concerning lubricants on ball bearings. The work was being co-funded by the UK defense group – British Aerospace – and the question that was asked, in that meeting, was 'What is the chance that this research will lead to development that would be used on a fighter plane that would subsequently be used to kill unarmed civilians?' Without our crystal balls we could not answer that question, but it was important that the question was asked. The proposal was given ethical clearance, I cannot say though if the research made its way into combat.

14.5   FUTURE DIRECTIONS FOR ETHICS AND HCI

Looming situations in HCI and ethics come from several directions. The first relates to the participation of individuals in research and design studies. Going forward there will need to be more conversations about the extent to which participants own the outcomes of their participation but also in terms of empowering them to ask their own questions. This will raise a huge problem for ethics boards as ethical problems will move as the studies mature. At this juncture, situational ethics (Munteanu et al., 2021) and in-action ethics (Frauenberger et al., 2017) become the norm. Situational ethics are proposed to deal with the way that ethical challenges keep on arising especially in field work with participants who may have several vulnerabilities. The claim from Munteanu et al. (2021) is that any static ethics process is inadequate for such situations. They propose suggestions for HCI researchers as they work:

- To look carefully for ethical triggers – vulnerable participants, sensitive settings – in the world deployment.
- To incorporate, into the research design, the chance to assess risks in the field - a situational ethics approach that can adapt.
- To open up discussions with university ethics boards about how HCI research can be considered.
- To work in multidisciplinary teams in order to get a broad perspective.
- To campaign to not have work assessed by mono-disciplinary ethics boards.
- To be active in dealing with ethical guidelines.

In-Action ethics (Frauenberger et al., 2017) also came from a critique of HCI work being done under a 'static' or 'formal' ethics agreement. This solution offers qualities that researchers can appropriate in their work which include being reflective in-action, co-constructing ethics with participants, being open and transparent, having a working culture that allows debate and disagreement and having shared responsibility for being ethical.

Ubiquitous and serendipitous interactions with new and emerging technologies will also provide new challenges for HCI as interactions may be in the wrong moment or at the wrong time for the human; theoretically causing harm which is against all our ethical codes. Drone technologies and other unmanned vehicles spring to mind but so too do social media applications and virtual experiences which are very difficult to predict.

Machine learning and AI, and machines that think for themselves, also bring huge challenges, and the ethics around AI design would need their own chapter – for human-centred AI. Xu (2019) proposed a framework where ethically aligned AI meets human factors meets technology enhancement – the challenge in this is, of course, what is ethically aligned AI. Longo et al. (2020) made the case that the way to explore ethical issues in the future might be by taking a Value Sensitive Design

(VSD) approach where human values shape design – would this be enough to determine if we then had 'ethically aligned AI' it is early days to know this.

14.6   CONCLUSION

In a short chapter, it is not possible to cover all the aspects of ethics as they pertain to the expanding world of HCI. HCI work will always raise ethical questions and researchers and practitioners will always face ethical dilemmas. In this chapter, some of the easier to reach ethical challenges around participation of users in our work have been highlighted; we can look to recruit from more diverse populations, we can be more critical about who we ask and why we ask them, we can aim for 'better' consent, and we can explore ways to make participation more meaningful and explainable.

Challenges around what we design, and our roles in this as innovators, are less easy to address; these require us to be constantly on our guard against becoming complacent in our design work, against designing for the sake of design, and against poorly thought-out persuasive design. We need to actively join the debates on the ethics of emerging technologies – the prospect of humanoid robots having their own feelings, of unexpected user actions in virtual realities, of out of control AI systems – all need the HCI community to input.

From this chapter it is hoped that HCI researchers will adopt one or more of the ideas to become more critical, more user-centred researchers, taking ethics out of the ethics board into everyday action. The HCI community has a long history of exploring the ethics of its own work, with core themes around design for good and participatory design, with feminist thinking and critical design; as future technologies challenge some of the static approaches to ethics, we need to continuously critique our own actions and be the innovators of user-centred context aware ethics for the changing world we inhabit.

REFERENCES

Anderson, R. E., Johnson, D. G., Gotterbarn, D., & Perrolle, J. (1993). Using the new ACM code of ethics in decision making. *Communications of the ACM*, *36*(2), 98–107.

Bates, O., Thomas, V., & Remy, C. (2017). Doing good in HCI: Can we broaden our agenda? *Interactions*, *24*(5), 80–82.

Beauchamp, T., & Childress, J. (2019). Principles of biomedical ethics: Marking its fortieth anniversary. *The American Journal of Bioethics*, *19*, 9–12.

Berdichevsky, D., & Neuenschwander, E. (1999). Toward an ethics of persuasive technology. *Communications of the ACM*, *42*(4), 51–58.

Bødker, S., & Kyng, M. (2018). Participatory design that matters-facing the big issues. *ACM Transactions on Computer-Human Interaction*, *25*(1), 4.

Boenink, M., Swierstra, T., & Stemerding, D. (2010). Anticipating the interaction between technology and morality: A scenario study of experimenting with humans in bionanotechnology. *Studies in Ethics, Law, and Technology*, *4*(2).

Böhme, R., & Köpsell, S. (2010). Trained to accept? A field experiment on consent dialogs. *Proceedings of the 2010 SIGCHI Conference on Human Factors in Computing Systems,* Atlanta, GA.

Brown, B., Weilenmann, A., McMillan, D., & Lampinen, A. (2016). Five provocations for ethical HCI research. *Proceedings of the 2016 SIGCHI Conference on Human Factors in Computing Systems,* San Jose, CA.

Brubaker, J., Kivran-Swaine, F., Taber, L., & Hayes, G. (2012). Grief-stricken in a crowd: The language of bereavement and distress in social media. *Proceedings of the 2012 International AAAI Conference on Web and Social Media,* Trinity College, Dublin.

Cary, J. (2017). *Design for good: A new era of architecture for everyone.* Island Press.

Coeckelbergh, M. (2021). How to use virtue ethics for thinking about the moral standing of social robots: A relational interpretation in terms of practices, habits, and performance. *International Journal of Social Robotics*, *13*(1), 31–40.

Druin, A., Montemayor, J., Hendler, J., McAlister, B., Boltman, A., Fiterman, E., & Plaisant, A. (1999). Designing PETS: A personal electronic teller of stories. *Proceedings of the 1999 SIGCHI Conference on Human Factors in Computing Systems,* Pittsburgh, PA.

Ehn, P. (2008). Participation in design things. In *Participatory Design Conference (PDC)* (pp. 92–101). Bloomington, IN: ACM Digital Library.

Fitton, D., & Read, J. (2023). Money from the Queen": Exploring children's ideas for monetization in free-to-play mobile games. *Proceedings of the 2023 IFIP Conference; Interact*, York, UK.

Fogg, B. J. (1998). Persuasive computers: Perspectives and research directions. *Proceedings of the 1998 SIGCHI Conference on Human Factors in Computing Systems,* Los Angeles, CA.

Frauenberger, C., Antle, A. N., Landoni, M., Read, J. C., & Fails, J. A. (2018). Ethics in interaction design and children: A panel and community dialogue. *Proceedings of the 2018 SIGCHI Conference on Interaction Design and Children,* Trondheim.

Frauenberger, C., Rauhala, M., & Fitzpatrick, G. (2017). In-action ethics. *Interacting with Computers*, 29(2), 220–236.

Gotterbarn, D., Bruckman, A., Flick, C., Miller, K., & Wolf, M. J. (2017). ACM code of ethics: A guide for positive action. *Communications of the ACM*, 61, 121–128.

Gram-Hansen, S. B. (2010). Persuasive everyware-possibilities and limitations. *Proceedings of the 2010 IIIS World Multi-Conference on Systemics, Cybernetics and Informatics,* Orlando, US

Grisso, T., & Vierling, L. (1978). Minors' consent to treatment: A developmental perspective. *Professional Psychology*, 9(3), 412.

Guha, M. L., Druin, A., Chipman, G., A, F. J., Simms, S., & Farber, A. (2004). Mixing ideas: A new technique for working with young children as design partners. *Proceedings of the 2004 SIGCHI Conference on Interaction Design and Children*, College Park, Maryland, US

Haimson, O. L., Brubaker, J. R., Dombrowski, L., & Hayes, G. R. (2016). Digital footprints and changing networks during online identity transitions. *Proceedings of the 2016 SIGCHI Conference on Human Factors in Computing Systems,* San Jose, CA.

Henrich, J., Heine, S. J., & Norenzayan, A. (2010). Most people are not WEIRD. *Nature*, 466(7302), 29–29.

Iversen, O. S., Smith, R. C., & Dindler, C. (2017). Child as protagonist: Expanding the role of children in participatory design. *Proceedings of the 2017 SIGCHI Conference on Interaction Design and Children*, Stanford, CA.

Kumar, P., Vitak, J., Chetty, M., Clegg, T. L., Yang, J., McNally, B., & Bonsignore, E. (2018). Co-designing online privacy-related games and stories with children. *Proceedings of the 2018 SIGCHI Conference on Interaction Design and Children,* Trondheim

Lemley, M. A., & Casey, B. (2019). Remedies for robots. *The University of Chicago Law Review*, 86(5), 1311–1396.

Linxen, S., Sturm, C., Brühlmann, F., Cassau, V., Opwis, K., & Reinecke, K. (2021). How weird is CHI? *Proceedings of the 2021 SIGCHI Conference on Human Factors in Computing Systems*

Longo, F., Padovano, A., & Umbrello, S. (2020). Value-oriented and ethical technology engineering in industry 5.0: A human-centric perspective for the design of the factory of the future. *Applied Sciences*, 10(12), 4182.

Luger, E., & Rodden, T. (2014). Sustaining consent through agency: A framework for future development. *Proceedings of the 2014 ACM International Joint Conference on Pervasive and Ubiquitous Computing,* Seattle.

Mazzone, E., Read, J. C., & Beale, R. (2008). Design with and for disaffected teenagers. *Proceedings of the 2008 Nordic Conference on Human-Computer Interaction*, Lund, Sweden.

Miller, V. A., Drotar, D., & Kodish, E. (2004). Children's competence for assent and consent: A review of empirical findings. *Ethics & Behavior*, 14(3), 255–295.

Moore, N. (2012). The politics and ethics of naming: Questioning anonymisation in (archival) research. *International Journal of Social Research Methodology*, 15(4), 331–340.

Morrison, A., McMillan, D., & Chalmers, M. (2014). Improving consent in large scale mobile HCI through personalised representations of data. *Proceedings of the 2014 Nordic Conference on Human-Computer Interaction: Fun, Fast, Foundational*, Helsinki.

Moser, M. R. (1988). Ethical conflict at work: A critique of the literature and recommendations for future research. *Journal of Business Ethics*, 7, 381–387.

Mthoko, H. L., & Pade-Khene, C. (2013). Towards a theoretical framework on ethical practice in ICT4D programmes. *Information Development*, 29(1), 36–53.

Munteanu, C., Waycott, J., & McNaney, R. (2021). Dealing with ethical challenges in HCI fieldwork. *Proceedings of the 2021 SIGCHI Conference on Human Factors in Computing Systems*

Offenwanger, A., Milligan, A. J., Chang, M., Bullard, J., & Yoon, D. (2021). Diagnosing bias in the gender representation of HCI research participants: How it happens and where we are. *Proceedings of the 2021 SIGCHI Conference on Human Factors in Computing Systems*

Palm, E., & Hansson, S. O. (2006). The case for ethical technology assessment (eTA). *Technological Forecasting and Social Change, 73*(5), 543–558.

Pennington, G. (2018). Global Code of Conduct-for Research in Resource-Poor Settings. www.globalcodeof-conduct.org.

Peters, D., Vold, K., Robinson, D., & Calvo, R. A. (2020). Responsible AI-two frameworks for ethical design practice. *IEEE Transactions on Technology and Society, 1*(1), 34–47.

Rashid, A.M., Ling, K., Tassone, R.D., Resnick, P., Kraut, R. & Riedl, J., 2006, April. Motivating participation by displaying the value of contribution. In *Proceedings of the SIGCHI Conference on Human Factors in Computing Systems* (pp. 955-958).

Read, J., Sim, G., Horton, M., & Fitton, D. (2022). Reporting Back in HCI Work with Children. *Proceedings of the 2022 SIGCHI Conference on Interaction Design and Children,* Braga.

Read, J. C., Fitton, D., & Horton, M. (2014). Giving ideas an equal chance: Inclusion and representation in participatory design with children. *Proceedings of the 2014 SIGCHI Conference on Interaction Design and Children,* Aarhus, Denmark.

Read, J. C., Fitton, D., Sim, G., & Horton, M. (2016). How ideas make it through to designs: Process and practice. *Proceedings of the 2016 Nordic Conference on Human-Computer Interaction,* Gothenburg, Sweden.

Read, J. C., Horton, M., Sim, G., Gregory, P., Fitton, D., & Cassidy, B. (2013). CHECk: A tool to inform and encourage ethical practice in participatory design with children. *Proceedings of the 2013 SIGCHI Conference on Human Factors in Computing Systems,* Paris, France.

Read, J. C., Mazzone, E., & Höysniemi, J. (2005). Wizard of Oz evaluations with children: Deception and discovery. *Proceedings of the 2005 SIGCHI Conference on Interaction Design and Children,* Boulder, Colorado.

Rode, J. A. (2009). *Digital parenting: Designing children's safety. Proceedings of the 2009 British Computer Society HCI Conference,* Cambridge, UK.

Russell, W. M. S., & Burch, R. L. (1959). *The principles of humane experimental technique.* Methuen.

Schneiderman, B. (2020). Bridging the gap between ethics and practice: Guidelines for reliable, safe, and trustworthy human-centered AI systems. *ACM Transactions on Interactive Intelligent Systems (TiiS), 10*(4), 1–31.

Sollie, P. (2007). Ethics, technology development and uncertainty: An outline for any future ethics of technology. *Journal of Information, Communication and Ethics in Society, 5*(4), 293–306.

Stahl, B. C., Heersmink, R., Goujon, P., Flick, C., Van Den Hoven, J., Wakunuma, K., Ikonen, V., & Rader, M. (2010). Identifying the ethics of emerging information and communication technologies: An essay on issues, concepts and method. *International Journal of Technoethics (IJT), 1*(4), 20–38.

Stibe, A., & Cugelman, B. (2016). Persuasive backfiring: When behavior change interventions trigger unintended negative outcomes. *Proceedings of the 2016 ACM PERSUASIVE Conference,* Salzburg, Austria.

Strengers, Y., Sadowski, J., Li, Z., Shimshak, A., & 'Floyd' Mueller, F. (2021). What can HCI learn from sexual consent? A feminist process of embodied consent for interactions with emerging technologies. *Proceedings of the 2021 SIGCHI Conference on Human Factors in Computing Systems.*

Sultana, S., Sultana, Z., & Ahmed, S. I. (2020). Parareligious-HCI: Designing for 'alternative' rationality in rural wellbeing in Bangladesh. *Proceedings of the 2020SIGCHI Conference on Human Factors in Computing Systems.*

Torning, K., & Oinas-Kukkonen, H. (2009). Persuasive system design: State of the art and future directions. *Proceedings of the 2009 ACM PERSUASIVE Conference* (pp. 1–8). Claremont, CA.

Vallor, S. (2016). *Technology and the virtues: A philosophical guide to a future worth wanting.* Oxford University Press.

Verbeek, P.-P. (2006). Materializing morality: Design ethics and technological mediation. *Science, Technology, & Human Values, 31*(3), 361–380.

Vines, J., Clarke, R., Wright, P., McCarthy, J., & Olivier, P. (2013). Configuring participation: On how we involve people in design. *Proceedings of the 2013 SIGCHI Conference on Human Factors in Computing Systems,* Paris.

Wiles, R., Coffey, A., Robinson, J., & Heath, S. (2012). Anonymisation and visual images: Issues of respect, 'voice' and protection. *International Journal of Social Research Methodology, 15*(1), 41–53.

Williamson, J. R., & Sundén, D. (2015). Deep cover HCI: A case for covert research in HCI. *Proceedings of the 2015 SIGCHI Conference on Human Factors in Computing Systems,* Seoul.

Wood, G., & Rimmer, M. (2003). Codes of ethics: What are they really and what should they be? *International Journal of Value-Based Management, 16,* 181–195.

Xu, W. (2019). Toward human-centered AI: A perspective from human-computer interaction. *Interactions, 26*(4), 42–46.

Index

Note: **Bold** page numbers refer to tables; *italic* page numbers refer to figures and page numbers followed by "n" denote endnotes.

For Product Safety Concerns and Information please contact our EU
representative GPSR@taylorandfrancis.com
Taylor & Francis Verlag GmbH, Kaufingerstraße 24, 80331 München, Germany